BIG IDEAS MATH®

Advanced 1

A Common Core Curriculum

CALIFORNIA EDITION

Ron Larson
Laurie Boswell

BIG IDEAS LEARNING®

Erie, Pennsylvania
BigIdeasLearning.com

Big Ideas Learning, LLC
1762 Norcross Road
Erie, PA 16510-3838
USA

For product information and customer support, contact Big Ideas Learning at **1-877-552-7766** or visit us at ***BigIdeasLearning.com***.

About the Cover
The cover images on the *Big Ideas Math* series illustrate the advancements in aviation from the hot-air balloon to spacecraft. This progression symbolizes the launch of a student's successful journey in mathematics. The sunrise in the background is representative of the dawn of the Common Core era in math education, while the cradle signifies the balanced instruction that is a pillar of the *Big Ideas Math* series.

Printed in the U.S.A.

ISBN 13: 978-1-60840-673-9
ISBN 10: 1-60840-673-3

4 5 6 7 8 9 10 WEB 17 16 15 14

AUTHORS

Ron Larson is a professor of mathematics at Penn State Erie, The Behrend College, where he has taught since receiving his Ph.D. in mathematics from the University of Colorado. Dr. Larson is well known as the lead author of a comprehensive program for mathematics that spans middle school, high school, and college courses. His high school and Advanced Placement books are published by Houghton Mifflin Harcourt. Ron's numerous professional activities keep him in constant touch with the needs of students, teachers, and supervisors. Ron and Laurie Boswell began writing together in 1992. Since that time, they have authored over two dozen textbooks. In their collaboration, Ron is primarily responsible for the pupil edition and Laurie is primarily responsible for the teaching edition of the text.

Laurie Boswell is the Head of School and a mathematics teacher at the Riverside School in Lyndonville, Vermont. Dr. Boswell received her Ed.D. from the University of Vermont in 2010. She is a recipient of the Presidential Award for Excellence in Mathematics Teaching. Laurie has taught math to students at all levels, elementary through college. In addition, Laurie was a Tandy Technology Scholar, and served on the NCTM Board of Directors from 2002 to 2005. She currently serves on the board of NCSM, and is a popular national speaker. Along with Ron, Laurie has co-authored numerous math programs.

ABOUT THE BOOK

The *Big Ideas Math Advanced* series allows students to complete the Common Core State Standards for grades 6, 7, and 8 in two years. After completing this series, students will be ready for Algebra 1 in the eighth grade. The *Big Ideas Math Advanced* series uses the same research-based strategy of a balanced approach to instruction that made the *Big Ideas Math* series so successful. This approach opens doors to abstract thought, reasoning, and inquiry as students persevere to answer the Essential Questions that introduce each section. The foundation of the program is the Common Core Standards for Mathematical Content and Standards for Mathematical Practice. Students are subtly introduced to "Habits of Mind" that help them internalize concepts for a greater depth of understanding. These habits serve students well not only in mathematics, but across all curricula throughout their academic careers.

Big Ideas Math exposes students to highly motivating and relevant problems. Woven throughout the series are the depth and rigor students need to prepare for career-readiness and other college-level courses. In addition, *Big Ideas Math* prepares students to meet the challenge of the new Common Core testing.

We consider *Big Ideas Math* to be the crowning jewel of 30 years of achievement in writing educational materials.

Ron Larson   Laurie Boswell

TEACHER REVIEWERS

- Lisa Amspacher
  Milton Hershey School
  Hershey, PA
- Mary Ballerina
  Orange County Public Schools
  Orlando, FL
- Lisa Bubello
  School District of Palm Beach County
  Lake Worth, FL
- Sam Coffman
  North East School District
  North East, PA
- Kristen Karbon
  Troy School District
  Rochester Hills, MI
- Laurie Mallis
  Westglades Middle School
  Coral Springs, FL
- Dave Morris
  Union City Area School District
  Union City, PA
- Bonnie Pendergast
  Tolleson Union High School District
  Tolleson, AZ
- Valerie Sullivan
  Lamoille South Supervisory Union
  Morrisville, VT
- Becky Walker
  Appleton Area School District
  Appleton, WI
- Zena Wiltshire
  Dade County Public Schools
  Miami, FL

STUDENT REVIEWERS

- Mike Carter
- Matthew Cauley
- Amelia Davis
- Wisdom Dowds
- John Flatley
- Nick Ganger
- Hannah Iadeluca
- Paige Lavine
- Emma Louie
- David Nichols
- Mikala Parnell
- Jordan Pashupathi
- Stephen Piglowski
- Robby Quinn
- Michael Rawlings
- Garrett Sample
- Andrew Samuels
- Addie Sedelmyer
- Tyler Steffy
- Erin Taylor
- Reid Wilson

CONSULTANTS

- Patsy Davis
  Educational Consultant
  Knoxville, Tennessee

- Bob Fulenwider
  Mathematics Consultant
  Bakersfield, California

- Linda Hall
  Mathematics Assessment Consultant
  Norman, Oklahoma

- Ryan Keating
  Special Education Advisor
  Gilbert, Arizona

- Michael McDowell
  Project-Based Instruction Specialist
  Fairfax, California

- Sean McKeighan
  Interdisciplinary Advisor
  Norman, Oklahoma

- Bonnie Spence
  Differentiated Instruction Consultant
  Missoula, Montana

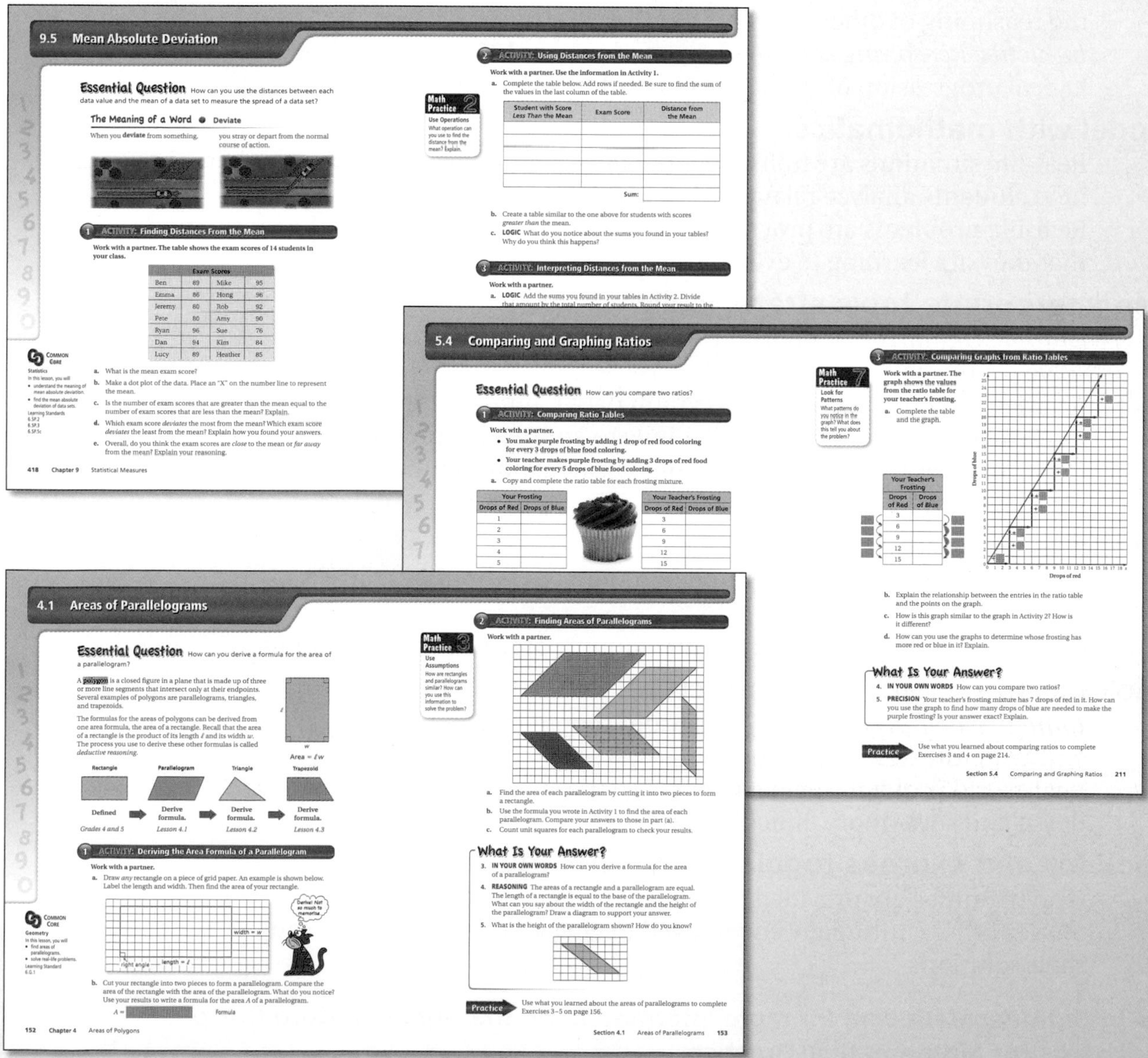

Common Core State Standards for Mathematical Practice

Make sense of problems and persevere in solving them.

- Multiple representations are presented to help students move from concrete to representative and into abstract thinking
- *Essential Questions* help students focus and analyze
- *In Your Own Words* provide opportunities for students to look for meaning and entry points to a problem

Reason abstractly and quantitatively.

- Visual problem solving models help students create a coherent representation of the problem
- Opportunities for students to decontextualize and contextualize problems are presented in every lesson

Construct viable arguments and critique the reasoning of others.

- *Error Analysis*; *Different Words, Same Question*; and *Which One Doesn't Belong* features provide students the opportunity to construct arguments and critique the reasoning of others
- *Inductive Reasoning* activities help students make conjectures and build a logical progression of statements to explore their conjecture

Model with mathematics.

- Real-life situations are translated into diagrams, tables, equations, and graphs to help students analyze relations and to draw conclusions
- Real-life problems are provided to help students learn to apply the mathematics that they are learning to everyday life

Use appropriate tools strategically.

- *Graphic Organizers* support the thought process of what, when, and how to solve problems
- A variety of tool papers, such as graph paper, number lines, and manipulatives, are available as students consider how to approach a problem
- Opportunities to use the web, graphing calculators, and spreadsheets support student learning

Attend to precision.

- *On Your Own* questions encourage students to formulate consistent and appropriate reasoning
- Cooperative learning opportunities support precise communication

Look for and make use of structure.

- *Inductive Reasoning* activities provide students the opportunity to see patterns and structure in mathematics
- Real-world problems help students use the structure of mathematics to break down and solve more difficult problems

Look for and express regularity in repeated reasoning.

- Opportunities are provided to help students make generalizations
- Students are continually encouraged to check for reasonableness in their solutions

Go to *BigIdeasMath.com* for more information on the Common Core State Standards for Mathematical Practice.

Common Core State Standards for Mathematical Content for Grade 6 Advanced

Chapter Coverage for Standards

1 · 2 · 3 · 4 · **5** · 6 · 7 · 8 · 9 · 10 · 11 · 12 · 13 · **14** · **15**

Domain Ratios and Proportional Relationships

- Understand ratio concepts and use ratio reasoning to solve problems.
- Analyze proportional relationships and use them to solve real-world and mathematical problems.

1 · **2** · 3 · 4 · 5 · **6** · 7 · 8 · 9 · 10 · **11** · **12** · 13 · 14 · 15

Domain The Number System

- Apply and extend previous understandings of multiplication and division to divide fractions by fractions.
- Compute fluently with multi-digit numbers and find common factors and multiples.
- Apply and extend previous understandings of numbers to the system of rational numbers.
- Apply and extend previous understandings of operations with fractions to add, subtract, multiply, and divide rational numbers.

1 · 2 · **3** · 4 · 5 · 6 · **7** · 8 · 9 · 10 · 11 · 12 · **13** · 14 · 15

Domain Expressions and Equations

- Apply and extend previous understandings of arithmetic to algebraic expressions.
- Reason about and solve one-variable equations and inequalities.
- Represent and analyze quantitative relationships between dependent and independent variables.
- Use properties of operations to generate equivalent expressions.
- Solve real-life and mathematical problems using numerical and algebraic expressions and equations.

1 · 2 · 3 · **4** · 5 · 6 · 7 · **8** · 9 · 10 · 11 · 12 · 13 · 14 · 15

Domain Geometry

- Solve real-world and mathematical problems involving area, surface area, and volume.

1 · 2 · 3 · 4 · 5 · 6 · 7 · 8 · **9** · **10** · 11 · 12 · 13 · 14 · 15

Domain Statistics and Probability

- Develop understanding of statistical variability.
- Summarize and describe distributions.

Go to *BigIdeasMath.com* for more information on the Common Core State Standards for Mathematical Content.

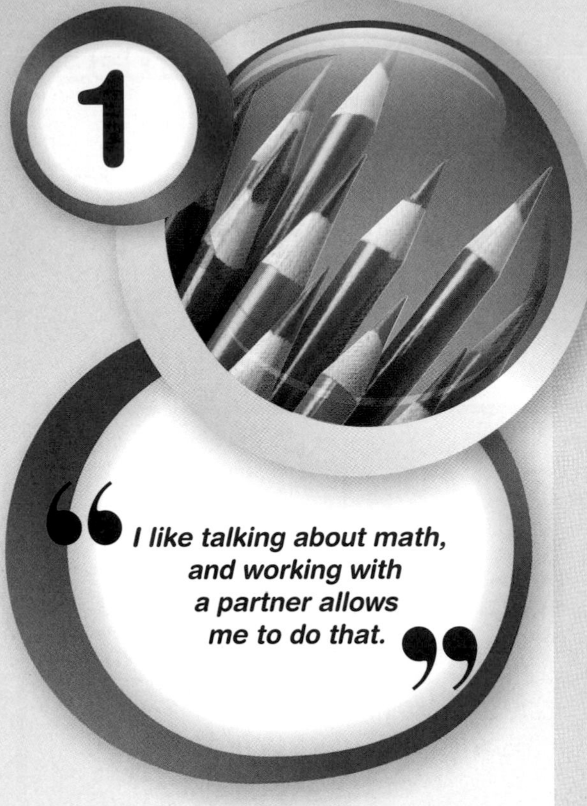
1
I like talking about math, and working with a partner allows me to do that.

Numerical Expressions and Factors

Fractions and Decimals

2

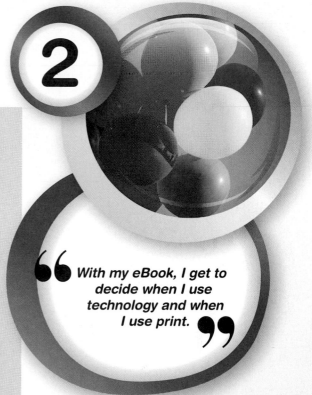

“With my eBook, I get to decide when I use technology and when I use print.”

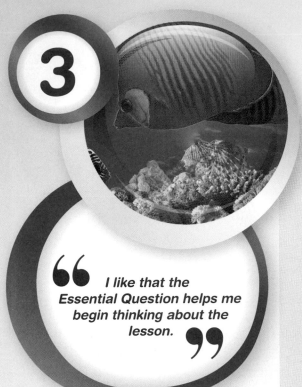

3 Algebraic Expressions and Properties

"I like that the Essential Question helps me begin thinking about the lesson."

Areas of Polygons

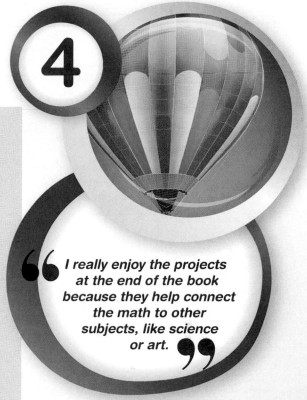

“I really enjoy the projects at the end of the book because they help connect the math to other subjects, like science or art.”

5

Ratios and Rates

“I like Newton and Descartes! The cartoons are funny and I like that they model the math that we are learning.”

Integers and the Coordinate Plane

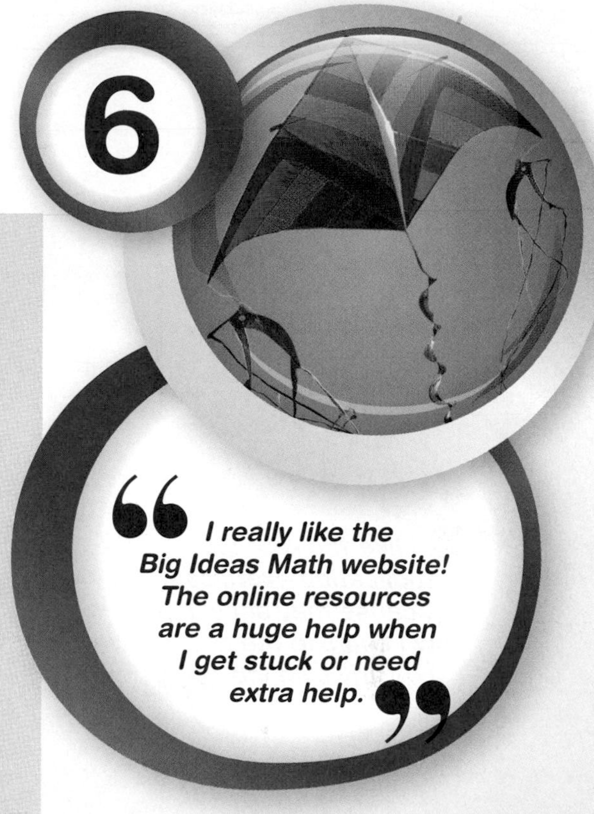
6
I really like the Big Ideas Math website! The online resources are a huge help when I get stuck or need extra help.

7

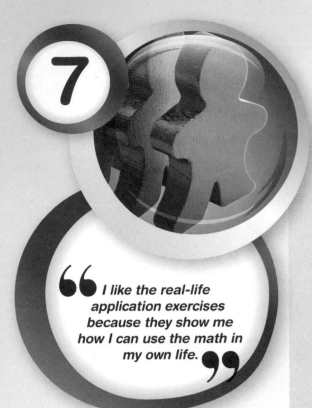

Equations and Inequalities

"I like the real-life application exercises because they show me how I can use the math in my own life."

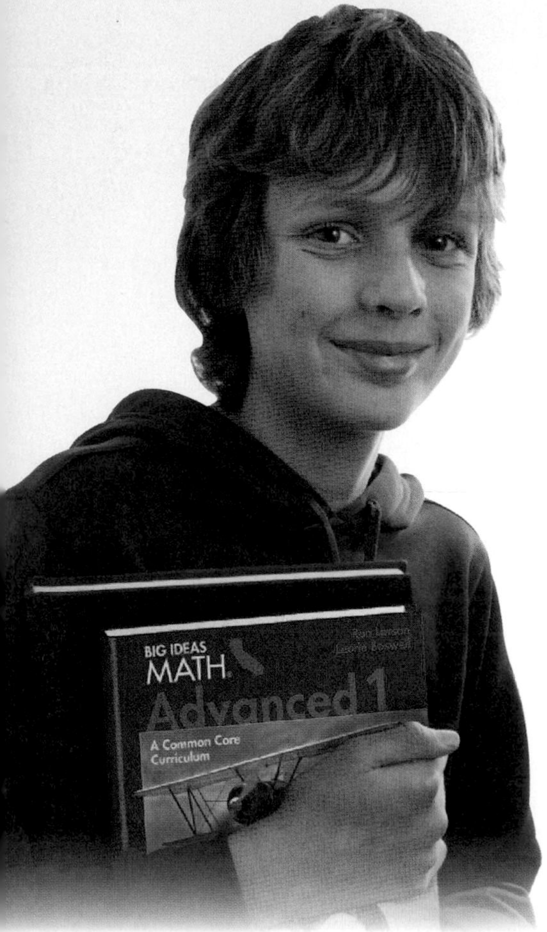

Surface Area and Volume

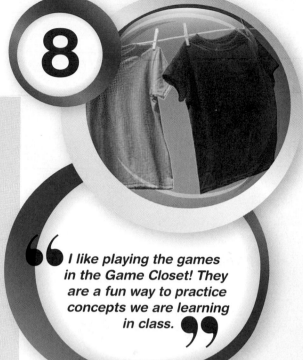

"I like playing the games in the Game Closet! They are a fun way to practice concepts we are learning in class."

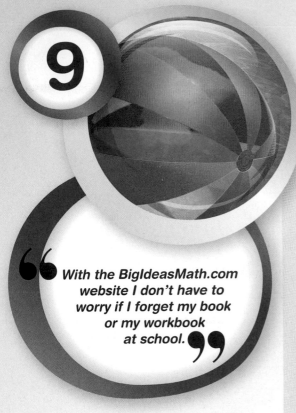

9 Statistical Measures

"With the BigIdeasMath.com website I don't have to worry if I forget my book or my workbook at school."

Data Displays

10

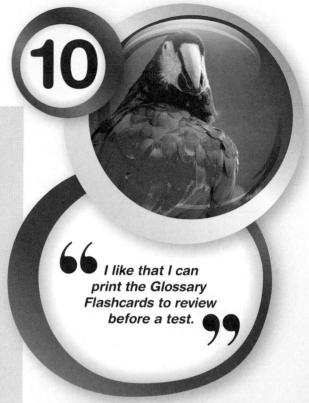

"I like that I can print the Glossary Flashcards to review before a test."

11

Integers

"Before my school had Big Ideas Math I would always lose test points because I left units off my answers. Now I see why they are so important."

Rational Numbers

12

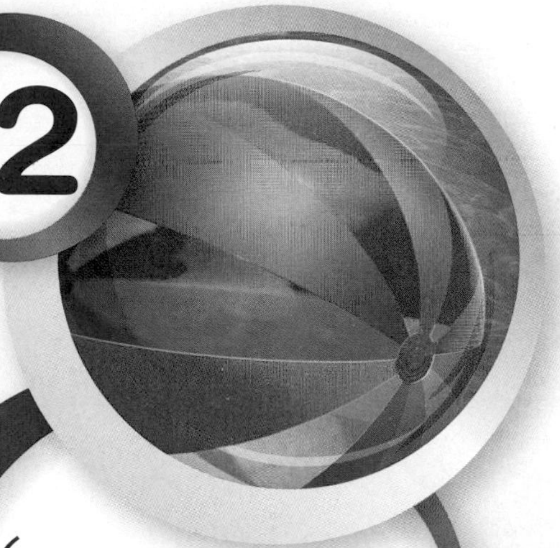

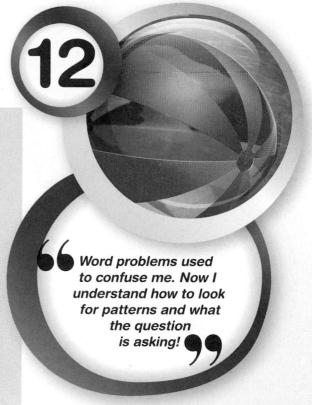

"Word problems used to confuse me. Now I understand how to look for patterns and what the question is asking!"

13 Expressions and Equations

"I like the Big Ideas Math Tutorials because they help explain the math when I am at home."

Ratios and Proportions

14

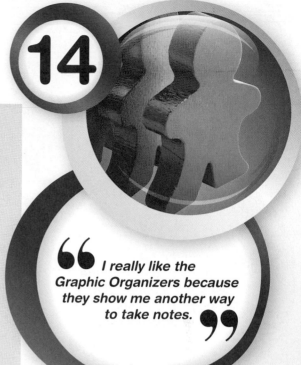

“I really like the Graphic Organizers because they show me another way to take notes.”

15

Percents

“Using the Interactive Manipulatives from the Dynamic Student Edition helps me to see the mathematics that I am learning.”

Appendix A: My Big Ideas Projects

"The Skills Review Handbook helps me review topics that I learned before."

How to Use Your Math Book

- Read the **Essential Question** in the activity.

  Discuss the **Math Practice** question with your partner.

  Work with a partner to decide **What Is Your Answer?**

  Now you are ready to do the **Practice** problems.

- Find the **Key Vocabulary** words, **highlighted in yellow**.

  Read their definitions. Study the concepts in each **Key Idea**.
  If you forget a definition, you can look it up online in the

  Multi-Language Glossary at BigIdeasMath.com.

- After you study each **EXAMPLE**, do the exercises in the **On Your Own**.

  Now You're Ready to do the exercises that correspond to the example.

  As you study, look for a **Study Tip** or a **Common Error**.

- The exercises are divided into 3 parts.

  Vocabulary and Concept Check

  Practice and Problem Solving

  Fair Game Review

  If an exercise has a **1** next to it, look back at Example 1 for help with that exercise.

  More help is available at **Check It Out Lesson Tutorials BigIdeasMath.com**.

- To help study for your test, use the following.

  Quiz **Study Help**

  Chapter Review **Chapter Test**

SCAVENGER HUNT

Use this *Scavenger Hunt* to find where things are in **Chapter 1**.

Newton

Find an **Essential Question**.

Find a **Math Practice**.

Find a **Practice**.

Find the definition of a **numerical expression**.

Find the 🔑 Order of Operations.

Use the index in the back of the book.

Find an **EXAMPLE** of greatest common factor.

Find an **On Your Own** about $6 + 2^4 - 1$.

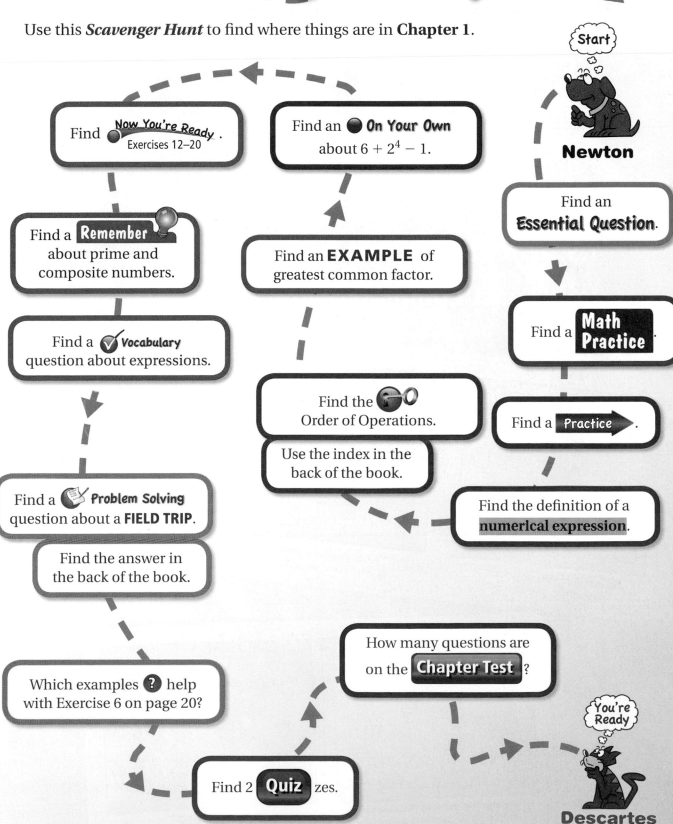

Find **Now You're Ready** Exercises 12–20.

Find a **Remember** about prime and composite numbers.

Find a **Vocabulary** question about expressions.

Find a **Problem Solving** question about a **FIELD TRIP**.

Find the answer in the back of the book.

Which examples **?** help with Exercise 6 on page 20?

Find 2 **Quiz**zes.

How many questions are on the **Chapter Test**?

Descartes

1 Numerical Expressions and Factors

1.1 Whole Number Operations

1.2 Powers and Exponents

1.3 Order of Operations

1.4 Prime Factorization

1.5 Greatest Common Factor

1.6 Least Common Multiple

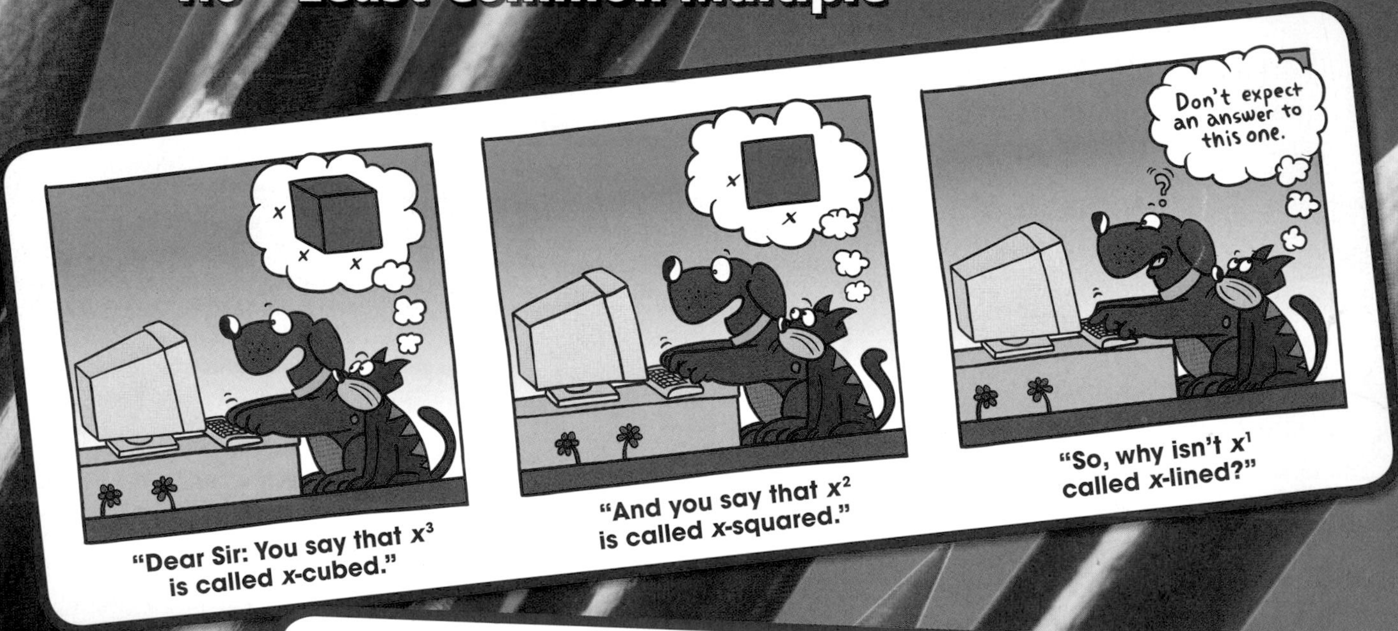

"Dear Sir: You say that x^3 is called x-cubed."
"And you say that x^2 is called x-squared."
Don't expect an answer to this one.
"So, why isn't x^1 called x-lined?"

$\frac{1}{2}+\frac{1}{3}=\frac{3}{6}+\frac{2}{6}$
$=\frac{5}{6}$
There aren't any hyenas around here.
"My sign on adding fractions with unlike denominators is keeping the hyenas away."
"See, it's working."

What You Learned Before

"Because 6 is composite, you can have 2 piles of 3 blocks, 3 piles of 2 blocks, or..."

Identifying Prime and Composite Numbers (4.OA.4)

Example 1 **Determine whether 26 is prime or composite.**
Because the factors of 26 are 1, 2, 13, and 26, it is composite.

Example 2 **Determine whether 37 is prime or composite.**
Because the only factors of 37 are 1 and 37, it is prime.

Try It Yourself

Determine whether the number is prime or composite.

1. 5 **2.** 14 **3.** 17

4. 23 **5.** 28 **6.** 33

7. 43 **8.** 57 **9.** 64

Adding and Subtracting Mixed Numbers with Like Denominators (4.NF.3c)

Example 3 **Find $2\frac{3}{5} + 4\frac{1}{5}$.**

$$2\frac{3}{5} + 4\frac{1}{5} = \frac{2 \cdot 5 + 3}{5} + \frac{4 \cdot 5 + 1}{5}$$ Rewrite the mixed numbers as improper fractions.

$$= \frac{13}{5} + \frac{21}{5}$$ Simplify.

$$= \frac{13 + 21}{5}$$ Add the numerators.

$$= \frac{34}{5}, \text{ or } 6\frac{4}{5}$$ Simplify.

Try It Yourself

Add or subtract.

10. $4\frac{1}{9} + 2\frac{7}{9}$ **11.** $6\frac{1}{11} + 3\frac{6}{11}$ **12.** $3\frac{7}{8} + 4\frac{3}{8}$

13. $5\frac{8}{13} - 1\frac{2}{13}$ **14.** $7\frac{1}{4} - 3\frac{3}{4}$ **15.** $4\frac{1}{6} - 2\frac{5}{6}$

1.1 Whole Number Operations

Essential Question How do you know which operation to choose when solving a real-life problem?

1 ACTIVITY: Choosing an Operation

Work with a partner. The double bar graph shows the history of a citywide cleanup day.

- **Copy each question below.**
- **Underline a key word or phrase that helps you know which operation to use to answer the question. State the operation. Why do you think the key word or phrase indicates the operation you chose?**
- **Write an expression you can use to answer the question.**
- **Find the value of your expression.**

a. What is the total amount of trash collected from 2010 to 2013?

b. How many more pounds of recyclables were collected in 2013 than in 2010?

c. How many times more recyclables were collected in 2012 than in 2010?

d. The amount of trash collected in 2014 is estimated to be twice the amount collected in 2011. What is that amount?

COMMON CORE

Whole Numbers

In this lesson, you will

- determine which operation to perform.
- divide multi-digit numbers.

Learning Standard
6.NS.2

2 ACTIVITY: Checking Answers

Math Practice 6

Communicate Precisely

What key words should you use so that your partner understands your explanation?

Work with a partner.

a. Explain how you can use estimation to check the reasonableness of the value of your expression in Activity 1(a).

b. Explain how you can use addition to check the value of your expression in Activity 1(b).

c. Explain how you can use estimation to check the reasonableness of the value of your expression in Activity 1(c).

d. Use mental math to check the value of your expression in Activity 1(d). Describe your strategy.

3 ACTIVITY: Using Estimation

Work with a partner. Use the map. Explain how you found each answer.

a. Which two lakes have a combined area of about 33,000 square miles?

b. Which lake covers an area about three times greater than the area of Lake Erie?

c. Which lake covers an area that is about 16,000 square miles greater than the area of Lake Ontario?

d. Estimate the total area covered by the Great Lakes.

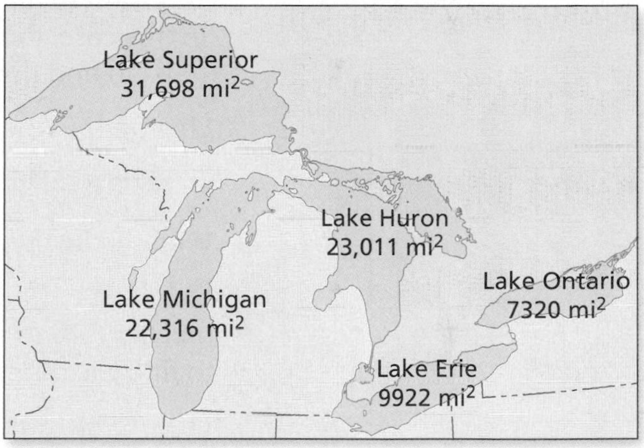

What Is Your Answer?

4. IN YOUR OWN WORDS How do you know which operation to choose when solving a real-life problem?

5. In a *magic square*, the sum of the numbers in each row, column, and diagonal is the same and each number from 1 to 9 is used only once. Complete the magic square. Explain how you found the missing numbers.

?	9	2
?	5	?
8	?	?

Practice Use what you learned about choosing operations to complete Exercises 8–11 on page 7.

1.1 Lesson

Check It Out
Lesson Tutorials
BigIdeasMath.com

Recall the four basic operations: addition, subtraction, multiplication, and division.

Operation	Words	Algebra
Addition	the *sum* of	$a + b$
Subtraction	the *difference* of	$a - b$
Multiplication	the *product* of	$a \times b \quad a \cdot b$
Division	the *quotient* of	$a \div b \quad \frac{a}{b} \quad b\overline{)a}$

EXAMPLE 1 Adding and Subtracting Whole Numbers

The bar graph shows the attendance at a three-day art festival.

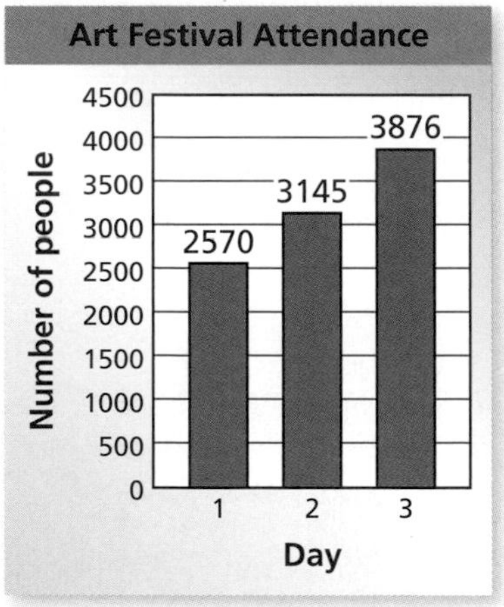

a. What is the total attendance for the art festival?

You want to find the total attendance for the three days. In this case, the phrase *total attendance* indicates you need to find the sum of the daily attendances.

Line up the numbers by their place values, then add.

$$\begin{array}{r} {}^{111} \\ 2570 \\ 3145 \\ +\ 3876 \\ \hline 9591 \end{array}$$

∴ The total attendance is 9591 people.

b. What is the increase in attendance from Day 1 to Day 2?

You want to find how many more people attended on Day 2 than on Day 1. In this case, the phrase *how many more* indicates you need to find the difference of the attendances on Day 2 and Day 1.

Line up the numbers by their place values, then subtract.

$$\begin{array}{r} {}^{\;10} \\ {}^{2\,\not{0}\,14} \\ \not{3}\not{1}\not{4}5 \\ -\ 2570 \\ \hline 575 \end{array}$$

∴ The increase in attendance from Day 1 to Day 2 is 575 people.

EXAMPLE 2 Multiplying Whole Numbers

A school lunch contains 12 chicken nuggets. Ninety-five students buy the lunch. What is the total number of chicken nuggets served?

You want to find the total number of chicken nuggets in 95 groups of 12 chicken nuggets. The phrase *95 groups of 12* indicates you need to find the product of 95 and 12.

$$\begin{array}{r} 12 \\ \times\ 95 \\ \hline 60 \\ 108 \\ \hline 1140 \end{array}$$

60 — Multiply 12 by the ones digit, 5.
108 — Multiply 12 by the tens digit, 9.
1140 — Add.

∴ There were 1140 chicken nuggets served.

Study Tip

In Example 2, you can use estimation to check the reasonableness of your answer.

$12 \times 95 \approx 12 \times 100$
$= 1200$

Because $1200 \approx 1140$, the answer is reasonable.

On Your Own

Now You're Ready Exercises 12–20

Find the value of the expression. Use estimation to check your answer.

1. $1745 + 682$
2. $912 - 799$
3. 42×118

EXAMPLE 3 Dividing Whole Numbers: No Remainder

You make 24 equal payments for a go-kart. You pay a total of \$840. How much is each payment?

You want to find the number of groups of 24 in \$840. The phrase *groups of 24 in \$840* indicates you need to find the quotient of 840 and 24.

Use long division to find the quotient.
Decide where to write the first digit of the quotient.

$$\begin{array}{r} ? \\ 24\overline{)840} \end{array}$$

Do not use the hundreds place because 24 is greater than 8.

$$\begin{array}{r} ? \\ 24\overline{)840} \end{array}$$

Use the tens place because 24 is less than 84.

So, divide the tens and write the first digit of the quotient in the tens place.

$$\begin{array}{r} 3 \\ 24\overline{)840} \\ \underline{-\,72} \\ 12 \end{array}$$

Divide 84 by 24: There are three groups of 24 in 84.
Multiply 3 and 24.
Subtract 72 from 84.

Next, bring down the 0 and divide the ones.

$$\begin{array}{r} 35 \\ 24\overline{)840} \\ \underline{-\,72}\downarrow \\ 120 \\ \underline{-\,120} \\ 0 \end{array}$$

Divide 120 by 24: There are five groups of 24 in 120.
Multiply 5 and 24.
Subtract 120 from 120.

Check Find the product of the quotient and the divisor.

$$\begin{array}{rl} 35 & \text{quotient} \\ \times\ 24 & \text{divisor} \\ \hline 140 & \\ 70 & \\ \hline 840 & \text{dividend } \checkmark \end{array}$$

The quotient of 840 and 24 is 35.

∴ So, each payment is \$35.

Remember

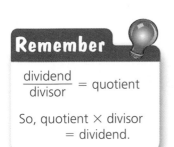

$\dfrac{\text{dividend}}{\text{divisor}} = \text{quotient}$

So, quotient × divisor = dividend.

On Your Own

Now You're Ready
Exercises 21–23

Find the value of the expression. Use estimation to check your answer.

4. $234 \div 9$ **5.** $\frac{986}{58}$ **6.** $840 \div 105$

7. Find the quotient of 9920 and 320.

When you use long division to divide whole numbers and you obtain a remainder, you can write the quotient as a mixed number using the rule

$$\text{dividend} \div \text{divisor} = \text{quotient} + \frac{\text{remainder}}{\text{divisor}}.$$

EXAMPLE 4 Real-Life Application

A 301-foot-high swing at an amusement park can take 64 people on each ride. A total of 8983 people ride the swing today. All the rides are full except for the last ride. How many rides are given? How many people are on the last ride?

To find the number of rides given, you need to find the number of groups of 64 people in 8983 people. The phrase *groups of 64 people in 8983 people* indicates you need to find the quotient of 8983 and 64.

Divide the place-value positions from left to right.

$$\begin{array}{r} 140 \text{ R}23 \\ 64\overline{)8983} \\ -\,64 \\ \hline 258 \\ -\,256 \\ \hline 23 \\ -\,0 \\ \hline 23 \end{array}$$

There is one group of 64 in 89.

There are four groups of 64 in 258.

There are no groups of 64 in 23.

The remainder is 23.

Do not stop here. You must write a 0 in the ones place of the quotient.

The quotient is $140\frac{23}{64}$. This indicates 140 groups of 64, with 23 remaining.

∴ So, 141 rides are given, with 23 people on the last ride.

On Your Own

Now You're Ready
Exercises 24–26

Find the value of the expression. Use estimation to check your answer.

8. $\frac{6096}{30}$ **9.** $45{,}691 \div 28$ **10.** $3215 \div 430$

11. WHAT IF? In Example 4, 9038 people ride the swing. What is the least number of rides possible?

1.1 Exercises

Check It Out
Help with Homework
BigIdeasMath.com

Vocabulary and Concept Check

VOCABULARY **Determine which operation the word or phrase represents.**

1. sum
2. times
3. the quotient of
4. decreased by
5. total of
6. minus
7. **VOCABULARY** Use the division problem shown to tell whether the number is the divisor, dividend, or quotient.

   a. 884 b. 26 c. 34

$$\begin{array}{r} 26 \\ 34\overline{)884} \end{array}$$

Practice and Problem Solving

The bar graph shows the attendance at a food festival. Write an expression you can use to answer the question. Then find the value of your expression.

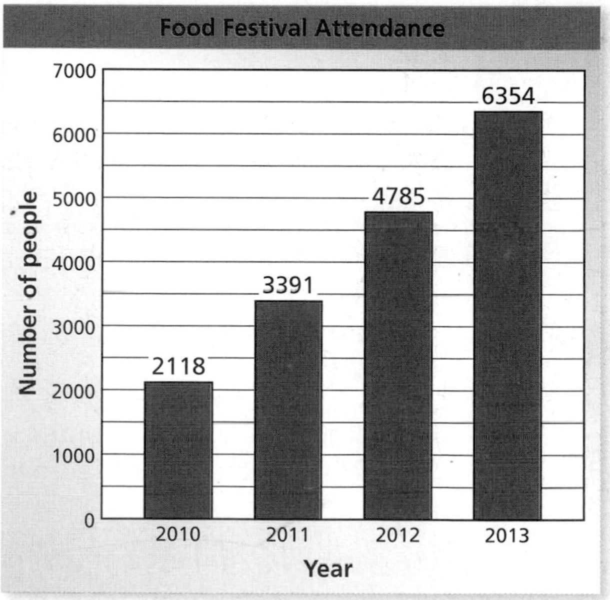

8. What is the total attendance at the food festival from 2010 to 2013?
9. How many more people attended the food festival in 2012 than in 2011?
10. How many times more people attended the food festival in 2013 than in 2010?
11. The festival projects that the total attendance for 2014 will be twice the attendance in 2012. What is the projected attendance for 2014?

Find the value of the expression. Use estimation to check your answer.

(1) 12. $2219 + 872$

13. $\begin{array}{r} 5351 \\ +\ 1730 \\ \hline \end{array}$

14. $3968 + 1879$

15. $7694 - 5232$

16. $9165 - 4729$

17. $\begin{array}{r} 2416 \\ -\ 1983 \\ \hline \end{array}$

(2) 18. $\begin{array}{r} 84 \\ \times\ 37 \\ \hline \end{array}$

19. 124×56

20. 419×236

(3) 21. $837 \div 27$

22. $\dfrac{588}{84}$

23. $7440 \div 124$

(4) 24. $6409 \div 61$

25. $8241 \div 173$

26. $\dfrac{33{,}505}{160}$

ERROR ANALYSIS **Describe and correct the error in finding the value of the expression.**

27. 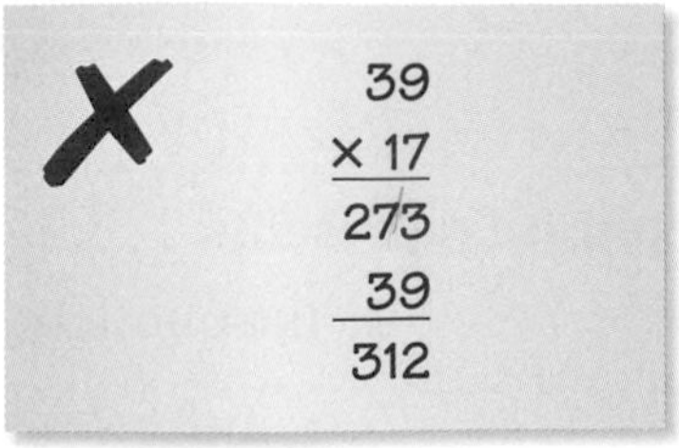

28. 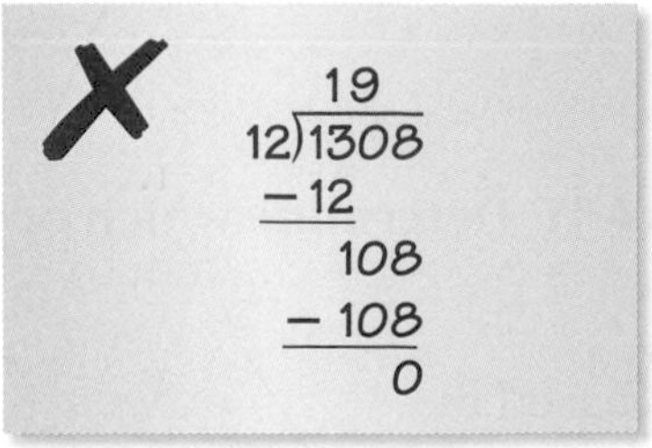

Determine the operation you would use to solve the problem. Do not answer the question.

29. Gymnastic lessons cost $30 per week. How much will 18 weeks of gymnastic lessons cost?

30. The scores on your first two tests were 82 and 93. By how many points did your score improve?

31. You are setting up tables for a banquet for 150 guests. Each table seats 12 people. What is the minimum number of tables you will need?

32. A store has 15 boxes of peaches. Each box contains 45 peaches. How many peaches does the store have?

33. Two shirts cost $18 and $25. What is the total cost of the shirts?

34. A gardener works for 14 hours during a week and charges $168. How much does the gardener charge for each hour?

Find the perimeter and area of the rectangle.

35. 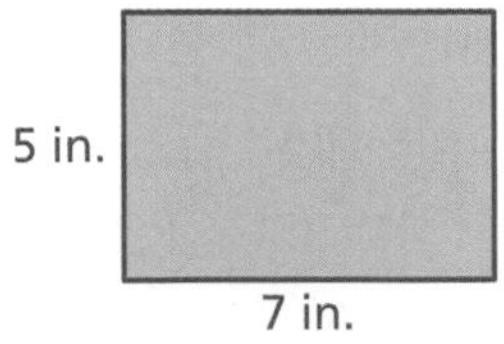

36. 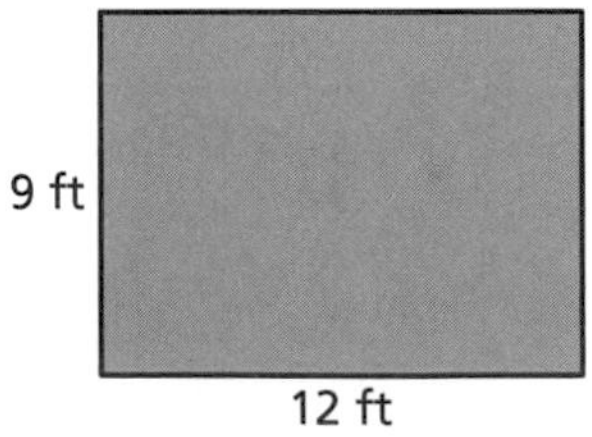

37. 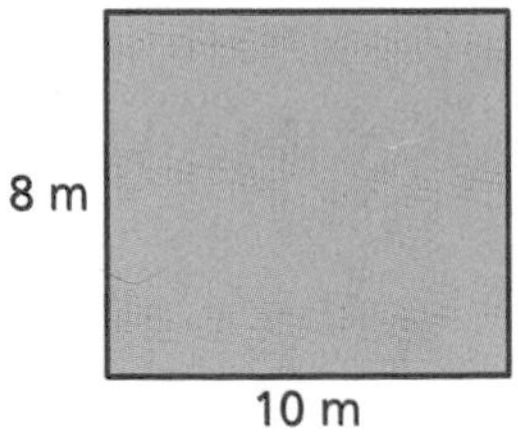

38. **BOX OFFICE** The number of tickets sold for the opening weekend of a movie is 879,575. The movie was shown in 755 theaters across the nation. What was the average number of tickets sold at each theater?

39. **LOGIC** You find that the product of 93 and 6 is 558. How can you use addition to check your answer? How can you use division to check your answer?

40. **NUMBER SENSE** Without calculating, decide which is greater: $3999 \div 129$ or $3834 \div 142$. Explain.

41. REASONING In a division problem, can the remainder be greater than the divisor? Explain.

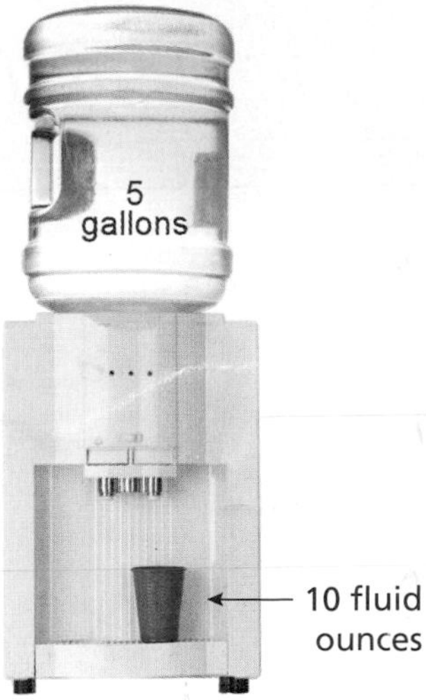

42. WATER COOLER You change the water jug on the water cooler. How many cups can be completely filled before you need to change the water jug again?

43. ARCADE You have \$9, one of your friends has \$10, and two of your other friends each have \$13. You combine your money to buy arcade tokens. You use a coupon to buy 8 tokens for \$1. The cost of the remaining tokens is four for \$1. You and your friends share the tokens evenly. How many tokens does each person get?

44. BOOK SALE You borrow bookcases like the one shown to display 943 books at a book sale. You plan to put 22 books on each shelf. No books will be on top of the bookcases.

a. How many bookcases must you borrow to display all the books?

b. You fill the shelves of each bookcase in order, starting with the top shelf. How many books are on the third shelf of the last bookcase?

45. MODELING The siding of a house is 2250 square feet. The siding needs two coats of paint. The table shows information about the paint.

Can Size	Cost	Coverage
1 quart	\$18	80 square feet
1 gallon	\$29	320 square feet

a. What is the minimum cost of the paint needed to complete the job?

b. How much paint is left over?

46. Critical Thinking Use the digits 3, 4, 6, and 9 to complete the division problem. Use each digit once.

$$\square\square,000 \div \square 00 = \square 0$$

Fair Game Review What you learned in previous grades & lessons

Plot the ordered pair in a coordinate plane. *(Skills Review Handbook)*

47. (1, 3) **48.** (0, 4) **49.** (6, 0) **50.** (4, 2)

51. MULTIPLE CHOICE Which of the following numbers is *not* prime? *(Skills Review Handbook)*

Ⓐ 1 Ⓑ 2 Ⓒ 3 Ⓓ 5

1.2 Powers and Exponents

Essential Question How can you use repeated factors in real-life situations?

As I was going to St. Ives
I met a man with seven wives
Each wife had seven sacks
Each sack had seven cats
Each cat had seven kits
Kits, cats, sacks, wives
How many were going to St. Ives? Nursery Rhyme, 1730

1 ACTIVITY: Analyzing a Math Poem

Work with a partner. Here is a "St. Ives" poem written by two students. Answer the question in the poem.

As I was walking into town
I met a ringmaster with five clowns
Each clown had five magicians
Each magician had five bunnies
Each bunny had five fleas
Fleas, bunnies, magicians, clowns
How many were going into town?

Number of clowns:	5	$=$	
Number of magicians:	5×5	$=$	
Number of bunnies:	$5 \times 5 \times 5$	$=$	
Number of fleas:	$5 \times 5 \times 5 \times 5$	$=$	

So, the number of fleas, bunnies, magicians, and clowns is ______.
Explain how you found your answer.

COMMON CORE
Numerical Expressions
In this lesson, you will
- write expressions as powers.
- find values of powers.

Preparing for Standard 6.EE.1

2 ACTIVITY: Writing Repeated Factors

Math Practice 8

Repeat Calculations

What patterns do you notice with each problem? How does this help you write exponents?

Work with a partner. Copy and complete the table.

Repeated Factors	Using an Exponent	Value
a. 4×4		
b. 6×6		
c. $10 \times 10 \times 10$		
d. $100 \times 100 \times 100$		
e. $3 \times 3 \times 3 \times 3$		
f. $4 \times 4 \times 4 \times 4 \times 4$		
g. $2 \times 2 \times 2 \times 2 \times 2 \times 2$		

h. In your own words, describe what the two numbers in the expression 3^5 mean.

3 ACTIVITY: Writing and Analyzing a Math Poem

Work with a partner.

a. Write your own "St. Ives" poem.

b. Draw pictures for your poem.

c. Answer the question in your poem.

d. Show how you can use exponents to write your answer.

What Is Your Answer?

4. IN YOUR OWN WORDS How can you use repeated factors in real-life situations? Give an example.

5. STRUCTURE Use exponents to complete the table. Describe the pattern.

10	100	1000	10,000	100,000	1,000,000
10^1	10^2				

Practice

Use what you learned about exponents to complete Exercises 4–6 on page 14.

1.2 Lesson

Check It Out
Lesson Tutorials
BigIdeasMath.com

Key Vocabulary
power, *p. 12*
base, *p. 12*
exponent, *p. 12*
perfect square, *p. 13*

A **power** is a product of repeated factors. The **base** of a power is the repeated factor. The **exponent** of a power indicates the number of times the base is used as a factor.

base → 3; exponent → 4

$$\underbrace{3^4}_{\text{power}} = \underbrace{3 \cdot 3 \cdot 3 \cdot 3}_{\text{3 is used as a factor 4 times.}}$$

Power	Words
3^2	Three *squared*, or three to the second
3^3	Three *cubed*, or three to the third
3^4	Three to the fourth

EXAMPLE 1 Writing Expressions as Powers

Math Practice 5
Choose Tools
Why are calculators more efficient when finding the values of expressions involving exponents?

Write each product as a power.

a. $4 \cdot 4 \cdot 4 \cdot 4 \cdot 4$

Because 4 is used as a factor 5 times, its exponent is 5.

So, $4 \cdot 4 \cdot 4 \cdot 4 \cdot 4 = 4^5$.

b. $12 \times 12 \times 12$

Because 12 is used as a factor 3 times, its exponent is 3.

So, $12 \times 12 \times 12 = 12^3$.

On Your Own

Now You're Ready
Exercises 4–12

Write the product as a power.

1. $6 \cdot 6 \cdot 6 \cdot 6 \cdot 6 \cdot 6$

2. $15 \times 15 \times 15 \times 15$

EXAMPLE 2 Finding Values of Powers

Find the value of each power.

a. 7^2

b. 5^3

a.		b.
$7^2 = 7 \cdot 7$	Write as repeated multiplication.	$5^3 = 5 \cdot 5 \cdot 5$
$= 49$	Simplify.	$= 125$

The square of a whole number is a **perfect square**.

EXAMPLE 3 Identifying Perfect Squares

Determine whether each number is a perfect square.

a. 64

Because $8^2 = 64$, 64 is a perfect square.

b. 20

No whole number squared equals 20. So, 20 is not a perfect square.

On Your Own

Now You're Ready
Exercises 14–21 and 25–32

Find the value of the power.

3. 6^3 **4.** 9^2 **5.** 3^4 **6.** 18^2

Determine whether the number is a perfect square.

7. 25 **8.** 2 **9.** 99 **10.** 100

Remember

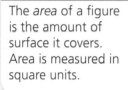

The *area* of a figure is the amount of surface it covers. Area is measured in square units.

The area of a square is equal to its side length squared.

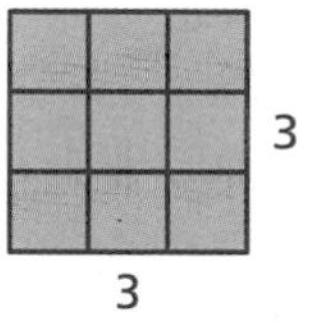

Area $= 3^2 = 9$ square units

EXAMPLE 4 Real-Life Application

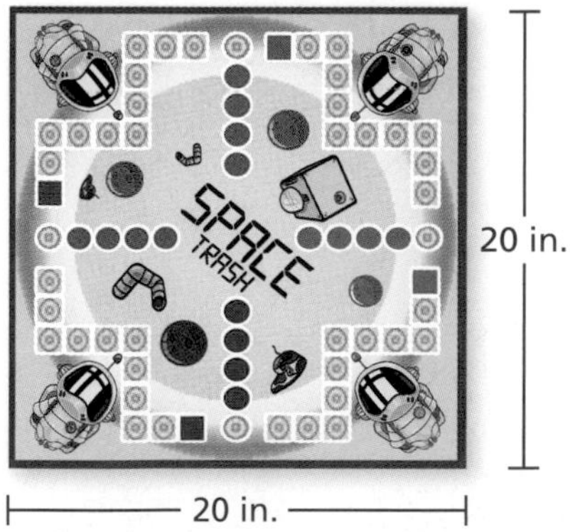

A game board is a square with a side length of 20 inches. What is the area of the game board?

Use a verbal model to solve the problem.

area of game board	=	$(\text{side length})^2$	
	=	20^2	Substitute 20 for side length.
	=	400	Multiply.

The area of the game board is 400 square inches.

On Your Own

11. What is the area of the square traffic sign in square inches? in square feet?

1.2 Exercises

Check It Out
Help with Homework
BigIdeasMath.com

Vocabulary and Concept Check

1. **VOCABULARY** How are exponents and powers different?
2. **VOCABULARY** Is 10 a perfect square? Is 100 a perfect square? Explain.
3. **WHICH ONE DOESN'T BELONG?** Which one does *not* belong with the other three? Explain your reasoning.

$2^4 = 2 \times 2 \times 2 \times 2$	$3 + 3 + 3 + 3 = 3(4)$	$3^2 = 3 \times 3$	$5 \cdot 5 \cdot 5 = 5^3$

Practice and Problem Solving

Write the product as a power.

1 4. 9×9 5. 13×13 6. $15 \times 15 \times 15$

7. $2 \cdot 2 \cdot 2 \cdot 2 \cdot 2$ 8. $14 \times 14 \times 14$ 9. $8 \cdot 8 \cdot 8 \cdot 8$

10. $11 \times 11 \times 11 \times 11 \times 11$ 11. $7 \cdot 7 \cdot 7 \cdot 7 \cdot 7 \cdot 7$ 12. $16 \cdot 16 \cdot 16 \cdot 16$

13. **ERROR ANALYSIS** Describe and correct the error in writing the product as a power.

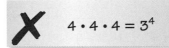

✗ $4 \cdot 4 \cdot 4 = 3^4$

Find the value of the power.

2 14. 5^2 15. 4^3 16. 2^5 17. 14^2

Use a calculator to find the value of the power.

18. 7^6 19. 4^8 20. 12^4 21. 17^5

22. **ERROR ANALYSIS** Describe and correct the error in finding the value of the power.

✗ $8^3 = 8 \cdot 3 = 24$

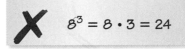

23. **POPULATION** The population of Virginia is about 8×10^6. About how many people live in Virginia?

24. **FIGURINES** The smallest figurine in a gift shop is 2 inches tall. The height of each figurine is twice the height of the previous figurine. Write a power to represent the height of the tallest figurine. Then find the height.

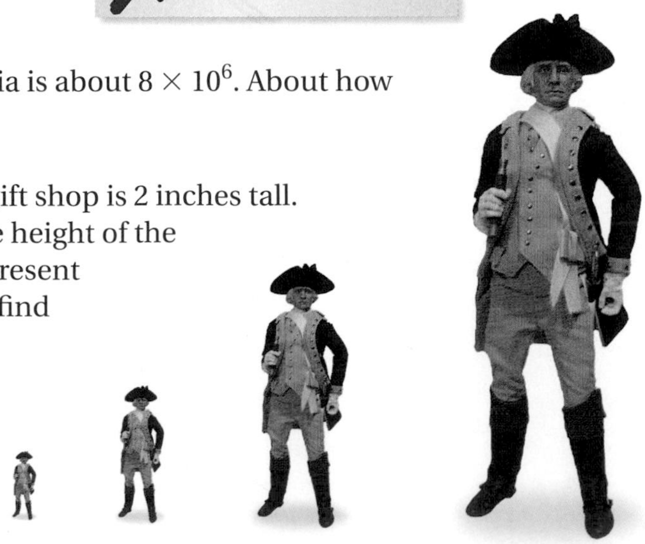

Determine whether the number is a perfect square.

3 **25.** 8 **26.** 4 **27.** 81 **28.** 44

29. 49 **30.** 125 **31.** 150 **32.** 144

33. PAINTING A square painting measures 2 meters on each side. What is the area of the painting in square centimeters?

34. NUMBER SENSE Write three powers that have values greater than 120 and less than 130.

35. CHECKERS A checkers board has 64 squares. How many squares are in each row?

36. PATIO A landscaper has 125 tiles to build a square patio. The patio must have an area of at least 80 square feet.

a. What are the possible arrangements for the patio?

b. How many tiles are not used in each arrangement?

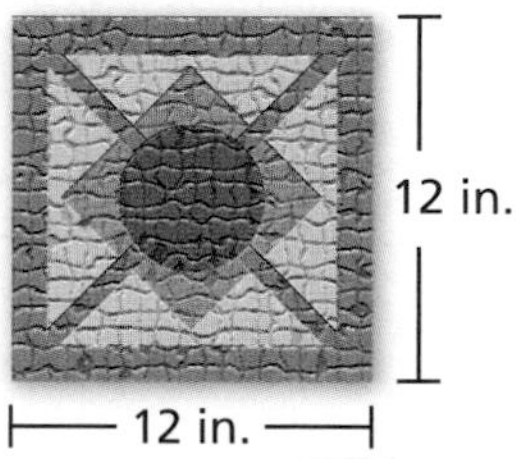

37. PATTERNS Copy and complete the table. Describe what happens to the value of the power as the exponent decreases. Use this pattern to find the value of 4^0.

Power	4^6	4^5	4^4	4^3	4^2	4^1
Value	4096	1024				

38. REASONING Consider the equation $56 = \blacksquare^2$. The missing number is between what two whole numbers?

39. Repeated Reasoning How many blocks do you need to add to Square 6 to get Square 7? to Square 9 to get Square 10? to Square 19 to get Square 20? Explain.

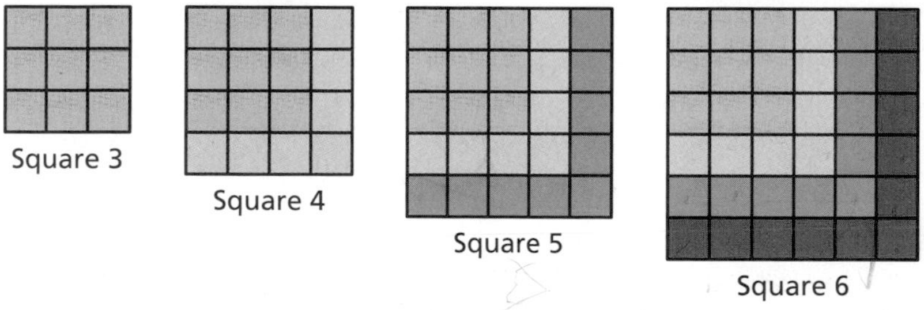

Fair Game Review What you learned in previous grades & lessons

Find the value of the expression. *(Skills Review Handbook)*

40. 6×14 **41.** 11×15 **42.** $56 \div 7$ **43.** $112 \div 16$

44. MULTIPLE CHOICE You buy a box of sugar-free gum that has 12 packs. Each pack has 5 pieces. Which expression represents the total number of pieces of gum? *(Skills Review Handbook)*

Ⓐ $12 + 5$ Ⓑ $12 - 5$ Ⓒ 12×5 Ⓓ $12 \div 5$

1.3 Order of Operations

Essential Question What is the effect of inserting parentheses into a numerical expression?

1 ACTIVITY: Comparing Different Orders

Work with a partner. Find the value of the expression by using different orders of operations. Are your answers the same? *(Circle yes or no.)*

a.	Add, then multiply. $3 + 4 \times 2 =$	Multiply, then add. $3 + 4 \times 2 =$	Same? Yes No
b.	Add, then subtract. $5 + 3 - 1 =$	Subtract, then add. $5 + 3 - 1 =$	Same? Yes No
c.	Divide, then multiply. $12 \div 3 \cdot 2 =$	Multiply, then divide. $12 \div 3 \cdot 2 =$	Same? Yes No
d.	Divide, then add. $16 \div 4 + 4 =$	Add, then divide. $16 \div 4 + 4 =$	Same? Yes No
e.	Multiply, then subtract. $8 \times 4 - 2 =$	Subtract, then multiply. $8 \times 4 - 2 =$	Same? Yes No
f.	Multiply, then divide. $8 \cdot 4 \div 2 =$	Divide, then multiply. $8 \cdot 4 \div 2 =$	Same? Yes No
g.	Subtract, then add. $13 - 4 + 6 =$	Add, then subtract. $13 - 4 + 6 =$	Same? Yes No
h.	Multiply, then add. $1 \times 2 + 3 =$	Add, then multiply. $1 \times 2 + 3 =$	Same? Yes No

COMMON CORE

Numerical Expressions

In this lesson, you will
- evaluate numerical expressions with whole-number exponents.

Learning Standard
6.EE.1

2 ACTIVITY: Using Parentheses

Work with a partner. Use all the symbols and numbers to write an expression that has the given value.

	Symbols and Numbers	*Value*	*Expression*
a.	( ), +, ÷, 3, 4, 5	3	
b.	( ), −, ×, 2, 5, 8	11	
c.	( ), ×, ÷, 4, 4, 16	16	
d.	( ), −, ÷, 3, 8, 11	1	
e.	( ), +, ×, 2, 5, 10	70	

3 ACTIVITY: Reviewing Fractions and Decimals

Math Practice 2

Use Operations

How do you know which operation to perform first?

Work with a partner. Evaluate the expression.

a. $\frac{3}{4} - \left(\frac{1}{4} + \frac{1}{2}\right)$ =

b. $\left(\frac{5}{6} - \frac{1}{6}\right) - \frac{1}{12}$ =

c. $7.4 - (3.5 - 3.1)$ =

d. $10.4 - (8.6 + 0.9)$ =

e. (\$7.23 + \$2.32) − \$5.40 =

f. \$124.60 − (\$72.41 + \$5.67) =

What Is Your Answer?

4. In an expression with two or more operations, why is it necessary to agree on an order of operations? Give examples to support your explanation.

5. **IN YOUR OWN WORDS** What is the effect of inserting parentheses into a numerical expression?

Practice Use what you learned about the order of operations to complete Exercises 3–5 on page 20.

1.3 Lesson

Check It Out
Lesson Tutorials
BigIdeasMath.com

Key Vocabulary
numerical expression, *p. 18*
evaluate, *p. 18*
order of operations, *p. 18*

A **numerical expression** is an expression that contains only numbers and operations. To **evaluate**, or find the value of, a numerical expression, use a set of rules called the **order of operations**.

Key Idea

Order of Operations

1. Perform operations in **P**arentheses.
2. Evaluate numbers with **E**xponents.
3. **M**ultiply or **D**ivide from left to right.
4. **A**dd or **S**ubtract from left to right.

EXAMPLE 1 Using Order of Operations

a. Evaluate $12 - 2 \times 4$.

$12 - 2 \times 4 = 12 - 8$ — Multiply 2 and 4.

$= 4$ — Subtract 8 from 12.

b. Evaluate $7 + 60 \div (3 \times 5)$.

$7 + 60 \div (3 \times 5) = 7 + 60 \div 15$ — Perform operation in parentheses.

$= 7 + 4$ — Divide 60 by 15.

$= 11$ — Add 7 and 4.

EXAMPLE 2 Using Order of Operations with Exponents

Study Tip
Remember to multiply and divide from left to right. In Example 2, you should divide before multiplying because the division symbol comes first when reading from left to right.

Evaluate $30 \div (7 + 2^3) \times 6$.

Evaluate the power in parentheses first.

$30 \div (7 + 2^3) \times 6 = 30 \div (7 + 8) \times 6$ — Evaluate 2^3.

$= 30 \div 15 \times 6$ — Perform operation in parentheses.

$= 2 \times 6$ — Divide 30 by 15.

$= 12$ — Multiply 2 and 6.

On Your Own

Now You're Ready
Exercises 6–14

Evaluate the expression.

1. $7 \cdot 5 + 3$
2. $(28 - 20) \div 4$
3. $6 \times 15 - 10 \div 2$
4. $6 + 2^4 - 1$
5. $4 \cdot 3^2 + 18 - 9$
6. $16 + (5^2 - 7) \div 3$

The symbols $\times$ and $\bullet$ are used to indicate multiplication. You can also use parentheses to indicate multiplication. For example, $3(2 + 7)$ is the same as $3 \times (2 + 7)$.

EXAMPLE 3 Using Order of Operations

a. Evaluate $9 + 7(5 - 2)$.

$9 + 7(5 - 2) = 9 + 7(3)$	Perform operation in parentheses.
$= 9 + 21$	Multiply 7 and 3.
$= 30$	Add 9 and 21.

b. Evaluate $15 - 4(6 + 1) \div 2^2$.

$15 - 4(6 + 1) \div 2^2 = 15 - 4(7) \div 2^2$	Perform operation in parentheses.
$= 15 - 4(7) \div 4$	Evaluate 2^2.
$= 15 - 28 \div 4$	Multiply 4 and 7.
$= 15 - 7$	Divide 28 by 4.
$= 8$	Subtract 7 from 15.

EXAMPLE 4 Real-Life Application

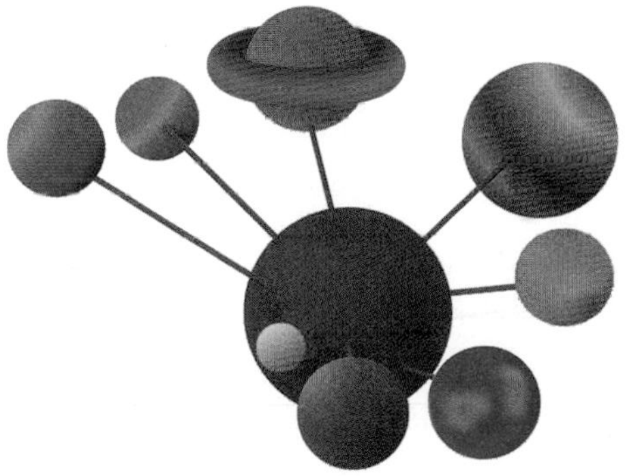

You buy foam spheres, paint bottles, and wooden rods to construct a model of our solar system. What is your total cost?

Item	Quantity	Cost per Item
Spheres	9	\$2
Paint	6	\$3
Rods	8	\$1

Use a verbal model to solve the problem.

cost of 9 spheres + cost of 6 paint bottles + cost of 8 rods

$9 \bullet 2 \quad + \quad 6 \bullet 3 \quad + \quad 8 \bullet 1$

$9 \bullet 2 + 6 \bullet 3 + 8 \bullet 1 = 18 + 18 + 8$	Multiply.
$= 44$	Add.

∴ Your total cost is \$44.

On Your Own

Now You're Ready
Exercises 18–23

Evaluate the expression.

7. $50 + 6(12 \div 4) - 8^2$ **8.** $5^2 - 5(10 - 5)$ **9.** $\dfrac{8(3 + 4)}{7}$

10. WHAT IF? In Example 4, you add the dwarf planet Pluto to your model. Use a verbal model to find your total cost assuming you do not need more paint. Explain.

1.3 Exercises

Check It Out
Help with Homework
BigIdeasMath.com

Vocabulary and Concept Check

1. **WRITING** Why does $12 - 8 \div 2 = 8$, but $(12 - 8) \div 2 = 2$?
2. **REASONING** Describe the steps in evaluating the expression $8 \div (6 - 4) + 3^2$.

Practice and Problem Solving

Find the value of the expression.

3. $(4 \times 15) - 3$
4. $10 - (7 + 1)$
5. $18 \div (6 + 3)$

Evaluate the expression.

1 2

6. $5 + 18 \div 6$
7. $(11 - 3) \div 2 + 1$
8. $45 \div 9 \times 12$
9. $6^2 - 3 \cdot 4$
10. $42 \div (15 - 2^3)$
11. $4^2 \cdot 2 + 8 \cdot 7$
12. $3^2 + 12 \div (6 - 3) \times 8$
13. $(10 + 4) \div (26 - 19)$
14. $(5^2 - 4) \cdot 2 - 18$

ERROR ANALYSIS **Describe and correct the error in evaluating the expression.**

15. ✗ $9 + 2 \times 3 = 11 \times 3$
$= 33$

16. ✗ $19 - 6 + 12 = 19 - 18$
$= 1$

17. **POETRY** You need to read 20 poems in 5 days for an English project. Each poem is 2 pages long. Evaluate the expression $20 \times 2 \div 5$ to find how many pages you need to read each day.

Evaluate the expression.

3

18. $9^2 - 8(6 + 2)$
19. $(3 - 1)^3 + 7(6) - 5^2$
20. $8\left(1\frac{1}{6} + \frac{5}{6}\right) \div 4$
21. $7^2 - 2\left(\frac{11}{8} - \frac{3}{8}\right)$
22. $8(7.3 + 3.7) - 14 \div 2$
23. $2^4(5.2 - 3.2) \div 4$

24. **MONEY** You have four \$10 bills and eighteen \$5 bills in your piggy bank. How much money do you have?

25. **THEATER** Before a show, there are 8 people in a theater. Five groups of 4 people enter, and then three groups of 2 people leave. Evaluate the expression $8 + 5(4) - 3(2)$ to find how many people are in the theater.

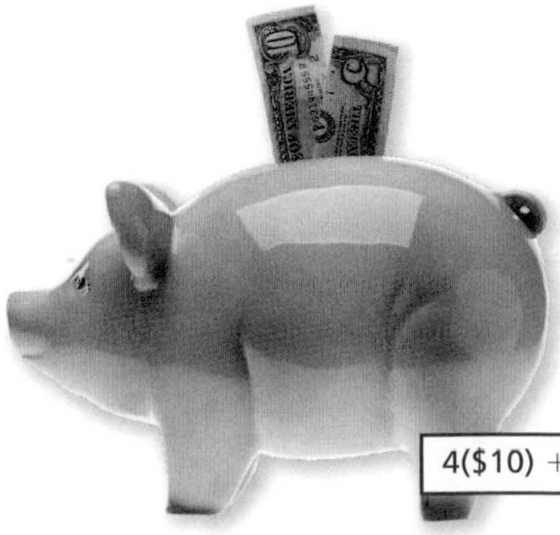

4(\$10) + 18(\$5)

Evaluate the expression.

26. $\frac{6(3+5)}{4}$ **27.** $\frac{12^2 - 4(6) + 1}{11^2}$ **28.** $\frac{26 \div 2 + 5}{3^2 - 3}$

29. FIELD TRIP Eighty students are going on a field trip to a history museum. The total cost includes

- 2 bus rentals and
- $10 per student for lunch.

What is the total cost per student?

30. OPEN-ENDED Use all four operations without parentheses to write an expression that has a value of 100.

31. SHOPPING You buy 6 notebooks, 10 folders, 1 pack of pencils, and 1 lunch box for school. After using a $10 gift card, how much do you owe? Explain how you solved the problem.

32. LITTER CLEANUP Two groups collect litter along the side of a road. It takes each group 5 minutes to clean up a 200-yard section. How long does it take to clean up 2 *miles*? Explain how you solved the problem.

33. Number Sense Copy each statement. Insert +, −, ×, or ÷ symbols to make each statement true.

a. 27 ▢ 3 ▢ 5 ▢ 2 = 19

b. 9^2 ▢ 11 ▢ 8 ▢ 4 ▢ 1 = 60

c. 5 ▢ 6 ▢ 15 ▢ 9 = 24

d. 14 ▢ 2 ▢ 7 ▢ 3 ▢ 9 = 10

Fair Game Review What you learned in previous grades & lessons

Add or subtract. *(Skills Review Handbook)*

34. 5.2 + 0.5 **35.** 8 − 1.9 **36.** 12.6 − 3 **37.** 0.7 + 0.2

38. MULTIPLE CHOICE You are making two recipes. One recipe calls for $2\frac{1}{3}$ cups of flour. The other recipe calls for $1\frac{1}{4}$ cups of flour. How much flour do you need to make both recipes? *(Skills Review Handbook)*

Ⓐ $1\frac{1}{12}$ cups Ⓑ $3\frac{1}{12}$ cups Ⓒ $3\frac{2}{7}$ cups Ⓓ $3\frac{7}{12}$ cups

1 Study Help

Check It Out
Graphic Organizer
BigIdeasMath.com

You can use an **information frame** to help you organize and remember concepts. Here is an example of an information frame for powers.

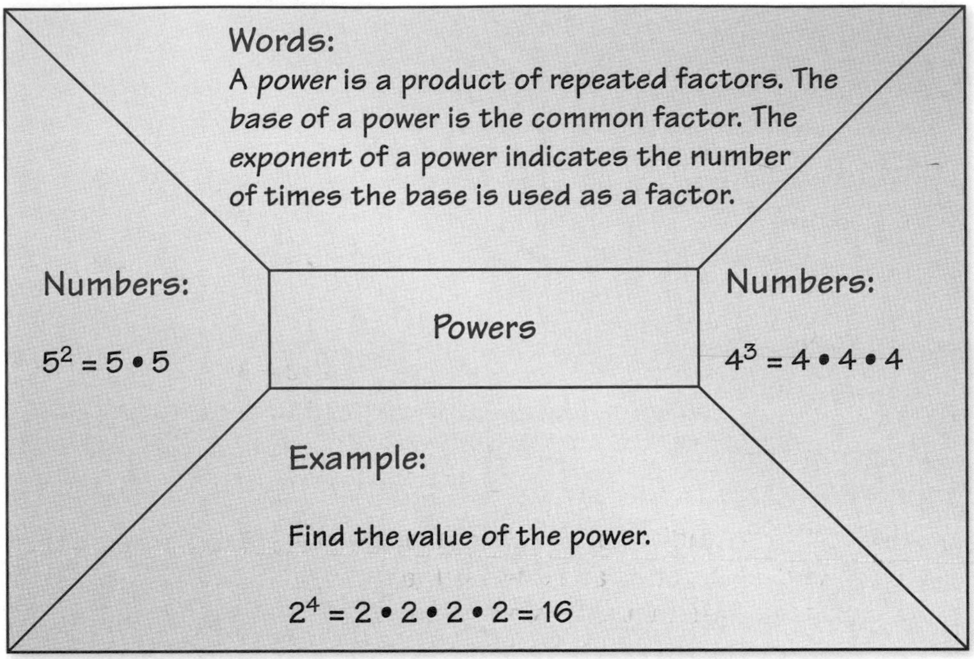

Words:
A *power* is a product of repeated factors. The *base* of a power is the common factor. The *exponent* of a power indicates the number of times the base is used as a factor.
Numbers:
$5^2 = 5 \bullet 5$
Powers
Numbers:
$4^3 = 4 \bullet 4 \bullet 4$
Example:
Find the value of the power.
$2^4 = 2 \bullet 2 \bullet 2 \bullet 2 = 16$

On Your Own

Make information frames to help you study these topics.

1. adding whole numbers
2. subtracting whole numbers
3. multiplying whole numbers
4. dividing whole numbers
5. order of operations

After you complete this chapter, make information frames for the following topics.

6. prime factorization
7. greatest common factor (GCF)
8. least common multiple (LCM)
9. least common denominator (LCD)

"Dear Mom, I am sending you an information frame card for Mother's Day!"

1.1–1.3 Quiz

Check It Out
Progress Check
BigIdeasMath.com

Find the value of the expression. Use estimation to check your answer. *(Section 1.1)*

1. $4265 + 3896$

2. $5327 - 2624$

3. 276×49

4. $648 \div 72$

Find the value of the power. *(Section 1.2)*

5. 3^3

6. 11^2

Determine whether the number is a perfect square. *(Section 1.2)*

7. 36

8. 15

Evaluate the expression. *(Section 1.3)*

9. $6 + 21 \div 7$

10. $\dfrac{4(12 - 3)}{12}$

11. $16 \div 2^3 + 6 - 2$

12. $2 \times 14 \div (3^2 - 2)$

13. AUDITORIUM An auditorium has a total of 592 seats. There are 37 rows of seats, and each row has the same number of seats. How many seats are there in a single row? *(Section 1.1)*

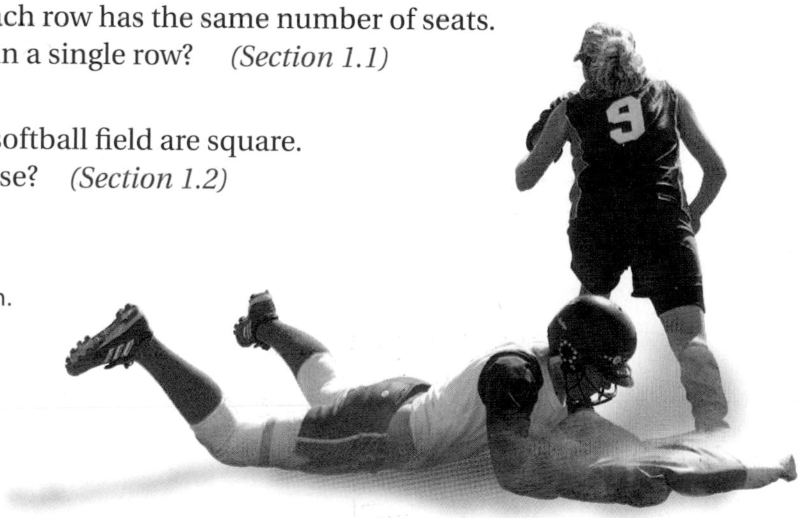

14. SOFTBALL The bases on a softball field are square. What is the area of each base? *(Section 1.2)*

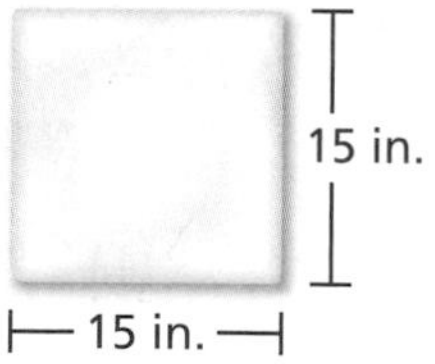

15. DUATHLON In an 18-mile duathlon, you run, then bike 12 miles, and then run again. The two runs are the same distance. Find the distance of each run. *(Section 1.3)*

16. AMUSEMENT PARK Tickets for an amusement park cost \$10 for adults and \$6 for children. Find the total cost for 2 adults and 3 children. *(Section 1.3)*

1.4 Prime Factorization

Essential Question Without dividing, how can you tell when a number is divisible by another number?

1 ACTIVITY: Finding Divisibility Rules for 2, 3, 5, and 10

Work with a partner. Copy the set of numbers (1–50) as shown.

1	2	3	4	5	6	7	8	9	10
11	12	13	14	15	16	17	18	19	20
21	22	23	24	25	26	27	28	29	30
31	32	33	34	35	36	37	38	39	40
41	42	43	44	45	46	47	48	49	50

a. Highlight all the numbers that are divisible by 2.

b. Put a box around the numbers that are divisible by 3.

c. Underline the numbers that are divisible by 5.

d. Circle the numbers that are divisible by 10.

e. **STRUCTURE** In parts (a)–(d), what patterns do you notice? Write four rules to determine when a number is divisible by 2, 3, 5, and 10.

COMMON CORE

Common Factors and Multiples

In this lesson, you will

- use divisibility rules to find prime factorizations of numbers.

Preparing for Standard 6.NS.4

2 ACTIVITY: Finding Divisibility Rules for 6 and 9

Work with a partner.

a. List ten numbers that are divisible by 6. Write a rule to determine when a number is divisible by 6. Use a calculator to check your rule with large numbers.

b. List ten numbers that are divisible by 9. Write a rule to determine when a number is divisible by 9. Use a calculator to check your rule with large numbers.

3 ACTIVITY: Rewriting a Number Using 2s, 3s, and 5s

Work with three other students. Use the following rules and only the prime factors 2, 3, and 5 to write each number below as a product.

- Your group should have four sets of cards: a set with all 2s, a set with all 3s, a set with all 5s, and a set of blank cards. Each person gets one set of cards.
- Begin by choosing two cards to represent the given number as a product of two factors. The person with the blank cards writes any factors that are not 2, 3, or 5.
- Use the cards again to represent any number written on a blank card as a product of two factors. Continue until you have represented each handwritten card as a product of two prime factors.
- You may use only one blank card for each step.

a. Sample: 108

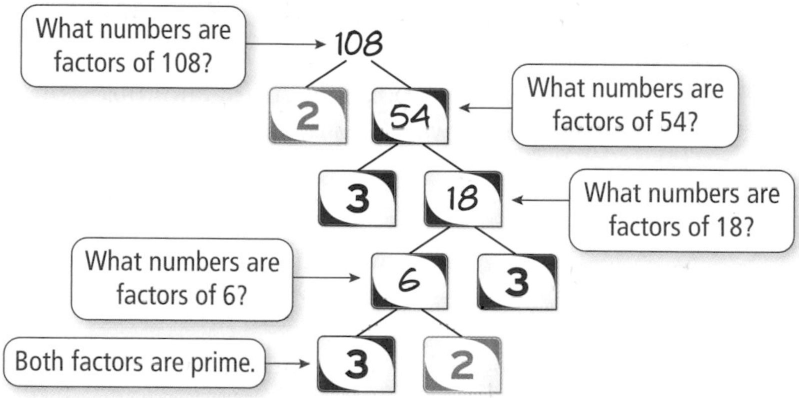

Math Practice 4

Interpret Results

How do you know your answer makes sense?

$108 = 2 \cdot 3 \cdot 3 \cdot 3 \cdot 2$

b. 80 **c.** 162 **d.** 300

e. Compare your results with those of other groups. Are your steps the same for each number? Is your final answer the same for each number?

What Is Your Answer?

4. IN YOUR OWN WORDS Without dividing, how can you tell when a number is divisible by another number? Give examples to support your explanation.

5. Explain how you can use your divisibility rules from Activities 1 and 2 to help with Activity 3.

Practice Use what you learned about divisibility rules to complete Exercises 4–7 on page 28.

1.4 Lesson

Check It Out
Lesson Tutorials
BigIdeasMath.com

Because 2 is factor of 10 and 2 • 5 = 10, 5 is also a factor of 10. The pair 2, 5 is called a **factor pair** of 10.

Key Vocabulary
factor pair, *p. 26*
prime factorization, *p. 26*
factor tree, *p. 26*

EXAMPLE 1 Finding Factor Pairs

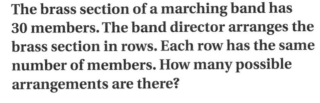

The brass section of a marching band has 30 members. The band director arranges the brass section in rows. Each row has the same number of members. How many possible arrangements are there?

Use the factor pairs of 30 to find the number of arrangements.

30 = 1 • 30	There could be 1 row of 30 or 30 rows of 1.
30 = 2 • 15	There could be 2 rows of 15 or 15 rows of 2.
30 = 3 • 10	There could be 3 rows of 10 or 10 rows of 3.
30 = 5 • 6	There could be 5 rows of 6 or 6 rows of 5.
30 = 6 • 5	The factors 5 and 6 are already listed.

∴ There are 8 possible arrangements: 1 row of 30, 30 rows of 1, 2 rows of 15, 15 rows of 2, 3 rows of 10, 10 rows of 3, 5 rows of 6, or 6 rows of 5.

Study Tip
When making an organized list of factor pairs, stop finding pairs when the factors begin to repeat.

On Your Own

Now You're Ready
Exercises 8–15

List the factor pairs of the number.

1. 18 **2.** 24 **3.** 51

4. WHAT IF? The woodwinds section of the marching band has 38 members. Which has more possible arrangements, the brass section or the woodwinds section? Explain.

Key Idea

Prime Factorization

The **prime factorization** of a composite number is the number written as a product of its prime factors.

You can use factor pairs and a **factor tree** to help find the prime factorization of a number. The factor tree is complete when only prime factors appear in the product. A factor tree for 60 is shown.

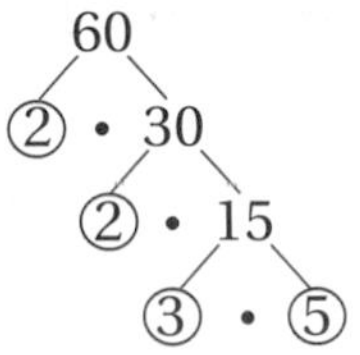

$60 = 2 \cdot 2 \cdot 3 \cdot 5$, or $2^2 \cdot 3 \cdot 5$

Remember
A *prime number* is a whole number greater than 1 with exactly two factors, 1 and itself. A *composite number* is a whole number greater than 1 with factors other than 1 and itself.

EXAMPLE 2 Writing a Prime Factorization

Write the prime factorization of 48.

Choose any factor pair of 48 to begin the factor tree.

> **Study Tip**
> Notice that beginning with different factor pairs results in the same prime factorization. Every composite number has only one prime factorization.

Tree 1

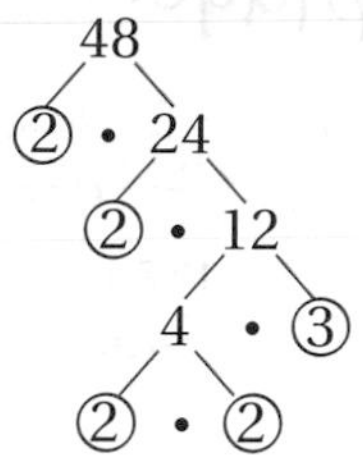

Find a factor pair and draw "branches."

Circle the prime factors as you find them.

Find factors until each branch ends at a prime factor.

Tree 2

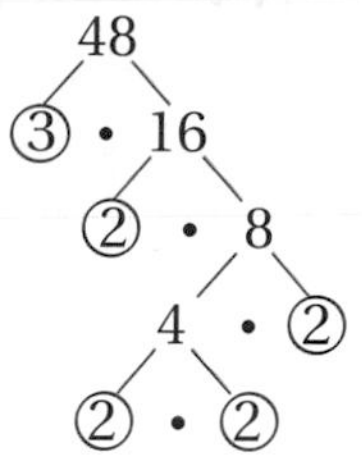

$48 = 2 \cdot 2 \cdot 3 \cdot 2 \cdot 2$

$48 = 3 \cdot 2 \cdot 2 \cdot 2 \cdot 2$

∴ The prime factorization of 48 is $2 \cdot 2 \cdot 2 \cdot 2 \cdot 3$, or $2^4 \cdot 3$.

EXAMPLE 3 Using a Prime Factorization

What is the greatest perfect square that is a factor of 1575?

Because 1575 has many factors, it is not efficient to list all of its factors and check for perfect squares. Use the prime factorization of 1575 to find any perfect squares that are factors.

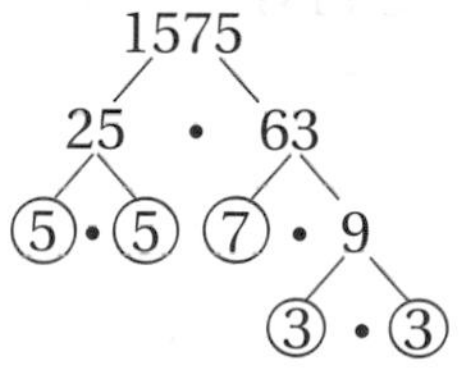

$1575 = 3 \cdot 3 \cdot 5 \cdot 5 \cdot 7$

The prime factorization shows that 1575 has three factors other than 1 that are perfect squares.

$3 \cdot 3 = 9$ $\quad$ $5 \cdot 5 = 25$ $\quad$ $(3 \cdot 5) \cdot (3 \cdot 5) = 15 \cdot 15 = 225$

∴ So, the greatest perfect square that is a factor of 1575 is 225.

On Your Own

Now You're Ready Exercises 16–23 and 29–32

Write the prime factorization of the number.

5. 20 **6.** 88 **7.** 90 **8.** 462

9. What is the greatest perfect square that is a factor of 396? Explain.

1.4 Exercises

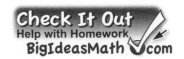

Check It Out
Help with Homework
BigIdeasMath.com

Vocabulary and Concept Check

1. **VOCABULARY** What is the prime factorization of a number?
2. **VOCABULARY** How can you use a factor tree to help you write the prime factorization of a number?
3. **WHICH ONE DOESN'T BELONG?** Which factor pair does not belong with the other three? Explain your reasoning.

2, 28	4, 14	6, 9	7, 8

Practice and Problem Solving

Use divisibility rules to determine whether the number is divisible by 2, 3, 5, 6, 9, and 10. Use a calculator to check your answer.

4. 1044   5. 1485   6. 1620   7. 1709

List the factor pairs of the number.

(1) 8. 15   9. 22   10. 34   11. 39

12. 45   13. 54   14. 59   15. 61

Write the prime factorization of the number.

(2) 16. 16   17. 25   18. 30   19. 26

20. 84   21. 54   22. 65   23. 77

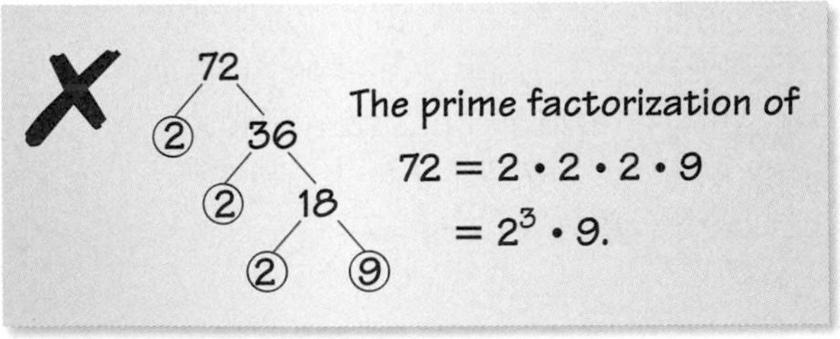

24. **ERROR ANALYSIS** Describe and correct the error in writing the prime factorization.

25. **FACTOR RAINBOW** You can use a factor rainbow to check whether a list of factors is correct. To create a factor rainbow, list the factors of a number in order from least to greatest. Then draw arches that link the factor pairs. For perfect squares, there is no connecting arch in the middle. So, just circle the middle number. A factor rainbow for 12 is shown. Create factor rainbows for 6, 24, 36, and 48.

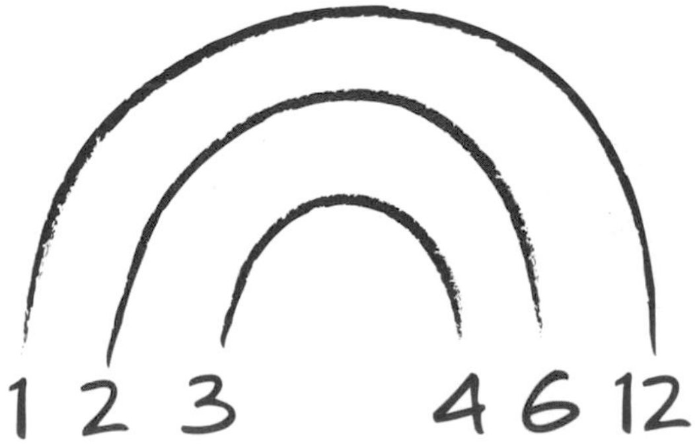

Find the number represented by the prime factorization.

26. $2^2 \cdot 3^2 \cdot 5$ **27.** $3^2 \cdot 5^2 \cdot 7$ **28.** $2^3 \cdot 11^2 \cdot 13$

Find the greatest perfect square that is a factor of the number.

3 **29.** 244 **30.** 650 **31.** 756 **32.** 1290

33. CRITICAL THINKING Is 2 the only even prime number? Explain.

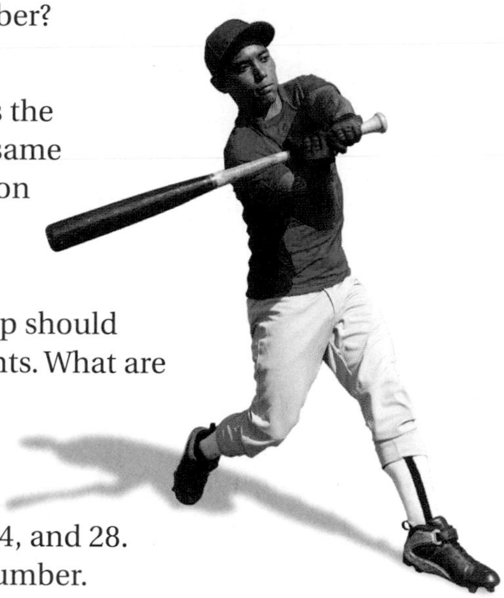

34. BASEBALL The coach of a baseball team separates the players into groups for drills. Each group has the same number of players. Is the total number of players on the baseball team *prime* or *composite*? Explain.

35. SCAVENGER HUNT A teacher divides 36 students into equal groups for a scavenger hunt. Each group should have at least 4 students but no more than 8 students. What are the possible group sizes?

36. PERFECT NUMBERS A *perfect number* is a number that equals the sum of its factors, not including itself. For example, the factors of 28 are 1, 2, 4, 7, 14, and 28. Because $1 + 2 + 4 + 7 + 14 = 28$, 28 is a perfect number. What are the perfect numbers between 1 and 28?

37. BAKE SALE One table at a bake sale has 75 oatmeal cookies. Another table has 60 lemon cupcakes. Which table allows for more rectangular arrangements when all the cookies and cupcakes are displayed? Explain.

38. MODELING The stage manager of a school play creates a rectangular acting area of 42 square yards. String lights will outline the acting area. To the nearest whole number, how many yards of string lights does the manager need to enclose this area?

Rectangular Prism

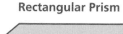

Volume = 40 cubic inches

39. **Volume** The volume of a rectangular prism can be found using the formula *volume* = *length* × *width* × *height*. Using only whole number dimensions, how many different prisms are possible? Explain.

Fair Game Review What you learned in previous grades & lessons

Find the difference. *(Skills Review Handbook)*

40. 192 − 47 **41.** 451 − 94 **42.** 3210 − 815 **43.** 4752 − 3504

44. MULTIPLE CHOICE You buy 168 pears. There are 28 pears in each bag. How many bags of pears do you buy? *(Skills Review Handbook)*

Ⓐ 5 Ⓑ 6 Ⓒ 7 Ⓓ 28

1.5 Greatest Common Factor

Essential Question How can you find the greatest common factor of two numbers?

A **Venn diagram** uses circles to describe relationships between two or more sets. The Venn diagram shows the names of students enrolled in two activities. Students enrolled in both activities are represented by the overlap of the two circles.

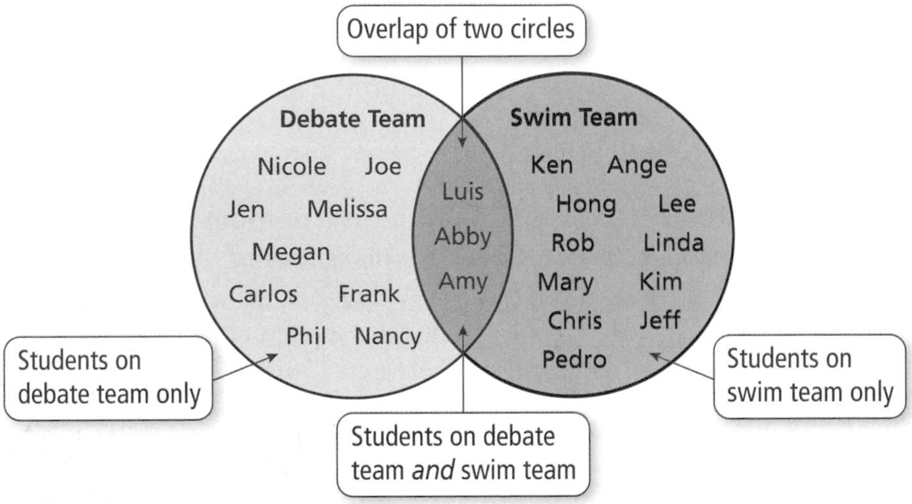

1 ACTIVITY: Identifying Common Factors

Work with a partner. Copy and complete the Venn diagram. Identify the *common factors* of the two numbers.

a. 36 and 48

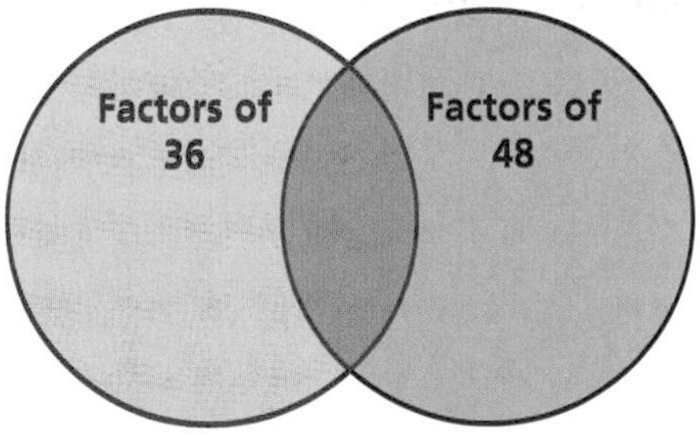

b. 16 and 56

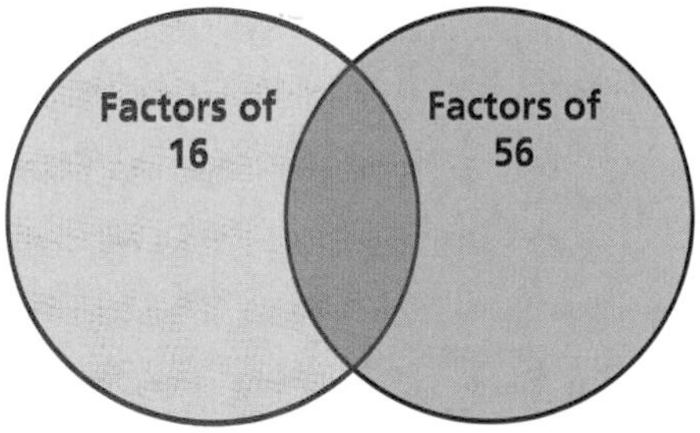

c. 30 and 75

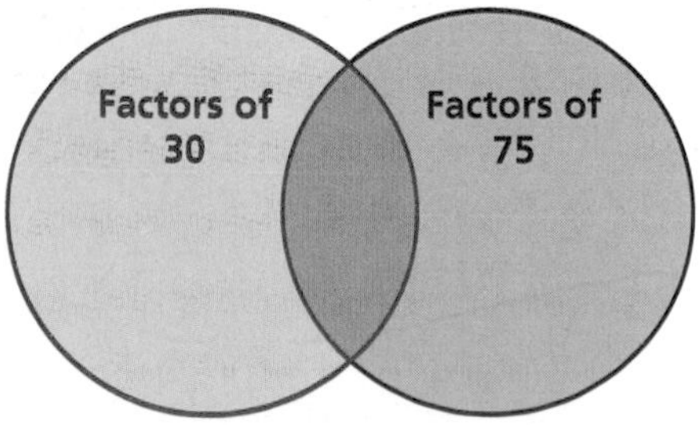

d. 54 and 90

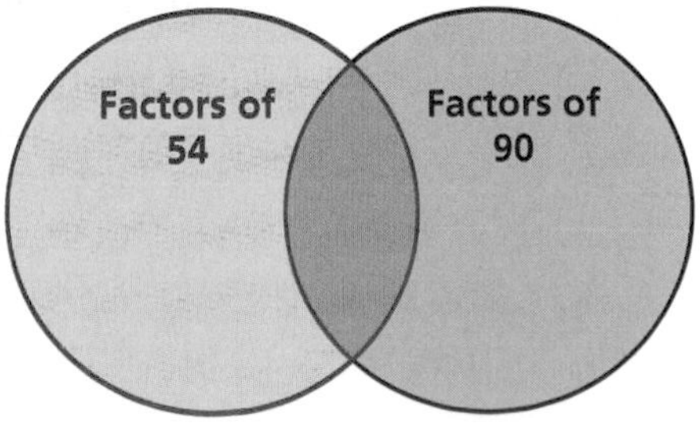

e. Look at the Venn diagrams in parts (a)–(d). Explain how to identify the *greatest common factor* of each pair of numbers. Then circle it in each diagram.

COMMON CORE

Common Factors

In this lesson, you will

- use diagrams to identify common factors.
- find greatest common factors.

Learning Standards
6.NS.4
6.EE.2b

2 ACTIVITY: Interpreting a Venn Diagram of Prime Factors

Work with a partner. The Venn diagram represents the prime factorization of two numbers. Identify the two numbers. Explain your reasoning.

a.

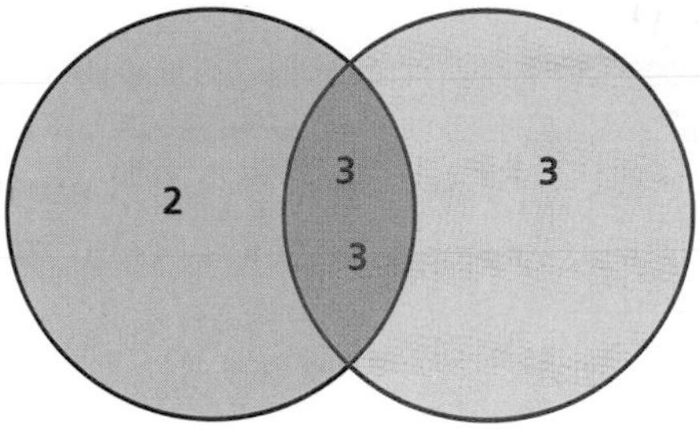

b.

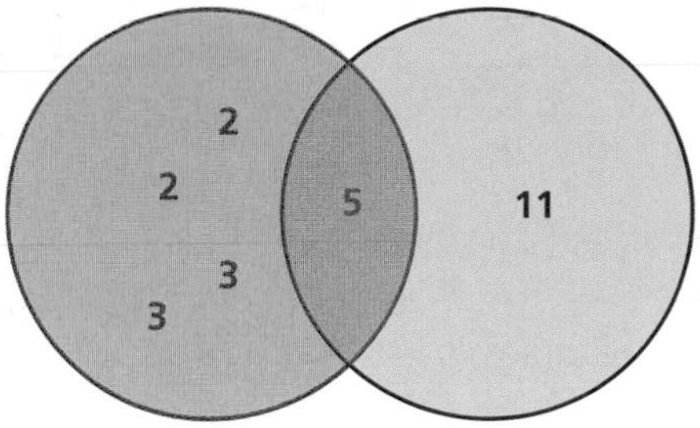

3 ACTIVITY: Identifying Common Prime Factors

Math Practice 1

Interpret a Solution

What does the diagram of the resulting prime factorization mean?

Work with a partner.

a. Write the prime factorizations of 36 and 48. Use the results to complete the Venn diagram.

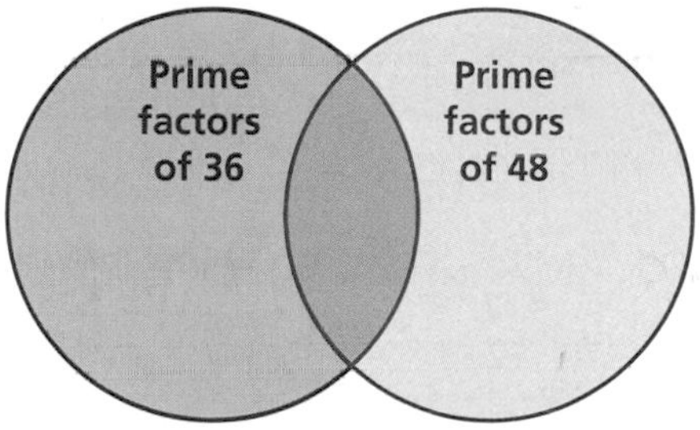

b. Repeat part (a) for the remaining number pairs in Activity 1.

c. **STRUCTURE** Compare the numbers in the overlap of the Venn diagrams to your results in Activity 1. What conjecture can you make about the relationship between these numbers and your results in Activity 1?

What Is Your Answer?

4. **IN YOUR OWN WORDS** How can you find the greatest common factor of two numbers? Give examples to support your explanation.

5. Can you think of another way to find the greatest common factor of two numbers? Explain.

Practice Use what you learned about greatest common factors to complete Exercises 4–6 on page 34.

1.5 Lesson

Check It Out
Lesson Tutorials
BigIdeasMath.com

Factors that are shared by two or more numbers are called **common factors**. The greatest of the common factors is called the **greatest common factor** (GCF). One way to find the GCF of two or more numbers is by listing factors.

Key Vocabulary
Venn diagram, *p. 30*
common factors, *p. 32*
greatest common factor, *p. 32*

EXAMPLE 1 Finding the GCF Using Lists of Factors

Find the GCF of 24 and 40.

List the factors of each number.

Factors of 24: ①, ②, 3, ④, 6, ⑧, 12, 24

Factors of 40: ①, ②, ④, 5, ⑧, 10, 20, 40

Circle the common factors.

The common factors of 24 and 40 are 1, 2, 4, and 8. The greatest of these common factors is 8.

∴ So, the GCF of 24 and 40 is 8.

Another way to find the GCF of two or more numbers is by using prime factors. The GCF is the product of the common prime factors of the numbers.

EXAMPLE 2 Finding the GCF Using Prime Factorizations

Find the GCF of 12 and 56.

Study Tip
Examples 1 and 2 show two different methods for finding the GCF. After solving with one method, you can use the other method to check your answer.

Make a factor tree for each number.

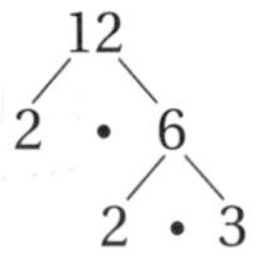

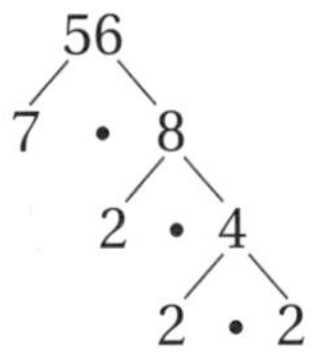

Write the prime factorization of each number.

$12 = (2) \cdot (2) \cdot 3$

$56 = (2) \cdot (2) \cdot 2 \cdot 7$

Circle the common prime factors.

$2 \cdot 2 = 4$ Find the product of the common prime factors.

∴ So, the GCF of 12 and 56 is 4.

On Your Own

Now You're Ready
Exercises 7–18

Find the GCF of the numbers using lists of factors.

1. 8, 36
2. 18, 72
3. 14, 28, 49

Find the GCF of the numbers using prime factorizations.

4. 20, 45
5. 32, 90
6. 45, 75, 120

EXAMPLE 3 Finding Two Numbers with a Given GCF

Which pair of numbers has a GCF of 15?

(A) 10, 15 (B) 30, 60 (C) 21, 45 (D) 45, 75

The number 15 cannot be a factor of the lesser number 10. So, you can eliminate Statement A.

The number 15 cannot be a factor of a number that does not have a 0 or 5 in the ones place. So, you can eliminate Statement C.

List the factors for Statements B and D. Then identify the GCF for each.

Choice B: Factors of 30: (1), (2), (3), (5), (6), (10), (15), (30)

Factors of 60: (1), (2), (3), 4, (5), (6), (10), 12, (15), 20, (30), 60

The GCF of 30 and 60 is 30.

Choice D: Factors of 45: (1), (3), (5), 9, (15), 45

Factors of 75: (1), (3), (5), (15), 25, 75

The GCF of 45 and 75 is 15.

∴ The correct answer is (D).

EXAMPLE 4 Real-Life Application

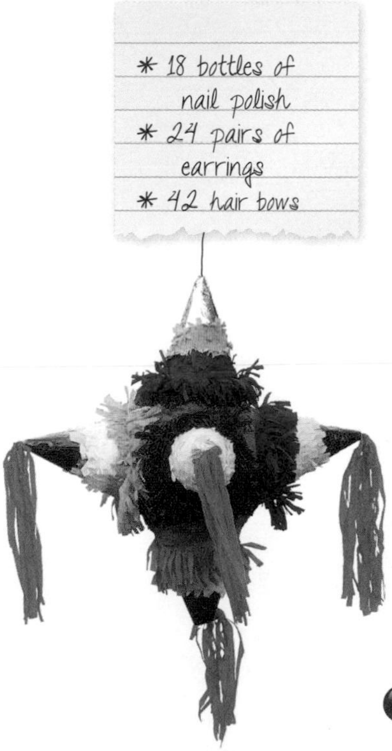

You are filling piñatas for your sister's birthday party. The list shows the gifts you are putting into the piñatas. You want identical groups of gifts in each piñata with no gifts left over. What is the greatest number of piñatas you can make?

The GCF of the numbers of gifts represents the greatest number of identical groups of gifts you can make with no gifts left over. So, to find the number of piñatas, find the GCF.

$$18 = 2 \cdot 3 \cdot 3$$
$$24 = 2 \cdot 3 \cdot 2 \cdot 2$$
$$42 = 2 \cdot 3 \cdot 7$$

$2 \cdot 3 = 6$ Find the product of the common prime factors.

The GCF of 18, 24, and 42 is 6.

∴ So, you can make at most 6 piñatas.

On Your Own

Now You're Ready
Exercises 23–25

7. Write a pair of numbers whose greatest common factor is 10.
8. **WHAT IF?** In Example 4, you add 6 more pairs of earrings. Does this change your answer? Explain your reasoning.

1.5 Exercises

Check It Out
Help with Homework
BigIdeasMath.com

Vocabulary and Concept Check

1. **VOCABULARY** What is the greatest common factor (GCF) of two numbers?
2. **WRITING** Describe how to find the GCF of two numbers by using prime factorization.
3. **DIFFERENT WORDS, SAME QUESTION** Which is different? Find "both" answers.

   What is the greatest common factor of 24 and 32?

   What is the greatest common divisor of 24 and 32?

   What is the greatest prime factor of 24 and 32?

   What is the product of the common prime factors of 24 and 32?

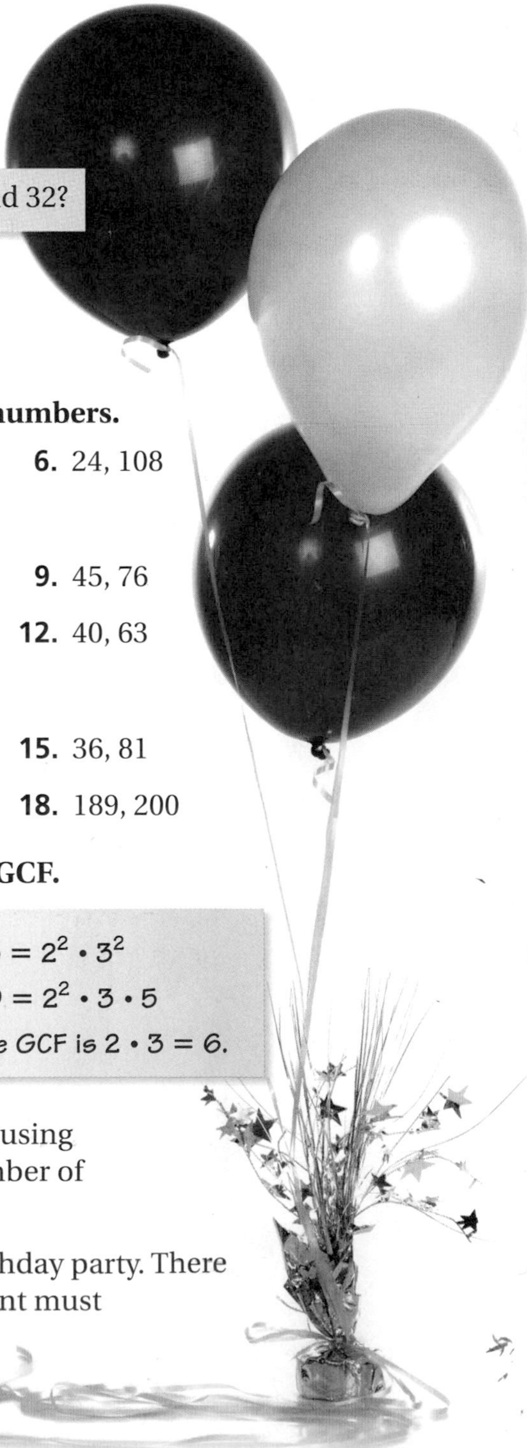

Practice and Problem Solving

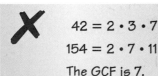

Use a Venn diagram to find the greatest common factor of the numbers.

4. 12, 30
5. 32, 54
6. 24, 108

Find the GCF of the numbers using lists of factors.

(1) 7. 6, 15
8. 14, 84
9. 45, 76
10. 39, 65
11. 51, 85
12. 40, 63

Find the GCF of the numbers using prime factorizations.

(2) 13. 45, 60
14. 27, 63
15. 36, 81
16. 72, 84
17. 61, 73
18. 189, 200

ERROR ANALYSIS Describe and correct the error in finding the GCF.

19. ✗

$42 = 2 \cdot 3 \cdot 7$

$154 = 2 \cdot 7 \cdot 11$

The GCF is 7.

20. ✗

$36 = 2^2 \cdot 3^2$

$60 = 2^2 \cdot 3 \cdot 5$

The GCF is $2 \cdot 3 = 6$.

21. **CLASSROOM** A teacher is making identical activity packets using 92 crayons and 23 sheets of paper. What is the greatest number of packets the teacher can make with no items left over?
22. **BALLOONS** You are making balloon arrangements for a birthday party. There are 16 white balloons and 24 red balloons. Each arrangement must be identical. What is the greatest number of arrangements you can make using every balloon?

Find the GCF of the numbers.

4 **23.** 35, 56, 63 **24.** 30, 60, 78 **25.** 42, 70, 84

26. OPEN-ENDED Write a set of three numbers that have a GCF of 16. What procedure did you use to find your answer?

27. REASONING You need to find the GCF of 256 and 400. Would you rather list their factors or use their prime factorizations? Explain.

CRITICAL THINKING Tell whether the statement is *always*, *sometimes*, or *never* true.

28. The GCF of two even numbers is 2.

29. The GCF of two prime numbers is 1.

30. When one number is a multiple of another, the GCF of the numbers is the greater of the numbers.

31. BOUQUETS A florist is making identical bouquets using 72 red roses, 60 pink roses, and 48 yellow roses. What is the greatest number of bouquets that the florist can make if no roses are left over? How many of each color are in each bouquet?

32. VENN DIAGRAM Consider the numbers 252, 270, and 300.

a. Create a Venn diagram using the prime factors of the numbers.

b. Use the Venn diagram to find the GCF of 252, 270, and 300.

c. What is the GCF of 252 and 270? 252 and 300? Explain how you found your answer.

33. FRUIT BASKETS You are making fruit baskets using 54 apples, 36 oranges, and 73 bananas.

a. Explain why you cannot make identical fruit baskets without leftover fruit.

b. What is the greatest number of identical fruit baskets you can make with the least amount of fruit left over? Explain how you found your answer.

34. Problem Solving Two rectangular, adjacent rooms share a wall. One-foot-by-one-foot tiles cover the floor of each room. Describe how the greatest possible length of the adjoining wall is related to the total number of tiles in each room. Draw a diagram that represents one possibility.

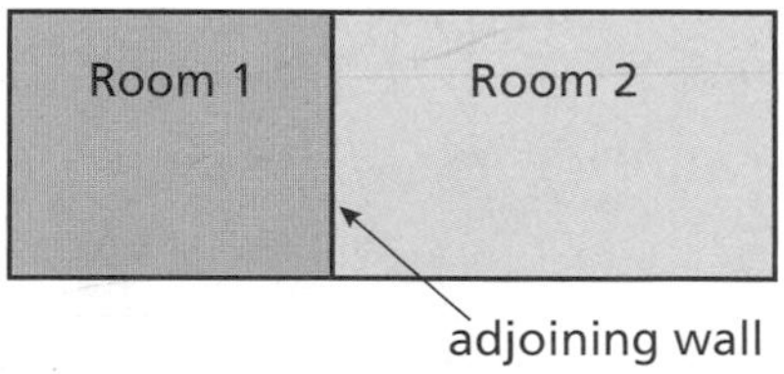

Fair Game Review What you learned in previous grades & lessons

Tell which property is being illustrated. *(Skills Review Handbook)*

35. $13 + (29 + 7) = 13 + (7 + 29)$ **36.** $13 + (7 + 29) = (13 + 7) + 29$

37. $(6 \times 37) \times 5 = (37 \times 6) \times 5$ **38.** $(37 \times 6) \times 5 = 37 \times (6 \times 5)$

39. MULTIPLE CHOICE In what order should you perform the operations in the expression $4 \times 3 - 12 \div 2 + 5$? *(Section 1.3)*

Ⓐ $\times, -, \div, +$ Ⓑ $\times, \div, -, +$ Ⓒ $\times, \div, +, -$ Ⓓ $\times, +, -, \div$

1.6 Least Common Multiple

Essential Question How can you find the least common multiple of two numbers?

1 ACTIVITY: Identifying Common Multiples

Work with a partner. Using the first several multiples of each number, copy and complete the Venn diagram. Identify any *common multiples* of the two numbers.

a. 8 and 12

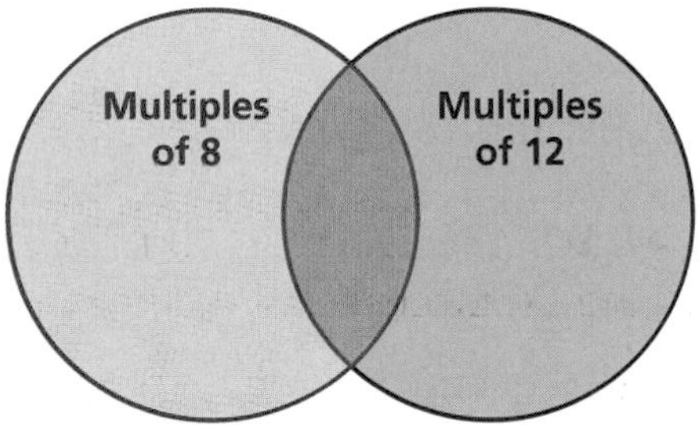

b. 4 and 14

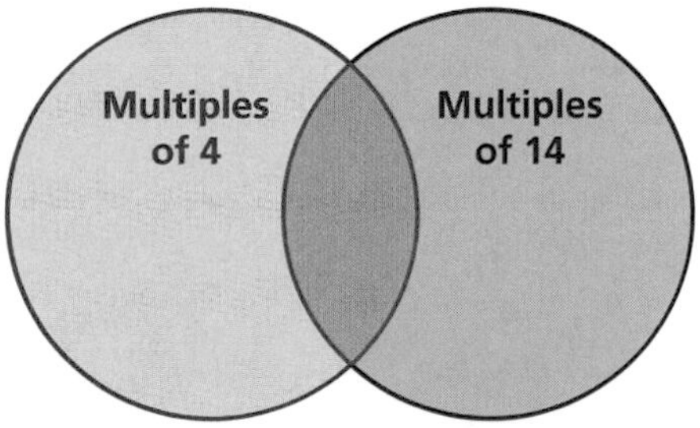

c. 10 and 15

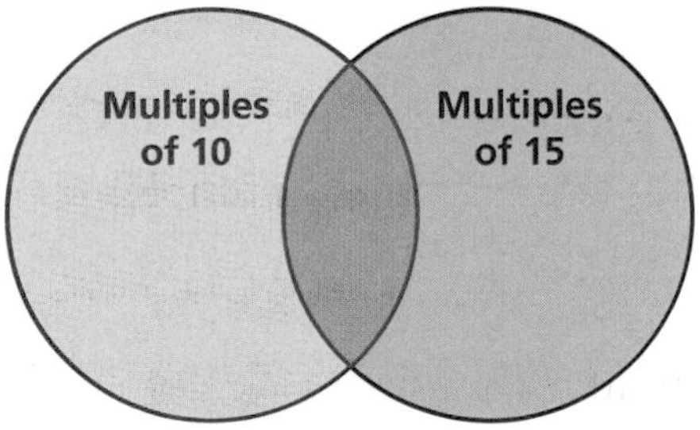

d. 20 and 35

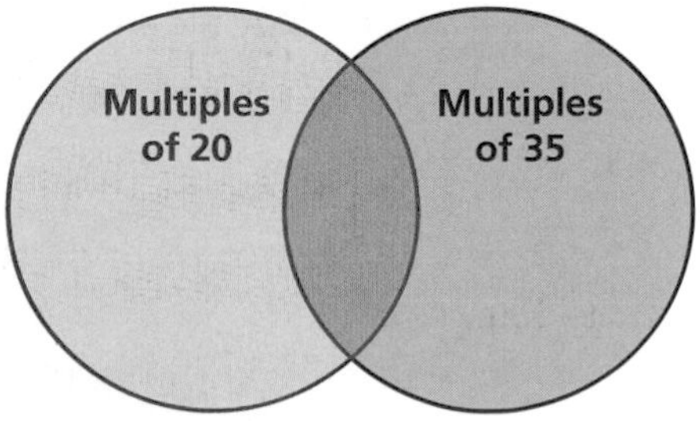

e. Look at the Venn diagrams in parts (a)–(d). Explain how to identify the *least common multiple* of each pair of numbers. Then circle it in each diagram.

COMMON CORE

Common Multiples

In this lesson, you will

- use diagrams to identify common multiples.
- find least common multiples.

Learning Standard 6.NS.4

2 ACTIVITY: Interpreting a Venn Diagram of Prime Factors

Work with a partner.

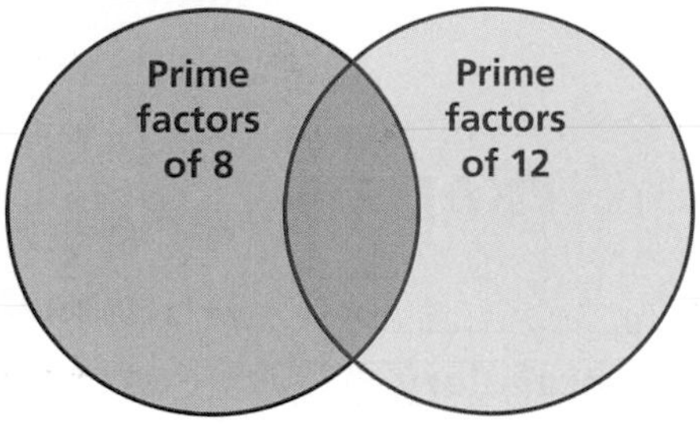

a. Write the prime factorizations of 8 and 12. Use the results to complete the Venn diagram.

b. Repeat part (a) for the remaining number pairs in Activity 1.

c. **STRUCTURE** Compare the numbers from each section of the Venn diagrams to your results in Activity 1. What conjecture can you make about the relationship between these numbers and your results in Activity 1?

What Is Your Answer?

Math Practice 3

Construct Arguments

How can you use diagrams to support your explanation?

3. **IN YOUR OWN WORDS** How can you find the least common multiple of two numbers? Give examples to support your explanation.

4. The Venn diagram shows the prime factors of two numbers.

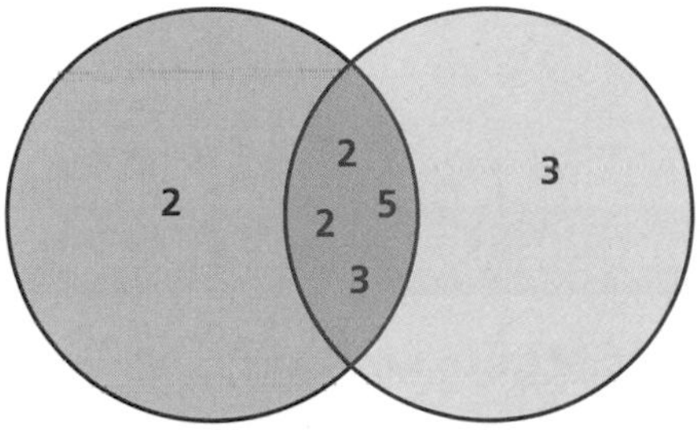

Use the diagram to do the following tasks.

a. Identify the two numbers.

b. Find the greatest common factor.

c. Find the least common multiple.

5. A student writes the prime factorizations of 8 and 12 in a table as shown. She claims she can use the table to find the greatest common factor and the least common multiple of 8 and 12. How is this possible?

8 =	2	2	2	
12 =	2	2		3

6. Can you think of another way to find the least common multiple of two or more numbers? Explain.

Practice

Use what you learned about least common multiples to complete Exercises 3–5 on page 40.

1.6 Lesson

Check It Out
Lesson Tutorials
BigIdeasMath.com

Multiples that are shared by two or more numbers are called **common multiples**. The least of the common multiples is called the **least common multiple** (LCM). You can find the LCM of two or more numbers by listing multiples or using prime factors.

Key Vocabulary
common multiples, *p. 38*
least common multiple, *p. 38*

EXAMPLE 1 Finding the LCM Using Lists of Multiples

Find the LCM of 4 and 6.

List the multiples of each number.

Multiples of 4: 4, 8, (12), 16, 20, (24), 28, 32, (36), . . .

Multiples of 6: 6, (12), 18, (24), 30, (36), . . .

Circle the common multiples.

Some common multiples of 4 and 6 are 12, 24, and 36. The least of these common multiples is 12.

So, the LCM of 4 and 6 is 12.

On Your Own

Now You're Ready
Exercises 6–11

Find the LCM of the numbers using lists of multiples.

1. 3, 8 **2.** 9, 12 **3.** 6, 10

EXAMPLE 2 Finding the LCM Using Prime Factorizations

Find the LCM of 16 and 20.

Make a factor tree for each number.

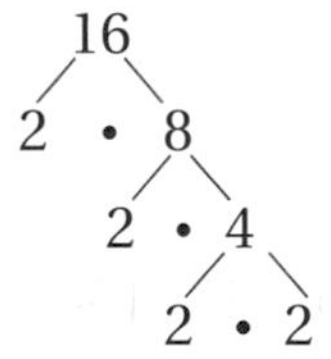

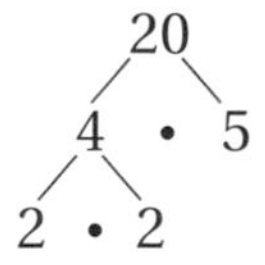

Write the prime factorization of each number. Circle each different factor where it appears the greater number of times.

$16 = (2) \cdot (2) \cdot (2) \cdot (2)$ — 2 appears more often here, so circle all 2s.

$20 = 2 \cdot 2 \cdot (5)$ — 5 appears once. Do not circle the 2s again.

$2 \cdot 2 \cdot 2 \cdot 2 \cdot 5 = 80$ — Find the product of the circled factors.

So, the LCM of 16 and 20 is 80.

On Your Own

Now You're Ready
Exercises 12–17

Find the LCM of the numbers using prime factorizations.

4. 14, 18 **5.** 28, 36 **6.** 24, 90

EXAMPLE 3 Finding the LCM of Three Numbers

Find the LCM of 4, 15, and 18.

Write the prime factorization of each number. Circle each different factor where it appears the greatest number of times.

$4 = ② \cdot ②$	2 appears most often here, so circle both 2s.
$15 = 3 \cdot ⑤$	5 appears here only, so circle 5.
$18 = 2 \cdot ③ \cdot ③$	3 appears most often here, so circle both 3s.
$2 \cdot 2 \cdot 5 \cdot 3 \cdot 3 = 180$	Find the product of the circled factors.

So, the LCM of 4, 15, and 18 is 180.

On Your Own

Now You're Ready
Exercises 22–27

Find the LCM of the numbers.

7. 2, 5, 8

8. 6, 10, 12

9. Write a set of numbers whose least common multiple is 100.

EXAMPLE 4 Real-Life Application

A traffic light changes every 30 seconds. Another traffic light changes every 40 seconds. Both lights just changed. After how many minutes will both lights change at the same time again?

Find the LCM of 30 and 40 by listing multiples of each number. Circle the least common multiple.

Multiples of 30: 30, 60, 90, (120), . . .

Multiples of 40: 40, 80, (120), 160, . . .

The LCM is 120. So, both lights will change again after 120 seconds.

Because there are 60 seconds in 1 minute, there are $120 \div 60 = 2$ minutes in 120 seconds.

Both lights will change at the same time again after 2 minutes.

On Your Own

10. WHAT IF? In Example 4, the traffic light that changes every 40 seconds is adjusted to change every 45 seconds. Both lights just changed. After how many minutes will both lights change at the same time again?

1.6 Exercises

Check It Out
Help with Homework
BigIdeasMath.com

Vocabulary and Concept Check

1. **VOCABULARY** What is the least common multiple (LCM) of two numbers?

2. **WRITING** Describe how to find the LCM of two numbers by using prime factorization.

Practice and Problem Solving

Use a Venn diagram to find the least common multiple of the numbers.

3. 3, 7
4. 6, 8
5. 12, 15

Find the LCM of the numbers using lists of multiples.

1
6. 2, 9
7. 3, 4
8. 8, 9
9. 5, 8
10. 15, 20
11. 12, 18

Find the LCM of the numbers using prime factorizations.

2
12. 9, 21
13. 12, 27
14. 18, 45
15. 22, 33
16. 36, 60
17. 35, 50

✗ $6 \times 9 = 54$
The LCM of 6 and 9 is 54.

18. **ERROR ANALYSIS** Describe and correct the error in finding the LCM.

19. **AQUATICS** You have diving lessons every fifth day and swimming lessons every third day. Today you have both lessons. In how many days will you have both lessons on the same day again?

20. **HOT DOGS** Hot dogs come in packs of 10, while buns come in packs of eight. What are the least numbers of packs you should buy in order to have the same numbers of hot dogs and buns?

21. **MODELING** Which model represents an LCM that is different from the other three? Explain your reasoning.

A.
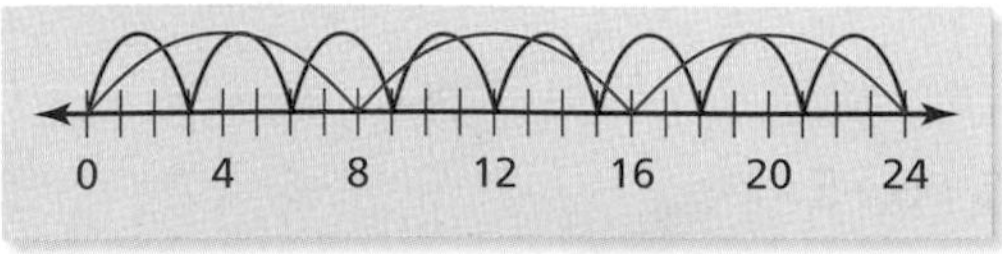

B.
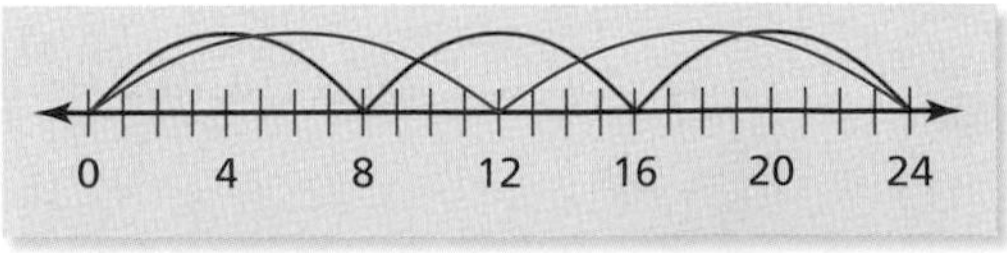

C.
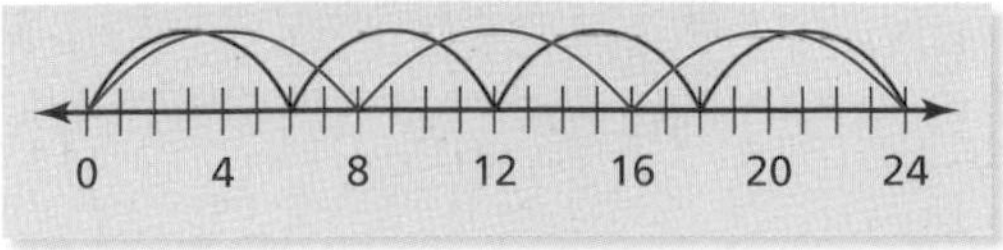

D.
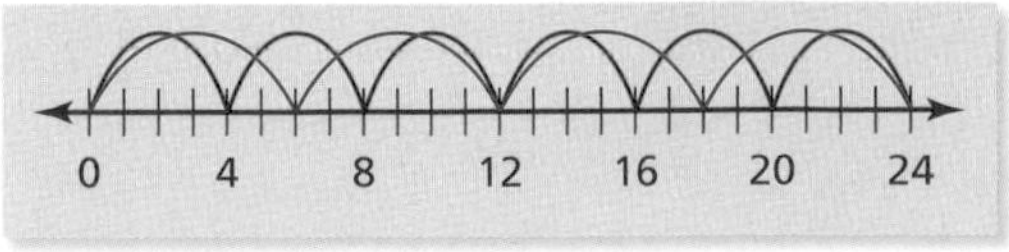

Find the LCM of the numbers.

3 **22.** 2, 3, 7 **23.** 3, 5, 11 **24.** 4, 9, 12

25. 6, 8, 15 **26.** 7, 18, 21 **27.** 9, 10, 28

28. **REASONING** You need to find the LCM of 13 and 14. Would you rather list their multiples or use their prime factorizations? Explain.

CRITICAL THINKING Tell whether the statement is *always*, *sometimes*, or *never* true.

29. The LCM of two different prime numbers is their product.

30. The LCM of a set of numbers is equal to one of the numbers in the set.

31. The GCF of two different numbers is the LCM of the numbers.

32. **SUBWAY** At Union Station, you notice that three subway lines just arrived at the same time. The table shows their arrival schedule. How long must you wait until all three lines arrive at Union Station at the same time again?

Subway Line	Arrival Time
A	every 10 min
B	every 12 min
C	every 15 min

33. **RADIO CONTEST** A radio station gives away $15 to every 15th caller, $25 to every 25th caller, and free concert tickets to every 100th caller. When will the station first give away *all* three prizes to one caller?

34. **TREADMILL** You and a friend are running on treadmills. You run 0.5 mile every 3 minutes, and your friend runs 2 miles every 14 minutes. You both start and stop running at the same time and run a whole number of miles. What is the least possible number of miles you and your friend can run?

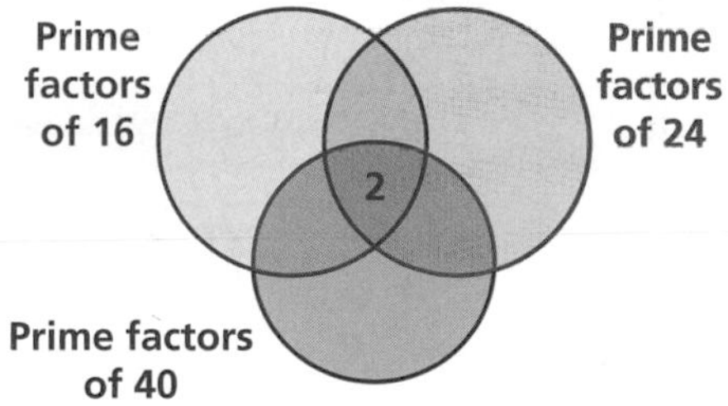

35. **VENN DIAGRAM** Refer to the Venn diagram.

a. Copy and complete the Venn diagram.

b. What is the LCM of 16, 24, and 40?

c. What is the LCM of 16 and 40? 24 and 40?

36. Number Sense When is the LCM of two numbers equal to their product?

Fair Game Review What you learned in previous grades & lessons

Write the product as a power. *(Section 1.2)*

37. 3×3 **38.** $5 \cdot 5 \cdot 5 \cdot 5$ **39.** $17 \times 17 \times 17 \times 17 \times 17$

40. **MULTIPLE CHOICE** Which two powers have the same value? *(Section 1.2)*

Ⓐ 1^3 and 3^1 Ⓑ 2^4 and 4^2 Ⓒ 3^2 and 2^3 Ⓓ 4^3 and 3^4

Extension 1.6 Adding and Subtracting Fractions

Check It Out
Lesson Tutorials
BigIdeasMath.com

Key Vocabulary
least common denominator, *p. 42*

Recall that you can add and subtract fractions with unlike denominators by writing equivalent fractions with a common denominator. One way to do this is by multiplying the numerator and the denominator of each fraction by the denominator of the other fraction.

EXAMPLE 1 Adding Fractions Using a Common Denominator

Find $\frac{5}{8} + \frac{1}{6}$.

Rewrite the fractions with a common denominator. Use the product of the denominators as the common denominator.

$\frac{5}{8} + \frac{1}{6} = \frac{5 \cdot 6}{8 \cdot 6} + \frac{1 \cdot 8}{6 \cdot 8}$ — Rewrite the fractions using a common denominator of $8 \cdot 6 = 48$.

$= \frac{30}{48} + \frac{8}{48}$ — Multiply.

$= \frac{38}{48}$ — Add the numerators.

$= \frac{\cancel{2}^{1} \cdot 19}{\cancel{2}_{1} \cdot 24}$ — Divide out the common factor 2.

$= \frac{19}{24}$ — Simplify.

Study Tip
A fraction is in *simplest form* when the numerator and the denominator have no common factors other than 1.

The **least common denominator** (LCD) of two or more fractions is the least common multiple (LCM) of the denominators. The LCD provides another method for adding and subtracting fractions with unlike denominators.

EXAMPLE 2 Adding Fractions Using the LCD

Find $\frac{5}{8} + \frac{1}{6}$.

Find the LCM of the denominators.

Multiples of 8: 8, 16, (24), 32, 40, (48), . . .

Multiples of 6: 6, 12, 18, (24), 30, 36, 42, (48), . . .

The LCM of 8 and 6 is 24. So, the LCD is 24.

$\frac{5}{8} + \frac{1}{6} = \frac{5 \cdot 3}{8 \cdot 3} + \frac{1 \cdot 4}{6 \cdot 4}$ — Rewrite the fractions using the LCD, 24.

$= \frac{15}{24} + \frac{4}{24}$ — Multiply.

$= \frac{19}{24}$ — Add the numerators.

COMMON CORE
Common Multiples
In this extension, you will
- use least common multiples to add and subtract fractions.

Applying Standard 6.NS.4

Multi-Language Glossary at BigIdeasMath.com

To add or subtract mixed numbers, first rewrite the numbers as improper fractions. Then find the common denominator.

EXAMPLE 3 Subtracting Mixed Numbers

Find $4\frac{3}{4} - 2\frac{3}{10}$.

Write the difference using improper fractions.

$$4\frac{3}{4} - 2\frac{3}{10} = \frac{19}{4} - \frac{23}{10}$$

> **Study Tip**
>
> Notice that Method 1 uses the same procedure shown in Example 1. You can generalize the procedure using the rule
>
> $\frac{a}{b} \pm \frac{c}{d} = \frac{ad \pm bc}{bd}$.

Method 1: Use the product of the denominators as the common denominator.

$$\frac{19}{4} - \frac{23}{10} = \frac{19 \cdot 10}{4 \cdot 10} - \frac{23 \cdot 4}{10 \cdot 4}$$ Rewrite the fractions using a common denominator of $4 \cdot 10 = 40$.

$$= \frac{190}{40} - \frac{92}{40}$$ Multiply.

$$= \frac{98}{40}$$ Subtract the numerators.

$$= \frac{49}{20}, \text{ or } 2\frac{9}{20}$$ Simplify.

Method 2: Use the LCD. The LCM of 4 and 10 is 20.

$$\frac{19}{4} - \frac{23}{10} = \frac{19 \cdot 5}{4 \cdot 5} - \frac{23 \cdot 2}{10 \cdot 2}$$ Rewrite the fractions using the LCD, 20.

$$= \frac{95}{20} - \frac{46}{20}$$ Multiply.

$$= \frac{49}{20}, \text{ or } 2\frac{9}{20}$$ Simplify.

Practice

Use the LCD to rewrite the fractions with the same denominator.

1. $\frac{1}{6}, \frac{3}{8}$
2. $\frac{4}{7}, \frac{3}{10}$
3. $\frac{5}{12}, \frac{2}{9}$
4. $\frac{3}{4}, \frac{5}{8}, \frac{1}{10}$

Copy and complete the statement using <, >, or =.

5. $\frac{4}{5} \ \square \ \frac{5}{6}$
6. $\frac{5}{14} \ \square \ \frac{3}{8}$
7. $2\frac{2}{5} \ \square \ \frac{24}{10}$
8. $4\frac{9}{25} \ \square \ 4\frac{7}{20}$

Add or subtract. Write the answer in simplest form.

9. $\frac{2}{3} + \frac{3}{4}$
10. $\frac{6}{7} + \frac{1}{2}$
11. $\frac{7}{10} - \frac{5}{12}$
12. $\frac{13}{18} - \frac{5}{8}$
13. $2\frac{1}{6} + 3\frac{4}{9}$
14. $4\frac{3}{16} + 1\frac{1}{10}$
15. $1\frac{5}{6} - \frac{3}{4}$
16. $3\frac{2}{3} - 2\frac{4}{11}$
17. **COMPARING METHODS** List some advantages and disadvantages of each method shown in the examples. Which method do you prefer? Why?

1.4–1.6 Quiz

Check It Out
Progress Check
BigIdeasMath.com

List the factor pairs of the number. *(Section 1.4)*

1. 48

2. 56

Write the prime factorization of the number. *(Section 1.4)*

3. 60

4. 72

Find the GCF of the numbers using lists of factors. *(Section 1.5)*

5. 18, 42

6. 24, 44, 52

Find the GCF of the numbers using prime factorizations. *(Section 1.5)*

7. 38, 68

8. 68, 76, 92

Find the LCM of the numbers using lists of multiples. *(Section 1.6)*

9. 8, 14

10. 3, 6, 16

Find the LCM of the numbers using prime factorizations. *(Section 1.6)*

11. 18, 30

12. 6, 24, 32

Add or subtract. Write the answer in simplest form. *(Section 1.6)*

13. $\frac{3}{5} + \frac{2}{3}$

14. $\frac{7}{8} - \frac{3}{4}$

15. PICNIC BASKETS You are creating identical picnic baskets using 30 sandwiches and 42 apples. What is the greatest number of baskets that you can fill using all of the food? *(Section 1.5)*

16. RIBBON You have 52 inches of yellow ribbon and 64 inches of red ribbon. You want to cut the ribbons into pieces of equal length with no leftovers. What is the greatest length of the pieces that you can make? *(Section 1.5)*

17. MUSIC LESSONS You have piano lessons every fourth day and guitar lessons every sixth day. Today you have both lessons. In how many days will you have both lessons on the same day again? Explain. *(Section 1.6)*

18. HAMBURGERS Hamburgers come in packs of 20, while buns come in packs of 12. What is the least number of packs you should buy in order to have the same numbers of hamburgers and buns? *(Section 1.6)*

1 Chapter Review

Check It Out
Vocabulary Help
BigIdeasMath.com

Review Key Vocabulary

power, *p. 12*
base, *p. 12*
exponent, *p. 12*
perfect square, *p. 13*
numerical expression, *p. 18*
evaluate, *p. 18*
order of operations, *p. 18*
factor pair, *p. 26*
prime factorization, *p. 26*
factor tree, *p. 26*
Venn diagram, *p. 30*
common factors, *p. 32*
greatest common factor (GCF), *p. 32*
common multiples, *p. 38*
least common multiple (LCM), *p. 38*
least common denominator (LCD), *p. 42*

Review Examples and Exercises

1.1 Whole Number Operations *(pp. 2–9)*

Use the tens place because 203 is less than 508.

$$\begin{array}{r} 2 \\ 203\overline{)\,5081} \\ -\,406 \\ \hline 102 \end{array}$$

Divide 508 by 203: There are two groups of 203 in 508.
Multiply 2 and 203.
Subtract 406 from 508.

Next, bring down the 1 and divide the ones.

$$\begin{array}{r} 25 \text{ R6} \\ 203\overline{)\,5081} \\ -\,406\downarrow \\ \hline 1021 \\ -\,1015 \\ \hline 6 \end{array}$$

Divide 1021 by 203: There are five groups of 203 in 1021.
Multiply 5 and 203.
Subtract 1015 from 1021.

∴ The quotient of 5081 and 203 is $25\frac{6}{203}$.

Exercises

Find the value of the expression. Use estimation to check your answer.

1. $4382 + 2899$
2. $8724 - 3568$
3. 192×38
4. $216 \div 31$

1.2 Powers and Exponents *(pp. 10–15)*

Evaluate 6^2.

$6^2 = 6 \cdot 6 = 36$ Write as repeated multiplication and simplify.

Exercises

Find the value of the power.

5. 7^3
6. 2^6
7. 4^4

1.3 Order of Operations (pp. 16–21)

Evaluate $4^3 - 15 \div 5$.

$4^3 - 15 \div 5 = 64 - 15 \div 5$	Evaluate 4^3.
$= 64 - 3$	Divide 15 by 5.
$= 61$	Subtract 3 from 64.

Exercises

Evaluate the expression.

8. $3 \times 6 - 12 \div 6$ **9.** $20 \times (3^2 - 4) \div 50$ **10.** $5 + (4^2 + 2) \div 6$

1.4 Prime Factorization (pp. 24–29)

Write the prime factorization of 18.

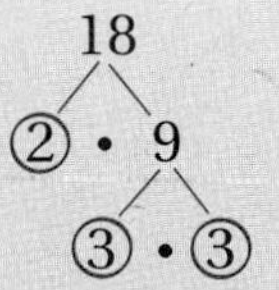

Find a factor pair and draw "branches."

Circle the prime factors as you find them.

Continue until each branch ends at a prime factor.

∴ The prime factorization of 18 is $2 \cdot 3 \cdot 3$, or $2 \cdot 3^2$.

Exercises

List the factor pairs of the number.

11. 28 **12.** 44 **13.** 63

Write the prime factorization of the number.

14. 42 **15.** 50 **16.** 66

1.5 Greatest Common Factor (pp. 30–35)

a. Find the GCF of 32 and 76.

Factors of 32: ①, ②, ④, 8, 16, 32
Factors of 76: ①, ②, ④, 19, 38, 76

The greatest of the common factors is 4.

∴ So, the GCF of 32 and 76 is 4.

b. Find the GCF of 45 and 63.

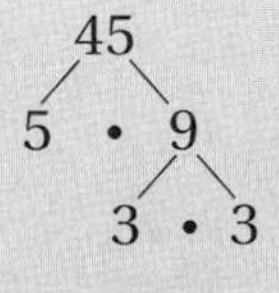

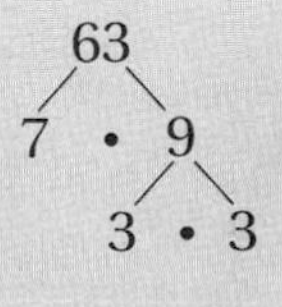

$45 = 3 \cdot 3 \cdot 5$

$63 = 3 \cdot 3 \cdot 7$

$3 \cdot 3 = 9$

∴ So, the GCF of 45 and 63 is 9.

Exercises

Find the GCF of the numbers using lists of factors.

17. 27, 45 **18.** 30, 48 **19.** 28, 48, 64

Find the GCF of the numbers using prime factorizations.

20. 24, 90 **21.** 52, 68 **22.** 32, 56, 96

1.6 Least Common Multiple *(pp. 36–43)*

a. Find the LCM of 8 and 12.

Make a factor tree for each number.

Write the prime factorization of each number. Circle each different factor where it appears the greater number of times.

$8 = (2) \cdot (2) \cdot (2)$ 2 appears more often here, so circle all 2s.

$12 = 2 \cdot 2 \cdot (3)$ 3 appears once. Do not circle the 2s again.

$2 \cdot 2 \cdot 2 \cdot 3 = 24$ Find the product of the circled factors.

∴ So, the LCM of 8 and 12 is 24.

b. Find $\frac{1}{2} + \frac{1}{3}$.

The LCM of 2 and 3 is 6. So, the LCD is 6.

$$\frac{1}{2} + \frac{1}{3} = \frac{1 \cdot 3}{2 \cdot 3} + \frac{1 \cdot 2}{3 \cdot 2} = \frac{3}{6} + \frac{2}{6} = \frac{5}{6}$$

Exercises

Find the LCM of the numbers using lists of multiples.

23. 4, 14 **24.** 6, 20 **25.** 12, 28

Find the LCM of the numbers using prime factorizations.

26. 6, 45 **27.** 10, 12 **28.** 18, 27

Add or subtract. Write the answer in simplest form.

29. $\frac{2}{7} + \frac{1}{4}$ **30.** $\frac{5}{9} + \frac{3}{8}$ **31.** $3\frac{5}{6} - 2\frac{7}{15}$

32. WATER PITCHER A water pitcher contains $\frac{2}{3}$ gallon of water. You add $\frac{5}{7}$ gallon of water to the pitcher. How much water does the pitcher contain?

1 Chapter Test

Check It Out
Test Practice
BigIdeasMath.com

Find the value of the expression. Use estimation to check your answer.

1. $3963 + 2379$

2. $6184 - 2348$

3. 184×26

4. $207 \div 23$

Find the value of the power.

5. 2^3

6. 15^2

7. 5^4

Evaluate the expression.

8. $11 \times 8 - 6 \div 2$

9. $5 + 2^3 \div 4 - 2$

10. $6 + 4(11 - 2) \div 3^2$

List the factor pairs of the number.

11. 52

12. 66

Write the prime factorization of the number.

13. 46

14. 28

Find the GCF of the numbers using lists of factors.

15. 24, 54

16. 16, 32, 72

Find the GCF of the numbers using prime factorizations.

17. 52, 65

18. 18, 45, 63

Find the LCM of the numbers using lists of multiples.

19. 14, 21

20. 9, 24

Find the LCM of the numbers using prime factorizations.

21. 26, 39

22. 6, 12, 14

23. **BRACELETS** You have 16 yellow beads, 20 red beads, and 24 orange beads to make identical bracelets. What is the greatest number of bracelets that you can make using all the beads?

24. **MARBLES** A bag contains equal numbers of green and blue marbles. You can divide all the green marbles into groups of 12 and all the blue marbles into groups of 16. What is the least number of each color of marble that can be in the bag?

25. **SCALE** You place a $3\frac{3}{8}$-pound weight on the left side of a balance scale and a $1\frac{1}{5}$-pound weight on the right side. How much weight do you need to add to the right side to balance the scale?

1 Standards Assessment

Test-Taking Strategy
Solve Directly or Eliminate Choices
How many hyenas are 5−2²−1 hyenas?
Ⓐ 0 Ⓑ 2 Ⓒ 8 Ⓓ 10
Survival strategy: A
"Which strategy would you use on this one: solve directly or eliminate choices?"

1. You are making identical bagel platters using 40 plain bagels, 30 raisin bagels, and 24 blueberry bagels. What is the greatest number of platters that you can make if there are no leftover bagels? *(6.NS.4)*

A. 2 **C.** 8

B. 6 **D.** 10

2. The top of an end table is a square with a side length of 16 inches. What is the area of the tabletop? *(6.EE.1)*

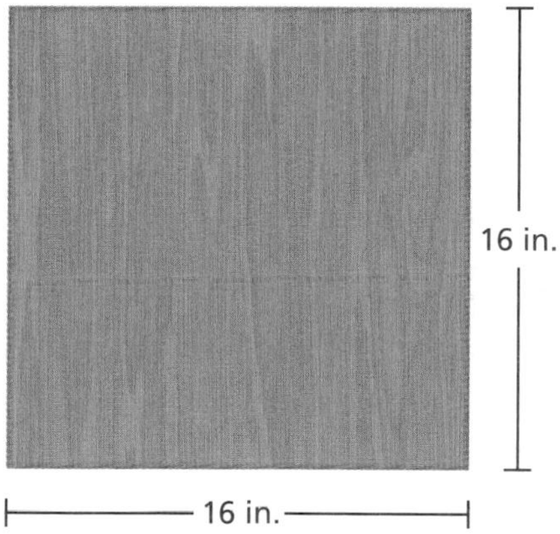
16 in.
16 in.

F. 16 in.^{2} **H.** 64 in.^{2}

G. 32 in.^{2} **I.** 256 in.^{2}

3. Which number is equivalent to the expression below? *(6.EE.1)*

$$3 \cdot 2^3 - 8 \div 4$$

A. 0 **C.** 22

B. 4 **D.** 214

4. What is the least common multiple of 14 and 49? *(6.NS.4)*

5. Which number is equivalent to the expression 7059 ÷ 301? *(6.NS.2)*

F. 23

H. $23\frac{136}{301}$

G. $23\frac{136}{7059}$

I. 136

6. You are building identical displays for the school fair using 65 blue boxes and 91 yellow boxes. What is the greatest number of displays you can build using all the boxes? *(6.NS.4)*

A. 13

C. 91

B. 35

D. 156

7. You hang the two strands of decorative lights shown below.

Strand 1: changes between red and blue every 15 seconds

Strand 2: changes between green and gold every 18 seconds

Both strands just changed color. After how many seconds will the strands change color at the same time again? *(6.NS.4)*

F. 3 seconds

H. 90 seconds

G. 30 seconds

I. 270 seconds

8. Which expression is equivalent to $\frac{29}{63}$? *(6.NS.4)*

A. $\frac{28}{60} + \frac{1}{3}$

C. $\frac{5}{21} + \frac{2}{9}$

B. $\frac{4}{27} + \frac{25}{36}$

D. $\frac{22}{47} + \frac{7}{16}$

9. Which expression is *not* equivalent to 32? *(6.EE.1)*

F. $6^2 - 8 \div 2$

H. $30 + 4^2 \div (2 + 6)$

G. $30 \div 2 + 5^2 - 8$

I. $8^2 \div 4 - 2$

10. Which number is equivalent to the expression 148 × 27? *(6.NS.2)*

A. 3696

C. 3946

B. 3896

D. 3996

11. You have 60 nickels, 48 dimes, and 42 quarters. You want to divide the coins into identical groups with no coins left over. What is the greatest number of groups that you can make? *(6.NS.4)*

12. Erica was evaluating the expression in the box below.

$$\begin{aligned} 56 \div (2^3 - 1) \times 4 &= 56 \div (8 - 1) \times 4 \\ &= 56 \div 7 \times 4 \\ &= 56 \div 28 \\ &= 2 \end{aligned}$$

What should Erica do to correct the error that she made? *(6.EE.1)*

F. Divide 56 by 8 because operations are performed left to right.

G. Multiply 1 by 4 because multiplication is done before subtraction.

H. Divide 56 by 7 because operations are performed left to right.

I. Divide 56 by 8 and multiply 1 by 4 because division and multiplication are performed before subtraction.

13. Find the greatest common factor for each pair of numbers.

Think Solve Explain

10 and 15   10 and 21   15 and 21

What can you conclude about the greatest common factor of 10, 15, and 21? Explain your reasoning. *(6.NS.4)*

14. Which number is *not* a perfect square? *(6.EE.1)*

A. 64

B. 81

C. 96

D. 100

15. Which number pair has a least common multiple of 48? *(6.NS.4)*

F. 4, 12

G. 6, 8

H. 8, 24

I. 16, 24

16. Which number is equivalent to the expression below? *(6.EE.1)*

$$\frac{3(6 + 2^2) + 2}{8}$$

A. 3

B. 4

C. 7

D. $24\frac{1}{4}$

2 Fractions and Decimals

2.1 Multiplying Fractions

2.2 Dividing Fractions

2.3 Dividing Mixed Numbers

2.4 Adding and Subtracting Decimals

2.5 Multiplying Decimals

2.6 Dividing Decimals

"Dear Sir: You say that MOST humans use only a fraction of their brain power."
My brain hurts.
"But, $\frac{3}{2}$ is a fraction. So, does that mean that SOME humans use one and a half of their brain power?"

What You Learned Before

"On a scale from 1 to 10, how do you like my painting?"

Estimating Whole Number Products and Quotients (5.NBT.4)

Example 1 **Estimate 32 × 88.**

32 is close to 30.

$$32 \times 88 \approx 30 \times 90 = 2700$$

88 is close to 90.

Example 2 **Estimate 176 ÷ 57.**

176 is close to 180.

$$176 \div 57 \approx 180 \div 60 = 3$$

57 is close to 60.

Try It Yourself

Estimate the product or the quotient.

1. 9×23 **2.** 19×22 **3.** 49×21 **4.** 38×61

5. $38 \div 9$ **6.** $63 \div 22$ **7.** $118 \div 19$ **8.** $245 \div 62$

Multiplying and Dividing Whole Numbers (5.NBT.5, 5.NBT.6)

Example 3 **Find 356 × 21.**

```
   11
  356
×  21
  356
+7120
 7476
```

Example 4 **Estimate 765 ÷ 3.**

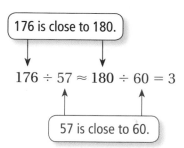

```
   255
 3)765
  -6↓
   16
  -15↓
    15
   -15
     0
```

Try It Yourself

Find the product or the quotient.

9. 425×9 **10.** 721×18 **11.** 599×29 **12.** 503×12

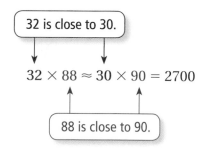

13. $7\overline{)280}$ **14.** $4\overline{)428}$ **15.** $14\overline{)532}$ **16.** $23\overline{)8303}$

2.1 Multiplying Fractions

Essential Question What does it mean to multiply fractions?

1 ACTIVITY: Multiplying Fractions

Work with a partner. A bottle of water is $\frac{1}{2}$ full. You drink $\frac{2}{3}$ of the water. How much of the bottle of water do you drink?

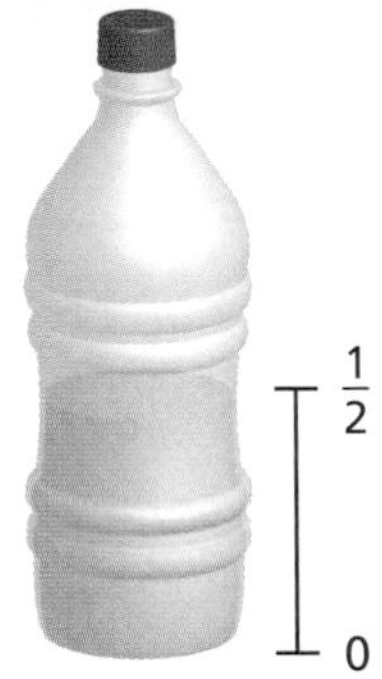

THINK ABOUT THE QUESTION: To help you think about this question, rewrite the question.

Words: What is $\frac{2}{3}$ of $\frac{1}{2}$? **Numbers:** $\frac{2}{3} \times \frac{1}{2} = ?$

Here is one way to get the answer.

- **Draw** a length of $\frac{1}{2}$.

Because you want to find $\frac{2}{3}$ of the length, divide it into 3 equal sections.

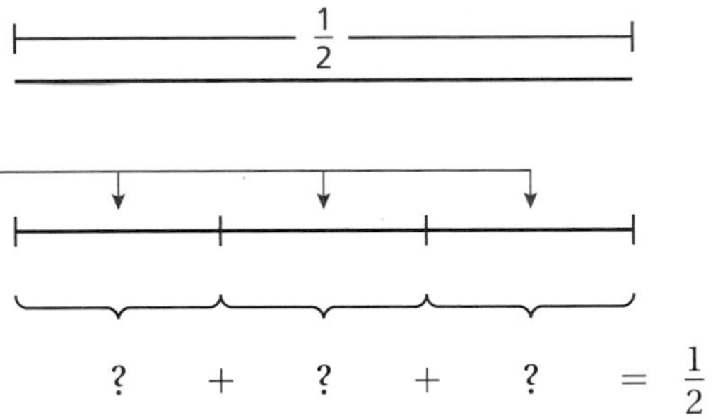

Now, you need to think of a way to divide $\frac{1}{2}$ into 3 equal parts.

- **Rewrite** $\frac{1}{2}$ as a fraction whose numerator is divisible by 3.

Because the length is divided into 3 equal sections, multiply the numerator and denominator by 3.

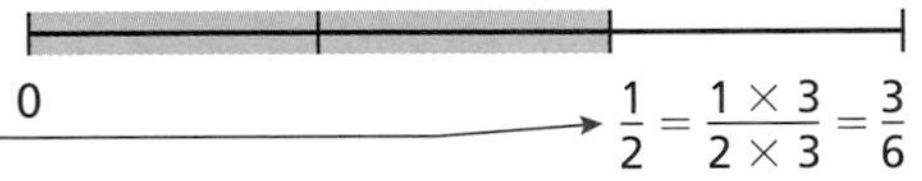

In this form, you see that $\frac{3}{6}$ can be divided into 3 equal parts of $\frac{1}{6}$.

- Each part is $\frac{1}{6}$ of the bottle of water, and you drank two of them. Written as multiplication, you have

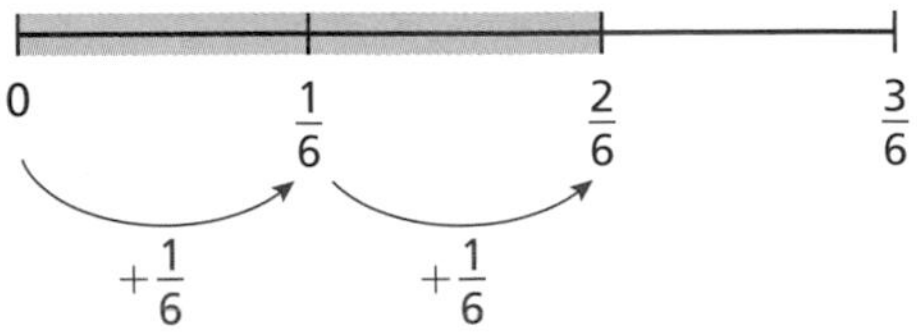

$$\frac{2}{3} \times \frac{1}{2} = \frac{2}{\square} = \frac{\square}{\square}.$$

So, you drank $\frac{\square}{\square}$ of the bottle of water.

COMMON CORE

Dividing Fractions

In this lesson, you will
- use models to multiply fractions.
- multiply fractions by fractions.

Preparing for Standard 6.NS.1

2 ACTIVITY: Multiplying Fractions

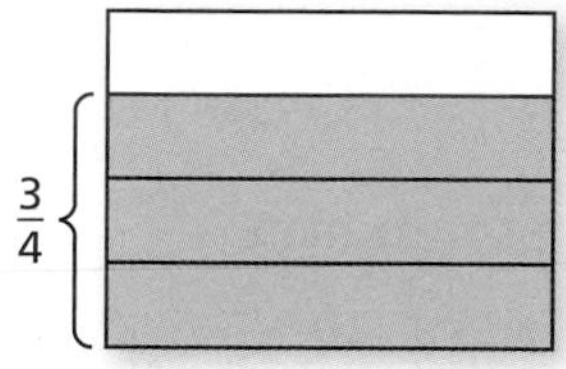

Work with a partner. A park has a playground that is $\frac{3}{4}$ of its width and $\frac{4}{5}$ of its length. What fraction of the park is covered by the playground?

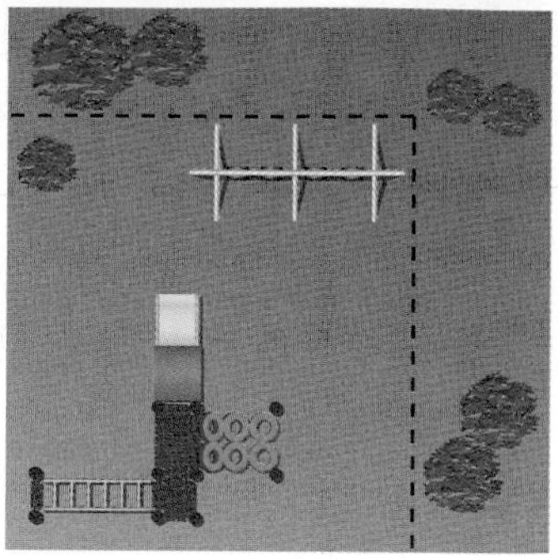

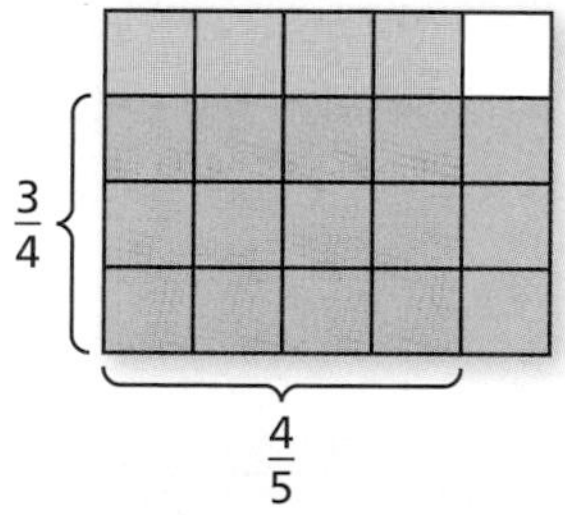

Fold a piece of paper horizontally into fourths and shade three of the fourths to represent $\frac{3}{4}$.

Fold the paper vertically into fifths and shade $\frac{4}{5}$ of the paper another color.

Count the total number of squares. This number is the denominator. The numerator is the number of squares shaded with both colors.

$\frac{3}{4} \times \frac{4}{5} = \frac{\square}{\square} = \frac{\square}{\square}$. So, $\frac{\square}{\square}$ of the park is covered by the playground.

Inductive Reasoning

Work with a partner. Complete the table by using a model or folding paper.

	Exercise	Verbal Expression	Answer
1	**3.** $\frac{2}{3} \times \frac{1}{2}$	$\frac{2}{3}$ of $\frac{1}{2}$	$\frac{\square}{\square}$
2	**4.** $\frac{3}{4} \times \frac{4}{5}$	$\frac{3}{4}$ of $\frac{4}{5}$	$\frac{\square}{\square}$
	5. $\frac{2}{3} \times \frac{5}{6}$		
	6. $\frac{1}{6} \times \frac{1}{4}$		
	7. $\frac{2}{5} \times \frac{1}{2}$		
	8. $\frac{5}{8} \times \frac{4}{5}$		

Math Practice 1

Consider Similar Problems

What are the similarities in constructing the models for each problem? What are the differences?

What Is Your Answer?

9. IN YOUR OWN WORDS What does it mean to multiply fractions?

10. STRUCTURE Write a general rule for multiplying fractions.

Practice Use what you learned about multiplying fractions to complete Exercises 4–11 on page 59.

2.1 Lesson

Check It Out
Lesson Tutorials
BigIdeasMath.com

Key Idea

Multiplying Fractions

Words Multiply the numerators and multiply the denominators.

Numbers $\frac{3}{7} \times \frac{1}{2} = \frac{3 \times 1}{7 \times 2} = \frac{3}{14}$

Algebra $\frac{a}{b} \cdot \frac{c}{d} = \frac{a \cdot c}{b \cdot d}$, where $b, d \neq 0$

EXAMPLE 1 Multiplying Fractions

Find $\frac{1}{5} \times \frac{1}{3}$.

$\frac{1}{5} \times \frac{1}{3} = \frac{1 \times 1}{5 \times 3}$ — Multiply the numerators. Multiply the denominators.

$= \frac{1}{15}$ — Simplify.

EXAMPLE 2 Multiplying Fractions with Common Factors

Find $\frac{8}{9} \times \frac{3}{4}$. **Estimate** $1 \times \frac{3}{4} = \frac{3}{4}$

$\frac{8}{9} \times \frac{3}{4} = \frac{8 \times 3}{9 \times 4}$ — Multiply the numerators. Multiply the denominators.

$= \frac{\overset{2}{\cancel{8}} \times \overset{1}{\cancel{3}}}{\underset{3}{\cancel{9}} \times \underset{1}{\cancel{4}}}$ — Divide out common factors.

$= \frac{2}{3}$ — Simplify.

∴ The product is $\frac{2}{3}$. **Reasonable?** $\frac{2}{3} \approx \frac{3}{4}$ ✓

Study Tip

When the numerator of one fraction is the same as the denominator of another fraction, you can use mental math to multiply. For example, $\frac{4}{5} \times \frac{5}{9} = \frac{4}{9}$ because you can divide out the common factor 5.

On Your Own

Now You're Ready
Exercises 4–19

Multiply. Write the answer in simplest form.

1. $\frac{1}{2} \times \frac{5}{6}$
2. $\frac{7}{8} \times \frac{1}{4}$
3. $\frac{3}{7} \times \frac{2}{3}$
4. $\frac{4}{9} \times \frac{3}{10}$

EXAMPLE 3 Real-Life Application

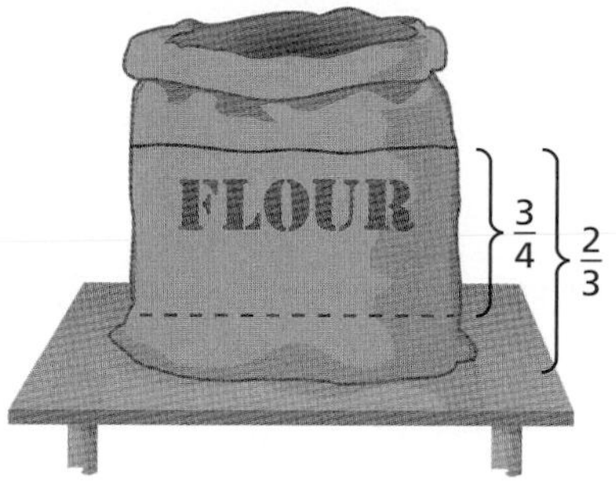

You have $\frac{2}{3}$ of a bag of flour. You use $\frac{3}{4}$ of the flour to make empanada dough. How much of the entire bag do you use to make the dough?

Method 1: Use a model. Six of the 12 squares have both types of shading.

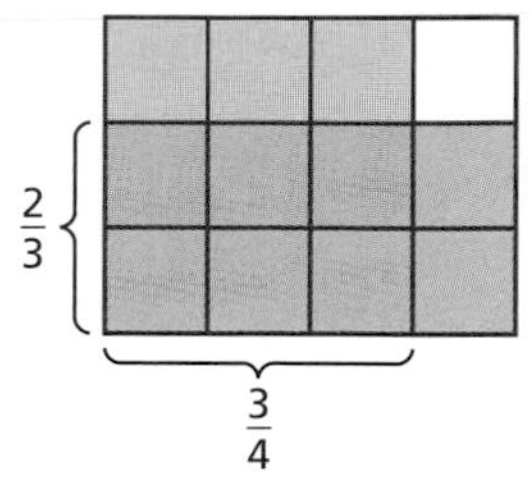

So, you use $\frac{6}{12} = \frac{1}{2}$ of the entire bag.

Method 2: To find $\frac{3}{4}$ of $\frac{2}{3}$, multiply.

$$\frac{3}{4} \times \frac{2}{3} = \frac{\overset{1}{\cancel{3}} \times \overset{1}{\cancel{2}}}{\underset{2}{\cancel{4}} \times \underset{1}{\cancel{3}}}$$ Multiply the numerators and the denominators. Divide out common factors.

$$= \frac{1}{2}$$ Simplify.

So, you use $\frac{1}{2}$ of the entire bag.

On Your Own

5. **WHAT IF?** In Example 3, you use $\frac{1}{4}$ of the flour to make the dough. How much of the entire bag do you use to make the dough?

Key Idea

Multiplying Mixed Numbers

Write each mixed number as an improper fraction. Then multiply as you would with fractions.

EXAMPLE 4 Multiplying a Fraction and a Mixed Number

Find $\frac{1}{2} \times 2\frac{3}{4}$. **Estimate $\frac{1}{2} \times 3 = 1\frac{1}{2}$**

$$\frac{1}{2} \times 2\frac{3}{4} = \frac{1}{2} \times \frac{11}{4}$$ Write $2\frac{3}{4}$ as the improper fraction $\frac{11}{4}$.

$$= \frac{1 \times 11}{2 \times 4}$$ Multiply the numerators and the denominators.

$$= \frac{11}{8}, \text{ or } 1\frac{3}{8}$$ Simplify.

The product is $1\frac{3}{8}$. **Reasonable?** $1\frac{3}{8} \approx 1\frac{1}{2}$ ✓

EXAMPLE 5 Multiplying Mixed Numbers

Find $1\frac{4}{5} \times 3\frac{2}{3}$. **Estimate $2 \times 4 = 8$**

$1\frac{4}{5} \times 3\frac{2}{3} = \frac{9}{5} \times \frac{11}{3}$ Write $1\frac{4}{5}$ and $3\frac{2}{3}$ as improper fractions.

$= \frac{\overset{3}{\cancel{9}} \times 11}{5 \times \underset{1}{\cancel{3}}}$ Multiply fractions. Divide out the common factor 3.

$= \frac{33}{5}$, or $6\frac{3}{5}$ Simplify.

∴ The product is $6\frac{3}{5}$. **Reasonable? $6\frac{3}{5} \approx 8$ ✓**

On Your Own

Now You're Ready
Exercises 26–41

Multiply. Write the answer in simplest form.

6. $\frac{1}{3} \times 1\frac{1}{6}$
7. $3\frac{1}{2} \times \frac{4}{9}$
8. $1\frac{7}{8} \times 2\frac{2}{5}$
9. $5\frac{5}{7} \times 2\frac{1}{10}$

EXAMPLE 6 Real-Life Application

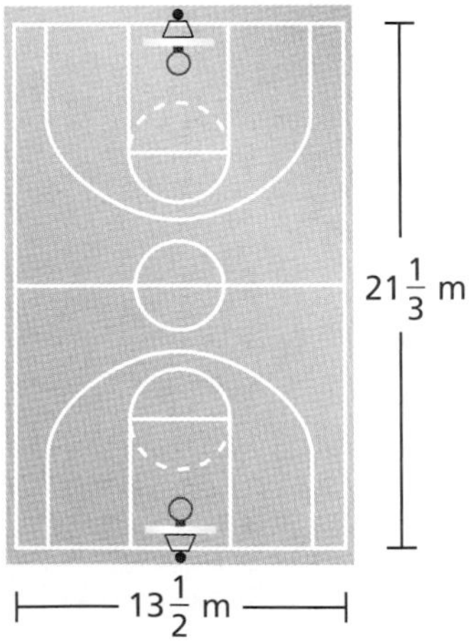

A city is resurfacing a basketball court. Find the area of the court.

Estimate $21 \times 14 = 294$

$A = \ell w$ Write the formula for the area of a rectangle.

$= 21\frac{1}{3} \times 13\frac{1}{2}$ Substitute $21\frac{1}{3}$ for ℓ and $13\frac{1}{2}$ for w.

$= \frac{64}{3} \times \frac{27}{2}$ Write $21\frac{1}{3}$ and $13\frac{1}{2}$ as improper fractions.

$= \frac{\overset{32}{\cancel{64}} \times \overset{9}{\cancel{27}}}{\underset{1}{\cancel{3}} \times \underset{1}{\cancel{2}}}$ Multiply fractions. Divide out common factors.

$= 288$ Simplify.

∴ So, the area of the court is 288 square meters.

Reasonable? $288 \approx 294$ ✓

On Your Own

10. Find the area of a rectangular air hockey table that is $8\frac{1}{4}$ feet by $4\frac{3}{8}$ feet.

2.1 Exercises

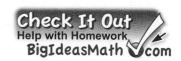

Check It Out
Help with Homework
BigIdeasMath.com

Vocabulary and Concept Check

1. **WRITING** Explain how to multiply two fractions.

2. **REASONING** Name the missing denominator.

$$\frac{3}{7} \times \frac{1}{\square} = \frac{3}{28}$$

3. **OPEN-ENDED** Write two mixed numbers between 3 and 4 that have a product between 9 and 12.

Practice and Problem Solving

Multiply. Write the answer in simplest form.

(1) (2)

4. $\frac{1}{7} \times \frac{2}{3}$
5. $\frac{5}{8} \times \frac{1}{2}$
6. $\frac{1}{4} \times \frac{2}{5}$
7. $\frac{3}{7} \times \frac{1}{4}$
8. $\frac{2}{3} \times \frac{4}{7}$
9. $\frac{5}{7} \times \frac{7}{8}$
10. $\frac{3}{8} \times \frac{1}{9}$
11. $\frac{5}{6} \times \frac{2}{5}$
12. $\frac{5}{12} \times 10$
13. $6 \times \frac{7}{8}$
14. $\frac{3}{4} \times \frac{8}{15}$
15. $\frac{4}{9} \times \frac{4}{5}$
16. $\frac{3}{7} \times \frac{3}{7}$
17. $\frac{5}{6} \times \frac{2}{9}$
18. $\frac{13}{18} \times \frac{6}{7}$
19. $\frac{7}{9} \times \frac{21}{10}$

20. **ERROR ANALYSIS** Describe and correct the error in finding the product.

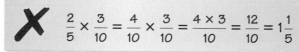

✗ $\frac{2}{5} \times \frac{3}{10} = \frac{4}{10} \times \frac{3}{10} = \frac{4 \times 3}{10} = \frac{12}{10} = 1\frac{1}{5}$

21. **AQUARIUM** In an aquarium, $\frac{2}{5}$ of the fish are surgeonfish. Of these, $\frac{3}{4}$ are yellow tangs. What fraction of all fish in the aquarium are yellow tangs?

22. **JUMP ROPE** You exercise for $\frac{3}{4}$ of an hour. You jump rope for $\frac{1}{3}$ of that time. What fraction of the hour do you spend jumping rope?

Without finding the product, copy and complete the statement using <, >, or =. Explain your reasoning.

23. $\frac{4}{7} \ \square \ \left(\frac{9}{10} \times \frac{4}{7}\right)$
24. $\left(\frac{5}{8} \times \frac{22}{15}\right) \ \square \ \frac{5}{8}$
25. $\frac{5}{6} \ \square \ \left(\frac{5}{6} \times \frac{7}{7}\right)$

Multiply. Write the answer in simplest form.

4 5 26. $1\frac{1}{3} \times \frac{2}{3}$ 27. $6\frac{2}{3} \times \frac{3}{10}$ 28. $2\frac{1}{2} \times \frac{4}{5}$ 29. $\frac{3}{5} \times 3\frac{1}{3}$

30. $7\frac{1}{2} \times \frac{2}{3}$ 31. $\frac{5}{9} \times 3\frac{3}{5}$ 32. $\frac{3}{4} \times 1\frac{1}{3}$ 33. $3\frac{3}{4} \times \frac{2}{5}$

34. $4\frac{3}{8} \times \frac{4}{5}$ 35. $\frac{3}{7} \times 2\frac{5}{6}$ 36. $1\frac{3}{10} \times 18$ 37. $15 \times 2\frac{4}{9}$

38. $1\frac{1}{6} \times 6\frac{3}{4}$ 39. $2\frac{5}{12} \times 2\frac{2}{3}$ 40. $5\frac{5}{7} \times 3\frac{1}{8}$ 41. $2\frac{4}{5} \times 4\frac{1}{16}$

ERROR ANALYSIS Describe and correct the error in finding the product.

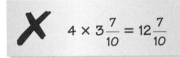

42. ✗ $4 \times 3\frac{7}{10} = 12\frac{7}{10}$

43. ✗ $2\frac{1}{2} \times 7\frac{4}{5} = (2 \times 7) + \left(\frac{1}{2} \times \frac{4}{5}\right)$

$= 14 + \frac{2}{5} = 14\frac{2}{5}$

44. **VITAMIN C** A vitamin C tablet contains $\frac{1}{40}$ of a gram of vitamin C. You take $1\frac{1}{2}$ tablets every day. How many grams of vitamin C do you take every day?

45. **SCHOOL BANNER** You make a banner for a football rally.

a. What is the area of the banner?

b. You add a $\frac{1}{4}$-foot border on each side. What is the new area of the banner?

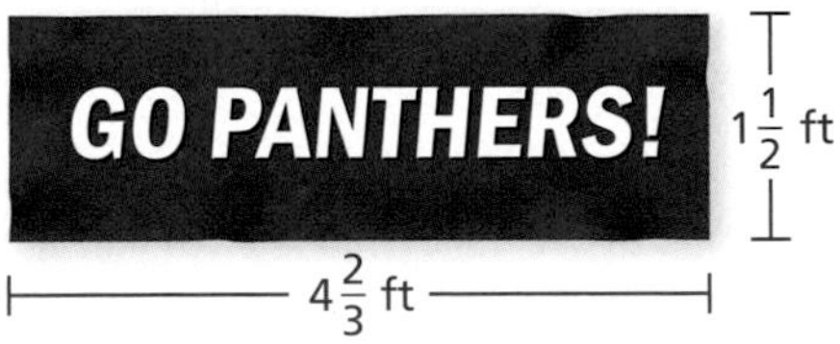

46. **NUMBER SENSE** Without calculating, is $1\frac{1}{6} \cdot \frac{4}{5}$ less than or greater than $1\frac{1}{6}$? Is the product less than or greater than $\frac{4}{5}$? Explain your reasoning.

Multiply. Write the answer in simplest form.

47. $\frac{1}{2} \times \frac{3}{5} \times \frac{4}{9}$ 48. $\frac{4}{7} \times 4\frac{3}{8} \times \frac{5}{6}$ 49. $1\frac{1}{15} \times 5\frac{2}{5} \times 4\frac{7}{12}$

50. $\left(\frac{3}{5}\right)^3$ 51. $\left(\frac{4}{5}\right)^2 \times \left(\frac{3}{4}\right)^2$ 52. $\left(\frac{5}{6}\right)^2 \times \left(1\frac{1}{10}\right)^2$

53. **PICTURES** Three pictures hang side by side on a wall. What is the total area of the wall that the pictures cover?

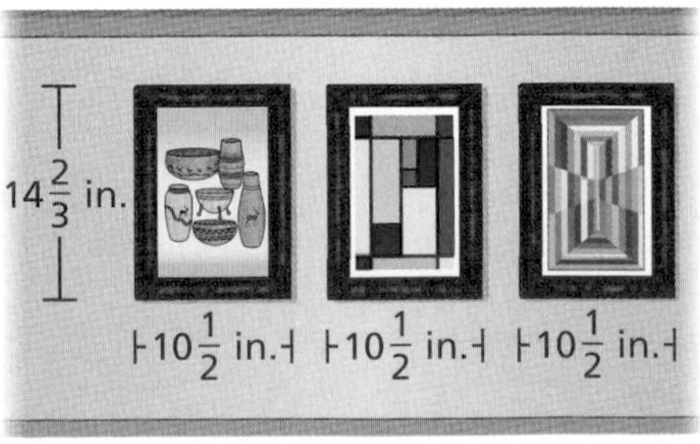

54. **OPEN-ENDED** Find a fraction that, when multiplied by $\frac{1}{2}$, is less than $\frac{1}{4}$.

55. **DISTANCES** You are in a bike race. When you get to the first checkpoint, you are $\frac{2}{5}$ of the distance to the second checkpoint. When you get to the second checkpoint, you are $\frac{1}{4}$ of the distance to the finish. What is the distance from the start to the first checkpoint?

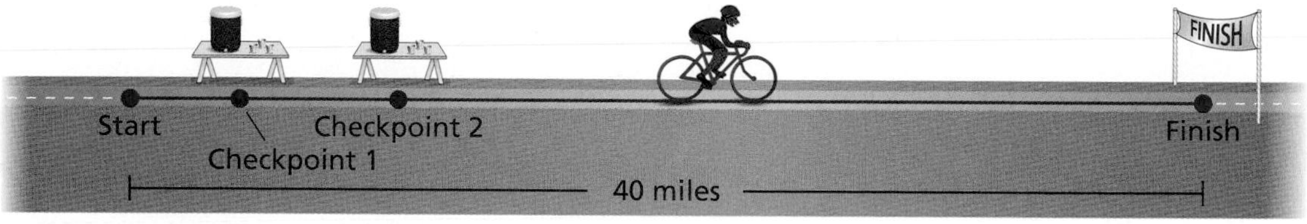

56. **NUMBER SENSE** Is the product of two positive mixed numbers ever less than 1? Explain.

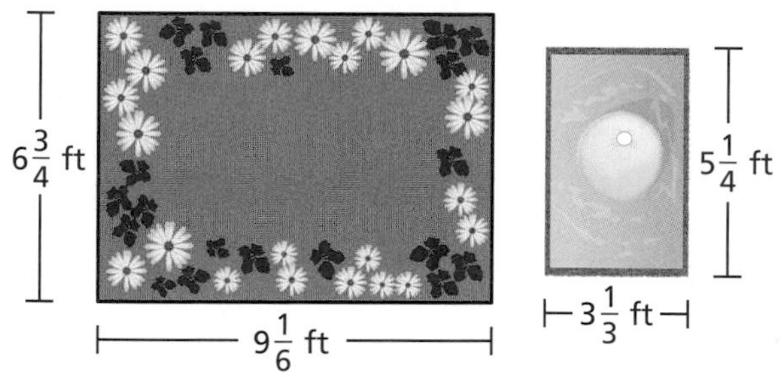

57. **MODELING** You plan to add a fountain to your garden.

   a. Draw a diagram of the fountain in the garden. Label the dimensions.

   b. Describe two methods for finding the area of the garden that surrounds the fountain.

   c. Find the area. Which method did you use, and why?

58. **COOKING** The cooking time for a ham is $\frac{2}{5}$ of an hour for each pound.

   a. How long should you cook a ham that weighs $12\frac{3}{4}$ pounds?

   b. Dinner time is 4:45 P.M. What time should you start cooking the ham?

59. **PETS** You ask 150 people about their pets. The results show that $\frac{9}{25}$ of the people own a dog. Of the people who own a dog, $\frac{1}{6}$ of them also own a cat.

   a. What fraction of the people own a dog and a cat?

   b. Reasoning How many people own a dog but not a cat? Explain.

Fair Game Review What you learned in previous grades & lessons

Find the prime factorization of the number. *(Section 1.4)*

60. 24 **61.** 45 **62.** 53 **63.** 60

64. **MULTIPLE CHOICE** A science experiment calls for $\frac{3}{4}$ cup of baking powder. You have $\frac{1}{3}$ cup of baking powder. How much more baking powder do you need? *(Section 1.6)*

Ⓐ $\frac{1}{4}$ cup Ⓑ $\frac{5}{12}$ cup Ⓒ $\frac{4}{7}$ cup Ⓓ $1\frac{1}{12}$ cups

2.2 Dividing Fractions

Essential Question How can you divide by a fraction?

1 ACTIVITY: Dividing by a Fraction

Work with a partner. Write the division problem and solve it using a model.

a. How many two-thirds are in three?

The division problem is $\square \div \dfrac{\square}{\square}$.

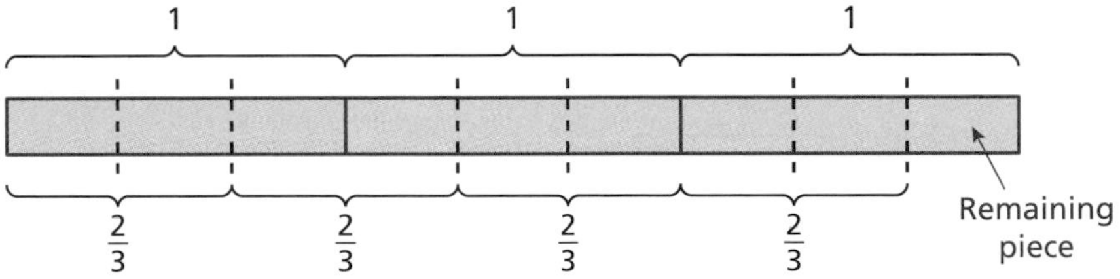

How many groups of $\dfrac{2}{3}$ are in 3? $\square$

The remaining piece represents $\dfrac{\square}{\square}$ of $\dfrac{2}{3}$.

So, there are $\square\dfrac{\square}{\square}$ groups of $\dfrac{2}{3}$ in 3.

So, $\square \div \dfrac{\square}{\square} = \square\dfrac{\square}{\square}$.

b. How many halves are in five halves?

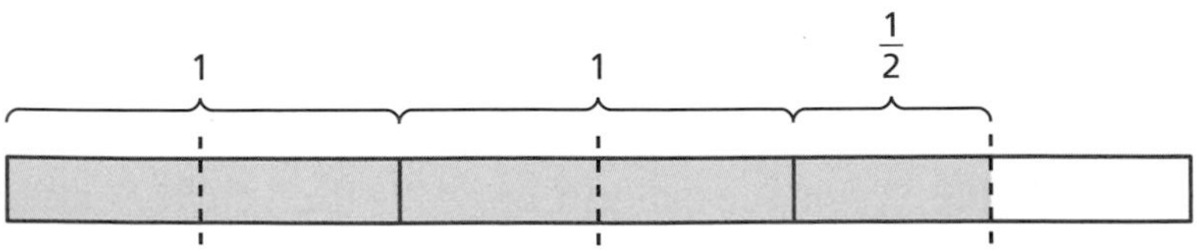

c. How many four-fifths are in eight?

d. How many one-thirds are in seven halves?

e. How many three-fourths are in five halves?

Common Core

Dividing Fractions

In this lesson, you will

- write reciprocals of numbers.
- use models to divide fractions.
- divide fractions by fractions.
- solve real-life problems.

Learning Standard 6.NS.1

2 ACTIVITY: Using Tables to Recognize a Pattern

Work with a partner.

a. Complete each table.

Division Table

$8 \div 16$	$\frac{1}{2}$
$8 \div 8$	1
$8 \div 4$	2
$8 \div 2$	4
$8 \div 1$	8
$8 \div \frac{1}{2}$	
$8 \div \frac{1}{4}$	
$8 \div \frac{1}{8}$	

Multiplication Table

$8 \times \frac{1}{16}$	$\frac{1}{2}$
$8 \times \frac{1}{8}$	1
$8 \times \frac{1}{4}$	2
$8 \times \frac{1}{2}$	4
8×1	8
8×2	
8×4	
8×8	

Math Practice 4

Analyze Relationships

How is multiplying numbers similar to dividing numbers?

b. Describe the relationship between the red numbers in the division table and the red numbers in the multiplication table.

c. Describe the relationship between the blue numbers in the division table and the blue numbers in the multiplication table.

d. **STRUCTURE** Make a conjecture about how you can use multiplication to divide by a fraction.

e. Test your conjecture using the problems in Activity 1.

What Is Your Answer?

3. **IN YOUR OWN WORDS** How can you divide by a fraction? Give an example.

4. How many halves are in a fourth? Explain how you found your answer.

Practice Use what you learned about dividing fractions to complete Exercises 11–18 on page 67.

2.2 Lesson

Check It Out
Lesson Tutorials
BigIdeasMath.com

Key Vocabulary
reciprocals, *p. 64*

Two numbers whose product is 1 are **reciprocals**. To write the reciprocal of a number, write the number as a fraction. Then invert the fraction. So, the reciprocal of a fraction $\frac{a}{b}$ is $\frac{b}{a}$, where a and $b \neq 0$.

The Meaning of a Word ● Invert

When you **invert** a glass, you turn it over.

Study Tip

The product of a nonzero number and its reciprocal is 1.

$$\frac{a}{b} \cdot \frac{b}{a} = 1$$

This is called the *Multiplicative Inverse Property*. You will learn more about this property in Chapter 7.

EXAMPLE 1 Writing Reciprocals

	Original Number	*Fraction*	*Reciprocal*	*Check*
a.	$\frac{3}{5}$	$\frac{3}{5}$	$\frac{5}{3}$	$\frac{3}{5} \times \frac{5}{3} = 1$
b.	$\frac{9}{5}$	$\frac{9}{5}$	$\frac{5}{9}$	$\frac{9}{5} \times \frac{5}{9} = 1$
c.	2	$\frac{2}{1}$	$\frac{1}{2}$	$\frac{2}{1} \times \frac{1}{2} = 1$

Study Tip

When any number is multiplied by 0, the product is 0. So, the number 0 does not have a reciprocal.

On Your Own

Now You're Ready
Exercises 7–10

Write the reciprocal of the number.

1. $\frac{3}{4}$ **2.** 5 **3.** $\frac{7}{2}$ **4.** $\frac{4}{9}$

Key Idea

Dividing Fractions

Words To divide a number by a fraction, multiply the number by the reciprocal of the fraction.

Numbers $\frac{1}{5} \div \frac{3}{4} = \frac{1}{5} \times \frac{4}{3} = \frac{1 \times 4}{5 \times 3}$

Algebra $\frac{a}{b} \div \frac{c}{d} = \frac{a}{b} \cdot \frac{d}{c} = \frac{a \cdot d}{b \cdot c}$, where b, c, and $d \neq 0$

Multi-Language Glossary at BigIdeasMath.com

EXAMPLE 2 Dividing a Fraction by a Fraction

Find $\frac{1}{6} \div \frac{2}{3}$.

$$\frac{1}{6} \div \frac{2}{3} = \frac{1}{6} \times \frac{3}{2}$$ Multiply by the reciprocal of $\frac{2}{3}$, which is $\frac{3}{2}$.

$$= \frac{1 \times \cancel{3}^{1}}{{}_{2}\cancel{6} \times 2}$$ Multiply fractions. Divide out the common factor 3.

$$= \frac{1}{4}$$ Simplify.

EXAMPLE 3 Dividing a Whole Number by a Fraction

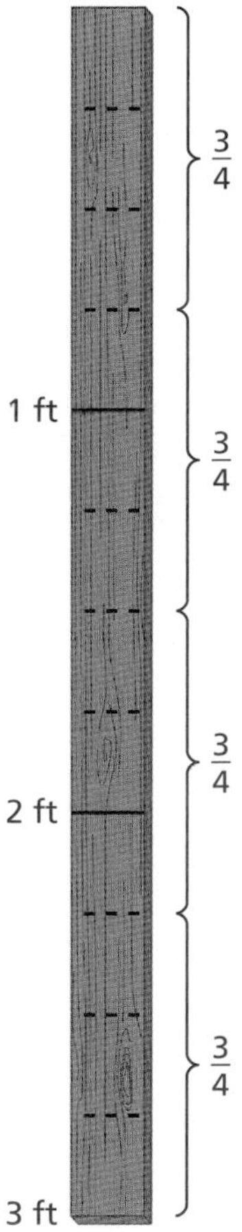

A piece of wood is 3 feet long. How many $\frac{3}{4}$-foot pieces can you cut from the piece of wood?

Method 1: Draw a diagram. Mark each foot on the diagram. Then divide each foot into $\frac{1}{4}$-foot sections.

Count the number of $\frac{3}{4}$-foot pieces of wood. There are four.

∴ So, you can cut four $\frac{3}{4}$-foot pieces from the piece of wood.

Method 2: Divide 3 by $\frac{3}{4}$ to find the number of $\frac{3}{4}$-foot pieces.

$$3 \div \frac{3}{4} = 3 \times \frac{4}{3}$$ Multiply by the reciprocal of $\frac{3}{4}$, which is $\frac{4}{3}$.

$$= \frac{{}^{1}\cancel{3} \times 4}{\cancel{3}_{1}}$$ Multiply. Divide out the common factor 3.

$$= 4$$ Simplify.

∴ So, you can cut four $\frac{3}{4}$-foot pieces from the piece of wood.

On Your Own

Divide. Write the answer in simplest form.

5. $\frac{2}{7} \div \frac{1}{3}$ **6.** $\frac{1}{2} \div \frac{1}{8}$ **7.** $\frac{3}{8} \div \frac{1}{4}$ **8.** $\frac{2}{5} \div \frac{3}{10}$

9. How many $\frac{1}{2}$-foot pieces can you cut from a 7-foot piece of wood?

EXAMPLE 4 Dividing a Fraction by a Whole Number

Find $\frac{4}{5} \div 2$.

$$\frac{4}{5} \div 2 = \frac{4}{5} \div \frac{2}{1}$$ Write 2 as an improper fraction.

$$= \frac{4}{5} \times \frac{1}{2}$$ Multiply by the reciprocal of $\frac{2}{1}$, which is $\frac{1}{2}$.

$$= \frac{\overset{2}{\cancel{4}} \times 1}{5 \times \underset{1}{\cancel{2}}}$$ Multiply fractions. Divide out the common factor 2.

$$= \frac{2}{5}$$ Simplify.

On Your Own

Now You're Ready
Exercises 11–26

Divide. Write the answer in simplest form.

10. $\frac{1}{2} \div 3$

11. $\frac{2}{3} \div 10$

12. $\frac{5}{8} \div 4$

13. $\frac{6}{7} \div 4$

EXAMPLE 5 Using Order of Operations

Evaluate $\frac{3}{8} + \frac{5}{6} \div 5$.

$$\frac{3}{8} + \frac{5}{6} \div 5 = \frac{3}{8} + \frac{5}{6} \times \frac{1}{5}$$ Multiply by the reciprocal of 5, which is $\frac{1}{5}$.

$$= \frac{3}{8} + \frac{\overset{1}{\cancel{5}} \times 1}{6 \times \underset{1}{\cancel{5}}}$$ Multiply $\frac{5}{6}$ and $\frac{1}{5}$. Divide out the common factor 5.

$$= \frac{3}{8} + \frac{1}{6}$$ Simplify.

$$= \frac{18}{48} + \frac{8}{48}$$ Rewrite fractions using a common denominator.

$$= \frac{26}{48}, \text{ or } \frac{13}{24}$$ Simplify.

Study Tip

You can use the LCD, 24, to add the fractions in Example 5.

$$\frac{3}{8} + \frac{1}{6} = \frac{9}{24} + \frac{4}{24} = \frac{13}{24}$$

On Your Own

Now You're Ready
Exercises 43–51

Evaluate the expression. Write the answer in simplest form.

14. $\frac{4}{5} + \frac{2}{5} \div 4$

15. $\frac{3}{8} \div \frac{3}{4} - \frac{1}{6}$

16. $\frac{8}{9} \div 2 \div 8$

2.2 Exercises

Check It Out
Help with Homework
BigIdeasMath.com

Vocabulary and Concept Check

1. **OPEN-ENDED** Write a fraction and its reciprocal.
2. **WHICH ONE DOESN'T BELONG?** Which of the following does *not* belong with the other three? Explain your reasoning.

$\frac{1}{3}$ $\frac{1}{6}$ $\frac{2}{9}$ $\frac{1}{8}$

MATCHING Match the expression with its value.

3. $\frac{2}{5} \div \frac{8}{15}$
4. $\frac{8}{15} \div \frac{2}{5}$
5. $\frac{2}{15} \div \frac{8}{5}$
6. $\frac{8}{5} \div \frac{2}{15}$

A. $\frac{1}{12}$
B. $\frac{3}{4}$
C. 12
D. $1\frac{1}{3}$

Practice and Problem Solving

Write the reciprocal of the number.

1 7. 8
8. $\frac{6}{7}$
9. $\frac{2}{5}$
10. $\frac{8}{11}$

Divide. Write the answer in simplest form.

2 3 4 11. $\frac{1}{8} \div \frac{1}{4}$
12. $\frac{5}{6} \div \frac{2}{7}$
13. $12 \div \frac{3}{4}$
14. $8 \div \frac{2}{5}$
15. $\frac{3}{7} \div 6$
16. $\frac{12}{25} \div 4$
17. $\frac{2}{9} \div \frac{2}{3}$
18. $\frac{8}{15} \div \frac{4}{5}$
19. $\frac{1}{3} \div \frac{1}{9}$
20. $\frac{7}{10} \div \frac{3}{8}$
21. $\frac{14}{27} \div 7$
22. $\frac{5}{8} \div 15$
23. $\frac{27}{32} \div \frac{7}{8}$
24. $\frac{4}{15} \div \frac{10}{13}$
25. $9 \div \frac{4}{9}$
26. $10 \div \frac{5}{12}$

ERROR ANALYSIS Describe and correct the error in finding the quotient.

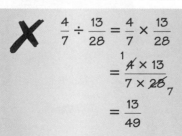

27. ✗

$$\frac{4}{7} \div \frac{13}{28} = \frac{4}{7} \times \frac{13}{28} = \frac{{}^{1}\cancel{4} \times 13}{7 \times \cancel{28}_{7}} = \frac{13}{49}$$

28. ✗

$$\frac{2}{5} \div \frac{8}{9} = \frac{5}{2} \times \frac{8}{9} = \frac{5 \times \cancel{8}^{4}}{{}_{1}\cancel{2} \times 9} = \frac{20}{9}$$

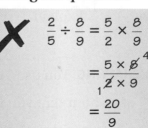

29. **REASONING** How can you use estimation to show that the quotient in Exercise 28 is incorrect?

30. **APPLE PIE** You have $\frac{3}{5}$ of an apple pie. You divide the remaining pie into 5 equal slices. What fraction of the original pie is each slice?

31. **ANIMALS** How many times longer is the baby alligator than the baby gecko?

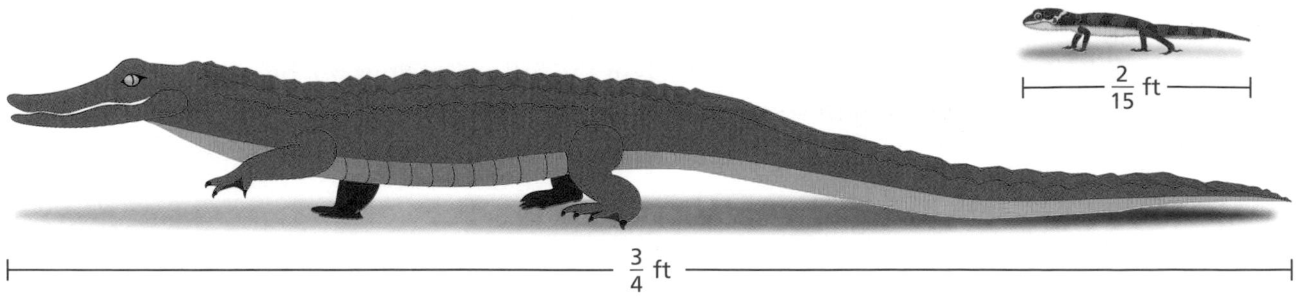

$\frac{2}{15}$ ft

$\frac{3}{4}$ ft

Determine whether the numbers are reciprocals. If not, write the reciprocal of each number.

32. $9, \frac{1}{9}$ 33. $\frac{4}{5}, \frac{10}{8}$ 34. $\frac{5}{6}, \frac{15}{18}$ 35. $\frac{6}{5}, \frac{5}{6}$

Copy and complete the statement.

36. $\frac{5}{12} \times \rule{1cm}{0.4cm} = 1$ 37. $3 \times \rule{1cm}{0.4cm} = 1$ 38. $7 \div \rule{1cm}{0.4cm} = 56$

Without finding the quotient, copy and complete the statement using <, >, or =. Explain your reasoning.

39. $5 \div \frac{7}{9} \ \rule{1cm}{0.4cm} \ 5$ 40. $\frac{3}{7} \div 1 \ \rule{1cm}{0.4cm} \ \frac{3}{7}$ 41. $8 \div \frac{3}{4} \ \rule{1cm}{0.4cm} \ 8$ 42. $\frac{5}{6} \div \frac{7}{8} \ \rule{1cm}{0.4cm} \ \frac{5}{6}$

Evaluate the expression. Write the answer in simplest form.

5 43. $\frac{1}{6} \div 6 \div 6$ 44. $\frac{7}{12} \div 14 \div 6$ 45. $\frac{3}{5} \div \frac{4}{7} \div \frac{9}{10}$

46. $4 \div \frac{8}{9} - \frac{1}{2}$ 47. $\frac{3}{4} + \frac{5}{6} \div \frac{2}{3}$ 48. $\frac{7}{8} - \frac{3}{8} \div 9$

49. $\frac{9}{16} \div \frac{3}{4} \cdot \frac{2}{13}$ 50. $\frac{3}{14} \cdot \frac{2}{5} \div \frac{6}{7}$ 51. $\frac{10}{27} \cdot \left(\frac{3}{8} \div \frac{5}{24} \right)$

52. **REASONING** Use a model to evaluate the quotient $\frac{1}{2} \div \frac{1}{6}$. Explain.

53. **VIDEO CHATTING** You use $\frac{1}{8}$ of your battery for every $\frac{2}{5}$ of an hour that you video chat. You use $\frac{3}{4}$ of your battery video chatting. How long did you video chat?

54. **NUMBER SENSE** When is the reciprocal of a fraction a whole number? Explain.

55. **BUDGETS** The table shows the portions of a family budget that are spent on several expenses.

Expense	Portion of Budget
Housing	$\frac{1}{4}$
Food	$\frac{1}{12}$
Automobiles	$\frac{1}{15}$
Recreation	$\frac{1}{40}$

a. How many times more is the expense for housing than for automobiles?

b. How many times more is the expense for food than for recreation?

c. The expense for automobile fuel is $\frac{1}{60}$ of the total expenses. What fraction of the automobile expense is spent on fuel?

56. **PROBLEM SOLVING** You have 6 pints of glaze. It takes $\frac{7}{8}$ of a pint to glaze a bowl and $\frac{9}{16}$ of a pint to glaze a plate.

a. How many bowls could you glaze? How many plates could you glaze?

b. You want to glaze 5 bowls, and then use the rest for plates. How many plates can you glaze? How much glaze will be left over?

c. How many of each object could you glaze so that there is no glaze left over? Explain how you found your answer.

57. **Reasoning** A water tank is $\frac{1}{8}$ full. The tank is $\frac{3}{4}$ full when 42 gallons of water are added to the tank.

a. How much water can the tank hold?

b. How much water was originally in the tank?

c. How much water is in the tank when it is $\frac{1}{2}$ full?

Fair Game Review What you learned in previous grades & lessons

Find the GCF of the numbers. *(Section 1.5)*

58. 8, 16 59. 24, 66 60. 48, 80 61. 15, 45, 100

62. **MULTIPLE CHOICE** How many inches are in $5\frac{1}{2}$ yards? *(Skills Review Handbook)*

Ⓐ $15\frac{1}{2}$ Ⓑ $16\frac{1}{2}$ Ⓒ 66 Ⓓ 198

2.3 Dividing Mixed Numbers

Essential Question How can you model division by a mixed number?

1 ACTIVITY: Writing a Story

Work with a partner. Think of a story that uses division by a mixed number.

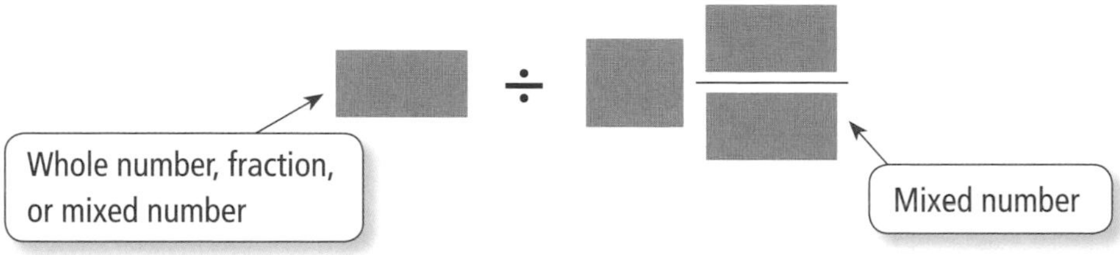

a. Write your story. Then draw pictures for your story.

b. Solve the division problem and use the answer in your story. Include a diagram of the division problem.

There are many possible stories. Here is one that uses $6 \div 1\frac{1}{2}$.

Joe goes on a camping trip with his aunt, his uncle, and three cousins. They leave at 5:00 P.M. and drive 2 hours to the campground.

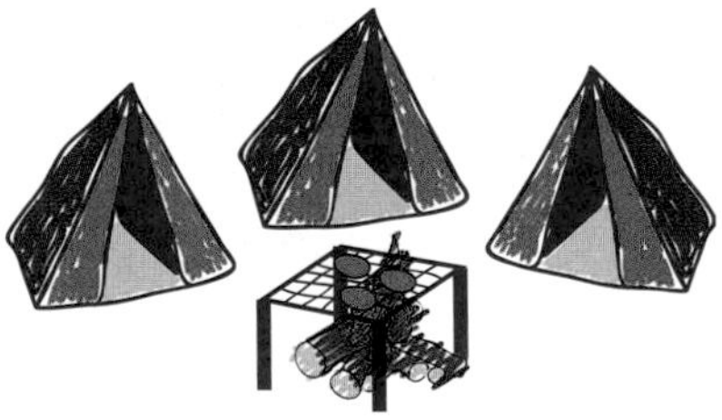

Joe helps his uncle put up three tents. His aunt cooks hamburgers on a grill that is over a fire.

In the morning, Joe tells his aunt that he is making pancakes. He decides to triple the recipe so there will be plenty of pancakes for everyone. A single recipe uses 2 cups of water, so he needs a total of 6 cups.

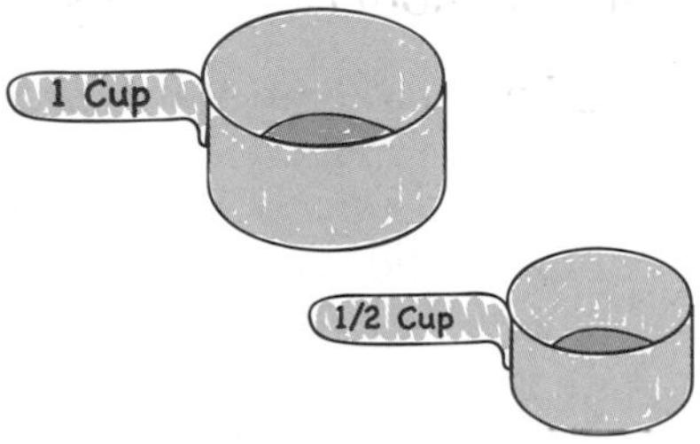

Joe's aunt has a 1-cup measuring cup and a ½-cup measuring cup. The water faucet is about 50 yards from the campsite. Joe tells his cousins that he can get 6 cups of water in only 4 trips.

When his cousins ask him how he knows that, he uses a stick to draw a diagram in the dirt. Joe says, "This diagram shows that there are four 1½s in 6." In other words, $6 \div 1\frac{1}{2} = 4$.

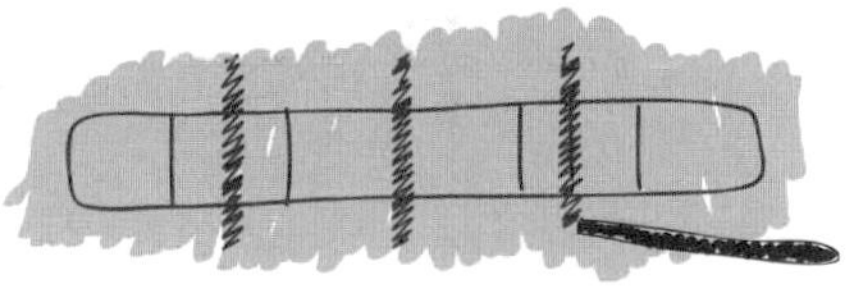

COMMON CORE
Dividing Fractions
In this lesson, you will
- use models to divide mixed numbers.
- divide mixed numbers.
- solve real-life problems.

Learning Standard
6.NS.1

2 ACTIVITY: Dividing Mixed Numbers

Work with a partner. Write the division problem and solve it using a model.

Math Practice 2
Make Sense of Quantities
What values do the parts of the model represent?

a. How many three-fourths are in four and one-half?

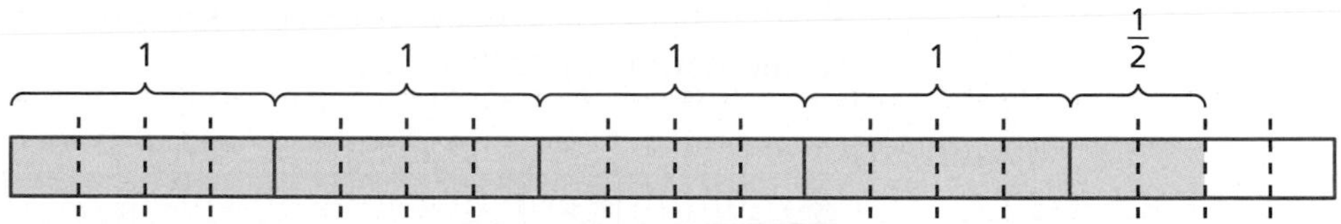

b. How many five-sixths are in three and one-third?

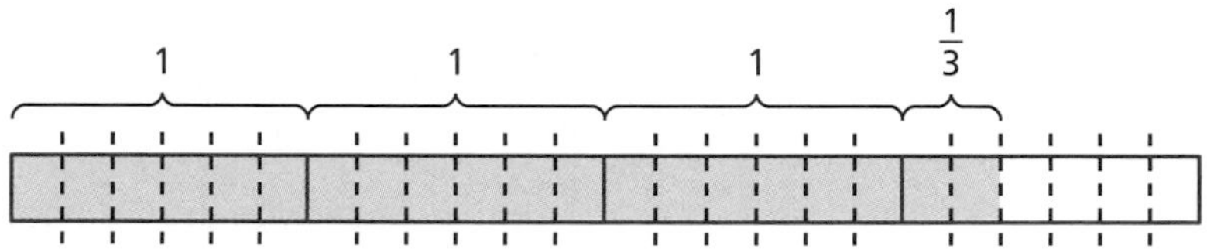

c. How many three-eighths are in three and three-fourths?

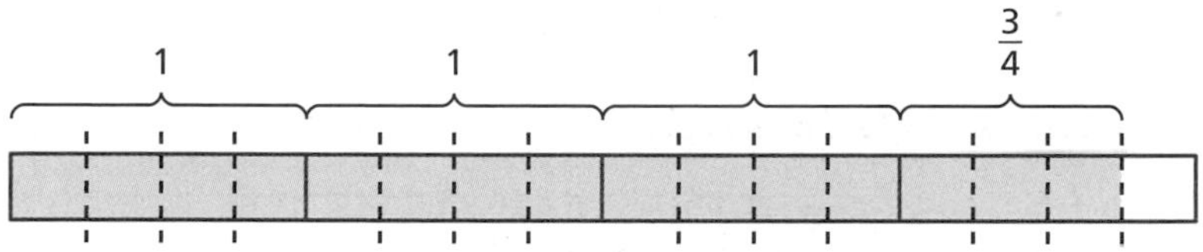

d. How many one and one-halves are in six?

e. How many one and one-fifths are in five?

f. How many one and one-fourths are in four and one-half?

g. How many two and one-thirds are in five and five-sixths?

What Is Your Answer?

3. IN YOUR OWN WORDS How can you model division by a mixed number?

4. Can you think of another method you can use to obtain your answers in Activity 2?

Practice Use what you learned about dividing mixed numbers to complete Exercises 5–12 on page 74.

Check It Out
Lesson Tutorials
BigIdeasMath.com

Key Idea

Dividing Mixed Numbers

Write each mixed number as an improper fraction. Then divide as you would with proper fractions.

EXAMPLE 1 Dividing a Mixed Number by a Fraction

Find $2\frac{1}{4} \div \frac{3}{8}$.

$2\frac{1}{4} \div \frac{3}{8} = \frac{9}{4} \div \frac{3}{8}$ — Write $2\frac{1}{4}$ as the improper fraction $\frac{9}{4}$.

$= \frac{9}{4} \times \frac{8}{3}$ — Multiply by the reciprocal of $\frac{3}{8}$, which is $\frac{8}{3}$.

$= \frac{\overset{3}{\cancel{9}} \times \overset{2}{\cancel{8}}}{\underset{1}{\cancel{4}} \times \underset{1}{\cancel{3}}}$ — Multiply fractions. Divide out common factors.

$= 6$ — Simplify.

Check

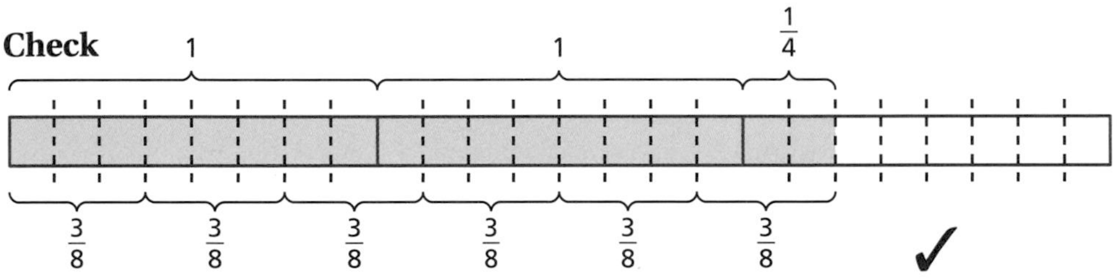

EXAMPLE 2 Dividing Mixed Numbers

Find $3\frac{5}{6} \div 1\frac{2}{3}$. **Estimate** $4 \div 2 = 2$

$3\frac{5}{6} \div 1\frac{2}{3} = \frac{23}{6} \div \frac{5}{3}$ — Write each mixed number as an improper fraction.

$= \frac{23}{6} \times \frac{3}{5}$ — Multiply by the reciprocal of $\frac{5}{3}$, which is $\frac{3}{5}$.

$= \frac{23 \times \overset{1}{\cancel{3}}}{\underset{2}{\cancel{6}} \times 5}$ — Multiply fractions. Divide out common factors.

$= \frac{23}{10}$, or $2\frac{3}{10}$ — Simplify.

So, the quotient is $2\frac{3}{10}$. **Reasonable?** $2\frac{3}{10} \approx 2$ ✓

On Your Own

Now You're Ready
Exercises 5–20

Divide. Write the answer in simplest form.

1. $1\frac{3}{7} \div \frac{2}{3}$
2. $2\frac{1}{6} \div \frac{3}{4}$
3. $8\frac{1}{4} \div 1\frac{1}{2}$
4. $6\frac{4}{5} \div 2\frac{1}{8}$

EXAMPLE 3 Using Order of Operations

Evaluate $5\frac{1}{4} \div 1\frac{1}{8} - \frac{2}{3}$.

Remember

Be sure to check your answers whenever possible. In Example 3, you can use estimation to check that your answer is reasonable.

$5\frac{1}{4} \div 1\frac{1}{8} - \frac{2}{3}$

$\approx 5 \div 1 - 1$

$= 5 - 1$

$= 4$ ✓

$5\frac{1}{4} \div 1\frac{1}{8} - \frac{2}{3} = \frac{21}{4} \div \frac{9}{8} - \frac{2}{3}$ — Write each mixed number as an improper fraction.

$= \frac{21}{4} \times \frac{8}{9} - \frac{2}{3}$ — Multiply by the reciprocal of $\frac{9}{8}$, which is $\frac{8}{9}$.

$= \frac{\overset{7}{\cancel{21}} \times \overset{2}{\cancel{8}}}{\underset{1}{\cancel{4}} \times \underset{3}{\cancel{9}}} - \frac{2}{3}$ — Multiply $\frac{21}{4}$ and $\frac{8}{9}$. Divide out common factors.

$= \frac{14}{3} - \frac{2}{3}$ — Simplify.

$= \frac{12}{3}$, or 4 — Subtract.

EXAMPLE 4 Real-Life Application

One serving of tortilla soup is $1\frac{2}{3}$ cups. A restaurant cook makes 50 cups of soup. Is there enough to serve 35 people? Explain.

Divide 50 by $1\frac{2}{3}$ to find the number of available servings.

$50 \div 1\frac{2}{3} = \frac{50}{1} \div \frac{5}{3}$ — Rewrite each number as an improper fraction.

$= \frac{50}{1} \cdot \frac{3}{5}$ — Multiply by the reciprocal of $\frac{5}{3}$, which is $\frac{3}{5}$.

$= \frac{\overset{10}{\cancel{50}} \cdot 3}{1 \cdot \underset{1}{\cancel{5}}}$ — Multiply fractions. Divide out common factors.

$= 30$ — Simplify.

No. Because 30 is less than 35, there is not enough soup to serve 35 people.

On Your Own

Now You're Ready
Exercises 26–37

Evaluate the expression. Write the answer in simplest form.

5. $1\frac{1}{2} \div \frac{1}{6} - \frac{7}{8}$

6. $3\frac{1}{3} \div \frac{5}{6} + \frac{8}{9}$

7. $\frac{2}{5} + 2\frac{4}{5} \div 1\frac{3}{4}$

8. $\frac{2}{3} - 1\frac{4}{7} \div 4\frac{5}{7}$

9. In Example 4, can 30 cups of tortilla soup serve 15 people? Explain.

2.3 Exercises

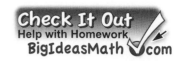

Check It Out
Help with Homework
BigIdeasMath.com

Vocabulary and Concept Check

1. **VOCABULARY** What is the reciprocal of $7\frac{1}{3}$?
2. **NUMBER SENSE** Is $5\frac{1}{4} \div 3\frac{1}{2}$ the same as $3\frac{1}{2} \div 5\frac{1}{4}$? Explain.
3. **NUMBER SENSE** Is the reciprocal of an improper fraction *sometimes, always,* or *never* a proper fraction? Explain.
4. **DIFFERENT WORDS, SAME QUESTION** Which is different? Find "both" answers.

What is $5\frac{1}{2}$ divided by $\frac{1}{8}$?	Find the quotient of $5\frac{1}{2}$ and $\frac{1}{8}$.
What is $5\frac{1}{2}$ times 8?	Find the product of $5\frac{1}{2}$ and $\frac{1}{8}$.

Practice and Problem Solving

Divide. Write the answer in simplest form.

1 2
5. $2\frac{1}{4} \div \frac{3}{4}$
6. $3\frac{4}{5} \div \frac{2}{5}$
7. $8\frac{1}{8} \div \frac{5}{6}$
8. $7\frac{5}{9} \div \frac{4}{7}$
9. $7\frac{1}{2} \div 1\frac{9}{10}$
10. $3\frac{3}{4} \div 2\frac{1}{12}$
11. $7\frac{1}{5} \div 8$
12. $8\frac{4}{7} \div 15$
13. $8\frac{1}{3} \div \frac{2}{3}$
14. $9\frac{1}{6} \div \frac{5}{6}$
15. $13 \div 10\frac{5}{6}$
16. $12 \div 5\frac{9}{11}$
17. $\frac{7}{8} \div 3\frac{1}{16}$
18. $\frac{4}{9} \div 1\frac{7}{15}$
19. $4\frac{5}{16} \div 3\frac{3}{8}$
20. $6\frac{2}{9} \div 5\frac{5}{6}$
21. **ERROR ANALYSIS** Describe and correct the error in finding the quotient.

$$✗\quad 3\frac{1}{2} \div 1\frac{2}{3} = 3\frac{1}{2} \times 1\frac{3}{2} = \frac{7}{2} \times \frac{5}{2} = \frac{35}{4} = 8\frac{3}{4}$$

22. **DOG FOOD** A bag contains 42 cups of dog food. Your dog eats $2\frac{1}{3}$ cups of dog food each day. How many days does the bag of dog food last?
23. **HAMBURGERS** How many $\frac{1}{4}$-pound hamburgers can you make from $3\frac{1}{2}$ pounds of ground beef?
24. **BOOKS** How many $1\frac{3}{5}$-inch-thick books can fit on a $14\frac{1}{2}$-inch-long bookshelf?

25. **LOGIC** Alexei uses the model shown to state that $2\frac{1}{2} \div 1\frac{1}{6} = 2\frac{1}{6}$. Is Alexei correct? Justify your answer using the model.

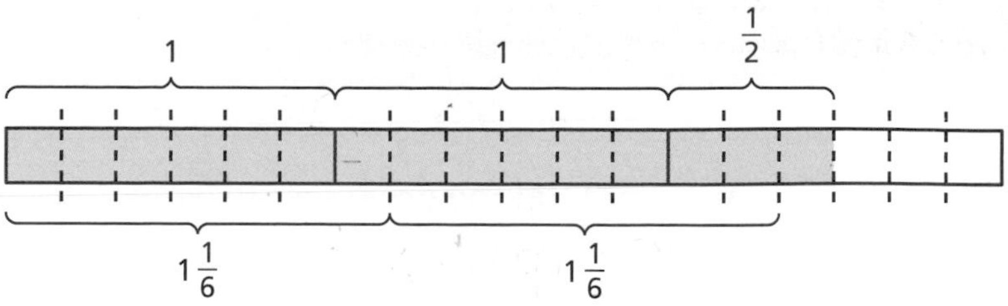

Evaluate the expression. Write the answer in simplest form.

3 26. $3 \div 1\frac{1}{5} + \frac{1}{2}$

27. $4\frac{2}{3} - 1\frac{1}{3} \div 2$

28. $\frac{2}{5} + 2\frac{1}{6} \div \frac{5}{6}$

29. $5\frac{5}{6} \div 3\frac{3}{4} - \frac{2}{9}$

30. $6\frac{1}{2} - \frac{7}{8} \div 5\frac{11}{16}$

31. $9\frac{1}{6} \div 5 + 3\frac{1}{3}$

32. $3\frac{3}{5} + 4\frac{4}{15} \div \frac{4}{9}$

33. $\frac{3}{5} \times \frac{7}{12} \div 2\frac{7}{10}$

34. $4\frac{3}{8} \div \frac{3}{4} \times \frac{4}{7}$

35. $1\frac{9}{11} \times 4\frac{7}{12} \div \frac{2}{3}$

36. $3\frac{4}{15} \div \left(8 \times 6\frac{3}{10}\right)$

37. $2\frac{5}{14} \div \left(2\frac{5}{8} \times 1\frac{3}{7}\right)$

38. **TRAIL MIX** You have 12 cups of granola and $8\frac{1}{2}$ cups of peanuts to make trail mix. What is the greatest number of full batches of trail mix you can make? Explain how you found your answer.

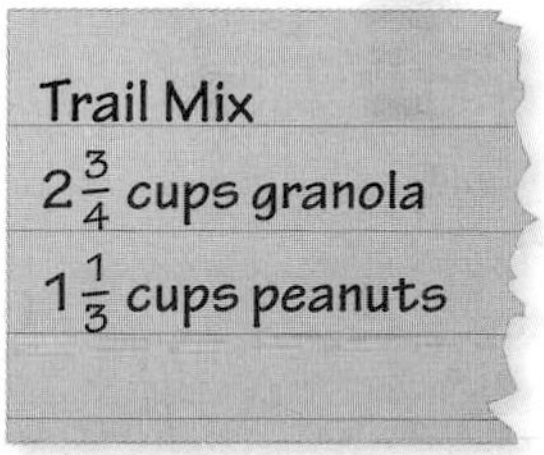

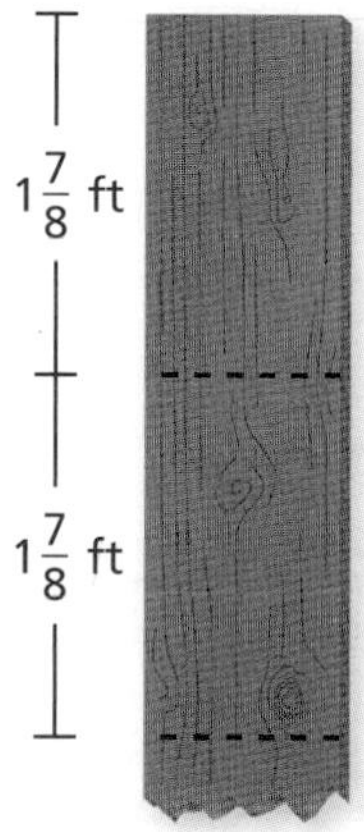

39. **RAMPS** You make skateboard ramps by cutting pieces from a board that is $12\frac{1}{2}$ feet long.

a. Estimate how many ramps you can cut from the board. Is your estimate reasonable? Explain.

b. How many ramps can you cut from the board? How much wood is left over?

40. **Reasoning** At a track meet, the longest shot-put throw by a boy is 25 feet 8 inches. The longest shot-put throw by a girl is 19 feet 3 inches. How many times greater is the longest shot-put throw by the boy than by the girl?

Fair Game Review What you learned in previous grades & lessons

Write the number as a decimal. *(Skills Review Handbook)*

41. forty-three hundredths

42. thirteen thousandths

43. three and eight tenths

44. seven and nine thousandths

45. **MULTIPLE CHOICE** The winner in a vote for class president received $\frac{3}{4}$ of the 240 votes. How many votes did the winner receive? *(Skills Review Handbook)*

Ⓐ 60 Ⓑ 150 Ⓒ 180 Ⓓ 320

2 Study Help

Check It Out
Graphic Organizer
BigIdeasMath.com

You can use a **notetaking organizer** to write notes, vocabulary, and questions about a topic. Here is an example of a notetaking organizer for dividing fractions.

Write important vocabulary or formulas in this space. →

$$\frac{a}{b} \div \frac{c}{d} = \frac{a}{b} \cdot \frac{d}{c} = \frac{a \cdot d}{b \cdot c}$$

(where b, c, and $d \neq 0$)

Dividing fractions

To divide a number by a fraction, multiply the number by the reciprocal of the fraction.

Example:

$$\frac{1}{5} \div \frac{3}{4} = \frac{1}{5} \times \frac{4}{3} = \frac{1 \times 4}{5 \times 3} = \frac{4}{15}$$

← Write your notes about the topic in this space.

How do you divide a mixed number by a fraction?

← Write your questions about the topic in this space.

On Your Own

Make notetaking organizers to help you study these topics.

1. multiplying fractions
2. multiplying mixed numbers
3. dividing mixed numbers

After you complete this chapter, make notetaking organizers for the following topics.

4. adding and subtracting decimals
5. multiplying decimals by whole numbers
6. multiplying decimals by decimals
7. dividing decimals by whole numbers
8. dividing decimals by decimals

"The notetaking organizer in my math class gave me an idea of how to organize my doggy biscuits."

2.1–2.3 Quiz

Check It Out
Progress Check
BigIdeasMath.com

Multiply. Write the answer in simplest form. *(Section 2.1)*

1. $\frac{3}{7} \times \frac{1}{4}$

2. $\frac{9}{10} \times \frac{2}{3}$

3. $1\frac{1}{6} \times \frac{2}{5}$

4. $3\frac{1}{2} \times 5\frac{7}{10}$

Divide. Write the answer in simplest form. *(Section 2.2 and Section 2.3)*

5. $\frac{1}{9} \div \frac{1}{3}$

6. $7 \div \frac{5}{8}$

7. $4\frac{7}{8} \div \frac{1}{8}$

8. $7\frac{2}{3} \div 1\frac{1}{9}$

Evaluate the expression. Write the answer in simplest form. *(Section 2.2 and Section 2.3)*

9. $6 \div \frac{2}{3} + \frac{1}{2}$

10. $\frac{7}{12} \div \frac{1}{4} \times \frac{9}{14}$

11. $3\frac{1}{3} \times 3\frac{3}{4} \div \frac{5}{6}$

12. $6\frac{2}{9} \div \left(4 \times 1\frac{1}{6}\right)$

13. MALL In a mall, $\frac{1}{15}$ of the stores sell shoes. There are 180 stores in the mall. How many of the stores sell shoes? *(Section 2.1)*

14. CONCERT FLOOR The floor of a concert venue is $100\frac{3}{4}$ feet by $75\frac{1}{2}$ feet. What is the area of the floor? *(Section 2.1)*

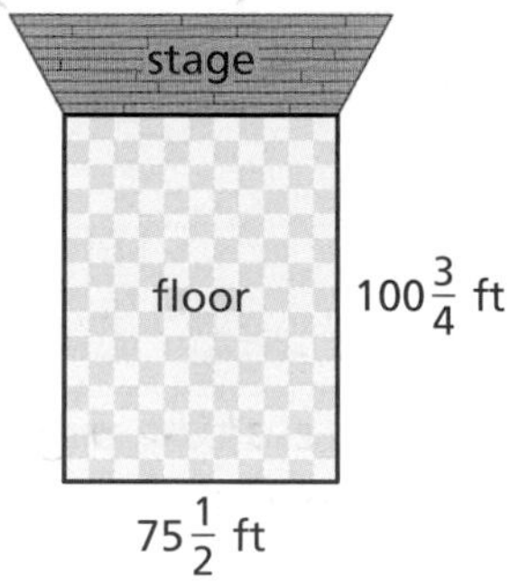

15. BAND Band members make $\frac{2}{3}$ of their profit from selling concert tickets. They make $\frac{1}{5}$ of their profit from selling band merchandise at the concerts. How many times more profit do they make from ticket sales than from merchandise sales? *(Section 2.2)*

16. SKATEBOARDS You are cutting as many $32\frac{1}{4}$-inch sections as you can out of the board to make skateboards. How many skateboards can you make? *(Section 2.3)*

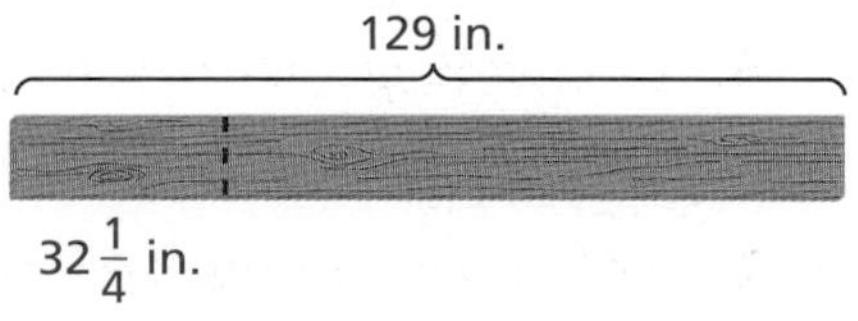

2.4 Adding and Subtracting Decimals

Essential Question How can you add and subtract decimals?

Base ten blocks can be used to model numbers.

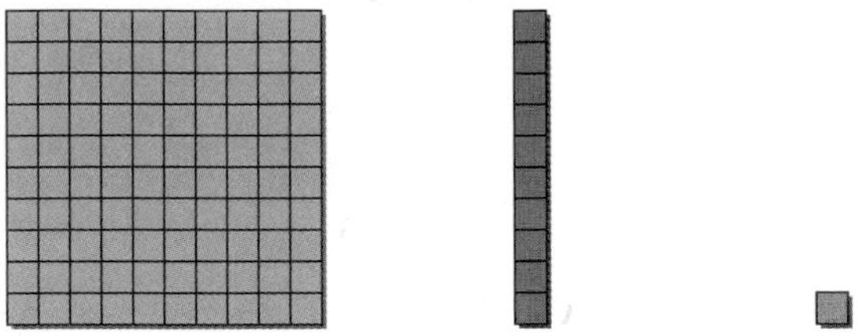

1 one 1 tenth 1 hundredth

1 ACTIVITY: Modeling a Sum

Work with a partner. Use base ten blocks to find the sum.

a. 1.23 + 0.87

Which base ten blocks do you need to model the numbers in the sum? How many of each do you need?

+

1.23 + 0.87

How many of each base ten block do you have when you combine the blocks?

____ ones ____ tenths ____ hundredths

How many of each base ten block do you have when you trade the blocks?

____ ones ____ tenths ____ hundredths

∴ So, 1.23 + 0.87 = ____.

b. 1.25 + 1.35 **c.** 2.14 + 0.92 **d.** 0.73 + 0.86

Common Core

Adding and Subtracting Decimals

In this lesson, you will

- use models to add and subtract decimals.
- add and subtract decimals.

Learning Standard
6.NS.3

2 ACTIVITY: Modeling a Difference

Work with a partner. Use base ten blocks to find the difference.

a. 2.43 − 0.73

Which number is shown by the model? ____

Circle the portion of the model that represents 0.73.

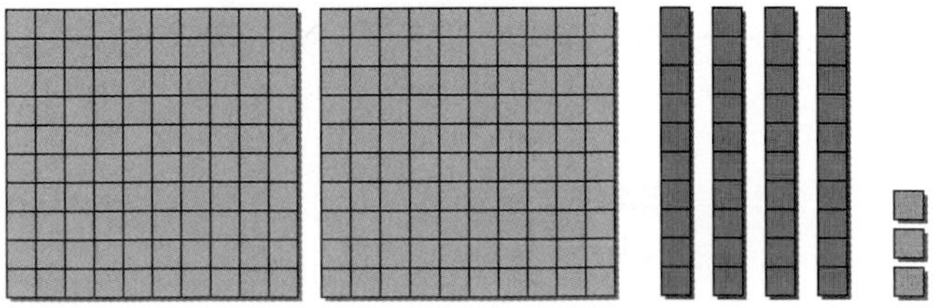

∴ So, 2.43 − 0.73 = ____.

b. 1.86 − 1.26 **c.** 3.72 − 0.5 **d.** 1.58 − 0.09

3 ACTIVITY: Making a Conjecture

Work with a partner.

a. Find each sum or difference.

$123 + 87$	$125 + 135$	$214 + 92$	$73 + 86$
$243 - 73$	$186 - 126$	$372 - 50$	$158 - 9$

b. How are the numerical expressions in part (a) related to the numerical expressions in Activities 1 and 2? How are the sums and differences related?

c. **STRUCTURE** There is a relationship between adding and subtracting decimals and adding and subtracting whole numbers. What conjecture can you make about this relationship?

4 ACTIVITY: Using a Place Value Chart

Work with a partner. Use the place value chart to find the sum or difference.

Math Practice 3

Analyze Conjectures

How can the conjecture you wrote in Activity 3 help you to solve these problems?

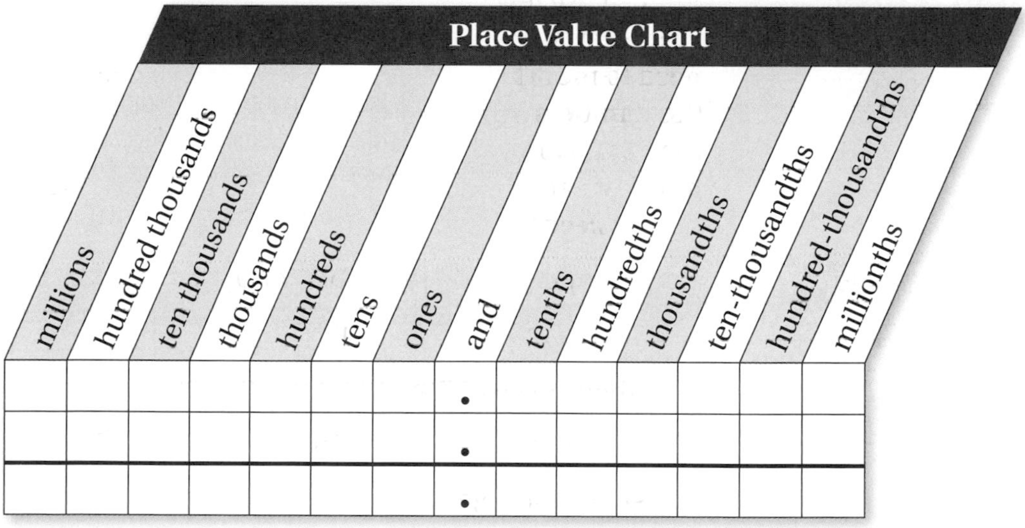
Place Value Chart
millions
hundred thousands
ten thousands
thousands
hundreds
tens
ones
and
tenths
hundredths
thousandths
ten-thousandths
hundred-thousandths
millionths

a. $16.05 + 2.94$

b. $7.421 + 92.55$

c. $38.72 - 8.61$

d. $64.968 - 51.167$

What Is Your Answer?

5. **MODELING** Describe two real-life examples of when you would need to add and subtract decimals.
6. **IN YOUR OWN WORDS** How can you add and subtract decimals?

Practice Use what you learned about adding and subtracting decimals to complete Exercises 3–4 on page 82.

Key Idea

Adding and Subtracting Decimals

To add or subtract decimals, write the numbers vertically and line up the decimal points. Then bring down the decimal point and add or subtract as you would with whole numbers.

EXAMPLE 1 Adding Decimals

a. Add 8.13 + 2.76. **Estimate** $8.13 + 2.76 \approx 8 + 3 = 11$

Line up the decimal points.

$$\begin{array}{r} 8.13 \\ +\ 2.76 \\ \hline 10.89 \end{array}$$

Add as you would with whole numbers.

Reasonable? $10.89 \approx 11$ ✓

Study Tip

Be sure to add or subtract only digits that have the same place value.

b. Add 1.459 + 23.7.

$$\begin{array}{r} {\scriptstyle 1} \\ 1.459 \\ +\ 23.700 \\ \hline 25.159 \end{array}$$

Insert zeros so that both numbers have the same number of decimal places.

EXAMPLE 2 Subtracting Decimals

a. Subtract 5.508 − 3.174. **Estimate** $5.508 - 3.174 \approx 6 - 3 = 3$

Line up the decimal points.

$$\begin{array}{r} {\scriptstyle 4\,10} \\ 5.\not{5}\not{0}8 \\ -\ 3.174 \\ \hline 2.334 \end{array}$$

Subtract as you would with whole numbers.

Reasonable? $2.334 \approx 3$ ✓

b. Subtract 21.9 − 1.605.

$$\begin{array}{r} {\scriptstyle 9} \\ {\scriptstyle 8\,\not{10}\,10} \\ 21.\not{9}\not{0}\not{0} \\ -\ 1.605 \\ \hline 20.295 \end{array}$$

Insert zeros so that both numbers have the same number of decimal places.

On Your Own

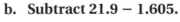

Now You're Ready
Exercises 5–16

Add or subtract.

1. 4.206 + 10.85
2. 15.5 + 8.229
3. 78.41 + 90.99
4. 6.34 − 5.33
5. 27.9 − 0.905
6. 18.626 − 13.88

EXAMPLE 3 Real-Life Application

Your meal at the school cafeteria costs $3.45. Your friend's meal costs $3.90. You pay for both meals with a $10 bill. How much change do you receive?

Use a verbal model to solve the problem.

amount of change = amount given − (cost of your meal + cost of friend's meal)

$= 10.00 - (3.45 + 3.90)$	Substitute values.
$= 10.00 - 7.35$	Add inside parentheses.
$= 2.65$	Subtract.

So, you receive $2.65.

EXAMPLE 4 Real-Life Application

The Lincoln Memorial Reflecting Pool is approximately rectangular. Its width is 50.9 meters, and its length is 618.44 meters. You walk the perimeter of the pool. About how many meters do you walk?

Draw a diagram and label the dimensions.

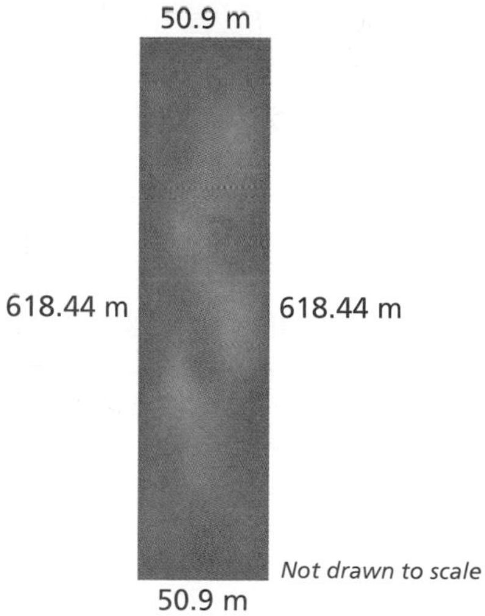

Find the sum of the side lengths.

$$\begin{array}{r} {\scriptstyle 112} \\ 618.44 \\ 50.90 \\ 618.44 \\ +\ \ 50.90 \\ \hline 1338.68 \end{array}$$

So, you walk about 1339 meters.

On Your Own

Now You're Ready
Exercises 21–26

7. **WHAT IF?** In Example 3, your meal costs $4.10 and your friend's meal costs $3.65. You pay for both meals with a $20 bill. How much change do you receive?
8. Find the perimeter of the triangle.

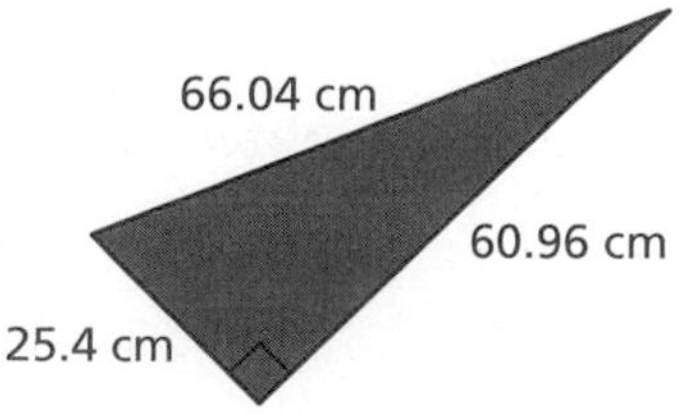

2.4 Exercises

Vocabulary and Concept Check

1. **CHOOSE TOOLS** Why is it helpful to estimate the answer before adding or subtracting decimals?

2. **WRITING** When adding or subtracting decimals, how can you be sure to add or subtract only digits that have the same place value?

Practice and Problem Solving

Write and evaluate the numerical expression modeled by the base ten blocks.

3. 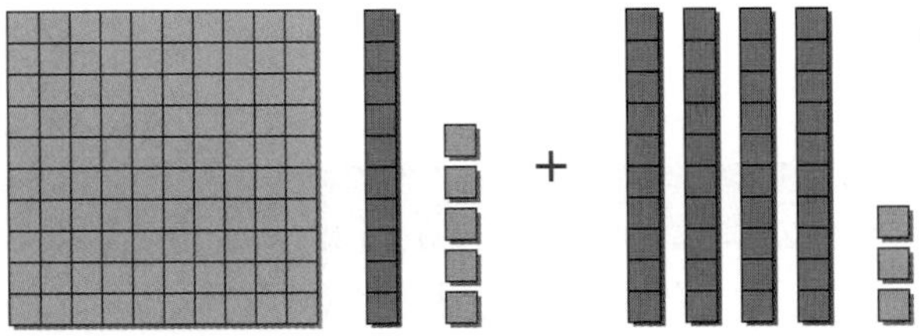

4. 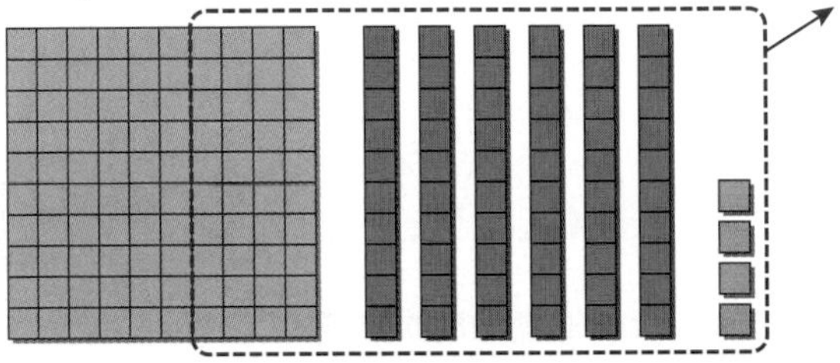

Add.

(1) 5. 7.82 + 3.209 6. 3.7 + 2.774 7. 12.829 + 10.07

8. 20.35 + 13.748 9. 17.440 + 12.497 10. 15.255 + 19.058

Subtract.

(2) 11. 4.58 − 3.12 12. 8.629 − 5.309 13. 6.98 − 2.614

14. 15.131 − 11.57 15. 13.5 − 10.856 16. 25.82 − 22.936

ERROR ANALYSIS **Describe and correct the error in the solution.**

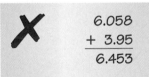

17. ✗
$$\begin{array}{r} 6.058 \\ +\ 3.95 \\ \hline 6.453 \end{array}$$

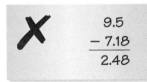

18. ✗
$$\begin{array}{r} 9.5 \\ -\ 7.18 \\ \hline 2.48 \end{array}$$

Breakfast Menu
7:30 A.M. to 11:00 A.M.

2 Eggs (any style)	\$2.95	Bacon & Eggs	\$3.95
Steak & Eggs	\$6.25	Cheese Omelet	\$3.55
Ham & Eggs	\$3.95	Ham Omelet	\$4.35
Sausage & Eggs	\$3.95	Ham & Cheese Omelet	\$4.95
Salami & Eggs	\$3.95		

19. **BREAKFAST** You order the sausage and eggs breakfast, and your friend orders the ham omelet. How much is the bill before taxes and tip?

20. **HAM & CHEESE** How much more does the ham and cheese omelet cost than the cheese omelet?

Evaluate the expression.

3 **21.** $6.105 + 10.4 + 3.075$

22. $22.6 - 12.286 - 3.542$

23. $15.35 + 7.604 - 12.954$

24. $16.5 - 13.45 + 7.293$

25. $25.92 - 18.478 + 8.164$

26. $23.45 + 17.75 - 19.618$

27. **STRUCTURE** When is the sum of two decimals equal to a whole number? When is the difference of two decimals equal to a whole number?

28. **OPEN-ENDED** Write three decimals that have a sum of 27.905.

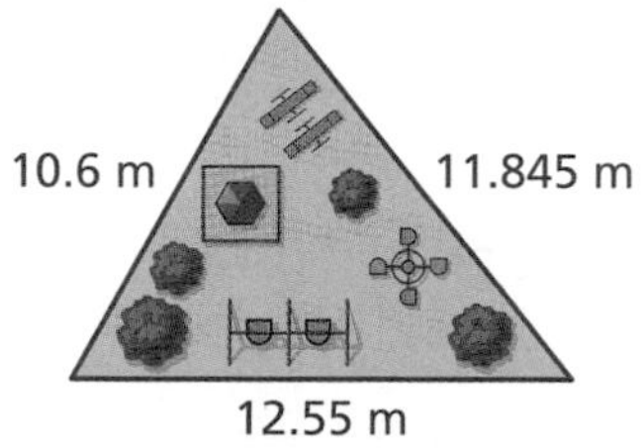

29. **DAY CARE** A day-care center is building a new outdoor play area. The diagram shows the dimensions in meters. How much fencing is needed to enclose the play area?

30. **HOMEWORK** You work 1.15 hours on English homework and 1.75 hours on math homework. Your science homework takes 1.05 hours less than your math homework. How many hours do you work on homework?

ASTRONOMY An astronomical unit (AU) is the average distance of Earth from the Sun. In Exercises 31–34, use the table that shows the average distance of each planet in our solar system from the Sun.

31. How much farther is Jupiter from the Sun than Mercury?

32. How much farther is Neptune from the Sun than Mars?

33. Estimate the greatest distance between Earth and Uranus.

34. Estimate the greatest distance between Venus and Saturn.

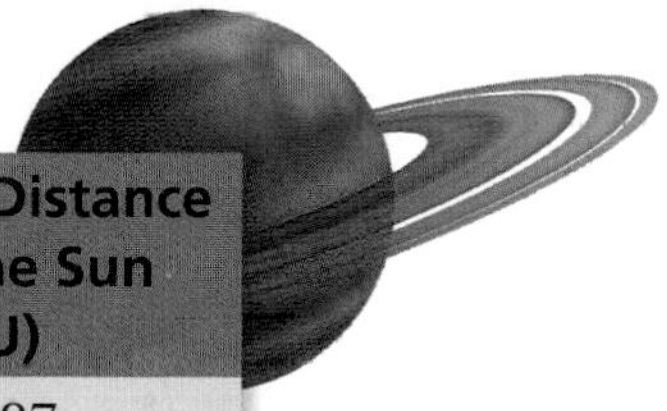

Planet	Average Distance from the Sun (AU)
Mercury	0.387
Venus	0.723
Earth	1.000
Mars	1.524
Jupiter	5.203
Saturn	9.537
Uranus	19.189
Neptune	30.07

35. Critical Thinking The length of a rectangle is twice the width. The perimeter of the rectangle can be expressed as $3 \cdot 13.7$. What is the width?

Fair Game Review What you learned in previous grades & lessons

Multiply. Write the answer in simplest form. *(Section 2.1)*

36. $\frac{7}{10} \times \frac{5}{7}$

37. $\frac{5}{6} \times \frac{3}{10}$

38. $\frac{3}{4} \times \frac{2}{9}$

39. $\frac{2}{5} \times \frac{1}{8}$

40. **MULTIPLE CHOICE** What is the LCM of 6, 12, and 18? *(Section 1.6)*

(A) 6 (B) 18 (C) 36 (D) 72

2.5 Multiplying Decimals

Essential Question How can you multiply decimals?

1 ACTIVITY: Multiplying Decimals Using a Rectangle

Work with a partner. Use a rectangle to find the product.

a. 2.7 • 1.3

Arrange base ten blocks to form a rectangle of length 2.7 units and width 1.3 units.

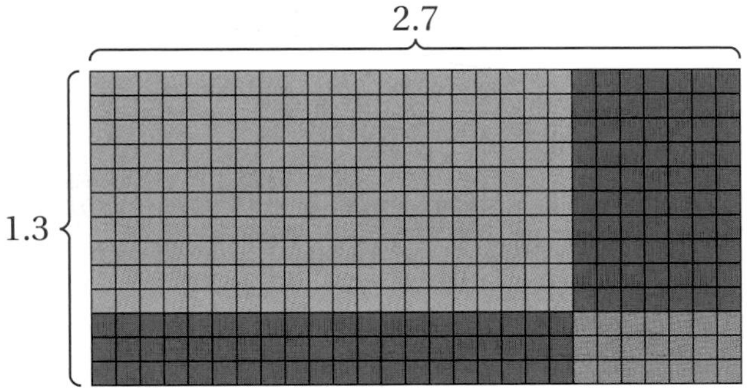

The area of the rectangle represents the product.

Find the total area represented by each grouping of base ten blocks.

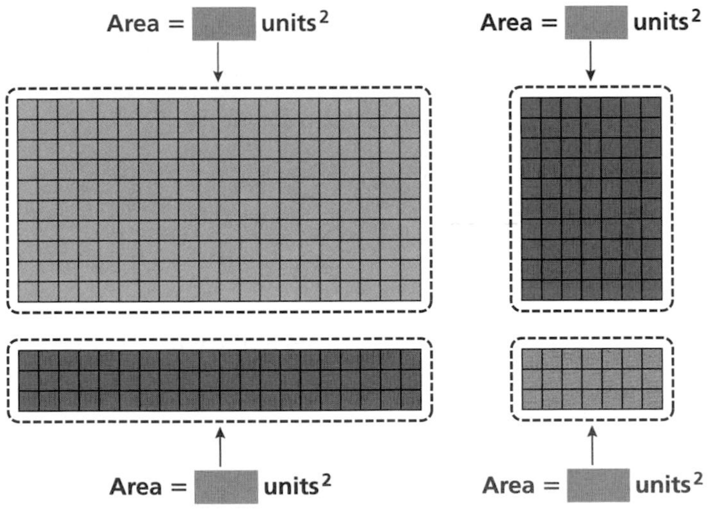

The area of the rectangle is:

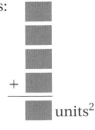

$$\begin{array}{r} \square \\ \square \\ \square \\ +\ \square \\ \hline \square \end{array} \text{ units}^2$$

So, 2.7 • 1.3 = ____.

b. 1.8 • 1.1 **c.** 4.6 • 1.2 **d.** 3.2 • 2.4

COMMON CORE

Multiplying Decimals

In this lesson, you will

- use models to multiply decimals.
- multiply decimals.

Learning Standard 6.NS.3

2 ACTIVITY: Multiplying Decimals Using an Area Model

Work with a partner. Use an area model to find the product. Explain your reasoning.

a. 0.8 • 0.5

Because ▮ hundredths are shaded with both colors, the product is

$\frac{\blacksquare}{100} = \blacksquare$.

∴ So, 0.8 • 0.5 = ▮.

b. 0.3 • 0.5 **c.** 0.7 • 0.6 **d.** 0.2 • 0.9

Math Practice 7

View as Components

How can you use an area model to find the product?

3 ACTIVITY: Making a Conjecture

Work with a partner.

a. Find each product.

27 • 13	18 • 11	46 • 12	32 • 24
8 • 5	3 • 5	7 • 6	2 • 9

b. How are the numerical expressions in part (a) related to the numerical expressions in Activities 1 and 2? How are the products related?

c. **STRUCTURE** What conjecture can you make about the relationship between multiplying decimals and multiplying whole numbers?

What Is Your Answer?

4. **IN YOUR OWN WORDS** How can you multiply decimals?

Practice Use what you learned about multiplying decimals to complete Exercises 9–12 on page 89.

Check It Out
Lesson Tutorials
BigIdeasMath.com

Key Idea

Multiplying Decimals by Whole Numbers

Words Multiply as you would with whole numbers. Then count the number of decimal places in the decimal factor. The product has the same number of decimal places.

Numbers

$$\begin{array}{r} 13.91 \\ \times \quad 7 \\ \hline 97.37 \end{array} \quad \text{2 decimal places} \qquad \begin{array}{r} 6.218 \\ \times \quad 4 \\ \hline 24.872 \end{array} \quad \text{3 decimal places}$$

EXAMPLE 1 Multiplying Decimals and Whole Numbers

a. Find 6×3.91.

Estimate $6 \times 4 = 24$

$$\begin{array}{r} {}^{5} \\ 3.91 \\ \times \quad 6 \\ \hline 23.46 \end{array}$$

3.91 ← 2 decimal places

23.46 ← Count 2 decimal places from right to left.

So, $6 \times 3.91 = 23.46$.

Reasonable? $23.46 \approx 24$ ✓

b. Find 3×0.016.

Estimate $3 \times 0 = 0$

$$\begin{array}{r} {}^{1} \\ 0.016 \\ \times \quad 3 \\ \hline 0.048 \end{array}$$

0.016 ← 3 decimal places

0.048 ← To have 3 decimal places, insert zeros to the left of 48.

So, $3 \times 0.016 = 0.048$.

Reasonable? $0.048 \approx 0$ ✓

EXAMPLE 2 Use Mental Math

How high is a stack of 100 dimes?

1.35 millimeters

Method 1: Multiply 1.35 by 100.

$$\begin{array}{r} 1.35 \\ \times 1\,00 \\ \hline 0\,00 \\ 00\,0 \\ 135 \\ \hline 135.00 \end{array}$$

2 decimal places

Method 2: You are multiplying by a power of 10. Use mental math.

There are two zeros in 100. So, move the decimal point in 1.35 two places to the right.

$1.35 \times 100 = 135. = 135$

So, a stack of 100 dimes is 135 millimeters high.

On Your Own

Now You're Ready
Exercises 13–24

Multiply. Use estimation to check your answer.

1. 12.3×8
2. 5×14.51
3. 0.88×9
4. 0.003×10
5. A quarter is 1.75 millimeters thick. How high is a stack of 1000 quarters? Solve using both methods.

The rule for multiplying two decimals is similar to the rule for multiplying a decimal by a whole number.

Key Idea

Multiplying Decimals by Decimals

Words Multiply as you would with whole numbers. Then add the number of decimal places in the factors. The sum is the number of decimal places in the product.

Numbers

$$\begin{array}{r} 4.716 \\ \times \quad 0.2 \\ \hline 0.9432 \end{array} \quad \begin{array}{l} \leftarrow \text{3 decimal places} \\ \leftarrow \underline{+\text{ 1 decimal place}} \\ \leftarrow \text{4 decimal places} \end{array}$$

EXAMPLE 3 Multiplying Decimals

a. Multiply 4.8 × 7.2. **Estimate** $5 \times 7 = 35$

$$\begin{array}{r} 4.8 \\ \times\ 7.2 \\ \hline 96 \\ 336 \\ \hline 34.56 \end{array} \quad \begin{array}{l} \leftarrow \text{1 decimal place} \\ \leftarrow \underline{+\text{ 1 decimal place}} \\ \\ \\ \leftarrow \text{2 decimal places} \end{array}$$

So, $4.8 \times 7.2 = 34.56$. **Reasonable?** $34.56 \approx 35$ ✓

b. Multiply 3.1 × 0.05. **Estimate** $3 \times 0 = 0$

$$\begin{array}{r} 3.1 \\ \times\ 0.05 \\ \hline 0.155 \end{array} \quad \begin{array}{l} \leftarrow \text{1 decimal place} \\ \leftarrow \underline{+\text{ 2 decimal places}} \\ \leftarrow \text{3 decimal places} \end{array}$$

So, $3.1 \times 0.05 = 0.155$. **Reasonable?** $0.155 \approx 0$ ✓

On Your Own

Now You're Ready
Exercises 30–45

Multiply. Use estimation to check your answer.

6. 8.1×5.6
7. 2.7×9.04
8. 6.32×0.09
9. 1.785×0.2

EXAMPLE 4 Evaluating an Expression

What is the value of 2.44(4.5 − 3.175)?

Ⓐ 3.233 Ⓑ 3.599 Ⓒ 7.805 Ⓓ 32.33

Step 1: Subtract first because the minus sign is in parentheses.

$$\begin{array}{r} \overset{4}{\cancel{5}}\overset{\overset{9}{\cancel{10}}}{\cancel{0}}\overset{10}{\cancel{0}} \\ 4.500 \\ -\ 3.175 \\ \hline 1.325 \end{array}$$

So, 2.44(4.5 − 3.175) = 2.44(1.325).

Step 2: Multiply the result from Step 1 by 2.44.

$$\begin{array}{r} 1.325 \\ \times\ 2.44 \\ \hline 5300 \\ 5300 \\ 2650 \\ \hline 3.23300 \end{array}$$

∴ The correct answer is Ⓐ.

On Your Own

Now You're Ready
Exercises 52–60

Evaluate the expression.

10. 12.67 + 8.2 • 1.9

11. 6.4(1.8 • 7.5)

EXAMPLE 5 Real-Life Application

You buy 2.75 pounds of tomatoes. You hand the cashier a $10 bill. How much change will you receive?

Step 1: Find the cost of the tomatoes. Multiply 1.89 by 2.75.

$$\begin{array}{rl} 1.89 & \leftarrow \text{ 2 decimal places} \\ \times\ 2.75 & \leftarrow \text{ + 2 decimal places} \\ \hline 945 & \\ 1323 & \\ 378 & \\ \hline 5.1975 & \leftarrow \text{ 4 decimal places} \end{array}$$

The cost of 2.75 pounds of tomatoes is $5.20.

Step 2: Subtract the cost of the tomatoes from the amount of money you hand the cashier.

10.00 − 5.20 = $4.80

∴ So, you will receive $4.80 in change.

On Your Own

12. **WHAT IF?** You buy 2.25 pounds of grapes. You hand the cashier a $5 bill. How much change will you receive?

2.5 Exercises

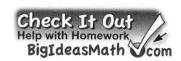

Check It Out
Help with Homework
BigIdeasMath.com

Vocabulary and Concept Check

1. **NUMBER SENSE** If you know $12 \times 24 = 288$, how can you find 1.2×2.4?
2. **NUMBER SENSE** Is the product 1.23×8 greater than or less than 8? Explain.

Copy the problem and place the decimal point in the product.

3. $\begin{array}{r} 1.78 \\ \times\ \ 4.9 \\ \hline 8722 \end{array}$

4. $\begin{array}{r} 9.24 \\ \times\ \ 0.68 \\ \hline 62832 \end{array}$

5. $\begin{array}{r} 3.75 \\ \times\ \ 5.22 \\ \hline 195750 \end{array}$

How many decimal places are in the product?

6. 6.17×8.2
7. 1.684×10.2
8. 0.053×2.78

Practice and Problem Solving

Use base ten blocks or an area model to find the product.

9. $\begin{array}{r} 2.1 \\ \times\ 1.5 \\ \hline \end{array}$

10. $\begin{array}{r} 0.6 \\ \times\ 0.4 \\ \hline \end{array}$

11. $\begin{array}{r} 0.7 \\ \times\ 0.3 \\ \hline \end{array}$

12. $\begin{array}{r} 2.7 \\ \times\ 2.3 \\ \hline \end{array}$

Multiply. Use estimation to check your answer.

1 2 13. $\begin{array}{r} 4.8 \\ \times\ \ 7 \\ \hline \end{array}$

14. $\begin{array}{r} 6.3 \\ \times\ \ 5 \\ \hline \end{array}$

15. $\begin{array}{r} 7.19 \\ \times\ \ 16 \\ \hline \end{array}$

16. $\begin{array}{r} 0.87 \\ \times\ \ 21 \\ \hline \end{array}$

17. $\begin{array}{r} 1.95 \\ \times\ \ 11 \\ \hline \end{array}$

18. $\begin{array}{r} 5.89 \\ \times\ \ 5 \\ \hline \end{array}$

19. $\begin{array}{r} 3.472 \\ \times\ \ 4 \\ \hline \end{array}$

20. $\begin{array}{r} 8.188 \\ \times\ \ 12 \\ \hline \end{array}$

21. 100×0.024
22. 19×0.004
23. 0.0038×9
24. 10×0.0093

ERROR ANALYSIS Describe and correct the error in the solution.

25. ✗ $\begin{array}{r} 0.0045 \\ \times\ \ \ 9 \\ \hline 4.05 \end{array}$

26. ✗ $\begin{array}{r} 0.32 \\ \times\ \ \ 5 \\ \hline 0.160 \end{array}$

27. **MOON** The weight of an object on the Moon is about 0.167 of its weight on Earth. How much does a 180-pound astronaut weigh on the Moon?
28. **BAMBOO** A bamboo plant grows about 1.25 feet each day. Find the growth in one week.
29. **NAILS** A fingernail grows about 0.1 millimeter each day. How much does a fingernail grow in 30 days? 90 days?

Multiply.

3 **30.** $\begin{array}{r} 0.7 \\ \times\ 0.2 \\ \hline \end{array}$ **31.** $\begin{array}{r} 0.08 \\ \times\ 0.3 \\ \hline \end{array}$ **32.** $\begin{array}{r} 0.007 \\ \times\ 0.03 \\ \hline \end{array}$ **33.** $\begin{array}{r} 0.0008 \\ \times\ \ 0.09 \\ \hline \end{array}$

34. $\begin{array}{r} 0.004 \\ \times\ \ 0.9 \\ \hline \end{array}$ **35.** $\begin{array}{r} 0.06 \\ \times\ 0.5 \\ \hline \end{array}$ **36.** $\begin{array}{r} 0.0008 \\ \times\ 0.004 \\ \hline \end{array}$ **37.** $\begin{array}{r} 0.0002 \\ \times\ \ 0.06 \\ \hline \end{array}$

38. 12.4×0.2 **39.** 18.6×5.9 **40.** 7.91×0.72 **41.** 1.16×3.35

42. 6.478×18.21 **43.** 1.9×7.216 **44.** 0.0021×18.2 **45.** 6.109×8.4

46. ERROR ANALYSIS Describe and correct the error in the solution.

X $\begin{array}{r} 4.9 \\ \times\ \ 3.8 \\ \hline 186.2 \end{array}$

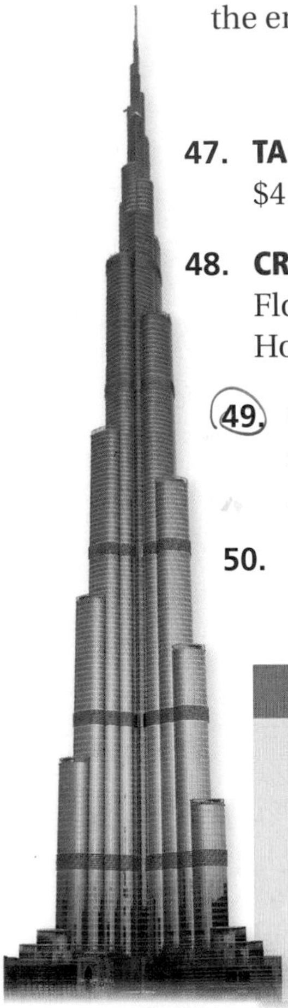

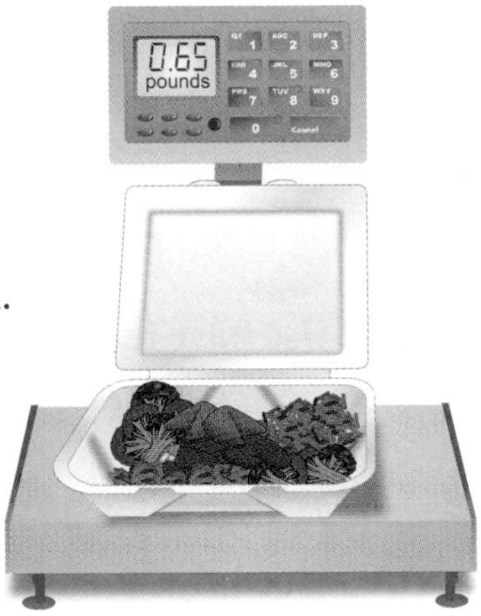

47. TAKEOUT A Chinese restaurant offers buffet takeout for \$4.99 per pound. How much does your takeout meal cost?

48. CROPLAND Alabama has about 2.51 million acres of cropland. Florida has about 1.15 times as much cropland as Alabama. How much cropland does Florida have?

49. GOLD On a tour of an old gold mine, you find a nugget containing 0.82 ounce of gold. Gold is worth \$1566.80 per ounce. How much is your nugget worth?

50. BUILDING HEIGHTS One meter is approximately 3.28 feet. Find the height of each building in feet by multiplying its height in meters by 3.28.

Continent	Tallest Building	Height (meters)
Africa	Carlton Centre Office Tower	223
Asia	Burj Khalifa	828
Australia	Q1 Tower	323
Europe	The Shard	310
North America	Willis Tower	442
South America	Gran Torre	300

51. REASONING Show how to evaluate $7.12 \times 8.22 \times 100$ without multiplying the two decimals.

ORDER OF OPERATIONS **Evaluate the expression.**

4 **52.** $2.4 \times 16 + 7$ **53.** $6.85 \times 2 \times 10$ **54.** $1.047 \times 5 - 0.88$

55. $4.32(3.7 + 1.65)$ **56.** $23.98 - 1.7^2 \cdot 7.6$ **57.** $12 \cdot 5.16 + 10.064$

58. $0.9(8.2 \cdot 20.35)$ **59.** $7.5^2(6.084 - 5.44)$ **60.** $6.8 \cdot 2.18 \cdot 3.95$

61. REASONING Without multiplying, how many decimal places does 3.4^2 have? 3.4^3? 3.4^4? Explain your reasoning.

REPEATED REASONING **Describe the pattern. Find the next three numbers.**

62. 1, 0.6, 0.36, 0.216, . . .

63. 15, 1.5, 0.15, 0.015, . . .

64. 0.04, 0.02, 0.01, 0.005, . . .

65. 5, 7.5, 11.25, 16.875, . . .

66. FOOD You buy 2.6 pounds of apples and 1.475 pounds of peaches. You hand the cashier a $20 bill. How much change will you receive?

67. MILEAGE A car can travel 22.36 miles on one gallon of gasoline.

a. How far can the car travel on 8.5 gallons of gasoline?

b. A hybrid car can travel 33.1 miles on one gallon of gasoline. How much farther can the hybrid car travel on 8.5 gallons of gasoline?

68. OPEN-ENDED You and four friends have dinner at a restaurant.

a. Draw a restaurant menu that has main items, desserts, and beverages, with their prices.

b. Write a guest check that shows what each of you ate. Find the subtotal.

c. Multiply by 0.07 to find the tax. Then find the total.

d. Round the total to the nearest whole number. Multiply by 0.20 to estimate a tip. Including the tip, how much did you spend?

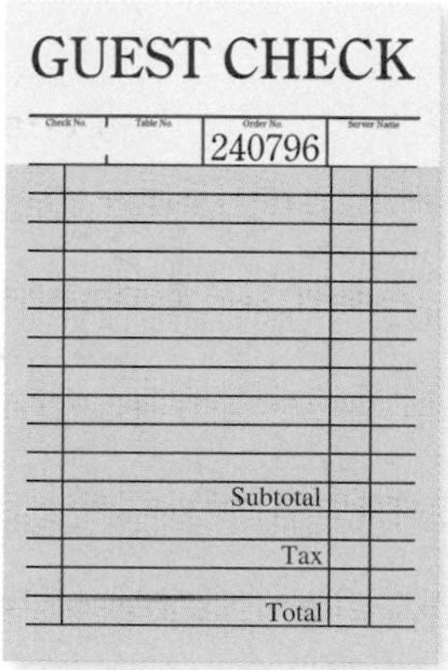

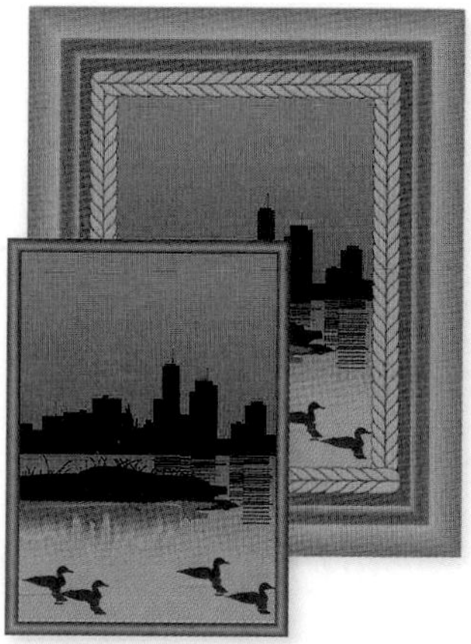

69. **Geometry** A rectangular painting has an area of 9.52 square feet.

a. Draw three different ways in which this can happen.

b. The cost of a frame depends on the perimeter of the painting. Which of your drawings from part (a) is the least expensive to frame? Explain your reasoning.

c. The thin, black framing costs $1 per foot. The fancy framing costs $5 per foot. Will the fancy framing cost five times as much as the black framing? Explain why or why not.

d. Suppose the cost of a frame depends on the outside perimeter of the frame. Does this change your answer to part (c)? Explain why or why not.

Fair Game Review *What you learned in previous grades & lessons*

Divide. *(Skills Review Handbook)*

70. $78 \div 3$

71. $65 \div 13$

72. $57 \div 19$

73. $84 \div 12$

74. MULTIPLE CHOICE How many edges does the rectangular prism at the right have? *(Skills Review Handbook)*

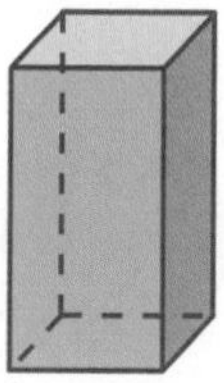

Ⓐ 4 Ⓑ 6

Ⓒ 8 Ⓓ 12

2.6 Dividing Decimals

Essential Question How can you use base ten blocks to model decimal division?

1 ACTIVITY: Dividing Decimals

Work with a partner. Use base ten blocks to model the division. Then find the quotient.

a. 2.4 ÷ 0.6

Begin by modeling 2.4.

2.4

How many of each base ten block did you use?

____ ones

____ tenths

____ hundredths

Next, think of the division problem 2.4 ÷ 0.6 as the question,

"How can you divide 2.4 into groups of 0.6?"

Rearrange the model for 2.4 into groups of 0.6. There are ____ groups of 0.6.

So, 2.4 ÷ 0.6 = ____.

b. 1.8 ÷ 2 **c.** 3.9 ÷ 3 **d.** 2.8 ÷ 0.7 **e.** 3.2 ÷ 0.4

f. Write and solve the division problem represented by the model.

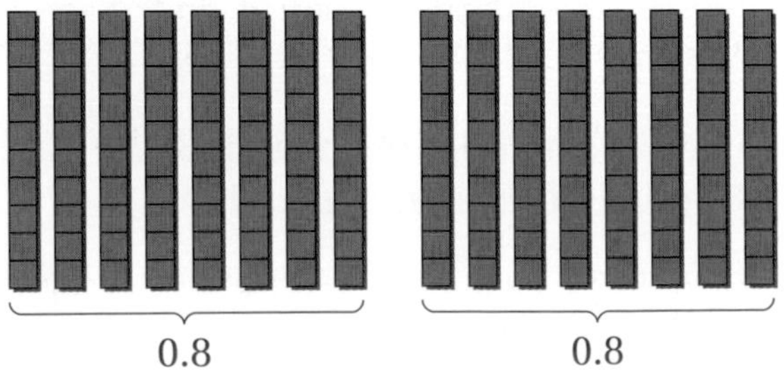

Common Core

Dividing Decimals

In this lesson, you will
- use models to divide decimals.
- divide decimals.

Learning Standard
6.NS.3

2 ACTIVITY: Dividing Decimals

Work with a partner. Use base ten blocks to model the division. Then find the quotient.

Math Practice 8

Evaluate Results

What can you do to check the reasonableness of your answer?

a. 0.3 ÷ 0.06

Model 0.3. | Replace tenths with hundredths. | How many 0.06s are in 0.3? Divide hundredths into groups of 0.06.

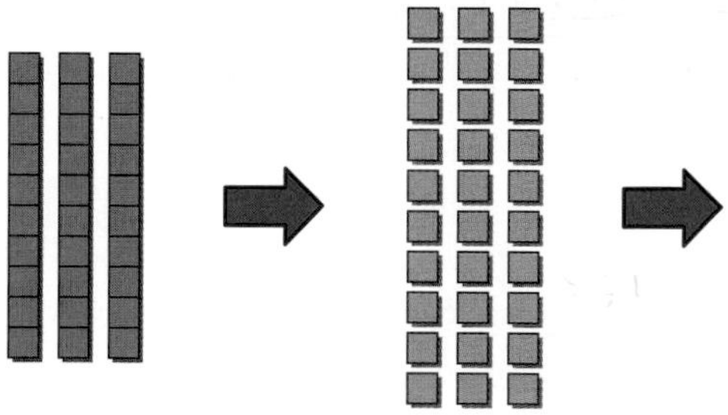

∴ There are ▭ groups of 0.06. So, 0.3 ÷ 0.06 = ▭.

b. 0.2 ÷ 0.04

c. 0.6 ÷ 0.01

d. 0.16 ÷ 0.08

e. 0.28 ÷ 0.07

What Is Your Answer?

3. **IN YOUR OWN WORDS** How can you use base ten blocks to model decimal division? Use examples from Activity 1 and Activity 2 as part of your answer.

4. **WRITING** Newton's poem is about dividing fractions. Write a poem about dividing decimals.

"When you must divide a fraction, do this very simple action: Flip what you're dividing BY, and then it's easy—multiply!"

5. Think of your own cartoon about dividing decimals. Draw your cartoon.

Practice Use what you learned about dividing decimals to complete Exercises 8–11 on page 97.

2.6 Lesson

Check It Out
Lesson Tutorials
BigIdeasMath.com

Key Idea

Dividing Decimals by Whole Numbers

Words Place the decimal point in the quotient above the decimal point in the dividend. Then divide as you would with whole numbers. Continue until there is no remainder.

Numbers

```
    1.83
 4)7.32
```

Place the decimal point in the quotient above the decimal point in the dividend.

EXAMPLE 1 Dividing Decimals by Whole Numbers

a. Find 7.6 ÷ 4. **Estimate 8 ÷ 4 = 2**

```
   1.9
 4)7.6
 − 4
   3 6
 − 3 6
     0
```

Place the decimal point in the quotient above the decimal point in the dividend.

So, 7.6 ÷ 4 = 1.9. **Reasonable? 1.9 ≈ 2** ✓

b. Find 4.38 ÷ 12.

```
    0.365
 12)4.380
  − 3 6
      78
    − 72
      60
    − 60
       0
```

Place the decimal point in the quotient above the decimal point in the dividend.

Insert a zero and continue to divide.

So, 4.38 ÷ 12 = 0.365. **Check 0.365 × 12 = 4.38** ✓

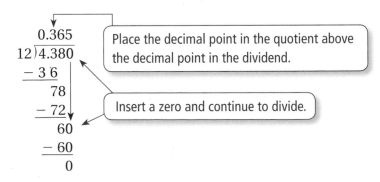

On Your Own

Now You're Ready
Exercises 12–23

Divide. Use estimation to check your answer.

1. 36.4 ÷ 2 **2.** 22.2 ÷ 6 **3.** 59.64 ÷ 7

4. 43.26 ÷ 14 **5.** 6.2 ÷ 4 **6.** 3.12 ÷ 16

Key Idea

Dividing Decimals by Decimals

Words Multiply the divisor *and* the dividend by a power of 10 to make the divisor a whole number. Then place the decimal point in the quotient and divide as you would with whole numbers. Continue until there is no remainder.

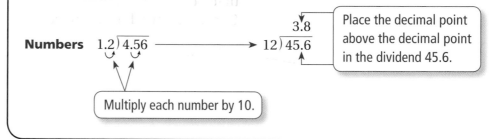

Numbers $1.2\overline{)4.56}$ → $12\overline{)45.6}$ with quotient 3.8

Multiply each number by 10.

Place the decimal point above the decimal point in the dividend 45.6.

EXAMPLE 2 Dividing Decimals

a. Find 18.2 ÷ 1.4.

$1.4\overline{)18.2}$ → $14\overline{)182.}$

Multiply each number by 10.

Place the decimal point above the decimal point in the dividend 182.

```
     13.
14)182.
   -14
     42
    -42
      0
```

So, 18.2 ÷ 1.4 = 13. **Check 13 × 1.4 = 18.2 ✓**

b. Find 0.273 ÷ 0.39.

$0.39\overline{)0.273}$ → $39\overline{)27.3}$

Multiply each number by 100.

```
     0.7
39)27.3
  -27.3
      0
```

So, 0.273 ÷ 0.39 = 0.7. **Check 0.7 × 0.39 = 0.273 ✓**

Study Tip

Multiplying the divisor and the dividend by a power of 10 does not change the quotient.

For example:

18.2 ÷ 1.4 = 13
182 ÷ 14 = 13
1820 ÷ 140 = 13

On Your Own

Now You're Ready
Exercises 36–39

Divide. Check your answer.

7. $1.2\overline{)9.6}$

8. $3.4\overline{)57.8}$

9. 21.643 ÷ 2.3

10. 0.459 ÷ 0.51

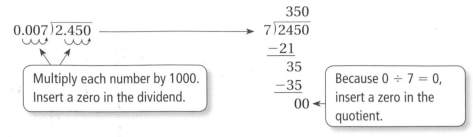

EXAMPLE 3 Inserting Zeros in the Dividend and the Quotient

Divide 2.45 ÷ 0.007.

> **Study Tip**
> Remember to check your answer by multiplying the quotient by the divisor.

$0.007\overline{)2.450}$ ⟶ $7\overline{)2450}$ with quotient 350

Multiply each number by 1000. Insert a zero in the dividend.

```
    350
 7)2450
  -21
    35
   -35
     00
```

Because $0 \div 7 = 0$, insert a zero in the quotient.

So, $2.45 \div 0.007 = 350$.

On Your Own

Now You're Ready Exercises 40–43

Divide. Check your answer.

11. $3.8 \div 0.16$

12. $15.6 \div 0.78$

13. $7.2 \div 0.048$

14. $0.18 \div 0.003$

EXAMPLE 4 Real-Life Application

How many times more cellular phone subscribers were there in 2011 than in 1991? Round to the nearest whole number.

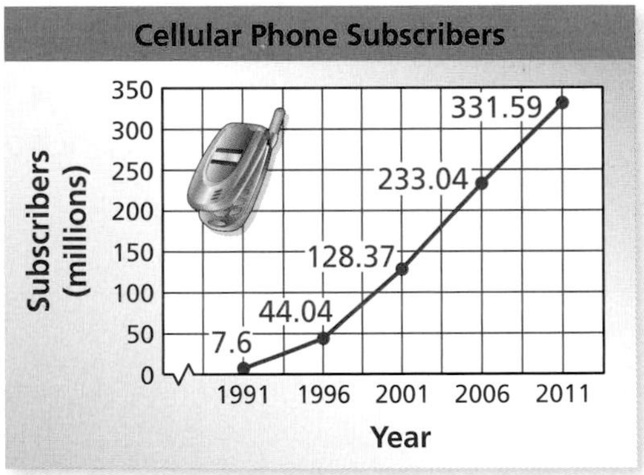

From the graph, there were 331.59 million subscribers in 2011 and 7.6 million in 1991. So, divide 331.59 by 7.6.

Estimate $320 \div 8 = 40$

$7.6\overline{)331.59}$ ⟶ $76\overline{)3315.9}$

```
      43.6   ← Rounds to 44.
 76)3315.9
   -304
     275
    -228
     47 9
    -45 6
      2 3
```

So, there were about 44 times more subscribers in 2011 than in 1991.

Reasonable? $44 \approx 40$ ✓

On Your Own

15. How many times more subscribers were there in 2006 than in 1996? Round to the nearest whole number.

2.6 Exercises

Check It Out
Help with Homework
BigIdeasMath.com

Vocabulary and Concept Check

1. **NUMBER SENSE** Fix the one that is not correct.

$$\begin{array}{r}6.1\\4\overline{)24.4}\end{array} \qquad \begin{array}{r}61\\4\overline{)244}\end{array} \qquad \begin{array}{r}6.1\\4\overline{)2.44}\end{array}$$

Copy the problem and place the decimal point in the correct location.

2. $18.6 \div 4 = 465$
3. $6.38 \div 11 = 58$
4. $88.27 \div 7 = 1261$

Rewrite the problem so that the divisor is a whole number.

5. $4.7\overline{)13.6}$
6. $0.21\overline{)17.66}$
7. $2.16\overline{)18.5}$

Practice and Problem Solving

Use base ten blocks to find the quotient.

8. $3.6 \div 0.3$
9. $2.6 \div 0.2$
10. $0.72 \div 0.06$
11. $0.36 \div 0.04$

Divide. Use estimation to check your answer.

(1) 12. $6\overline{)25.2}$
13. $5\overline{)33.5}$
14. $7\overline{)3.5}$
15. $8\overline{)10.4}$
16. $38.7 \div 9$
17. $37.6 \div 4$
18. $43.4 \div 7$
19. $25.6 \div 8$
20. $44.64 \div 8$
21. $0.294 \div 3$
22. $3.6 \div 24$
23. $64.26 \div 18$

ERROR ANALYSIS **Describe and correct the error in finding the quotient.**

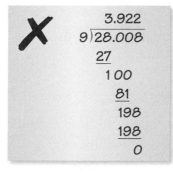

24. ✗
$$\begin{array}{r}3.922\\9\overline{)28.008}\\\underline{27}\\1\,00\\\underline{81}\\198\\\underline{198}\\0\end{array}$$

25. ✗
$$\begin{array}{r}0.86\\6\overline{)0.516}\\\underline{48}\\36\\\underline{36}\\0\end{array}$$

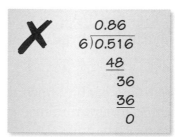

26. **TEXT MESSAGING** You send 40 text messages in one month. The total cost is \$4.80. How much does each text message cost?

27. **SUNBLOCK** Of the two bottles of sunblock shown, which is the better buy? Explain.

4-ounce bottle \$8.49 5-ounce bottle \$10.29

ORDER OF OPERATIONS Evaluate the expression.

28. $7.68 + 3.18 \div 12$ **29.** $10.56 \div 3 - 1.9$ **30.** $19.6 \div 7 \times 9$

31. $5.5 \times 16.56 \div 9$ **32.** $35.25 \div 5 \div 3$ **33.** $13.41 \times (5.4 \div 9)$

34. FRUIT PUNCH Which pack of fruit punch is the best buy? Explain.

35. SALE You buy 3 pairs of jeans for \$35.95 each and get a fourth pair for free. What is your cost per pair of jeans?

Divide. Check your answer.

② **36.** $2.1\overline{)25.2}$ **37.** $3.8\overline{)34.2}$ **38.** $36.47 \div 0.7$ **39.** $0.984 \div 12.3$

③ **40.** $4.23 \div 0.012$ **41.** $0.52 \div 0.0013$ **42.** $95.04 \div 0.0132$ **43.** $32.2 \div 0.07$

Divide. Round to the nearest hundredth if necessary.

44. $80.88 \div 8.425$ **45.** $0.8 \div 0.6$ **46.** $38.9 \div 6.44$ **47.** $11.6 \div 0.95$

48. ERROR ANALYSIS Describe and correct the error in rewriting the problem.

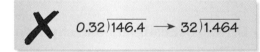

✗ $0.32\overline{)146.4} \rightarrow 32\overline{)1.464}$

49. TICKETS Tickets to the school musical cost \$6.25. The amount received from ticket sales is \$706.25. How many tickets were sold?

50. HEIGHT A person's running stride is about 1.14 times the person's height. Your friend's stride is 5.472 feet. How tall is your friend?

51. MP3 PLAYER You have 3.4 gigabytes available on your MP3 player. Each song is about 0.004 gigabyte. How many more songs can you download onto your MP3 player?

52. SWIMMING The table shows the top three times in a swimming event at the Summer Olympics. The event consists of a team of four women swimming 100 meters each.

Women's 4 × 100 Freestyle Relay		
Medal	**Country**	**Time (seconds)**
Gold	Australia	215.94
Silver	United States	216.39
Bronze	Netherlands	217.59

a. Suppose the times of all four swimmers on each team were the same. For each team, how much time does it take a swimmer to swim 100 meters?

b. Suppose each U.S. swimmer completed 100 meters a quarter second faster. Would the U.S. team have won the gold medal? Explain your reasoning.

Without finding the quotient, copy and complete the statement using <, >, or =.

53. 6.66 ÷ 0.74 ▢ 66.6 ÷ 7.4

54. 32.2 ÷ 0.7 ▢ 3.22 ÷ 7

55. 160.72 ÷ 16.4 ▢ 160.72 ÷ 1.64

56. 75.6 ÷ 63 ▢ 7.56 ÷ 0.63

57. BEES To approximate the number of bees in a hive, multiply the number of bees that leave the hive in one minute by 3 and divide by 0.014. You count 25 bees leaving a hive in one minute. How many bees are in the hive?

58. PROBLEM SOLVING You are saving money to buy a new bicycle that costs $155.75. You have $30 and plan to save $5 each week. Your aunt decides to give you an additional $10 each week.

a. How many weeks will you have to save until you have enough money to buy the bicycle?

b. How many more weeks would you have to save to buy a new bicycle that costs $203.89? Explain how you found your answer.

Applesauce	
3.9-ounce bowl	$0.52
24-ounce jar	$2.63

59. PRECISION A store sells applesauce in two sizes.

a. How many *bowls* of applesauce fit in a *jar*? Round your answer to the nearest hundredth.

b. Explain two ways to find the better buy.

c. What is the better buy?

60. Geometry The large rectangle's dimensions are three times the dimensions of the small rectangle.

23.1 ft

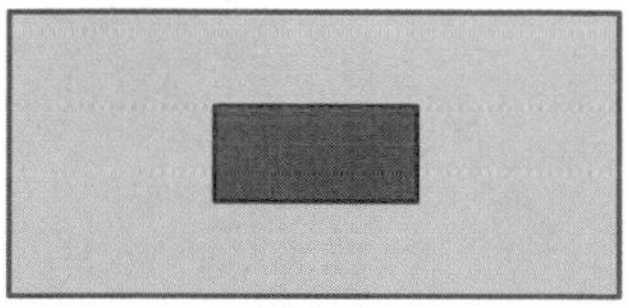

49.2 ft

a. How many times greater is the perimeter of the large rectangle compared to the perimeter of the small rectangle?

b. How many times greater is the area of the large rectangle compared to the area of the small rectangle?

c. Are the answers to parts (a) and (b) the same? *Explain* why or why not.

d. What happens in parts (a) and (b) if the dimensions of the large rectangle are two times the dimensions of the small rectangle?

Fair Game Review What you learned in previous grades & lessons

Add or subtract. Write your answer in simplest form. *(Section 1.6)*

61. $\frac{1}{2} + \frac{2}{3}$

62. $\frac{2}{5} + \frac{3}{4}$

63. $\frac{3}{10} - \frac{1}{4}$

64. $\frac{11}{12} - \frac{7}{8}$

65. MULTIPLE CHOICE Melissa earns $7.40 an hour working at a grocery store. She works 14.25 hours this week. How much does she earn? *(Section 2.5)*

Ⓐ $83.13 Ⓑ $105.45 Ⓒ $156.75 Ⓓ $1054.50

2.4–2.6 Quiz

Check It Out
Progress Check
BigIdeasMath.com

Add or subtract. *(Section 2.4)*

1. $6.329 + 14.38$

2. $43.56 + 41.82$

3. $85.8 - 2.354$

4. $26.782 - 14.96$

Multiply. Use estimation to check your answer. *(Section 2.5)*

5. 7.6×5

6. 0.62×17

7. 0.54×0.9

8. 4.16×0.7

Divide. Use estimation to check your answer. *(Section 2.6)*

9. $5\overline{)8.4}$

10. $6\overline{)6.48}$

11. $5.6 \div 0.7$

12. $1.8 \div 0.03$

13. FIELD HOCKEY A field hockey field is rectangular. Its width is 54.88 meters, and its length is 91.46 meters. Find the perimeter of the field. *(Section 2.4)*

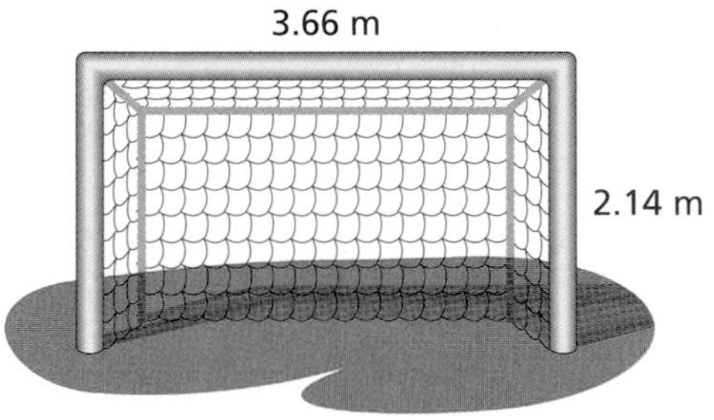

14. GEOMETRY Find the area of the mouth of the field hockey goal. *(Section 2.5)*

15. BROADWAY The bar graph shows the yearly attendance at traveling Broadway shows. *(Section 2.6)*

a. Suppose the attendance was the same each month in 2008. How many people attended each month?

b. How many times more people attended shows in 2006 than in 2009? Round your answer to the nearest tenth.

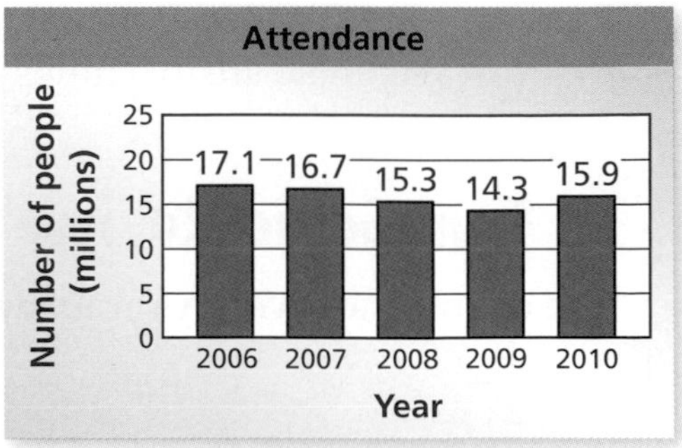

2 Chapter Review

Check It Out
Vocabulary Help
BigIdeasMath.com

Review Key Vocabulary

reciprocals, *p. 64*

Review Examples and Exercises

2.1 Multiplying Fractions *(pp. 54–61)*

a. Find $\frac{1}{4} \times \frac{3}{5}$.

$\frac{1}{4} \times \frac{3}{5} = \frac{1 \times 3}{4 \times 5} = \frac{3}{20}$ Multiply the numerators and the denominators.

b. Find $\frac{3}{5} \times 1\frac{1}{8}$.

$\frac{3}{5} \times 1\frac{1}{8} = \frac{3}{5} \times \frac{9}{8}$ Write $1\frac{1}{8}$ as the improper fraction $\frac{9}{8}$.

$= \frac{3 \times 9}{5 \times 8} = \frac{27}{40}$ Multiply the numerators and the denominators.

Exercises

Multiply. Write the answer in simplest form.

1. $\frac{1}{8} \times \frac{5}{7}$
2. $\frac{3}{5} \times \frac{1}{2}$
3. $\frac{2}{9} \times \frac{3}{4}$
4. $\frac{3}{10} \times \frac{4}{5}$
5. $2\frac{2}{3} \times \frac{4}{5}$
6. $\frac{2}{7} \times 4\frac{4}{9}$
7. $1\frac{5}{6} \times 2\frac{3}{8}$
8. $2\frac{3}{10} \times 5\frac{1}{3}$

2.2 Dividing Fractions *(pp. 62–69)*

Find $\frac{3}{7} \div \frac{5}{8}$.

$\frac{3}{7} \div \frac{5}{8} = \frac{3}{7} \times \frac{8}{5}$ Multiply by the reciprocal of $\frac{5}{8}$, which is $\frac{8}{5}$.

$= \frac{3 \times 8}{7 \times 5} = \frac{24}{35}$ Multiply fractions and simplify.

Exercises

Divide. Write the answer in simplest form.

9. $\frac{1}{9} \div \frac{2}{5}$
10. $\frac{3}{4} \div \frac{5}{6}$
11. $5 \div \frac{1}{3}$
12. $\frac{8}{9} \div \frac{3}{10}$

2.3 Dividing Mixed Numbers (pp. 70–75)

Find $3\frac{3}{4} \div 1\frac{1}{2}$.

$3\frac{3}{4} \div 1\frac{1}{2} = \frac{15}{4} \div \frac{3}{2}$ Write each mixed number as an improper fraction.

$= \frac{15}{4} \times \frac{2}{3}$ Multiply by the reciprocal of $\frac{3}{2}$, which is $\frac{2}{3}$.

$= \frac{\overset{5}{\cancel{15}} \times \overset{1}{\cancel{2}}}{\underset{2}{\cancel{4}} \times \underset{1}{\cancel{3}}}$ Multiply fractions. Divide out common factors.

$= \frac{5}{2}$, or $2\frac{1}{2}$ Simplify.

Exercises

Divide. Write the answer in simplest form.

13. $1\frac{2}{5} \div \frac{4}{7}$ **14.** $2\frac{3}{8} \div \frac{3}{5}$ **15.** $4\frac{1}{8} \div 2\frac{1}{4}$ **16.** $5\frac{5}{8} \div 1\frac{2}{9}$

17. PANCAKES A box contains 10 cups of pancake mix. You use $\frac{2}{3}$ cup each time you make pancakes. How many times can you make pancakes?

2.4 Adding and Subtracting Decimals (pp. 78–83)

a. Add 7.36 + 2.22.

Line up the decimal points.

$$\begin{array}{r} 7.36 \\ +\ 2.22 \\ \hline 9.58 \end{array}$$

Add as you would with whole numbers.

b. Subtract 5.467 − 2.736.

Line up the decimal points.

$$\begin{array}{r} \overset{4\ 14}{\cancel{5}.\cancel{4}67} \\ -\ 2.736 \\ \hline 2.731 \end{array}$$

Subtract as you would with whole numbers.

Exercises

Add or subtract.

18. 3.78 + 8.94 **19.** 19.89 + 4.372

20. 7.638 − 2.365 **21.** 14.21 − 4.103

2.5 Multiplying Decimals *(pp. 84–91)*

Find 7.5 × 5.3.

```
      7.5  ←——   1 decimal place
    × 5.3  ←——  + 1 decimal place
    2 2 5
  + 3 7 5
   3 9.7 5 ←——  2 decimal places
```

∴ So, $7.5 \times 5.3 = 39.75$.

Exercises

Multiply. Use estimation to check your answer.

22. 5.3×8 **23.** 6.1×7 **24.** 4.68×3

25. 9.475×8.03 **26.** 0.27×4.42 **27.** 0.051×0.244

28. AREA Find the area of the computer screen.

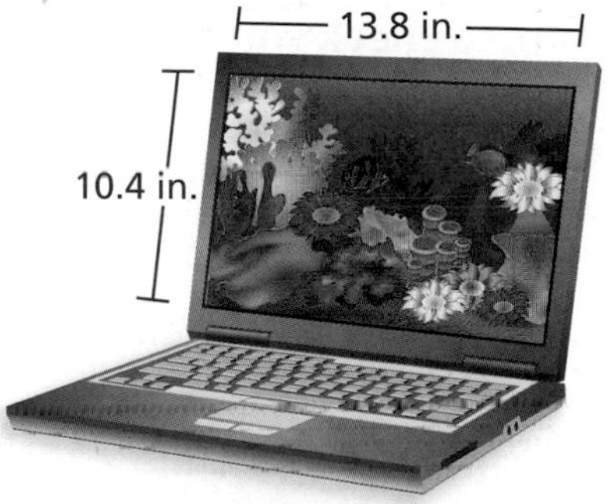

2.6 Dividing Decimals *(pp. 92–99)*

Find 22.8 ÷ 1.2.

Multiply 1.2 by 10.

Multiply 22.8 by 10.

Place the decimal point above the decimal point in the dividend 228.

```
1.2)22.8  ————→     19.
                 12)228.
                  − 12
                    108
                  − 108
                      0
```

∴ So, $22.8 \div 1.2 = 19$.

Exercises

Divide. Use estimation to check your answer.

29. $6.8 \div 4$ **30.** $13.2 \div 6 + 4$ **31.** $49.7 \div 7$

32. $0.12\overline{)3.6}$ **33.** $2.5\overline{)0.125}$ **34.** $3.9\overline{)22.23}$

2 Chapter Test

Check It Out
Test Practice
BigIdeasMath.com

Multiply. Write the answer in simplest form.

1. $\frac{9}{16} \times \frac{2}{3}$

2. $\frac{1}{10} \times \frac{5}{6}$

3. $1\frac{3}{7} \times 6\frac{7}{10}$

Divide. Write the answer in simplest form.

4. $\frac{1}{6} \div \frac{1}{3}$

5. $10 \div \frac{2}{5}$

6. $8\frac{3}{4} \div 2\frac{7}{8}$

Add or subtract.

7. $4.92 + 3.79$

8. $5.138 + 2.624$

9. $5.316 - 1.942$

Multiply. Use estimation to check your answer.

10. 6.7×8

11. 0.4×0.7

12. 4.87×7.23

Divide. Use estimation to check your answer.

13. $5.6 \div 7$

14. $2.6 \div 0.02$

15. $4\overline{)9.32}$

16. $0.25\overline{)5.46}$

17. DVD SALE Which deal is the better buy?

18. BLOG You spend $2\frac{1}{2}$ hours online. You spend $\frac{1}{5}$ of that time writing a blog. How long do you spend writing your blog?

19. GRAPES A grocery store sells grapes for $1.99 per pound. You buy 2.34 pounds of the grapes. How much do you pay?

20. PHOTOGRAPHY A motocross rider is in the air for 2.5 seconds. Your camera can take a picture every 0.125 second. Your friend's camera can take a picture every 0.15 second.

a. How many times faster is your camera than your friend's camera?

b. How many more pictures can you take while the rider is in the air?

2 Standards Assessment

Test-Taking Strategy
Estimate the Answer

5½ treats are divided evenly between you and Fluffy. How many do you get?
(A) 1½ (B) 2¾ (C) 5½ (D) 11
Fluffy: 1/2
Me: Five

"Using estimation you can see that the answer is about 3. So, you should choose B."

1. At a party, 10 people equally shared $2\frac{1}{2}$ gallons of ice cream. How much ice cream did each person eat? *(6.NS.1)*

A. $\frac{1}{5}$ gal **C.** $\frac{2}{5}$ gal

B. $\frac{1}{4}$ gal **D.** $\frac{3}{4}$ gal

2. What is the value of the expression below? *(6.NS.3)*

$$4.643 + 11.02 \div 2.32$$

3. Which number is equivalent to the expression below? *(6.EE.1)*

$$2 \cdot 4^2 + 3(6 \div 2)$$

F. 25 **H.** 73

G. 41 **I.** 105

4. Your friend divided two decimal numbers. Her work is shown in the box below. What should your friend change in order to divide the two decimal numbers correctly? *(6.NS.3)*

$$0.07\overline{)14.56} \rightarrow \overset{2.08}{7\overline{)14.56}}$$

A. Rewrite the problem as $0.07\overline{)0.1456}$. **C.** Rewrite the problem as $7\overline{)0.1456}$.

B. Rewrite the problem as $0.07\overline{)1456}$. **D.** Rewrite the problem as $7\overline{)1456}$.

5. You bought some grapes at a farm stand. You paid $2.48 per pound.

What was the total amount that you paid for the grapes? *(6.NS.3)*

6. The steps your friend took to divide two mixed numbers are shown below.

$$4\frac{2}{3} \div 2\frac{1}{4} = \frac{14}{3} \times \frac{9}{4}$$
$$= \frac{21}{2}$$
$$= 10\frac{1}{2}$$

What should your friend change in order to divide the two mixed numbers correctly? *(6.NS.1)*

F. Find a common denominator of 3 and 4.

G. Multiply by the reciprocal of $\frac{14}{3}$.

H. Multiply by the reciprocal of $\frac{9}{4}$.

I. Rename $4\frac{2}{3}$ as $3\frac{5}{3}$.

7. Which pair of numbers does *not* have a least common multiple less than 100? *(6.NS.4)*

A. 10, 15

B. 12, 16

C. 16, 18

D. 18, 24

8. You are making identical snack bags. You have 18 fruit-chew snacks and 24 granola snacks. What is the greatest number of snack bags that you can make with no snacks left over? *(6.NS.4)*

F. 1

G. 2

H. 3

I. 6

9. Which expression is *not* equivalent to $\frac{2}{3}$? *(6.NS.1)*

A. $\frac{1}{4} + \frac{1}{3} \div \frac{4}{5}$

B. $\frac{13}{30} + \frac{1}{5} \div \frac{6}{7}$

C. $\frac{5}{6} - \frac{1}{8} \div \frac{1}{2}$

D. $\frac{13}{18} - \frac{1}{26} \div \frac{9}{13}$

10. Which number is equivalent to $5.139 - 2.64$? *(6.NS.3)*

F. 2.499

G. 2.599

H. 3.519

I. 3.599

11. Which expression is equivalent to $\frac{4}{9} \div \frac{5}{7}$? *(6.NS.1)*

A. $\frac{20}{63}$

B. $\frac{28}{45}$

C. $\frac{45}{28}$

D. $\frac{63}{20}$

12. Which of the following expressions is equivalent to a perfect square? *(6.EE.1)*

F. $3 + 2^2 \times 7$

G. $34 + 18 \div 3^2$

H. $(80 + 4) \div 4$

I. $3^2 + 6 \times 5 \div 3$

13. You are filling baskets using 18 green eggs, 36 red eggs, and 54 blue eggs. What is the greatest number of baskets that you can fill so that the baskets are identical and there are no eggs left over? *(6.NS.4)*

A. 3

B. 6

C. 9

D. 18

14. A walkway was built using identical concrete blocks. *(6.NS.1)*

Think
Solve
Explain

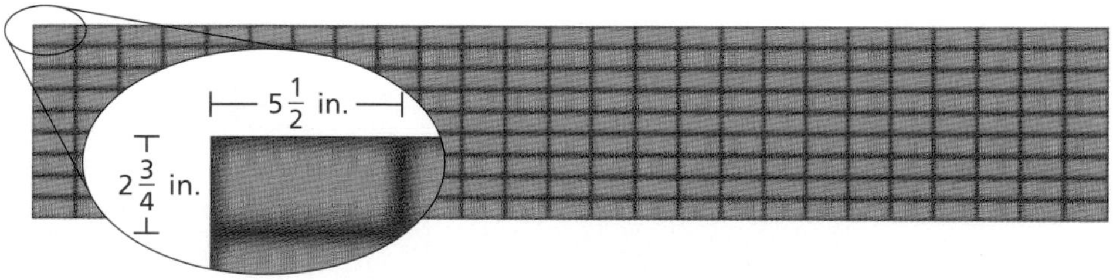

Part A How much longer, in inches, is the length of the walkway than the width of the walkway? Show your work and explain your reasoning.

Part B How many times longer is the length of the walkway than the width of the walkway? Show your work and explain your reasoning.

3 Algebraic Expressions and Properties

"Did you know that $5 \times 6 = 6 \times 5$, but $5 \div 6 \neq 6 \div 5$?"

"Only certain operations like addition and multiplication preserve equality when you switch the numbers around."

"Descartes, evaluate this expression when $x = 2$ to determine the number of cat treats you are going to eat today."

"Remember that you evaluate an algebraic expression by substituting the value of x into the expression."

What You Learned Before

"Great! You're up to $x = 2$. Let's keep going."

Interpreting Numerical Expressions (5.OA.2)

Example 1 **Write a sentence interpreting the expression $3 \times (19{,}762 + 418)$.**

$3 \times (19{,}762 + 418)$ is 3 times as large as $19{,}762 + 418$.

Example 2 **Write a sentence interpreting the expression $(316 + 43{,}449) + 5$.**

$(316 + 43{,}449) + 5$ is 5 more than $316 + 43{,}449$.

Example 3 **Write a sentence interpreting the expression $(20{,}008 - 752) \div 2$.**

$(20{,}008 - 752) \div 2$ is half as large as $20{,}008 - 752$.

Try It Yourself

Write a sentence interpreting the expression.

1. $3 \times (372 + 20{,}967)$
2. $2 \times (432 + 346{,}322)$
3. $4 \times (6722 + 4086)$
4. $(115 + 36{,}372) + 6$
5. $(392 + 75{,}325) + 78$
6. $(352 + 46{,}795) + 100$
7. $(30{,}929 + 425) \div 2$
8. $(58{,}742 - 721) \div 2$
9. $(96{,}792 + 564) \div 3$

Using Order of Operations (5.OA.1, 6.EE.1)

Example 4 **Simplify $4^2 \div 2 + 3(9 - 5)$.**

First:	Parentheses	$4^2 \div 2 + 3(9 - 5) = 4^2 \div 2 + 3 \cdot 4$
Second:	Exponents	$= 16 \div 2 + 3 \cdot 4$
Third:	Multiplication and Division (from left to right)	$= 8 + 12$
Fourth:	Addition and Subtraction (from left to right)	$= 20$

Try It Yourself

Simplify the expression.

10. $3^2 + 5(4 - 2)$
11. $3 + 4 \div 2$
12. $10 \div 5 \cdot 3$
13. $4(3^3 - 8) \div 2$
14. $3 \cdot 6 - 4 \div 2$
15. $12 + 7 \cdot 3 - 24$

3.1 Algebraic Expressions

Essential Question How can you write and evaluate an expression that represents a real-life problem?

1 ACTIVITY: Reading and Re-Reading

Work with a partner.

a. You babysit for 3 hours. You receive $12. What is your hourly wage?

- Write the problem. Underline the important numbers and units you need to solve the problem.
- Read the problem carefully a second time. Circle the key word for the question.

You babysit for 3 hours. You receive $12.
What is your hourly wage?

- Write each important number or word, with its units, on a piece of paper. Write +, −, ×, ÷, and = on five other pieces of paper.

hourly wage ($ per hour)

- Arrange the pieces of paper to answer the key word question, "What is your hourly wage?"
- Evaluate the expression that represents the hourly wage.

hourly wage = ___ ÷ ___ Write.

= ___ Evaluate.

So, your hourly wage is $___ per hour.

b. How can you use your hourly wage to find how much you will receive for any number of hours worked?

COMMON CORE

Algebraic Expressions

In this lesson, you will

- use order of operations to evaluate algebraic expressions.
- solve real-life problems.

Learning Standard 6.EE.2c

2 ACTIVITY: Reading and Re-Reading

Math Practice 2

Make Sense of Quantities

What are the units in the problem? How does this help you write an expression?

Work with a partner. Use the strategy shown in Activity 1 to write an expression for each problem. After you have written the expression, evaluate it using mental math or some other method.

a. You wash cars for 2 hours. You receive \$6. How much do you earn per hour?

b. You have \$60. You buy a pair of jeans and a shirt. The pair of jeans costs \$27. You come home with \$15. How much did you spend on the shirt?

c. For lunch, you buy 5 sandwiches that cost \$3 each. How much do you spend?

d. You are running a 4500-foot race. How much farther do you have to go after running 2000 feet?

e. A young rattlesnake grows at a rate of about 20 centimeters per year. How much does a young rattlesnake grow in 2 years?

What Is Your Answer?

3. **IN YOUR OWN WORDS** How can you write and evaluate an expression that represents a real-life problem? Give one example with addition, one with subtraction, one with multiplication, and one with division.

Practice Use what you learned about evaluating expressions to complete Exercises 4–7 on page 115.

3.1 Lesson

Check It Out
Lesson Tutorials
BigIdeasMath.com

Key Vocabulary
- algebraic expression, *p. 112*
- terms, *p. 112*
- variable, *p. 112*
- coefficient, *p. 112*
- constant, *p. 112*

An **algebraic expression** is an expression that may contain numbers, operations, and one or more symbols. Parts of an algebraic expression are called **terms**.

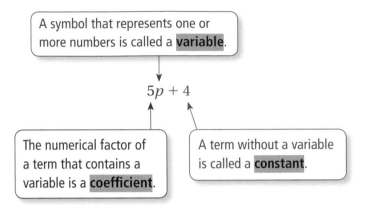

A symbol that represents one or more numbers is called a **variable**.

$$5p + 4$$

The numerical factor of a term that contains a variable is a **coefficient**.

A term without a variable is called a **constant**.

EXAMPLE 1 Identifying Parts of an Algebraic Expression

Identify the terms, coefficients, and constants in each expression.

a. $5x + 13$

$5x + 13$

Terms: $5x$, 13

Coefficient: 5

Constant: 13

b. $2z^2 + y + 3$

$2z^2 + y + 3$

Terms: $2z^2$, $1y$, 3

Coefficients: 2, 1

Constant: 3

Study Tip

A variable by itself has a coefficient of 1. So, the term y in Example 1(b) has a coefficient of 1.

On Your Own

Now You're Ready
Exercises 8–13

Identify the terms, coefficients, and constants in the expression.

1. $12 + 10c$ **2.** $15 + 3w + \frac{1}{2}$ **3.** $z^2 + 9z$

EXAMPLE 2 Writing Algebraic Expressions Using Exponents

Write each expression using exponents.

a. $d \cdot d \cdot d \cdot d$

Because d is used as a factor 4 times, its exponent is 4.

So, $d \cdot d \cdot d \cdot d = d^4$.

b. $1.5 \cdot h \cdot h \cdot h$

Because h is used as a factor 3 times, its exponent is 3.

So, $1.5 \cdot h \cdot h \cdot h = 1.5h^3$.

On Your Own

Now You're Ready
Exercises 16–21

Write the expression using exponents.

4. $j \cdot j \cdot j \cdot j \cdot j \cdot j$

5. $9 \cdot k \cdot k \cdot k \cdot k \cdot k$

To evaluate an algebraic expression, substitute a number for each variable. Then use the order of operations to find the value of the numerical expression.

EXAMPLE 3 Evaluating Algebraic Expressions

a. Evaluate $k + 10$ when $k = 25$.

$k + 10 = 25 + 10$ Substitute 25 for k.

$= 35$ Add 25 and 10.

b. Evaluate $4 \cdot n$ when $n = 12$.

$4 \cdot n = 4 \cdot 12$ Substitute 12 for n.

$= 48$ Multiply 4 and 12.

Study Tip

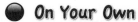

You can write the product of 4 and n in several ways.

$4 \cdot n$

$4n$

$4(n)$

On Your Own

Now You're Ready
Exercises 25–32

6. Evaluate $24 + c$ when $c = 9$.

7. Evaluate $d - 17$ when $d = 30$.

EXAMPLE 4 Evaluating an Expression with Two Variables

Evaluate $a \div b$ when $a = 16$ and $b = \frac{2}{3}$.

$a \div b = 16 \div \frac{2}{3}$ Substitute 16 for a and $\frac{2}{3}$ for b.

$= 16 \cdot \frac{3}{2}$ Multiply by the reciprocal of $\frac{2}{3}$, which is $\frac{3}{2}$.

$= 24$ Multiply.

On Your Own

Now You're Ready
Exercises 33–36

Evaluate the expression when $p = 24$ and $q = 8$.

8. $p \div q$

9. $q + p$

10. $p - q$

11. pq

EXAMPLE 5 Evaluating Expressions with Two Operations

a. Evaluate $3x - 14$ when $x = 5$.

$3x - 14$	$= 3(5) - 14$	Substitute 5 for *x*.
	$= 15 - 14$	Using order of operations, multiply 3 and 5.
	$= 1$	Subtract 14 from 15.

b. Evaluate $z^2 + 8.5$ when $z = 2$.

$z^2 + 8.5$	$= 2^2 + 8.5$	Substitute 2 for *z*.
	$= 4 + 8.5$	Using order of operations, evaluate 2^2.
	$= 12.5$	Add 4 and 8.5.

On Your Own

Now You're Ready
Exercises 43–51

Evaluate the expression when $y = 6$.

12. $5y + 1$ **13.** $30 - 24 \div y$ **14.** $y^2 - 7$ **15.** $1.5 + y^2$

EXAMPLE 6 Real-Life Application

You are saving money to buy a skateboard. You begin with $45 and you save $3 each week. The expression $45 + 3w$ gives the amount of money you save after *w* weeks.

a. How much will you have after 4 weeks, 10 weeks, and 20 weeks?

b. After 20 weeks, can you buy the skateboard? Explain.

Substitute the given number of weeks for *w*.

a.

Number of Weeks, *w*	$45 + 3w$	Amount Saved
4	$45 + 3(4)$	$45 + 12 = \$57$
10	$45 + 3(10)$	$45 + 30 = \$75$
20	$45 + 3(20)$	$45 + 60 = \$105$

b. After 20 weeks, you have $105. So, you cannot buy the $125 skateboard.

On Your Own

16. WHAT IF? In Example 6, the expression for how much money you have after *w* weeks is $45 + 4w$. Can you buy the skateboard after 20 weeks? Explain.

3.1 Exercises

Check It Out
Help with Homework
BigIdeasMath.com

Vocabulary and Concept Check

1. **WHICH ONE DOESN'T BELONG?** Which expression does *not* belong with the other three? Explain your reasoning.

$2x + 1$	$5w \cdot c$	$3(4) + 5$	$y \div z$

2. **NUMBER SENSE** Which step in the order of operations is first? second? third? fourth?

Add or subtract from left to right.	Multiply or divide from left to right.
Evaluate terms with exponents.	Perform operations in parentheses.

3. **NUMBER SENSE** Will the value of the expression $20 - x$ *increase, decrease,* or *stay the same* as x increases? Explain.

Practice and Problem Solving

Write and evaluate an expression for the problem.

4. You receive \$8 for raking leaves for 2 hours. What is your hourly wage?
5. Music lessons cost \$20 per week. How much do 6 weeks of lessons cost?
6. The scores on your first two history tests were 82 and 95. By how many points did you improve on your second test?
7. You buy a hat for \$12 and give the cashier a \$20 bill. How much change do you receive?

Identify the terms, coefficients, and constants in the expression.

1 8. $7h + 3$

9. $g + 12 + 9g$

10. $5c^2 + 7d$

11. $2m^2 + 15 + 2p^2$

12. $6 + n^2 + \frac{1}{2}d$

13. $8x + \frac{x^2}{3}$

14. **ERROR ANALYSIS** Describe and correct the error in identifying the terms, coefficients, and constants in the algebraic expression $2x^2y$.

> ✗
> Terms: 2, x^2, y
> Coefficient: 2
> Constant: none

15. **PERIMETER** You can use the expression $2\ell + 2w$ to find the perimeter of a rectangle where ℓ is the length and w is the width.

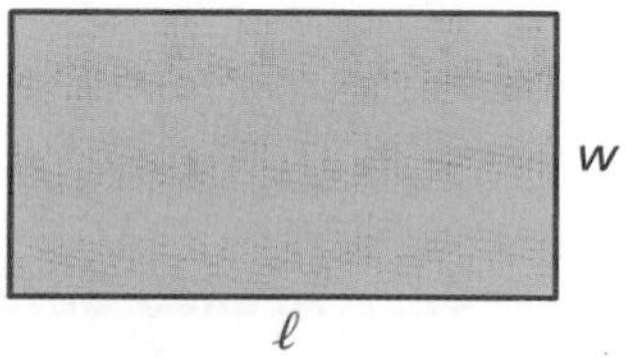

a. Identify the terms, coefficients, and constants in the expression.

b. Interpret the coefficients of the terms.

Write each expression using exponents.

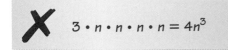

2 **16.** $b \cdot b \cdot b$ **17.** $g \cdot g \cdot g \cdot g \cdot g$ **18.** $8 \cdot w \cdot w \cdot w \cdot w$

19. $5.2 \cdot y \cdot y \cdot y$ **20.** $a \cdot a \cdot c \cdot c$ **21.** $2.1 \cdot x \cdot z \cdot z \cdot z \cdot z$

✗ $3 \cdot n \cdot n \cdot n \cdot n = 4n^3$

22. ERROR ANALYSIS Describe and correct the error in writing the product using exponents.

23. AREA Write an expression using exponents that represents the area of the square.

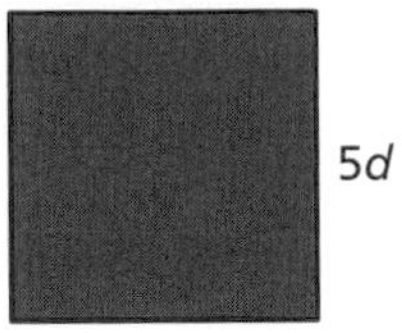

As I was going to St. Ives
I met a man with seven wives
Each wife had seven sacks
Each sack had seven cats
Each cat had seven kits
Kits, cats, sacks, wives
How many were going to St. Ives?

24. ST. IVES Suppose the man in the St. Ives poem has x wives, each wife has x sacks, each sack has x cats, and each cat has x kits. Write an expression using exponents that represents the total number of kits, cats, sacks, and wives going to St. Ives.

ALGEBRA Evaluate the expression when $a = 3$, $b = 2$, and $c = 12$.

3 **25.** $6 + a$ **26.** $b \cdot 5$ **27.** $c - 1$ **28.** $27 \div a$

29. $12 - b$ **30.** $c + 5$ **31.** $2a$ **32.** $c \div 6$

4 **33.** $a + b$ **34.** $c - a$ **35.** $\frac{c}{a}$ **36.** $b \cdot c$

37. ERROR ANALYSIS Describe and correct the error in evaluating the expression when $m = 8$.

✗
$$5m + 3 = 5 \cdot 8 + 3$$
$$= 5 \cdot 11$$
$$= 55$$

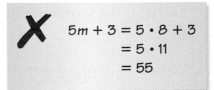

38. LAWNS You earn $15n$ dollars for mowing n lawns. How much do you earn for mowing one lawn? seven lawns?

39. PLANT After m months, the height of a plant is $10 + 3m$ millimeters. How tall is the plant after eight months? three years?

Copy and complete the table.

40.

x	3	6	9
$x \cdot 8$			

41.

x	2	4	8
$64 \div x$			

42. FALLING OBJECT An object falls $16t^2$ feet in t seconds. You drop a rock from a bridge that is 75 feet above the water. Will the rock hit the water in 2 seconds? Explain.

ALGEBRA **Evaluate the expression when $a = 10$, $b = 9$, and $c = 4$.**

5 **43.** $2a + 3$ **44.** $4c - 7.8$ **45.** $\frac{a}{4} + \frac{1}{3}$

46. $\frac{24}{b} + 8$ **47.** $c^2 + 6$ **48.** $a^2 - 18$

49. $a + 9c$ **50.** $bc + 12.3$ **51.** $3a + 2b - 6c$

Standard Rentals
$3

New Releases
$4

52. **MOVIES** You rent x new releases and y standard rentals. Which expression tells you how much money you will need?

$3x + 4y$ $4x + 3y$ $7(x + y)$

53. **WATER PARK** You float 2000 feet along a "Lazy River" water ride. The ride takes less than 10 minutes. Give two examples of possible times and speeds. Illustrate the water ride with a drawing.

54. **SCIENCE CENTER** The expression $20a + 13c$ is the cost (in dollars) for a adults and c students to enter a science center.

a. How much does it cost for an adult? a student? Explain your reasoning.

b. Find the total cost for 4 adults and 24 students.

c. You find the cost for a group. Then the numbers of adults and students in the group both double. Does the cost double? Explain your answer using an example.

d. In part (b), the number of adults is cut in half, but the number of students doubles. Is the cost the same? Explain your answer.

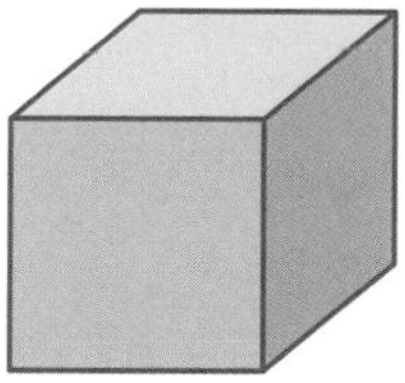

x in.

55. **Reasoning** The volume of the cube is equal to four times the area of one of its faces. What is the volume of the cube?

Fair Game Review What you learned in previous grades & lessons

Find the value of the power. *(Section 1.2)*

56. 3^5 **57.** 8^3 **58.** 7^4 **59.** 2^8

60. **MULTIPLE CHOICE** Which numbers have a least common multiple of 24? *(Section 1.6)*

Ⓐ 4, 6 Ⓑ 2, 22 Ⓒ 3, 8 Ⓓ 6, 12

3.2 Writing Expressions

Essential Question How can you write an expression that represents an unknown quantity?

1 ACTIVITY: Ordering Lunch

Work with a partner. You use a $20 bill to buy lunch at a café. You order a sandwich from the menu board shown.

a. Complete the table. In the last column, write a numerical expression for the amount of change received.

Sandwich	Price (dollars)	Change Received (dollars)
Reuben		
BLT		
Egg salad		
Roast beef		

b. **REPEATED REASONING** Write an expression for the amount of change you receive when you order any sandwich from the menu board.

c. Compare the expression you wrote in part (b) with the expressions in the last column of the table in part (a).

d. The café offers several side dishes, each at the same price. You order a chicken salad sandwich and two side dishes. Write an expression for the total amount of money you spend. Explain how you wrote your expression.

e. The expression $20 - 4.65s$ represents the amount of change one customer receives after ordering from the menu board. Explain what each part of the expression represents. Do you know what the customer ordered? Explain your reasoning.

COMMON CORE

Algebraic Expressions

In this lesson, you will

- use variables to represent numbers in algebraic expressions.
- write algebraic expressions.

Learning Standard
6.EE.2a

2 ACTIVITY: Words That Imply Addition or Subtraction

Math Practice 2

Use Expressions

How do the key words in the phrase help you write the given relationship as an expression?

Work with a partner.

a. Complete the table.

Variable	Phrase	Expression
n	4 **more than** a number	
m	the **difference** of a number and 3	
x	the **sum** of a number and 8	
p	10 **less than** a number	
n	7 units **farther** away	
t	8 minutes **sooner**	
w	12 minutes **later**	
y	a number **increased** by 9	

b. Here is a word problem that uses one of the expressions in the table.

You arrive at the café 8 minutes sooner than your friend. Your friend arrives at 6:42 P.M. *When did you arrive?*

Which expression from the table can you use to solve the problem?

c. Write a problem that uses a different expression from the table.

3 ACTIVITY: Words That Imply Multiplication or Division

Work with a partner. Match each phrase with an expression.

the product of a number and 3	$n \div 3$
the quotient of 3 and a number	$4p$
4 times a number	$n \cdot 3$
a number divided by 3	$2m$
twice a number	$3 \div n$

What Is Your Answer?

4. **IN YOUR OWN WORDS** How can you write an expression that represents an unknown quantity? Give examples to support your explanation.

Practice Use what you learned about writing expressions to complete Exercises 9–12 on page 122.

3.2 Lesson

Check It Out
Lesson Tutorials
BigIdeasMath.com

Some words imply math operations.

Operation	Addition	Subtraction	Multiplication	Division
Key Words and Phrases	added to plus sum of more than increased by total of and	subtracted from minus difference of less than decreased by fewer than take away	multiplied by times product of twice of	divided by quotient of

EXAMPLE 1 Writing Numerical Expressions

Write the phrase as an expression.

a. 8 fewer than 21

$21 - 8$ The phrase *fewer than* means *subtraction.*

b. the product of 30 and 9

30×9, or $30 \cdot 9$ The phrase *product of* means *multiplication.*

EXAMPLE 2 Writing Algebraic Expressions

Write the phrase as an expression.

a. 14 more than a number x

$x + 14$ The phrase *more than* means *addition.*

b. a number y minus 75

$y - 75$ The word *minus* means *subtraction.*

c. the quotient of 3 and a number z

$3 \div z$, or $\frac{3}{z}$ The phrase *quotient of* means *division.*

Common Error

When writing expressions involving subtraction or division, order is important. For example, the quotient of a number x and 2 means

$x \div 2$, not $2 \div x$.

On Your Own

Now You're Ready
Exercises 3–18

Write the phrase as an expression.

1. the sum of 18 and 35
2. 6 times 50
3. 25 less than a number b
4. a number x divided by 4
5. the total of a number t and 11
6. 100 decreased by a number k

EXAMPLE 3 Writing an Algebraic Expression

The length of Interstate 90 from the West Coast to the East Coast is 153.5 miles more than 2 times the length of Interstate 15 from southern California to northern Montana. Let m be the length of Interstate 15. Which expression can you use to represent the length of Interstate 90?

Ⓐ $2m + 153.5$ Ⓑ $2m - 153.5$ Ⓒ $153.5 - 2m$ Ⓓ $153.5m + 2$

The word *times* means *multiplication*. So, multiply 2 and m.

$$2m + 153.5$$

The phrase *more than* means *addition*. So, add $2m$ and 153.5.

∴ The correct answer is Ⓐ.

EXAMPLE 4 Real-Life Application

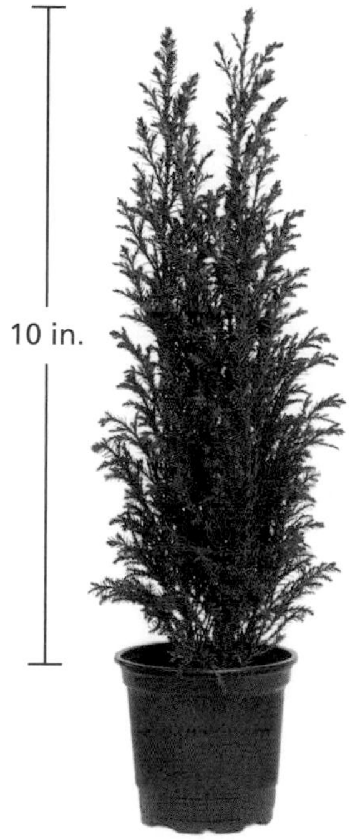

You plant a cypress tree that is 10 inches tall. Each year, its height increases by 15 inches.

a. Make a table that shows the height of the tree for 4 years. Then write an expression for the height after t years.

b. What is the height after 9 years?

a. The height is *increasing*, so *add* 15 each year as shown in the table.

Year, t	Height (inches)
0	10
1	$10 + 15(1) = 25$
2	$10 + 15(2) = 40$
3	$10 + 15(3) = 55$
4	$10 + 15(4) = 70$

When t is 0, the height is 10 inches.

You can see that an expression is $10 + 15t$.

∴ So, the height after year t is $10 + 15t$.

b. Evaluate $10 + 15t$ when $t = 9$.

$$10 + 15t = 10 + 15(9) = 145$$

∴ After 9 years, the height of the tree is 145 inches.

Study Tip

Sometimes, like in Example 3, a variable represents a single value. Other times, like in Example 4, a variable can represent more than one value.

On Your Own

7. Your friend has 5 more than twice as many game tokens as your sister. Let t be the number of game tokens your sister has. Write an expression for the number of game tokens your friend has.

8. **WHAT IF?** In Example 4, what is the height of the cypress tree after 16 years?

Now You're Ready
Exercises 27–30

3.2 Exercises

Check It Out
Help with Homework
BigIdeasMath.com

Vocabulary and Concept Check

1. **DIFFERENT WORDS, SAME QUESTION** Which is different? Write "both" expressions.

12 more than x	x increased by 12	x take away 12	the sum of x and 12

2. **REASONING** You pay $0.25p$ dollars to print p photos. What does the coefficient represent?

Practice and Problem Solving

Write the phrase as an expression.

1 2 3. 5 less than 8
4. the product of 3 and 12
5. 28 divided by 7
6. the total of 6 and 10
7. 3 fewer than 18
8. 17 added to 15
9. 13 subtracted from a number x
10. 5 times a number d
11. the quotient of 18 and a number a
12. the difference of a number s and 6
13. 7 increased by a number w
14. a number b squared
15. the sum of a number y and 4
16. the difference of 12 and a number x
17. twice a number z
18. a number t cubed

ERROR ANALYSIS Describe and correct the error in writing the phrase as an expression.

19. the quotient of 8 and a number y

✗ $\frac{y}{8}$

20. 16 decreased by a number x

✗ $x - 16$

21. **DINNER** Five friends share the cost of a dinner equally.
    a. Write an expression for the cost per person.
    b. Make up a total cost and test your expression. Is the result reasonable?

22. **TV SHOW** A television show has 19 episodes per season.
    a. Copy and complete the table.
    b. Write an expression for the number of episodes in n seasons.

Seasons	1	2	3	4	5
Episodes					

Give two ways to write the expression as a phrase.

23. $n + 6$
24. $4w$
25. $15 - b$
26. $14 - 3z$

③ ④ Write the phrase as an expression. Then evaluate when $x = 5$ and $y = 20$.

27. 3 less than the quotient of a number y and 4

28. the sum of a number x and 4, all divided by 3

29. 6 more than the product of 8 and a number x

30. the quotient of 40 and the difference of a number y and 16

31. MODELING It costs $3 to bowl a game and $2 for shoe rental.

a. Make a table for the cost of up to 5 games.

b. Write an expression for the cost of g games.

c. Use your expression to find the cost of 8 games.

32. PUZZLE Florida has 8 less than 5 times the number of counties in Arizona. Georgia has 25 more than twice the number of counties in Florida.

a. Write an expression for the number of counties in Florida.

b. Write an expression for the number of counties in Georgia.

c. Arizona has 15 counties. How many do Florida and Georgia have?

33. PATTERNS There are 140 people in a singing competition. The graph shows the results for the first five rounds.

a. Write an expression for the number of people after each round.

b. How many people compete in the ninth round? Explain your reasoning.

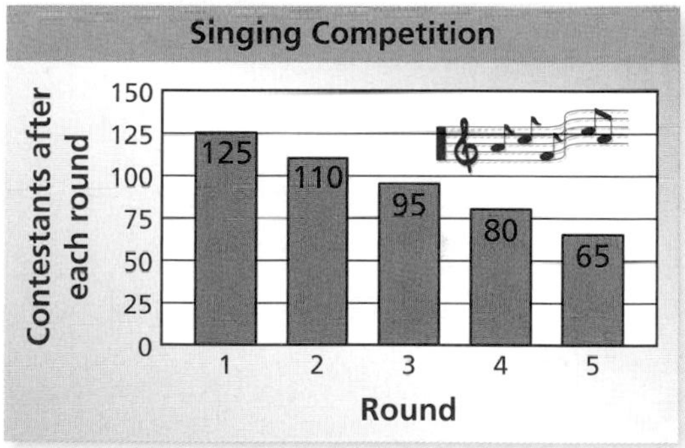

34. NUMBER SENSE The difference between two numbers is 8. The lesser number is a. Write an expression for the greater number.

35. Reasoning One number is four times another. The greater number is x. Write an expression for the lesser number.

Fair Game Review What you learned in previous grades & lessons

Evaluate the expression. *(Skills Review Handbook)*

36. $8 + (22 + 15)$ **37.** $(13 + 9) + 37$ **38.** $(13 \times 6) \times 5$ **39.** $4 \times (7 \times 5)$

40. MULTIPLE CHOICE A grocery store is making fruit baskets using 144 apples, 108 oranges, and 90 pears. Each basket will be identical. What is the greatest number of fruit baskets the store can make using all the fruit? *(Section 1.5)*

Ⓐ 6 Ⓑ 9 Ⓒ 16 Ⓓ 18

3 Study Help

Check It Out
Graphic Organizer
BigIdeasMath.com

You can use an **information wheel** to organize information about a topic. Here is an example of an information wheel for identifying parts of an algebraic expression.

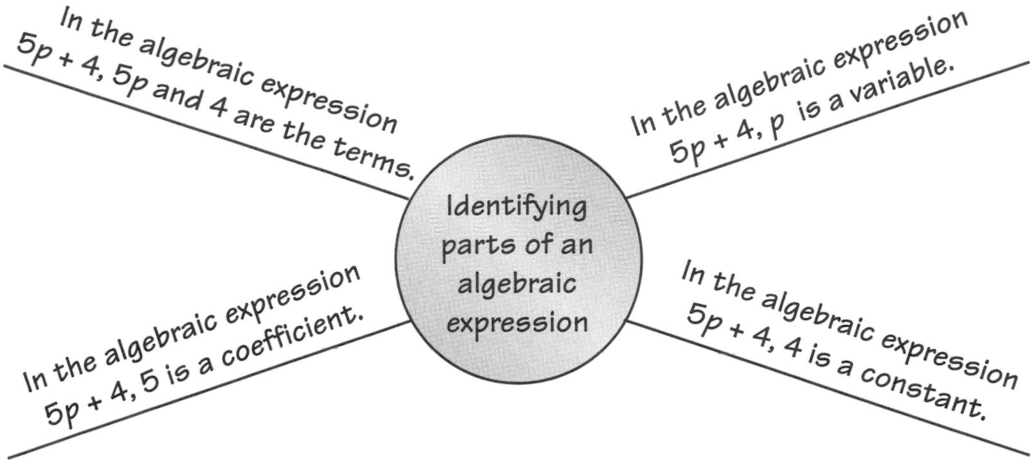

Identifying parts of an algebraic expression
In the algebraic expression 5p + 4, 5p and 4 are the terms.
In the algebraic expression 5p + 4, p is a variable.
In the algebraic expression 5p + 4, 5 is a coefficient.
In the algebraic expression 5p + 4, 4 is a constant.

On Your Own

Make information wheels to help you study these topics.

1. evaluating algebraic expressions
2. writing algebraic expressions

After you complete this chapter, make information wheels for the following topics.

3. Commutative Properties of Addition and Multiplication
4. Associative Properties of Addition and Multiplication
5. Addition Property of Zero
6. Multiplication Properties of Zero and One
7. Distributive Property
8. factoring expressions

"My information wheel for Fluffy has matching adjectives and nouns."

3.1–3.2 Quiz

Check It Out
Progress Check
BigIdeasMath.com

Identify the terms, coefficients, and constants of the expression. *(Section 3.1)*

1. $6q + 1$

2. $3r^2 + 4r + 8$

Write the expression using exponents. *(Section 3.1)*

3. $s \cdot s \cdot s \cdot s$

4. $2 \cdot t \cdot t \cdot t \cdot t \cdot t$

Evaluate the expression when $a = 8$ and $b = 2$. *(Section 3.1)*

5. $a + 5$

6. ab

7. $a^2 - 6$

Copy and complete the table. *(Section 3.1)*

8.

x	$x + 6$
1	
2	
3	

9.

x	$3x - 5$
3	
6	
9	

Write the phrase as an expression. *(Section 3.2)*

10. the sum of 28 and 35

11. a number x divided by 2

12. the product of a number m and 23

13. 10 less than a number a

14. COUPON The expression $p - 15$ is the amount you pay after using the coupon on a purchase of p dollars. How much do you pay for a purchase of $83? *(Section 3.1)*

Coupon
Good for $15 off any
purchase of $75 or more

15. AMUSEMENT PARK The expression $15a + 12c$ is the cost (in dollars) of admission at an amusement park for a adults and c children. Find the total cost for 5 adults and 10 children. *(Section 3.1)*

MOVING?
Call Now & Let Us Help You

16. MOVING TRUCK To rent a moving truck for the day, it costs $33 plus $1 for each mile driven. *(Section 3.2)*

a. Write an expression for the cost to rent the truck.

b. You drive the truck 300 miles. How much do you pay?

3.3 Properties of Addition and Multiplication

Essential Question Does the order in which you perform an operation matter?

1 ACTIVITY: Does Order Matter?

Work with a partner. Place each statement in the correct oval.

a. Fasten 5 shirt buttons.
b. Put on a shirt and tie.
c. Fill and seal an envelope.
d. Floss your teeth.
e. Put on your shoes.
f. Chew and swallow.

Order Matters | **Order Doesn't Matter**

Think of three math problems using the four operations where order matters and three where order doesn't matter.

The Meaning of a Word ● Commute

When you **commute** the positions of two stuffed animals on a shelf,

you switch their positions.

COMMON CORE

Equivalent Expressions

In this lesson, you will

- use properties of operations to generate equivalent expressions.

Learning Standards
6.EE.3
6.EE.4

2 ACTIVITY: Commutative Properties

Work with a partner.

a. Which of the following are true?

$3 + 5 \stackrel{?}{=} 5 + 3$ $\qquad$ $3 - 5 \stackrel{?}{=} 5 - 3$

$9 \times 3 \stackrel{?}{=} 3 \times 9$ $\qquad$ $9 \div 3 \stackrel{?}{=} 3 \div 9$

b. The true equations show the Commutative Properties of Addition and Multiplication. Why do you think they are called *commutative*?

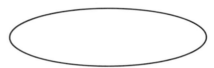

The Meaning of a Word ● Associate

You have two best friends. Sometimes you **associate** with one of them.

And sometimes you **associate** with the other.

3 ACTIVITY: Associative Properties

Math Practice 3

Use Counterexamples

What do the false equations tell you about the Associative Properties?

Work with a partner.

a. Which of the following are true?

$8 + (3 + 1) \stackrel{?}{=} (8 + 3) + 1$ $\qquad$ $8 - (3 - 1) \stackrel{?}{=} (8 - 3) - 1$

$12 \times (6 \times 2) \stackrel{?}{=} (12 \times 6) \times 2$ $\qquad$ $12 \div (6 \div 2) \stackrel{?}{=} (12 \div 6) \div 2$

b. The true equations show the Associative Properties of Addition and Multiplication. Why do you think they are called *associative*?

What Is Your Answer?

I know a trick for adding this.

4. **IN YOUR OWN WORDS** Does the order in which you perform an operation matter? Give examples to support your explanation.

5. **MENTAL MATH** Explain how you can add the sum in your head.

$$11 + 7 + 12 + 13 + 8 + 9$$

6. **SECRET CODE** The creatures on a distant planet use the symbols ■, ◆, ★, and ● for the four operations.

   a. Use the codes to decide which symbol represents addition and which symbol represents multiplication. Explain your reasoning.

   3 ● 4 = 4 ● 3

   3 ★ 4 = 4 ★ 3

   2 ● (5 ● 3) = (2 ● 5) ● 3

   2 ★ (5 ★ 3) = (2 ★ 5) ★ 3

   0 ● 4 = 0

   0 ★ 4 = 4

   b. Make up your own symbols for addition and multiplication. Write codes using your symbols. Trade codes with a classmate. Decide which symbol represents addition and which symbol represents multiplication.

Practice Use what you learned about the properties of addition and multiplication to complete Exercises 5–8 on page 130.

3.3 Lesson

Check It Out
Lesson Tutorials
BigIdeasMath.com

Key Vocabulary
equivalent expressions, *p. 128*

Expressions with the same value, like $12 + 7$ and $7 + 12$, are **equivalent expressions**. You can use the Commutative and Associative Properties to write equivalent expressions.

Key Ideas

Commutative Properties

Words Changing the order of addends or factors does not change the sum or product.

Numbers $5 + 8 = 8 + 5$ **Algebra** $a + b = b + a$

$5 \cdot 8 = 8 \cdot 5$ $a \cdot b = b \cdot a$

Associative Properties

Words Changing the grouping of addends or factors does not change the sum or product.

Numbers $(7 + 4) + 2 = 7 + (4 + 2)$

$(7 \cdot 4) \cdot 2 = 7 \cdot (4 \cdot 2)$

Algebra $(a + b) + c = a + (b + c)$

$(a \cdot b) \cdot c = a \cdot (b \cdot c)$

EXAMPLE 1 Using Properties to Write Equivalent Expressions

a. Simplify the expression $7 + (12 + x)$.

$7 + (12 + x) = (7 + 12) + x$ Associative Property of Addition

$= 19 + x$ Add 7 and 12.

b. Simplify the expression $(6.1 + x) + 8.4$.

$(6.1 + x) + 8.4 = (x + 6.1) + 8.4$ Commutative Property of Addition

$= x + (6.1 + 8.4)$ Associative Property of Addition

$= x + 14.5$ Add 6.1 and 8.4.

c. Simplify the expression $5(11y)$.

$5(11y) = (5 \cdot 11)y$ Associative Property of Multiplication

$= 55y$ Multiply 5 and 11.

Study Tip

One way to check whether expressions are equivalent is to evaluate each expression for any value of the variable. In Example 1(a), use $x = 2$.

$7 + (12 + x) = 19 + x$

$7 + (12 + 2) \stackrel{?}{=} 19 + 2$

$21 = 21$ ✓

On Your Own

Now You're Ready
Exercises 5–8

Simplify the expression. Explain each step.

1. $10 + (a + 9)$ **2.** $\left(c + \frac{2}{3}\right) + \frac{1}{2}$ **3.** $5(4n)$

Key Ideas

Addition Property of Zero

Words The sum of any number and 0 is that number.

Numbers $7 + 0 = 7$ **Algebra** $a + 0 = a$

Multiplication Properties of Zero and One

Words The product of any number and 0 is 0.

The product of any number and 1 is that number.

Numbers $9 \cdot 0 = 0$ **Algebra** $a \cdot 0 = 0$

$4 \cdot 1 = 4$ $a \cdot 1 = a$

EXAMPLE 2 Using Properties to Write Equivalent Expressions

a. Simplify the expression $9 \cdot 0 \cdot p$.

$9 \cdot 0 \cdot p = (9 \cdot 0) \cdot p$	Associative Property of Multiplication
$= 0 \cdot p = 0$	Multiplication Property of Zero

b. Simplify the expression $4.5 \cdot r \cdot 1$.

$4.5 \cdot r \cdot 1 = 4.5 \cdot (r \cdot 1)$	Associative Property of Multiplication
$= 4.5 \cdot r$	Multiplication Property of One
$= 4.5r$	

EXAMPLE 3 Real-Life Application

You and six friends play on a basketball team. A sponsor paid \$100 for the league fee, x dollars for each player's T-shirt, and \$68.25 for trophies. Write an expression for the total amount the sponsor paid.

Common Error

You **and** six friends are on the team, so use the expression $7x$, not $6x$, to represent the cost of the T-shirts.

Add the league fee, the cost of the T-shirts, and the cost of the trophies.

$100 + 7x + 68.25 = 7x + 100 + 68.25$	Commutative Property of Addition
$= 7x + 168.25$	Add 100 and 68.25.

An expression for the total amount is $7x + 168.25$.

On Your Own

Now You're Ready
Exercises 9–23

Simplify the expression. Explain each step.

4. $12 \cdot b \cdot 0$
5. $1 \cdot m \cdot 24$
6. $(t + 15) + 0$
7. **WHAT IF?** In Example 3, your sponsor paid \$54.75 for trophies. Write an expression for the total amount the sponsor paid.

3.3 Exercises

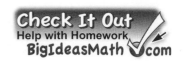

Check It Out
Help with Homework
BigIdeasMath.com

Vocabulary and Concept Check

1. **NUMBER SENSE** Write an example of a sum of fractions. Show that the Commutative Property of Addition is true for the sum.
2. **OPEN-ENDED** Write an algebraic expression that can be simplified using the Associative Property of Addition.
3. **OPEN-ENDED** Write an algebraic expression that can be simplified using the Associative Property of Multiplication and the Multiplication Property of One.
4. **WHICH ONE DOESN'T BELONG?** Which statement does *not* belong with the other three? Explain your reasoning.

$7 + (x + 4) = 7 + (4 + x)$	$(3 + b) + 2 = (b + 3) + 2$
$9 + (7 + w) = (9 + 7) + w$	$(4 + n) + 6 = (n + 4) + 6$

Practice and Problem Solving

Tell which property the statement illustrates.

1 5. $5 \cdot p = p \cdot 5$

6. $2 + (12 + r) = (2 + 12) + r$

7. $4 \cdot (x \cdot 10) = (4 \cdot x) \cdot 10$

8. $x + 7.5 = 7.5 + x$

2 9. $(c + 2) + 0 = c + 2$

10. $a \cdot 1 = a$

11. **ERROR ANALYSIS** Describe and correct the error in stating the property that the statement illustrates.

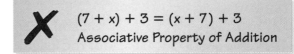

✗ $(7 + x) + 3 = (x + 7) + 3$
Associative Property of Addition

Simplify the expression. Explain each step.

12. $6 + (5 + x)$
13. $(14 + y) + 3$
14. $6(2b)$
15. $7(9w)$
16. $3.2 + (x + 5.1)$
17. $(0 + a) + 8$
18. $9 \cdot c \cdot 4$
19. $(18.6 \cdot d) \cdot 1$
20. $\left(3k + 4\frac{1}{5}\right) + 8\frac{3}{5}$
21. $(2.4 + 4n) + 9$
22. $(3s) \cdot 8$
23. $z \cdot 0 \cdot 12$

24. **GEOMETRY** The expression $12 + x + 4$ represents the perimeter of a triangle. Simplify the expression.

25. **RAISIN COOKIES** A case of raisin cookies has 10 cartons. A carton has 12 boxes. The amount you earn on a whole case is $10(12x)$ dollars.

 a. What does x represent?

 b. Simplify the expression.

26. STRUCTURE The volume of the rectangular prism is $12.5 \cdot x \cdot 1$.

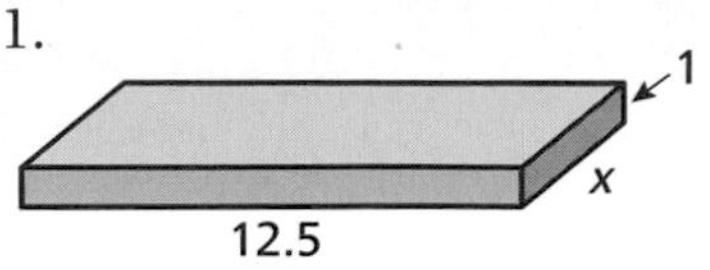

a. Simplify the expression.

b. Match $x = 0.25$, 12.5, and 144 with the object. Explain.

A. siding for a house **B.** ruler **C.** square floor tile

Write the phrase as an expression. Then simplify the expression.

27. 7 plus the sum of a number x and 5

28. the product of 8 and a number y multiplied by 9

Copy and complete the statement using the specified property.

	Property	Statement
29.	Associative Property of Multiplication	$7(2y) =$
30.	Commutative Property of Multiplication	$13.2 \cdot (x \cdot 1) =$
31.	Associative Property of Addition	$17 + (6 + 2x) =$
32.	Addition Property of Zero	$2 + (c + 0) =$
33.	Multiplication Property of One	$1 \cdot w \cdot 16 =$

34. HATS You and a friend sell hats at a fair booth. You sell 16 hats on the first shift and 21 hats on the third shift. Your friend sells x hats on the second shift.

a. Write an expression for the number of hats sold.

b. The expression $37(14) + 10x$ represents the amount that you both earned. How can you tell that your friend was selling the hats for a discounted price?

c. **Reasoning** You earned more money than your friend. What can you say about the value of x?

Fair Game Review What you learned in previous grades & lessons

Evaluate the expression. *(Section 1.3)*

35. $7(10 + 4)$ **36.** $12(10 - 1)$ **37.** $6(5 + 10)$ **38.** $8(30 - 5)$

Find the prime factorization of the number. *(Section 1.4)*

39. 37 **40.** 144 **41.** 147 **42.** 205

43. MULTIPLE CHOICE A bag has 16 blue, 20 red, and 24 green marbles. What fraction of the marbles in the bag are blue? *(Skills Review Handbook)*

Ⓐ $\frac{1}{5}$ Ⓑ $\frac{4}{15}$ Ⓒ $\frac{4}{11}$ Ⓓ $\frac{11}{15}$

3.4 The Distributive Property

Essential Question How do you use mental math to multiply two numbers?

The Meaning of a Word ● **Distribute**

When you **distribute** something to each person in a group,

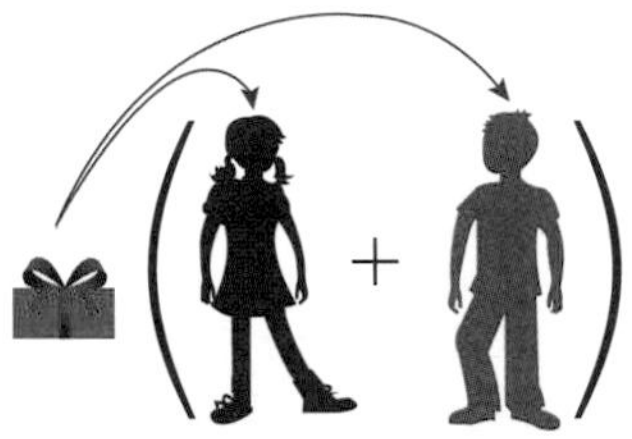

you give that thing to each person in the group.

1 ACTIVITY: Modeling a Property

Work with a partner.

a. **MODELING** Draw two rectangles of the same width but with different lengths on a piece of grid paper. Label the dimensions.

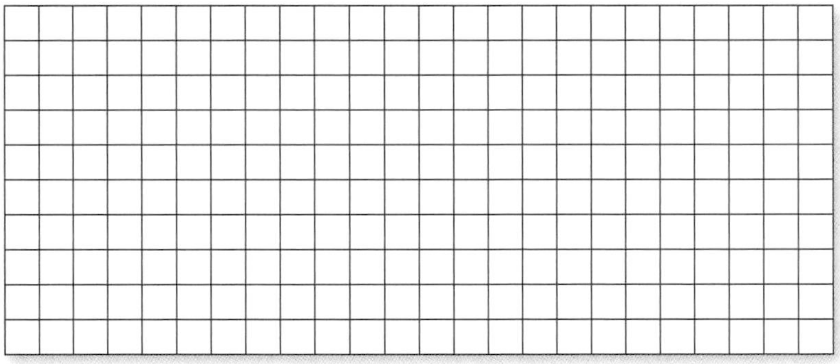

b. Write an expression for the total area of the rectangles.

$$(\square \times \square) + (\square \times \square)$$

c. Rearrange the rectangles by aligning the shortest sides to form one rectangle. Label the dimensions. Write an expression for the area.

$$\square \times (\square + \square)$$

d. Can the expressions from parts (b) and (c) be set equal to each other? Explain.

e. **REPEATED REASONING** Repeat this activity using different rectangles. Explain how this illustrates the Distributive Property. Write a rule for the Distributive Property.

COMMON CORE

Equivalent Expressions

In this lesson, you will
- use the Distributive Property to find products.
- use the Distributive Property to simplify algebraic expressions.

Learning Standards
6.NS.4
6.EE.2b
6.EE.3
6.EE.4

2 ACTIVITY: Using Mental Math

Work with a partner. Use the method shown to find the product.

a. Sample: 23×6

$$\begin{array}{r} 23 \\ \times \ \ 6 \\ \hline 120 \\ +\ 18 \\ \hline 138 \end{array}$$

23 is 20 + 3.

Multiply 20 and 6.
Multiply 3 and 6.
Add.

So, $23 \times 6 = 138$.

b. 33×7 **c.** 47×9

d. 28×5 **e.** 17×4

Math Practice 1

Find Entry Points

How can you rewrite the larger number as the sum of two numbers so that you can use mental math?

3 ACTIVITY: Using Mental Math

Work with a partner. Use the Distributive Property and mental math to find the product.

Hmmm. Which method is easier?

a. Sample: 6×23

$6 \times 23 = 6 \times (20 + 3)$	Write 23 as the sum of 20 and 3.
$= (6 \times 20) + (6 \times 3)$	Distribute the 6 over the sum.
$= 120 + 18$	Find the products.
$= 138$	Add.

So, $6 \times 23 = 138$.

b. 5×17 **c.** 8×26

d. 20×19 **e.** 40×29

f. 25×39 **g.** 15×47

What Is Your Answer?

4. Compare the methods in Activities 2 and 3.

5. **IN YOUR OWN WORDS** How do you use mental math to multiply two numbers? Give examples to support your explanation.

Practice Use what you learned about the Distributive Property to complete Exercises 5–8 on page 137.

3.4 Lesson

Check It Out
Lesson Tutorials
BigIdeasMath.com

Key Vocabulary
like terms, *p. 136*

Key Idea

Distributive Property

Words To multiply a sum or difference by a number, multiply each number in the sum or difference by the number outside the parentheses. Then evaluate.

Numbers $3(7 + 2) = 3 \times 7 + 3 \times 2$ **Algebra** $a(b + c) = ab + ac$

$3(7 - 2) = 3 \times 7 - 3 \times 2$ $a(b - c) = ab - ac$

EXAMPLE 1 Using Mental Math

Use the Distributive Property and mental math to find 8 × 53.

$8 \times 53 = 8(50 + 3)$ Write 53 as 50 + 3.

$= 8(50) + 8(3)$ Distributive Property

$= 400 + 24$ Multiply.

$= 424$ Add.

EXAMPLE 2 Using the Distributive Property

Use the Distributive Property to find $\frac{1}{2} \times 2\frac{3}{4}$.

$\frac{1}{2} \times 2\frac{3}{4} = \frac{1}{2} \times \left(2 + \frac{3}{4}\right)$ Rewrite $2\frac{3}{4}$ as the sum $2 + \frac{3}{4}$.

$= \left(\frac{1}{2} \times 2\right) + \left(\frac{1}{2} \times \frac{3}{4}\right)$ Distributive Property

$= 1 + \frac{3}{8}$ Multiply.

$= 1\frac{3}{8}$ Add.

On Your Own

Now You're Ready
Exercises 5–16

Use the Distributive Property to find the product.

1. 5×41
2. 9×19
3. $6(37)$
4. $\frac{2}{3} \times 1\frac{1}{2}$
5. $\frac{1}{4} \times 4\frac{1}{5}$
6. $\frac{2}{7} \times 3\frac{3}{4}$

EXAMPLE 3 Simplifying Algebraic Expressions

Use the Distributive Property to simplify the expression.

a. $4(n + 5)$

$4(n + 5) = 4(n) + 4(5)$	Distributive Property
$= 4n + 20$	Multiply.

b. $12(2y - 3)$

$12(2y - 3) = 12(2y) - 12(3)$	Distributive Property
$= 24y - 36$	Multiply.

c. $9(6 + x + 2)$

$9(6 + x + 2) = 9(6) + 9(x) + 9(2)$	Distributive Property
$= 54 + 9x + 18$	Multiply.
$= 9x + 54 + 18$	Commutative Property of Addition
$= 9x + 72$	Add 54 and 18.

Study Tip

You can use the Distributive Property when there are more than two terms in the sum or difference.

On Your Own

Now You're Ready
Exercises 17–32

Use the Distributive Property to simplify the expression.

7. $7(a + 2)$ **8.** $3(d - 11)$ **9.** $7(2 + 6 - 4d)$

EXAMPLE 4 Real-Life Application

José is *x* years old. His brother, Felipe, is 2 years older than José. Their aunt, Maria, is three times as old as Felipe. Write and simplify an expression that represents Maria's age in years.

Name	Description	Expression
José	He is x years old.	x
Felipe	He is 2 years *older* than José. So, *add* 2 to x.	$x + 2$
Maria	She is three *times* as old as Felipe. So, *multiply* 3 and $(x + 2)$.	$3(x + 2)$

$3(x + 2) = 3(x) + 3(2)$	Distributive Property
$= 3x + 6$	Multiply.

∴ Maria's age in years is represented by the expression $3x + 6$.

On Your Own

10. Alexis is x years old. Her sister, Gloria, is 7 years older than Alexis. Their grandfather is five times as old as Gloria. Write and simplify an expression that represents their grandfather's age in years.

In an algebraic expression, **like terms** are terms that have the same variables raised to the same exponents. Constant terms are also like terms.

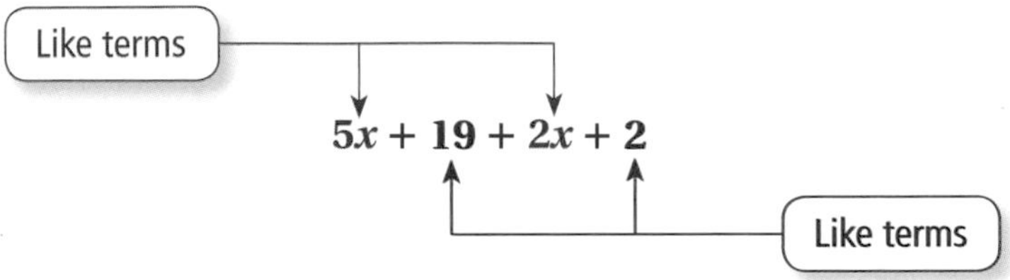

Use the Distributive Property to *combine* like terms.

EXAMPLE 5 Combining Like Terms

Simplify each expression.

a. $3x + 9 + 2x - 5$

$3x + 9 + 2x - 5 = 3x + 2x + 9 - 5$	Commutative Property of Addition
$= (3 + 2)x + 9 - 5$	Distributive Property
$= 5x + 4$	Simplify.

b. $y + y + y$

$y + y + y = 1y + 1y + 1y$	Multiplication Property of One
$= (1 + 1 + 1)y$	Distributive Property
$= 3y$	Add coefficients.

c. $7z + 2(z - 5y)$

$7z + 2(z - 5y) = 7z + 2(z) - 2(5y)$	Distributive Property
$= 7z + 2z - 10y$	Multiply.
$= (7 + 2)z - 10y$	Distributive Property
$= 9z - 10y$	Add coefficients.

On Your Own

Now You're Ready
Exercises 39–53

Simplify the expression.

11. $8 + 3z - z$

12. $3(b + 5) + b + 2$

3.4 Exercises

Vocabulary and Concept Check

1. **WRITING** One meaning of the word *distribute* is "to give something to each member of a group." How can this help you remember the Distributive Property?
2. **OPEN-ENDED** Write an algebraic expression in which you use the Distributive Property and then the Associative Property of Addition to simplify.
3. **WHICH ONE DOESN'T BELONG?** Which expression does *not* belong with the other three? Explain your reasoning.

   $2(x + 2)$ $\quad$ $5(x - 8)$ $\quad$ $4 + (x \cdot 4)$ $\quad$ $8(9 - x)$

4. Identify the like terms in the expression $8x + 1 + 7x + 4$.

Practice and Problem Solving

Use the Distributive Property and mental math to find the product.

(1) **5.** 3×21 **6.** 9×76 **7.** $12(43)$ **8.** $5(88)$

9. 18×52 **10.** 8×27 **11.** $8(63)$ **12.** $7(28)$

Use the Distributive Property to find the product.

(2) **13.** $\frac{1}{4} \times 2\frac{2}{7}$ **14.** $\frac{5}{6} \times 2\frac{2}{5}$ **15.** $\frac{5}{9} \times 4\frac{1}{2}$ **16.** $\frac{2}{15} \times 5\frac{5}{8}$

Use the Distributive Property to simplify the expression.

(3) **17.** $3(x + 4)$ **18.** $10(b - 6)$ **19.** $6(s - 9)$ **20.** $7(8 + y)$

21. $8(12 + a)$ **22.** $9(2n + 1)$ **23.** $12(6 - k)$ **24.** $18(5 - 3w)$

25. $9(3 + c + 4)$ **26.** $7(8 + x + 2)$ **27.** $8(5g + 5 - 2)$ **28.** $6(10 + z + 3)$

29. $4(x + y)$ **30.** $25(x - y)$ **31.** $7(p + q + 9)$ **32.** $13(n + 4 + 7m)$

33. ERROR ANALYSIS Describe and correct the error in rewriting the expression.

✗ $6(y + 8) = 6y + 8$

34. ART MUSEUM A class of 30 students visits an art museum and a special exhibit while there.

a. Use the Distributive Property to write and simplify an expression for the cost.

b. Estimate a reasonable value for x. Explain.

c. Use your estimate for x to evaluate the original expression and the simplified expression in part (a). Are the values the same?

PRICES

	Museum	Exhibit
Child (under 5)	Free	Free
Student	\$8	\$$x$
Regular	\$12	\$4
Senior	\$10	\$3

35. **FITNESS** Each day, you run on a treadmill for r minutes and lift weights for 15 minutes. Which expressions can you use to find how many minutes of exercise you do in 5 days? Explain your reasoning.

$5(r + 15)$ $\quad$ $5r + 5 \cdot 15$ $\quad$ $5r + 15$ $\quad$ $r(5 + 15)$

36. **SPEED** A cheetah can run 103 feet per second. A zebra can run x feet per second. Use the Distributive Property to write and simplify an expression for how much farther the cheetah can run in 10 seconds.

UNIFORMS Your baseball team has 16 players. Use the Distributive Property to write and simplify an expression for the total cost of buying the items shown for all the players.

37.

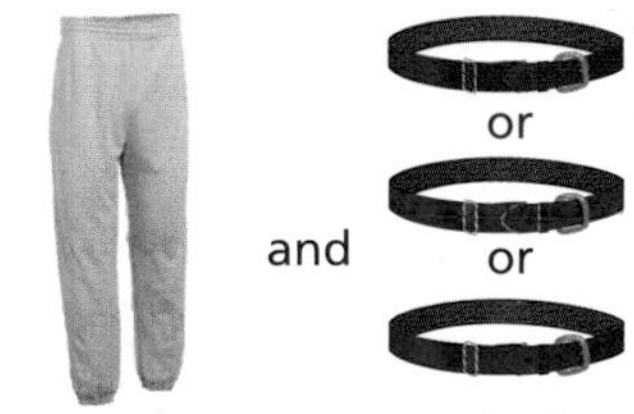

Pants: \$10 $\quad$ Belt: \$$x$

38.

Jersey: \$12 $\quad$ Socks: \$4 $\quad$ Hat: \$$x$

5 **Simplify the expression.**

39. $6(x + 4) + 1$
40. $5 + 8(3 + x)$
41. $7(8 + 4k) + 12$
42. $x + 3 + 5x$
43. $7y + 6 - 1 + 12y$
44. $w + w + 5w$
45. $4d + 9 - d - 8$
46. $n + 3(n - 1)$
47. $2v + 8v - 5v$
48. $5(z + 4) + 5(2 - z)$
49. $2.7(w - 5.2)$
50. $\frac{2}{3}y + \frac{1}{6}y + y$
51. $\frac{3}{4}\left(z + \frac{2}{5}\right) + 2z$
52. $7(x + y) - 7x$
53. $4x + 9y + 3(x + y)$

54. **ERROR ANALYSIS** Describe and correct the error in simplifying the expression.

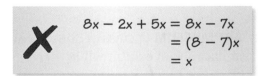

✗
$$\begin{aligned} 8x - 2x + 5x &= 8x - 7x \\ &= (8 - 7)x \\ &= x \end{aligned}$$

ALGEBRA Find the value of x that makes the expressions equivalent.

55. $4(x - 5);\ 32 - 20$
56. $2(x + 9);\ 30 + 18$
57. $7(8 - x);\ 56 - 21$

58. **REASONING** Simplify the expressions and compare. What do you notice? Explain.

$4(x + 6)$ $\quad$ $(x + 6) + (x + 6) + (x + 6) + (x + 6)$

GEOMETRY **Write and simplify expressions for the area and perimeter of the rectangle.**

59.

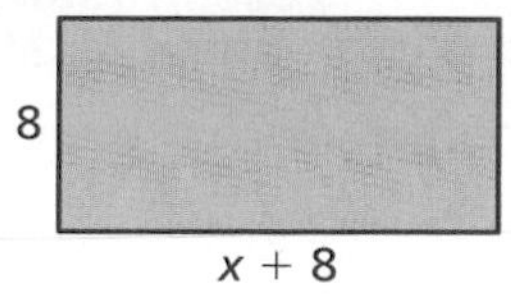

60.

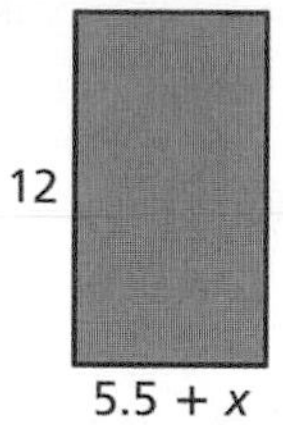

61.

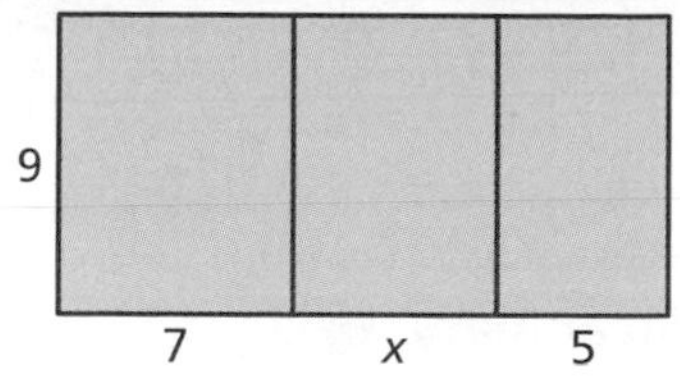

62. **FUNDRAISER** An art club sells 42 large candles and 56 small candles.

a. Use the Distributive Property to write and simplify an expression for the profit.

b. A large candle costs \$5, and a small candle costs \$3. What is the club's profit?

Profit = Price − Cost

63. **REASONING** Evaluate each expression by (1) using the Distributive Property and (2) evaluating inside the parentheses first. Which method do you prefer? Is your preference the same for both expressions? Explain your reasoning.

a. $2(3.22 - 0.12)$ **b.** $12\left(\frac{1}{2} + \frac{2}{3}\right)$

64. **REASONING** Write and simplify an expression for the difference between the perimeters of the rectangle and the hexagon. Interpret your answer.

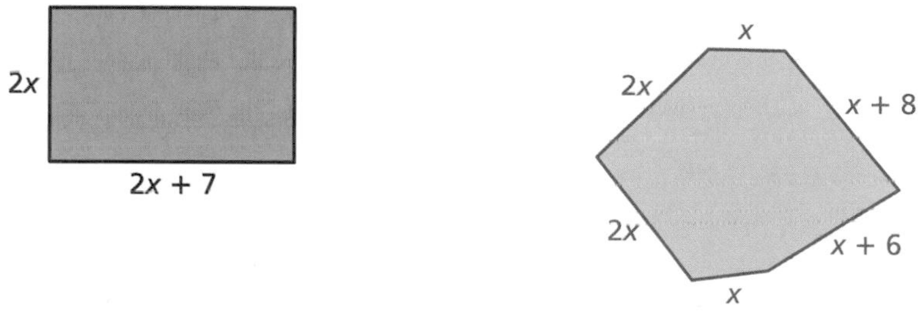

65. **Puzzle** Add one set of parentheses to the expression $7 \cdot x + 3 + 8 \cdot x + 3 \cdot x + 8 - 9$ so that it is equivalent to $2(9x + 10)$.

Fair Game Review What you learned in previous grades & lessons

Evaluate the expression. *(Section 2.4, Section 2.5, and Section 2.6)*

66. $4.871 + 7.4 - 1.63$

67. $25.06 - 0.049 + 8.995$

68. $15.3 \cdot 9.1 - 4.017$

69. $29.24 \div 3.4 \cdot 0.045$

70. **MULTIPLE CHOICE** What is the GCF of 48, 80, and 96? *(Section 1.5)*

Ⓐ 12 Ⓑ 16

Ⓒ 24 Ⓓ 480

Extension 3.4 Factoring Expressions

Check It Out
Lesson Tutorials
BigIdeasMath.com

Key Idea

Key Vocabulary
factoring an expression, p. 140

Factoring an Expression

Words Writing a numerical expression or algebraic expression as a product of factors is called **factoring the expression**. You can use the Distributive Property to factor expressions.

Numbers $3 \cdot 7 + 3 \cdot 2 = 3(7 + 2)$

$3 \cdot 7 - 3 \cdot 2 = 3(7 - 2)$

Algebra $ab + ac = a(b + c)$

$ab - ac = a(b - c)$

EXAMPLE 1 Factoring a Numerical Expression

Study Tip
When you factor an expression, you can *factor out* any common factor.

Factor 20 − 12 using the GCF.

Find the GCF of 20 and 12 by listing their factors.

Factors of 20: ①,②,④, 5, 10, 20

Factors of 12: ①,②, 3,④, 6, 12

Circle the common factors.

The GCF of 20 and 12 is 4.

Write each term of the expression as a product of the GCF and the remaining factor. Then use the Distributive Property to factor the expression.

$20 - 12 = 4(5) - 4(3)$ Rewrite using GCF.

$= 4(5 - 3)$ Distributive Property

EXAMPLE 2 Identifying Equivalent Expressions

Which expression is not equivalent to 16*x* + 24?

Ⓐ $2(8x + 12)$ Ⓑ $4(4x + 6)$ Ⓒ $6(3x + 4)$ Ⓓ $(2x + 3)8$

Each choice is a product of two factors in which one is a whole number and the other is the sum of two terms. For an expression to be equivalent to $16x + 24$, its whole number factor must be a common factor of 16 and 24.

Factors of 16: ①,②,④,⑧, 16

Factors of 24: ①,②, 3,④, 6,⑧, 12, 24

Circle the common factors.

The common factors of 16 and 24 are 1, 2, 4, and 8. Because 6 is not a common factor of 16 and 24, Choice C cannot be equivalent to $16x + 24$.

Check: $6(3x + 4) = 6(3x) + 6(4) = 18x + 24 \neq 16x + 24$ ✗

So, the correct answer is Ⓒ.

Common Core
Equivalent Expressions
In this extension, you will
- use the Distributive Property to produce equivalent expressions.

Learning Standards
6.NS.4
6.EE.3
6.EE.4

Multi-Language Glossary at BigIdeasMath.com

EXAMPLE 3 Factoring an Algebraic Expression

You receive a discount on each book you buy for your electronic reader. The original price of each book is x dollars. You buy 5 books for a total of $(5x - 15)$ dollars. Factor the expression. What can you conclude about the discount?

Find the GCF of $5x$ and 15 by writing their prime factorizations.

$$5x = 5 \cdot x$$
$$15 = 5 \cdot 3$$

Circle the common prime factor.

So, the GCF of $5x$ and 15 is 5. Use the GCF to factor the expression.

$$5x - 15 = 5(x) - 5(3)$$ Rewrite using GCF.

$$= 5(x - 3)$$ Distributive Property

The factor 5 represents the number of books purchased. The factor $(x - 3)$ represents the price of each book. This factor is a difference of two terms, showing that the price x of each book is decreased by \$3.

∴ So, the factored expression shows a \$3 discount for every book you buy. The original expression shows a total savings of \$15.

Practice

Factor the expression using the GCF.

1. $7 + 14$
2. $44 - 11$
3. $18 - 12$
4. $70 + 95$
5. $60 - 36$
6. $100 - 80$
7. $84 + 28$
8. $48 + 80$
9. $2x + 10$
10. $15x + 6$
11. $26x - 13$
12. $50x - 60$
13. $36x + 9$
14. $14x - 98$
15. $10x - 25y$
16. $24y + 88x$

17. **REASONING** The whole numbers a and b are divisible by c. Is $a + b$ divisible by c? Is $b - a$ divisible by c? Explain your reasoning.

18. **OPEN-ENDED** Write five expressions that are equivalent to $8x + 16$.

19. **GEOMETRY** The area of the parallelogram is $(4x + 16)$ square feet. Write an expression for the base.

20. **STRUCTURE** You buy 37 concert tickets for \$8 each, and then sell all 37 tickets for \$11 each. The work below shows two ways you can determine your profit. Describe each solution method. Which do you prefer? Explain your reasoning.

profit = 37(11) − (37)8
= 407 − 296
= \$111

profit = 37(11) − (37)8
= 37(11 − 8)
= 37(3)
= \$111

3.3–3.4 Quiz

Check It Out
Progress Check
BigIdeasMath.com

Tell which property the statement illustrates. *(Section 3.3)*

1. $3.5 \cdot z = z \cdot 3.5$

2. $14 + (35 + w) = (14 + 35) + w$

Simplify the expression. Explain each step. *(Section 3.3)*

3. $3.2 + (b + 5.7)$

4. $6 \cdot (10 \cdot k)$

Use the Distributive Property and mental math to find the product. *(Section 3.4)*

5. 6×49

6. 7×86

Use the Distributive Property to simplify the expression. *(Section 3.4)*

7. $5(x - 8)$

8. $7(y + 3)$

Simplify the expression. *(Section 3.4)*

9. $6q + 2 + 3q + 5$

10. $4r + 3(r - 2)$

Factor the expression using the GCF. *(Section 3.4)*

11. $12 + 21$

12. $16x - 36$

13. GEOMETRY The expression $18 + 7 + (18 + 2x) + 7$ represents the perimeter of the trapezoid. Simplify the expression. *(Section 3.3)*

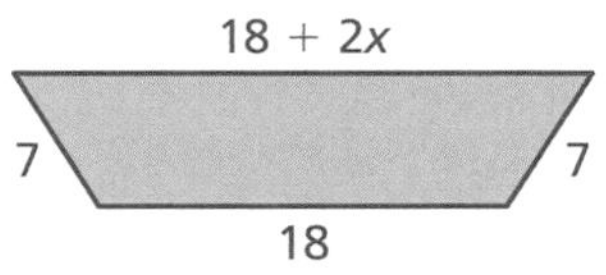

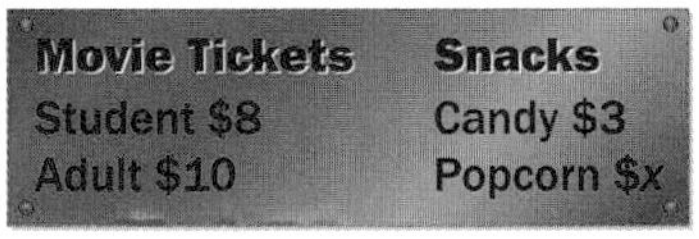

14. MOVIES You and four of your friends go to a movie and each buy popcorn. *(Section 3.4)*

a. Use the Distributive Property to write an expression for the total cost to buy movie tickets and popcorn. Simplify the expression.

b. Choose a reasonable value for x. Evaluate the expression.

15. GEOMETRY The length of a rectangle is 16 inches, and its area is $(32x + 48)$ square inches. Factor the expression for the area. Write an expression for the width. *(Section 3.4)*

3 Chapter Review

Check It Out
Vocabulary Help
BigIdeasMath.com

Review Key Vocabulary

algebraic expression, *p. 112*
terms, *p. 112*
variable, *p. 112*
coefficient, *p. 112*
constant, *p. 112*
equivalent expressions, *p. 128*
like terms, *p. 136*
factoring an expression, *p. 140*

Review Examples and Exercises

3.1 Algebraic Expressions *(pp. 110–117)*

a. Evaluate $a \div b$ when $a = 48$ and $b = 8$.

$a \div b = 48 \div 8$ — Substitute 48 for *a* and 8 for *b*.

$= 6$ — Divide 48 by 8.

b. Evaluate $y^2 - 14$ when $y = 5$.

$y^2 - 14 = 5^2 - 14$ — Substitute 5 for *y*.

$= 25 - 14$ — Using order of operations, evaluate 5^2.

$= 11$ — Subtract 14 from 25.

Exercises

Evaluate the expression when $x = 20$ and $y = 4$.

1. $x \div 5$
2. $y + x$
3. $8y - x$

4. **GAMING** In a video game, you score p game points and b triple bonus points. An expression for your score is $p + 3b$. What is your score when you earn 245 game points and 20 triple bonus points?

3.2 Writing Expressions *(pp. 118–123)*

Write the phrase as an expression.

a. a number z decreased by 18

$z - 18$ — The phrase *decreased by* means *subtraction*.

b. the sum of 7 and the product of a number x and 12

$7 + 12x$ — The phrase *sum of* means *addition*.
The phrase *product of* means *multiplication*.

Exercises

Write the phrase as an expression.

5. 11 fewer than a number b
6. the product of a number d and 32
7. 18 added to a number n
8. a number t decreased by 17

9. **BASKETBALL** Your basketball team scored 4 fewer than twice as many points as the other team.
   - **a.** Write an expression for the number of points your team scored.
   - **b.** The other team scored 24 points. How many points did your team score?

3.3 Properties of Addition and Multiplication *(pp. 126–131)*

a. Simplify the expression $(x + 18) + 4$.

$(x + 18) + 4 = x + (18 + 4)$ — Associative Property of Addition

$= x + 22$ — Add 18 and 4.

b. Simplify the expression $(5.2 + a) + 0$.

$(5.2 + a) + 0 = 5.2 + (a + 0)$ — Associative Property of Addition

$= 5.2 + a$ — Addition Property of Zero

c. Simplify the expression $36 \cdot r \cdot 1$.

$36 \cdot r \cdot 1 = 36 \cdot (r \cdot 1)$ — Associative Property of Multiplication

$= 36 \cdot r$ — Multiplication Property of One

$= 36r$

Exercises

Simplify the expression. Explain each step.

10. $10 + (2 + y)$
11. $(21 + b) + 1$
12. $3(7x)$
13. $1(3.2w)$
14. $5.3 + (w + 1.2)$
15. $(0 + t) + 9$

16. **GEOMETRY** The expression $7 + 3x + 4$ represents the perimeter of the triangle. Simplify the expression.

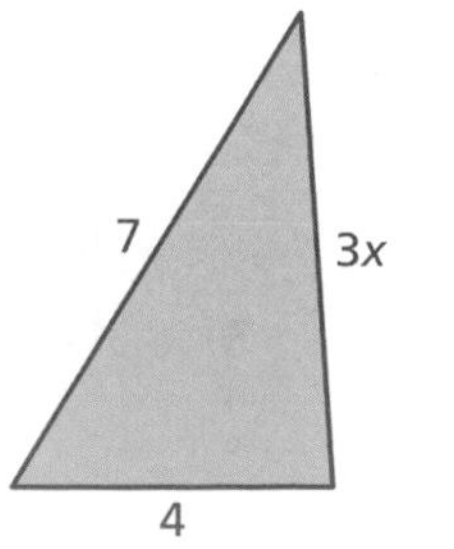

3.4 The Distributive Property (pp. 132–141)

a. Use the Distributive Property to simplify $3(n + 9)$.

$3(n + 9) = 3(n) + 3(9)$ Distributive Property

$= 3n + 27$ Multiply.

b. Simplify $5x + 7 + 3x - 2$.

$5x + 7 + 3x - 2 = 5x + 3x + 7 - 2$ Commutative Property of Addition

$= (5 + 3)x + 7 - 2$ Distributive Property

$= 8x + 5$ Simplify.

c. Factor $14x - 49$ using the GCF.

Find the GCF of $14x$ and 49 by writing their prime factorizations.

$14x = 2 \cdot 7 \cdot x$ Circle the common prime factor.

$49 = 7 \cdot 7$

So, the GCF of $14x$ and 49 is 7. Use the GCF to factor the expression.

$14x - 49 = 7(2x) - 7(7)$ Rewrite using GCF.

$= 7(2x - 7)$ Distributive Property

Exercises

Use the Distributive Property to find the product.

17. $\frac{3}{4} \times 2\frac{1}{3}$ **18.** $\frac{4}{7} \times 4\frac{5}{8}$ **19.** $\frac{1}{5} \times 5\frac{10}{11}$

Use the Distributive Property to simplify the expression.

20. $2(x + 12)$ **21.** $11(b - 3)$ **22.** $8(s - 1)$

23. $6(6 + y)$ **24.** $25(z - 4)$ **25.** $35(w - 2)$

26. HAIRCUT A family of four goes to a salon for haircuts. The cost of each haircut is $13. Use the Distributive Property and mental math to find the product 4×13 for the total cost.

Simplify the expression.

27. $5(n + 3) + 4n$ **28.** $t + 2 + 6t$ **29.** $3z + 4 + 5z - 9$

Factor the expression using the GCF.

30. $15 + 35$ **31.** $36x - 28$ **32.** $16x + 56y$

3 Chapter Test

Check It Out
Test Practice
BigIdeasMath.com

Evaluate the expression when $a = 6$ and $b = 8$.

1. $4 + a$
2. $a - 6$
3. ab

Write the phrase as an expression.

4. twice a number x
5. 25 more than 50
6. 40 divided by 5

Simplify the expression. Explain each step.

7. $3.1 + (8.6 + m)$
8. $(10 \cdot n) \cdot 7$
9. $3(15w)$

Use the Distributive Property to simplify the expression.

10. $4(x + 8)$
11. $12(y - 5)$

Simplify the expression.

12. $4(q + 2) - 6$
13. $3(2 + 5r) + 11$
14. $s + 3s + 4s$
15. $4t - 2 - 2t + 7$

Factor the expression using the GCF.

16. $18 + 24$
17. $40 - 16$
18. $15x + 20$
19. $32x - 40y$

20. **SOCCER GAME** Playing time is added at the end of a soccer game to make up for stoppages. An expression for the length of a 90-minute soccer game with x minutes of stoppage time is $90 + x$. How long is a game with 4 minutes of stoppage time?

21. **GEOMETRY** The expression $15 \cdot x \cdot 6$ represents the volume of a rectangular prism with a length of 15, a width of x, and a height of 6. Simplify the expression.

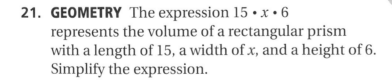

22. **PARTY FAVORS** You make party favors for an event. You tie 9 inches of ribbon around each party favor. Write an expression for the amount of ribbon you need for n party favors. The ribbon costs \$3 for each *yard*. Write an expression for the total cost of the ribbon.

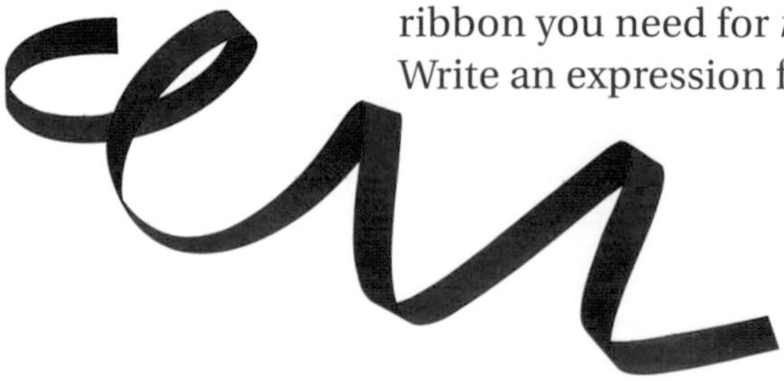

3 Standards Assessment

Test-Taking Strategy
After Answering Easy Questions, Relax

In x days you eat 3x cans of tuna. How many cans of tuna do you eat in 4 days?
(A) 10 (B) 12 (C) 14 (D) 16
I come equipped with natural can openers.

"After answering easy questions, relax and try the harder ones. For this, 3(4) = 12. So, it is B."

1. The student council is organizing a school fair. Council members are making signs to show the prices for admission and for each game a person can play.

SCHOOL FAIR
Admission $2.00
Price per game $0.25

Let x represent the number of games. Which expression can you use to determine the total amount, in dollars, a person pays for admission and playing x games? *(6.EE.2a)*

A. 2.25

B. $2.25x$

C. $2 + 0.25x$

D. $2x + 0.25$

2. Which property does the equation below represent? *(6.EE.3)*

$$17 \cdot 44 + 17 \cdot 56 = 17 \cdot 100$$

F. Distributive Property

G. Multiplication Property of One

H. Associative Property of Multiplication

I. Commutative Property of Multiplication

3. At a used book store, you can purchase two types of books.

You can use the expression $3h + 2p$ to find the total cost for h hardcover books and p paperback books. What is the total cost, in dollars, for 6 hardcover books and 4 paperback books? *(6.EE.2c)*

Hardcover Books - $3
Paperback Books - $2

4. What is the value of 9.6×12.643? *(6.NS.3)*

A. 12.13728

B. 121.3728

C. 1213.728

D. 12,137.28

5. What is the value of $4.391 + 5.954$? *(6.NS.3)*

F. 9.12145

H. 9.345

G. 9.245

I. 10.345

6. Which number pair has a greatest common factor of 6? *(6.NS.4)*

A. 18, 54

C. 30, 60

B. 30, 42

D. 36, 60

7. Properties of Addition and Multiplication are used to simplify an expression.

$$
\begin{aligned}
36 \cdot 23 + 33 \cdot 64 &= 36 \cdot 23 + 64 \cdot 33 \\
&= 36 \cdot 23 + 64 \cdot (23 + 10) \\
&= 36 \cdot 23 + 64 \cdot 23 + 64 \cdot 10 \\
&= x \cdot 23 + 64 \cdot 10 \\
&= 2300 + 640 \\
&= 2940
\end{aligned}
$$

What number belongs in place of the x? *(6.EE.3)*

8. Which property was used to simplify the expression? *(6.EE.3)*

$$
\begin{aligned}
(47 \times 125) \times 8 &= 47 \times (125 \times 8) \\
&= 47 \times 1000 \\
&= 47{,}000
\end{aligned}
$$

F. Distributive Property

G. Multiplication Property of One

H. Associative Property of Multiplication

I. Commutative Property of Multiplication

9. What is the value of the expression below when $a = 5$, $b = 7$, and $c = 6$? *(6.EE.2c)*

$$9b - 4a + 2c$$

A. 29

C. 55

B. 31

D. 78

10. Which equation correctly demonstrates the Distributive Property? *(6.EE.4)*

F. $a(b + c) = ab + c$

G. $a(b + c) = ab + ac$

H. $a + (b + c) = (a + b) + (a + c)$

I. $a + (b + c) = (a + b) \cdot (a + c)$

11. Which expression is equivalent to $3\frac{3}{5} \div 6\frac{1}{2}$? *(6.NS.1)*

A. $\frac{5}{18} \times \frac{13}{2}$

B. $\frac{18}{5} \times \frac{2}{13}$

C. $\frac{9}{5} \div \frac{6}{2}$

D. $\frac{18}{5} \div \frac{2}{13}$

12. Which number pair does *not* have a least common multiple of 24? *(6.NS.4)*

F. 2, 12

G. 3, 8

H. 6, 8

I. 12, 24

13. Use the Properties of Multiplication to simplify the expression in an efficient way. Show your work and explain how you used the Properties of Multiplication. *(6.EE.3)*

Think Solve Explain

$(25 \times 18) \times 4$

14. You evaluated an expression using $x = 6$ and $y = 9$. You correctly got an answer of 105. Which expression did you evaluate? *(6.EE.2c)*

A. $3x + 6y$

B. $5x + 10y$

C. $6x + 9y$

D. $10x + 5y$

15. Which number is equivalent to the expression below? *(6.EE.1)*

$$2 \times 12 - 8 \div 2^2$$

F. 2

G. 4

H. 8

I. 22

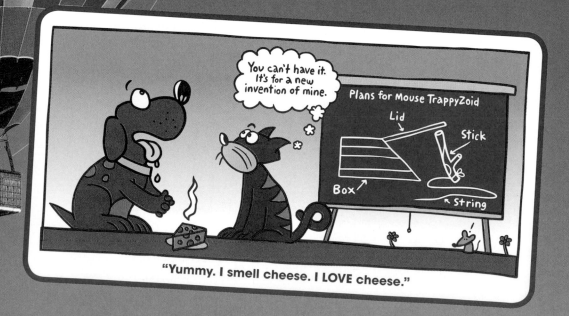

4 Areas of Polygons

4.1 Areas of Parallelograms

4.2 Areas of Triangles

4.3 Areas of Trapezoids

4.4 Polygons in the Coordinate Plane

COOL!
"Remember, Descartes, you don't have to measure area in standard units like square inches or square centimeters."
"You can also use nonstandard units ... like the length of your paw squared."

"Yummy. I smell cheese. I LOVE cheese."

What You Learned Before

Finding Areas of Squares and Rectangles (4.MD.3)

Example 1 **Find the area of the square or rectangle.**

a.

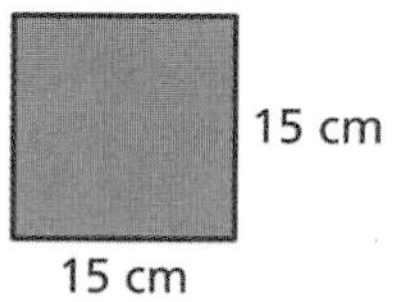

$A = s^2$ Write formula.

$= 15^2$ Substitute.

$= 225$ Simplify.

∴ The area of the square is 225 square centimeters.

b.

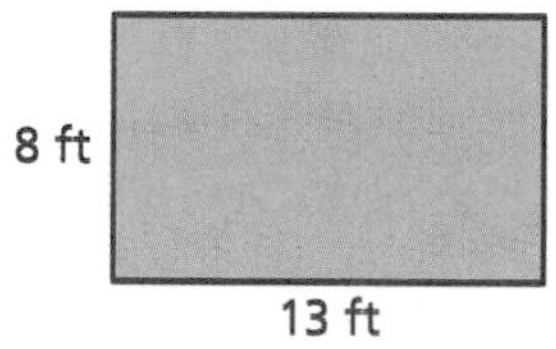

$A = \ell w$

$= 13(8)$

$= 104$

∴ The area of the rectangle is 104 square feet.

Try It Yourself

Find the area of the square or rectangle.

1.

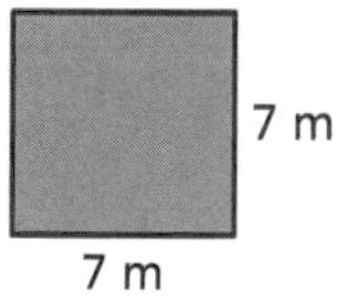

2.

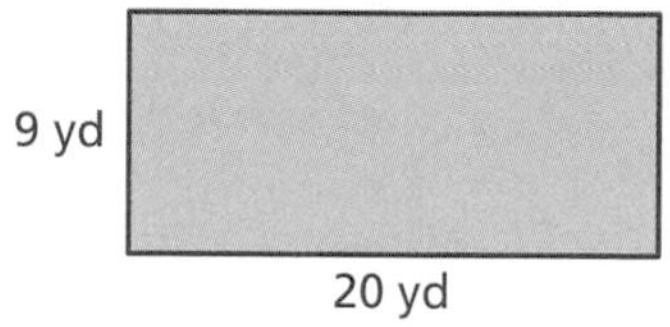

3.

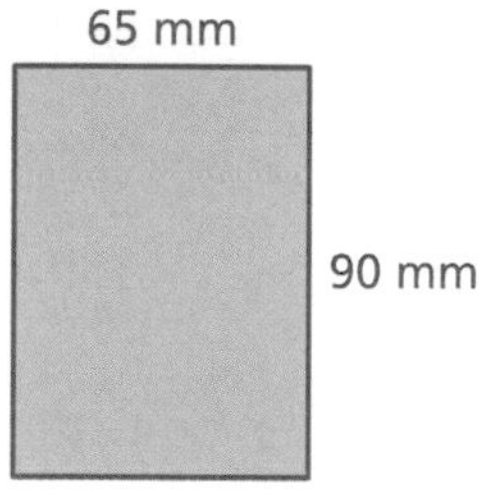

Plotting Ordered Pairs (5.G.1)

Example 2 **Plot (2, 3) in a coordinate plane.**

Start at the origin. Move 2 units right and 3 units up. Then plot the point.

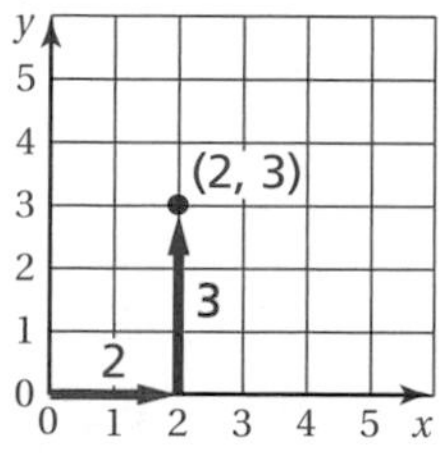

Try It Yourself

Plot the ordered pair in a coordinate plane.

4. (1, 4) **5.** (3, 2) **6.** (5, 1)

4.1 Areas of Parallelograms

Essential Question How can you derive a formula for the area of a parallelogram?

A **polygon** is a closed figure in a plane that is made up of three or more line segments that intersect only at their endpoints. Several examples of polygons are parallelograms, triangles, and trapezoids.

The formulas for the areas of polygons can be derived from one area formula, the area of a rectangle. Recall that the area of a rectangle is the product of its length ℓ and its width w. The process you use to derive these other formulas is called *deductive reasoning*.

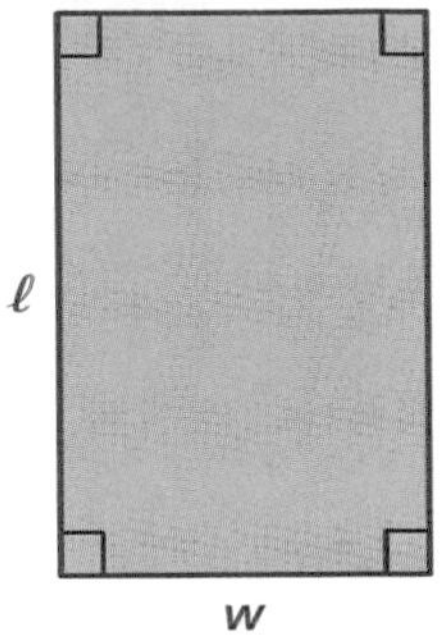

Area $= \ell w$

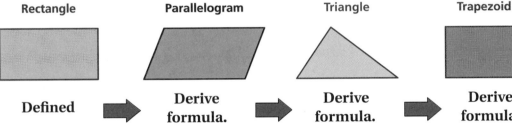

Rectangle	Parallelogram	Triangle	Trapezoid

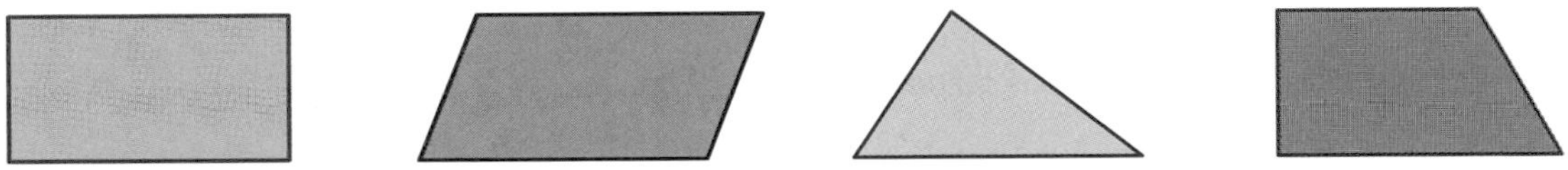

Defined →	**Derive formula.** →	**Derive formula.** →	**Derive formula.**
Grades 4 and 5	*Lesson 4.1*	*Lesson 4.2*	*Lesson 4.3*

1 ACTIVITY: Deriving the Area Formula of a Parallelogram

Work with a partner.

a. Draw *any* rectangle on a piece of grid paper. An example is shown below. Label the length and width. Then find the area of your rectangle.

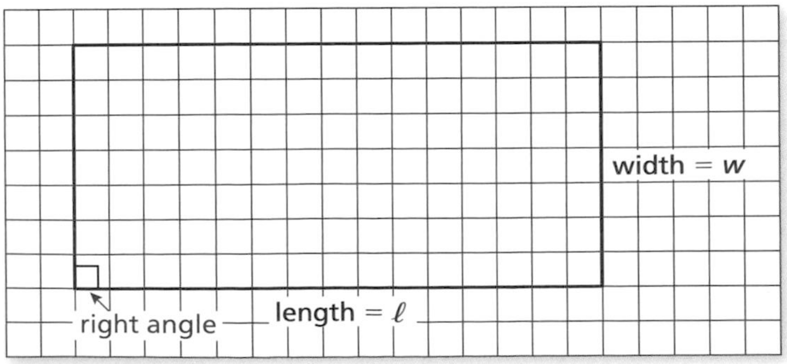

b. Cut your rectangle into two pieces to form a parallelogram. Compare the area of the rectangle with the area of the parallelogram. What do you notice? Use your results to write a formula for the area A of a parallelogram.

$A =$ ________ Formula

Common Core

Geometry

In this lesson, you will
- find areas of parallelograms.
- solve real-life problems.

Learning Standard 6.G.1

2 ACTIVITY: Finding Areas of Parallelograms

Math Practice 3

Use Assumptions

How are rectangles and parallelograms similar? How can you use this information to solve the problem?

Work with a partner.

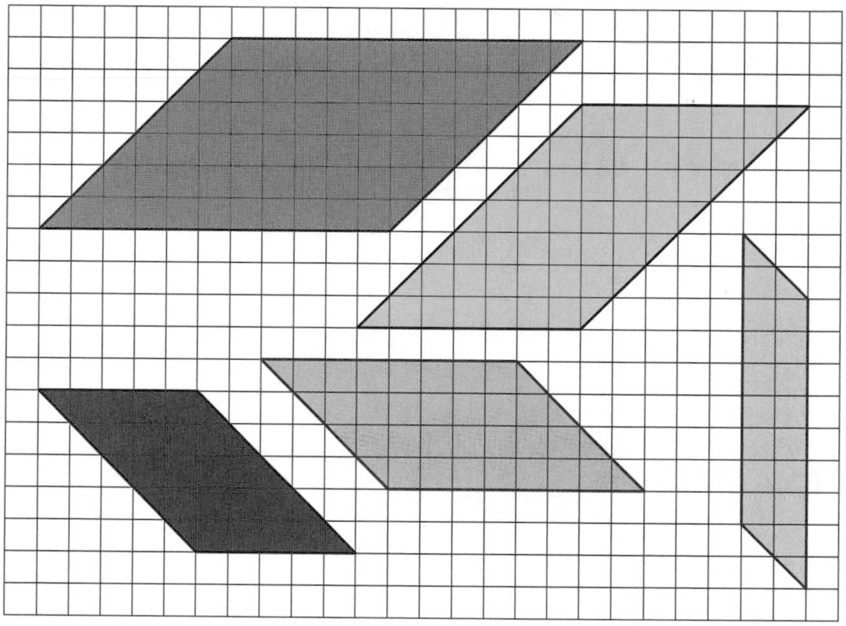

a. Find the area of each parallelogram by cutting it into two pieces to form a rectangle.

b. Use the formula you wrote in Activity 1 to find the area of each parallelogram. Compare your answers to those in part (a).

c. Count unit squares for each parallelogram to check your results.

What Is Your Answer?

3. IN YOUR OWN WORDS How can you derive a formula for the area of a parallelogram?

4. REASONING The areas of a rectangle and a parallelogram are equal. The length of a rectangle is equal to the base of the parallelogram. What can you say about the width of the rectangle and the height of the parallelogram? Draw a diagram to support your answer.

5. What is the height of the parallelogram shown? How do you know?

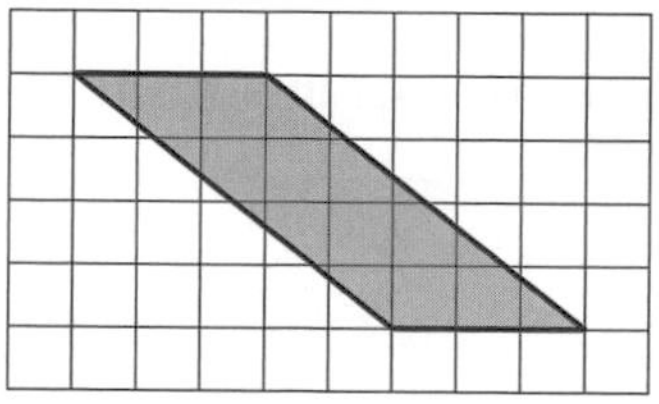

Practice Use what you learned about the areas of parallelograms to complete Exercises 3–5 on page 156.

4.1 Lesson

Check It Out
Lesson Tutorials
BigIdeasMath.com

Key Vocabulary
polygon, *p. 152*

The *area* of a polygon is the amount of surface it covers. You can find the area of a parallelogram in much the same way as you can find the area of a rectangle.

Key Idea

Area of a Parallelogram

Words The area A of a parallelogram is the product of its base b and its height h.

Algebra $A = bh$

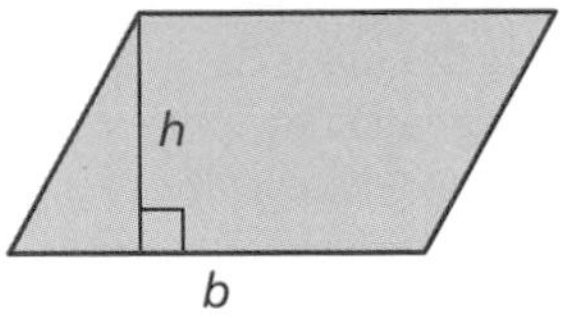

EXAMPLE 1 Finding Areas of Parallelograms

Find the area of each parallelogram.

a.

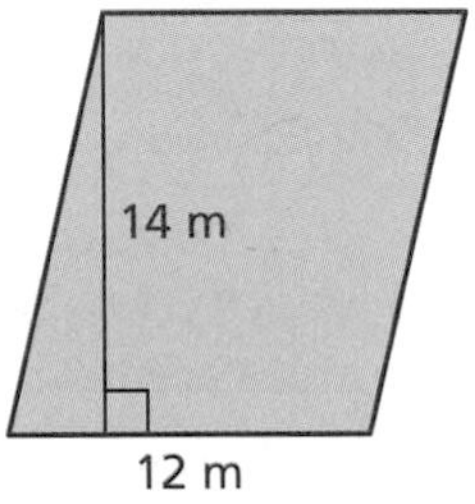

$A = bh$ Write formula.

$= 12(14)$ Substitute values.

$= 168$ Multiply.

The area of the parallelogram is 168 square meters.

b.

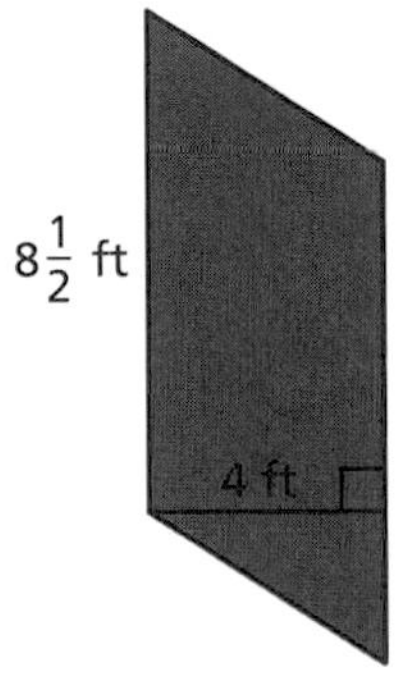

$A = bh$

$= 8\frac{1}{2}(4)$

$= 34$

The area of the parallelogram is 34 square feet.

Remember

Area is measured in square units.

On Your Own

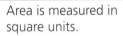

Now You're Ready
Exercises 3–8

Find the area of the parallelogram.

1.

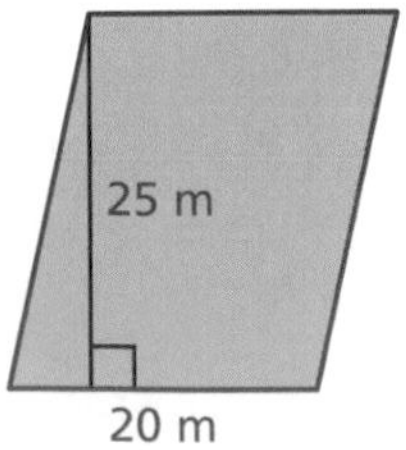

2.

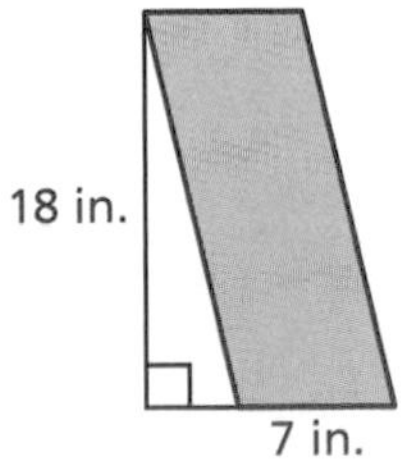

3.

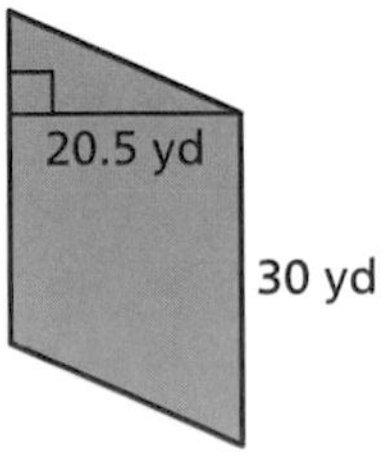

EXAMPLE 2 Finding the Area of a Parallelogram on a Grid

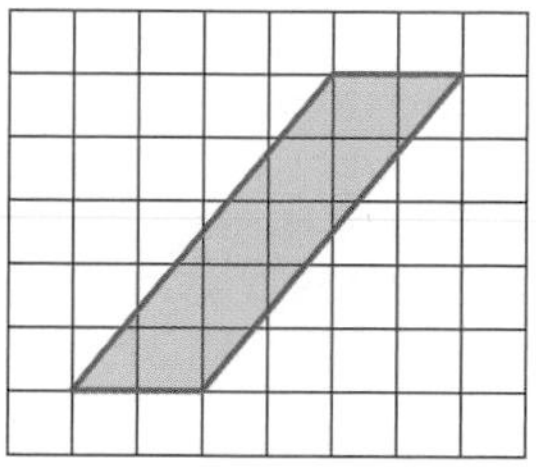

Find the area of the parallelogram.

Count grid lines to find the dimensions.
The base b is 2 units, and the height h is 5 units.

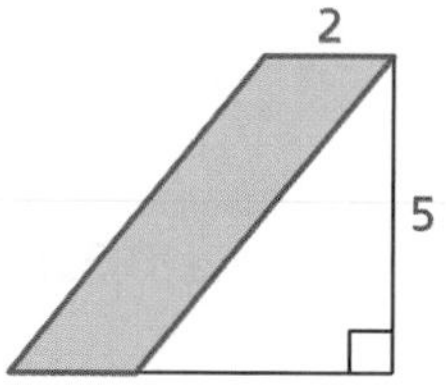

$A = bh$	Write formula.
$= 2(5)$	Substitute values.
$= 10$	Multiply.

The area of the parallelogram is 10 square units.

On Your Own

Now You're Ready
Exercises 11–13

4. Find the area of the parallelogram.

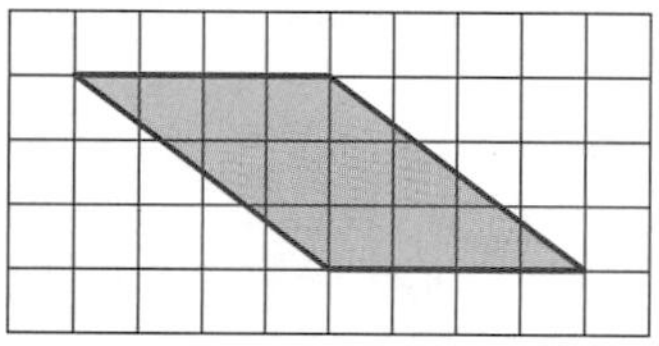

EXAMPLE 3 Real-Life Application

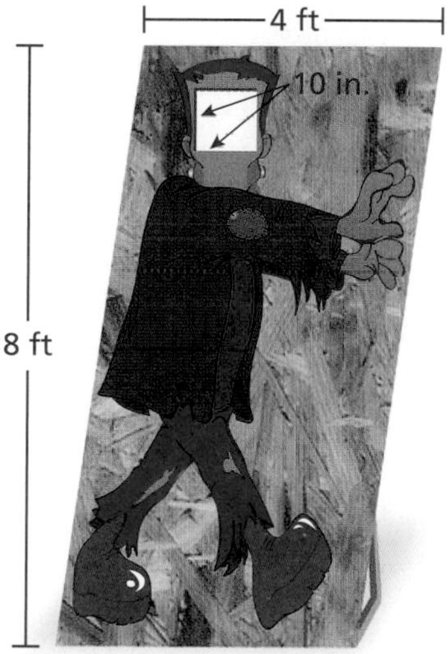

You make a photo prop for a school fair. You cut a 10-inch square out of a parallelogram-shaped piece of wood. What is the area of the photo prop?

Convert the dimensions of the piece of wood to inches.

There are 12 inches in 1 foot, so the base is $4 \cdot 12 = 48$ inches and the height is $8 \cdot 12 = 96$ inches.

Use a verbal model to solve the problem.

area of photo prop = area of wood − area of square

=	96(48)	−	10^2	Substitute.
=	96(48)	−	100	Evaluate 10^2.
=	4608	−	100	Multiply 96 and 48.
=	4508			Subtract 100 from 4608.

The area of the photo prop is 4508 square inches.

On Your Own

Now You're Ready
Exercises 14–16

5. Find the area of the shaded region.

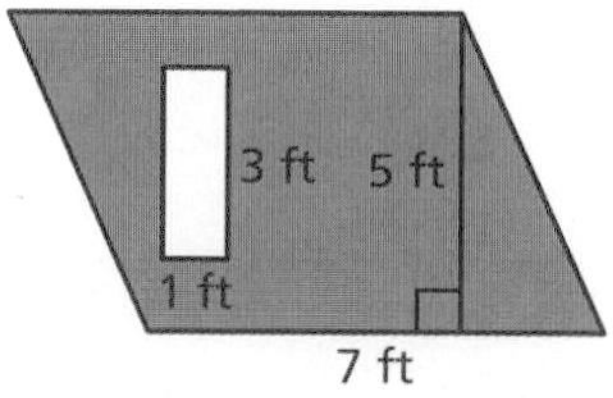

6. WHAT IF? In Example 3, you cut a 12-inch square out of the piece of wood. What is the area of the photo prop?

4.1 Exercises

Check It Out
Help with Homework
BigIdeasMath.com

Vocabulary and Concept Check

1. **WRITING** What is the area of a polygon? Explain how the perimeter and the area of the polygon are different.
2. **CHOOSE TOOLS** Construct a parallelogram that has an area of 24 square inches. Explain your method.

Practice and Problem Solving

Find the area of the parallelogram.

1 3.

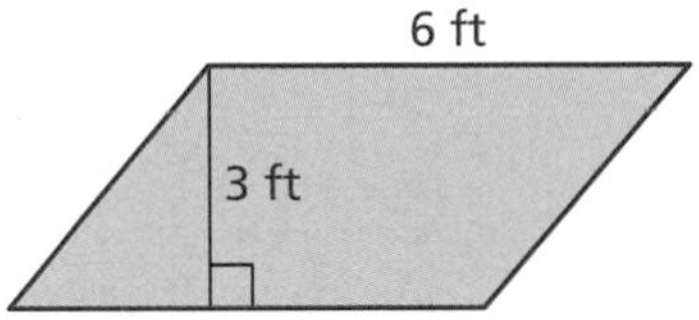

4.

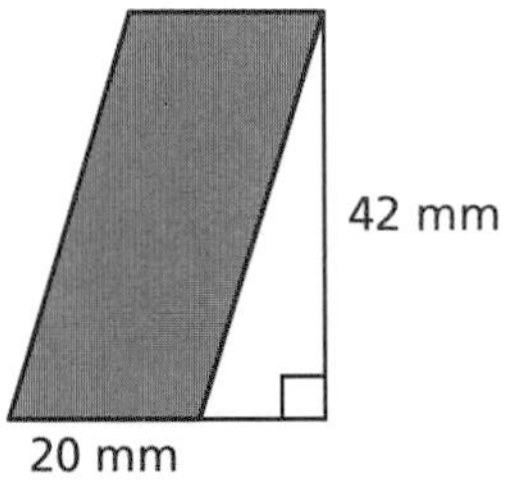

5.

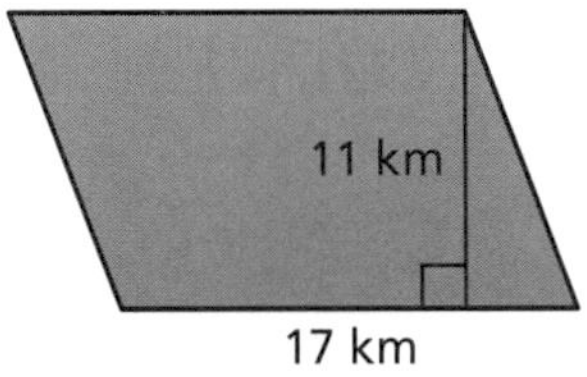

6.

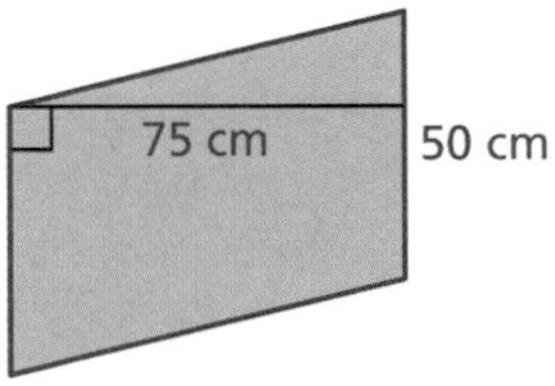

7.

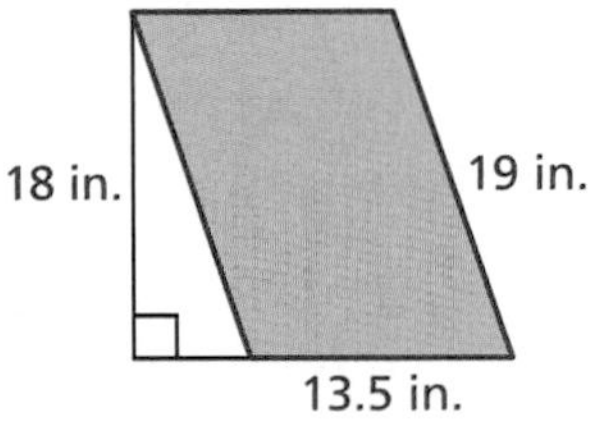

8.

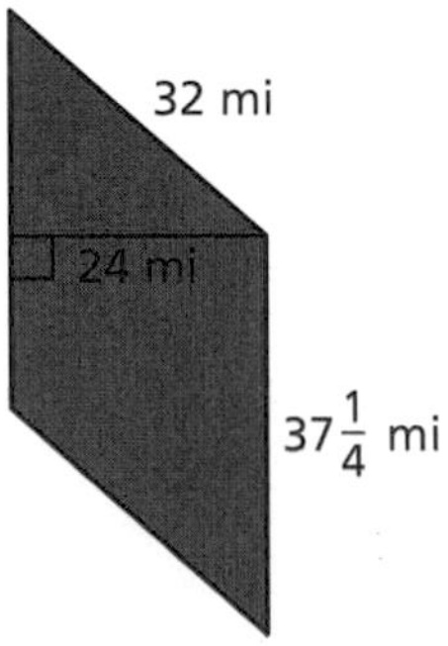

9. **ERROR ANALYSIS** Describe and correct the error in finding the area of the parallelogram.

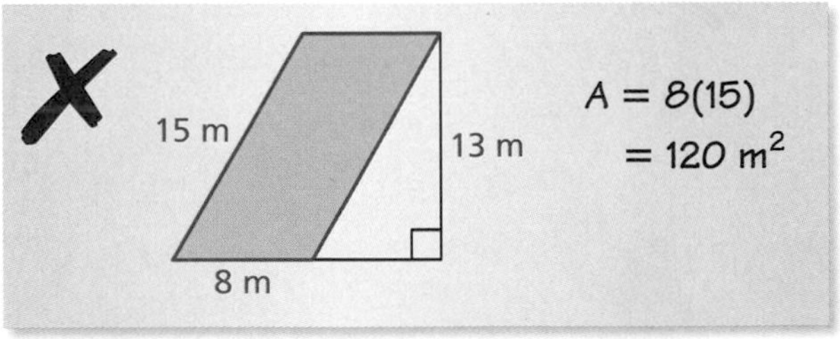

10. **CERAMIC TILE** A ceramic tile in the shape of a parallelogram has a base of 4 inches and a height of 1.5 inches. What is the area of the tile?

Find the area of the parallelogram.

2 11.

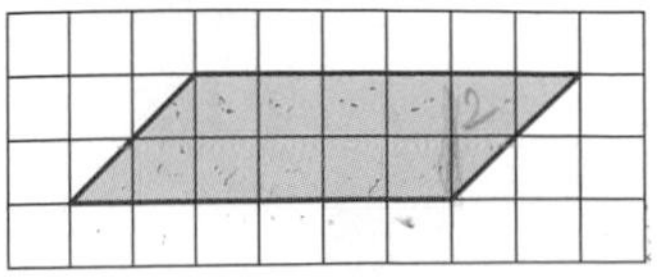

12.

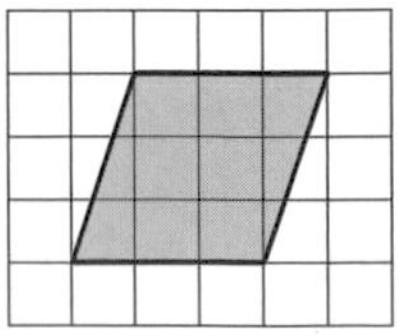

13.

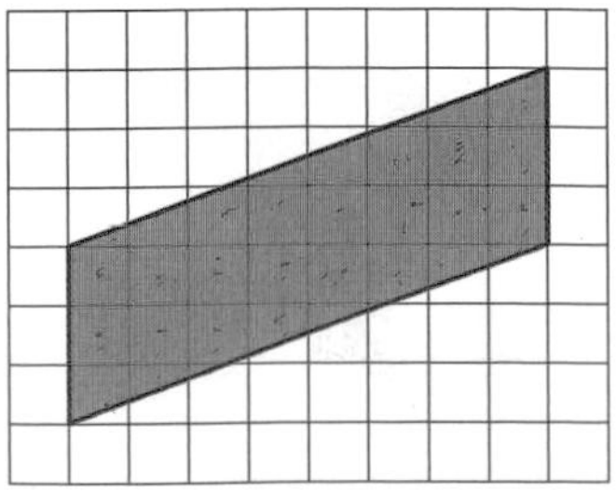

Find the area of the shaded region.

3 **14.**

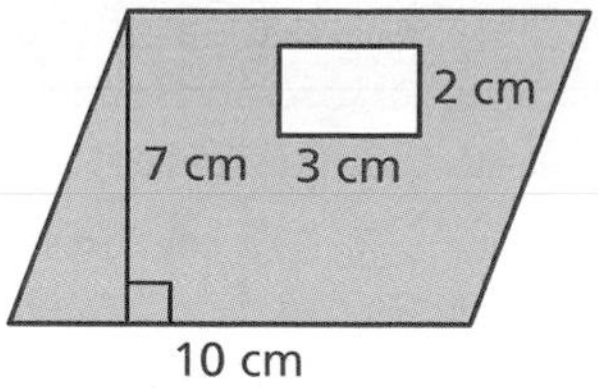

15.

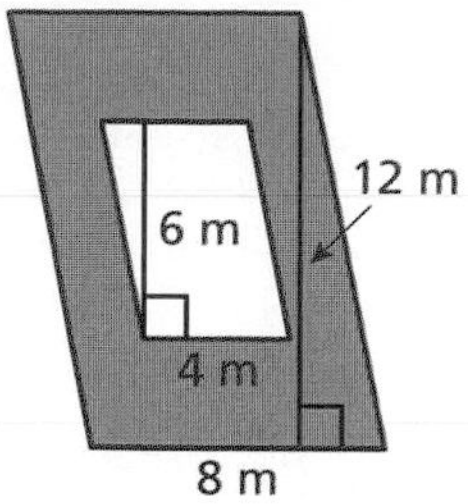

16.

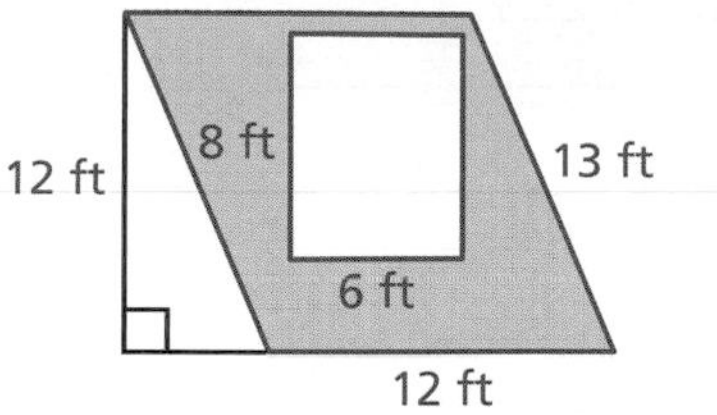

17. **DECK** Your deck has an area of 128 square feet. After adding a section, the area will be $s^2 + 128$ square feet. Draw a diagram of how this can happen.

18. **T-SHIRT DESIGN** You use the parallelogram-shaped sponge to create the T-shirt design. The area of the design is 66 square inches. How many times do you use the sponge to create the design? Draw a diagram to support your answer.

19. **STAIRCASE** The staircase has three parallelogram-shaped panels that are the same size. The horizontal distance between each panel is 4.25 inches. What is the area of one panel?

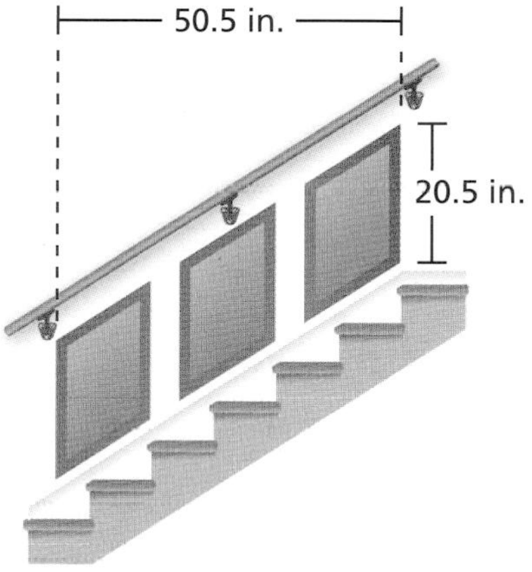

20. **REASONING** Find the missing dimensions in the table.

Parallelogram	Base	Height	Area
A	$x + 4$		$5x + 20$
B		8	$8x - 24$
C	6		$12x + 6y$

21. **Logic** Each dimension of a parallelogram is multiplied by a positive number n. Write an expression for the area of the new parallelogram.

Fair Game Review What you learned in previous grades & lessons

Use mental math to multiply. *(Skills Review Handbook)*

22. $\frac{1}{2} \times 26$ **23.** 82×20 **24.** 16×30 **25.** $\frac{1}{2} \times 236$

26. **MULTIPLE CHOICE** Which of the following describes angle B? *(Skills Review Handbook)*

Ⓐ acute Ⓑ obtuse

Ⓒ right Ⓓ isosceles

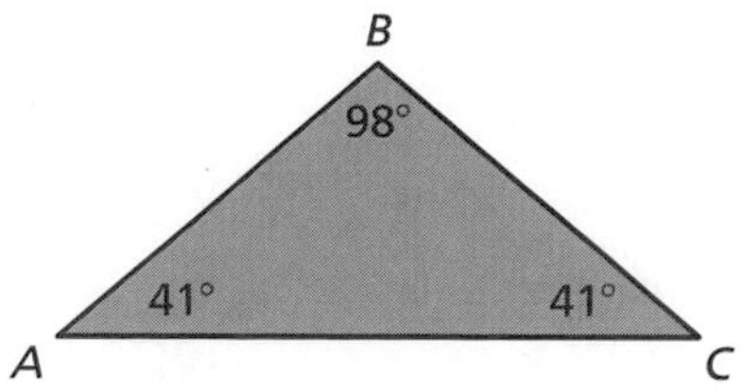

4.2 Areas of Triangles

Essential Question How can you derive a formula for the area of a triangle?

1 ACTIVITY: Deriving the Area Formula of a Triangle

Work with a partner.

a. Draw *any* rectangle on a piece of grid paper. An example is shown below. Label the length and width. Then find the area of your rectangle.

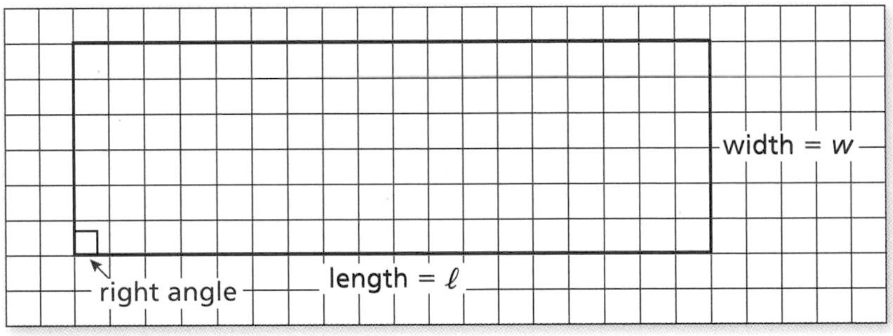

b. Draw a diagonal from one corner of your rectangle to the opposite corner. Cut along the diagonal. Compare the area of the rectangle with the area of the two pieces you cut. What do you notice? Use your results to write a formula for the area A of a triangle.

$A =$ ______ Formula

2 ACTIVITY: Deriving the Area Formula of a Triangle

Work with a partner.

a. Fold a piece of grid paper in half. Draw a triangle so that its base lies on one of the horizontal lines of the paper. Do not use a right triangle. Label the height and the base *inside* the triangle.

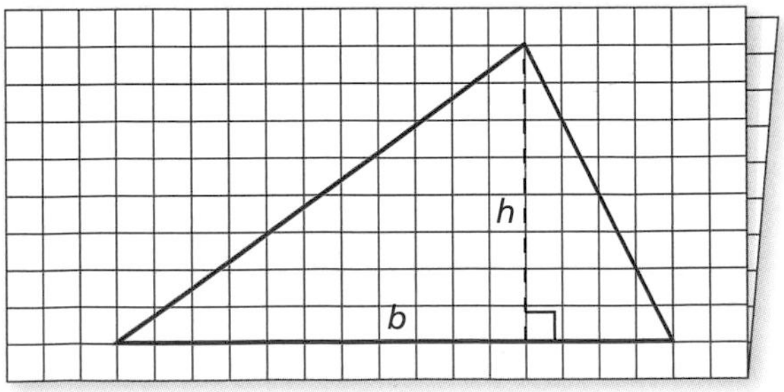

fold

b. Estimate the area of your triangle by counting unit squares.

Area ≈ ______ Estimate

c. Cut out the triangle so that you end up with two identical triangles. Form a quadrilateral whose area you know. What type of quadrilateral is it? Explain how you *know* it is this type.

d. Use your results to write a formula for the area of a triangle. Then use your formula to find the exact area of your triangle. Compare this area with your estimate in part (b).

COMMON CORE

Geometry

In this lesson, you will
- find areas of triangles.
- solve real-life problems.

Learning Standard
6.G.1

Math Practice 6

Calculate Accurately

How can you estimate the area of each triangle so that the answer is close to the exact area?

3 ACTIVITY: Estimating and Finding the Area of a Triangle

Work with a partner. Each grid square represents 1 square centimeter.

- **Use estimation to match each triangle with its area.**
- **Then check your work by finding the exact area of each triangle.**

	Area	*Estimate Match*	*Exact Match*
a.	15 cm^2		
b.	20 cm^2		
c.	9 cm^2		
d.	12 cm^2		
e.	60 cm^2		
f.	$12\frac{1}{2} \text{ cm}^2$		
g.	$24\frac{1}{2} \text{ cm}^2$		
h.	8 cm^2		

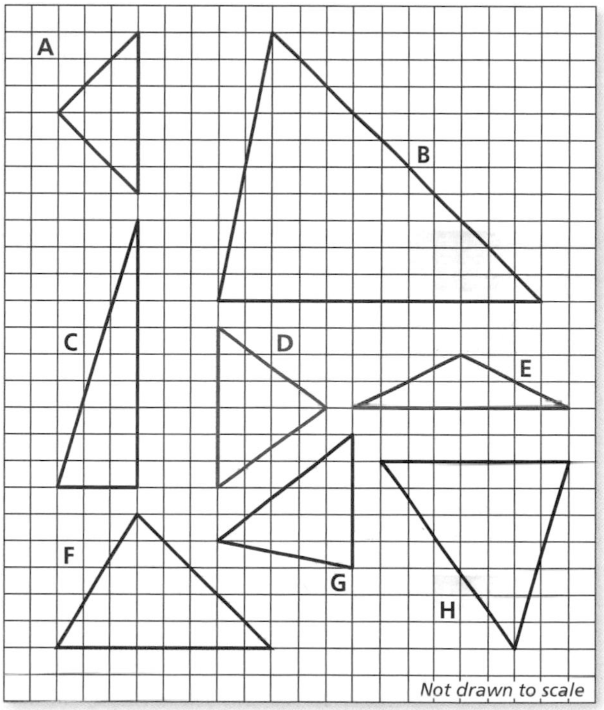

What Is Your Answer?

4. **PARTNER ACTIVITY** Use a piece of centimeter grid paper to create your own "triangle matching activity." Trade with your partner and solve each other's matching activity.
5. **IN YOUR OWN WORDS** How can you derive a formula for the area of a triangle?

Practice Use what you learned about the areas of triangles to complete Exercises 3–5 on page 162.

4.2 Lesson

Check It Out
Lesson Tutorials
BigIdeasMath.com

Key Idea

Area of a Triangle

Words The area A of a triangle is one-half the product of its base b and its height h.

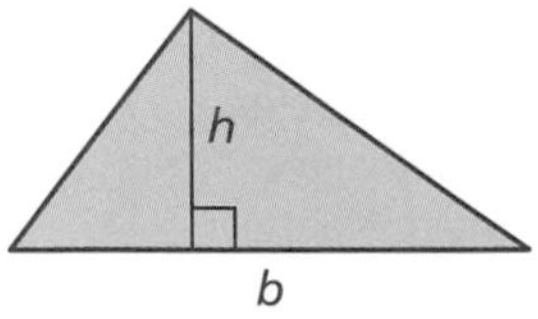

Algebra $A = \frac{1}{2}bh$

EXAMPLE 1 Finding the Area of a Triangle

Find the area of the triangle.

$A = \frac{1}{2}bh$	Write formula.
$= \frac{1}{2}(5)(8)$	Substitute 5 for b and 8 for h.
$= \frac{1}{2}(40)$	Multiply 5 and 8.
$= 20$	Multiply $\frac{1}{2}$ and 40.

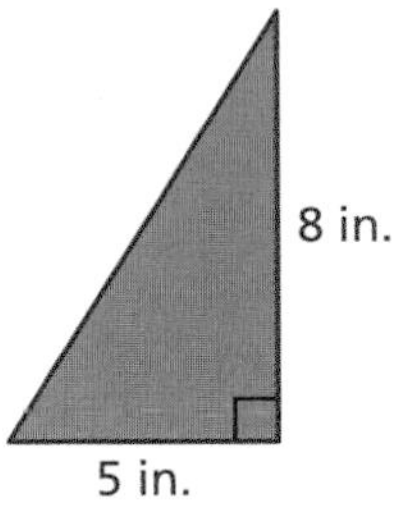

Remember

In Example 1, use the Associative Property of Multiplication to multiply 5 and 8 first.

The area of the triangle is 20 square inches.

Reasonable? Draw the triangle on grid paper and count unit squares. Each square in the grid represents 1 square inch.

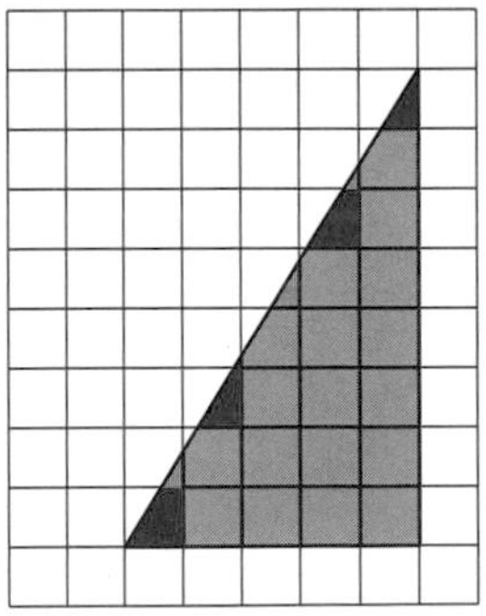

Squares full or nearly full: 18

Squares about half full: 4

The area is $18(1) + 4\left(\frac{1}{2}\right) = 20$ square inches.

So, the answer is reasonable. ✓

On Your Own

Now You're Ready
Exercises 3–8

Find the area of the triangle.

1. 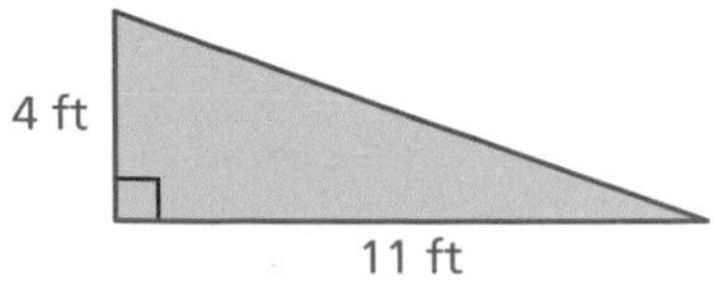

2. 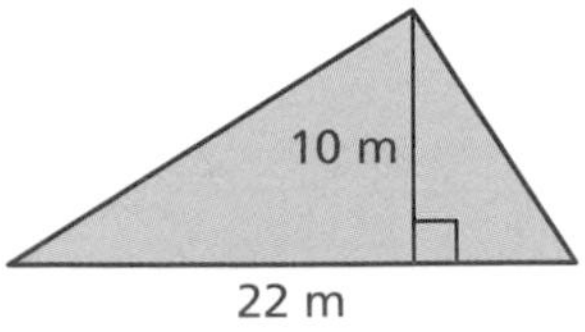

EXAMPLE 2 Finding the Area of a Triangle

Find the area of the triangle.

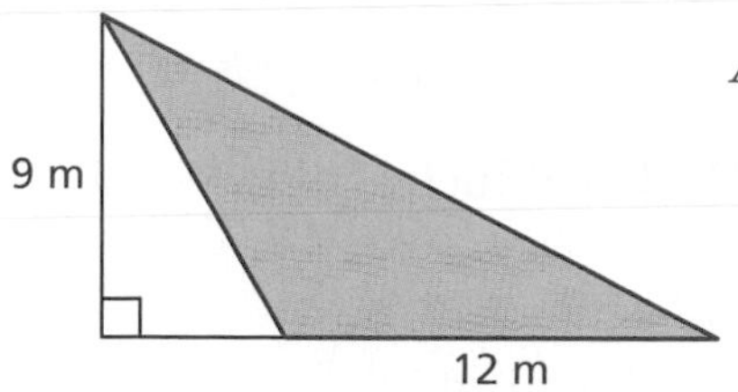

$$A = \frac{1}{2}bh \quad \text{Write formula.}$$

$$= \frac{1}{2}(12)(9) \quad \text{Substitute 12 for } b \text{ and 9 for } h.$$

$$= 54 \quad \text{Multiply.}$$

∴ The area of the triangle is 54 square meters.

EXAMPLE 3 Real-Life Application

The base and height of the red butterfly wing are two times greater than the base and height of the blue butterfly wing. How many times greater is the area of the red wing than the area of the blue wing?

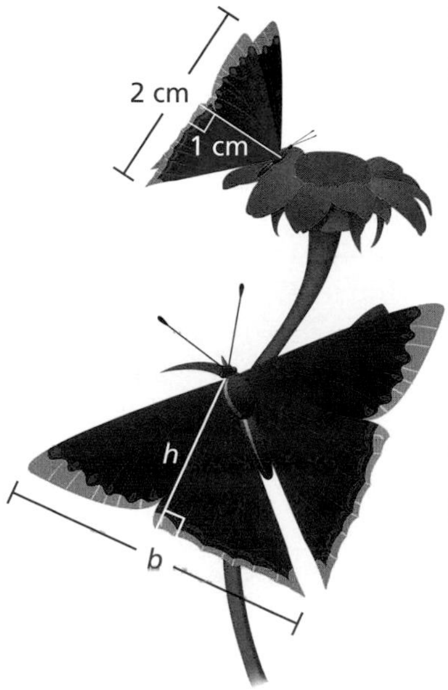

Find the area of the blue wing.

$$A = \frac{1}{2}bh \quad \text{Write formula.}$$

$$= \frac{1}{2}(2)(1) \quad \text{Substitute 2 for } b \text{ and 1 for } h.$$

$$= 1 \text{ cm}^2 \quad \text{Multiply.}$$

The red wing dimensions are 2 times greater, so the base is $2 \times 2 = 4$ cm and the height is $2 \times 1 = 2$ cm. Find the area of the red wing.

$$A = \frac{1}{2}bh \quad \text{Write formula.}$$

$$= \frac{1}{2}(4)(2) \quad \text{Substitute 4 for } b \text{ and 2 for } h.$$

$$= 4 \text{ cm}^2 \quad \text{Multiply.}$$

∴ Because $\frac{4 \cancel{\text{cm}^2}}{1 \cancel{\text{cm}^2}} = 4$, the area of the red wing is 4 times greater.

On Your Own

Now You're Ready
Exercises 12–14

3. Find the area of the triangle.

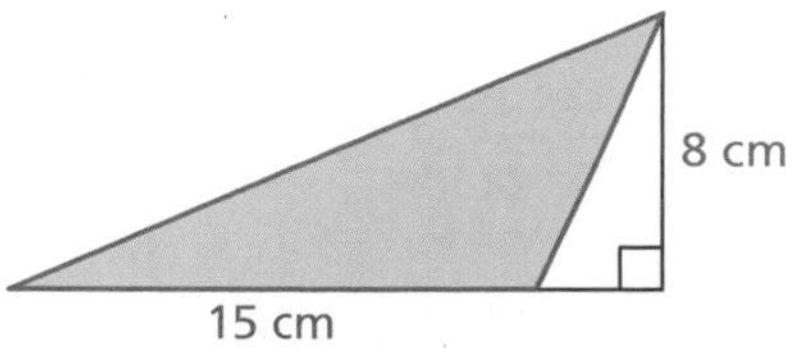

4. **WHAT IF?** In Example 3, the base and the height of the red butterfly wing are three times greater than those of the blue wing. How many times greater is the area of the red wing?

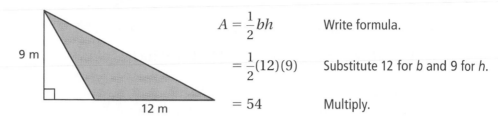

4.2 Exercises

Vocabulary and Concept Check

1. **CRITICAL THINKING** Can *any* side of a triangle be labeled as its base? Explain.

2. **DIFFERENT WORDS, SAME QUESTION** Which is different? Find "both" answers.

What is the area of the triangle?	What is the distance around the triangle?
How many unit squares fit in the triangle?	What is one-half the product of the base and the height?

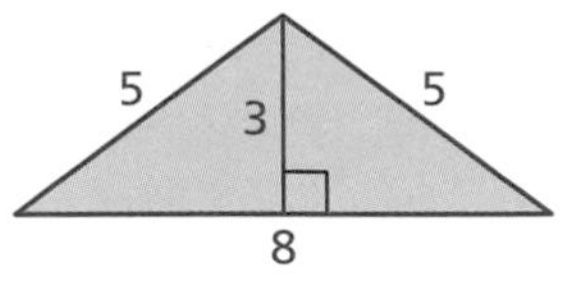

Practice and Problem Solving

Find the area of the triangle.

1 3.

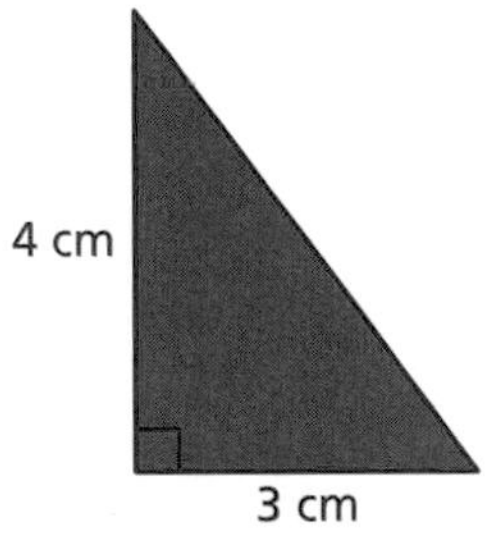

4.

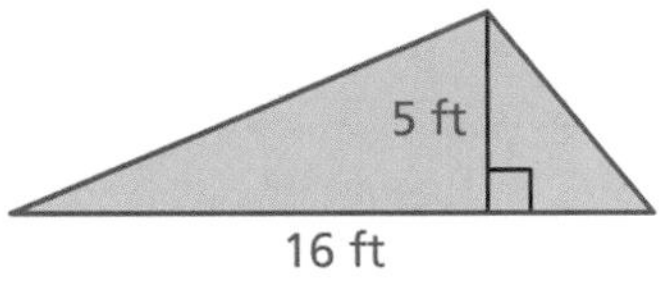

5.

6.

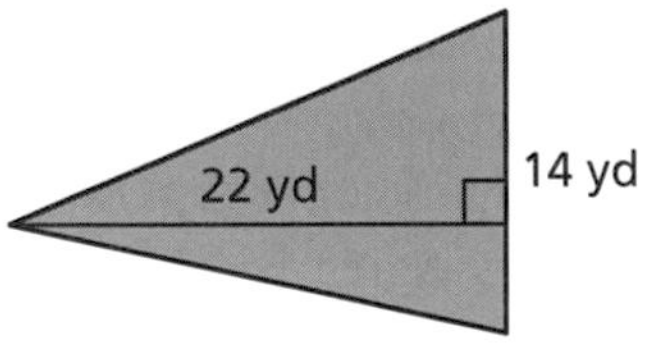

7.

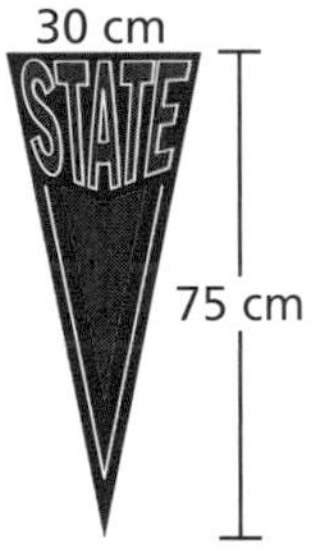

8.

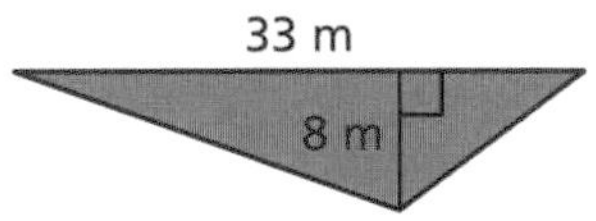

9. **ERROR ANALYSIS** Describe and correct the error in finding the area of the triangle.

✗

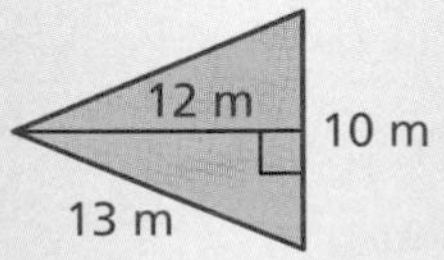

$$A = \frac{1}{2}(10)(13)$$
$$= 65 \text{ m}^2$$

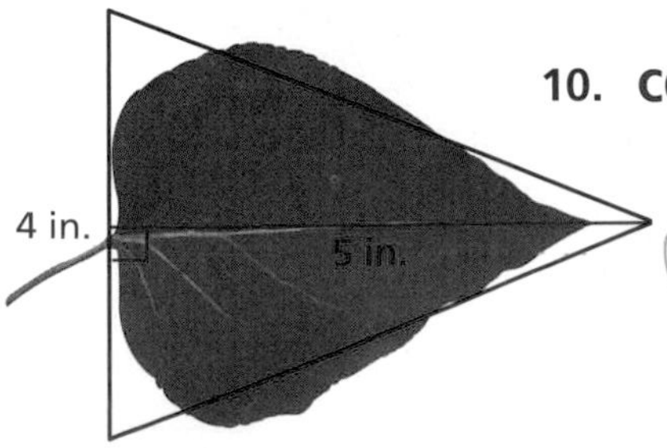

10. **COTTONWOOD LEAF** Estimate the area of the cottonwood leaf.

11. **CORNER SHELF** A shelf has the shape of a triangle. The base of the shelf is 36 centimeters, and the height is 18 centimeters. Find the area of the shelf.

Find the area of the triangle.

12.

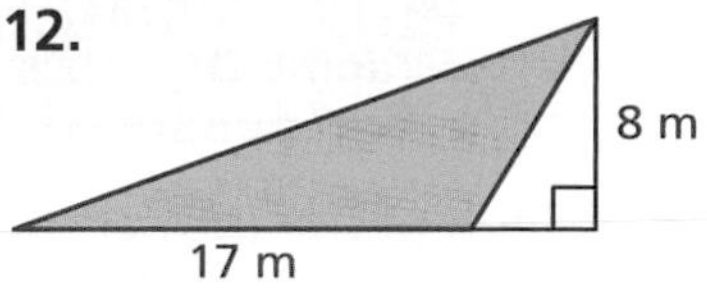

13.

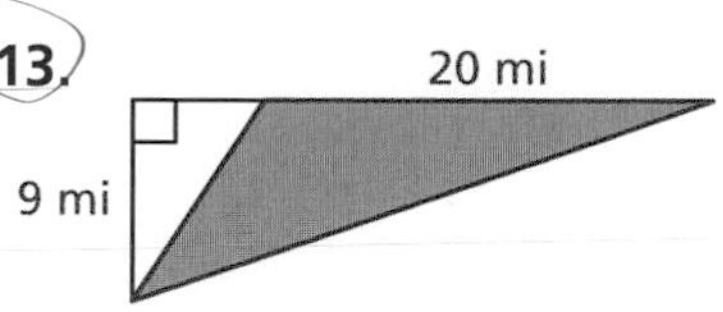

14.

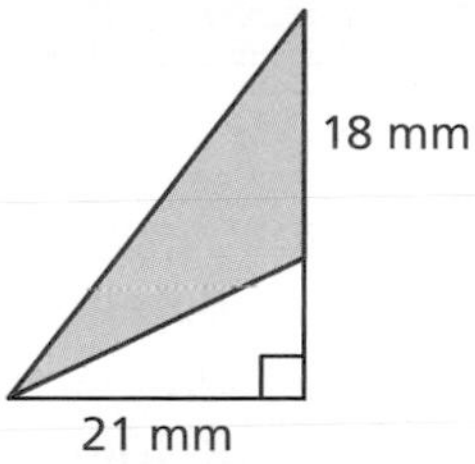

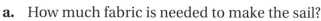

15. **OPEN-ENDED** Draw and label two triangles that each have an area of 24 square feet.

16. **HANG GLIDING** The wingspan of the triangular hang glider is 30 feet.

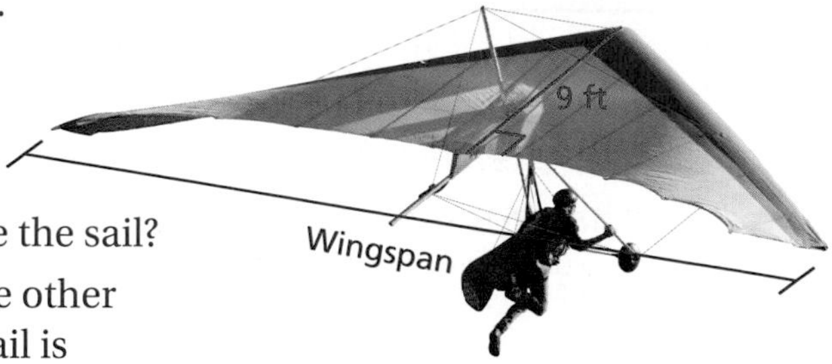

a. How much fabric is needed to make the sail?

b. **RESEARCH** Use the Internet or some other source to find how the area of the sail is related to the weight limit of the pilot.

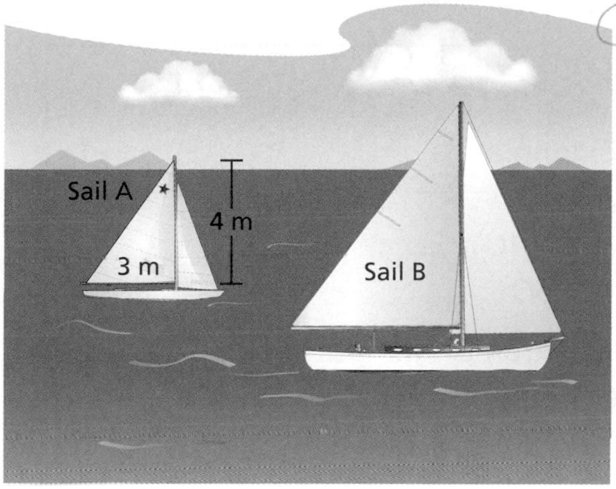

17. **SAILBOATS** The base and the height of Sail B are x times greater than the base and the height of Sail A. How many times greater is the area of Sail B? Write your answer as a power.

18. **WRITING** You know the height and the perimeter of an equilateral triangle. Explain how to find the area of the triangle. Draw a diagram to support your reasoning.

19. **REASONING** The base and the height of Triangle A are half the base and the height of Triangle B. How many times greater is the area of Triangle B?

20. **Critical Thinking** The total area of the polygon is 176 square feet. Find the value of x.

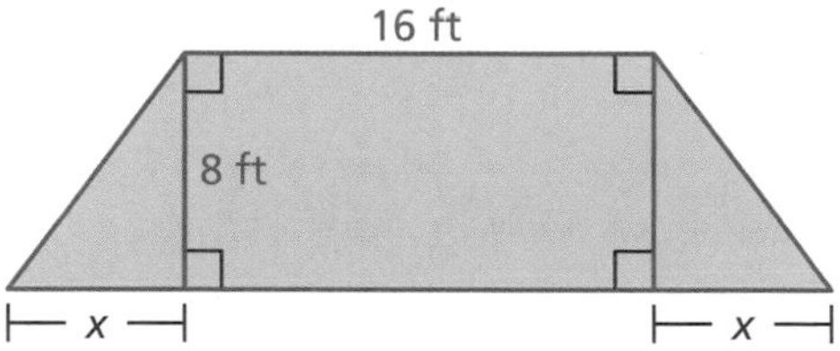

Fair Game Review What you learned in previous grades & lessons

Tell which property is illustrated by the statement. *(Section 3.3)*

21. $n \cdot 1 = n$

22. $4 \cdot m = m \cdot 4$

23. $(x + 2) + 5 = x + (2 + 5)$

24. **MULTIPLE CHOICE** What is the first step when using order of operations? *(Section 1.3)*

(A) Multiply or divide from left to right.
(B) Add or subtract from left to right.
(C) Perform operations in parentheses.
(D) Evaluate numbers with exponents.

4 Study Help

Check It Out
Graphic Organizer
BigIdeasMath.com

You can use a **four square** to organize information about a topic. Each of the four squares can be a category, such as *definition, vocabulary, example, non-example, words, algebra, table, numbers, visual, graph,* or *equation*. Here is an example of a four square for the area of a parallelogram.

Area of a parallelogram

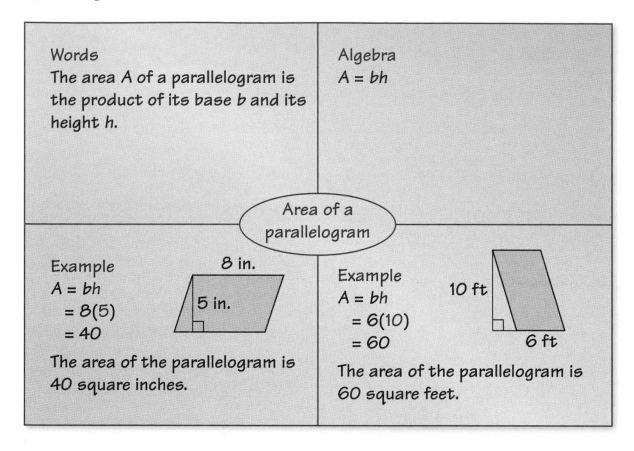

Words
The area A of a parallelogram is the product of its base b and its height h.

Algebra
$A = bh$

Example
$A = bh$
$= 8(5)$
$= 40$

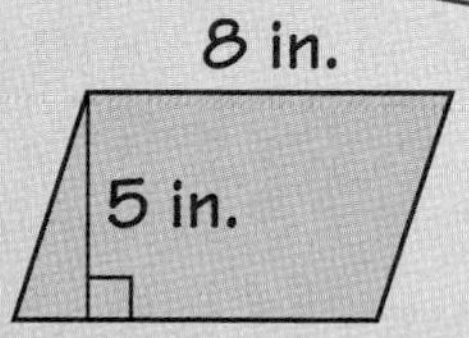

The area of the parallelogram is 40 square inches.

Example
$A = bh$
$= 6(10)$
$= 60$

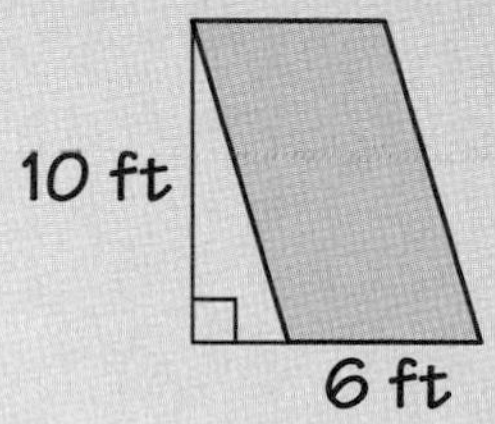

The area of the parallelogram is 60 square feet.

On Your Own

Make a four square to help you study the topic.

1. area of a triangle

After you complete this chapter, make four squares for the following topics.

2. area of a trapezoid
3. area of a composite figure
4. drawing a polygon in a coordinate plane
5. finding distances in the first quadrant

"Sorry, but I have limited space in my four square. I needed pet names with only three letters."

4.1–4.2 Quiz

Find the area of the parallelogram. *(Section 4.1)*

1. 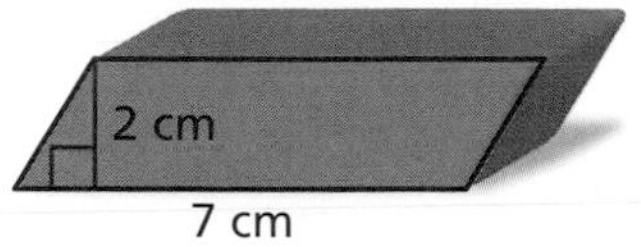

2. 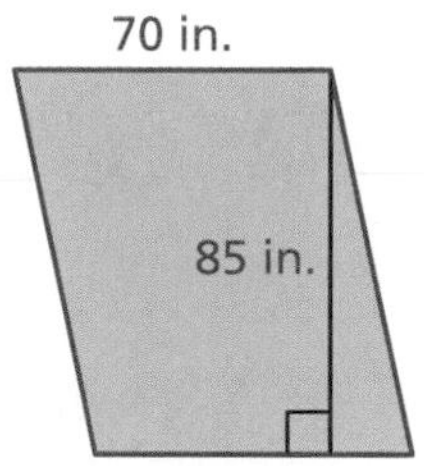

3. 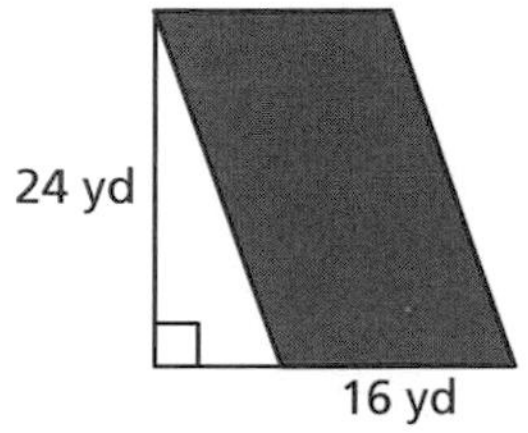

4. 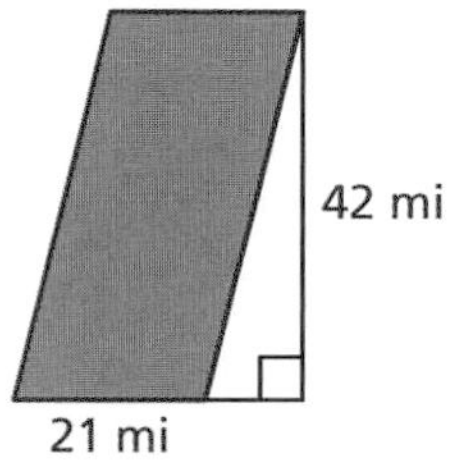

Find the area of the triangle. *(Section 4.2)*

5. 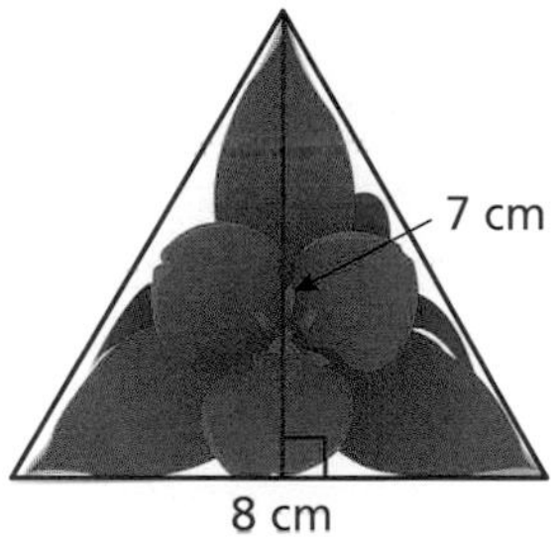

6. 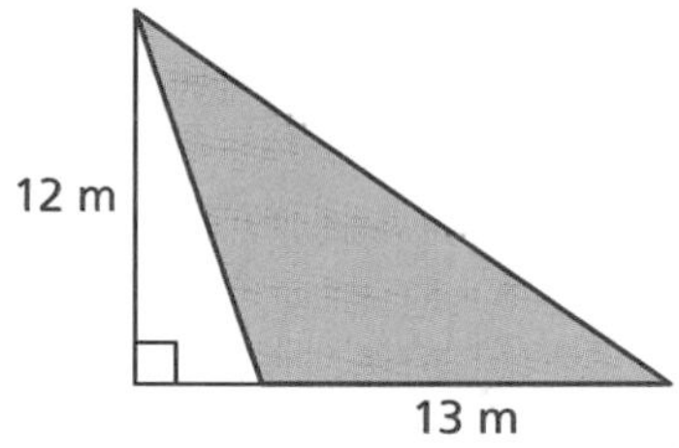

7. **LAND** A wildlife conservation group buys a plot of land. How much land does it buy? *(Section 4.2)*

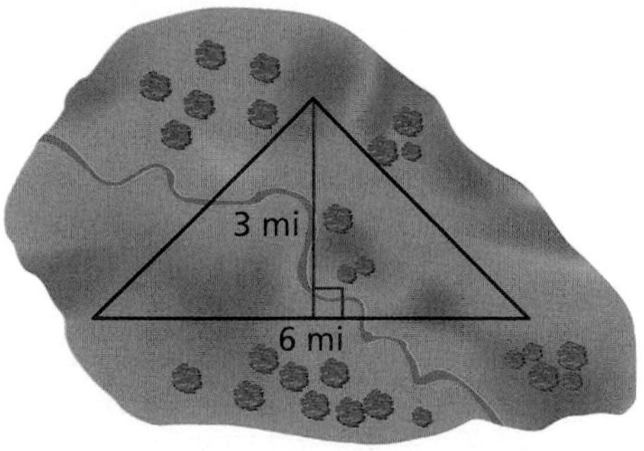

8. **FRAMING** A sheet of plywood is 4 feet wide by 8 feet long. What is the minimum number of sheets of plywood needed to cover the frame? Justify your answer. *(Section 4.2)*

4.3 Areas of Trapezoids

Essential Question How can you derive a formula for the area of a trapezoid?

1 ACTIVITY: Deriving the Area Formula of a Trapezoid

Work with a partner. Use a piece of centimeter grid paper.

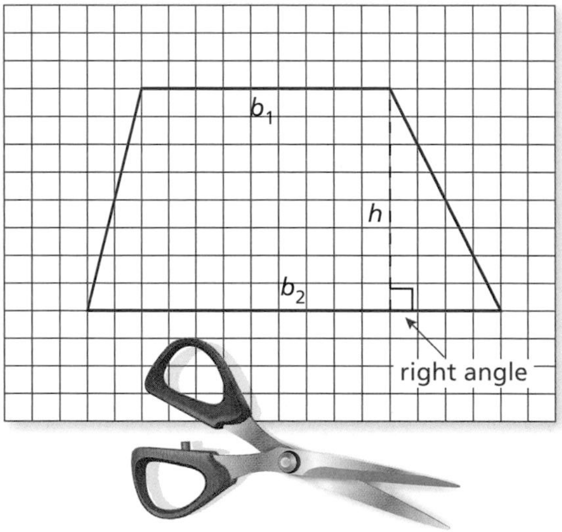

a. Draw *any* trapezoid so that its base lies on one of the horizontal lines of the paper.

b. Estimate the area of your trapezoid (in square centimeters) by counting unit squares.

Area ≈ ______ Estimate

c. Label the height and the bases *inside* the trapezoid.

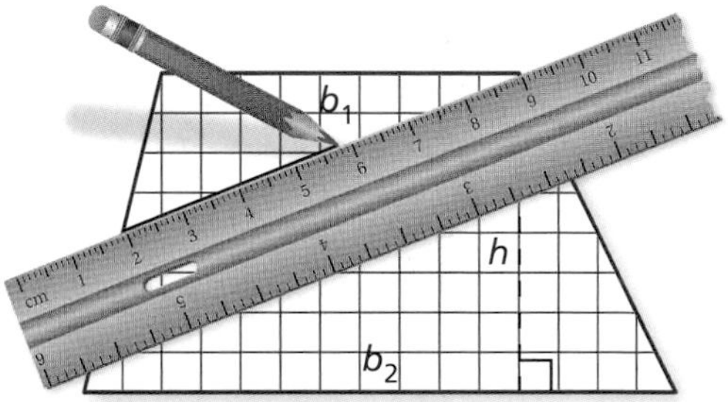

d. Cut out the trapezoid. Mark the midpoint of the side opposite the height. Draw a line from the midpoint to the opposite upper vertex.

e. Cut along the line. You will end up with a triangle and a quadrilateral. Arrange these two figures to form a figure whose area you know.

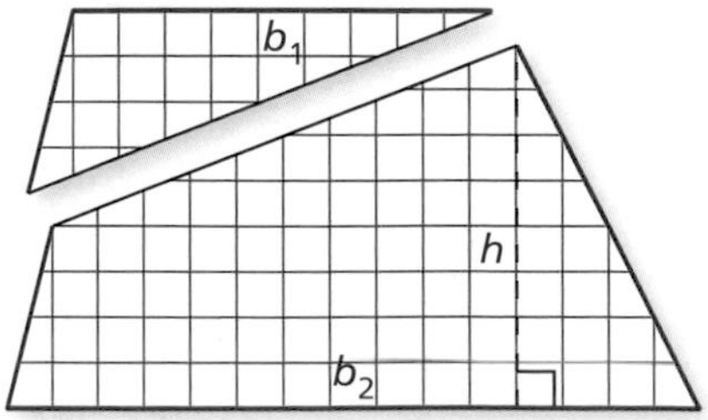

f. Use your result to write a *formula* for the area of a trapezoid.

Area = ______ Formula

g. Use your formula to find the area of your trapezoid (in square centimeters).

Area = ______ Exact Area

h. Compare this area with your estimate in part (b).

COMMON CORE

Geometry

In this lesson, you will
- find areas of trapezoids.
- solve real-life problems.

Learning Standard
6.G.1

2 ACTIVITY: Writing a Math Lesson

Work with a partner. Use your results from Activity 1 to write a lesson on finding the area of a trapezoid.

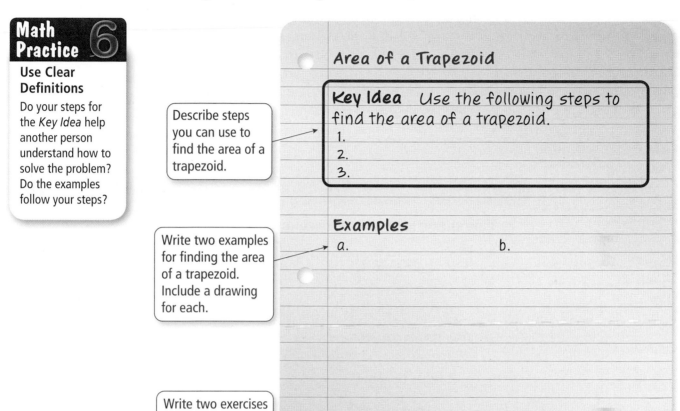

Math Practice 6

Use Clear Definitions

Do your steps for the *Key Idea* help another person understand how to solve the problem? Do the examples follow your steps?

Describe steps you can use to find the area of a trapezoid.

Write two examples for finding the area of a trapezoid. Include a drawing for each.

Write two exercises for finding the area of a trapezoid. Include an answer sheet.

Area of a Trapezoid

Key Idea Use the following steps to find the area of a trapezoid.

1.

2.

3.

Examples

a. b.

Exercises

Find the area.

1. 2.

What Is Your Answer?

3. **IN YOUR OWN WORDS** How can you derive a formula for the area of a trapezoid?

4. In this chapter, you used deductive reasoning to derive new area formulas from area formulas you have already learned. Describe a real-life career in which deductive reasoning is important.

Practice Use what you learned about the areas of trapezoids to complete Exercises 4–6 on page 170.

4.3 Lesson

Check It Out
Lesson Tutorials
BigIdeasMath.com

Key Idea

Area of a Trapezoid

Words The area A of a trapezoid is one-half the product of its height h and the sum of its bases b_1 and b_2.

Algebra $A = \frac{1}{2}h(b_1 + b_2)$

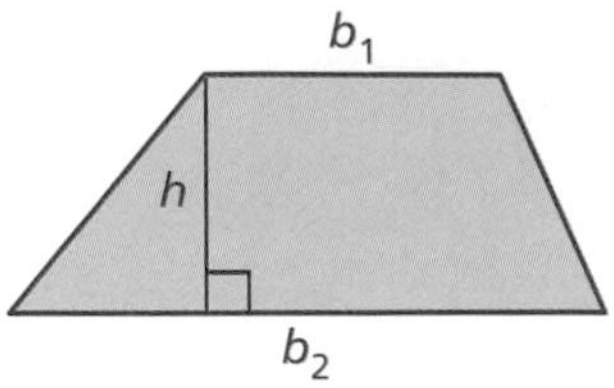

EXAMPLE 1 Finding Areas of Trapezoids

Find the area of each trapezoid.

a.

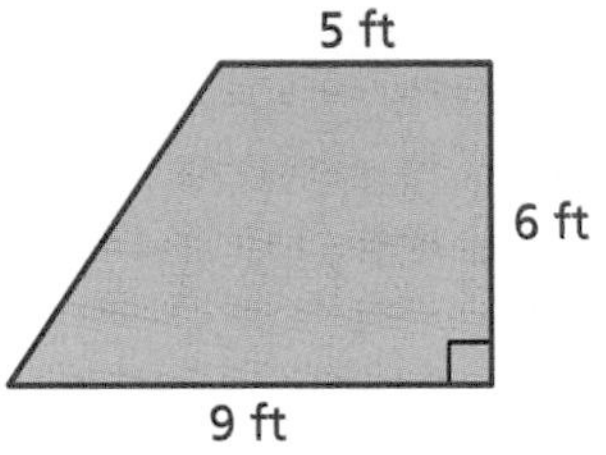

$A = \frac{1}{2}h(b_1 + b_2)$ Write formula.

$= \frac{1}{2}(6)(5 + 9)$ Substitute.

$= \frac{1}{2}(6)(14)$ Add.

$= 42$ Multiply.

∴ The area of the trapezoid is 42 square feet.

b.

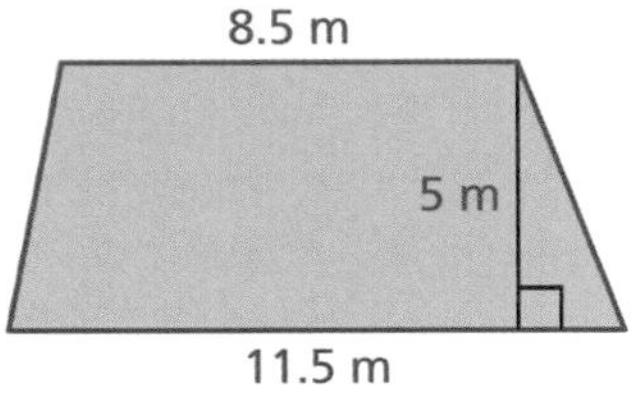

$A = \frac{1}{2}h(b_1 + b_2)$

$= \frac{1}{2}(5)(8.5 + 11.5)$

$= \frac{1}{2}(5)(20)$

$= 50$

∴ The area of the trapezoid is 50 square meters.

On Your Own

Now You're Ready
Exercises 7–9

Find the area of the trapezoid.

1.

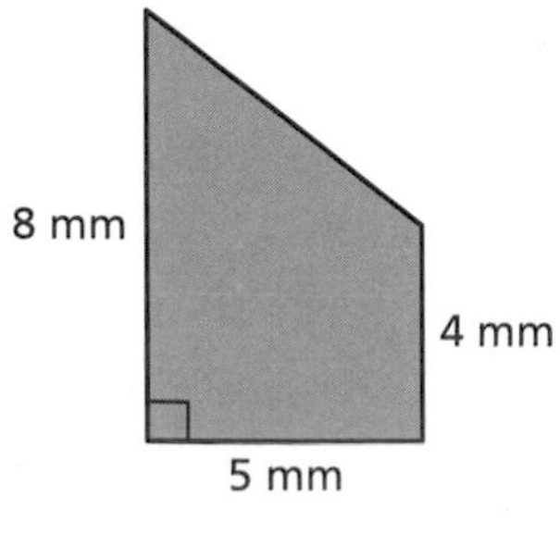

2.

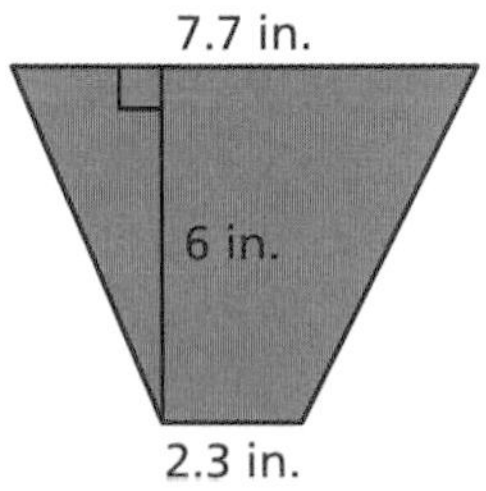

EXAMPLE 2 Finding the Area of a Trapezoid on a Grid

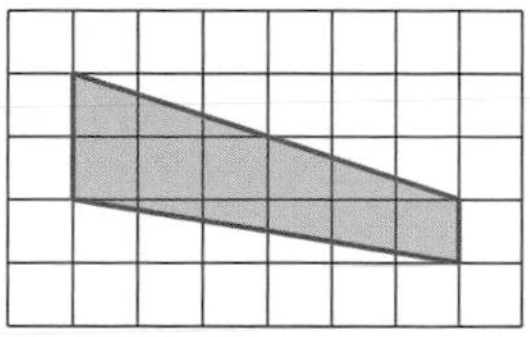

What is the area of the trapezoid?

(A) 6 units^2 (B) 7 units^2 (C) 9 units^2 (D) 12 units^2

Count grid lines to find the dimensions. The height h is 6 units, base b_1 is 1 unit, and base b_2 is 2 units.

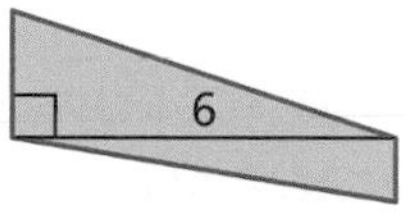

$$A = \frac{1}{2}h(b_1 + b_2) \qquad \text{Write formula.}$$

$$= \frac{1}{2}(6)(1 + 2) \qquad \text{Substitute values.}$$

$$= \frac{1}{2}(6)(3) \qquad \text{Add.}$$

$$= 9 \qquad \text{Multiply.}$$

∴ The area of the trapezoid is 9 square units.
The correct answer is (C).

EXAMPLE 3 Real-Life Application

You can use a trapezoid to approximate the shape of Scott County, Virginia. The population is about 23,200. About how many people are there per square mile?

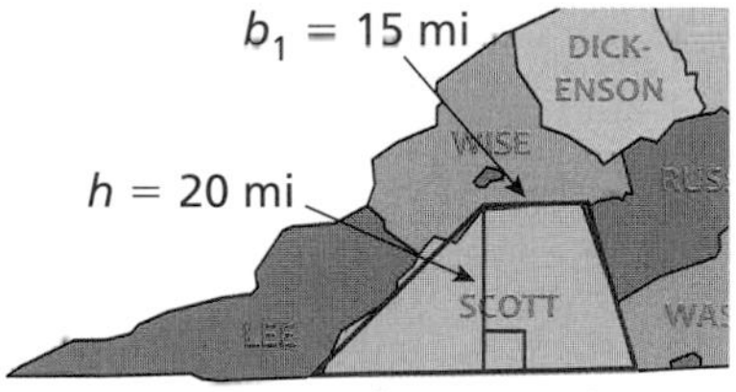

Find the area of Scott County.

$$A = \frac{1}{2}h(b_1 + b_2) \qquad \text{Write formula for area of a trapezoid.}$$

$$= \frac{1}{2}(20)(15 + 38) \qquad \text{Substitute 20 for } h\text{, 15 for } b_1\text{, and 38 for } b_2.$$

$$= \frac{1}{2}(20)(53) = 530 \qquad \text{Simplify.}$$

The area of Scott County is about 530 square miles. Divide the population by the area to find the number of people per square mile.

∴ So, there are about $\frac{23{,}200 \text{ people}}{530 \text{ mi}^2} \approx 44$ people per square mile.

On Your Own

Now You're Ready
Exercises 11–13

3. Find the area of the trapezoid.

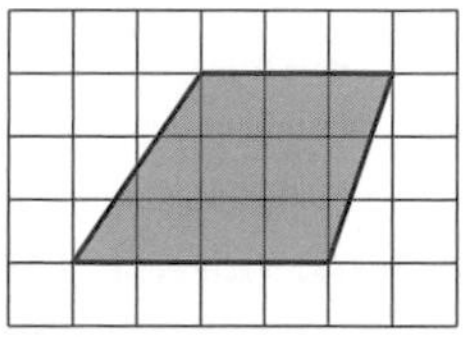

4. **WHAT IF?** In Example 3, the population of Scott County decreases by 550. By how much does the number of people per square mile change? Explain.

4.3 Exercises

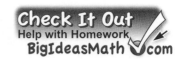

Check It Out
Help with Homework
BigIdeasMath.com

Vocabulary and Concept Check

1. **VOCABULARY** Identify the bases and the height of the trapezoid.

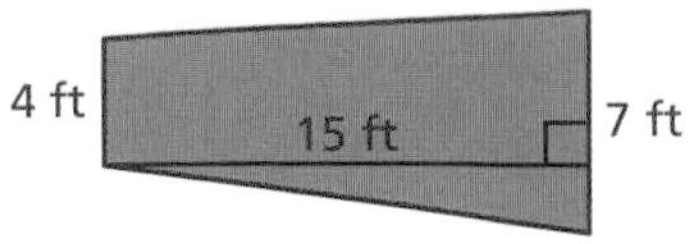

2. **REASONING** What measures do you need to find the area of a trapezoid?

3. **WHICH ONE DOESN'T BELONG?** Which one does *not* belong with the other three? Explain your reasoning.

$\frac{1}{2}bh$ $\qquad$ ℓw $\qquad$ $2\ell + 2w$ $\qquad$ $\frac{1}{2}h(b_1 + b_2)$

Practice and Problem Solving

Find the area of the trapezoid.

4. $b_1 = 4, b_2 = 8, h = 2$

5. $b_1 = 5, b_2 = 7, h = 4$

6. $b_1 = 12, b_2 = 6, h = 3$

1 7.

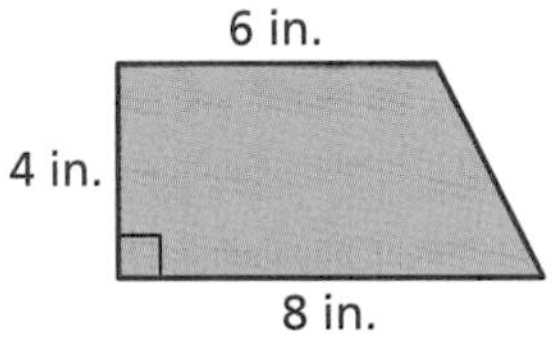

8.

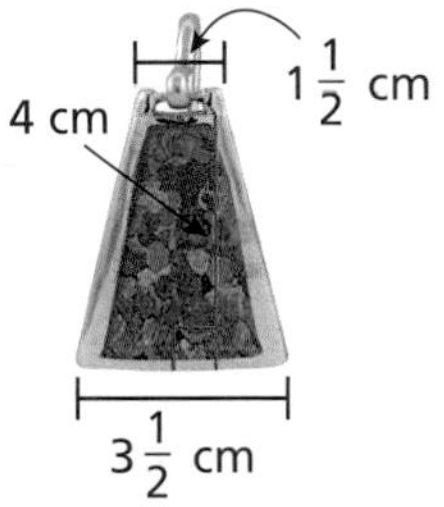

9.

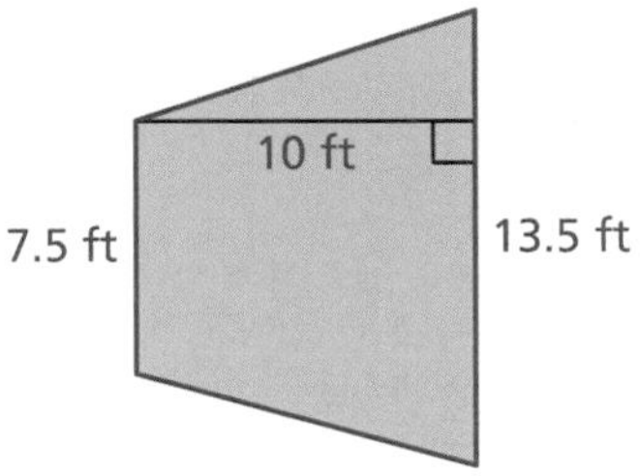

10. **ERROR ANALYSIS** Describe and correct the error in finding the area of the trapezoid.

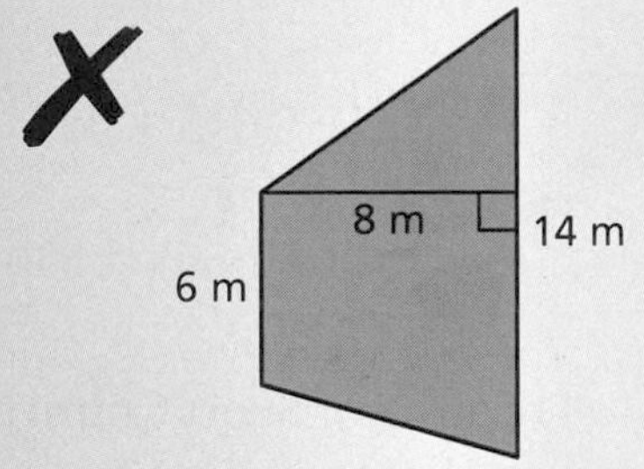

$$\text{Area} = \frac{1}{2}(6 + 14)$$
$$= 10 \text{ m}^2$$

Find the area of the trapezoid.

2 11.

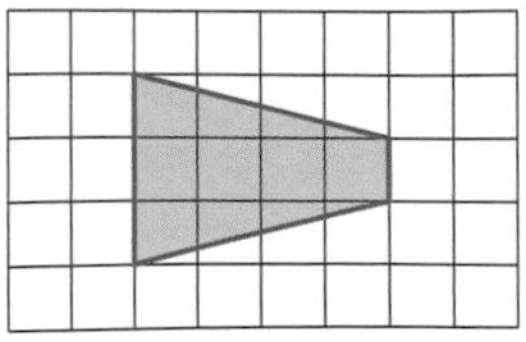

12.

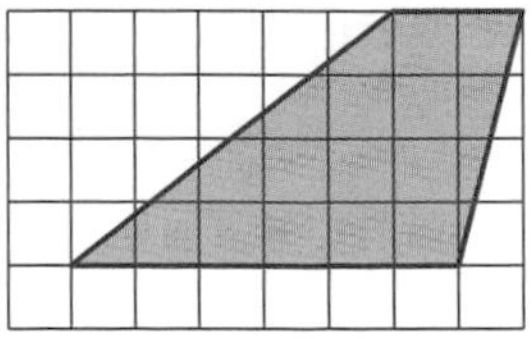

13.

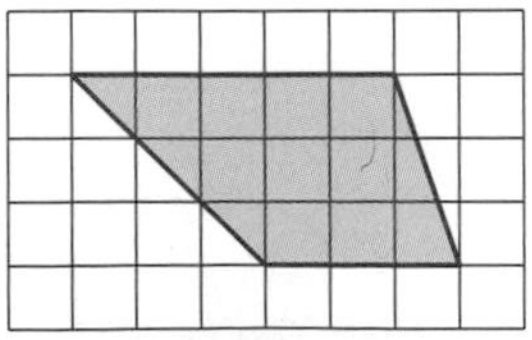

14. **LIGHT** Light shines through a window. What is the area of the trapezoid-shaped region created by the light?

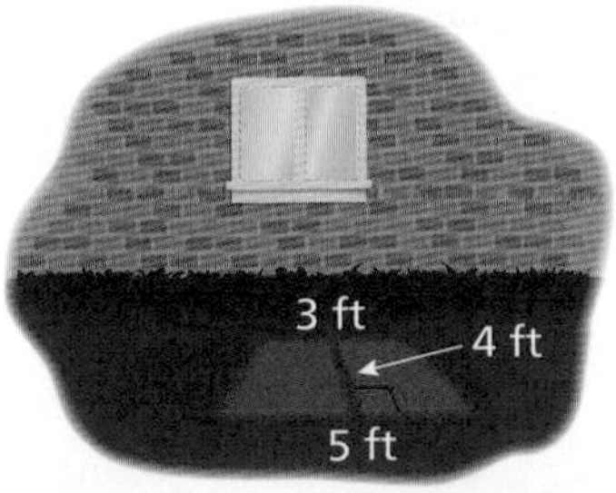

Find the area of a trapezoid with height h and bases b_1 and b_2.

15. $h = 6$ in.
$b_1 = 9$ in.
$b_2 = 11$ in.

16. $h = 22$ cm
$b_1 = 10.5$ cm
$b_2 = 12.5$ cm

17. $h = 12$ mi
$b_1 = 5.6$ mi
$b_2 = 7.4$ mi

18. $h = 14$ m
$b_1 = 21$ m
$b_2 = 22$ m

19. REASONING The rectangle and the trapezoid have the same area. What is the length ℓ of the rectangle?

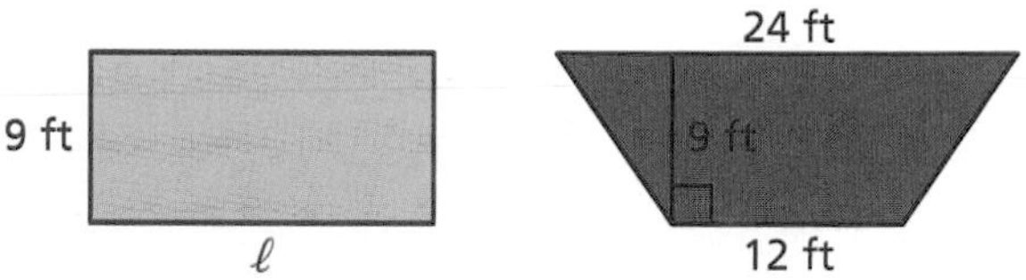

20. OPEN-ENDED The area of the trapezoidal student election sign is 5 square feet. Find two possible values for each base length.

21. AUDIO How many times greater is the area of the floor covered by the larger speaker than by the smaller speaker?

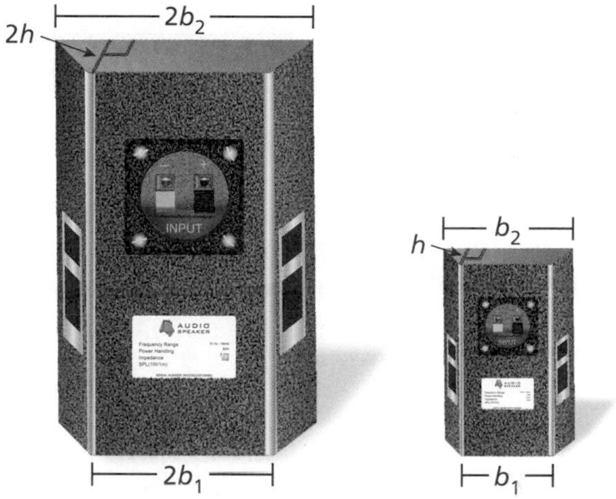

22. Critical Thinking The triangle and the trapezoid share a 15-inch base and a height of 10 inches.

a. The area of the trapezoid is less than twice the area of the triangle. Find the values of x. Explain your reasoning.

b. Can the area of the *trapezoid* be exactly twice the area of the triangle? Explain your reasoning.

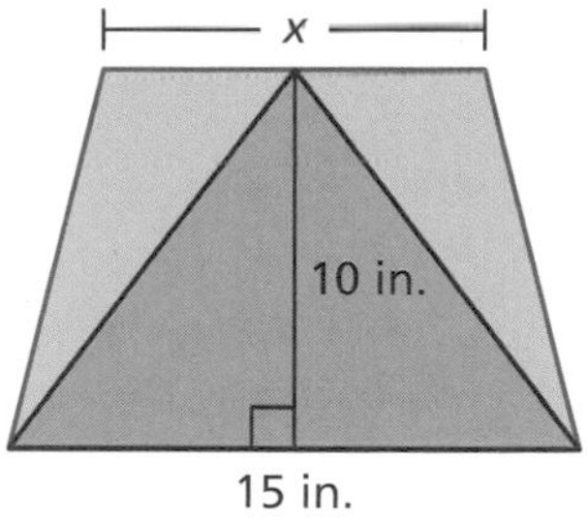

Fair Game Review What you learned in previous grades & lessons

Plot the ordered pair in a coordinate plane. *(Skills Review Handbook)*

23. (5, 0) **24.** (2, 4) **25.** (0, 3) **26.** (6, 1)

27. MULTIPLE CHOICE Which expression represents "6 more than x"? *(Section 3.2)*

Ⓐ $6 - x$ Ⓑ $6x$ Ⓒ $x + 6$ Ⓓ $\frac{6}{x}$

Extension 4.3

Areas of Composite Figures

Check It Out Lesson Tutorials BigIdeasMath.com

Key Vocabulary
composite figure, *p. 172*

A **composite figure** is made up of triangles, squares, rectangles, and other two-dimensional figures. Here are two examples.

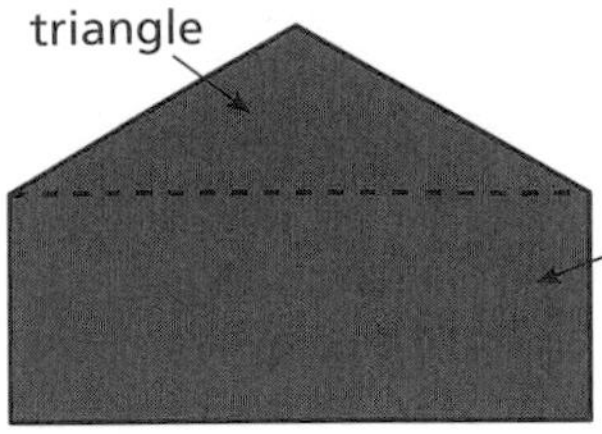

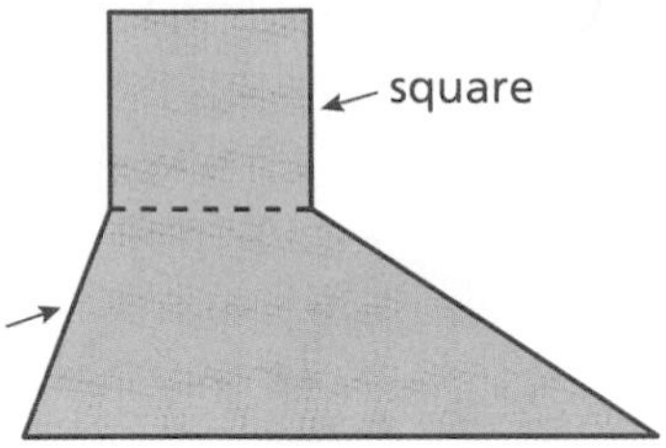

To find the area of a composite figure, separate it into figures with areas you know how to find. Then find the sum of the areas of those figures.

EXAMPLE 1 Finding the Area of a Composite Figure

Find the area of the purple figure.

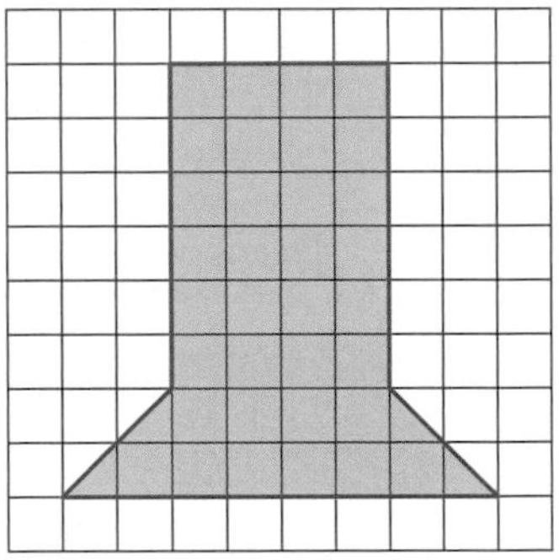

You can separate the figure into a rectangle and a trapezoid. Count grid lines to find the dimensions of each figure. Then find the area of each figure.

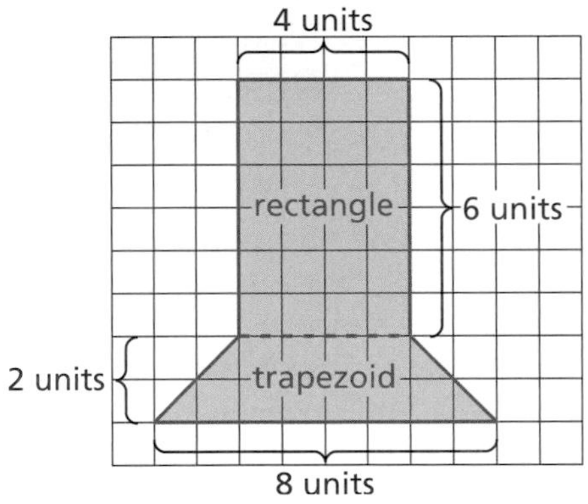

Area of Rectangle

$A = \ell w$

$= 6(4)$

$= 24$

Area of Trapezoid

$A = \frac{1}{2}h(b_1 + b_2)$

$= \frac{1}{2}(2)(4 + 8)$

$= 12$

So, the area of the purple figure is $24 + 12 = 36$ square units.

Reasonable? You can check your result by counting unit squares.

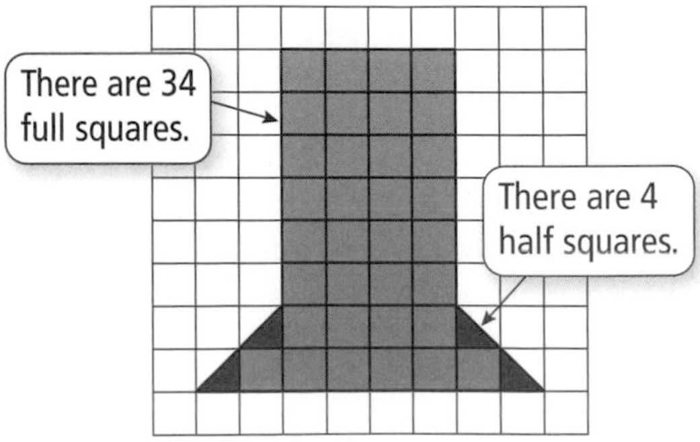

Full squares: 34

Half squares: 4

The area is

$$34(1) + 4\left(\frac{1}{2}\right) = 36 \text{ square units.}$$

So, the answer is reasonable. ✓

Study Tip
There is often more than one way to separate composite figures. In Example 1, you can separate the figure into one rectangle and two triangles.

Common Core
Geometry
In this extension, you will
- find areas of composite figures.
- solve real-life problems.

Applying Standard 6.G.1

EXAMPLE 2 Real-Life Application

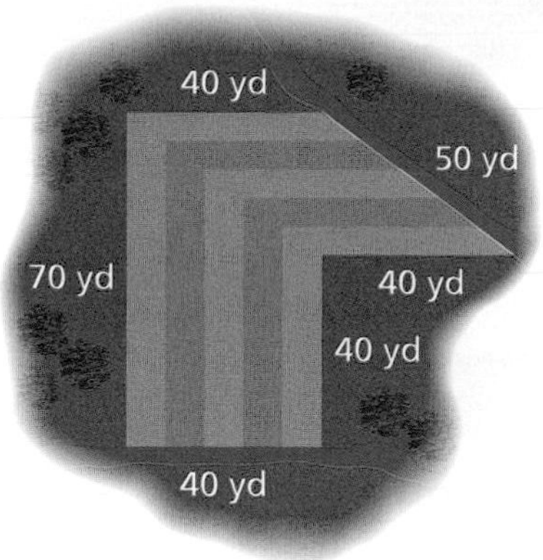

Find the area of the fairway between two streams on a golf course.

There are several ways to separate the fairway into figures whose areas you can find using formulas. It appears that one way is to separate it into a right triangle and a rectangle.

Identify each shape and find any missing dimensions.

The height of the right triangle is $70 - 40 = 30$ yards.

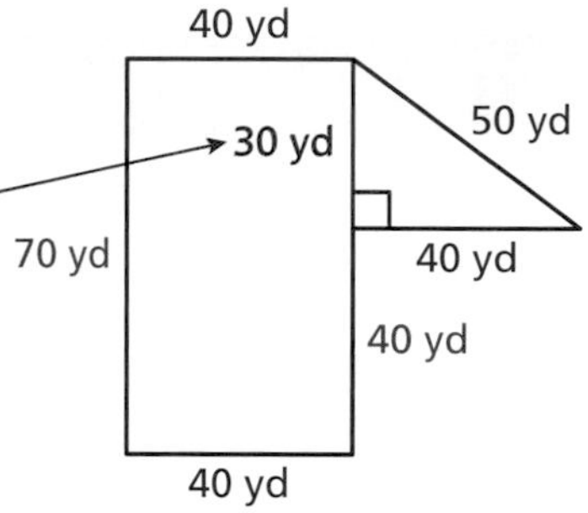

Area of Rectangle	***Area of Right Triangle***
$A = \ell w$	$A = \frac{1}{2}bh$
$= 70(40)$	$= \frac{1}{2}(40)(30)$
$= 2800$	$= 600$

So, the area of the fairway is $2800 + 600 = 3400$ square yards.

Practice

Find the area of the shaded figure.

1.

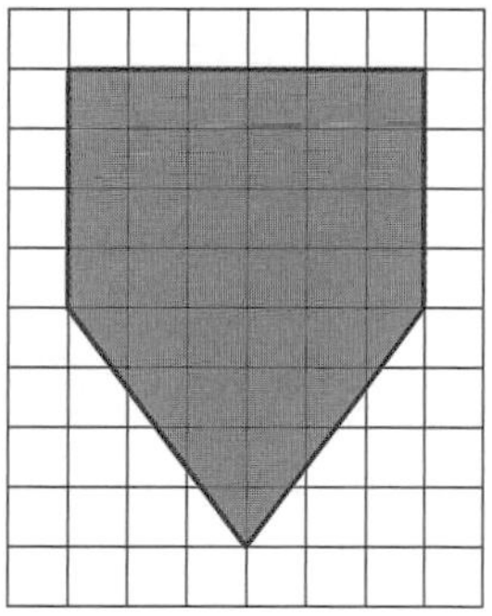

2.

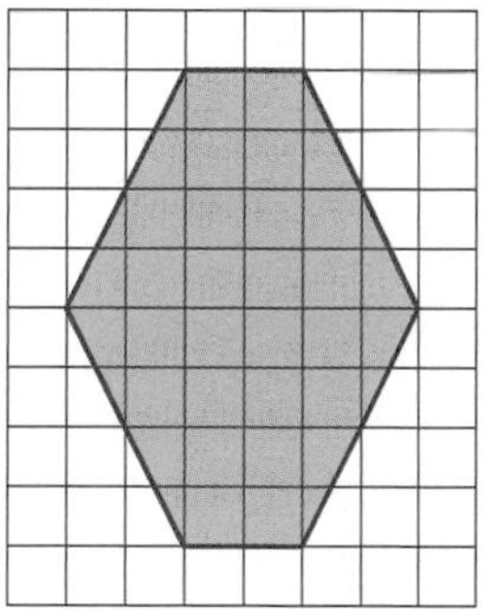

3.

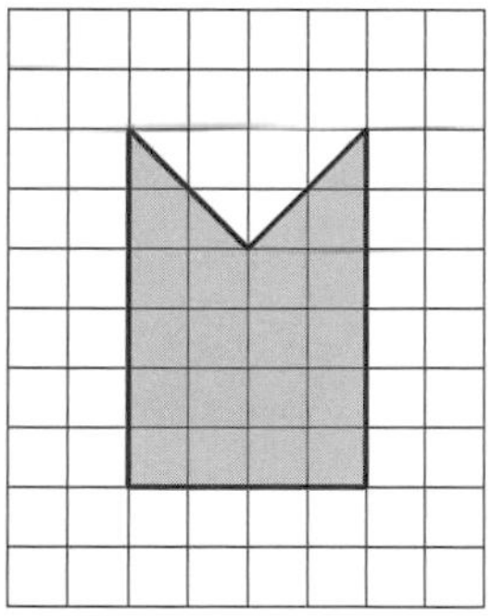

Find the area of the figure.

4.

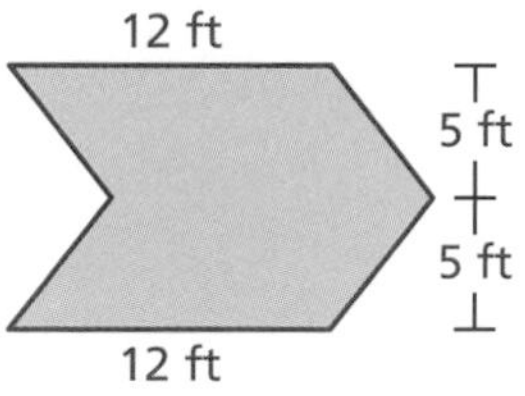

5.

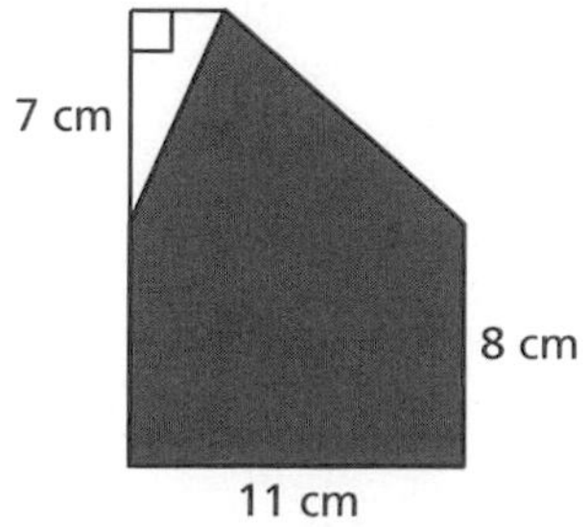

6.

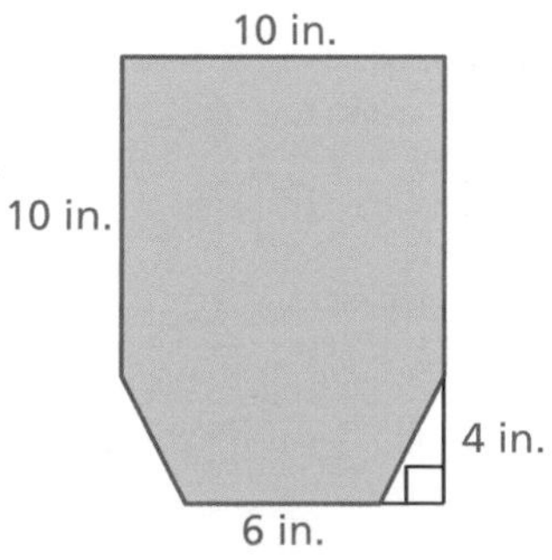

7. ANOTHER METHOD Find the area in Example 2 using a different method.

4.4 Polygons in the Coordinate Plane

Essential Question How can you find the lengths of line segments in a coordinate plane?

1 ACTIVITY: Finding Distances on a Map

Work with a partner. The coordinate grid shows a portion of a city. Each square on the grid represents one square mile.

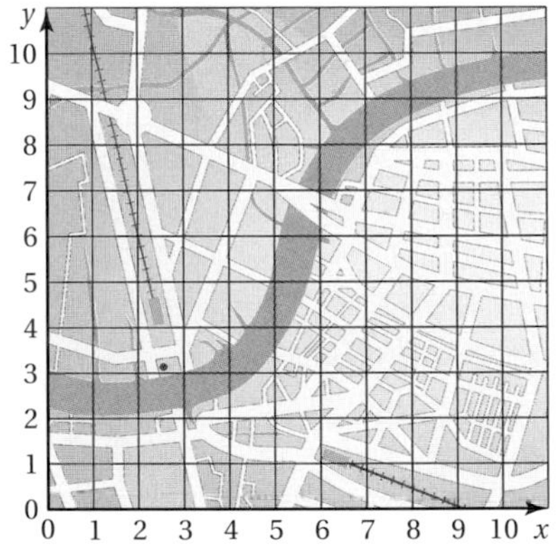

a. A public library is located at (4, 5). City Hall is located at (7, 5). Plot and label these points.

b. How far is the public library from City Hall?

c. A stadium is located 4 miles from the public library. Give the coordinates of several possible locations of the stadium. Justify your answers by graphing.

d. Connect the three locations of the public library, City Hall, and the stadium using your answers in part (c). What shapes are formed?

2 ACTIVITY: Graphing Polygons

Work with a partner. Plot and label each set of points in the coordinate plane. Then connect each set of points to form a polygon.

Rectangle: $A(2, 3)$, $B(2, 10)$, $C(6, 10)$, $D(6, 3)$

Triangle: $E(8, 3)$, $F(14, 8)$, $G(14, 3)$

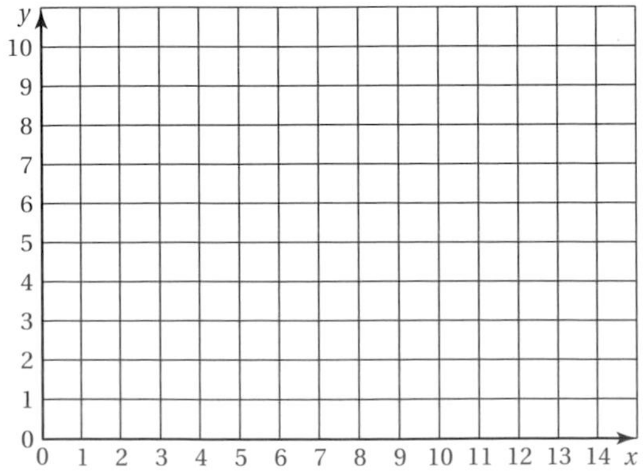

COMMON CORE

Geometry

In this lesson, you will
- draw polygons in the coordinate plane.
- find distances in the coordinate plane.
- solve real-life problems.

Learning Standard
6.G.3

3 ACTIVITY: Finding Distances in a Coordinate Plane

Work with a partner.

a. Find the length of each horizontal line segment in Activity 2.

b. **STRUCTURE** What relationship do you notice between the lengths of the line segments in part (a) and the coordinates of their endpoints? Explain.

c. Find the length of each vertical line segment in Activity 2.

d. **STRUCTURE** What relationship do you notice between the lengths of the line segments in part (c) and the coordinates of their endpoints? Explain.

e. Plot and label the points below in the coordinate plane. Then connect each pair of points with a line segment. Use the relationships you discovered in parts (b) and (d) above to find the length of each line segment. Show your work.

$S(3, 1)$ and $T(14, 1)$ $\quad$ $U(9, 8)$ and $V(9, 0)$

$W(0, 7)$ and $X(0, 10)$ $\quad$ $Y(1, 9)$ and $Z(7, 9)$

f. Check your answers in part (e) by counting grid lines.

Math Practice 8

Repeat Calculations

What calculations are repeated? How can you use this information to write a rule about the length of a line segment?

What Is Your Answer?

4. **IN YOUR OWN WORDS** How can you find the lengths of line segments in a coordinate plane? Give examples to support your explanation.

5. Do the methods you used in Activity 3 work for diagonal line segments? Explain why or why not.

6. Use the Internet or some other reference to find an example of how "finding distances in a coordinate plane" is helpful in each of the following careers.

a.
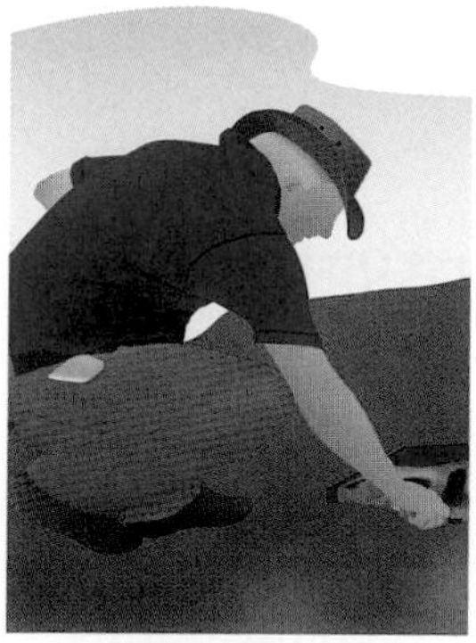
Archaeologist

b.

Surveyor

c.

Pilot

Practice Use what you learned about finding the lengths of line segments to complete Exercises 3–5 on page 178.

4.4 Lesson

Check It Out
Lesson Tutorials
BigIdeasMath.com

You can use ordered pairs to represent vertices of polygons. To draw a polygon in a coordinate plane, plot and connect the ordered pairs.

EXAMPLE 1 Drawing a Polygon in a Coordinate Plane

The vertices of a quadrilateral are $A(2, 4)$, $B(3, 9)$, $C(7, 8)$, and $D(8, 1)$. Draw the quadrilateral in a coordinate plane.

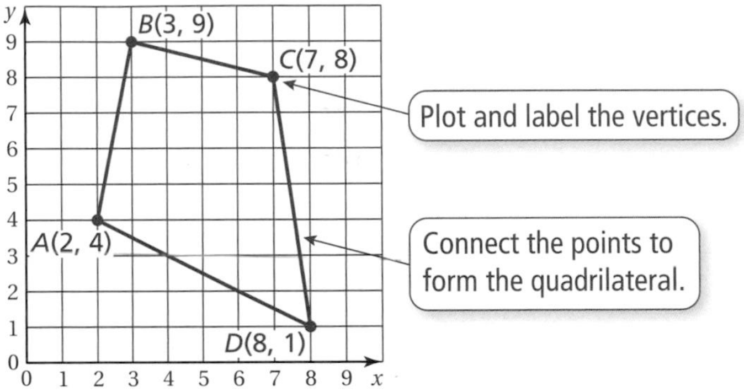

Study Tip

After you plot the vertices, connect them *in order* to draw the polygon.

On Your Own

Now You're Ready
Exercises 6–11

Draw the polygon with the given vertices in a coordinate plane.

1. $A(0, 0)$, $B(5, 7)$, $C(7, 4)$
2. $W(4, 4)$, $X(7, 4)$, $Y(7, 1)$, $Z(4, 1)$
3. $F(1, 3)$, $G(3, 6)$, $H(5, 6)$, $J(3, 3)$
4. $P(1, 4)$, $Q(3, 5)$, $R(7, 3)$, $S\left(6, \frac{1}{2}\right)$, $T\left(2, \frac{1}{2}\right)$

Key Idea

Finding Distances in the First Quadrant

You can find the length of a horizontal or vertical line segment in a coordinate plane by using the coordinates of the endpoints.

- When the x-coordinates are the same, the vertical distance between the points is the difference of the y-coordinates.
- When the y-coordinates are the same, the horizontal distance between the points is the difference of the x-coordinates.

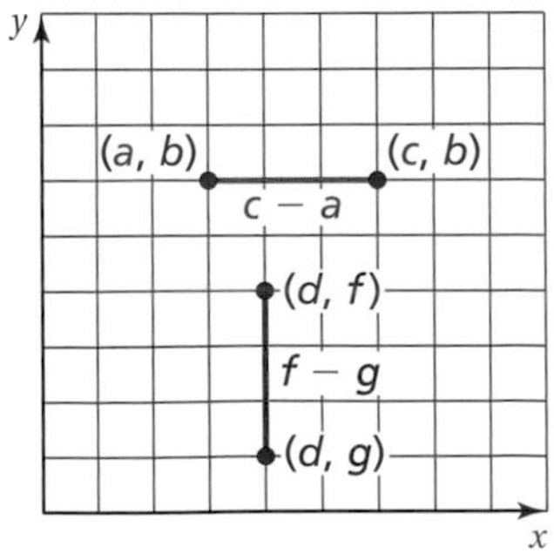

Be sure to subtract the lesser coordinate from the greater coordinate.

EXAMPLE 2 Finding a Perimeter

The vertices of a rectangle are $F(1, 6)$, $G(7, 6)$, $H(7, 2)$, and $J(1, 2)$. Draw the rectangle in a coordinate plane and find its perimeter.

Draw the rectangle and use the vertices to find its dimensions.

Study Tip

You can also find the length using vertices H and J. You can find the width using vertices F and J.

The length is the horizontal distance between $F(1, 6)$ and $G(7, 6)$, which is the difference of the x-coordinates.

length $= 7 - 1 = 6$ units

The width is the vertical distance between $G(7, 6)$ and $H(7, 2)$, which is the difference of the y-coordinates.

width $= 6 - 2 = 4$ units

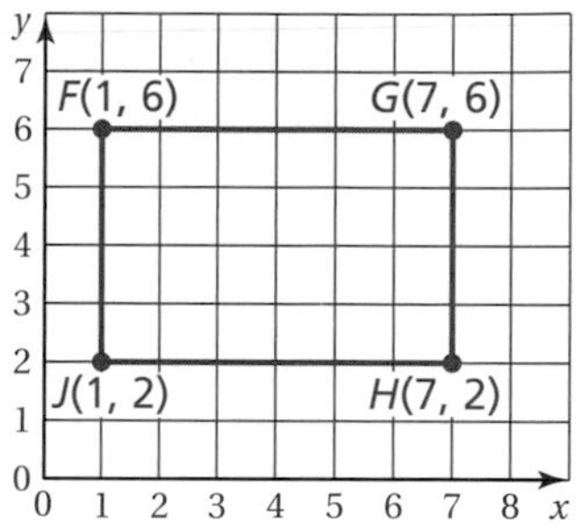

So, the perimeter of the rectangle is $2(6) + 2(4) = 20$ units.

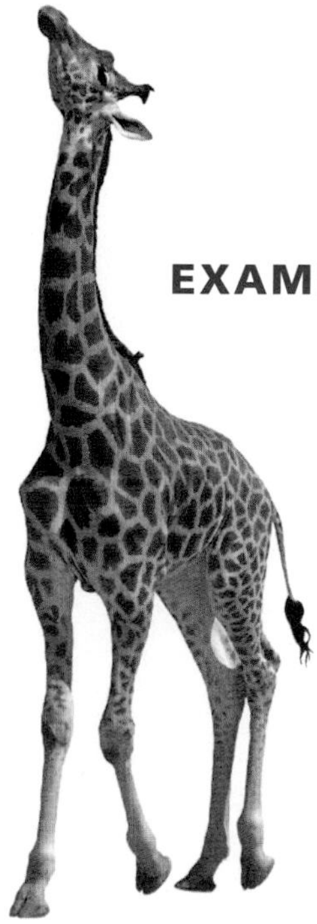

EXAMPLE 3 Real-Life Application

In a grid of the exhibits at a zoo, the vertices of the giraffe exhibit are $E(0, 90)$, $F(60, 90)$, $G(100, 30)$, and $H(0, 30)$. The coordinates are measured in feet. What is the area of the giraffe exhibit?

Plot and connect the vertices using a coordinate grid to form a trapezoid. Use the coordinates to find the lengths of the bases and the height.

$b_1 = 60 - 0 = 60$

$b_2 = 100 - 0 = 100$

$h = 90 - 30 = 60$

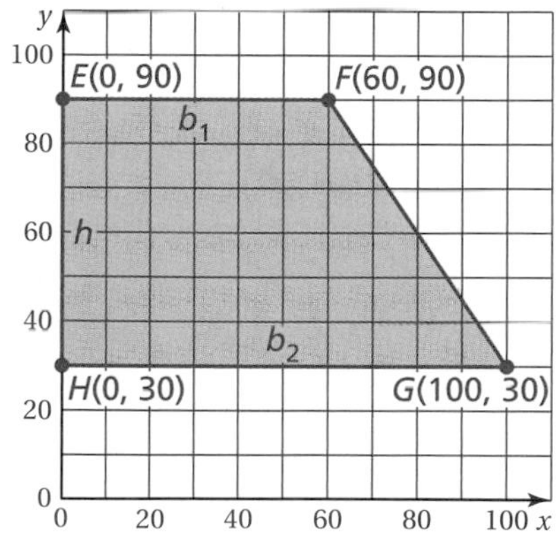

Use the formula for the area of a trapezoid.

$$A = \frac{1}{2}(60)(60 + 100)$$

$$= \frac{1}{2}(60)(160) = 4800$$

The area of the giraffe exhibit is 4800 square feet.

Common Error

You can count grid lines to find the dimensions, but make sure you consider the scale of the axes.

On Your Own

Now You're Ready
Exercises 12–15

5. The vertices of a rectangle are $J(2, 7)$, $K(4, 7)$, $L(4, 1.5)$, and $M(2, 1.5)$. Find the perimeter and the area of the rectangle.
6. **WHAT IF?** In Example 3, the giraffe exhibit is enlarged by moving vertex F to $(80, 90)$. How does this affect the area? Explain.

4.4 Exercises

Vocabulary and Concept Check

1. **WRITING** How can you use a coordinate plane to draw a polygon?
2. **WRITING** How can you find the perimeter of a rectangle in a coordinate plane?

Practice and Problem Solving

Plot and label each pair of points in a coordinate plane. Find the length of the line segment connecting the points.

3. $C(0, 1), D(8, 1)$
4. $K(5, 2), L(5, 6)$
5. $Q(3, 4), R(3, 9)$

Draw the polygon with the given vertices in a coordinate plane.

1 6. $A(4, 7), B(6, 2), C(0, 0)$

7. $D\left(\frac{1}{2}, 2\right), E(5, 5), F(4, 1)$

8. $G\left(1\frac{1}{2}, 4\right), H\left(1\frac{1}{2}, 8\right), J(5, 8), K(5, 4)$

9. $L(3, 2), M(3, 5), N(9, 5), P(9, 2)$

10. $Q(0, 4), R(10, 8), S(7, 4), T(10, 2), U(5, 0)$

11. $V(2, 2), W\left(3, 7\frac{1}{2}\right), X\left(8, 7\frac{1}{2}\right), Y(10, 4), Z(7, 0)$

Find the perimeter and the area of the polygon with the given vertices.

2 12. $C(1, 1), D(1, 4), E(4, 4), F(4, 1)$

13. $J(1, 2), K(7, 2), L(7, 8), M(1, 8)$

14. $N(0, 2), P(5, 2), Q(5, 5), R(0, 5)$

15. $S(3, 0), T(3, 9), U(8, 9), V(8, 0)$

16. **ERROR ANALYSIS** Describe and correct the error in drawing a triangle with vertices $A(5, 1)$, $B(7, 6)$, and $C(1, 3)$.

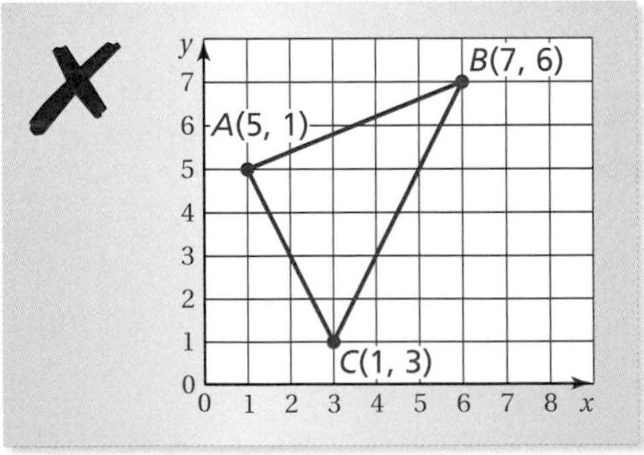

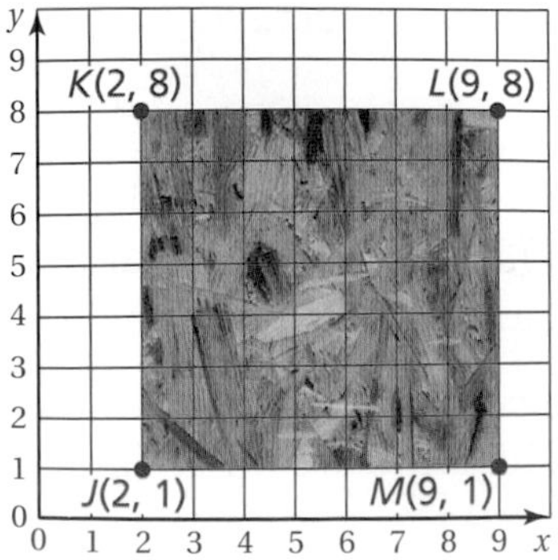

17. **TREE HOUSE** You design a tree house using a coordinate plane. You plot the vertices of the floor at $J(2, 1)$, $K(2, 8)$, $L(9, 8)$, and $M(9, 1)$. The coordinates are measured in feet.
    a. What is the shape of the floor?
    b. What are the perimeter and the area of the floor?

OPEN-ENDED Draw a polygon with the given conditions in a coordinate plane.

18. a square with a perimeter of 20 units

19. a rectangle with a perimeter of 18 units

20. a rectangle with an area of 24 units^2

21. a triangle with an area of 15 units^2

22. **STRUCTURE** The coordinate plane shows three vertices of a parallelogram. Find two possible points that could represent the fourth vertex.

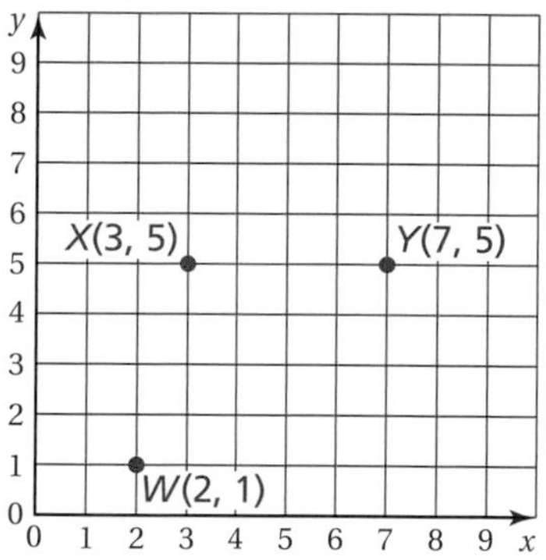

23. **BUS ROUTE** Polygon *JKLMNP* represents a bus route. Each grid square represents 9 square miles. What is the shortest distance, in miles, from station *P* to station *L* using the bus route? Explain your reasoning.

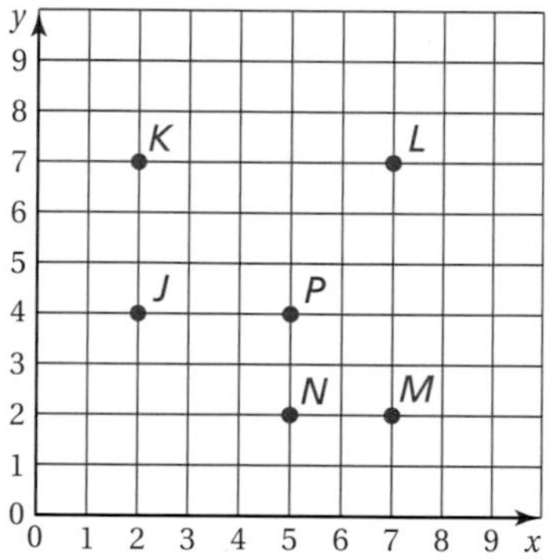

24. **CITY LIMITS** In a topographical map of a city, the vertices of the city limits are $A(10, 9)$, $B(18, 9)$, $C(18, 2)$, $D(14, 4.5)$, and $E(10, 4.5)$. The coordinates are measured in miles. What is the area of the city?

25. **BACKYARD** The vertices of a backyard are $W(10, 30)$, $X(10, 100)$, $Y(110, 100)$, and $Z(50, 30)$. The coordinates are measured in feet. The line segment XZ separates the backyard into a lawn and a garden. The area of the lawn is greater than the area of the garden. How many times larger is the lawn than the garden?

26. **Precision** The vertices of a rectangle are $(1, 0)$, $(1, a)$, $(5, a)$, and $(5, 0)$. The vertices of a parallelogram are $(1, 0)$, $(2, b)$, $(6, b)$, and $(5, 0)$. The value of a is greater than the value of b. Which polygon has a greater area? Explain your reasoning.

Fair Game Review What you learned in previous grades & lessons

Divide. Write the answer in simplest form. *(Section 2.3)*

27. $1\frac{1}{3} \div \frac{2}{3}$

28. $6\frac{3}{5} \div \frac{3}{4}$

29. $2\frac{1}{2} \div 8$

30. $4\frac{1}{6} \div 1\frac{1}{8}$

31. **MULTIPLE CHOICE** You are filling bottles from 5 gallons of lemonade. How many bottles can you fill when each bottle is $\frac{3}{8}$ of a gallon? *(Section 2.2)*

Ⓐ $1\frac{7}{8}$ Ⓑ 3 Ⓒ 8 Ⓓ $13\frac{1}{3}$

4.3–4.4 Quiz

Check It Out
Progress Check
BigIdeasMath.com

Find the area of the trapezoid. *(Section 4.3)*

1.

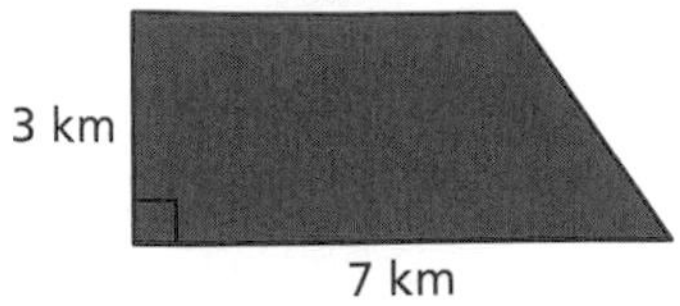

2.

3.

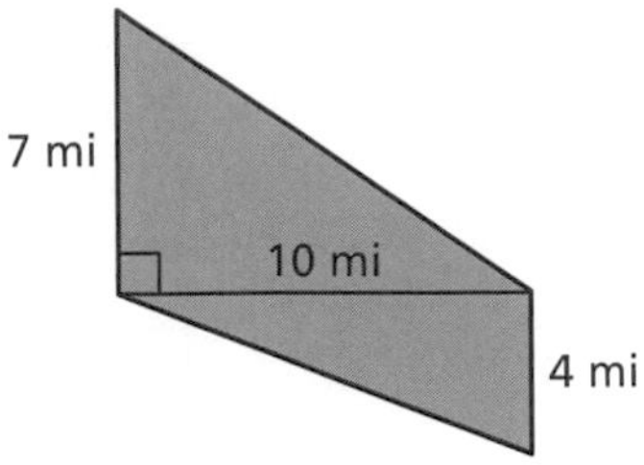

Find the area of the figure. *(Section 4.3)*

4.

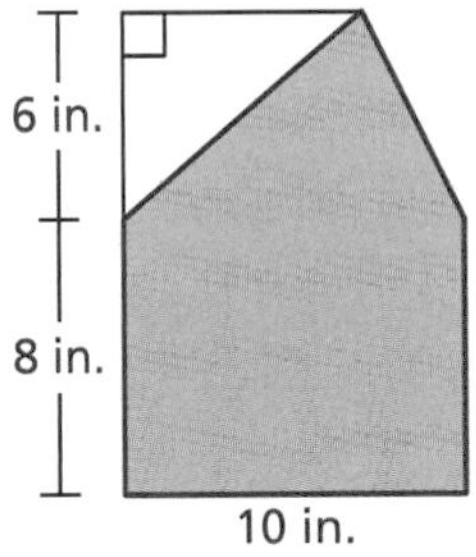

5.

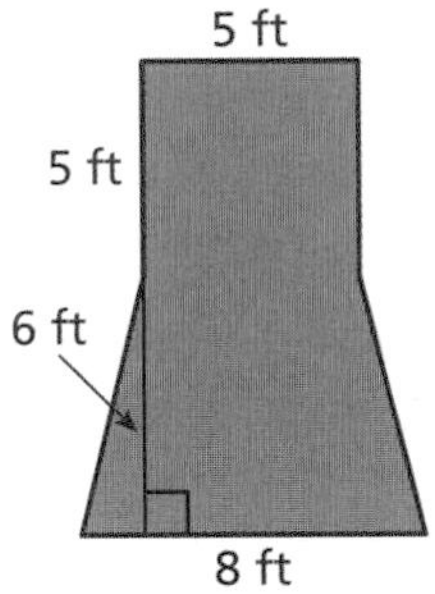

6.

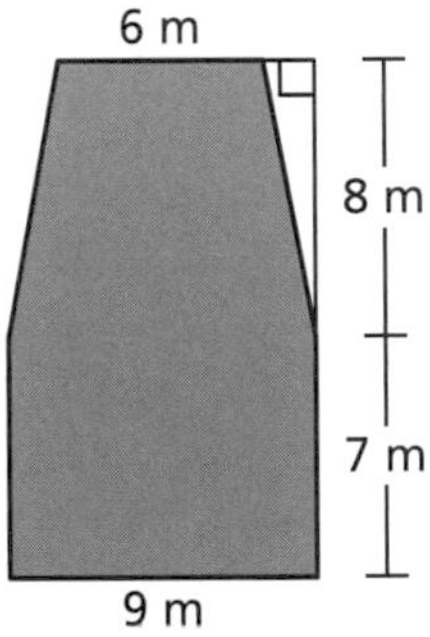

Draw the polygon with the given vertices in a coordinate plane. *(Section 4.4)*

7. $A(1, 2)$, $B(3, 5)$, $C(6, 1)$

8. $E(1, 2)$, $F(3, 6)$, $G(8, 6)$, $H(6, 2)$

Find the perimeter and the area of the polygon with the given vertices. *(Section 4.4)*

9. $J(1, 3)$, $K(1, 8)$, $L(5, 8)$, $M(5, 3)$

10. $P(1, 2)$, $Q(1, 7)$, $R(7, 7)$, $S(7, 2)$

11. BACK POCKET How much material do you need to make two back pockets? *(Section 4.3)*

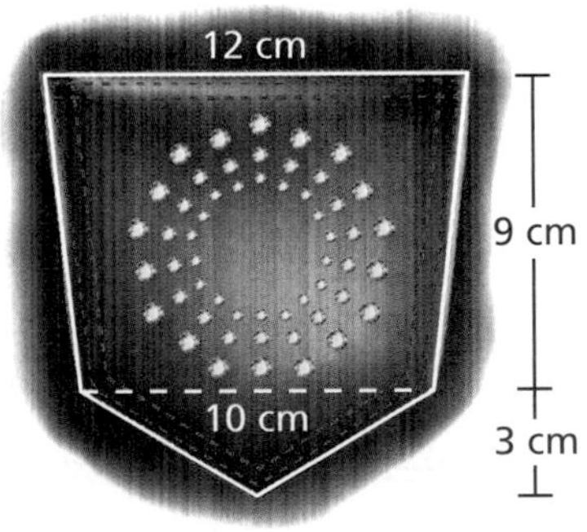

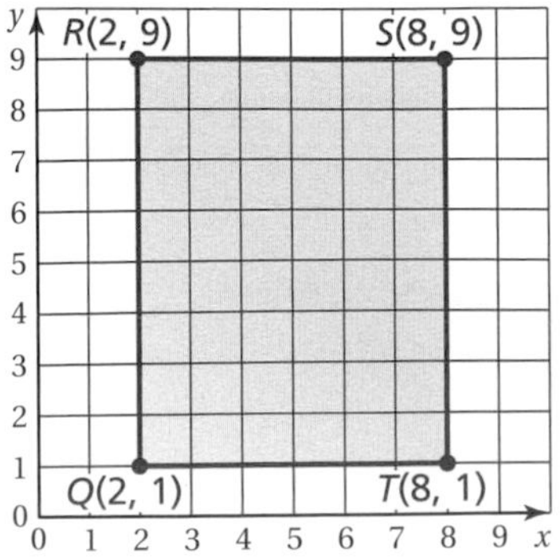

12. PATIO Plans for a patio are shown in the coordinate plane at the left. The coordinates are measured in feet. Find the perimeter and the area of the patio. *(Section 4.4)*

4 Chapter Review

Check It Out
Vocabulary Help
BigIdeasMath.com

Review Key Vocabulary

polygon, *p. 152* composite figure, *p. 172*

Review Examples and Exercises

4.1 Areas of Parallelograms *(pp. 152–157)*

Find the area of the parallelogram.

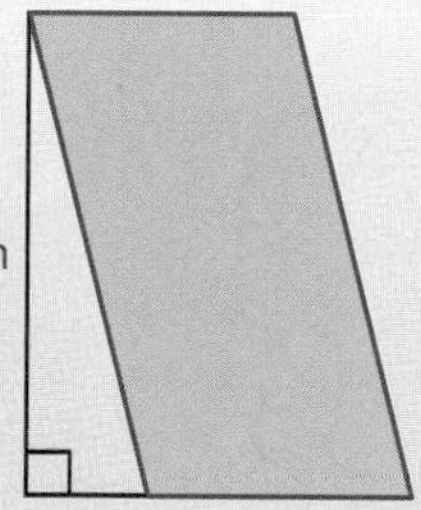

$A = bh$ Write formula.

$= 5(9)$ Substitute 5 for *b* and 9 for *h*.

$= 45$ Multiply.

∴ The area of the parallelogram is 45 square centimeters.

Exercises

Find the area of the parallelogram.

1.

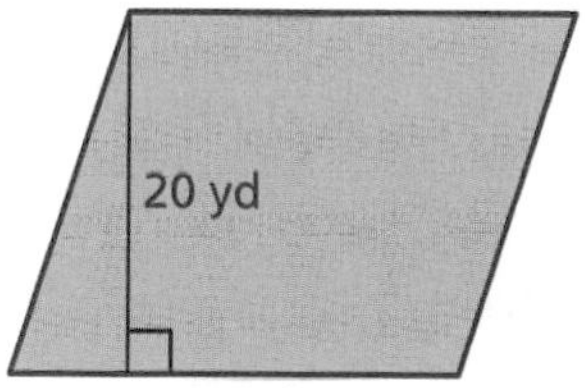

2.

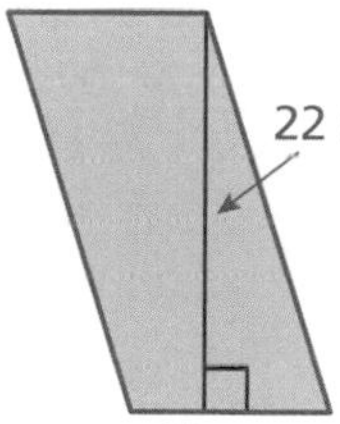

4.2 Areas of Triangles *(pp. 158–163)*

Find the area of the triangle.

$A = \frac{1}{2}bh$ Write formula.

$= \frac{1}{2}(10)(7)$ Substitute.

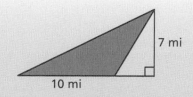

$= 35$ Multiply.

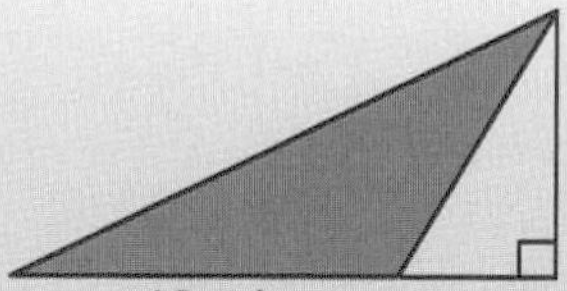

∴ The area of the triangle is 35 square miles.

Exercises

Find the area of the triangle.

3.

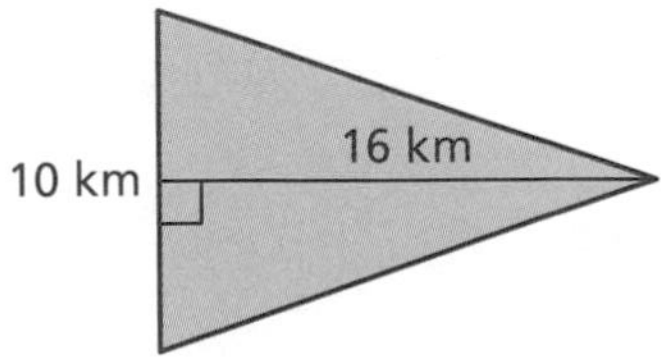

4.

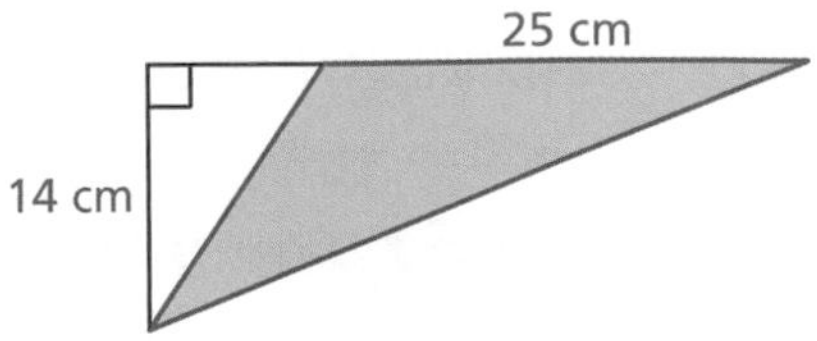

4.3 Areas of Trapezoids *(pp. 166–173)*

Find the area of the trapezoid.

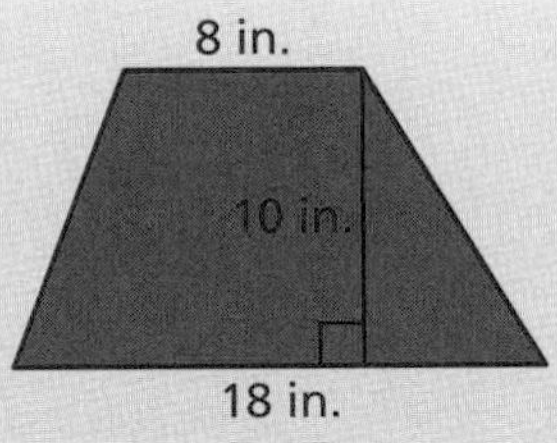

$A = \frac{1}{2}h(b_1 + b_2)$ Write formula.

$= \frac{1}{2}(10)(8 + 18)$ Substitute.

$= \frac{1}{2}(10)(26) = 130$ Multiply.

∴ The area of the trapezoid is 130 square inches.

Exercises

Find the area of the trapezoid.

5.

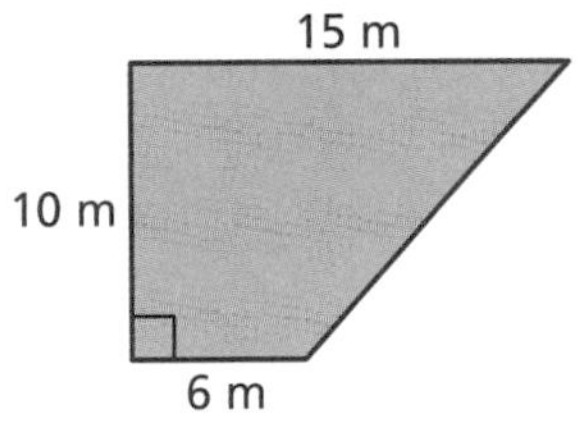

6.

7.

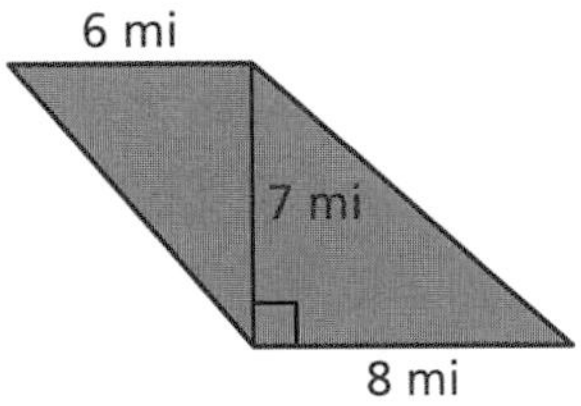

Find the area of the figure.

8.

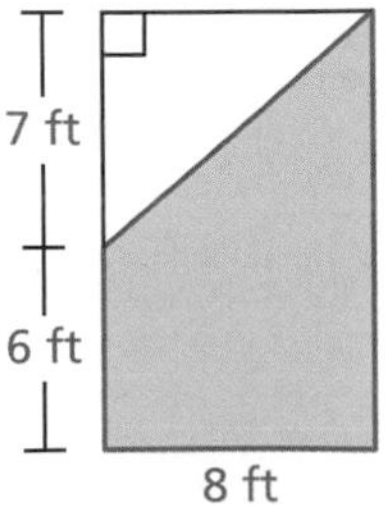

9.

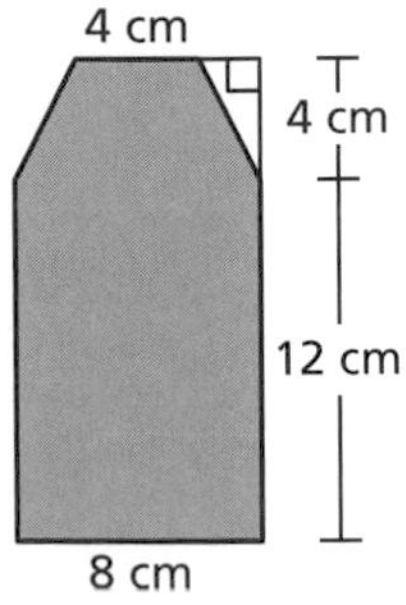

10.

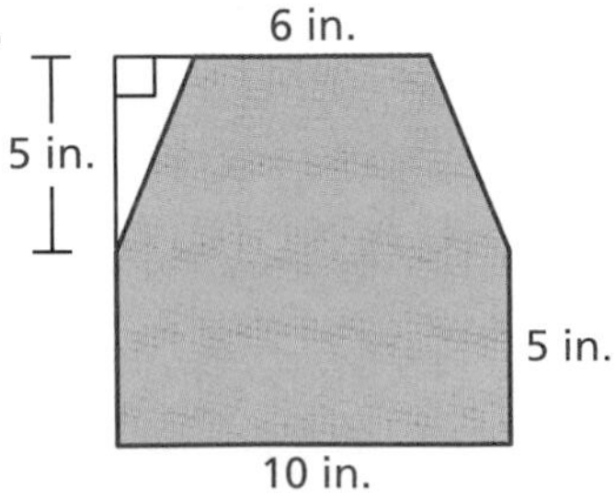

4.4 Polygons in the Coordinate Plane *(pp. 174–179)*

a. The vertices of a triangle are $A(1, 3)$, $B(5, 9)$, and $C(8, 2)$. Draw the triangle in a coordinate plane.

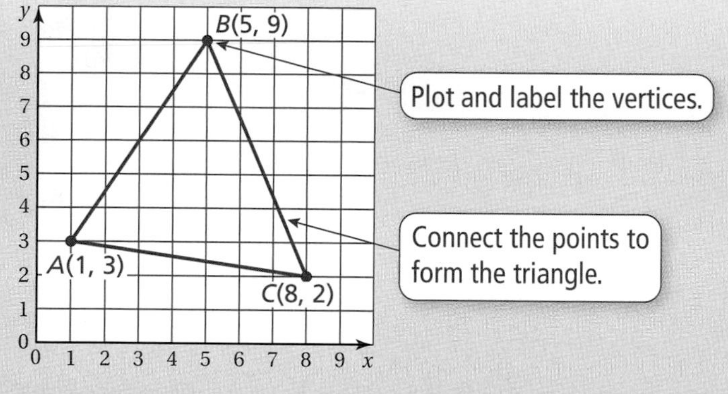

b. The vertices of a rectangle are $F(2, 6)$, $G(8, 6)$, $H(8, 1)$, and $J(2, 1)$. Draw the rectangle in a coordinate plane and find its perimeter.

Draw the rectangle and use the vertices to find its dimensions.

The length is the horizontal distance between $F(2, 6)$ and $G(8, 6)$, which is the difference of the x-coordinates.

length = 8 − 2 = 6 units

The width is the vertical distance between $G(8, 6)$ and $H(8, 1)$, which is the difference of the y-coordinates.

width = 6 − 1 = 5 units

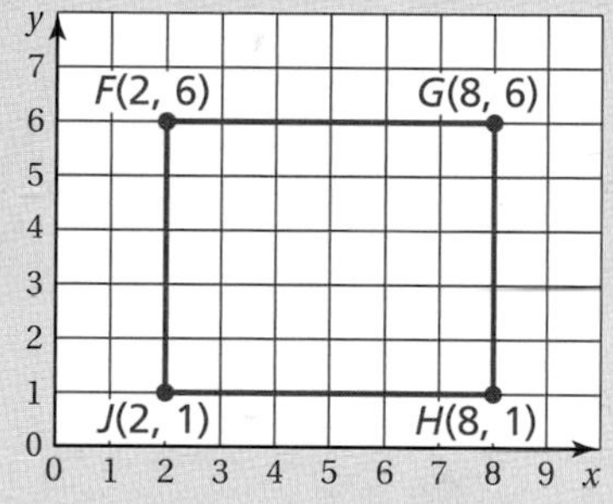

∴ So, the perimeter of the rectangle is 2(6) + 2(5) = 22 units.

Exercises

Draw the polygon with the given vertices in a coordinate plane.

11. $A(3, 2)$, $B(4, 7)$, $C(6, 0)$

12. $D(1, 1)$, $E(1, 5)$, $F(4, 5)$, $G(4, 1)$

13. $J(1, 2)$, $K(1, 7)$, $L(5, 7)$, $M(8, 2)$

14. $K\left(3, 3\frac{1}{2}\right)$, $L(5, 7)$, $M(8, 7)$, $N\left(6, 3\frac{1}{2}\right)$

Find the perimeter and the area of the polygon with the given vertices.

15. $P(4, 3)$, $Q(4, 7)$, $R(9, 7)$, $S(9, 3)$

16. $T(2, 7)$, $U(2, 9)$, $V(5, 9)$, $W(5, 7)$

17. $W(11, 2)$, $X(11, 8)$, $Y(14, 8)$, $Z(14, 2)$

18. $A(12, 2)$, $B(12, 13)$, $C(15, 13)$, $D(15, 2)$

4 Chapter Test

Check It Out
Test Practice
BigIdeasMath.com

Find the area of the parallelogram, triangle, or trapezoid.

1.

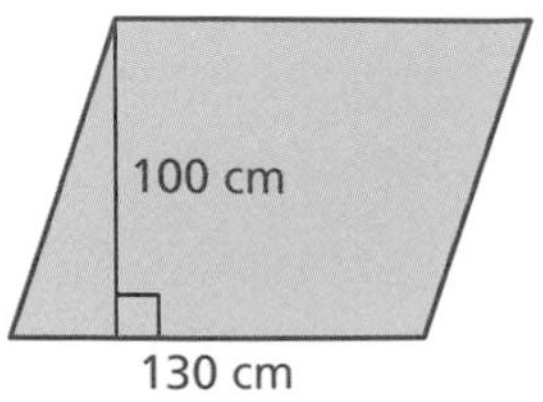

2.

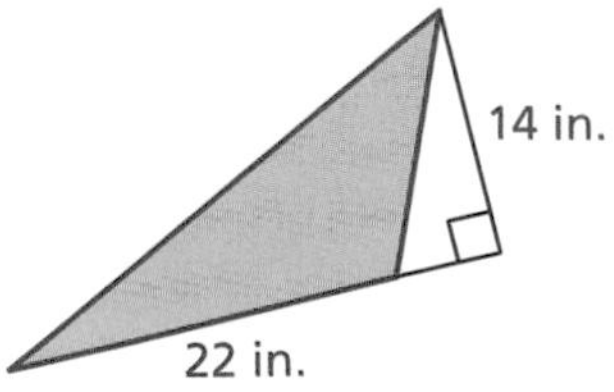

3.

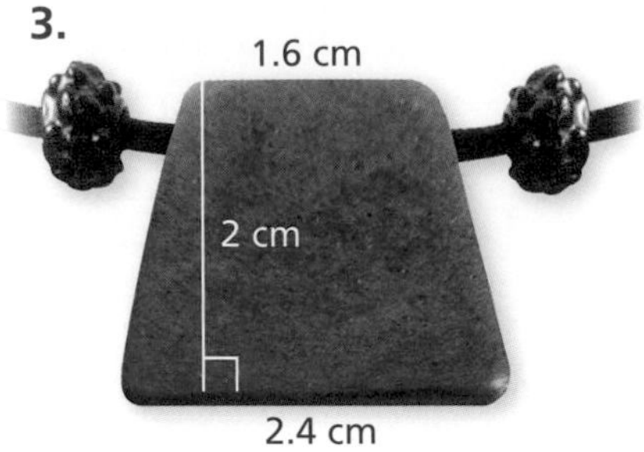

Find the area of the figure.

4.

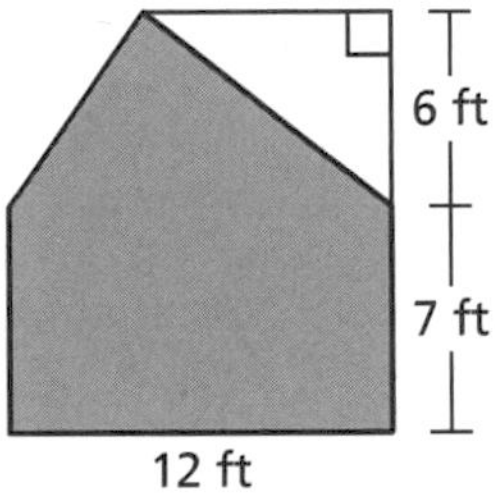

5.

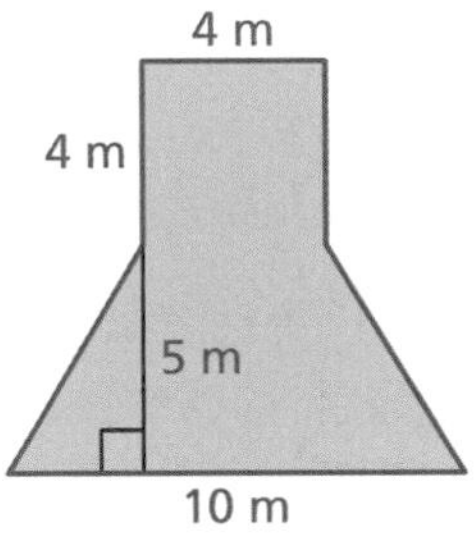

6.

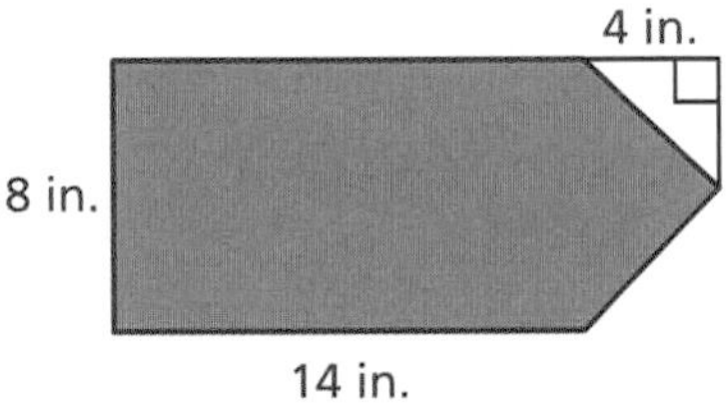

Draw the polygon with the given vertices in a coordinate plane.

7. $A(4, 2)$, $B(5, 6)$, $C(7, 4)$

8. $D(3, 4)$, $E(5, 8)$, $F(8, 8)$, $G(6, 4)$

Find the perimeter and the area of the polygon with the given vertices.

9. $Q(5, 6)$, $R(5, 10)$, $S(9, 10)$, $T(9, 6)$

10. $W(2, 8)$, $X(2, 16)$, $Y(8, 16)$, $Z(8, 8)$

11. TABLETOP The base lengths of a trapezoidal tabletop are 6 feet and 8 feet. The height is 5 feet. What is the area of the tabletop?

12. PENTAGON The Pentagon in Arlington, Virginia, is the headquarters of the U.S. Department of Defense.

a. Find the perimeter of the Pentagon.

b. A pentagon is made of a triangle and a trapezoid. The height of the triangle shown is about 541 feet, and the height of the trapezoid shown is about 876 feet. Estimate the land area of the Pentagon.

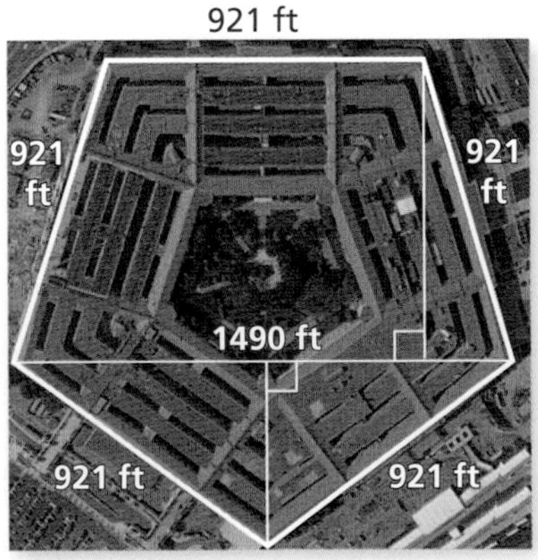

13. CAMPING The vertices of a campsite are (25, 15), (25, 30), (55, 30), and (55, 15). The vertices of your tent are (30, 20), (30, 25), (40, 25), and (40, 20). The coordinates are measured in feet. What is the area of the campsite not covered by your tent?

4 Standards Assessment

1. What is the area of the shaded figure shown below? *(6.G.1)*

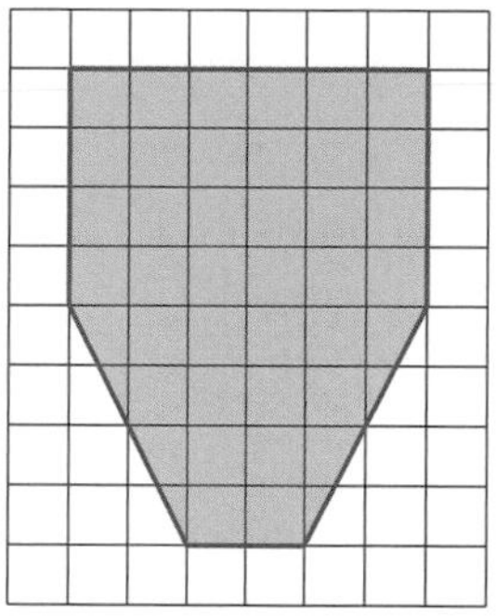

A. 32 $units^2$

B. 40 $units^2$

C. 44 $units^2$

D. 56 $units^2$

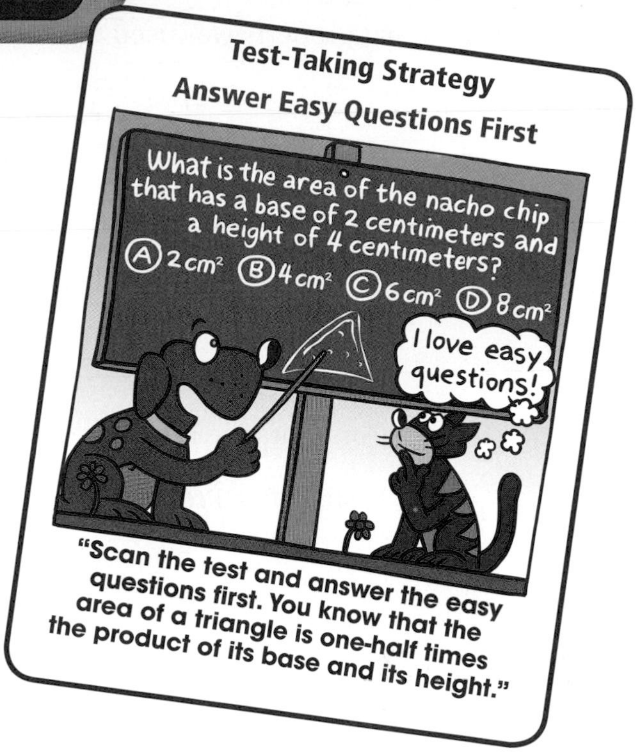

Test-Taking Strategy
Answer Easy Questions First
What is the area of the nacho chip that has a base of 2 centimeters and a height of 4 centimeters?
(A) 2 cm² (B) 4 cm² (C) 6 cm² (D) 8 cm²
I love easy questions!
"Scan the test and answer the easy questions first. You know that the area of a triangle is one-half times the product of its base and its height."

2. What is the value of the expression below? *(6.EE.1)*

$$18^3$$

3. You have 36 red apples and 42 green apples. What is the greatest number of identical fruit baskets you can make with no apples left over? *(6.NS.4)*

F. 6

G. 9

H. 12

I. 18

4. What is the perimeter of the rectangle with the vertices shown below? *(6.G.3)*

$$A(4, 7), B(4, 15), C(9, 15), D(9,7)$$

A. 8 units

B. 13 units

C. 26 units

D. 70 units

5. What property was used to simplify the expression? *(6.EE.3)*

$$\begin{aligned} 5 \times 78 &= 5(70 + 8) \\ &= 5(70) + 5(8) \\ &= 350 + 40 \\ &= 390 \end{aligned}$$

F. Associative Property of Multiplication

G. Commutative Property of Addition

H. Distributive Property

I. Multiplication Property of One

6. What is the area, in square yards, of the triangle below? *(6.G.1)*

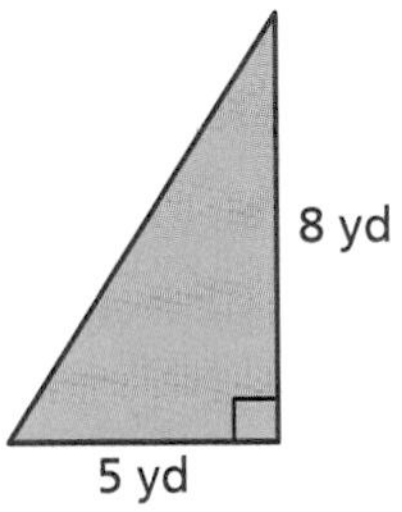

7. Which of the following is equivalent to $\frac{12}{35}$? *(6.NS.1)*

A. $\frac{5}{6} \div \frac{2}{7}$

B. $\frac{2}{7} \div \frac{6}{5}$

C. $\frac{2}{7} \div \frac{5}{6}$

D. $\frac{5}{6} \div \frac{7}{2}$

8. The description below represents the area of which polygon? *(6.G.1)*

"one-half the product of its height and the sum of the lengths of its bases"

F. rectangle

G. square

H. trapezoid

I. triangle

9. Edward was evaluating the expression in the box.

$$\begin{aligned} 180 \div 9 + 3^4 - 1 &= 180 \div 9 + 81 - 1 \\ &= 180 \div 90 - 1 \\ &= 2 - 1 \\ &= 1 \end{aligned}$$

What should Edward do to correct the error that he made? *(6.EE.1)*

A. Add 9 and 81 then subtract 1 before dividing.

B. Divide 180 by 9 before adding or subtracting.

C. Divide 180 by 9 then subtract 1 before adding 3^4.

D. Subtract 1 from 90 before dividing.

10. You have 3 times as many guitar picks as your cousin. Let v be the number of guitar picks that your cousin has. Which expression represents the number of guitar picks you have? *(6.EE.2a)*

F. $3v$

G. $v + 3$

H. $3 - v$

I. $\frac{v}{3}$

11. Your family hires a company to install invisible fencing around your yard. *(6.G.1)*

Think
Solve
Explain

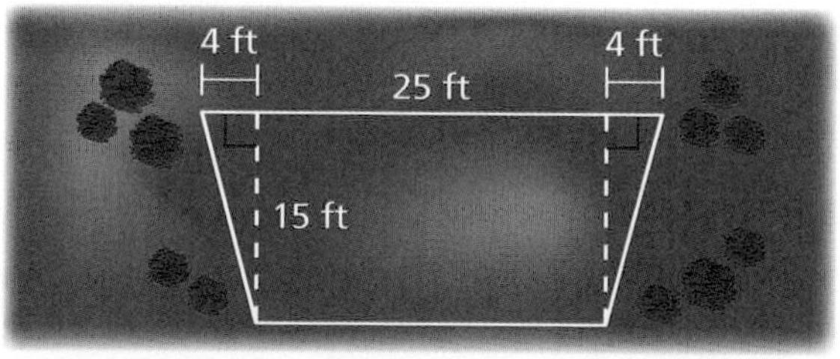

Part A Find the area of the yard using only the area formulas for rectangles and triangles. Show your work.

Part B Find the area of the yard using the area formula for trapezoids.

Part C Explain why the two methods of finding the area of the yard give the same result. Describe the advantages of each method.

5 Ratios and Rates

It's good to keep records.
"By my records, I ate 1460 dog biscuits last year."
But, this is pushing it.
"So I calculated that my unit rate is 4 biscuits per day."

"It says 75% tomatoes, 15% sugar, 5% vinegar, 4% water, and 1% salt."
That's a relief!
"See... no cats in catsup."

What You Learned Before

AromaTherapy for Dogs
Bacon Cheese Chicken Roses
No kidding.
"I like seventy-five percent of them."

Identifying Patterns (5.OA.3)

Example 1 Using the numbers from the table, find and state the rule in words. Then find the missing value.

x	y
1	6
2	12
3	18
4	

Each y-value is 6 times the x-value.

∴ The x-value times 6 equals the y-value. The missing value is $6(4) = 24$.

Try It Yourself

Using the numbers from the table, find and state the rule in words. Then find the missing value.

1.

x	y
1	2
3	6
5	10
7	

2.

x	y
2	8
4	16
6	24
8	

3.

x	y
1	5
2	10
3	15
4	

Multiplying and Dividing by Fractions (5.NF.4a, 6.NS.1)

Example 2 Find $\frac{5}{6} \cdot \frac{3}{4}$.

$$\frac{5}{6} \cdot \frac{3}{4} = \frac{5 \cdot \overset{1}{\cancel{3}}}{\underset{2}{\cancel{6}} \cdot 4}$$

$$= \frac{5}{8}$$

Example 3 Find $2 \div \frac{9}{10}$.

$$2 \div \frac{9}{10} = 2 \cdot \frac{10}{9}$$

Multiply by the reciprocal of the divisor.

$$= \frac{2 \cdot 10}{9}$$

$$= \frac{20}{9}$$

Try It Yourself

Evaluate the expression. Write the answer in simplest form.

4. $\frac{1}{5} \cdot \frac{13}{20}$

5. $\frac{3}{4} \cdot \frac{13}{25}$

6. $7 \div \frac{9}{10}$

7. $4 \div \frac{16}{17}$

5.1 Ratios

Essential Question How can you represent a relationship between two quantities?

1 ACTIVITY: Comparing Quantities

Work with a partner. Use the collection of objects to complete each statement.

There are ____ graphing calculators to ____ protractors.

There are ____ protractors to ____ graphing calculators.

There are ____ compasses to ____ protractors.

There are ____ graphing calculators to ____ compasses.

There are ____ protractors to ____ total objects.

The number of graphing calculators is $\frac{\square}{\square}$ of the total number of objects.

2 ACTIVITY: Playing Garbage Basketball

Work with a partner.

- **Take turns shooting a ball or other object into a wastebasket from a reasonable distance.**
- **Organize the numbers of shots you made and shots you missed in a chart.**

a. Write a statement similar to those in Activity 1 that describes the relationship between the number of shots you made and the number of shots you missed.

b. Write a statement similar to those in Activity 1 that describes the relationship between the number of shots you made and the total number of shots.

c. What fraction of your shots did you make? What fraction did you miss?

COMMON CORE

Ratios

In this lesson, you will
- understand the concept of a ratio.
- use ratios to describe the relationship between two quantities.

Learning Standard
6.RP.1

3 ACTIVITY: Reading a Diagram

Work with a partner. You mix different amounts of paint to create new colors. Write a statement that describes the relationship between the amounts of paint shown in each diagram.

a. Blue 

Green 

There are ___ parts blue for every ___ parts green.

b. Orange 

Yellow 

There are ___ for every ___.

c. Red 

Blue 

___.

d. White 

Purple

___.

Math Practice 4

Use a Table or Diagram

What are the quantities in this problem? How does a table or diagram represent the relationship between the quantities?

4 ACTIVITY: Describing Relationships

Work with a partner. Use a table or a diagram to represent the relationship between the two quantities.

a. For every 3 boys standing in a line, there are 4 girls.

b. For each vote Brian received, Sasha received 6 votes.

c. A class counts the number of vehicles that pass by its school from 1:00 to 2:00 P.M. There are 3 times as many cars as trucks.

d. A hand sanitizer contains 5 parts aloe for every 2 parts distilled water.

What Is Your Answer?

5. **IN YOUR OWN WORDS** How can you represent a relationship between two quantities? Give examples to support your explanation.

6. **MODELING** You make 48 pints of pink paint by using 5 pints of red paint for every 3 pints of white paint. Use a diagram to find the number of pints of red paint and white paint in your mixture. Explain.

Practice Use what you learned about comparing two quantities to complete Exercises 4 and 5 on page 194.

5.1 Lesson

Check It Out
Lesson Tutorials
BigIdeasMath.com

Key Vocabulary
ratio, *p. 192*

Key Idea

Ratio

Words A **ratio** is a comparison of two quantities. Ratios can be part-to-part, part-to-whole, or whole-to-part comparisons.

Examples
2 red crayons *to* 6 blue crayons
1 red crayon *for every* 3 blue crayons
3 blue crayons *per* 1 red crayon
3 blue crayons *for each* red crayon
3 blue crayons *out of every* 4 crayons
2 red crayons *out of* 8 crayons

Algebra The ratio of a to b can be written as $a : b$.

EXAMPLE 1 Writing Ratios

You have the coins shown.

a. Write the ratio of pennies to quarters.

6 pennies → 6 to 7 ← 7 quarters

∴ So, the ratio of pennies to quarters is 6 to 7, or 6 : 7.

b. Write the ratio of quarters to dimes.

7 quarters → 7 to 3 ← 3 dimes

∴ So, the ratio of quarters to dimes is 7 to 3, or 7 : 3

c. Write the ratio of dimes to the total number of coins.

3 dimes → 3 to 16 ← 16 coins

∴ So, the ratio of dimes to the total number of coins is 3 to 16, or 3 : 16.

Remember

Part-to-whole relationships compare a part of a whole to the whole. Fractions represent part-to-whole relationships. Part-to-part relationships compare a part of a whole to another part of the whole.

On Your Own

Now You're Ready
Exercises 6–13

1. In Example 1, write the ratio of dimes to pennies.
2. The circle graph shows the favorite ice-cream toppings of several students. Use ratio language to compare the number of students who favor peanuts to the total number of students.

Favorite Toppings

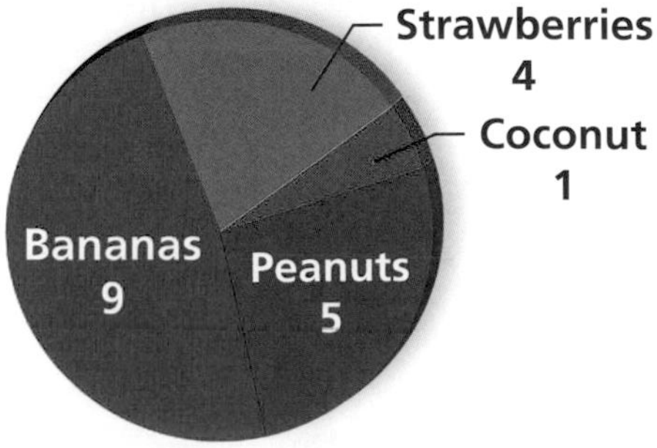

A *tape diagram* is a diagram that looks like a segment of tape. It shows the relationship between two quantities.

EXAMPLE 2 Using a Tape Diagram

The ratio of your monthly allowance to your friend's monthly allowance is 5 : 3. The monthly allowances total $40. How much is each allowance?

To help visualize the problem, express the ratio 5 : 3 using a tape diagram.

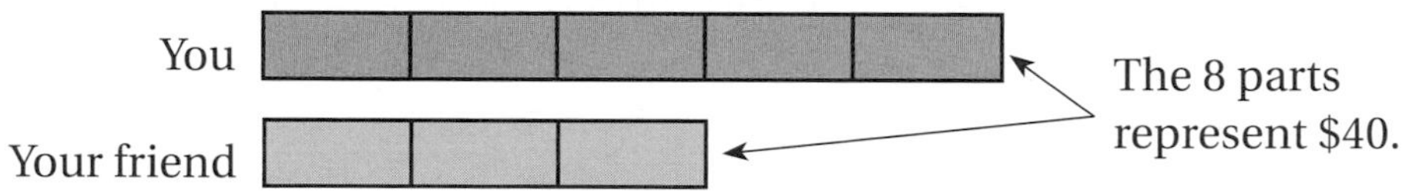

Because there are 8 parts, you know that 1 part represents $\$40 \div 8 = \$5$.

5 parts represent $\$5 \cdot 5 = \$25$.

3 parts represent $\$5 \cdot 3 = \$15$.

So, your monthly allowance is $25, and your friend's monthly allowance is $15.

EXAMPLE 3 Using a Tape Diagram

You separate 42 bulbs of garlic into two groups: one for planting and one for cooking. You will plant 3 bulbs for every 4 bulbs that you will use for cooking. Each bulb has about 8 cloves. About how many cloves will you plant?

To help visualize the problem, express the ratio *3 for every 4* using a tape diagram.

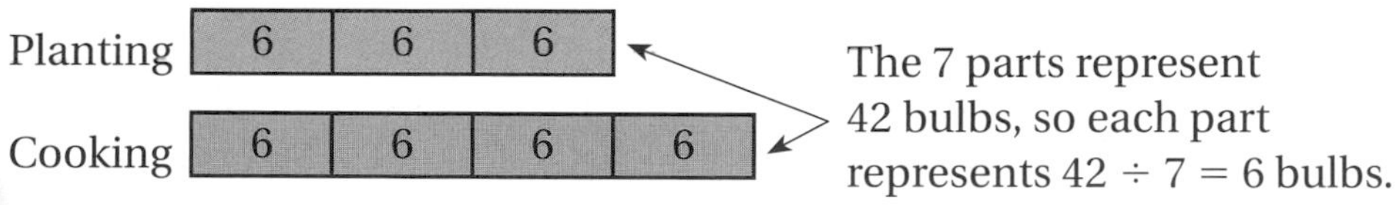

There are $3 \cdot 6 = 18$ bulbs for planting and $4 \cdot 6 = 24$ bulbs for cooking. The group of 18 bulbs has about $18 \cdot 8 = 144$ cloves.

So, you will plant about 144 cloves.

On Your Own

Now You're Ready
Exercises 15 and 16

3. **WHAT IF?** In Example 2, the ratio is 2 to 3. How much is each allowance?

4. **WHAT IF?** In Example 3, you will plant 1 bulb for every 2 bulbs that you will use for cooking. Will you plant more or fewer cloves than originally planned? Explain your reasoning.

5.1 Exercises

Check It Out
Help with Homework
BigIdeasMath.com

Vocabulary and Concept Check

1. **VOCABULARY** The ratio of vowels to consonants in a word is 5 to 7. Are there more vowels or consonants in the word? Explain.
2. **NUMBER SENSE** You are comparing apples to oranges in a fruit bowl. Is the ratio 2 : 3 the same as the ratio 3 : 2? Explain.
3. **WHICH ONE DOESN'T BELONG?** Which ratio does *not* belong with the other three? Explain your reasoning.

2 parts to 5 parts	2 out of every 5	2 for each 5	2 for every 5

Practice and Problem Solving

Use a table or a diagram to represent the relationship between the two quantities.

4. For each lion, there are 7 giraffes.
5. For every 5 seats, there are 4 fans.

Write the ratio. Explain what the ratio means.

1 6. frogs to turtles

7. basketballs to soccer balls

8. calculators : pencils

9. shirts : pants

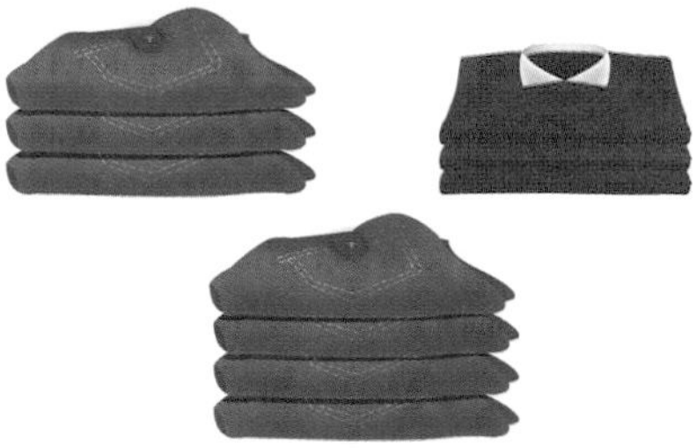

Use the table to write the ratio. Explain what the ratio means.

10. dramas to movies
11. comedies to movies
12. movies : action
13. movies : dramas

Movie	Number
Drama	3
Comedy	8
Action	4

Topic	Stamps
Birds	7
Celebrity	14
Horses	5
Ships	9

14. **STAMP COLLECTING** The table shows the numbers of stamps in a new stamp collection. Use ratio language to compare the number of celebrity stamps to the total number of stamps.

You and a friend tutor for a total of 12 hours. Use the tape diagram to find how many hours you tutor.

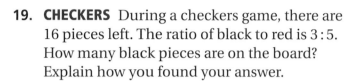

2 **15.** 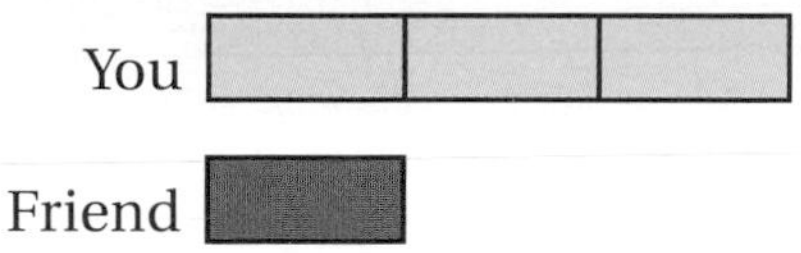

16. 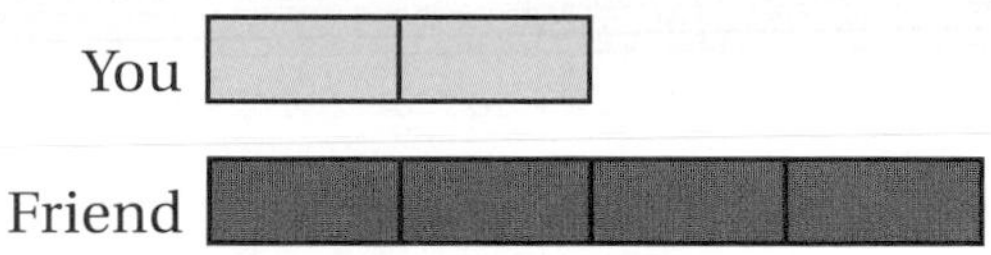

17. REASONING Twelve of the 28 students in a class have a dog. What is the ratio of students who have a dog to students who do not?

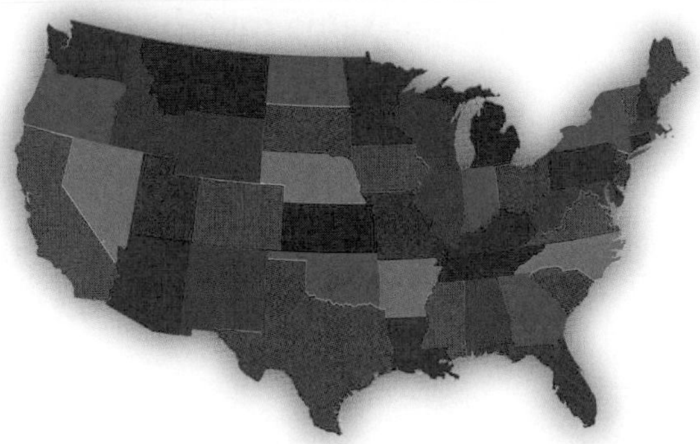

18. GEOGRAPHY In the continental United States, the ratio of states that border an ocean to states that do not border an ocean is 7 : 9. How many of the states border an ocean?

19. CHECKERS During a checkers game, there are 16 pieces left. The ratio of black to red is 3 : 5. How many black pieces are on the board? Explain how you found your answer.

20. SCHOOL PLAY There are 48 students in a school play. The ratio of boys to girls is 5 : 7. How many more girls than boys are in the play? Explain how you found your answer.

21. GEOMETRY Use the blue and green rectangles.

a. Find the ratio of the length of the blue rectangle to the length of the green rectangle. Repeat this for width, perimeter, and area.

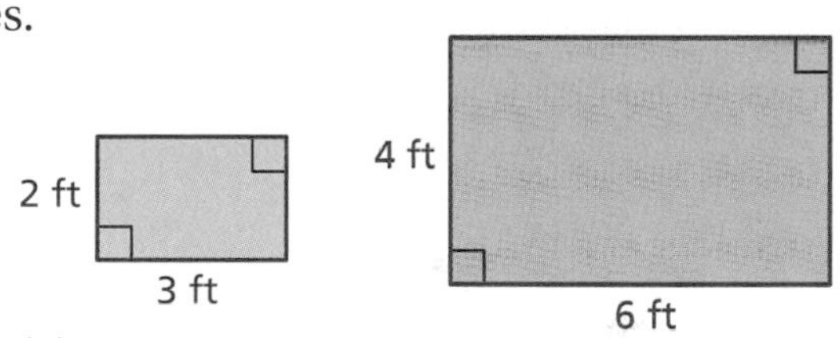

b. Compare and contrast your ratios in part (a).

22. PERIMETER The ratio of the side lengths of a triangle is 2 : 3 : 4. The shortest side is 15 inches. What is the perimeter? Explain.

23. PRECISION You mix soda water, fruit punch concentrate, and ginger ale in the ratio of 1 : 2 : 5 to make fruit punch. How many pints of each ingredient should you use to make 4 gallons of fruit punch? Is your answer reasonable? Explain.

24. **Reasoning** There are 12 boys and 10 girls in your gym class. If 6 boys joined the class, how many girls would need to join for the ratio of boys to girls to remain the same? Justify your answer.

Fair Game Review What you learned in previous grades & lessons

Divide. *(Section 2.6)*

25. $13.8 \div 3$ **26.** $16.45 \div 5$ **27.** $53.13 \div 21$ **28.** $19.214 \div 13$

29. MULTIPLE CHOICE What is the value of the expression $x \div y$ when $x = 30$ and $y = 18$? *(Section 3.1)*

Ⓐ $\frac{3}{5}$ Ⓑ $1\frac{2}{3}$ Ⓒ 12 Ⓓ 48

5.2 Ratio Tables

Essential Question How can you find two ratios that describe the same relationship?

1 ACTIVITY: Making a Mixture

Work with a partner. A mixture calls for 1 cup of lemonade and 3 cups of iced tea.

a. How many total cups does the mixture contain? ▢ cups

For every ▢ cup of lemonade, there are ▢ cups of iced tea.

b. How do you make a larger batch of this mixture? Describe your procedure and use the table below to organize your results. Add more columns to the table if needed.

Cups of Lemonade						
Cups of Iced Tea						
Total Cups						

c. Which operations did you use to complete your table? Do you think there is more than one way to complete the table? Explain.

d. How many total cups are in your final mixture? How many of those cups are lemonade? How many are iced tea? Compare your results with those of other groups in your class.

e. Suppose you take a sip from every group's final mixture. Do you think all the mixtures should taste the same? Do you think the color of all the mixtures should be the same? Explain your reasoning.

f. Why do you think it is useful to use a table when organizing your results in this activity? Explain.

COMMON CORE

Ratios

In this lesson, you will
- use ratio tables to find equivalent ratios.
- solve real-life problems.

Learning Standards
6.RP.1
6.RP.3a

2 ACTIVITY: Using a Multiplication Table

Math Practice 2

Use Operations

For each part of this problem, how do you know which operation to use?

Work with a partner. Use the information in Activity 1 and the multiplication table below.

	1	2	3	4	5	6	7	8	9	10	11	12
1	1	2	3	4	5	6	7	8	9	10	11	12
2	2	4	6	8	10	12	14	16	18	20	22	24
3	3	6	9	12	15	18	21	24	27	30	33	36
4	4	8	12	16	20	24	28	32	36	40	44	48

a. A mixture contains 8 cups of lemonade. How many cups of iced tea are in the mixture?

b. A mixture contains 21 cups of iced tea. How many cups of lemonade are in the mixture?

c. A mixture has a total of 40 cups. How many cups are lemonade? How many are iced tea?

d. **REPEATED REASONING** Explain how a multiplication table may have helped you in Activity 1.

3 ACTIVITY: Using More than One Ratio to Describe a Quantity

Work with a partner.

a. Find the ratio of pitchers of lemonade to pitchers of iced tea.

b. How can you divide the pitchers into equal groups? Is there more than one way? Use your results to describe the entire collection of pitchers.

c. Three more pitchers of lemonade are added. Is there more than one way to divide the pitchers into equal groups? Explain.

d. The number of pitchers of lemonade and iced tea are doubled. Can you use the ratio in part (b) to describe the entire collection of pitchers? Explain.

What Is Your Answer?

4. **IN YOUR OWN WORDS** How can you find two ratios that describe the same relationship? Give examples to support your explanation.

Practice Use what you learned about ratios to complete Exercises 4 and 5 on page 201.

5.2 Lesson

Check It Out
Lesson Tutorials
BigIdeasMath.com

Key Vocabulary
equivalent ratios, *p. 198*
ratio table, *p. 198*

Two ratios that describe the same relationship are **equivalent ratios**. You can find equivalent ratios by:

- adding or subtracting quantities in equivalent ratios.
- multiplying or dividing each quantity in a ratio by the same number.

You can find and organize equivalent ratios in a **ratio table**.

EXAMPLE 1 Completing Ratio Tables

Find the missing value(s) in each ratio table. Then write the equivalent ratios.

a.

Pens	1	2	
Pencils	3		9

b.

Dogs	4		24
Cats	6	12	

a. You can use repeated addition with the original ratio to find the missing values.

+1 +1

Pens	1	2	3
Pencils	3	6	9

+3 +3

∴ The equivalent ratios are 1 : 3, 2 : 6, and 3 : 9.

b. You can use multiplication to find the missing values.

×2 ×3

Dogs	4	8	24
Cats	6	12	36

×2 ×3

∴ The equivalent ratios are 4 : 6, 8 : 12, and 24 : 36.

On Your Own

Now You're Ready
Exercises 6–11

Find the missing value(s) in the ratio table. Then write the equivalent ratios.

1.

Plantains	4		12
Bananas	3	6	

2.

Euros	5	10	
Dollars	4		32

EXAMPLE 2 Making a Ratio Table

You are making sugar water for your hummingbird feeder. A website indicates to use 4 parts of water for every 1 part of sugar. You use 20 cups of water. How much sugar do you need?

You can solve this problem by using equivalent ratios. The ratio of water to sugar is 4 parts to 1 part. So, for every 4 cups of water, you need 1 cup of sugar. Find an equivalent ratio with 20 parts water.

Method 1: Use a ratio table and addition.

You can think of making a larger batch of sugar water as combining several batches of 4 to 1 mixtures. Use addition to obtain 20 in the water column.

+4 +4 +4 +4

Water (cups)	4	8	12	16	20
Sugar (cups)	1	2	3	4	5

+1 +1 +1 +1

The ratio 20 to 5 is equivalent to 4 to 1.

∴ So, you need 5 cups of sugar.

Method 2: Use a ratio table and multiplication.

You multiplied the amount of water in the recipe by 5 because $20 \div 4 = 5$. So, you need to multiply the amount of sugar by 5. Multiply each part of the ratio in the original recipe by 5.

×5

Water (cups)	4	20
Sugar (cups)	1	5

×5

The ratio 20 to 5 is equivalent to 4 to 1.

∴ So, you need 5 cups of sugar.

Study Tip

In Example 2, Method 1, notice that you can eliminate a step by adding columns 2 and 3 to obtain $8 + 12 = 20$ cups of water for $2 + 3 = 5$ cups of sugar.

On Your Own

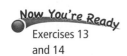

Now You're Ready Exercises 13 and 14

3. **WHAT IF?** You use 24 cups of water. How much sugar do you need?
4. You make a sweeter mixture of sugar water for your hummingbird feeder using 3 parts of water for every 1 part of sugar. You use 9 quarts of water. How much sugar do you need?

EXAMPLE 3 Using a Ratio Table

The nutrition facts label on a box of crackers shows that there are 240 milligrams of sodium in every 36 crackers.

a. You eat 15 crackers. How much sodium do you consume?

The ratio of sodium to crackers is 240 to 36. Use a ratio table to find an equivalent ratio with 15 crackers.

$\div 2 \quad \div 6 \quad \times 5$

Sodium (milligrams)	240	120	20	100
Crackers	36	18	3	15

$\div 2 \quad \div 6 \quad \times 5$

Study Tip

In Example 3, notice that you could use one step in the ratio table: multiply by

$\frac{1}{2} \cdot \frac{1}{6} \cdot 5 = \frac{5}{12}.$

The ratio 100 to 15 is equivalent to 240 to 36.

∴ So, you consume 100 milligrams of sodium.

b. You eat 21 crackers. How much sodium do you consume?

Notice that you can add the two middle columns in the table above.

∴ So, you consume $120 + 20 = 140$ milligrams of sodium in $18 + 3 = 21$ crackers.

EXAMPLE 4 Using a Ratio Table

You mix 3 pints of yellow paint for every 4 pints of blue paint to make green paint. You use 10 pints of blue paint. How much green paint do you make?

The ratio of yellow paint to blue paint is 3 to 4. Use a ratio table to find an equivalent ratio with 10 parts blue paint.

$\div 2 \quad \times 5$

Yellow (pints)	3	$\frac{3}{2}$	$7\frac{1}{2}$
Blue (pints)	4	2	10

$\div 2 \quad \times 5$

Study Tip

In Example 4, notice that you could use one step in the ratio table: multiply by

$\frac{1}{2} \cdot 5 = \frac{5}{2}.$

You use $7\frac{1}{2}$ pints of yellow paint and 10 pints of blue paint.

∴ So, you make $7\frac{1}{2} + 10 = 17\frac{1}{2}$ pints of green paint.

On Your Own

Now You're Ready Exercises 15 and 16

5. **WHAT IF?** In Example 3, you eat 24 crackers. How much sodium do you consume?
6. **WHAT IF?** In Example 4, you mix 2 pints of yellow paint for every 3 pints of blue paint. You use 5 pints of yellow paint. How much green paint do you make?

5.2 Exercises

Check It Out
Help with Homework
BigIdeasMath.com

Vocabulary and Concept Check

1. **VOCABULARY** How can you tell whether two ratios are equivalent?

2. **NUMBER SENSE** Consider the ratio 3 : 5. Can you create an equivalent ratio by adding the same number to each quantity in the ratio? Explain.

3. **WHICH ONE DOESN'T BELONG?** Which ratio does *not* belong with the other three? Explain your reasoning.

3 : 4   9 : 12   12 : 15   12 : 16

Practice and Problem Solving

Write several ratios that describe the collection.

4. baseballs to gloves

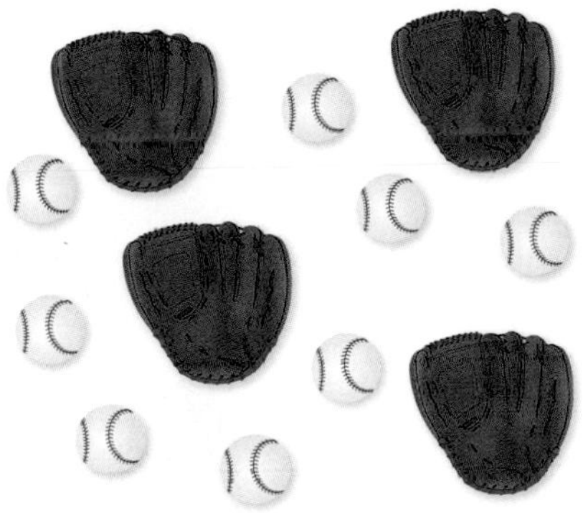

5. ladybugs to bees

Find the missing value(s) in the ratio table. Then write the equivalent ratios.

1 6.

Boys	1	
Girls	5	10

7.

Violins	8	24
Cellos	3	

8.

Taxis	6		36
Buses	5	15	

9.

Burgers	3		9
Hot Dogs	5	10	

10.

Towels	14	7	
Blankets	8		16

11.

Forks	16	8	
Spoons	10		30

12. **WORK** Your neighbor pays you \$17 for every 2 hours you work. You work for 8 hours on Saturday. How much does your neighbor owe you?

Complete the ratio table to solve the problem.

2 **13.** For every 3 tickets you sell, your friend sells 4. You sell a total of 12 tickets. How many does your friend sell?

You	3			12
Friend	4			

14. A store sells 2 printers for every 5 computers. The store sells 40 computers. How many printers does the store sell?

Printers	2		8	
Computers	5	10		40

3 **15.** First and second place in a contest use a ratio to share a cash prize. When first place pays \$100, second place pays \$60. How much does first place pay when second place pays \$36?

First	100		
Second	60		36

16. A grade has 81 girls and 72 boys. The grade is split into groups that have the same ratio of girls to boys as the whole grade. How many girls are in a group that has 16 boys?

Girls	81		
Boys	72		16

ERROR ANALYSIS Describe and correct the error in making the ratio table.

17. ✗

A	3	8	13
B	7	12	17

18. ✗

A	5	25	125
B	3	9	27

19. DONATION A sports store donates basketballs and soccer balls to the boys and girls club. The ratio of basketballs to soccer balls is 7 : 6. The store donates 24 soccer balls. How many basketballs does the store donate?

20. DOWNLOAD You are downloading songs to your MP3 player. The ratio of pop songs to rock songs is 5 : 4. You download 40 pop songs. How many rock songs do you download?

SCRAMBLED EGGS In Exercises 21–25, use the ratio table showing different batches of the same recipe for scrambled eggs.

Recipe	A	B	C	D	E	F
Servings	4	2	6	3	5	9
Eggs	8	4	12	6	10	18
Milk (cups)	$\frac{1}{2}$	$\frac{1}{4}$	$\frac{3}{4}$	$\frac{3}{8}$	$\frac{5}{8}$	$1\frac{1}{8}$

21. How can you use Recipes B and D to create Recipe E?

22. How can you use Recipes C and D to create Recipe F?

23. How can you use Recipes B and C to create Recipe A?

24. How can you use Recipes C and F to create Recipe D?

25. Describe one way to use the recipes to create a batch with 11 servings.

Two whole numbers A and B satisfy the following conditions. Find A and B.

26. $A + B = 30$
$A : B$ is equivalent to $2 : 3$.

27. $A + B = 44$
$A : B$ is equivalent to $4 : 7$.

28. $A - B = 18$
$A : B$ is equivalent to $11 : 5$.

29. $A - B = 25$
$A : B$ is equivalent to $13 : 8$.

Nutrition Facts
Serving Size: 1 ounce (28g)

Amount Per Serving	
Calories 161	Calories from Fat 109
	% Daily Value*
Total Fat 13g	20%
Saturated Fat 3g	13%
Trans Fat	
Cholesterol 0mg	0%
Sodium 4mg	0%
Total Carbohydrate 9g	3%
Dietary Fiber 1g	3%
Sugars 1g	
Protein 4g	
Vitamin A 0% • Vitamin C	0%
Calcium 1% • Iron	9%

*Percent Daily Values are based on a 2,000 calorie diet. Your daily values may be higher or lower depending on your calorie needs.

30. CASHEWS The nutrition facts label on a container of dry roasted cashews indicates there are 161 calories in 28 grams. You eat 9 cashews totaling 12 grams.

a. How many calories do you consume?

b. How many cashews are in one serving?

31. REASONING The ratio of three numbers is $4 : 3 : 1$. The sum of the numbers is 64. What is the greatest number?

32. SURVEY Seven out of every 8 students surveyed owns a bike. The difference between the number of students who own a bike and those who do not is 72. How many students were surveyed?

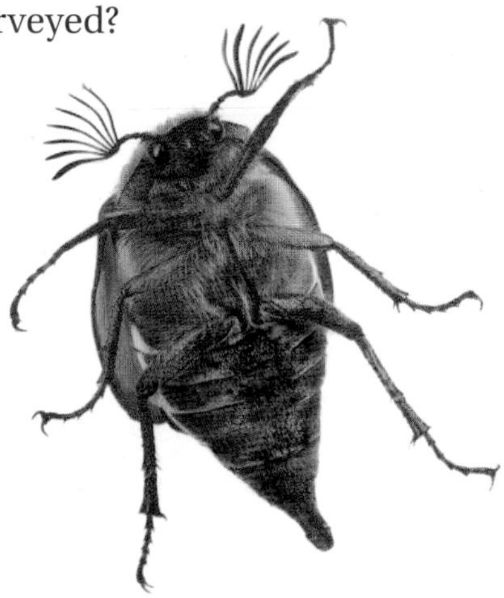

33. BUG COLLECTION You and a classmate have a bug collection for science class. You find 5 out of every 9 bugs in the collection. You find 4 more bugs than your classmate. How many bugs are in the collection?

34. Problem Solving You and a friend each have a collection of tokens. Initially, for every 8 tokens you had, your friend had 3. After you give half of your tokens to your friend, your friend now has 18 more tokens than you. Initially, how many more tokens did you have than your friend?

Fair Game Review What you learned in previous grades & lessons

Factor the expression using the GCF. *(Section 3.4)*

35. $54 + 27$

36. $60x - 84$

37. $42x + 28y$

38. MULTIPLE CHOICE Which expression does *not* give the area of the shaded figure? *(Section 4.3)*

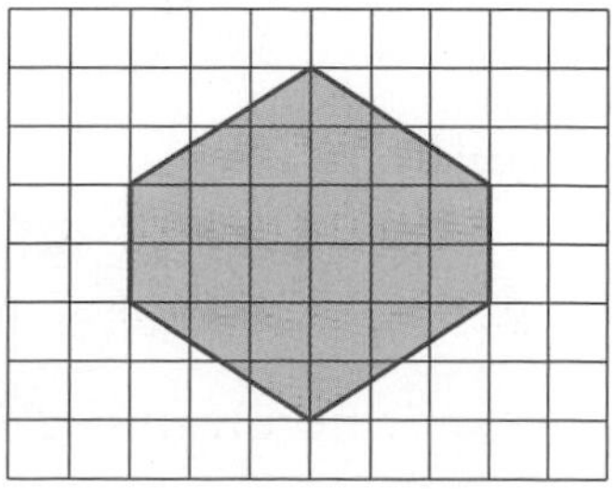

Ⓐ $2(6) + 2\left(\frac{1}{2}(6)(2)\right)$

Ⓑ $2\left(\frac{1}{2}(3)(2 + 6)\right)$

Ⓒ $6(6) - 4\left(\frac{1}{2}(3)(2)\right)$

Ⓓ $6(6) - \frac{1}{2}(6)(2)$

5.3 Rates

Essential Question How can you use rates to describe changes in real-life problems?

1 ACTIVITY: Stories Without Words

Work with a partner. Each diagram shows a story problem.

- **Describe the story problem in your own words.**
- **Write the rate indicated by the diagram. What are the units?**

a.

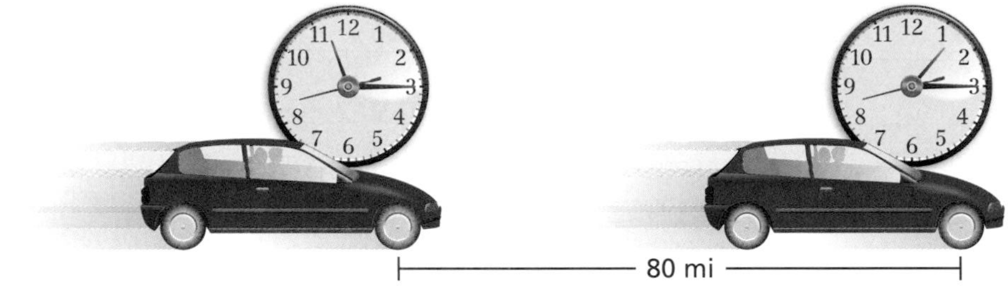

b.

c.

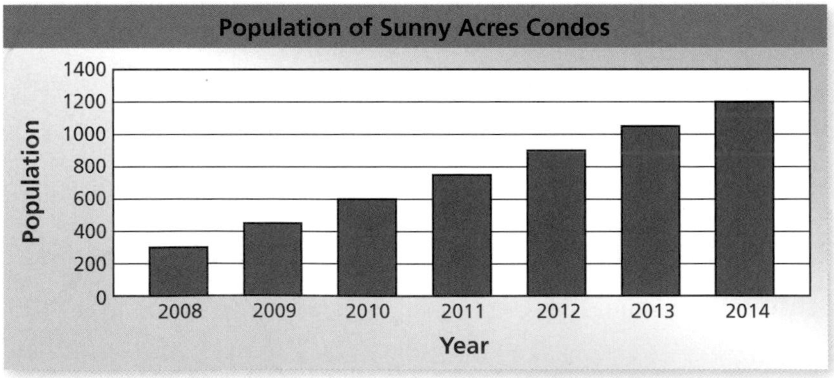

d.

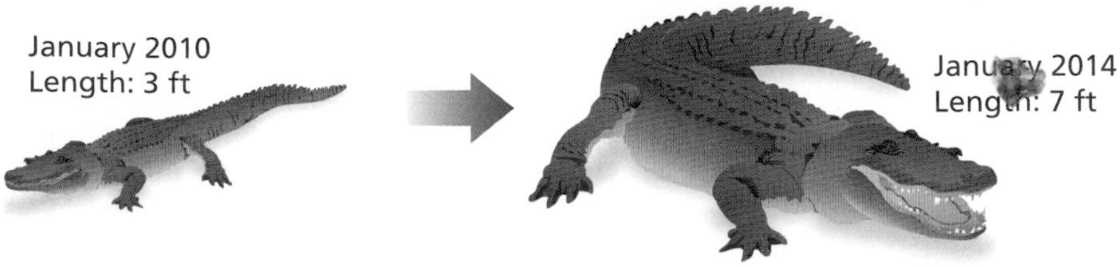

COMMON CORE

Rates

In this lesson, you will
- understand the concepts of rates and unit rates.
- write unit rates.
- solve real-life problems.

Learning Standards
6.RP.2
6.RP.3a
6.RP.3b

2 ACTIVITY: Finding Equivalent Rates

Math Practice 6

Specify Units

How do the given units help you find the units for your answer?

Work with a partner. Use the diagrams in Activity 1. Explain how you found each answer.

a. How many miles does the car travel in 1 hour?

b. How much money does the person earn every hour?

c. How much does the population of Sunny Acres Condos increase each year?

d. How many feet does the alligator grow per year?

3 ACTIVITY: Using a Double Number Line

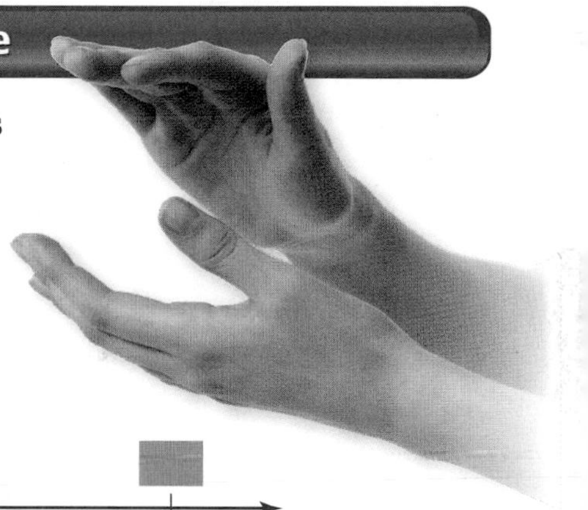

Work with a partner. Count the number of times you can clap your hands in 12 seconds. Have your partner keep track of the time and record your results.

a. Use the results to complete the double number line.

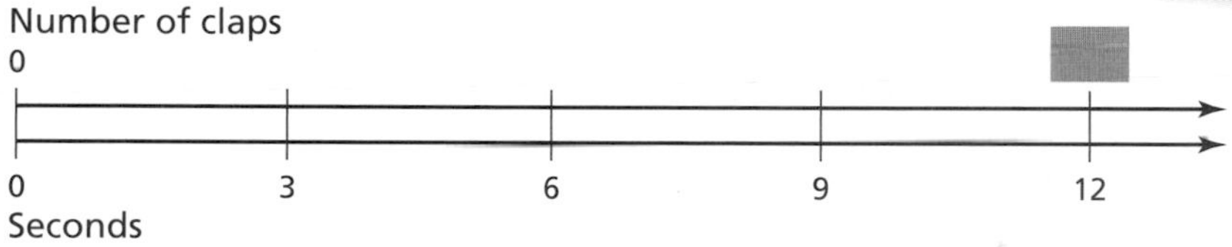

b. Explain how to use the double number line to find the number of times you clap your hands in 6 seconds and in 4 seconds.

c. Find the number of times you can clap your hands in 1 minute. Explain how you found your answer.

d. How can you find the number of times you can clap your hands in 2 minutes? 3 minutes? Explain.

What Is Your Answer?

4. IN YOUR OWN WORDS How can you use rates to describe changes in real-life problems? Give examples to support your explanation.

5. MODELING Use a double number line to model each story in Activity 1. Show how to use the double number line to answer each question in Activity 2. Why is a double number line a good problem-solving tool for these types of problems?

Practice Use what you learned about rates to complete Exercises 3 and 4 on page 208.

5.3 Lesson

Check It Out
Lesson Tutorials
BigIdeasMath.com

Key Vocabulary
rate, *p. 206*
unit rate, *p. 206*
equivalent rates, *p. 206*

Key Idea

Rate and Unit Rate

Words A **rate** is a ratio of two quantities using different units. A **unit rate** compares a quantity to one unit of another quantity. **Equivalent rates** have the same unit rate.

Numbers You pay $27 for 3 pizzas.

Rate: $27 : 3 pizzas

Unit rate: $9 : 1 pizza

Algebra Rate: a units : b units Unit rate: $\frac{a}{b}$ units : 1 unit

Study Tip

In a rate $a : b$, you can divide both a and b by b to find the unit rate.

EXAMPLE 1 Writing a Rate

The double number line shows the rate at which you earn points for successfully hitting notes in a music video game. Write a rate that represents this situation.

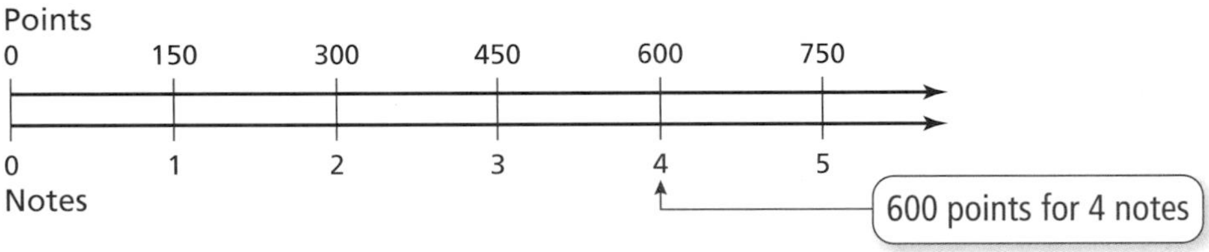

∴ One possible rate is 600 points for every 4 notes.

EXAMPLE 2 Finding a Unit Rate

A piece of space junk travels 5 miles in 8 seconds. How far does it travel per second?

Use a ratio table and divide by 8 to write an equivalent rate in which the time is 1 second.

	÷ 8 →	
Distance (miles)	5	$\frac{5}{8}$
Time (seconds)	8	1

The rate 5 miles : 8 seconds is equivalent to $\frac{5}{8}$ mile : 1 second.

∴ So, the space junk travels $\frac{5}{8}$ mile per second.

On Your Own

Now You're Ready
Exercises 3–14

1. Write another rate that represents the situation in Example 1.
2. A Japanese bullet train travels 558 miles in 3 hours. How far does it travel every hour?
3. You pay $8 for 16 ounces of sliced turkey. Write a rate that gives the price for each ounce of turkey.

EXAMPLE 3 Finding Equivalent Rates

a. A chef buys 6 pounds of salmon fillets for $51. How much will the chef pay for 9 more pounds of salmon fillets?

Using a ratio table, divide to find the unit rate and then multiply to find the cost for 9 pounds of salmon fillets.

$\div 6$ $\times 9$

Cost (dollars)	51	8.5	76.5
Salmon (pounds)	6	1	9

$\div 6$ $\times 9$

So, the chef will pay $76.50 for 9 more pounds of salmon fillets.

Study Tip

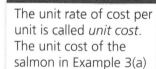

The unit rate of cost per unit is called *unit cost*. The unit cost of the salmon in Example 3(a) is $8.50 per pound.

b. You buy 2 pounds of tilapia fillets for $16. What is the cost for 7 pounds of tilapia fillets?

Because $16 is easily divided into halves, fourths, and eighths, it is appropriate to model the rate using a double number line.

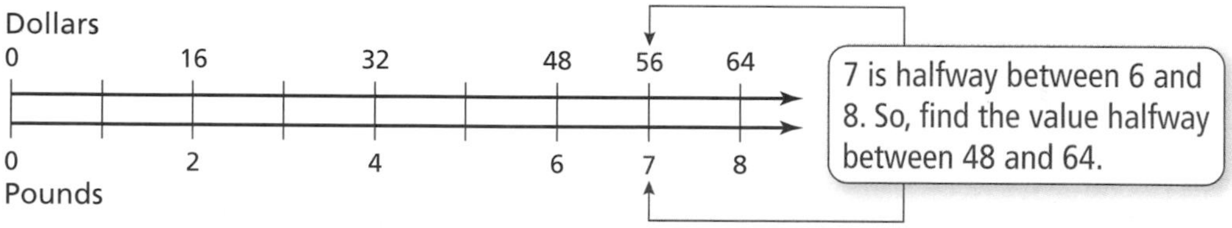

So, the cost for 7 pounds of tilapia fillets is $56.

On Your Own

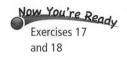

Now You're Ready
Exercises 17 and 18

4. Your download speed is 3 megabytes every 4 seconds.
   a. How many megabytes can you download in 1 minute?
   b. Construct a double number line that represents the situation. How many megabytes can you download in 10 seconds?

5.3 Exercises

Vocabulary and Concept Check

1. **WRITING** Describe a unit rate that you use in real life.
2. **DIFFERENT WORDS, SAME QUESTION** Which is different? Find "both" answers.

What is the cost per bagel?	What is the cost per dozen bagels?
What is the unit cost of a bagel?	How much does each bagel cost?

Practice and Problem Solving

Write a rate that represents the situation.

1 3.

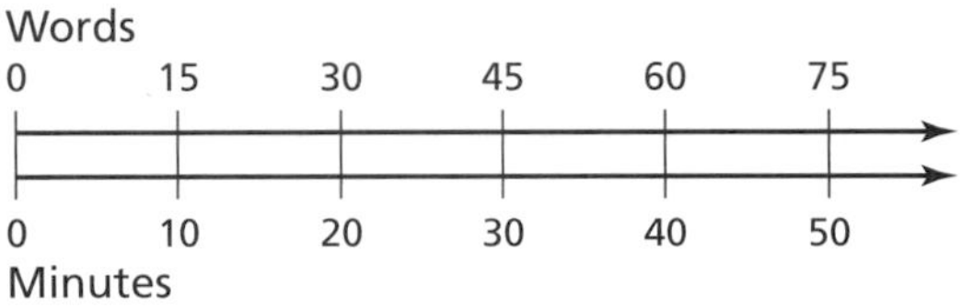

4.

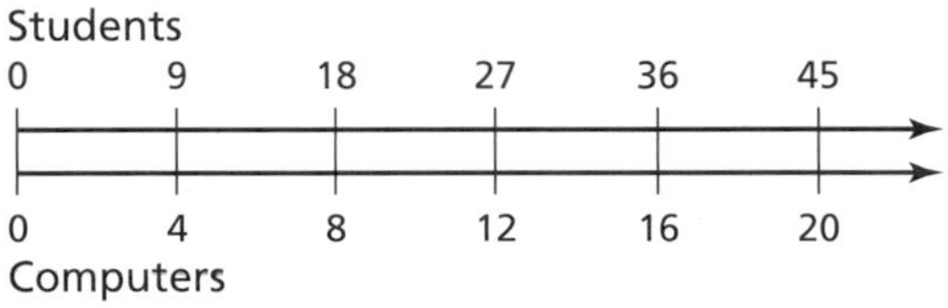

5.

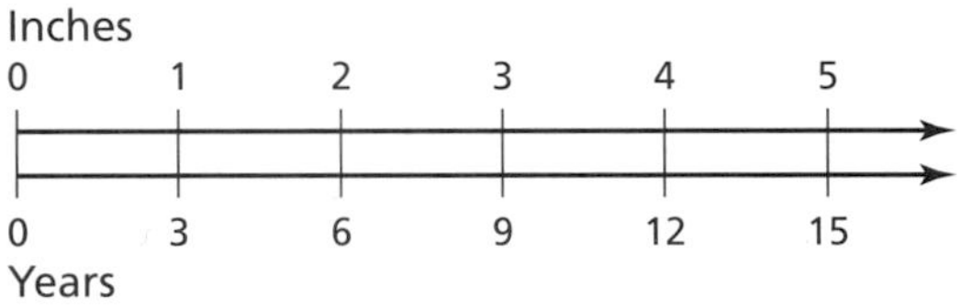

6.

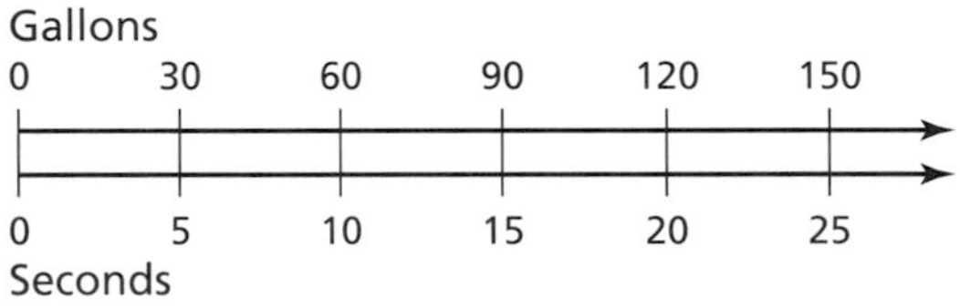

Write a unit rate for the situation.

2 7. \$28 saved in 4 weeks

8. 18 necklaces made in 3 hours

9. 270 miles in 6 hours

10. 228 students in 12 classes

11. 2520 kilobytes in 18 seconds

12. 880 calories in 8 servings

13. 1080 miles on 15 gallons

14. \$12.50 for 5 ounces

15. **LIGHTNING** Lightning strikes Earth 1000 times in 10 seconds. How many times does lightning strike per second?

16. **HEART RATE** Your heart beats 240 times in 4 minutes. How many times does your heart beat each minute?

3 17. **CAR WASH** You earn \$35 for washing 7 cars. How much do you earn for washing 4 cars?

18. **5K RACE** You jog 2 kilometers in 12 minutes. At this rate, how long will it take you to complete a 5-kilometer race?

Decide whether the rates are equivalent.

19. 24 laps in 6 minutes
72 laps in 18 minutes

20. 126 points every 3 games
210 points every 5 games

21. 15 breaths every 36 seconds
90 breaths every 3 minutes

22. \$16 for 4 pounds
\$1 for 4 ounces

23. **PRINTER** A printer prints 28 photos in 8 minutes.
   a. How many minutes does it take to print 21 more photos?
   b. Construct a double number line diagram that represents the situation. How many minutes does it take to print 35 more photos?

24. **SUN VISOR** An athletic director pays \$90 for 12 sun visors for the softball team.
   a. How much will the athletic director pay to buy 15 more sun visors?
   b. Construct a double number line diagram that represents the situation. What is the cost of 16 sun visors?

25. **FOOD DRIVE** The table shows the amounts of food collected by two homerooms. Homeroom A collects 21 additional items of food. How many more items does Homeroom B need to collect to have more items per student?

	Homeroom A	Homeroom B
Students	24	16
Canned Food	30	22
Dry Food	42	24

26. **MARATHON** A runner completed a 26.2-mile marathon in 210 minutes.
   a. Estimate the unit rate, in miles per minute.
   b. Estimate the unit rate, in minutes per mile.
   c. Another runner says, "I averaged 10-minute miles in the marathon." Is this runner talking about the kind of rate described in part (a) or in part (b)? Explain your reasoning.

27. **Logic** You can do one-half of a job in an hour. Your friend can do one-third of the same job in an hour. How long will it take to do the job if you work together?

Fair Game Review What you learned in previous grades & lessons

Write two fractions that are equivalent to the given fraction. *(Skills Review Handbook)*

28. $\frac{1}{3}$ 29. $\frac{5}{6}$ 30. $\frac{2}{5}$ 31. $\frac{4}{9}$

32. **MULTIPLE CHOICE** Which expression is equivalent to $6(x) - 6(2)$? *(Section 3.4)*

Ⓐ $2(x - 6)$ Ⓑ $6(x - 2)$ Ⓒ $12(x - 1)$ Ⓓ $36(x - 2)$

5.4 Comparing and Graphing Ratios

Essential Question How can you compare two ratios?

1 ACTIVITY: Comparing Ratio Tables

Work with a partner.

- **You make purple frosting by adding 1 drop of red food coloring for every 3 drops of blue food coloring.**
- **Your teacher makes purple frosting by adding 3 drops of red food coloring for every 5 drops of blue food coloring.**

a. Copy and complete the ratio table for each frosting mixture.

Your Frosting	
Drops of Red	Drops of Blue
1	
2	
3	
4	
5	

Your Teacher's Frosting	
Drops of Red	Drops of Blue
3	
6	
9	
12	
15	

b. Whose frosting is bluer? Whose frosting is redder? Justify your answers.

c. **STRUCTURE** Insert and complete a new column for each ratio table above that shows the total number of drops. How can you use this column to answer part (b)?

2 ACTIVITY: Graphing from a Ratio Table

Work with a partner.

a. Explain how you can use the values from the ratio table for your frosting to create a graph in the coordinate plane.

b. Use the values in the table to plot the points. Then connect the points and describe the graph. What do you notice?

c. What does the line represent?

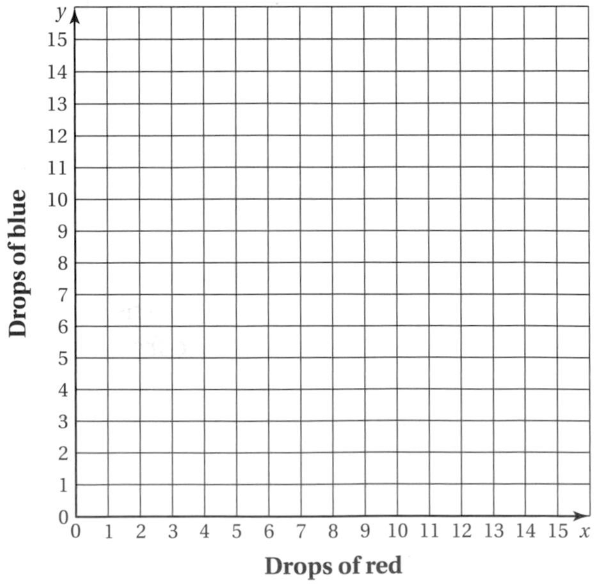

COMMON CORE

Ratios and Rates

In this lesson, you will
- compare ratios.
- compare unit rates.
- graph ordered pairs to compare ratios and rates.

Learning Standards
6.RP.2
6.RP.3a

3 ACTIVITY: Comparing Graphs from Ratio Tables

Math Practice 7
Look for Patterns
What patterns do you notice in the graph? What does this tell you about the problem?

Work with a partner. The graph shows the values from the ratio table for your teacher's frosting.

a. Complete the table and the graph.

Your Teacher's Frosting	
Drops of Red	Drops of Blue
3	
6	
9	
12	
15	

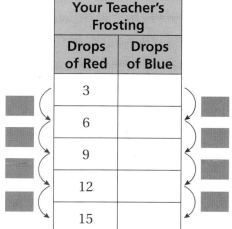

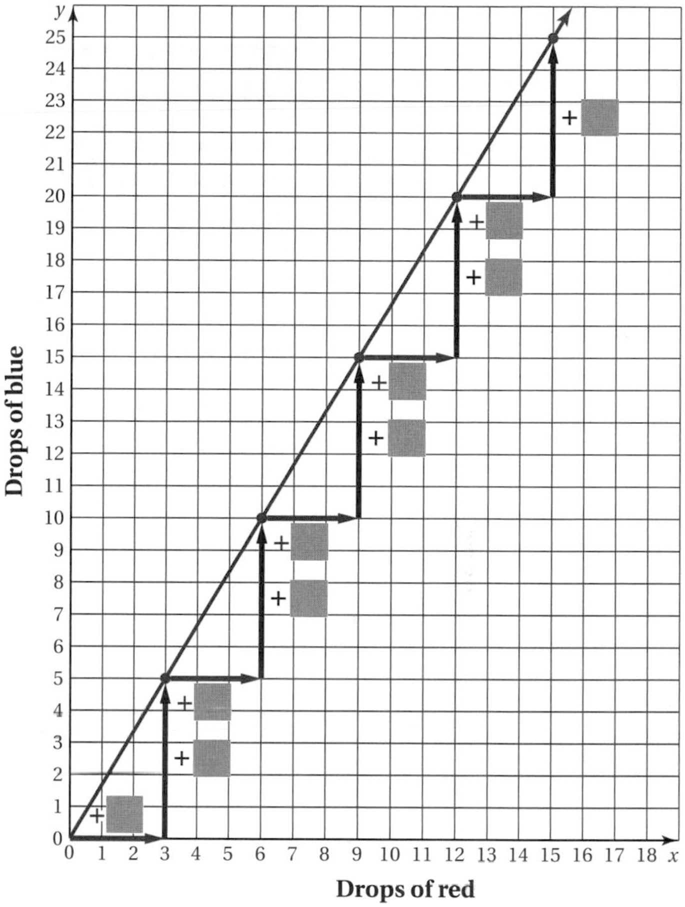

b. Explain the relationship between the entries in the ratio table and the points on the graph.

c. How is this graph similar to the graph in Activity 2? How is it different?

d. How can you use the graphs to determine whose frosting has more red or blue in it? Explain.

What Is Your Answer?

4. **IN YOUR OWN WORDS** How can you compare two ratios?

5. **PRECISION** Your teacher's frosting mixture has 7 drops of red in it. How can you use the graph to find how many drops of blue are needed to make the purple frosting? Is your answer exact? Explain.

Practice Use what you learned about comparing ratios to complete Exercises 3 and 4 on page 214.

5.4 Lesson

Check It Out
Lesson Tutorials
BigIdeasMath.com

One way to compare ratios is by using ratio tables.

EXAMPLE 1 Comparing Ratios

You mix 8 tablespoons of hot sauce and 3 cups of salsa in a green bowl. You mix 12 tablespoons of hot sauce and 4 cups of salsa in an orange bowl. Which mixture is hotter?

Use ratio tables to compare the mixtures. Find a larger batch of each mixture in which the amount of hot sauce or salsa is the same.

Green Bowl

$\times 4$

Hot Sauce (tablespoons)	8	32
Salsa (cups)	3	12

$\times 4$

Orange Bowl

$\times 3$

Hot Sauce (tablespoons)	12	36
Salsa (cups)	4	12

$\times 3$

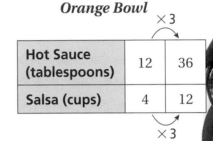

The tables show that for a larger batch of each mixture using 12 cups of salsa, the orange bowl would have $36 - 32 = 4$ more tablespoons of hot sauce.

So, the mixture in the orange bowl is hotter.

EXAMPLE 2 Comparing Unit Rates

Which bag of dog food is the better buy?

Use ratio tables to find and compare the unit costs.

20-Pound Bag

$\div 20$

Cost (dollars)	17.20	0.86
Food (pounds)	20	1

$\div 20$

30-Pound Bag

$\div 30$

Cost (dollars)	25.20	0.84
Food (pounds)	30	1

$\div 30$

The 20-pound bag costs \$0.86 per pound, and the 30-pound bag costs \$0.84 per pound.

Because \$0.84 is less than \$0.86, the 30-pound bag is the better buy.

On Your Own

Now You're Ready
Exercises 3–10

1. In Example 1, you mix 10 tablespoons of hot sauce and 3 cups of salsa in a red bowl. Which mixture is the mildest? Explain.
2. A 30-pack of paper towels costs \$48.30. A 32-pack costs \$49.60. Which is the better buy? Explain.

EXAMPLE 3 Graphing Values from Ratio Tables

A hot-air balloon rises 9 meters every 3 seconds. A blimp rises 7 meters every 2 seconds.

a. Complete the ratio table for each aircraft. Which rises faster?

Rises 9 meters every 3 seconds.

Balloon	
Time (seconds)	Height (meters)
3	9
6	**18**
9	**27**
12	**36**

×2, ×3, ×4

Blimp	
Time (seconds)	Height (meters)
2	7
4	14
6	21
8	28

×2, ×3, ×4

Every 6 seconds, the balloon rises 18 meters and the blimp rises 21 meters.

So, the blimp rises faster.

b. Graph the ordered pairs (time, height) from the tables in part (a). What can you conclude?

Write the ordered pairs.

Balloon: (3, 9), (6, 18), (9, 27), (12, 36)

Blimp: (2, 7), (4, 14), (6, 21), (8, 28)

Study Tip

When graphing speed, you often place time on the horizontal axis and distance on the vertical axis.

Plot and label each set of ordered pairs. Then draw a line through each set of points.

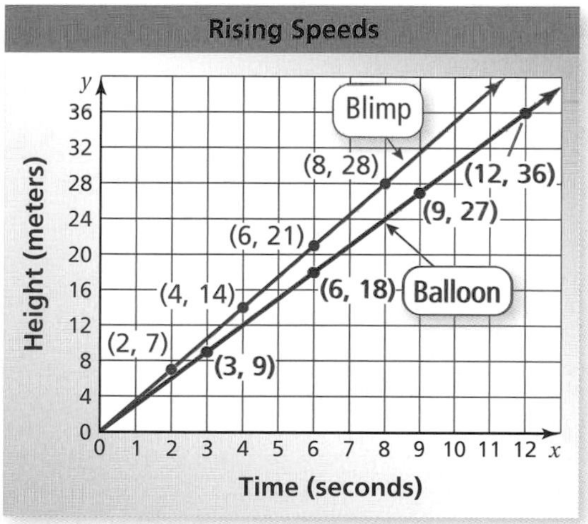

Both graphs begin at (0, 0). The graph for the blimp is steeper, so the blimp rises faster than the hot-air balloon.

On Your Own

Now You're Ready
Exercises 12 and 13

3. **WHAT IF?** The blimp rises 6 meters every 2 seconds. How does this affect your conclusion?

5.4 Exercises

Check It Out
Help with Homework
BigIdeasMath.com

Vocabulary and Concept Check

1. **WRITING** Explain how to use tables to compare ratios.

2. **NUMBER SENSE** Just by looking at the graph, determine who earns a greater hourly wage. Explain.

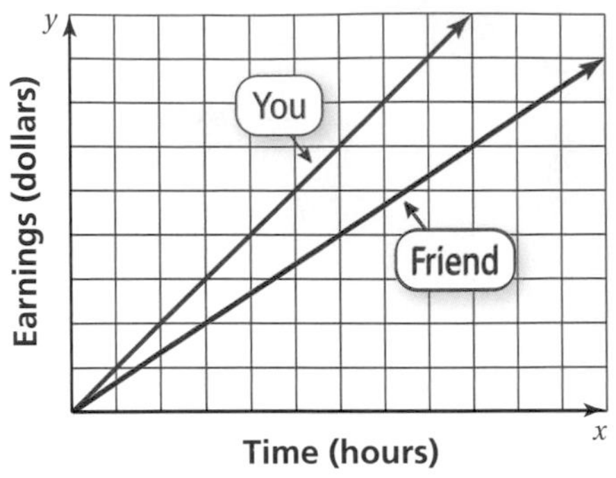

Practice and Problem Solving

Determine which car gets the better gas mileage.

1 3.

Car	A	B
Distance (miles)	125	120
Gallons Used	5	6

4.

Car	A	B
Distance (miles)	300	320
Gallons Used	8	10

5.

Car	A	B
Distance (miles)	450	405
Gallons Used	15	12

6.

Car	A	B
Distance (miles)	360	270
Gallons Used	20	18

Determine which is the better buy.

2 7.

Air Freshener	A	B
Cost (dollars)	6	12
Refills	2	3

8.

Kitten Food	A	B
Cost (dollars)	15	9
Cans	18	12

9.

Ham	A	B
Cost (dollars)	5.70	8.75
Pounds	3	5

10.

Cheese	A	B
Cost (dollars)	3.59	5.12
Slices	10	16

11. **SALT WATER GARGLE** Salt water gargle can temporarily relieve a sore throat. One recipe calls for $\frac{3}{4}$ teaspoon of salt in 1 cup of water. A second recipe calls for 1 teaspoon of salt in 2 cups of water. Which recipe will taste saltier?

Complete the ratio tables and graph the ordered pairs from the tables. What can you conclude?

3 **12.**

Water Tank	
Time (min)	Liters Leaked
2	4
4	
6	
8	

Swimming Pool	
Time (min)	Liters Leaked
3	2
6	
9	
12	

13.

Zoo	
People	Cost (dollars)
4	60
8	
12	
16	

Museum	
People	Cost (dollars)
5	95
10	
15	
20	

14. **MILK** In whole milk, 13 parts out of 400 are milk fat. In 2% milk, 1 part out of 50 is milk fat. Which type of milk has more milk fat per cup?

15. **HEART RATE** A horse's heart beats 440 times in 10 minutes. A cow's heart beats 390 times in 6 minutes. Which animal has a greater heart rate?

16. **CHOOSE TOOLS** A chemist prepares two acid solutions.

a. Use a ratio table to determine which solution is more acidic.

b. Use a graph to determine which solution is more acidic.

c. Which method do you prefer? Explain.

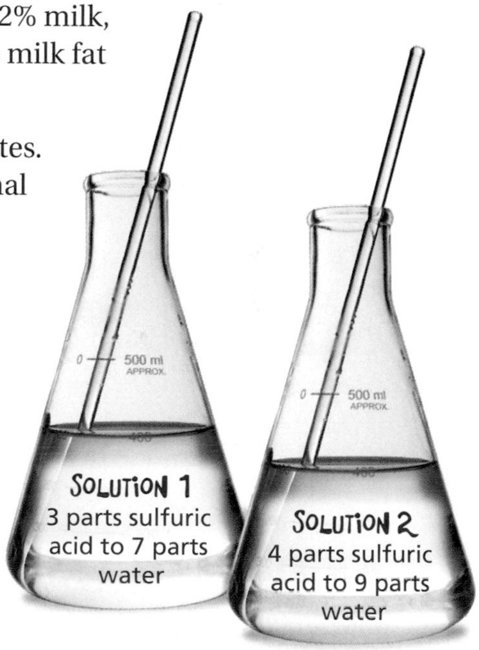

17. **NUT MIXTURE** A company offers a nut mixture with 7 peanuts for every 4 almonds. The company changes the mixture to have 8 peanuts for every 5 almonds, but the number of nuts per container does not change.

a. Create a ratio table for each mixture. How many nuts are in the smallest possible container?

b. Graph the ordered pairs from the tables. What can you conclude?

c. Almonds cost more than peanuts. Should the company charge more or less for the new mixture? Explain your reasoning.

18. **Structure** The point (p, q) is on the graph of values from a ratio table. What is another point on the graph?

Fair Game Review What you learned in previous grades & lessons

Divide. *(Section 1.1)*

19. $544 \div 34$ **20.** $1520 \div 83$ **21.** $8439 \div 245$

22. **MULTIPLE CHOICE** Which of the following numbers is equal to 9.32 when you increase it by 4.65? *(Section 2.4)*

Ⓐ 4.33 Ⓑ 4.67 Ⓒ 5.67 Ⓓ 13.97

5 Study Help

Check It Out
Graphic Organizer
BigIdeasMath.com

You can use a **definition and example chart** to organize information about a concept. Here is an example of a definition and example chart for ratio.

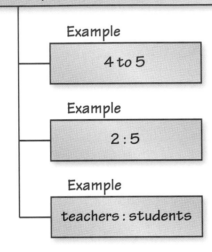

Ratio: a comparison of two quantities. Ratios can be part-to-part, part-to-whole, or whole-to-part comparisons.

- Example: 4 to 5
- Example: 2 : 5
- Example: teachers : students

On Your Own

Make definition and example charts to help you study these topics.

1. equivalent ratios
2. ratio table
3. rate
4. unit rate
5. equivalent rates

After you complete this chapter, make definition and example charts for the following topics.

6. percent
7. U.S. customary system
8. metric system
9. conversion factor
10. unit analysis

"My math teacher taught us how to make a definition and example chart."

5.1–5.4 Quiz

Write the ratio. Explain what the ratio means. *(Section 5.1)*

1. tulips to lilies

2. crayons to markers

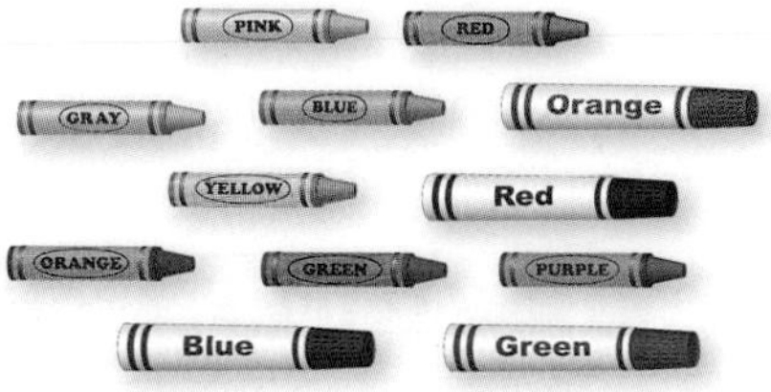

Find the missing values in the ratio table. Then write the equivalent ratios. *(Section 5.2)*

3.

Shoes	7		49
Boots	2	8	

4.

Trains	3	12	
Airplanes	8		48

Write a rate that represents the situation. *(Section 5.3)*

5.

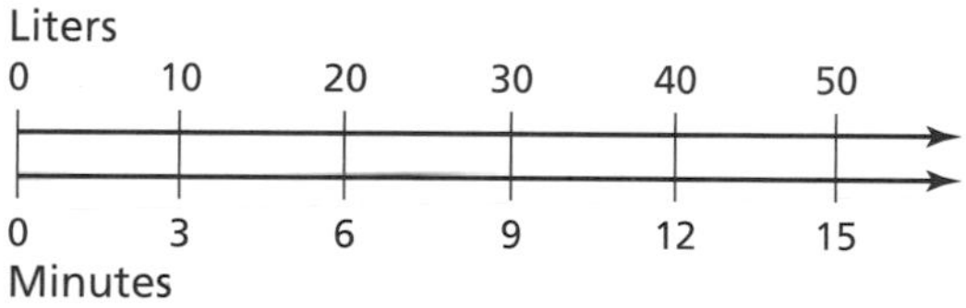

6.

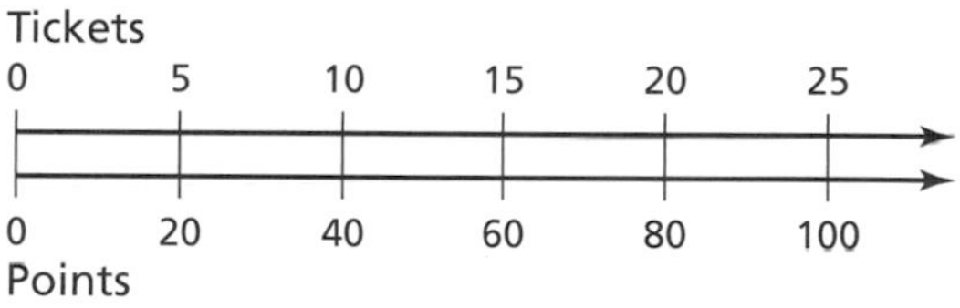

Write a unit rate for the situation. *(Section 5.3)*

7. 12 touchdowns in 6 games

8. 15 text messages in 5 minutes

9. 80 entries in 4 contests

10. 75 questions in 25 minutes

11. DOWNLOADS Three album downloads cost \$36. How much do 5 album downloads cost? *(Section 5.3)*

12. SHAMPOO You can buy 20 fluid ounces of shampoo for \$4.40 or 24 fluid ounces for \$4.80. Which is the better buy? Explain. *(Section 5.4)*

13. NBA CHAMPIONSHIPS Write each ratio. Explain what the ratio means. *(Section 5.1)*

a. Celtics championships to Lakers championships

b. Pistons championships to Spurs championships

c. Bulls championships to Lakers championships

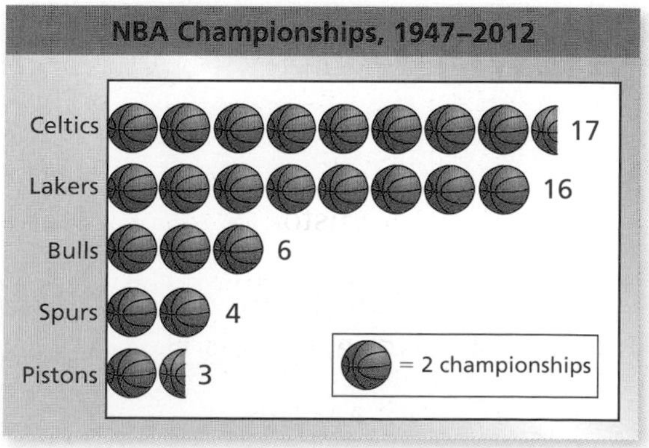

5.5 Percents

Essential Question What is the connection between ratios, fractions, and percents?

1 ACTIVITY: Writing Ratios

Work with a partner.

- **Write the fraction of the squares that are shaded.**
- **Write the ratio of the number of shaded squares to the total number of squares.**
- **How are the ratios and the fractions related?**
- **When can you write ratios as fractions?**

a.

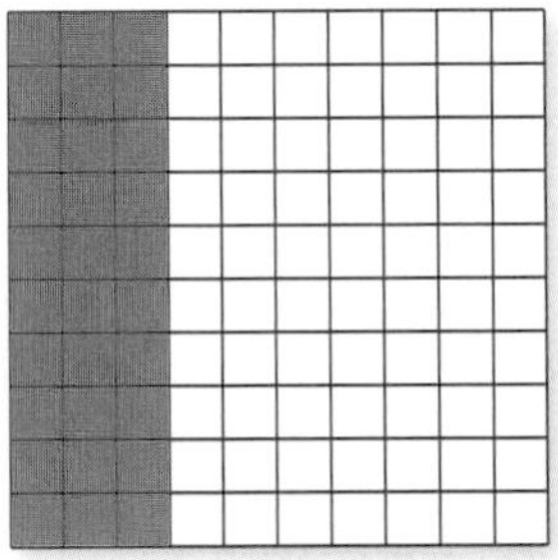

b.

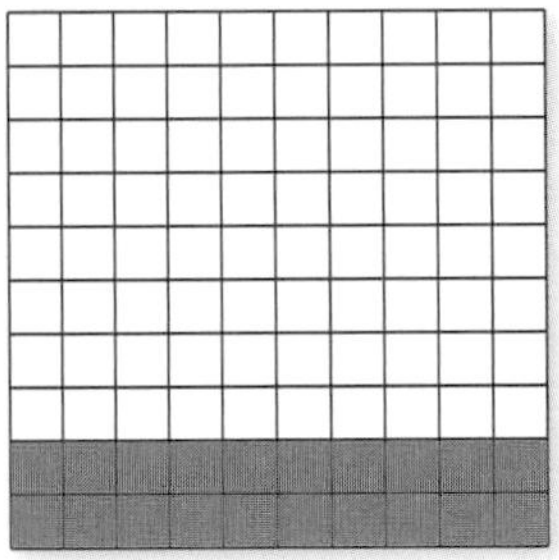

c.

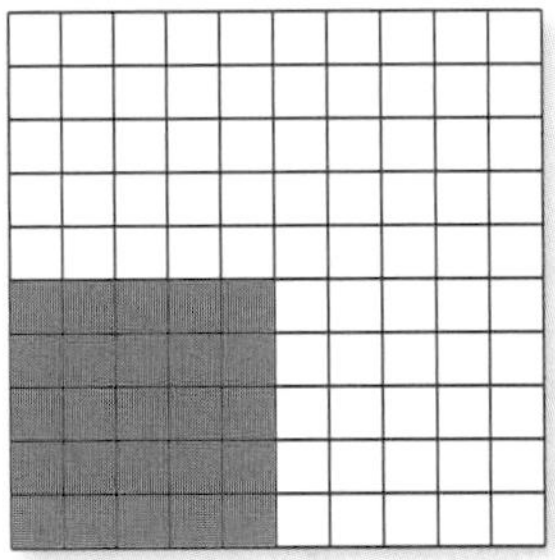

The Meaning of a Word ● Percent

A century is 100 years.

A cent is one hundredth of a dollar.

In Mexico, a centavo is one hundredth of a peso.

COMMON CORE

Percents

In this lesson, you will
- write percents as fractions with denominators of 100.
- write fractions as percents.

Learning Standard 6.RP.3c

Cent means *one hundred*, so **percent** means *per one hundred*. The symbol for percent is %.

2 ACTIVITY: Writing Percents as Fractions

Work with a partner.

- **What percent of each diagram in Activity 1 is shaded?**
- **What percent of each diagram below is shaded? Write each percent as a fraction in simplest form.**

a. 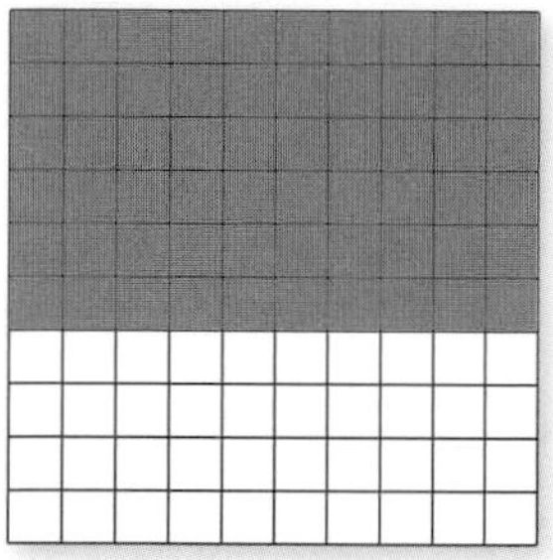

b. 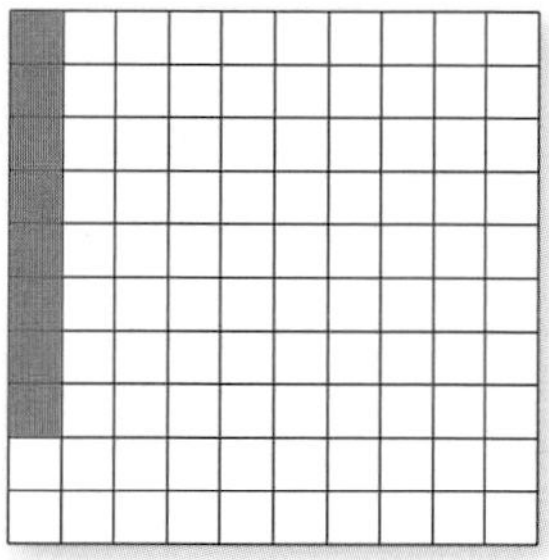

c. 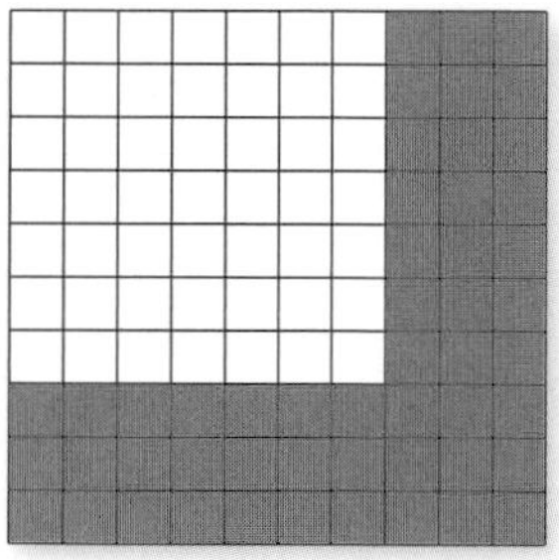

3 ACTIVITY: Writing Fractions as Percents

Work with a partner. Draw a model to represent the fraction. How can you write the fraction as a percent?

a. $\frac{2}{5} = \frac{\square}{100} = \square\%$

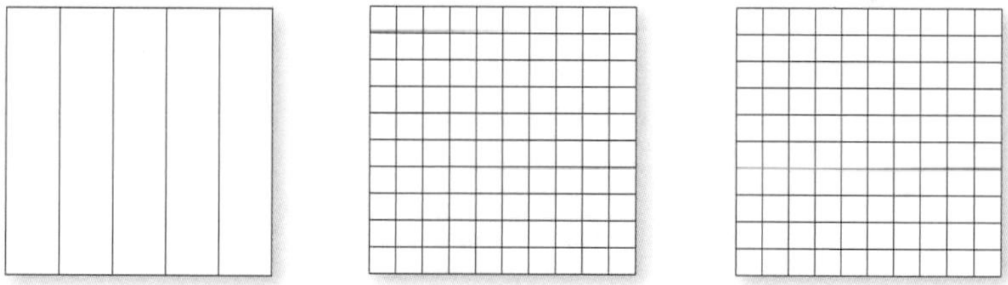

b. $\frac{7}{10}$

c. $\frac{3}{5}$

d. $\frac{3}{4}$

e. $\frac{3}{25}$

Math Practice 1

Consider Similar Problems

How is this problem similar to ones you have seen before? How does this help you find the solution?

What Is Your Answer?

4. **IN YOUR OWN WORDS** What is the connection between ratios, fractions, and percents? Give an example with your answer.
5. **REASONING** Your score on a test is 110%. What does this mean?

Practice Use what you learned about percents to complete Exercises 5–7 on page 222.

Key Vocabulary
percent, *p. 220*

Key Idea

Writing Percents as Fractions

Words A **percent** is a part-to-whole ratio where the whole is 100. So, you can write a percent as a fraction with a denominator of 100.

Numbers $60\% = 60 \text{ out of } 100 = \frac{60}{100}$ (60 ← part; fraction bar ← per; 100 ← one hundred (whole))

Algebra $n\% = \frac{n}{100}$

EXAMPLE 1 Writing Percents as Fractions

Study Tip
Equivalent fractions and percents represent the same number using different notations.

a. Write 35% as a fraction in simplest form.

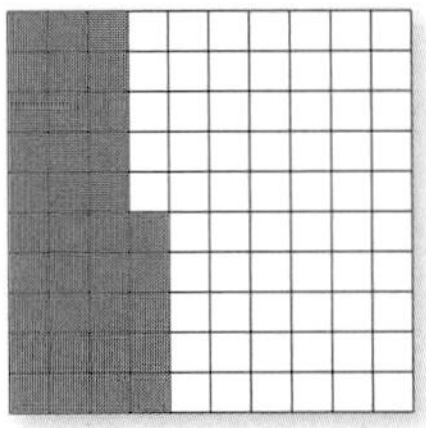

$35\% = \frac{35}{100}$ Write as a fraction with a denominator of 100.

$= \frac{7}{20}$ Simplify.

∴ So, $35\% = \frac{7}{20}$.

b. Write 100% as a fraction in simplest form.

$100\% = \frac{100}{100}$ Write as a fraction with a denominator of 100.

$= 1$ Simplify.

∴ So, $100\% = 1$.

c. Write 174% as a mixed number in simplest form.

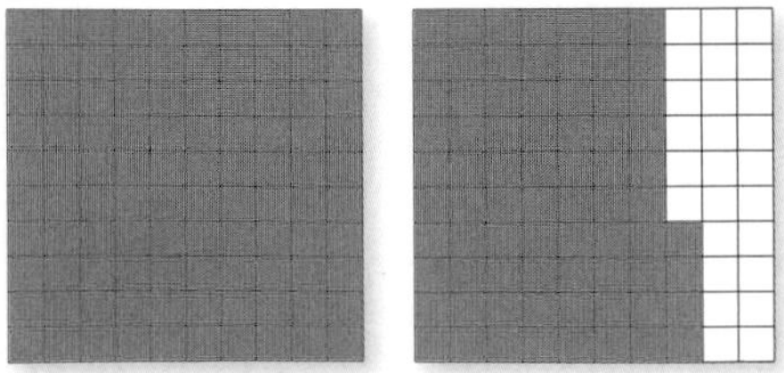

$174\% = \frac{174}{100}$ Write as a fraction with a denominator of 100.

$= \frac{87}{50}$, or $1\frac{37}{50}$ Simplify.

∴ So, $174\% = 1\frac{37}{50}$.

On Your Own

Now You're Ready
Exercises 8–19

Write the percent as a fraction or mixed number in simplest form.

1. 5% **2.** 168% **3.** 36% **4.** 83%

Key Idea

Writing Fractions as Percents

Words Write an equivalent fraction with a denominator of 100. Then write the numerator with the percent symbol.

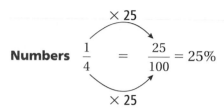

Numbers

$$\frac{1}{4} \overset{\times 25}{=} \frac{25}{100} = 25\%$$

(× 25 on the numerator and × 25 on the denominator)

EXAMPLE 2 Writing a Fraction as a Percent

Write $\frac{3}{50}$ as a percent.

$$\frac{3}{50} = \frac{6}{100} = 6\%$$

(× 2 on the numerator and × 2 on the denominator)

Because $50 \times 2 = 100$, multiply the numerator and denominator by 2. Write the numerator with a percent symbol.

EXAMPLE 3 Real-Life Application

A drought affects 9 out of 12 midwestern states. What percent of the midwestern states are affected by the drought?

Midwestern United States

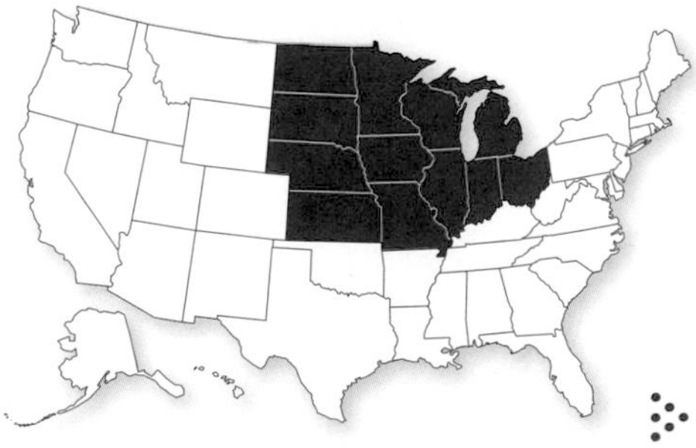

$\frac{9}{12} = \frac{3}{4}$ Simplify.

$= \frac{75}{100}$ ← $\frac{3 \times 25}{4 \times 25} = \frac{75}{100}$

$= 75\%$ Write the numerator with a percent symbol.

∴ So, 75% of the midwestern states are affected by the drought.

On Your Own

Now You're Ready
Exercises 21–28

Write the fraction or mixed number as a percent.

5. $\frac{31}{50}$ **6.** $\frac{7}{25}$ **7.** $\frac{19}{20}$ **8.** $1\frac{1}{2}$

9. **WHAT IF?** In Example 3, it rains in all the midwestern states. In what percent of the states affected by drought does it rain?

5.5 Exercises

Check It Out
Help with Homework
BigIdeasMath.com

Vocabulary and Concept Check

1. **WRITING** Explain how you can use a 10-by-10 grid to model 42%.
2. **WHICH ONE DOESN'T BELONG?** Which one does *not* have the same value as the other three? Explain your reasoning.

$\frac{10}{100}$	10%	$\frac{1}{10}$	0.01

3. **OPEN-ENDED** Write three different fractions that are less than 40%.
4. **NUMBER SENSE** Can $1\frac{1}{4}$ be written as a percent? Explain.

Practice and Problem Solving

Use a 10-by-10 grid to model the percent.

5. 10%
6. 55%
7. 35%

Write the percent as a fraction or mixed number in simplest form.

(1) 8. 45%
9. 90%
10. 15%
11. 7%
12. 34%
13. 79%
14. 77.5%
15. 188%
16. 8%
17. 224%
18. 0.25%
19. 0.4%

20. **ERROR ANALYSIS** Describe and correct the error in writing 225% as a fraction.

✗ $225\% = \frac{225}{1000} = \frac{9}{40}$

Write the fraction or mixed number as a percent.

(2) 21. $\frac{1}{10}$
22. $\frac{1}{5}$
23. $\frac{11}{20}$
24. $\frac{2}{25}$
25. $\frac{27}{50}$
26. $\frac{18}{25}$
27. $1\frac{17}{20}$
28. $2\frac{41}{50}$

29. **ERROR ANALYSIS** Describe and correct the error in writing $\frac{14}{25}$ as a percent.

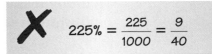

✗ $\frac{14}{25} = \frac{14 \times 4}{25 \times 4} = \frac{56}{100} = 0.56\%$

30. **LEFT-HANDED** Of the students in your class, 12% are left-handed. What *fraction* of the students are left-handed? Are there more right-handed or left-handed students? Explain.
31. **ARCADE** You have 125% of the tickets required for a souvenir. What *fraction* of the required tickets do you have? Do you need more tickets for the souvenir? Explain.

Find the percent.

32. 3 is what percent of 8?

33. 13 is what percent of 16?

34. 9 is what percent of 16?

35. 33 is what percent of 40?

36. **SOCIAL NETWORKING** A survey asked students to choose their favorite social networking website. The results are shown in the table.

Social Networking Website	Number of Students
Website A	35
Website B	13
Website C	22
Website D	10

a. What fraction of the students chose Website A?

b. What percent of the students chose Website C?

37. **GEOGRAPHY** The percent of the total area of the United States that is in each of four states is shown.

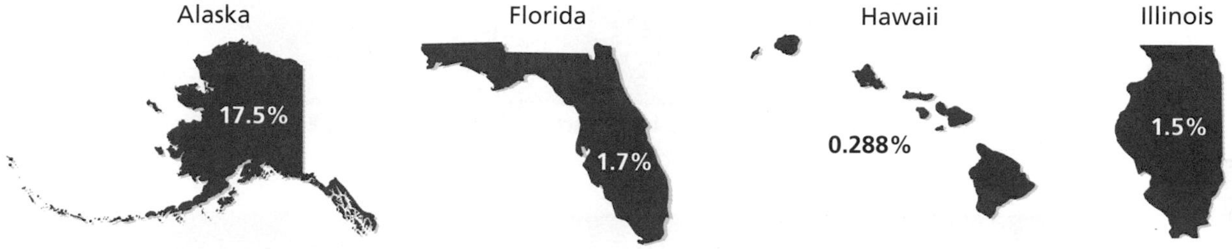

a. Write the percents as fractions in simplest form.

b. How many times larger is Illinois than Hawaii?

c. Compared to the map of Florida, is the map of Alaska the correct size? Explain your reasoning.

d. **RESEARCH** Which of the 50 states are larger than Illinois?

38. **CRITICAL THINKING** A school fundraiser raised 120% of its goal last year and 125% of its goal this year. Did the fundraiser raise more money this year? Explain your reasoning.

39. **CRITICAL THINKING** How can you use a 10-by-10 grid to model $\frac{1}{2}$%?

40. **Reasoning** Write $\frac{1}{12}$ as a percent. Explain how you found your answer.

Fair Game Review What you learned in previous grades & lessons

Divide. Write the answer in simplest form. *(Section 2.2)*

41. $\frac{1}{6} \div \frac{1}{3}$

42. $9 \div \frac{3}{4}$

43. $10 \div \frac{5}{8}$

44. $\frac{1}{6} \div 2$

45. **MULTIPLE CHOICE** Which of the following is *not* equal to 15? *(Section 2.1)*

Ⓐ $\frac{3}{4} \cdot 20$ Ⓑ $\frac{5}{9} \cdot 27$ Ⓒ $35 \cdot \frac{3}{7}$ Ⓓ $28 \cdot \frac{5}{7}$

5.6 Solving Percent Problems

Essential Question How can you use mental math to find the percent of a number?

"I have a secret way for finding 21% of 80." "10% is 8, and 1% is 0.8." "So, 21% is 8 + 8 + 0.8 = 16.8."

1 ACTIVITY: Finding 10% of a Number

Work with a partner.

a. How did Newton know that 10% of 80 is 8?

Write 10% as a fraction. $10\% = \frac{\square}{\square} = \frac{1}{\square}$ (10, per, cent)

Method 1: Use a model.

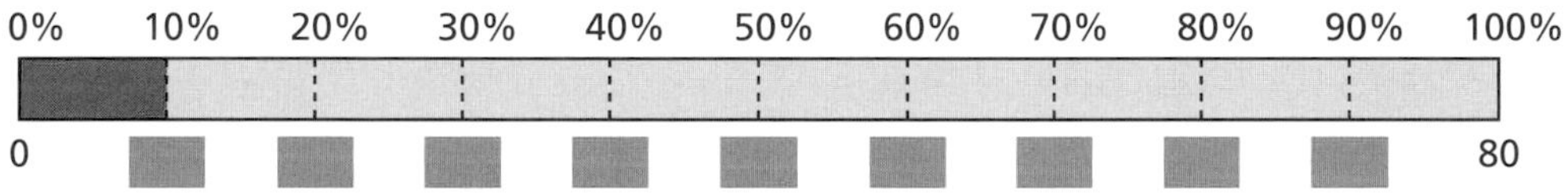

Method 2: Use multiplication.

$$10\% \text{ of } 80 = \frac{\square}{10} \text{ of } 80 = \frac{\square}{10} \times \square = \frac{\square}{10} = \square$$

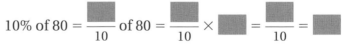

b. How do you move the decimal point to find 10% of a number?
Move the decimal point one place to the ☐. **10% of 80. =** ☐

COMMON CORE
Percents
In this lesson, you will
- find percents of numbers.
- find the whole given the part and the percent.

Learning Standard
6.RP.3c

2 ACTIVITY: Finding 1% of a Number

Work with a partner.

a. How did Newton know that 1% of 80 is 0.8?

b. How do you move the decimal point to find 1% of a number?

3 ACTIVITY: Using Mental Math

> **Math Practice 8**
>
> **Evaluate Results**
>
> Does your answer seem reasonable? How can you check your answer?

Work with a partner. Use mental math to find each percent of a number.

a. 12% of 40

Think: 12% = 10% + 1% + 1%

10% of 40 = ☐ 1% of 40 = ☐

☐ + ☐ + ☐ = ☐

b. 19% of 50

Think: 19% = 10% + 10% − 1%

10% of 50 = ☐ 1% of 50 = ☐

☐ + ☐ − ☐ = ☐

4 ACTIVITY: Using Mental Math

Work with a partner. Use mental math to find each percent of a number.

a. 20% tip for a $30 meal

b. 18% tip for a $30 meal

c. 6% sales tax on a $20 shirt

d. 9% sales tax on a $20 shirt

e. 6% service charge for a $200 boxing ticket

f. 2% delivery fee for a $200 boxing ticket

g. 21% bonus on a total of 40,000 points

h. 38% bonus on a total of 80,000 points

What Is Your Answer?

5. **IN YOUR OWN WORDS** How can you use mental math to find the percent of a number?
6. Describe two real-life examples of finding a percent of a number.
7. How can you use 10% of a number to find 20% of the number? 30%? Explain your reasoning.

Practice Use what you learned about finding the percent of a number to complete Exercises 3–10 on page 229.

Check It Out
Lesson Tutorials
BigIdeasMath.com

Key Idea

Finding the Percent of a Number

Words Write the percent as a fraction. Then multiply by the whole. The percent times the whole equals the part.

Numbers 20% of 60 is 12.

$$\frac{1}{5} \times 60 = 12$$

Model

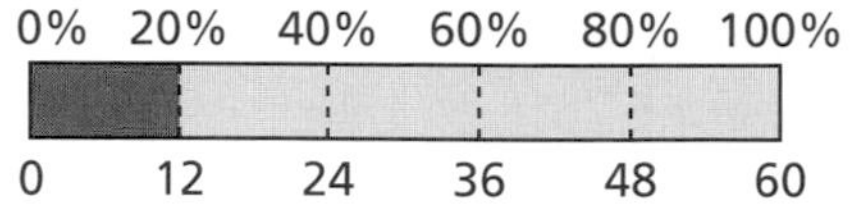

EXAMPLE 1 Finding the Percent of a Number

25% of 40 is what number?

$$25\% \text{ of } 40 = \frac{1}{4} \cdot 40$$ Write the percent as a fraction and multiply.

$$= \frac{1 \cdot \cancel{40}^{10}}{\cancel{4}_1}$$ Divide out the common factor.

$$= 10$$ Simplify.

So, 25% of 40 is 10.

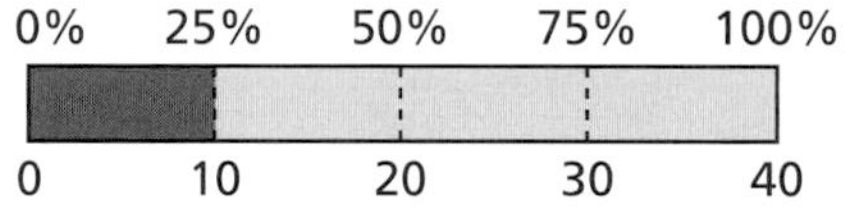

Study Tip

You can use mental math to check your answer in Example 1.

10% of 40 = 4

5% of 40 = 2

So, 25% of 40 is 4 + 4 + 2 = 10.

You can also use a ratio table to find the percent of a number.

EXAMPLE 2 Finding the Percent of a Number Using a Ratio Table

60% of 150 is what number?

Use a ratio table to find the part. Let one row be the *part,* and let the other be the *whole.* Find an equivalent ratio with 150 as the whole.

The first column represents the percent.

$$\frac{\text{part}}{\text{whole}} = \frac{60}{100} = 60\%$$

÷ 2 ×3

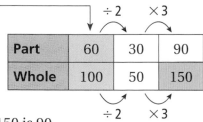

Part	60	30	90
Whole	100	50	150

÷ 2 ×3

So, 60% of 150 is 90.

On Your Own

Now You're Ready
Exercises 3–22

Find the percent of the number. Explain your method.

1. 90% of 20 **2.** 75% of 32 **3.** 10% of 110 **4.** 30% of 75

You can use a related division equation to find the whole given the part and the percent.

Key Idea

Finding the Whole

Write the percent as a fraction. Then divide the part by the fraction.

Words The part divided by the percent equals the whole.

Numbers 20% of 60 is 12.

$$\frac{1}{5} \times 60 = 12 \longrightarrow 12 \div \frac{1}{5} = 60$$

Multiplication equation → Related division equation

EXAMPLE 3 Finding the Whole

75% of what number is 48?

$48 \div 75\% = 48 \div \frac{3}{4}$ — Write the percent as a fraction and divide.

$= 48 \cdot \frac{4}{3}$ — Multiply by the reciprocal.

$= 64$ — Simplify.

So, 75% of 64 is 48.

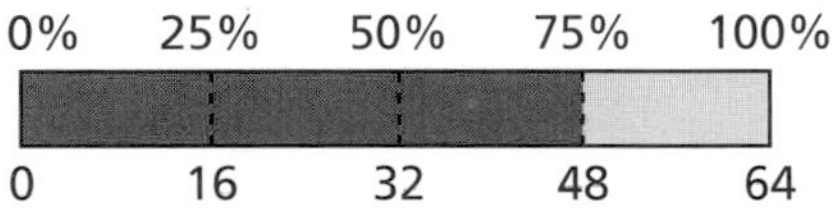

EXAMPLE 4 Finding the Whole Using a Ratio Table

120% of what number is 72?

Use a ratio table to find the whole. Find an equivalent ratio with 72 as the part.

The first column represents the percent.

$\frac{\text{part}}{\text{whole}} = \frac{120}{100} = 120\%$

	÷ 20 →	× 12 →	
Part	120	6	72
Whole	100	5	60

(÷ 20, × 12 applied to both rows)

So, 120% of 60 is 72.

On Your Own

Now You're Ready
Exercises 27–36

Find the whole. Explain your method.

5. 5% of what number is 10?

6. 62% of what number is 31?

EXAMPLE 5 Real-Life Application

The width of a rectangular room is 80% of its length. What is the area of the room?

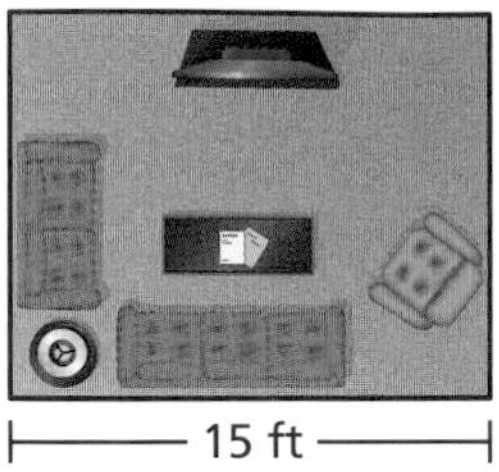

Find 80% of 15 feet.

$$80\% \text{ of } 15 = \frac{4}{5} \times 15$$

$$= \frac{4 \times \overset{3}{\cancel{15}}}{\underset{1}{\cancel{5}}}$$

$$= 12$$

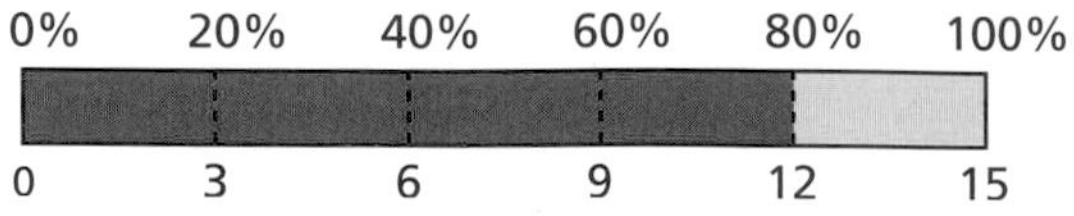

The width is 12 feet.

Use the formula for the area A of a rectangle.

$$A = 15 \times 12 = 180$$

∴ So, the area of the room is 180 square feet.

On Your Own

7. The width of a rectangular stage is 55% of its length. The stage is 120 feet long. What is the area?

EXAMPLE 6 Real-Life Application

You win an online auction for concert tickets. Your winning bid is 60% of your maximum bid. How much more were you willing to pay for the tickets than you actually paid?

Ⓐ \$72 Ⓑ \$80 Ⓒ \$120 Ⓓ \$200

Your maximum bid is the *whole*, and your winning bid is the *part*. Find your maximum bid by dividing the part by the percent.

$$120 \div 60\% = 120 \div \frac{3}{5}$$ Divide the part by the percent.

$$= 120 \cdot \frac{5}{3}$$ Multiply by the reciprocal.

$$= 200$$ Simplify.

Your maximum bid is \$200, and your winning bid is \$120. So, you were willing to pay $200 - 120 =$ \$80 more for the tickets.

∴ The correct answer is Ⓑ.

On Your Own

8. **WHAT IF?** Your winning bid is 96% of your maximum bid. How much more were you willing to pay for the tickets than you actually paid?

5.6 Exercises

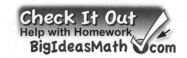

Check It Out
Help with Homework
BigIdeasMath.com

Vocabulary and Concept Check

1. **DIFFERENT WORDS, SAME QUESTION** Which is different? Find "both" answers.

What is twenty percent of 30?	What is one-fifth of 30?
Twenty percent of what number is 30?	What is two-tenths of 30?

2. **NUMBER SENSE** If 52 is 130% of a number, is the number greater or less than 52? Explain.

Practice and Problem Solving

Find the percent of the number. Explain your method.

(1) (2) **3.** 20% of 60 **4.** 10% of 40 **5.** 50% of 70 **6.** 30% of 30

7. 10% of 90 **8.** 15% of 20 **9.** 25% of 50 **10.** 5% of 60

11. 30% of 70 **12.** 75% of 48 **13.** 45% of 45 **14.** 92% of 19

15. 40% of 60 **16.** 38% of 22 **17.** 70% of 20 **18.** 87% of 55

19. 140% of 60 **20.** 120% of 33 **21.** 175% of 54 **22.** 250% of 146

23. **ERROR ANALYSIS** Describe and correct the error in finding 40% of 75.

✗ $40\% \text{ of } 75 = 40\% \times 75 = 3000$

24. **PINE TREES** A town had about 2120 acres of pine trees 40 years ago. Only about 13% of the pine trees remain. How many acres of pine trees remain?

25. **SPIDER MONKEY** The tail of the spider monkey is 64% of the length shown. What is the length of its tail?

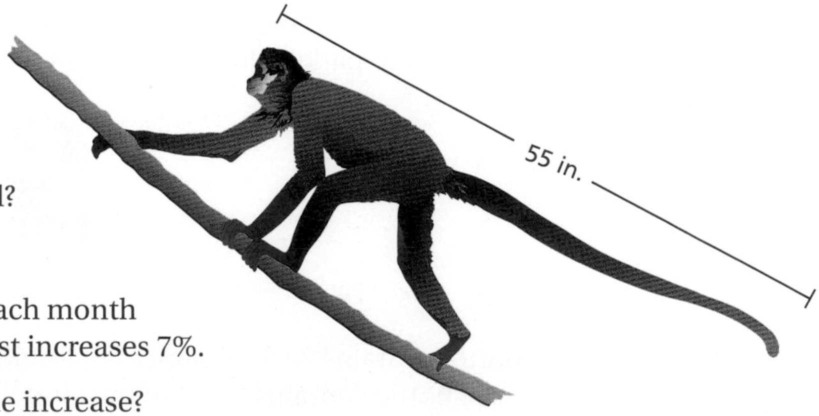

26. **CABLE** A family pays \$45 each month for cable television. The cost increases 7%.
   - **a.** How many dollars is the increase?
   - **b.** What is the new monthly cost?

Find the whole. Explain your method.

3 4 **27.** 10% of what number is 14? **28.** 20% of what number is 18?

29. 25% of what number is 21? **30.** 75% of what number is 27?

31. 15% of what number is 12? **32.** 85% of what number is 17?

33. 140% of what number is 35? **34.** 160% of what number is 32?

35. 125% of what number is 25? **36.** 175% of what number is 42?

37. ERROR ANALYSIS Describe and correct the error in finding the whole.

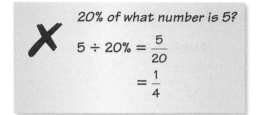

✗ *20% of what number is 5?*

$$5 \div 20\% = \frac{5}{20} = \frac{1}{4}$$

38. COUPON You have a coupon for a restaurant. You save \$3 on a meal. What was the original cost of the meal?

39. SURVEY The results of a survey are shown at the right. In the survey, 12 students said that they would like to learn French.

a. How many students were surveyed?

b. How many of the students surveyed would like to learn Spanish?

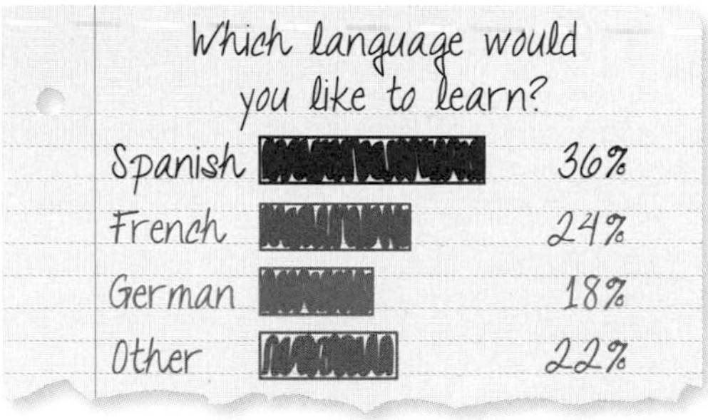

40. WEIGHT A sixth grader weighs 90 pounds, which is 120% of what he weighed in fourth grade. How much did he weigh in fourth grade?

41. PARKING LOT In a parking lot, 16% of the cars are blue. There are 4 blue cars in the parking lot. How many cars in the parking lot are *not* blue?

42. LOTION A bottle contains 20 fluid ounces of lotion and sells for \$5.80. The 20-fluid-ounce bottle contains 125% of the lotion in the next smallest size, which sells for \$5.12. Which is the better buy? Explain.

Copy and complete the statement using <, >, or =.

43. 80% of 60 $\square$ 60% of 80

44. 20% of 30 $\square$ 30% of 40

45. 120% of 5 $\square$ 0.8% of 250

46. 85% of 40 $\square$ 25% of 136

47. **TIME** How many minutes is 40% of 2 hours?

48. **LENGTH** How many inches is 78% of 3 feet?

49. **GEOMETRY** The width of the rectangle is 75% of its length.

- **a.** What is the area of the rectangle?
- **b.** The length of the rectangle is doubled. What percent of the length is the width now? Explain your reasoning.

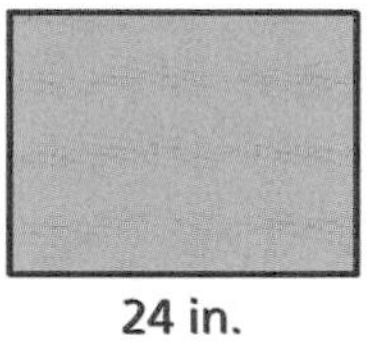

50. **BASKETBALL** To pass inspection, a new basketball should bounce back to between 68% and 75% of the starting height. A new ball is dropped from 6 feet and bounces back 4 feet 1 inch. Does the ball pass inspection? Explain.

51. **REASONING** You know that 15% of a number n is 12. How can you use this to find 30% of n? 45% of n? Explain.

52. **SURFBOARD** You have a coupon for 10% off the sale price of a surfboard. Which is the better buy? Explain your reasoning.

- 40% off the regular price
- 30% off the regular price and then 10% off the sale price

53. **Number Sense** On three 150-point geography tests, you earned grades of 88%, 94%, and 90%. The final test is worth 250 points. What *percent* do you need on the final to earn 93% of the total points on all tests?

Fair Game Review What you learned in previous grades & lessons

Multiply. *(Section 2.5)*

54. 0.6×8

55. 3.3×5

56. 0.74×9

57. 2.19×12

58. **MULTIPLE CHOICE** What is the quotient of 75 and 2.4? *(Section 2.6)*

Ⓐ 0.032 Ⓑ 0.3125 Ⓒ 3.2 Ⓓ 31.25

5.7 Converting Measures

Essential Question How can you compare lengths between the customary and metric systems?

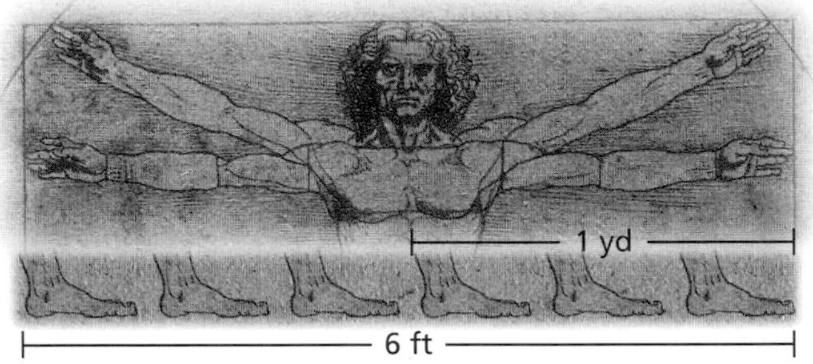

1 ACTIVITY: Customary Measure History

Work with a partner.

a. Match the measure of length with its historical beginning.

Length	*Historical Beginning*
Inch	The length of a human foot
Foot	The width of a human thumb
Yard	The distance a human can walk in 1000 paces (1 pace = 2 steps)
Mile	The distance from a human nose to the end of an outstretched human arm

b. Use a ruler to measure your thumb, arm, and foot. How do your measurements compare to your answers from part (a)? Are they close to the historical measures?

You know how to convert measures within the customary and metric systems.

Equivalent Customary Lengths

1 ft = 12 in. 1 yd = 3 ft 1 mi = 5280 ft

Equivalent Metric Lengths

1 m = 1000 mm 1 m = 100 cm 1 km = 1000 m

You will learn how to convert between the two systems.

Converting Between Systems

1 in. = 2.54 cm

1 mi ≈ 1.61 km

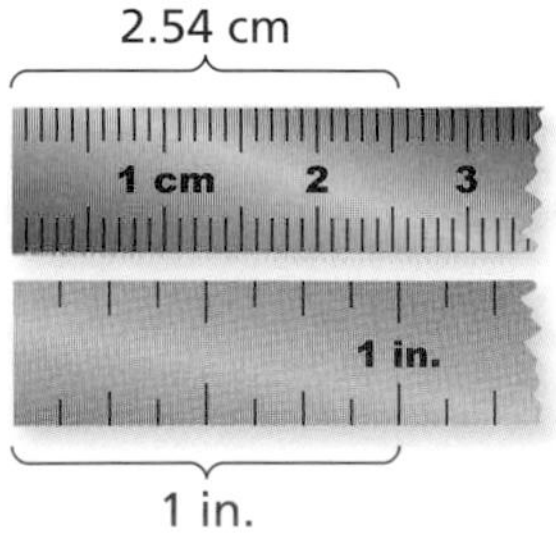

Common Core

Converting Measures

In this lesson, you will
- use conversion factors (rates) to convert units of measurement.

Learning Standard 6.RP.3d

2 ACTIVITY: Comparing Measures

> **Math Practice**
>
> **Analyze Givens**
>
> What is the relationship between the given quantities? What are you trying to find?

Work with a partner. Answer each question. Explain your answer. Use a diagram in your explanation.

		Metric	*Customary*
a.	**Car Speed:** Which is faster?	80 km/h	60 mi/h
b.	**Trip Distance:** Which is farther?	200 km	200 mi
c.	**Human Height:** Who is taller?	180 cm	5 ft 8 in.
d.	**Wrench Width:** Which is wider?	8 mm	5/16 in.
e.	**Swimming Pool Depth:** Which is deeper?	1.4 m	4 ft

3 ACTIVITY: Changing Units in a Rate

Work with a partner. Change the units of the rate by multiplying by a "Magic One." Write your answer as a unit rate. Show your work.

	Original Rate		*Magic One*		*New Units*		*Unit Rate*
a.	**Sample:**						
	$\frac{\$120}{\cancel{h}}$	$\times$	$\frac{1\cancel{h}}{60\text{ min}}$	$=$	$\frac{\$120}{60\text{ min}}$	$=$	$\frac{\$2}{1\text{ min}}$
b.	$\frac{\$3}{\text{min}}$	$\times$	1			$=$	$\$\frac{\square}{1\text{ h}}$
c.	$\frac{12\text{ in.}}{\text{ft}}$	$\times$	1			$=$	$\frac{\square\text{ in.}}{1\text{ yd}}$
d.	$\frac{2\text{ ft}}{\text{week}}$	$\times$	1			$=$	$\frac{\square\text{ ft}}{1\text{ yr}}$

What Is Your Answer?

4. One problem-solving strategy is called *Working Backwards*. What does this mean? How can you use this strategy to find the rates in Activity 3?

5. **IN YOUR OWN WORDS** How can you compare lengths between the customary and the metric systems? Give examples with your description.

Practice Use what you learned about converting measures between systems to complete Exercises 4 and 5 on page 236.

5.7 Lesson

Check It Out
Lesson Tutorials
BigIdeasMath.com

Key Vocabulary
U.S. customary system, *p. 234*
metric system, *p. 234*
conversion factor, *p. 234*
unit analysis, *p. 234*

The **U.S. customary system** is a system of measurement that contains units for length, capacity, and weight. The **metric system** is a decimal system of measurement, based on powers of 10, that contains units for length, capacity, and mass.

To convert from one unit of measure to another, multiply by one or more *conversion factors*. A conversion factor can be written using fraction notation.

Key Idea

Conversion Factor

A **conversion factor** is a rate that equals 1.

	Relationship	*Conversion Factors*
Example	1 m ≈ 3.28 ft	$\frac{1\text{ m}}{3.28\text{ ft}}$ and $\frac{3.28\text{ ft}}{1\text{ m}}$

You can use **unit analysis** to decide which conversion factor will produce the appropriate units.

EXAMPLE 1 Converting Units

a. Convert 36 quarts to gallons.

Use a conversion factor.

1 gal = 4 qt

$$36\text{ qt} \cdot \frac{1\text{ gal}}{4\text{ qt}} = \frac{36 \cdot 1\text{ gal}}{4}$$

$$= 9\text{ gal}$$

So, 36 quarts is 9 gallons.

b. Convert 20 centimeters to inches.

Use a conversion factor.

1 in. = 2.54 cm

$$20\text{ cm} \cdot \frac{1\text{ in.}}{2.54\text{ cm}} \approx 7.87\text{ in.}$$

So, 20 centimeters is about 7.87 inches.

On Your Own

Now You're Ready
Exercises 6–17

Copy and complete the statement. Round to the nearest hundredth if necessary.

1. 48 ft = ___ yd
2. 7 lb = ___ oz
3. 5 g = ___ mg
4. 7 mi ≈ ___ km
5. 12 qt ≈ ___ L
6. 25 kg ≈ ___ lb

EXAMPLE 2 Comparing Units

Copy and complete the statement using < or >: 25 oz $\square$ 2 kg.

Convert 25 ounces to kilograms.

1 lb = 16 oz

1 lb ≈ 0.45 kg

$$25\ \cancel{\text{oz}} \times \frac{1\ \cancel{\text{lb}}}{16\ \cancel{\text{oz}}} \times \frac{0.45\ \text{kg}}{1\ \cancel{\text{lb}}} = \frac{25 \cdot 1 \cdot 0.45\ \text{kg}}{16 \cdot 1} \approx 0.70\ \text{kg}$$

Because 0.70 kilogram is less than 2 kilograms, 25 oz < 2 kg.

EXAMPLE 3 Converting a Rate: Changing One Unit

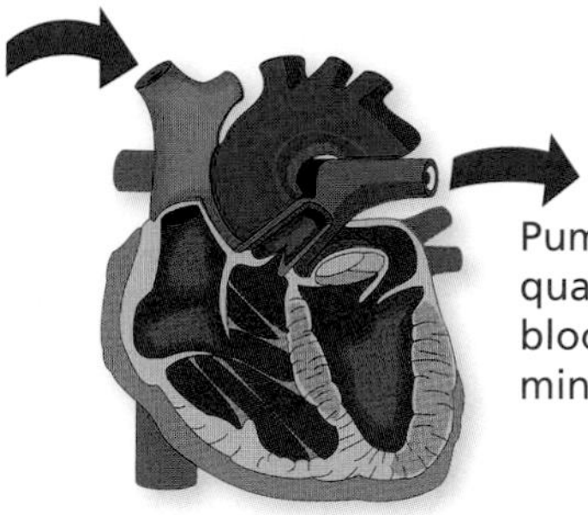

Pumps 5 quarts of blood per minute

How many liters does the human heart pump per minute?

1 qt ≈ 0.95 L

$$\frac{5\ \cancel{\text{qt}}}{1\ \text{min}} \cdot \frac{0.95\ \text{L}}{1\ \cancel{\text{qt}}} \approx \frac{4.75\ \text{L}}{1\ \text{min}}$$

The rate of 5 quarts per minute is about 4.75 liters per minute.

EXAMPLE 4 Converting a Speed: Changing Both Units

You are riding on a zip line. Your speed is 15 miles per hour. What is your speed in feet per second?

$$\frac{15\ \cancel{\text{mi}}}{1\ \cancel{\text{h}}}\left(\frac{5280\ \text{ft}}{1\ \cancel{\text{mi}}}\right)\left(\frac{1\ \cancel{\text{h}}}{3600\ \text{sec}}\right) = \frac{15 \cdot 5280\ \text{ft}}{3600\ \text{sec}}$$

$$= \frac{79{,}200\ \text{ft}}{3600\ \text{sec}}$$

$$= \frac{22\ \text{ft}}{1\ \text{sec}}$$

1 mi = 5280 ft

1 h = 3600 sec

Your speed is 22 feet per second.

On Your Own

Now You're Ready
Exercises 20–31

Copy and complete the statement using < or >.

7. 7 cm $\square$ 3 in.   8. 8 c $\square$ 2 L   9. 3 oz $\square$ 70 g

10. An oil tanker is leaking oil at a rate of 300 gallons per minute. What is this rate in gallons per second?

11. A tennis ball travels at a speed of 120 miles per hour. What is this rate in feet per second?

5.7 Exercises

Check It Out
Help with Homework
BigIdeasMath.com

Vocabulary and Concept Check

1. **VOCABULARY** Is $\frac{10\text{ mm}}{1\text{ cm}}$ a conversion factor? Explain.
2. **WRITING** Describe how to convert 2 liters per hour to milliliters per second.
3. **DIFFERENT WORDS, SAME QUESTION** Which is different? Find "both" answers.

Convert 5 inches to centimeters.	Find the number of inches in 5 centimeters.
How many centimeters are in 5 inches?	Five inches equals how many centimeters?

Practice and Problem Solving

Answer the question. Explain your answer.

4. Which juice container is larger: 2 L or 1 gal?
5. Which person is heavier: 75 kg or 110 lb?

Copy and complete the statement. Round to the nearest hundredth if necessary.

(1)
6. 3 pt = ____ c
7. 1500 mL = ____ L
8. 40 oz = ____ lb
9. 12 L ≈ ____ qt
10. 14 m ≈ ____ ft
11. 4 ft ≈ ____ m
12. 64 lb ≈ ____ kg
13. 0.3 km ≈ ____ mi
14. 75.2 in. ≈ ____ cm
15. 17 kg ≈ ____ lb
16. 15 cm ≈ ____ in.
17. 9 mi ≈ ____ km

18. **ERROR ANALYSIS** Describe and correct the error in converting the units.

✗
$$8\text{ L} \approx 8\text{ L} \cdot \frac{0.95\text{ qt}}{1\text{ L}} = 8\cancel{\text{L}} \cdot \frac{0.95\text{ qt}}{1\cancel{\text{L}}} = 7.6\text{ qt}$$

19. **BRIDGE** The Mackinac Bridge in Michigan is the third-longest suspension bridge in the United States.
    a. How high above the water is the roadway in meters?
    b. The bridge has a length of 26,372 feet. What is the length in kilometers?

Copy and complete the statement using < or >.

2 **20.** 8 kg $\square$ 30 oz　**21.** 6 ft $\square$ 300 cm　**22.** 3 gal $\square$ 6 L

23. 10 in. $\square$ 200 mm　**24.** 1200 g $\square$ 5 lb　**25.** 1500 m $\square$ 3000 ft

Copy and complete the statement.

3 4 **26.** $\frac{13\text{ km}}{\text{h}} \approx \frac{\square\text{ mi}}{\text{h}}$　**27.** $\frac{22\text{ L}}{\text{min}} = \frac{\square\text{ L}}{\text{h}}$　**28.** $\frac{63\text{ mi}}{\text{h}} = \frac{\square\text{ mi}}{\text{sec}}$

29. $\frac{3\text{ km}}{\text{min}} \approx \frac{\square\text{ mi}}{\text{h}}$　**30.** $\frac{17\text{ gal}}{\text{h}} \approx \frac{\square\text{ qt}}{\text{min}}$　**31.** $\frac{6\text{ cm}}{\text{min}} = \frac{\square\text{ m}}{\text{sec}}$

32. BOTTLE Can you pour the water from a full 2-liter bottle into a 2-quart pitcher without spilling any? Explain.

33. AUTOBAHN Germany suggests a speed limit of 130 kilometers per hour on highways.

a. Is the speed shown greater than the suggested limit?

b. Suppose the speed shown drops 30 miles per hour. Is the new speed below the suggested limit?

34. BIRDS The table shows the flying speeds of several birds.

a. Which bird is the fastest? Which is the slowest?

b. The peregrine falcon has a dive speed of 322 kilometers per hour. Is the dive speed of the peregrine falcon faster than the flying speed of any of the birds? Explain.

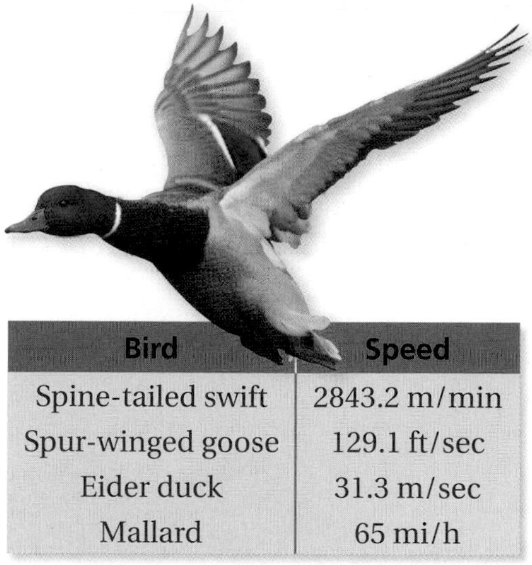

Bird	Speed
Spine-tailed swift	2843.2 m/min
Spur-winged goose	129.1 ft/sec
Eider duck	31.3 m/sec
Mallard	65 mi/h

35. SPEED OF LIGHT The speed of light is about 300,000 kilometers per second. Convert the speed to miles per hour.

36. Critical Thinking One liter of paint covers 100 square feet. How many gallons of paint does it take to cover a room whose walls have an area of 800 square meters?

Fair Game Review What you learned in previous grades & lessons

Find the percent of the number. *(Section 5.6)*

37. 25% of 120　**38.** 65% of 47　**39.** 120% of 15　**40.** 3.2% of 80

41. MULTIPLE CHOICE What is the area of a parallelogram with a base of 15 centimeters and a height of 12 centimeters? *(Section 4.1)*

Ⓐ 90 cm^2　Ⓑ 175 cm^2　Ⓒ 180 cm^2　Ⓓ 205 cm^2

5.5–5.7 Quiz

Check It Out
Progress Check
BigIdeasMath.com

Write the percent as a fraction or mixed number in simplest form. *(Section 5.5)*

1. 14%

2. 124%

Write the fraction or mixed number as a percent. *(Section 5.5)*

3. $\frac{13}{20}$

4. $1\frac{1}{4}$

Find the percent of the number. Explain your method. *(Section 5.6)*

5. 25% of 64

6. 120% of 50

Find the whole. Explain your method. *(Section 5.6)*

7. 60% of what number is 24?

8. 160% of what number is 80?

Copy and complete the statement. Round to the nearest hundredth if necessary. *(Section 5.7)*

9. 6.4 in. ≈ ____ cm

10. 4 qt ≈ ____ L

11. 10 kg ≈ ____ lb

12. ANATOMY About 62% of the human body is composed of water. Write this percent as a fraction in simplest form. *(Section 5.5)*

13. SAVES A goalie's saves (•) and goals scored against (×) are shown. What percent of shots did the goalie save? Explain. *(Section 5.5)*

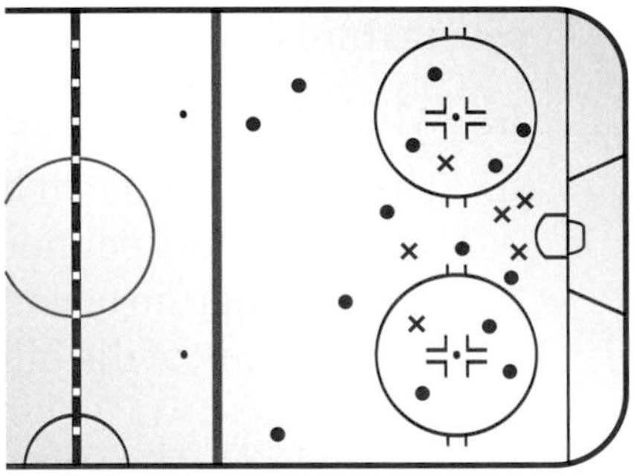

14. SHOPPING You went to the mall with $80. You spent 25% of your money on a pair of shorts and 65% of the remainder on sandals. How much did you spend on the sandals? Explain how you found your answer. *(Section 5.6)*

15. WINDSURFING Determine which windsurfer is traveling faster. Explain your reasoning. *(Section 5.7)*

Speed:
5 meters per second

Speed:
720 feet per minute

5 Chapter Review

Check It Out
Vocabulary Help
BigIdeasMath .com

Review Key Vocabulary

ratio, *p. 192*
equivalent ratios, *p. 198*
ratio table, *p. 198*
rate, *p. 206*
unit rate, *p. 206*
equivalent rates, *p. 206*
percent, *p. 220*
U.S. customary system, *p. 234*
metric system, *p. 234*
conversion factor, *p. 234*
unit analysis, *p. 234*

Review Examples and Exercises

5.1 Ratios *(pp. 190–195)*

Write the ratio of apples to oranges. Explain what the ratio means.

3 apples → 3 to 5 ← 5 oranges

So, the ratio of apples to oranges is 3 to 5, or 3 : 5. That means that for every 3 apples, there are 5 oranges.

Exercises

Write the ratio. Explain what the ratio means.

1. butterflies : caterpillars

2. saxophones : trumpets

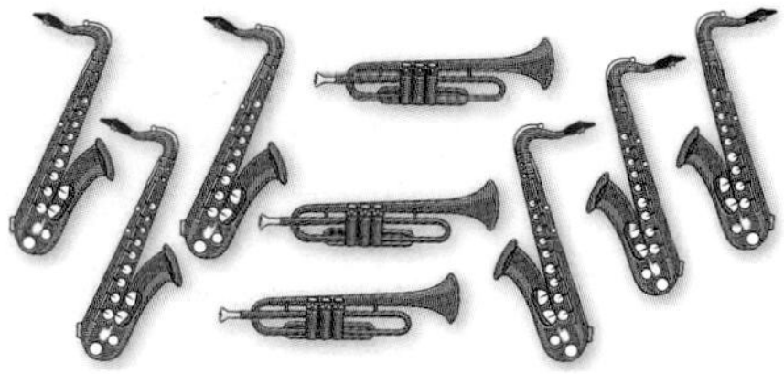

5.2 Ratio Tables *(pp. 196–203)*

Find the missing values in the ratio table. Then write the equivalent ratios.

Trees	2	6	
Birds	5		30

You can use multiplication to find the missing values.

× 3, × 2

Trees	2	6	12
Birds	5	15	30

× 3, × 2

The equivalent ratios are 2 : 5, 6 : 15, and 12 : 30.

Exercises

Find the missing values in the ratio table. Then write the equivalent ratios.

3.

Levers	6		18
Pulleys	3	6	

4.

Cars	3	6	
Trucks	4		24

5.3 Rates *(pp. 204–209)*

A horse can run 165 feet in 3 seconds. At this rate, how far can the horse run in 5 seconds?

Using a ratio table, divide to find the unit rate. Then multiply to find the distance that the horse can run in 5 seconds.

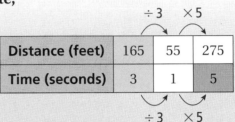

	÷ 3	× 5	
Distance (feet)	165	55	275
Time (seconds)	3	1	5
	÷ 3	× 5	

So, the horse can run 275 feet in 5 seconds.

Exercises

Write a unit rate for the situation.

5. 12 stunts in 4 movies

6. 3600 stitches in 3 minutes

7. **MUSIC** A song has 28 beats in 4 seconds. At this rate, how many beats are there in 30 seconds?

5.4 Comparing and Graphing Ratios *(pp. 210–215)*

There are 24 grams of sugar in 6 fluid ounces of Soft Drink A, and there are 15 grams of sugar in 4 fluid ounces of Soft Drink B. Which soft drink contains more sugar in a 12-ounce can?

Use ratio tables to compare the soft drinks.

Soft Drink A

	× 2	
Sugar (grams)	24	48
Volume (fluid ounces)	6	12
	× 2	

Soft Drink B

	× 3	
Sugar (grams)	15	45
Volume (fluid ounces)	4	12
	× 3	

The tables show that a 12-ounce can of Soft Drink A has $48 - 45 = 3$ more grams of sugar than Soft Drink B.

So, a 12-ounce can of Soft Drink A has more sugar.

Exercises

8. **TUNA** A 5-ounce can of tuna costs $0.90. A 12-ounce can of tuna costs $2.40. Which is the better buy?

5.5 Percents *(pp. 218–223)*

Write $\frac{3}{20}$ as a percent.

$$\frac{3}{20} \overset{\times 5}{=} \frac{15}{100} = 15\%$$

Because $20 \times 5 = 100$, multiply the numerator and denominator by 5. Write the numerator with a percent symbol.

Exercises

Write the percent as a fraction or mixed number in simplest form.

9. 12% **10.** 88% **11.** 0.8%

Write the fraction or mixed number as a percent.

12. $\frac{3}{5}$ **13.** $\frac{43}{25}$ **14.** $1\frac{21}{50}$

5.6 Solving Percent Problems *(pp. 224–231)*

a. 75% of 80 is what number?

$$75\% \text{ of } 80 = \frac{3}{4} \times 80 = \frac{3 \times \cancel{80}^{20}}{\cancel{4}_1} = 60$$

∴ So, 75% of 80 is 60.

b. 30% of what number is 27?

$$27 \div 30\% = 27 \div \frac{3}{10} = \cancel{27}^{9} \cdot \frac{10}{\cancel{3}_1} = 90$$

∴ So, 30% of 90 is 27.

Exercises

Find the percent of the number. Explain your method.

15. 60% of 80 **16.** 80% of 55 **17.** 150% of 48

Find the whole. Explain your method.

18. 70% of what number is 35? **19.** 140% of what number is 56?

5.7 Converting Measures *(pp. 232–237)*

Convert 8 kilometers to miles.

$$8 \text{ km} \times \frac{1 \text{ mi}}{1.6 \text{ km}} \approx 5 \text{ mi}$$

Because 1 mi ≈ 1.6 km, use the ratio $\frac{1 \text{ mi}}{1.6 \text{ km}}$.

Exercises

Copy and complete the statement. Round to the nearest hundredth if necessary.

20. 3 L ≈ ___ qt **21.** 9.2 in. ≈ ___ cm **22.** 15 lb ≈ ___ kg

5 Chapter Test

Check It Out
Test Practice
BigIdeasMath.com

Write the ratio. Explain what the ratio means.

1. scooters : bikes

2. starfish : seashells

Find the missing values in the ratio table. Then write the equivalent ratios.

3.

Lemons	4		36
Limes	2	6	

4.

Rabbits	2	4	
Hamsters	9		54

Write a unit rate for the situation.

5. \$54.00 for 3 tickets

6. 210 miles in 3 hours

Write the fraction or mixed number as a percent.

7. $\frac{21}{25}$

8. $\frac{17}{20}$

9. $1\frac{2}{5}$

Find the percent of the number. Explain your method.

10. 80% of 90

11. 30% of 50

12. 120% of 75

Find the whole. Explain your method.

13. 34 is 40% of what number?

14. 52 is 130% of what number?

Copy and complete the statement. Round to the nearest hundredth if necessary.

15. 5 L ≈ ▒ qt

16. 56 lb ≈ ▒ kg

17. SOUP There are 600 milligrams of sodium in 4 ounces of Soup A, and there are 720 milligrams of sodium in 6 ounces of Soup B. You prepare an 18-ounce bowl of each soup. Which bowl of soup contains more sodium?

18. ORANGE JUICE A 48-fluid-ounce container of orange juice costs \$2.40. A 60-fluid-ounce container of orange juice costs \$3.60. Which is the better buy?

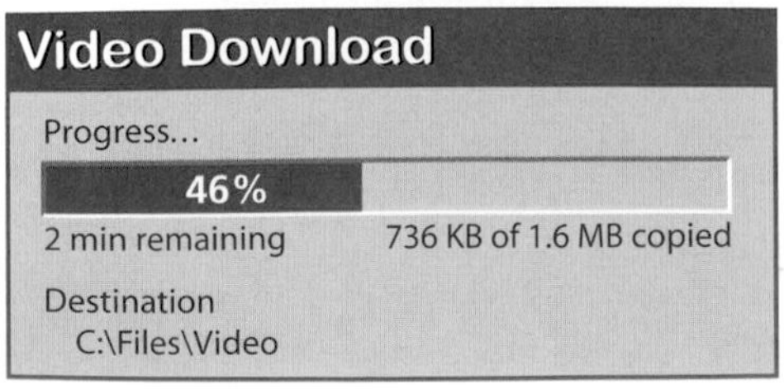

19. DOWNLOAD Your computer displays the progress of a downloading video. What fraction of the video is downloaded? Write your answer in simplest form.

20. GLASSES In a class of 20 students, 40% are boys. Twenty-five percent of the boys and 50% of the girls wear glasses. How many students wear glasses?

5 Standards Assessment

1. What is the value of the expression below? *(6.NS.1)*

$$8\frac{4}{9} \div 4\frac{2}{3}$$

A. $1\frac{17}{21}$ **C.** $32\frac{8}{27}$

B. $2\frac{2}{3}$ **D.** $39\frac{11}{27}$

Test-Taking Strategy
Solve Problem Before Looking at Choices

At a speed of 20 miles per hour, how far can a hyena run in 15 minutes?
(A) 20 mi (B) 5 mi (C) 80 mi (D) 10 mi
Toward me or away from me?

"Solve the problem before looking at the choices. You know one-fourth of 20 is 5, so the answer is 5 miles."

2. Which fraction is *not* equivalent to 25%? *(6.RP.3c)*

F. $\frac{1}{4}$ **H.** $\frac{5}{20}$

G. $\frac{2}{5}$ **I.** $\frac{25}{100}$

3. The school store sells 4 pencils for \$0.50. At that rate, what would be the cost of 10 pencils? *(6.RP.3b)*

A. \$1.10 **C.** \$2.00

B. \$1.25 **D.** \$5.00

4. Which expression is equivalent to the expression below? *(6.EE.4)*

$$2(m + n)$$

F. $2m \times 2n$ **H.** $(2 + m) \times (2 + n)$

G. $2m + 2n$ **I.** $(2 + m) + (2 + n)$

5. A service club wants to buy tickets to a baseball game. Tickets are available for the grandstand and for the bleachers.

Grandstand Ticket \$25	Bleachers Ticket \$15

Which expression represents the total cost, in dollars, for g grandstand tickets and b bleachers tickets? *(6.EE.2a)*

A. $375(g + b)$ **C.** $25g + 15b$

B. $40(g \times b)$ **D.** $25g \times 15b$

6. What property was used to simplify the expression? *(6.EE.3)*

$$\begin{aligned} 12 \times 47 &= 12 \times (40 + 7) \\ &= 12 \times 40 + 12 \times 7 \\ &= 480 + 84 \\ &= 564 \end{aligned}$$

F. Distributive Property

G. Identity Property of Addition

H. Commutative Property of Addition

I. Associative Property of Multiplication

7. What is 15% of 36? *(6.RP.3c)*

8. If 5 dogs share equally a bag of dog treats, each dog gets 24 treats. Suppose 8 dogs share equally the bag of treats. How many treats does each dog get? *(6.RP.3b)*

A. 3

B. 15

C. 21

D. 38

9. The figure below consists of a rectangle and a right triangle. *(6.G.1)*

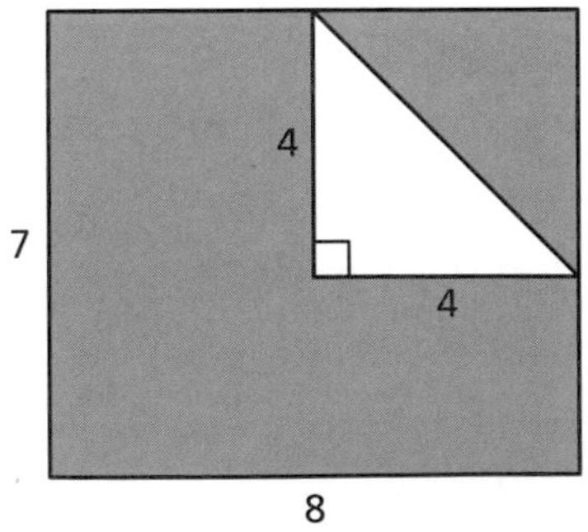

What is the area of the shaded region?

F. 23 units^2

G. 40 units^2

H. 48 units^2

I. 60 units^2

10. What is the area, in square inches, of the trapezoid-shaped award? *(6.G.1)*

11. Your friend evaluated an expression using $k = 0.5$ and $p = 1.6$ and got an answer of 12. Which expression did your friend evaluate? *(6.EE.2c)*

A. $5p + 8k$

B. $8p + 5k$

C. $0.5k + 1.6p$

D. $0.8k + 0.5p$

12. For a party, you made a gelatin dessert in a rectangular pan and cut the dessert into equal-sized pieces as shown below.

Think Solve Explain

The dessert consisted of 5 layers of equal height. Each layer was a different flavor, as shown below by a side view of the pan. *(6.RP.3c)*

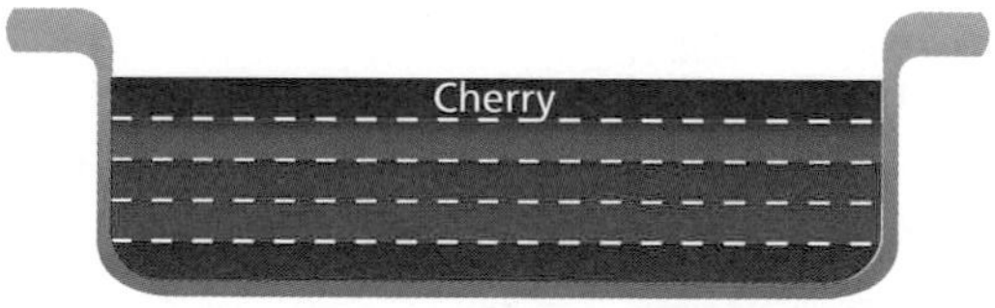

Your guests ate $\frac{3}{5}$ of the pieces of the dessert.

Part A Write the amount of cherry gelatin eaten by your guests as a fraction of the total dessert. Justify your answer.

Part B Write the amount of cherry gelatin eaten by your guests as a percent of the total dessert. Justify your answer.

6 Integers and the Coordinate Plane

"Don't worry. At negative 20 miles per hour, we're still under the speed limit."

"Dear Sir: You asked me to 'find' the opposite of −1."

"I didn't know it was missing."

What You Learned Before

"The Dog License Department wrote and said that it has run out of positive license numbers. So, they assigned me negative 3."

Ordering Decimals (4.NF.7)

Example 1 Use a number line to order 0.25, 1.15, 0.2, and 0.34 from least to greatest.

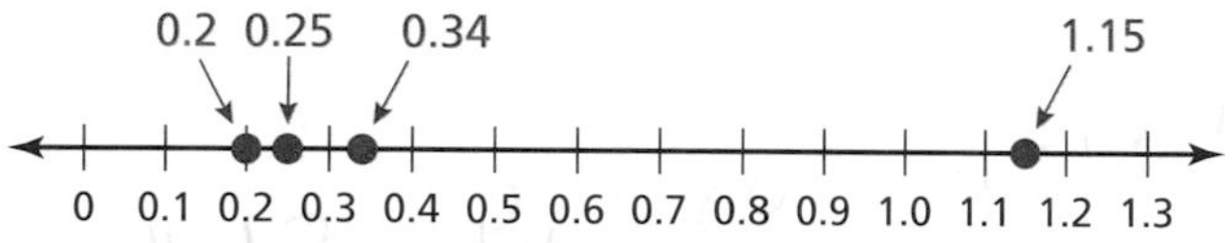

Try It Yourself

Use a number line to order the numbers from least to greatest.

1. 0.01, 0.42, 0.2, 0.5

2. 1.05, 0.95, 0.75, 1.01

Comparing Numbers (5.NBT.3b)

Complete the number sentence with <, >, or =.

Example 2 10 $\square$ 15

On a number line, 10 is closer to zero than 15.

∴ So, $10 < 15$.

Example 3 $0.875 \ \square \ \frac{7}{8}$

$$0.875 = \frac{875}{1000} = \frac{875 \div 125}{1000 \div 125} = \frac{7}{8}$$

∴ So, $0.875 = \frac{7}{8}$.

Example 4 Find three numbers that make the number sentence $1\frac{2}{5} \leq \square$ true.

∴ *Sample answer:* $1\frac{3}{5}, \frac{5}{2}, 2$

Try It Yourself

Complete the number sentence with <, >, or =.

3. 2.01 $\square$ 2.001

4. $4.5 \ \square \ \frac{9}{2}$

5. 3.18 $\square$ 3.2

Find three numbers that make the number sentence true.

6. $\frac{17}{2} \leq \square$

7. $1\frac{1}{2} > \square$

8. $0.75 \geq \square$

6.1 Integers

Essential Question How can you represent numbers that are less than 0?

1 ACTIVITY: Reading Thermometers

Work with a partner. The thermometers show the temperatures in four cities.

Honolulu, Hawaii | *Anchorage, Alaska*
Death Valley, California | *Seattle, Washington*

Write each temperature. Then match each temperature with its most appropriate location.

a.

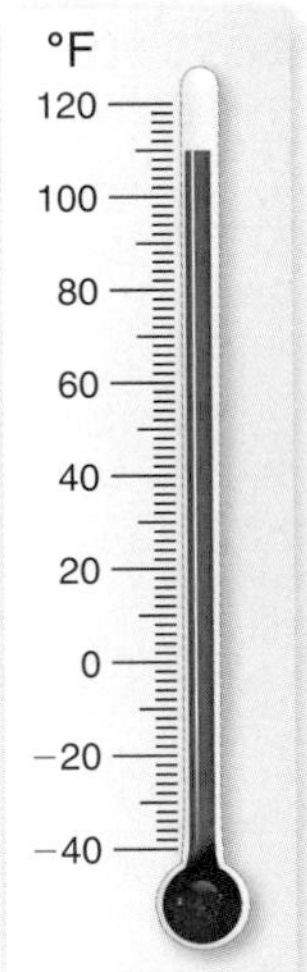

b.

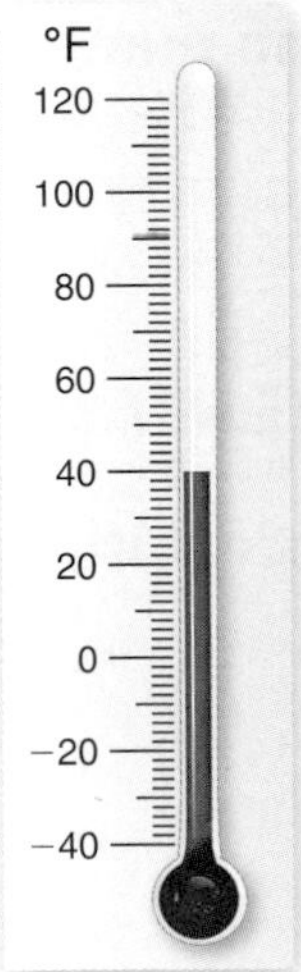

c.

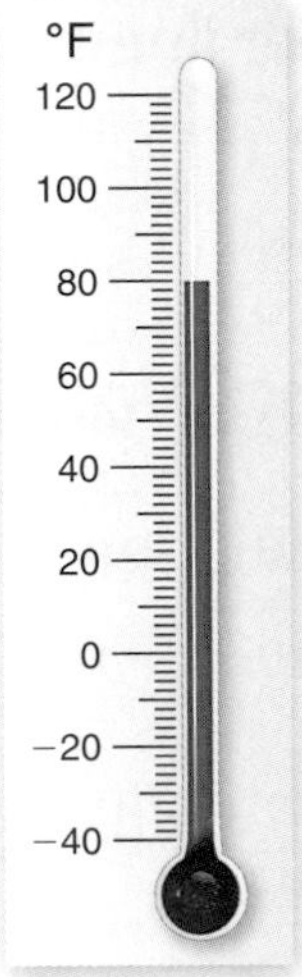

d.

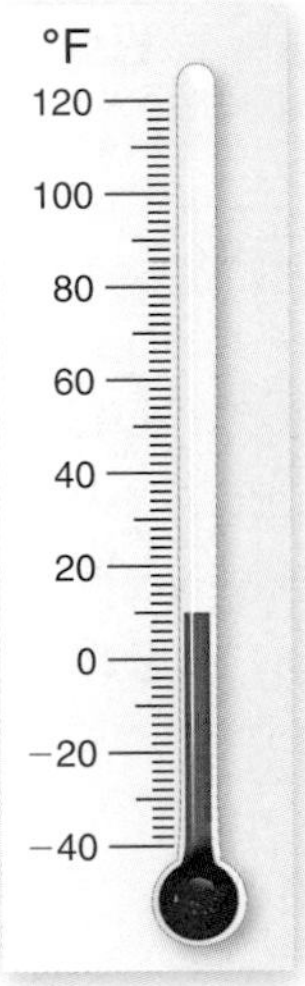

e. How would you describe all the temperatures in relation to 0°F?

2 ACTIVITY: Describing a Temperature

Work with a partner. The thermometer shows the coldest temperature ever recorded in Seattle, Washington.

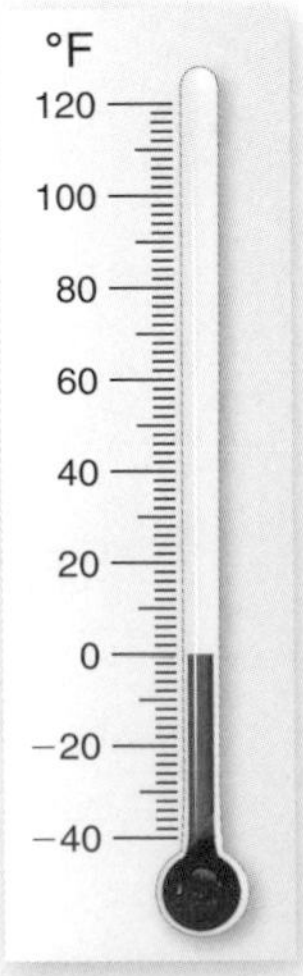

a. What is the temperature?

b. How do you write temperatures that are colder than this?

c. Suppose the record for the coldest temperature in Seattle is broken by 10 degrees. What is the new coldest temperature? Draw a thermometer that shows the new coldest temperature.

d. How is the new coldest temperature different from the temperatures in Activity 1?

COMMON CORE

Integers

In this lesson, you will
- understand positive and negative integers and use them to describe real-life situations.
- graph integers on a number line.

Learning Standards
6.NS.5
6.NS.6a
6.NS.6c

3 ACTIVITY: Extending the System of Whole Numbers

Math Practice 8

Maintain Oversight

How does this activity help you represent numbers less than 0?

Work with a partner.

a. Copy and complete the number line using whole numbers only.

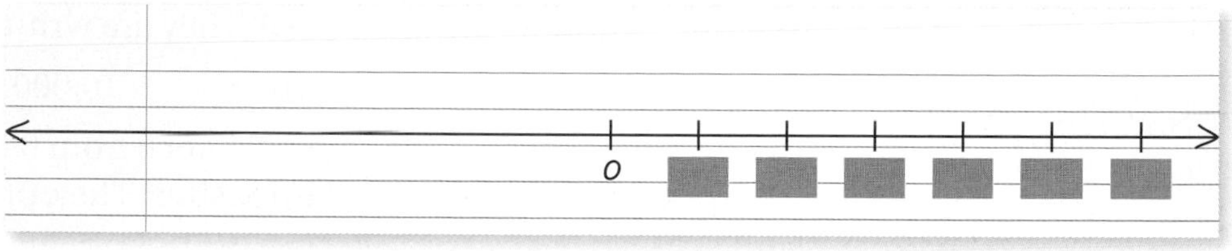

b. Fold the paper with your number line around 0 so that the lines overlap. Make tick marks on the other side of the number line to match the tick marks for the whole numbers.

c. **STRUCTURE** Compare this number line to the thermometers from Activities 1 and 2. What do you think the new tick marks represent? How would you label them?

What Is Your Answer?

4. **IN YOUR OWN WORDS** How can you represent numbers that are less than 0?

5. Describe another real-life example that uses numbers that are less than 0.

6. **REASONING** How are the temperatures shown by the thermometers at the right similar? How are they different?

7. **WRITING** The temperature in a town on Thursday evening is 25°F. On Sunday morning, the temperature drops below 0°F. Write a story to describe what may have happened in the town. Be sure to include the temperatures for each day.

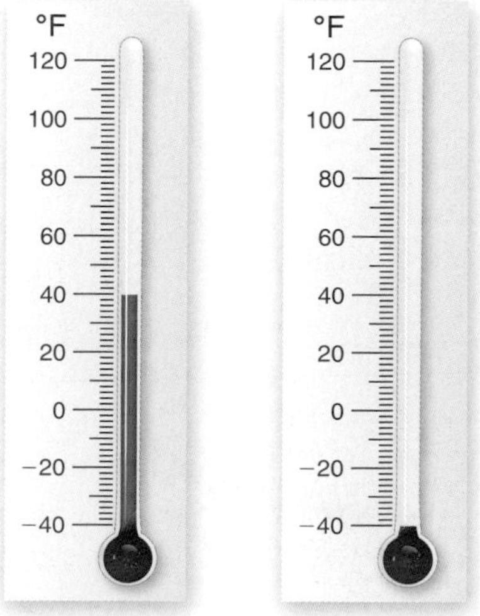

Practice Use what you learned about positive and negative numbers to complete Exercises 4–7 on page 252.

6.1 Lesson

Check It Out
Lesson Tutorials
BigIdeasMath.com

Key Vocabulary
positive numbers, *p. 250*
negative numbers, *p. 250*
opposites, *p. 250*
integers, *p. 250*

Positive numbers are greater than 0. They can be written with or without a positive sign (+).

+1   5   +20   10,000

Negative numbers are less than 0. They are written with a negative sign (−).

−1   −5   −20   −10,000

Two numbers that are the same distance from 0 on a number line, but on opposite sides of 0, are called **opposites**. The opposite of 0 is 0.

Key Idea

Integers

Words **Integers** are the set of whole numbers and their opposites.

Graph

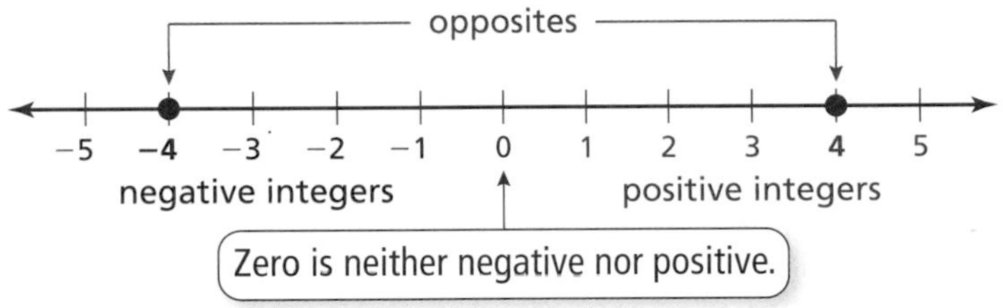

The Meaning of a Word

Opposite

When you sit across from your friend at the lunch table, you sit **opposite** your friend.

EXAMPLE 1 Writing Positive and Negative Integers

Write a positive or negative integer that represents the situation.

a. A contestant gains 250 points on a game show.

Gains indicates a number greater than 0. So, use a positive integer.

∴ +250, or 250

b. Gasoline freezes at 40 degrees below zero.

Below zero indicates a number less than 0. So, use a negative integer.

∴ −40

On Your Own

Now You're Ready
Exercises 8–13

Write a positive or negative integer that represents the situation.

1. A hiker climbs 900 feet up a mountain.
2. You have a debt of $24.
3. A student loses 5 points for being late to class.
4. A savings account earns $10.

EXAMPLE 2 Graphing Integers

Graph each integer and its opposite.

a. 3

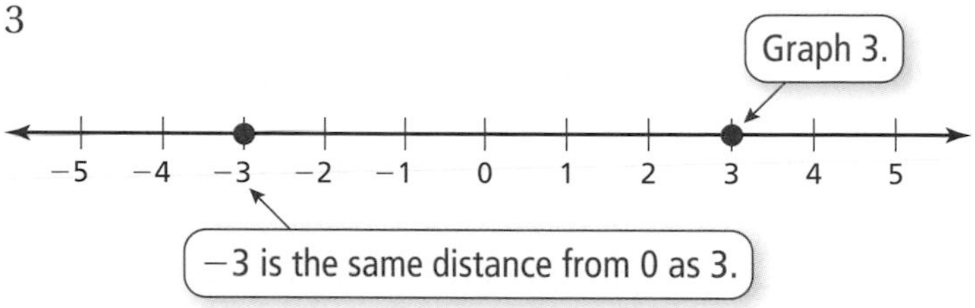

Graph 3.
−5 −4 −3 −2 −1 0 1 2 3 4 5
−3 is the same distance from 0 as 3.

b. −2

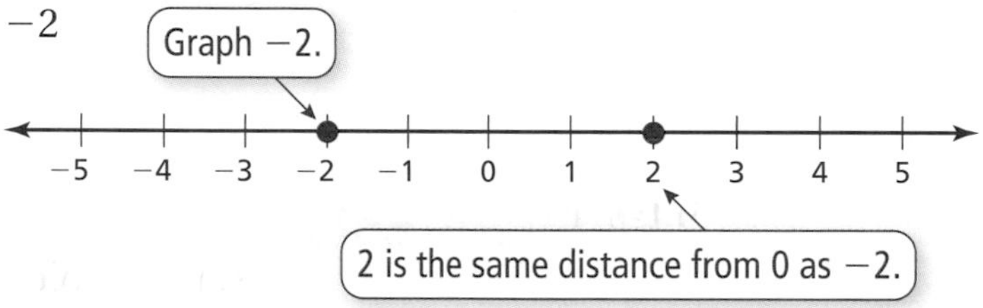

Graph −2.
−5 −4 −3 −2 −1 0 1 2 3 4 5
2 is the same distance from 0 as −2.

Reading

You can think of the negative sign (−) as referring to the opposite of a number. In Example 2(b), you can read −2 as "the opposite of 2."

EXAMPLE 3 Real-Life Application

You deliver flowers to an office building. You enter at ground level and go down 2 floors to make the first delivery. Then you go up 7 floors to make the second delivery.

a. Write an integer that represents each position.

Position	*Integer*
You enter at ground level.	0
You go down 2 floors.	−2
You go up 7 floors.	+7

b. Write an integer that represents how you return to ground level.

Use a number line to model your movement, as shown.

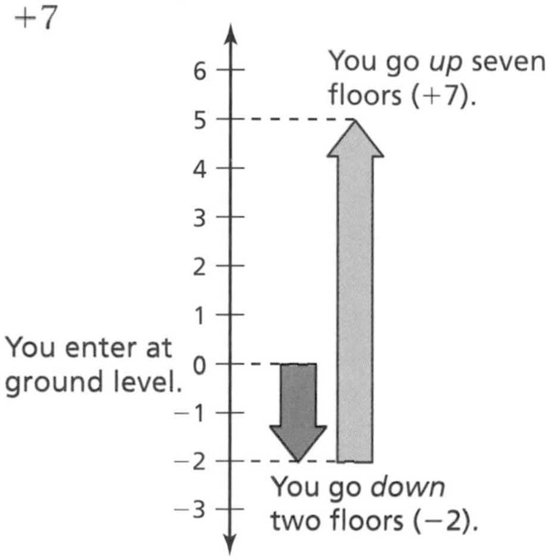

6
5
4
3
2
1
0
−1
−2
−3
You go *up* seven floors (+7).
You enter at ground level.
You go *down* two floors (−2).

The second delivery is on the fifth floor. You must go down 5 floors to return to ground level.

∴ The integer representing "down 5 floors" is −5.

On Your Own

Now You're Ready Exercises 16–23

Graph the integer and its opposite.

5. 6 **6.** −4 **7.** −12 **8.** 1

9. WHAT IF? In Example 3, you go up 9 floors to make the second delivery. Write an integer that represents how you return to ground level.

6.1 Exercises

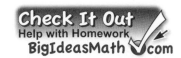

Check It Out
Help with Homework
BigIdeasMath.com

Vocabulary and Concept Check

1. **VOCABULARY** Which of the following numbers are integers?

$$8,\ -4.1,\ -9,\ \frac{1}{6},\ 1.75,\ 22$$

2. **OPEN-ENDED** Describe a real-life example that you can represent by -1200.
3. **VOCABULARY** List three words or phrases used in real life that indicate negative numbers.

Practice and Problem Solving

Graph the number that represents the situation on a number line.

4. A football team loses 3 yards.
5. The temperature is 6 degrees below zero.
6. A person climbs 600 feet up a mountain.
7. You earn $15 raking leaves.

Write a positive or negative integer that represents the situation.

(1) 8. You withdraw $42 from an account.
9. An airplane climbs to 37,500 feet.
10. The temperature rises 17 degrees.
11. You lose 56 points in a video game.
12. A ball falls 350 centimeters.
13. You receive 5 bonus points in class.

14. **STOCK MARKET** A stock market gains 83 points. The next day, the stock market loses 47 points. Write each amount as an integer.
15. **SCUBA DIVING** The world record for scuba diving is 318 meters below sea level. Write this as an integer.

Graph the integer and its opposite.

(2) 16. -5
17. -8
18. 14
19. 9
20. 30
21. -150
22. -32
23. 400

24. **ERROR ANALYSIS** Describe and correct the error in describing positive integers.

> ✗ The positive integers are 0, 1, 2, 3, ...

25. **TEMPERATURE** The highest temperature in February is 25°F. The lowest temperature in February is the opposite of the highest temperature. Graph both temperatures.

Identify the integer represented by the point on the number line.

26. A **27.** B **28.** C **29.** D

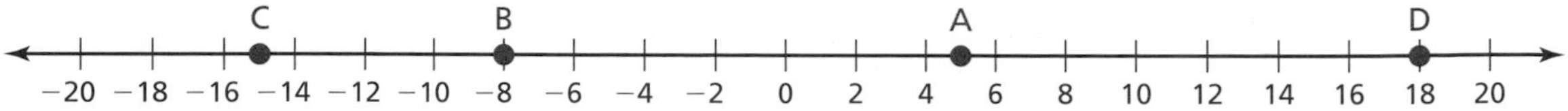

30. TIDES Use the information below.

- Low tide is 1 foot below the average water level.
- High tide is 5 feet higher than low tide.

Write an integer that represents the average water level relative to high tide.

31. REPEATED REASONING Choose any positive integer.

a. Find the opposite of the integer. **b.** Find the opposite of the integer in part (a).

c. What can you conclude about the opposite of the opposite of the integer? Is this true for all integers? Use a number line to justify your answer.

d. Describe the meaning of $-(-(-6))$. Find its value.

32. **Number Sense** In a game of tug-of-war, a team wins by pulling the flag over its goal line. The flag begins at 0. During a game, the flag moves 8 feet to the right, 12 feet to the left, and 13 feet back to the right. Did a team win? Explain.

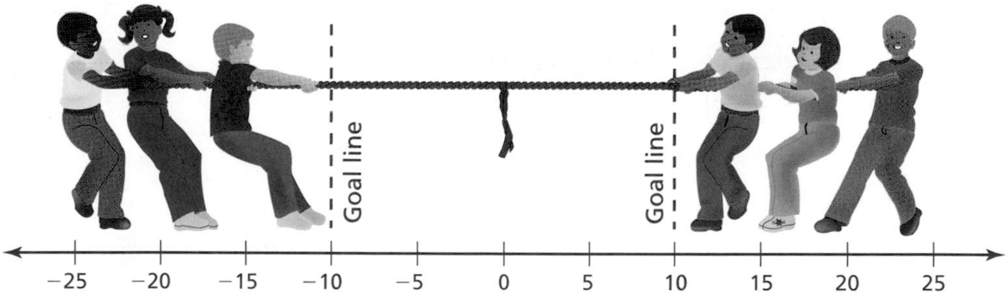

Fair Game Review What you learned in previous grades & lessons

Order the numbers from least to greatest. *(Skills Review Handbook)*

33. $\frac{7}{8}, \frac{1}{2}, \frac{3}{8}, \frac{3}{4}$ **34.** 4.5, 4.316, 4.32, 4.312

35. MULTIPLE CHOICE The height of a statue is 276 inches. What is the height of the statue in meters? Round your answer to the nearest hundredth. *(Section 5.7)*

Ⓐ 1.09 m Ⓑ 7.01 m Ⓒ 108.66 m Ⓓ 701.04 m

6.2 Comparing and Ordering Integers

Essential Question How can you use a number line to order real-life events?

1 ACTIVITY: Seconds to Takeoff

Work with a partner. You are listening to a command center before the liftoff of a rocket.

You hear the following:

> **"T minus 10 seconds . . . go for main engine start . . . T minus 9 . . . 8 . . . 7 . . . 6 . . . 5 . . . 4 . . . 3 . . . 2 . . . 1 . . . we have liftoff."**

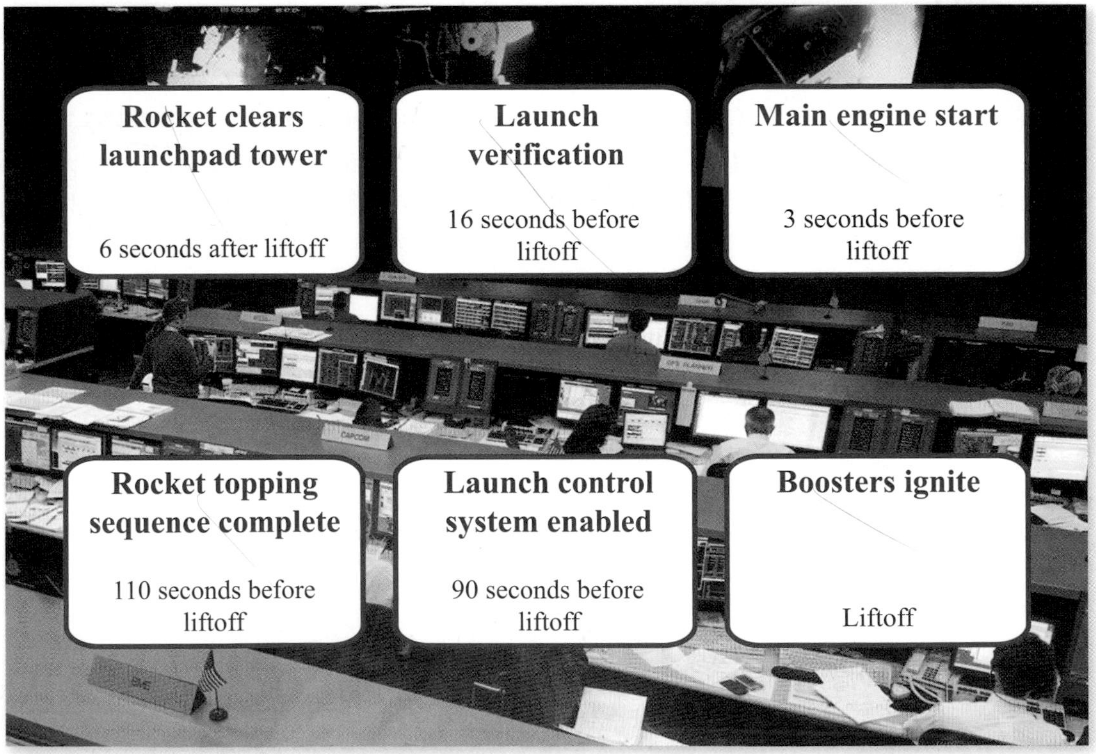

COMMON CORE

Integers

In this lesson, you will
- use a number line to compare positive and negative integers.
- use a number line to order positive and negative integers for real-life situations.

Learning Standards
6.NS.6c
6.NS.7a
6.NS.7b

a. Draw a number line. Then locate the events shown above at appropriate points on the number line.

b. Which event occurs at zero on your number line? Explain.

c. Which of the events occurs first? Which of the events occurs last? How do you know?

d. List the events in the order they occurred.

2 ACTIVITY: Being Careful with Terminology

Work with a partner.

a. Use a number line to show that the phrase "3 seconds away from liftoff" can have two meanings.

b. Reword the phrase "3 seconds away from liftoff" in two ways so that each meaning is absolutely clear.

c. Explain why you must be very careful with terminology if you are working in the command center for a rocket launch.

3 ACTIVITY: A Day in the Life of an Astronaut

Math Practice 5

Recognize the Usefulness of Tools

Which sources would give you the most accurate information? How do you know you can trust the information you find?

Make a time line that shows a day in the life of an astronaut. Use the Internet or another reference source to gather information.

- Use a number line with units representing hours. Start at 12 hours before liftoff and end at 12 hours after liftoff. Locate the liftoff at 0. Assume liftoff occurs at noon.
- Include at least five events before liftoff, such as when the astronauts suit up.
- Include at least five events after liftoff, such as when the rocket enters Earth's orbit.
- How do you determine where each event occurs on the number line?

What Is Your Answer?

4. **IN YOUR OWN WORDS** How can you use a number line to order real-life events?
5. Describe how you can use a number line to create a time line.

Practice Use what you learned about number lines to complete Exercises 4–7 on page 258.

6.2 Lesson

Check It Out
Lesson Tutorials
BigIdeasMath.com

On a horizontal number line, numbers to the left are less than numbers to the right. Numbers to the right are greater than numbers to the left.

EXAMPLE 1 Comparing Integers on a Horizontal Number Line

Compare 2 and −6.

∴ 2 is to the right of −6. So, $2 > -6$.

On a vertical number line, numbers below are less than numbers above. Numbers above are greater than numbers below.

EXAMPLE 2 Comparing Integers on a Vertical Number Line

Compare −5 and −3.

∴ −5 is below −3. So, $-5 < -3$.

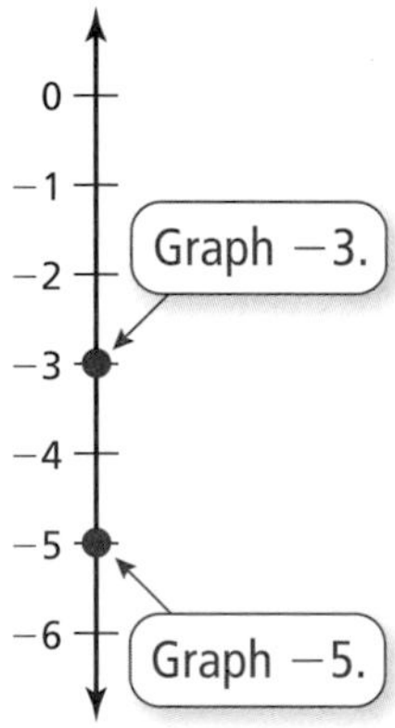

On Your Own

Now You're Ready
Exercises 4–11

Copy and complete the statement using < or >.

1. $0 \ \square \ -4$ **2.** $-5 \ \square \ 5$ **3.** $-8 \ \square \ -7$

EXAMPLE 3 Ordering Integers

Order −4, 3, 0, −1, −2 from least to greatest.

Graph each integer on a number line.

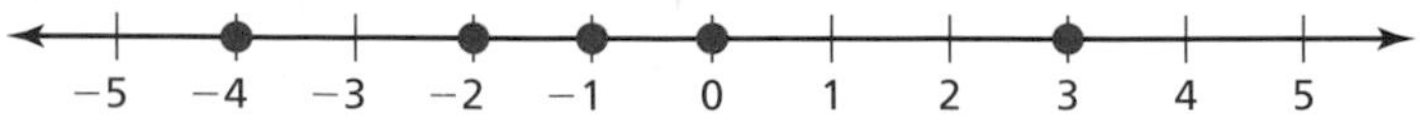

Write the integers as they appear on the number line from left to right.

∴ So, the order from least to greatest is −4, −2, −1, 0, 3.

EXAMPLE 4 Reasoning with Integers

A number is greater than −8 and less than 0. What is the greatest possible integer value of this number?

Ⓐ −10 Ⓑ −7 Ⓒ −1 Ⓓ 2

> **Study Tip**
> In Example 4, you can eliminate Choices A and D because −10 is to the left of −8 and 2 is to the right of 0.

The number is greater than −8 and less than 0. So, the number must be to the right of −8 and to the left of 0 on a horizontal number line.

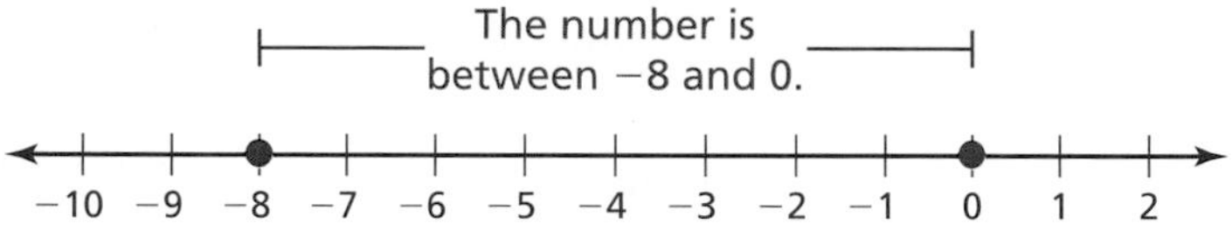

The greatest possible integer value between −8 and 0 is the integer farthest to the right on the number line between these values, which is −1.

So, the correct answer is Ⓒ.

EXAMPLE 5 Real-Life Application

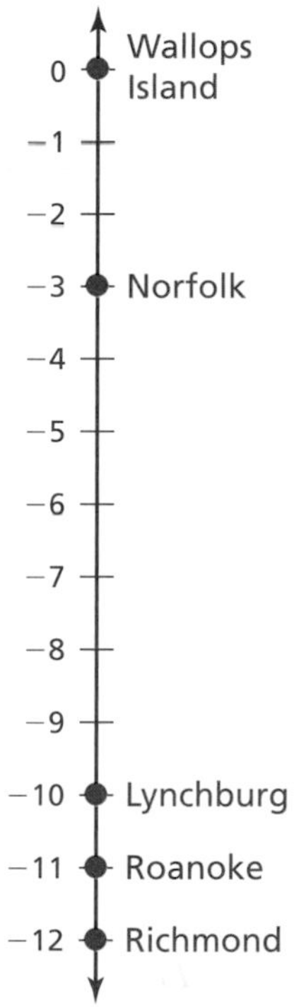

The diagram shows the coldest recorded temperatures for several cities in Virginia.

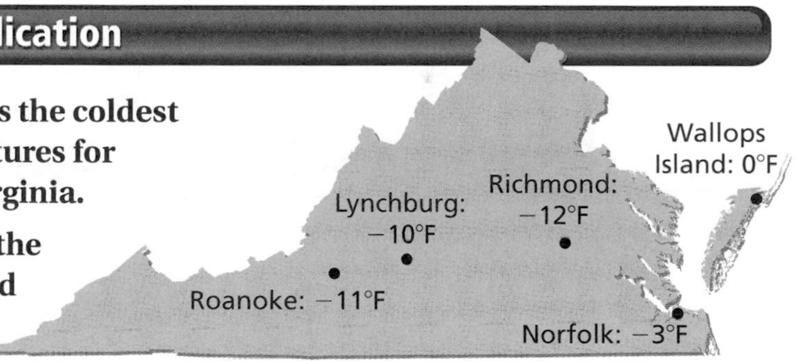

a. Which city has the coldest recorded temperature?

Graph each integer on a vertical number line.

−12 is the lowest on the number line. So, Richmond has the coldest recorded temperature.

b. Has a negative Fahrenheit temperature ever been recorded on Wallops Island? Explain.

The coldest recorded temperature on Wallops Island is 0°F, which is greater than every negative temperature. So, a negative temperature has never been recorded on Wallops Island.

On Your Own

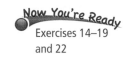

Now You're Ready
Exercises 14–19 and 22

Order the integers from least to greatest.

4. −2, −3, 3, 1, −1
5. 4, −7, −8, 6, 1
6. In Example 4, what is the least possible integer value of the number?
7. In Example 5, Norfolk recorded a new record low last night. The new record low is greater than the record low in Lynchburg. What integers can represent the new record low in Norfolk?

6.2 Exercises

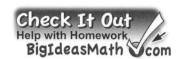

Check It Out
Help with Homework
BigIdeasMath.com

Vocabulary and Concept Check

1. **WRITING** Explain how to use a number line to compare two integers.
2. **REASONING** The positions of four fish are shown.
   a. Use red, blue, yellow, and green dots to graph the positions of the fish on a horizontal number line and a vertical number line.
   b. Explain how to use the number lines from part (a) to order the positions from least to greatest.

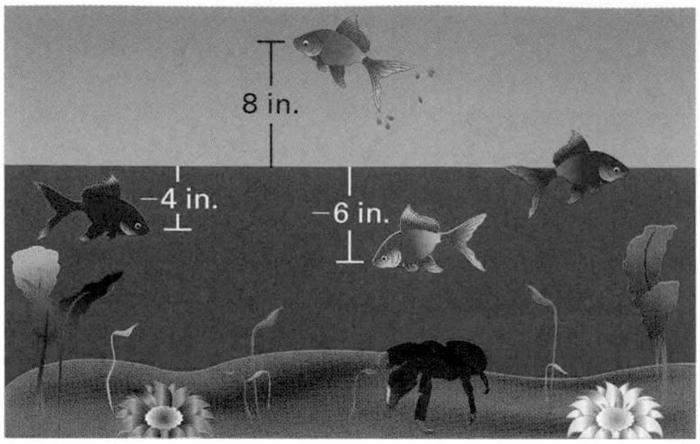

3. **NUMBER SENSE** a and b are negative integers. Compare a and b. Explain your reasoning.

Practice and Problem Solving

Copy and complete the statement using < or >.

1 2 **4.** $3 \ \square \ 0$ **5.** $-2 \ \square \ 0$ **6.** $6 \ \square \ -6$ **7.** $3 \ \square \ -4$

8. $-1 \ \square \ 4$ **9.** $-7 \ \square \ -8$ **10.** $-3 \ \square \ -2$ **11.** $-5 \ \square \ -10$

ERROR ANALYSIS Describe and correct the error in comparing the negative numbers.

12. ✗ Compare −3 and −1.
$3 > 1$. So, $-3 > -1$.

13. ✗ Compare −7 and −3.
Because $-7 < -3$, −7 is to the right of −3 on a number line.

Order the integers from least to greatest.

3 **14.** 0, −1, 2, 3, −3 **15.** −4, −2, −3, 2, 1 **16.** −2, 3, −3, −4, 4

17. −7, 2, 6, −4, 3 **18.** 10, −10, 30, −30, −50 **19.** −5, 15, −10, −20, 25

20. ARCHAEOLOGY An archaeologist discovers the two artifacts shown.
   a. What integer represents ground level?
   b. A dinosaur bone is found 42 centimeters below ground level. Is it deeper than both of the artifacts?

21. TEMPERATURE The freezing temperature of nitrogen is −210°C, and the freezing temperature of oxygen is −223°C. Which temperature is colder?

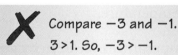

4 22. **REASONING** A number is between −2 and −10. What is the least possible integer value of this number? What is the greatest possible integer value of this number?

Tell whether the statement is *always*, *sometimes*, or *never* true. Explain.

23. A positive integer is greater than its opposite.

24. An integer is less than its opposite and greater than 0.

25. **ELEVATION** The table shows the highest and lowest elevations for five states.

    a. Order the states by their highest elevations, from least to greatest.

    b. Order the states by their lowest elevations, from least to greatest.

    c. What does the lowest elevation for Florida represent?

State	Highest Elevation (feet)	Lowest Elevation (feet)
Arkansas	2,753	55
California	14,494	−282
Florida	345	0
Louisiana	535	−8
Tennessee	6,643	178

26. **NUMBER LINE** Point A is on a number line halfway between −17 and 5. Point B is halfway between point A and 0. What integer does point B represent?

27. **TEMPERATURE** Eleven Fahrenheit temperatures are shown on a map during a weather report. When the temperatures are ordered from least to greatest, the middle temperature is below 0°F. Do you know exactly how many of the temperatures are represented by negative numbers? Explain.

28. **Puzzle** Nine students choose integers. Here are seven of them:

    5, −8, 10, −1, −12, −20, and 1.

    a. When all nine integers are ordered from least to greatest, the middle integer is 1. Describe the integers chosen by the other two students.

    b. When all nine integers are ordered from least to greatest, the middle integer is –3. Describe the integers chosen by the other two students.

Fair Game Review What you learned in previous grades & lessons

Graph the decimal on a number line. *(Skills Review Handbook)*

29. 2.4 30. 1.3 31. 0.65 32. 2.45

33. **MULTIPLE CHOICE** What is the area of the trapezoid? *(Section 4.3)*

Ⓐ 6.3 ft^2 Ⓑ 44.1 ft^2

Ⓒ 50.4 ft^2 Ⓓ 88.2 ft^2

6.3 Fractions and Decimals on the Number Line

Essential Question How can you use a number line to compare positive and negative fractions and decimals?

1 ACTIVITY: Locating Fractions on a Number Line

On your time line for "A Day in the Life of an Astronaut" from Activity 3 in Section 6.2, include the following events. Represent each using a fraction or a mixed number.

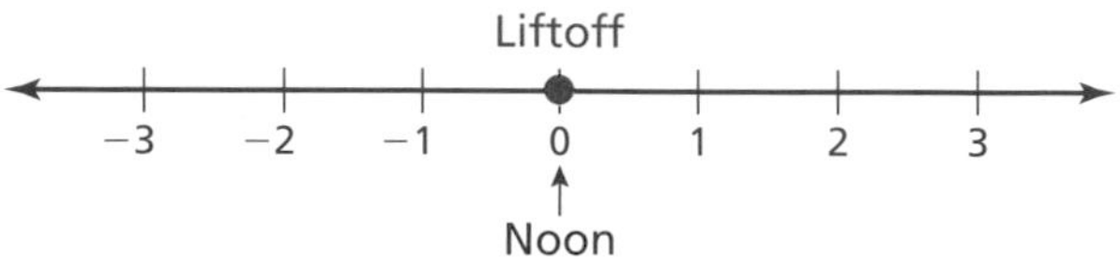

a. Radio Transmission: 10:30 A.M.

b. Space Walk: 7:30 P.M.

c. Physical Exam: 4:45 A.M.

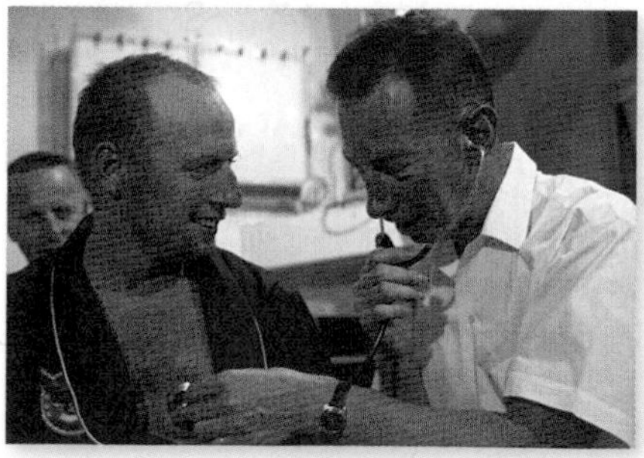

d. Photograph Taken: 3:15 A.M.

e. Float in the Cabin: 6:20 P.M.

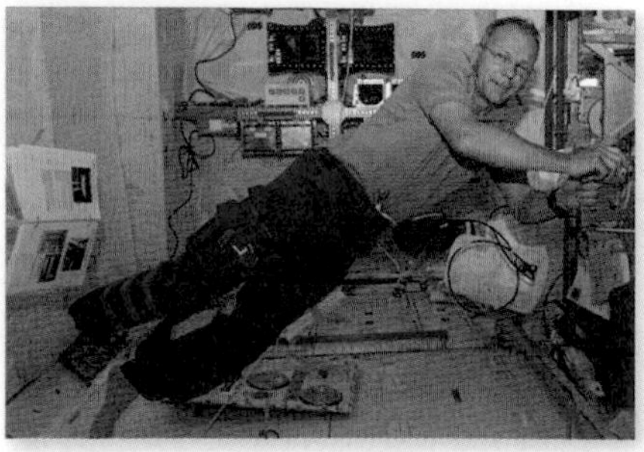

f. Eat Dinner: 8:40 P.M.

COMMON CORE

Fractions and Decimals

In this lesson, you will
- understand positive and negative numbers and use them to describe real-life situations.
- graph numbers on a number line.

Learning Standards
6.NS.5
6.NS.6a
6.NS.6c
6.NS.7a
6.NS.7b

2 ACTIVITY: Fractions and Decimals on a Number Line

Math Practice 1

Make a Plan

How can you find a number between two given numbers?

Work with a partner. Find a number that is between the two numbers. The number must be greater than the green number *and* less than the blue number.

a.

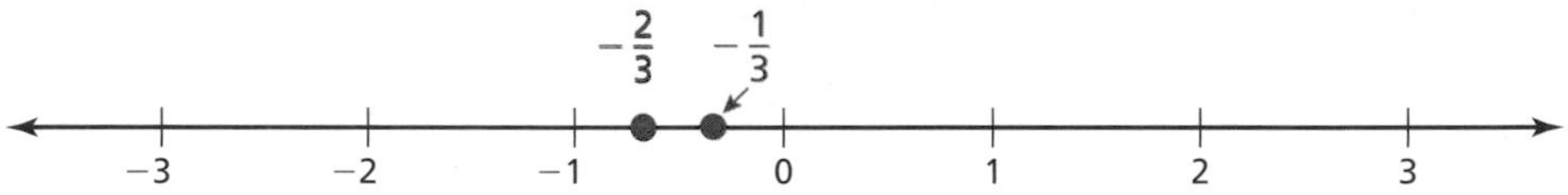

b.

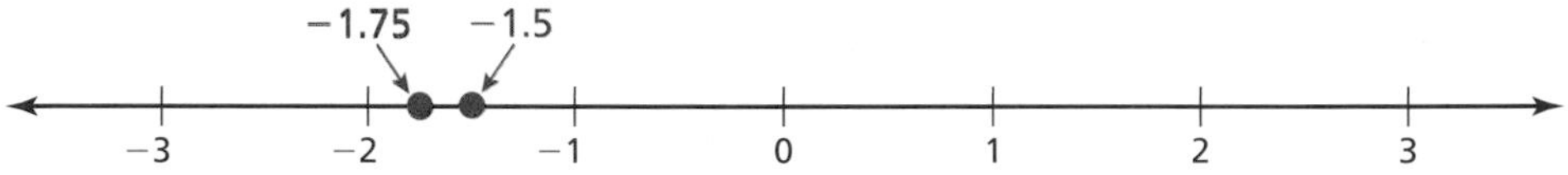

c.

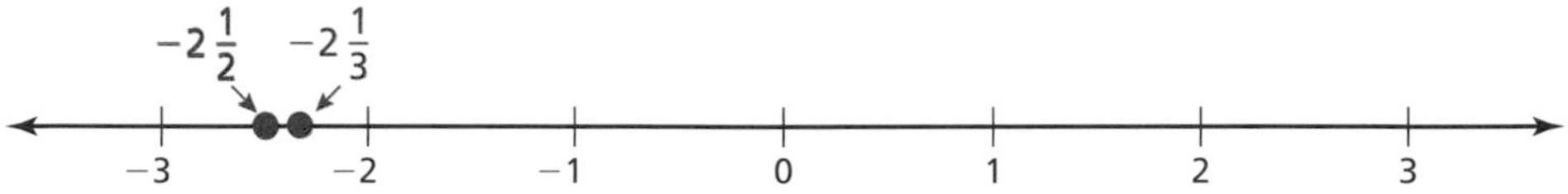

3 ACTIVITY: Decimals on a Number Line

Work with a partner.

Snorkeling: −5 meters

Scuba diving: −50 meters

Deep-sea diving: −700 meters

a. Write the position of each diver in kilometers.

b. **CHOOSE TOOLS** Would a horizontal or a vertical number line be more appropriate for representing these data? Why?

c. Use a number line to order the positions from deepest to shallowest.

What Is Your Answer?

4. **IN YOUR OWN WORDS** How can you use a number line to compare positive and negative fractions and decimals?

5. Draw a number line. Graph and label three values between −2 and −1.

Practice Use what you learned about fractions and decimals on a number line to complete Exercises 4 and 5 on page 264.

6.3 Lesson

Check It Out
Lesson Tutorials
BigIdeasMath.com

In Section 6.1, you learned that integers can be negative. Fractions and decimals can also be negative.

EXAMPLE 1 Graphing Negative Fractions and Decimals

Graph each number and its opposite.

a. $\frac{3}{4}$

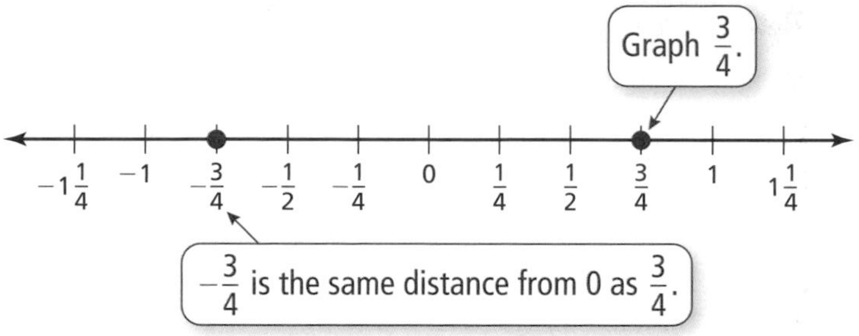

b. -1.6

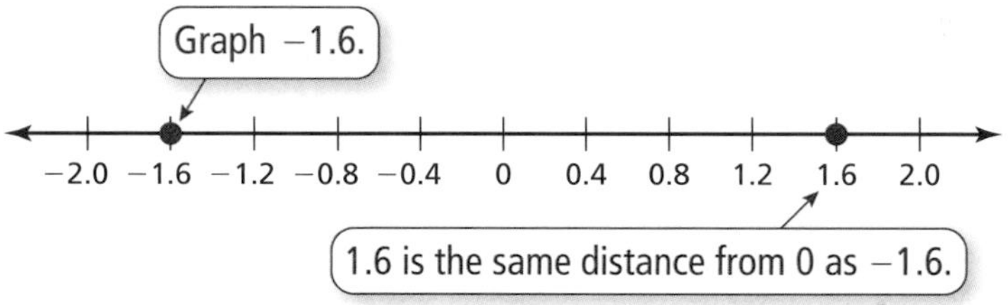

On Your Own

Now You're Ready
Exercises 6–9

Graph the number and its opposite.

1. $2\frac{1}{2}$ **2.** $-\frac{4}{5}$ **3.** -3.5 **4.** 5.25

EXAMPLE 2 Comparing Fractions and Mixed Numbers

a. Compare $-\frac{1}{2}$ and $-\frac{3}{4}$.

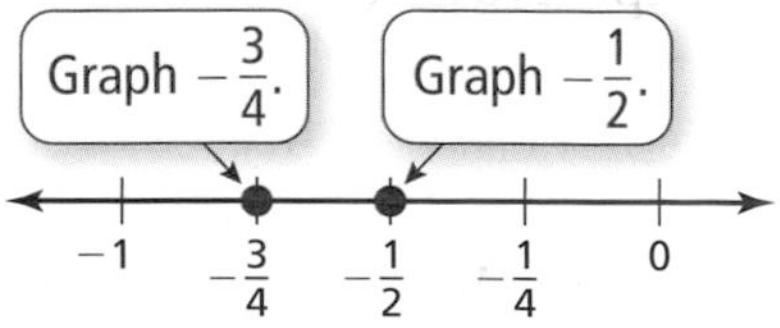

$-\frac{1}{2}$ is to the right of $-\frac{3}{4}$.

So, $-\frac{1}{2} > -\frac{3}{4}$.

b. Compare $-4\frac{5}{6}$ and $-4\frac{1}{6}$.

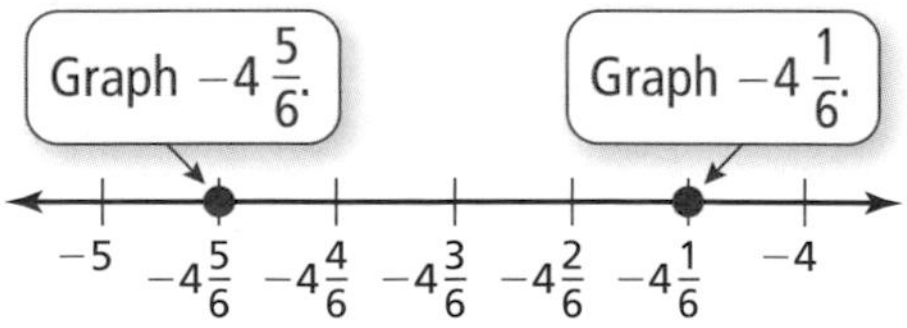

$-4\frac{5}{6}$ is to the left of $-4\frac{1}{6}$.

So, $-4\frac{5}{6} < -4\frac{1}{6}$.

EXAMPLE 3 Comparing Decimals

Compare −3.08 and −3.8.

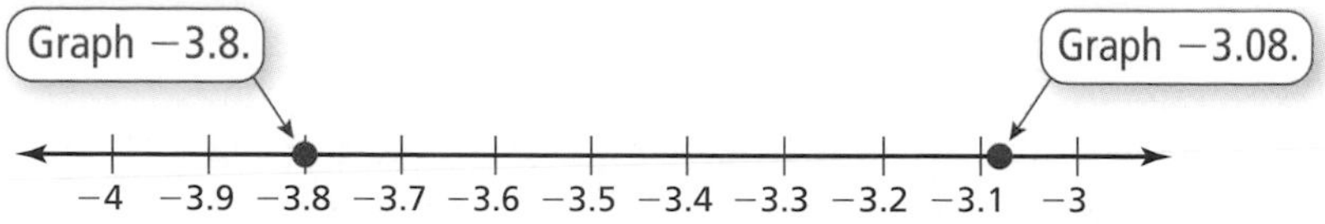

−3.08 is to the right of −3.8.

So, $-3.08 > -3.8$.

EXAMPLE 4 Real-Life Application

A *Chinook wind* is a warm mountain wind that can cause rapid temperature changes. The table shows three of the greatest temperature drops ever recorded after a Chinook wind occurred. On which date did the temperature drop the fastest? Explain.

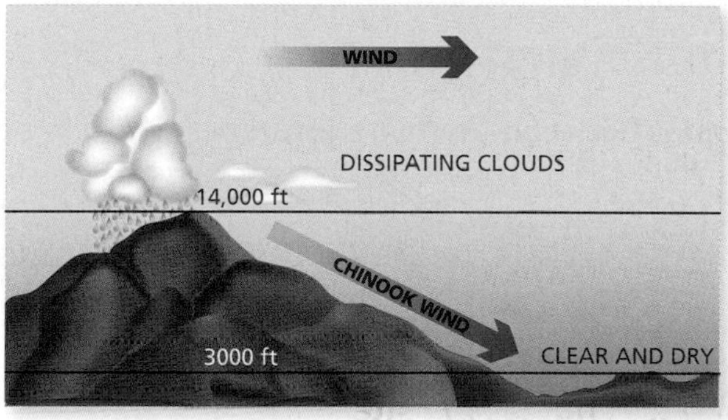

Date	Temperature Change
January 10, 1911	$-3\frac{1}{10}$°F per minute
November 10, 1911	$-\frac{5}{8}$°F per minute
January 22, 1943	$-2\frac{1}{5}$°F per minute

Graph the numbers on a number line.

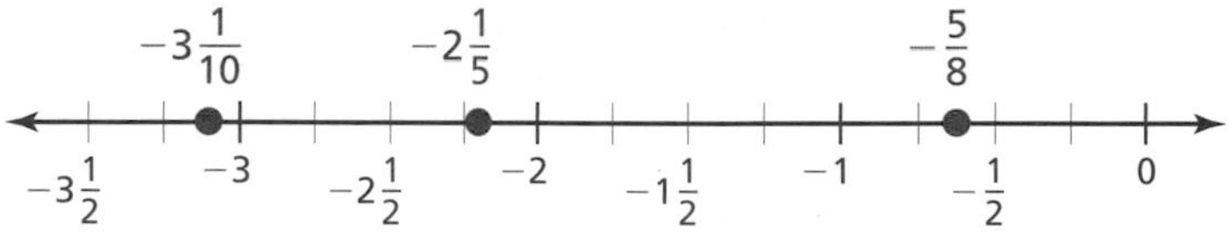

$-3\frac{1}{10}$ is farthest to the left.

So, the temperature dropped the fastest on January 10, 1911.

On Your Own

Now You're Ready
Exercises 10–18 and 20–23

Copy and complete the statement using < or >.

5. $-\frac{4}{7}$ ▒ $-\frac{1}{7}$

6. $-1\frac{2}{3}$ ▒ $-1\frac{5}{6}$

7. -0.5 ▒ 0.3

8. **WHAT IF?** In Example 4, a temperature change of $-3\frac{2}{5}$°F per minute is recorded. How does this temperature change compare with the other temperature changes? Explain.

6.3 Exercises

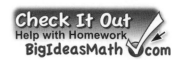

Check It Out
Help with Homework
BigIdeasMath.com

Vocabulary and Concept Check

1. **NUMBER SENSE** Which statement is *not* true?
   - **a.** On a number line, $-2\frac{1}{6}$ is to the left of $-2\frac{2}{3}$.
   - **b.** $-2\frac{2}{3}$ is less than $-2\frac{1}{6}$.
   - **c.** $-2\frac{1}{6}$ is greater than $-2\frac{2}{3}$.
   - **d.** On a number line, $-2\frac{2}{3}$ is to the left of $-2\frac{1}{6}$.
2. **NUMBER SENSE** Is a negative decimal *always*, *sometimes*, or *never* equal to a positive decimal? Explain.
3. **NUMBER SENSE** On a number line, is -2.06 or -2.6 farther to the left?

Practice and Problem Solving

Find a fraction or mixed number that is between the two numbers.

4.
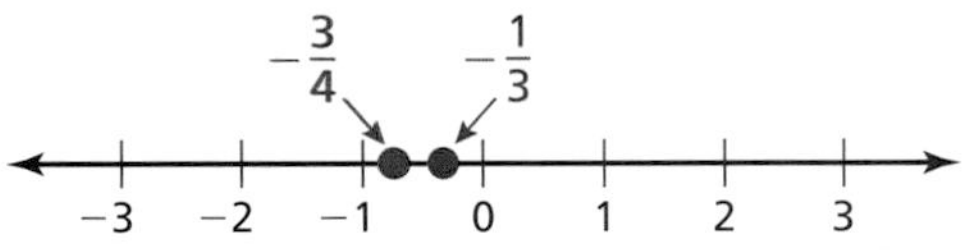

5.
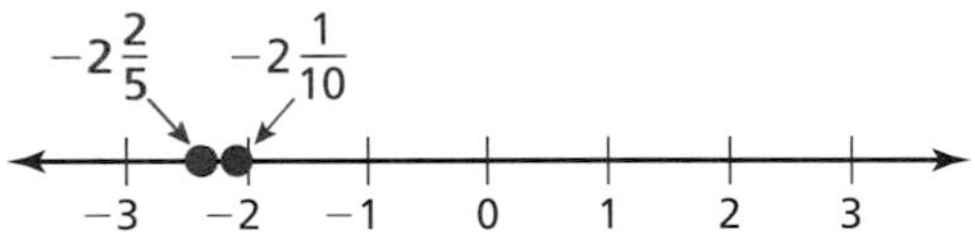

Graph the number and its opposite.

1 6. $\frac{2}{3}$ 7. $-2\frac{1}{4}$ 8. -3.8 9. 2.15

Copy and complete the statement using < or >.

2 3 10. $-3\frac{1}{3}$ ■ $-3\frac{2}{3}$ 11. $-\frac{1}{2}$ ■ $-\frac{1}{6}$ 12. $-\frac{3}{4}$ ■ $\frac{5}{8}$

13. $-2\frac{2}{3}$ ■ $-2\frac{1}{2}$ 14. $-1\frac{5}{6}$ ■ $-1\frac{3}{4}$ 15. -4.6 ■ -4.8

16. -0.12 ■ -0.05 17. 2.41 ■ -3.16 18. -3.524 ■ -3.542

19. **SAND DOLLARS** In rough water, a small sand dollar burrows $-\frac{1}{2}$ centimeter into the sand. A larger sand dollar burrows $-1\frac{1}{4}$ centimeters into the sand. Which sand dollar burrowed farther?

Order the numbers from least to greatest.

4 **20.** $-2\frac{3}{10}, -2\frac{2}{5}, -2, -2\frac{1}{2}, -3$

21. $-\frac{1}{20}, -\frac{5}{8}, 0, -1, -\frac{3}{4}$

22. 1.3, −2, −1.8, 0, −1.75

23. −4, −4.35, −4.9, −5, −4.3

24. STARS The *apparent magnitude* of a star measures how bright the star appears as seen from Earth. The brighter the star, the lesser the number. Which star is the brightest?

Star	Alpha Centauri	Antares	Canopus	Deneb	Sirius
Apparent Magnitude	−0.27	0.96	−0.72	1.25	−1.46

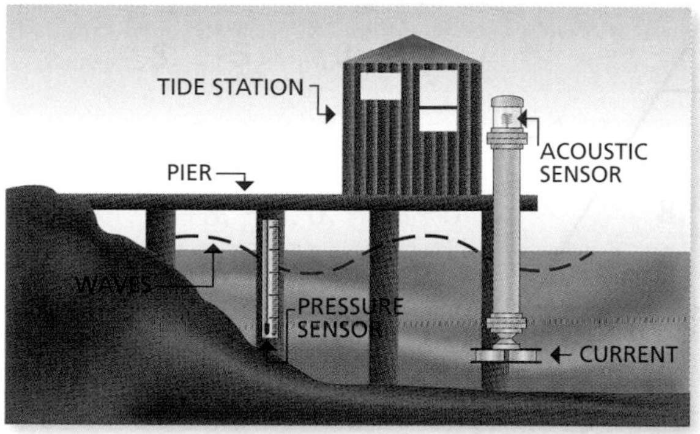

25. LOW TIDE The daily water level is recorded for seven straight days at a tide station on the Big Marco River in Florida. On which days is the water level higher than on the previous day? On which days is it lower?

Day	Sun.	Mon.	Tues.	Wed.	Thurs.	Fri.	Sat.
Water Level of the Day (feet)	$-\frac{3}{25}$	$-\frac{7}{20}$	$-\frac{27}{50}$	$-\frac{13}{20}$	$-\frac{16}{25}$	$-\frac{53}{100}$	$-\frac{1}{3}$

26. PROBLEM SOLVING A guitar tuner allows you to tune a guitar string to its correct pitch. The units on a tuner are measured in *cents*. The units tell you how far the string tone is above or below the correct pitch.

Guitar String	6	5	4	3	2	1
Number of Cents Away from the Correct Pitch	−0.3	1.6	−2.3	2.8	2.4	−3.6

a. What number on the tuner represents a correctly tuned guitar string?

b. Which strings have a pitch below the correct pitch?

c. Which string has a pitch closest to its correct pitch?

d. Which string has a pitch farthest from its correct pitch?

e. The tuner is rated to be accurate to within 0.5 cent of the true pitch. Which string could possibly be correct?

27. Number Sense What integer values of x make the statement $-\frac{3}{x} < -\frac{x}{3}$ true?

Fair Game Review What you learned in previous grades & lessons

Graph the integer and its opposite. *(Section 6.1)*

28. −7 **29.** 40 **30.** 100 **31.** −15

32. MULTIPLE CHOICE You pay \$48 for 8 pounds of chicken. Which is an equivalent rate? *(Section 5.3)*

(A) \$44 for 4 pounds (B) \$28 for 4 pounds

(C) \$15 for 3 pounds (D) \$30 for 5 pounds

6.4 Absolute Value

Essential Question How can you describe how far an object is from sea level?

1 ACTIVITY: Sea Level

Work with a partner. Write an integer that represents the elevation of each object. How far is each object from sea level? Explain your reasoning.

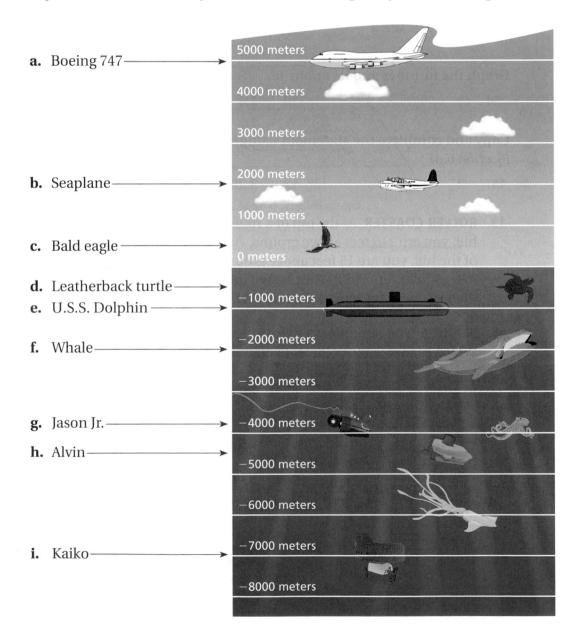

a. Boeing 747 →

b. Seaplane →

c. Bald eagle →

d. Leatherback turtle →

e. U.S.S. Dolphin →

f. Whale →

g. Jason Jr. →

h. Alvin →

i. Kaiko →

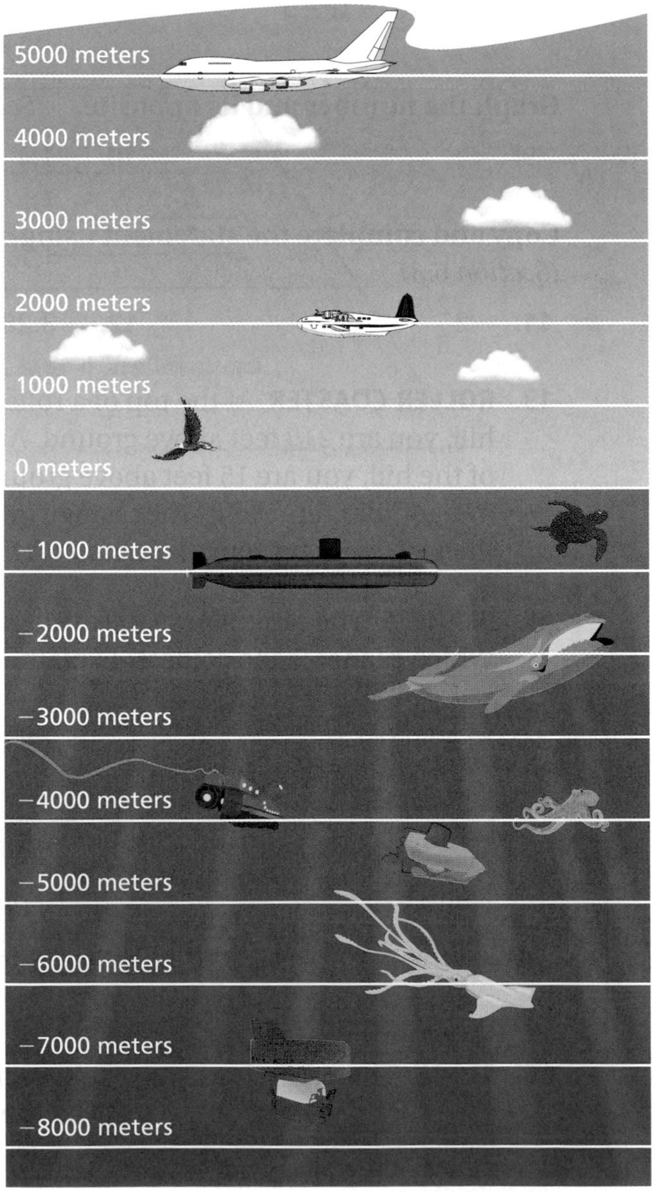

COMMON CORE

Absolute Value

In this lesson, you will
- find the absolute value of numbers.
- use absolute value to compare numbers in real-life situations.

Learning Standards
6.NS.7c
6.NS.7d

2 ACTIVITY: Finding a Distance

Work with a partner. Use the diagram in Activity 1.

a. What integer represents sea level?

b. The vessel *Kaiko* ascends to the same depth as the U.S.S. *Dolphin*. About how many meters did *Kaiko* travel? Explain how you found your answer.

c. The vessel *Jason Jr.* descends to the same depth as the *Alvin*. About how many meters did *Jason Jr.* travel? Explain how you found your answer.

d. **REASONING** Which pairs of objects are the same distance from sea level? How do you know?

e. **REASONING** An airplane is the same distance from sea level as the *Kaiko*. How far is the airplane from sea level?

3 ACTIVITY: Oceanography Project

Math Practice 5

Use Technology to Explore

How can you find more information on oceanography? What information is useful to your report?

Work with a partner. Use the Internet or some other resource to write a report that describes two ways in which mathematics is used in oceanography.

Here are two possible ideas. You can use one or both of these, or you can use other ideas.

Diving Bell

Mine Neutralization Vehicle

What Is Your Answer?

4. **IN YOUR OWN WORDS** How can you describe how far an object is from sea level?

5. **PRECISION** In Activity 1, an object has an elevation of -7500 meters. Is -7500 greater than or less than -7000? Does this object have a depth greater than or less than 7000 meters? Explain your reasoning.

Practice Use what you learned about elevation and sea level to complete Exercises 4–6 on page 272.

Check It Out
Lesson Tutorials
BigIdeasMath.com

Key Vocabulary
absolute value, *p. 270*

Key Idea

Absolute Value

Words The **absolute value** of a number is the distance between the number and 0 on a number line. The absolute value of a number a is written as $|a|$.

Numbers $|-2| = 2$ $|2| = 2$

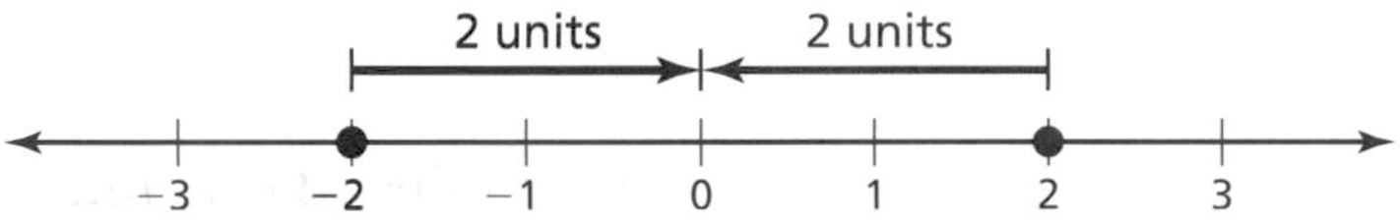

EXAMPLE 1 Finding Absolute Value

a. Find the absolute value of 3.

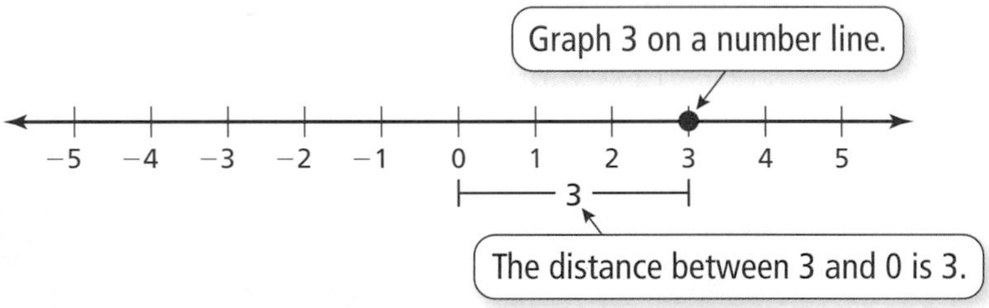

So, $|3| = 3$.

b. Find the absolute value of $-2\frac{1}{2}$.

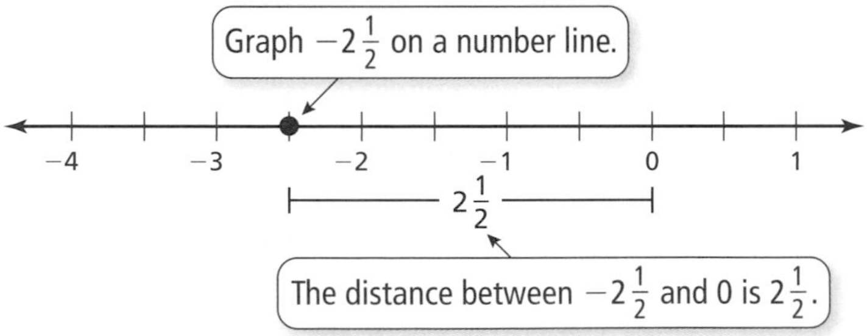

So, $\left|-2\frac{1}{2}\right| = 2\frac{1}{2}$.

On Your Own

Now You're Ready
Exercises 7–14

Find the absolute value.

1. $|8|$ **2.** $|-6|$ **3.** $|0|$

4. $\left|\frac{1}{4}\right|$ **5.** $\left|-7\frac{1}{3}\right|$ **6.** $|-12.9|$

Multi-Language Glossary at BigIdeasMath.com

EXAMPLE 2 Comparing Values

Compare 2 and $|-5|$.

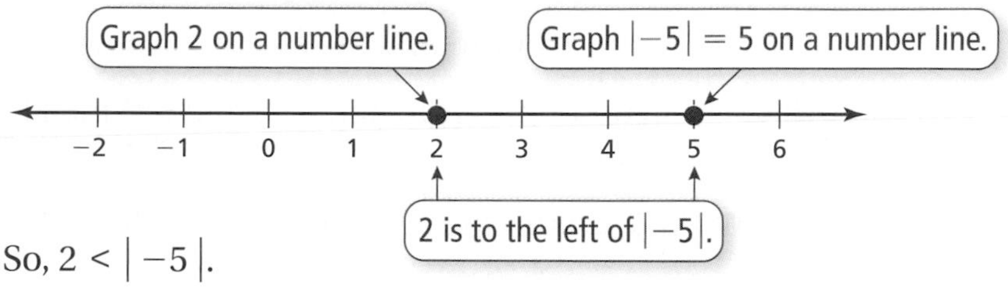

So, $2 < |-5|$.

On Your Own

Now You're Ready
Exercises 17–22

Copy and complete the statement using <, >, or =.

7. $|-4|$ ▢ -2

8. -5 ▢ $|5|$

9. $|9|$ ▢ 10

10. 3.9 ▢ $|-3.9|$

EXAMPLE 3 Real-Life Application

Animal	Elevation (ft)
Shark	−4
Sea lion	5
Seagull	56
Shrimp	−65
Turtle	−22

The table shows the elevations of several animals.

a. Which animal is the deepest? Explain.

Graph each elevation.

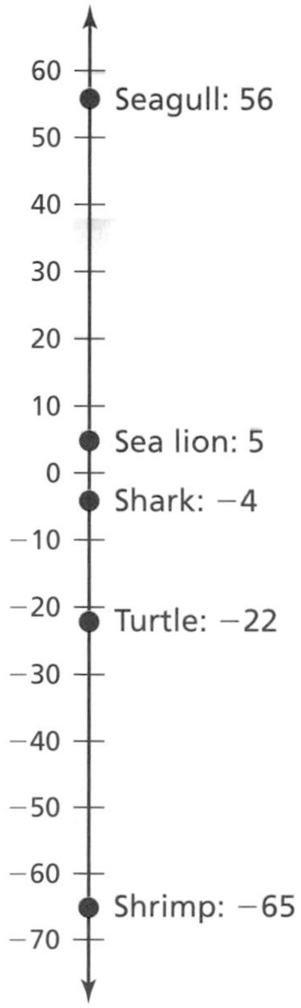

The lowest elevation represents the animal that is the deepest. The integer that is lowest on the number line is −65.

So, the shrimp is the deepest.

b. Is the shark or the sea lion closer to sea level?

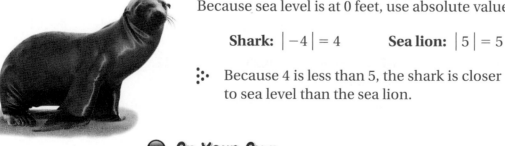

Because sea level is at 0 feet, use absolute values.

Shark: $|-4| = 4$ **Sea lion:** $|5| = 5$

Because 4 is less than 5, the shark is closer to sea level than the sea lion.

On Your Own

11. Is the seagull or the shrimp closer to sea level? Explain your reasoning.

6.4 Exercises

Check It Out
Help with Homework
BigIdeasMath.com

Vocabulary and Concept Check

1. **VOCABULARY** Explain how to find the absolute value of an integer.
2. **REASONING** Which integer is greater, -50 or 25? Which has the greater absolute value? Explain.
3. **DIFFERENT WORDS, SAME QUESTION** Which is different? Find "both" answers.

How far is -3 from 0?	What integer is 3 units to the left of 0?
What is the absolute value of -3?	What is the distance between -3 and 0?

Practice and Problem Solving

Use a vertical number line to graph the location of each object. Then tell which object is farther from sea level.

4. Scuba diver: -15 m
   Dolphin: -22 m
5. Seagull: 12 m
   School of fish: -4 m
6. Shark: -40 m
   Flag on a ship: 32 m

Find the absolute value.

1 7. $|-2|$ 8. $|23|$ 9. $|-8.35|$ 10. $\left|\frac{1}{6}\right|$

11. $\left|-3\frac{2}{5}\right|$ 12. $|11|$ 13. $|14.06|$ 14. $|-68|$

15. **REASONING** Write two integers that have an absolute value of 10.

16. **ERROR ANALYSIS** Describe and correct the error in finding the absolute value.

✗ $|14| = -14$

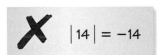

Copy and complete the statement using <, >, or =.

2 17. $6 \;\blacksquare\; |-8|$ 18. $|-3| \;\blacksquare\; 3$ 19. $|-5.5| \;\blacksquare\; |-3.1|$

20. $\frac{3}{4} \;\blacksquare\; \left|-\frac{2}{5}\right|$ 21. $|-6.8| \;\blacksquare\; |8.25|$ 22. $-12 \;\blacksquare\; |12|$

23. **CAVES** Three scientists explore a cave. Which scientist is farthest underground?

Scientist A: -48 ft Scientist B: -62 ft Scientist C: -53 ft

MATCHING Match the account balance with the debt that it represents. Explain your reasoning.

24. account balance $= -\$25$
25. account balance $< -\$25$
26. account balance $> -\$25$

A. debt $> \$25$
B. debt $= \$25$
C. debt $< \$25$

Order the values from least to greatest.

27. $5, 0, |-1|, |4|, -2$

28. $|-3|, |5|, -3, -4, |-4|$

29. $10, |-6|, 9, |3|, -11, 0$

30. $-18, |30|, -19, |-22|, -20, |-18|$

Simplify the expression.

31. $|0|$

32. $-|6|$

33. $-|-1|$

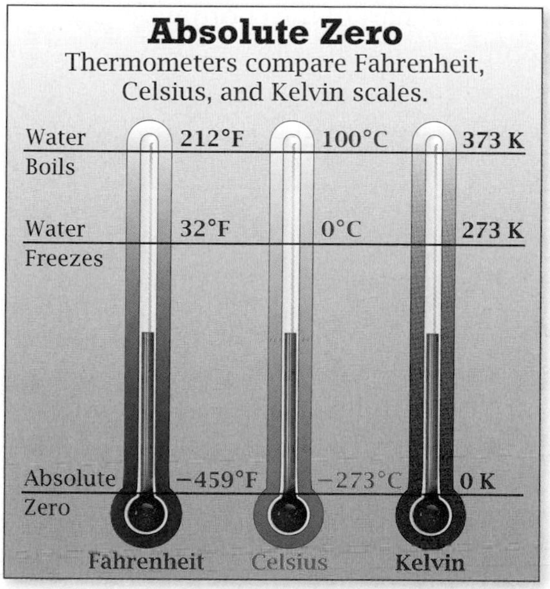

34. **ABSOLUTE ZERO** The coldest possible temperature is called *absolute zero*. It is represented by 0 K on the Kelvin temperature scale.

a. Which temperature is closer to 0 K: 32°F or −50°C?

b. What do absolute values and temperatures on the Kelvin scale have in common?

Tell whether the statement is *always*, *sometimes*, or *never* true. Explain.

35. The absolute value of a number is greater than the number.

36. The absolute value of a negative number is positive.

37. The absolute value of a positive number is its opposite.

38. **PALINDROME** A *palindrome* is a word or sentence that reads the same forward as it does backward.

a. Graph and label the following points on a number line: $A = -2$, $C = -1$, $E = 0$, $R = -3$. Then graph and label the absolute value of each point on the *same* number line.

b. What word do the letters spell? Is this a palindrome?

c. Make up your own palindrome.

39. Critical Thinking Find values of x and y so that $|x| < |y|$ and $x > y$.

Fair Game Review What you learned in previous grades & lessons

Draw the polygon with the given vertices in a coordinate plane. *(Section 4.4)*

40. $A(1, 1), B(3, 5), C(5, 0)$

41. $D(0, 6), E(2, 1), F(6, 3)$

42. $P(2, 1), Q(4, 4), R(8, 4), S(6, 1)$

43. $W(1, 6), X(9, 6), Y(9, 1), Z(4, 1)$

44. **MULTIPLE CHOICE** Which expression represents "6 less than the product of 4 and a number x"? *(Section 3.2)*

Ⓐ $(6-4)x$ Ⓑ $6-4x$ Ⓒ $\frac{6}{4x}$ Ⓓ $4x-6$

6.5 The Coordinate Plane

Essential Question How can you graph and locate points that contain negative numbers in a coordinate plane?

You have already graphed points and polygons in one part of the coordinate plane. In Activity 1, you will form the *entire* coordinate plane.

1 ACTIVITY: Forming the Entire Coordinate Plane

Work with a partner.

a. In the middle of a sheet of grid paper, construct a horizontal number line as shown. Label the tick marks. On a different sheet of grid paper, construct and label a similar vertical number line.

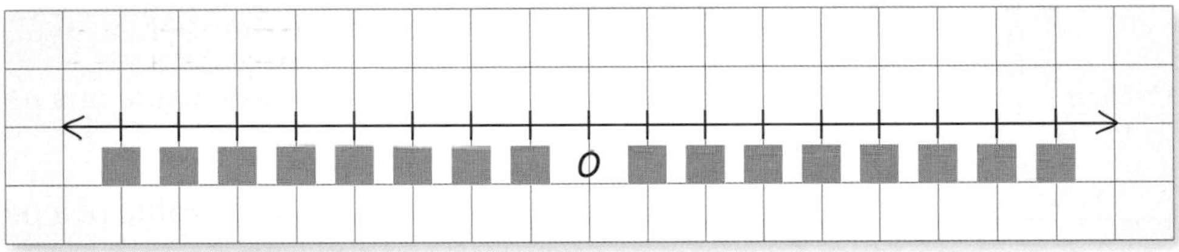

b. Cut out the vertical number line and tape it on top of the horizontal number line so that the zeros overlap. Make sure the number lines are perpendicular to one another. How many regions did you form by doing this?

c. **REASONING** What ordered pair represents the point where the number lines intersect? Why do you think this point is called the *origin*? Explain.

2 ACTIVITY: Describing Points in the Coordinate Plane

Work with a partner. Use your perpendicular number lines from Activity 1.

a. Plot and label (3, 2) on your coordinate plane. Shade this region in your coordinate plane. What do you notice about the integers along the number lines that surround (3, 2)?

b. Can you plot a point in your coordinate plane so that it is surrounded by negative numbers on the axes? If so, where is this point? Use a different color to shade this region in your coordinate plane.

c. What do you notice about the integers along the number lines for points in the regions that are not shaded?

d. **STRUCTURE** Describe how you would plot $(-3, -2)$. How is plotting this point similar to plotting (3, 2)? Plot $(-3, -2)$ in your coordinate plane.

e. **REASONING** Where in your coordinate plane do you plot $(2, -4)$? Where do you plot $(-2, 4)$? Explain your reasoning.

COMMON CORE

Coordinate Plane

In this lesson, you will
- describe the locations of points in the coordinate plane.
- plot points in the coordinate plane given ordered pairs.
- find distances between points in the coordinate plane.

Learning Standards
6.NS.6b
6.NS.6c
6.NS.8

3 ACTIVITY: Plotting Points in a Coordinate Plane

Math Practice 1

Check Progress

How can you check your progress to make sure you are accurately drawing the picture?

Work with a partner. Plot and connect the points to make a picture. Describe and color the picture when you are done.

1(6, 9)	**2**(4, 11)	**3**(2, 12)	**4**(0, 11)	**5**(−2, 9)
6(−6, 2)	**7**(−9, 1)	**8**(−11, −3)	**9**(−7, 0)	**10**(−5, −1)
11(−5, −5)	**12**(−4, −8)	**13**(−6, −10)	**14**(−3, −9)	**15**(−3, −10)
16(−4, −11)	**17**(−4, −12)	**18**(−3, −11)	**19**(−2, −12)	**20**(−2, −11)
21(−1, −12)	**22**(−1, −11)	**23**(−2, −10)	**24**(−2, −9)	**25**(1, −9)
26(2, −8)	**27**(2, −10)	**28**(1, −11)	**29**(1, −12)	**30**(2, −11)
31(3, −12)	**32**(3, −11)	**33**(4, −12)	**34**(4, −11)	**35**(3, −10)
36(3, −8)	**37**(4, −6)	**38**(6, 0)	**39**(9, −3)	**40**(9, −1)
41(8, 1)	**42**(5, 3)	**43**(3, 6)	**44**(3, 7)	**45**(4, 8)

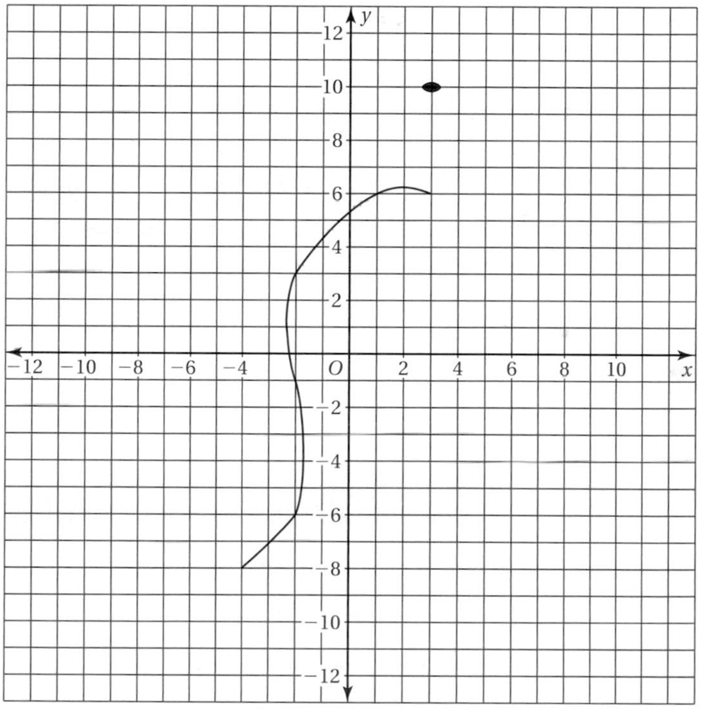

What Is Your Answer?

4. **IN YOUR OWN WORDS** How can you graph and locate points that contain negative numbers in a coordinate plane?

5. Make up your own "dot-to-dot" picture. Use at least 20 points. Your picture should have at least two points in each region of the coordinate plane.

Practice Use what you learned about the coordinate plane to complete Exercise 4 on page 279.

6.5 Lesson

Check It Out
Lesson Tutorials
BigIdeasMath.com

Key Vocabulary
coordinate plane, *p. 276*
origin, *p. 276*
quadrants, *p. 276*

Previously, you plotted points with positive coordinates. Now you will plot points with positive and negative coordinates.

Key Idea

The Coordinate Plane

A **coordinate plane** is formed by the intersection of a horizontal number line and a vertical number line. The number lines intersect at the **origin** and separate the coordinate plane into four regions called **quadrants**.

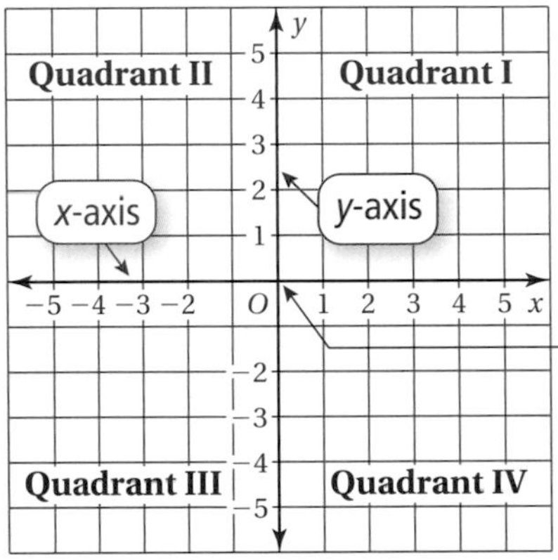

An *ordered pair* is used to locate a point in a coordinate plane.

ordered pair: $(4, -2)$

x-coordinate → 4; y-coordinate → −2

EXAMPLE 1 Identifying an Ordered Pair

Which ordered pair corresponds to point T?

Ⓐ $(-3, -3)$ Ⓑ $(-3, 3)$

Ⓒ $(3, -3)$ Ⓓ $(3, 3)$

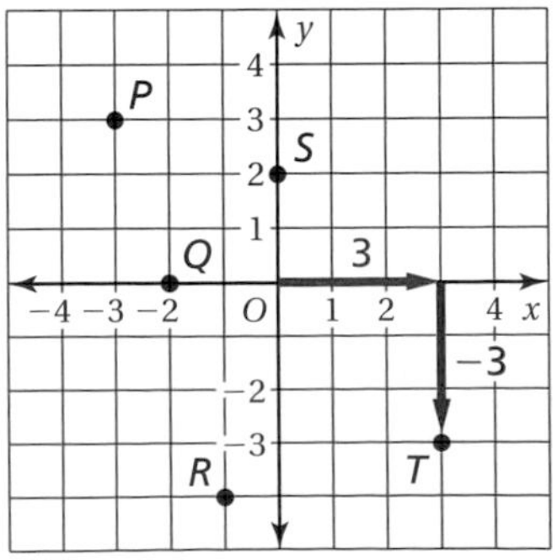

Point T is 3 units to the right of the origin and 3 units down. So, the x-coordinate is 3 and the y-coordinate is -3.

∴ The ordered pair $(3, -3)$ corresponds to point T. The correct answer is Ⓒ.

On Your Own

Now You're Ready
Exercises 5–14

Use the graph in Example 1 to write an ordered pair corresponding to the point.

1. Point P 2. Point Q 3. Point R 4. Point S

Multi-Language Glossary at BigIdeasMath.com

EXAMPLE 2 Plotting Ordered Pairs

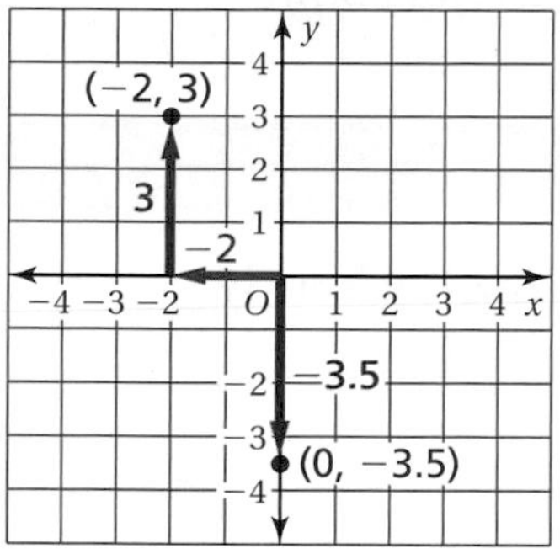

Plot (a) (−2, 3) and (b) (0, −3.5) in a coordinate plane. Describe the location of each point.

a. Start at the origin. Move 2 units left and 3 units **up**. Then plot the point.

∴ The point is in Quadrant II.

b. Start at the origin. Move 3.5 units **down**. Then plot the point.

∴ The point is on the y-axis.

On Your Own

Now You're Ready
Exercises 15–22

Plot the ordered pair in a coordinate plane. Describe the location of the point.

5. (3, −1) **6.** (−5, 0) **7.** (−2.5, −1) **8.** $\left(-1\frac{1}{2}, \frac{1}{2}\right)$

EXAMPLE 3 Finding Distances in the Coordinate Plane

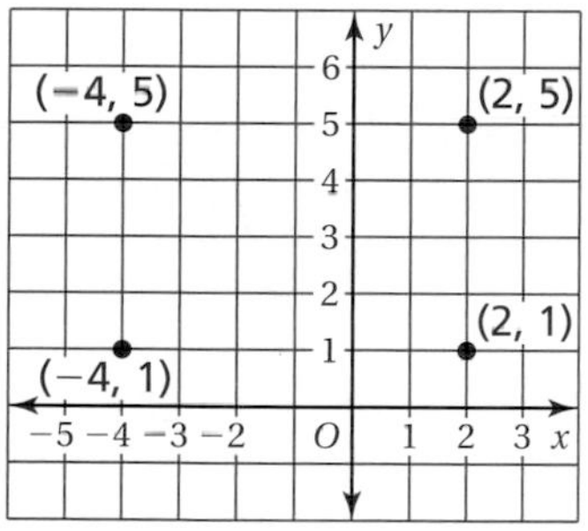

An *archaeologist* divides an area using a coordinate plane in which each unit represents 1 meter. The corners of a secret chamber are shown in the graph. What are the dimensions of the secret chamber?

The length of the chamber is the distance between (−4, 5) and (2, 5). The width of the chamber is the distance between (2, 5) and (2, 1).

You can use absolute values to find the distances between the points.

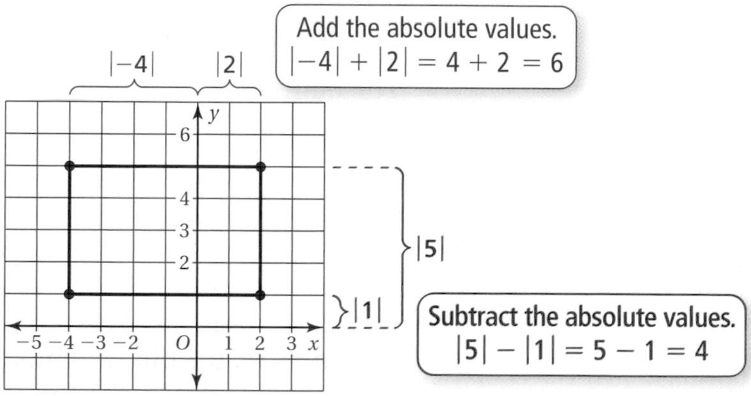

∴ The secret chamber is 6 meters long and 4 meters wide.

Reading

An **archaeologist** studies ancient ruins and objects to learn about people and cultures.

On Your Own

Now You're Ready
Exercises 25–30

9. In Example 3, the archaeologist finds a gold coin at (−1, 4), a silver coin at (−4, 2), and pottery at (−4, 4). How much closer is the pottery to the silver coin than to the gold coin?

You can use line graphs to display data that is collected over a period of time. Graphing and connecting the ordered pairs can show patterns or trends in the data. This type of line graph is also called a *time series graph*.

EXAMPLE 4 Real-Life Application

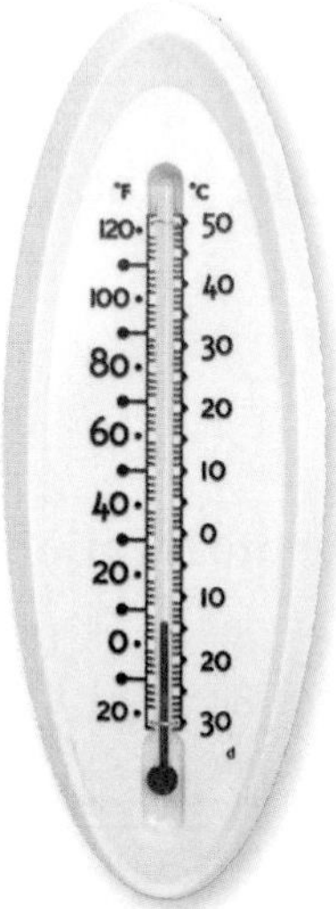

A blizzard hits a town at midnight. The table shows the hourly temperatures from midnight to 8:00 A.M.

Hours after Midnight, x	0	1	2	3	4	5	6	7	8
Temperature, y	7°F	5°F	3°F	0°F	−1°F	−4°F	−5°F	−2°F	2°F

a. Display the data in a line graph.

Write the ordered pairs.

(0, 7) (1, 5) (2, 3)

(3, 0) (4, −1) (5, −4)

(6, −5) (7, −2) (8, 2)

Plot and label the ordered pairs. Then connect the ordered pairs with line segments.

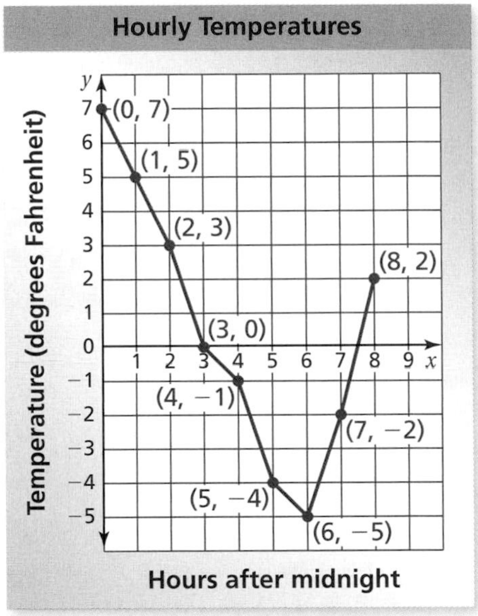

b. Make three observations from the graph.

Three possible observations follow:

- The hourly temperatures decrease from midnight to 6:00 A.M.
- The hourly temperatures increase from 6:00 A.M. to 8:00 A.M.
- The greatest decrease in hourly temperatures from one hour to the next is 3°F. This happens twice: from 2:00 A.M. to 3:00 A.M. and from 4:00 A.M. to 5:00 A.M.

Study Tip

The observations given in Example 4(b) are sample answers. You can make many other correct observations.

On Your Own

10. In Example 4, the blizzard hits another town at noon. The table shows the hourly temperatures from noon to 6:00 P.M.

Hours after Noon	0	1	2	3	4	5	6
Temperature	6°F	7°F	5°F	1°F	1°F	0°F	−3°F

a. Display the data in a line graph.

b. Make three observations from the graph.

6.5 Exercises

Check It Out
Help with Homework
BigIdeasMath.com

Vocabulary and Concept Check

1. **VOCABULARY** How many quadrants are in a coordinate plane?
2. **VOCABULARY** Is the point (0, −7) on the x-axis or the y-axis?
3. **WHICH ONE DOESN'T BELONG?** Which point does *not* belong with the other three? Explain your reasoning.

(−2, 1) (−4, 5) (2, −3) (−1, 3)

Practice and Problem Solving

4. Plot and connect the points to make a picture.

1(5, 0) **2**(2, −3) **3**(2, −2) **4**(0, −2) **5**(−3, −2)
6(−3, 0) **7**(−3, 2) **8**(0, 2) **9**(2, 2) **10**(2, 3)

Write an ordered pair corresponding to the point.

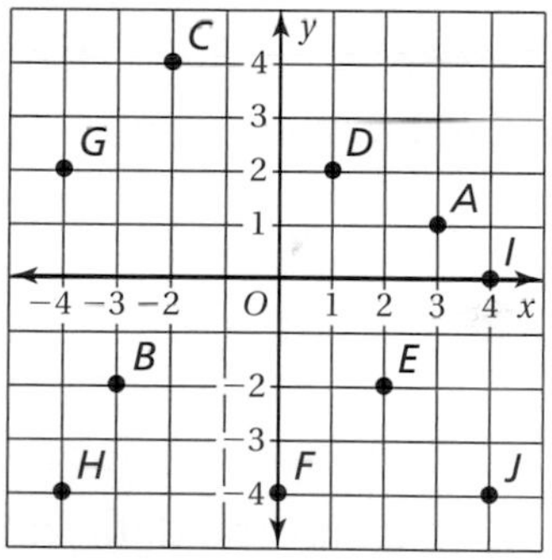

1 5. Point A
6. Point B
7. Point C
8. Point D
9. Point E
10. Point F
11. Point G
12. Point H
13. Point I
14. Point J

Plot the ordered pair in a coordinate plane. Describe the location of the point.

2 15. $K(4, 3)$
16. $L(-1, 2)$
17. $M(0, -6)$
18. $N(3.5, -1.5)$
19. $P(2, -4)$
20. $R(-4, 1)$
21. $S\left(2\frac{1}{2}, 0\right)$
22. $T(-4, -5)$

ERROR ANALYSIS Describe and correct the error in the solution.

23. ✗ To plot (4, 5), start at (0, 0) and move 5 units right and 4 units up.

24. ✗ To plot (−6, 3), start at (0, 0) and move 6 units right and 3 units down.

Plot the points and find the distance between the points.

3 25. (2, −3), (6, −3)
26. (4, 2), (4, −1)
27. (−1, 1), (−1, 7)
28. (−5, −2), (4, −2)
29. (−3, 4), (5, 4)
30. (−2, −4), (−2, 1)

31. **REASONING** The coordinates of three vertices of a square are shown in the figure. What are the coordinates of the fourth vertex?

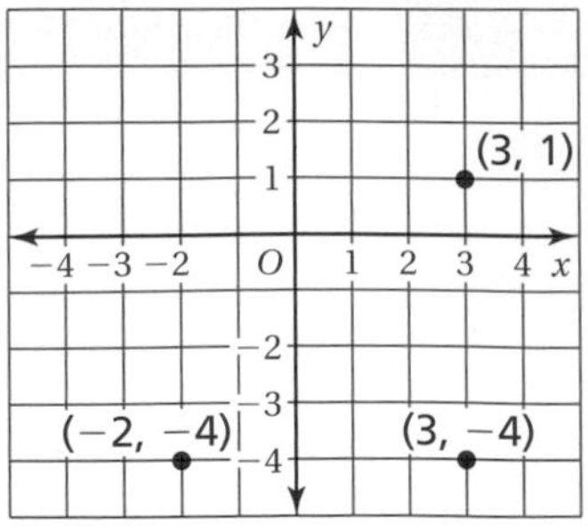

Draw the figure with the given vertices in a coordinate plane. Find the perimeter and the area of the figure.

32. $D(1, 1), E(1, -2), F(-2, -2), G(-2, 1)$

33. $P(-2, 3), Q(5, 3), R(5, -1), S(-2, -1)$

34. $W(-3, 2), X(2, 2), Y(2, -7), Z(-3, -7)$

35. **POPULATION** The line graph shows the population of a city from 2005 to 2013.

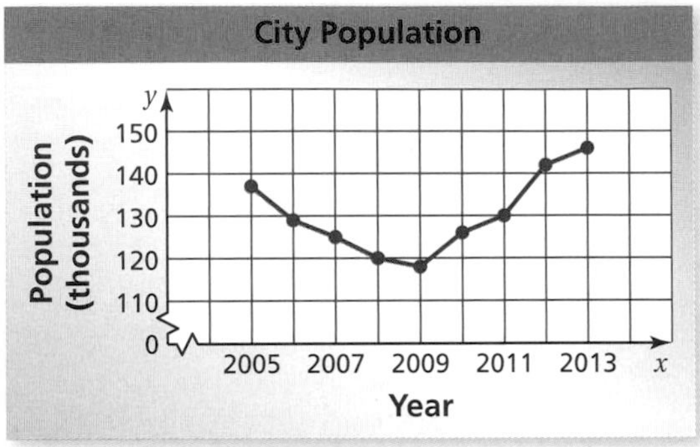

a. Estimate the population of the city in 2012.

b. Between which two years did the population increase the most?

c. Estimate the total change in population from 2005 to 2013.

36. **MODELING** The table shows the total miles run through 18 weeks for a marathon training program.

Week	1	2	3	4	5	6	7	8	9
Total Miles	22	46	72	96	124	151	181	211	244

Week	10	11	12	13	14	15	16	17	18
Total Miles	279	317	357	397	437	473	506	530	544

a. Create a table for the distance run during each week of training.

b. Display the data from part (a) in a line graph.

c. Make three observations from the graph.

d. Explain the pattern shown in the graph.

37. **PROFITS** The table shows the profits of a company from 2007 to 2013.

Years since 2000, x	7	8	9	10	11	12	13
Profit (millions of dollars), y	0.6	−0.2	−1.2	1.2	0.8	1	−0.6

a. Display the data in a line graph.

b. Make three observations from the graph.

c. What was the total profit from 2007 to 2013?

d. How could you include profits from the years 1990 to 2006 on your graph? Explain.

Describe the possible location(s) of the point (x, y).

38. $x > 0, y > 0$

39. $x < 0, y < 0$

40. $x > 0, y < 0$

41. $x > 0$

42. $y < 0$

43. $x = 0, y = 0$

Tell whether the statement is *sometimes*, *always*, or *never* true. Explain your reasoning.

44. The x-coordinate of a point on the x-axis is zero.

45. The y-coordinates of points in Quadrant III are positive.

46. The x-coordinate of a point in Quadrant II has the same sign as the y-coordinate of a point in Quadrant IV.

ZOO In Exercises 47–51, use the map of the zoo.

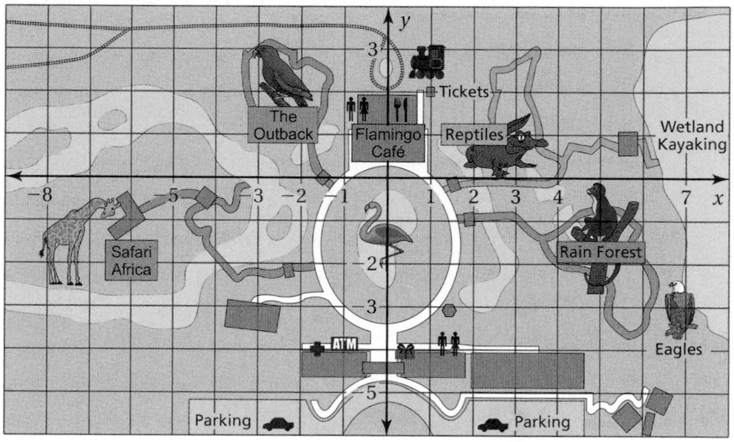

47. Which exhibit is located at (2, 1)?

48. Name an attraction on the positive y-axis.

49. Is parking available in Quadrant II? If not, name a quadrant in which you can park.

50. Write two different ordered pairs that represent the location of the Rain Forest.

51. Which exhibit is closest to $(-8, -3)$?

52. **NUMBER SENSE** Name the ordered pair that is 5 units right and 2 units down from $(-3, 4)$.

53. **OPEN-ENDED** The vertices of triangle ABC are $A(-6, -3)$ and $B(2, -3)$. List four possible coordinates of the third vertex so that the triangle has an area of 24 square units.

54. **Reasoning** Your school is located at $(2, -1)$, which is 2 blocks east and 1 block south of the center of town. To get from your house to the school, you walk 5 blocks west and 2 blocks north.

a. What ordered pair corresponds to the location of your house?

b. Is your house or your school closer to the center of town? Explain.

c. You can only walk along streets that are north and south or streets that are east and west. You are at the center of town and decide to take the shortest path home that passes by the school. When you are at the school, what percent of the walk home remains?

Fair Game Review What you learned in previous grades & lessons

Write the phrase as an expression. *(Section 3.2)*

55. 4 less than a number y

56. the product of 18 and a number b

57. a number x increased by 9

58. a number w divided by 3

59. **MULTIPLE CHOICE** What is the ratio of ducks to swans? *(Section 5.1)*

Ⓐ 4 : 9 Ⓑ 4 : 5

Ⓒ 5 : 4 Ⓓ 5 : 9

Extension 6.5 Reflecting Points in the Coordinate Plane

Check It Out
Lesson Tutorials
BigIdeasMath.com

You can *reflect* a point in the *x*-axis, in the *y*-axis, or in both axes.

The red points are mirror images of each other in the *x*-axis because the *x*-coordinates are the same and the *y*-coordinates are opposites. So, the red points are 3 units from the *x*-axis in opposite directions. The red points represent a *reflection in the x-axis.*

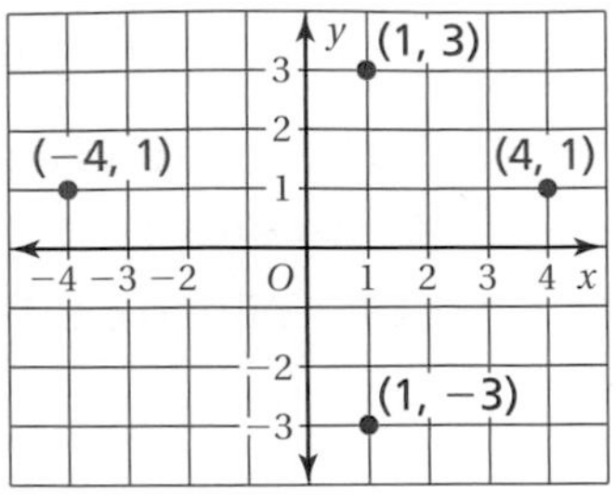

The blue points are mirror images of each other in the *y*-axis because the *y*-coordinates are the same and the *x*-coordinates are opposites. So, the blue points are 4 units from the *y*-axis in opposite directions. The blue points represent a *reflection in the y-axis.*

Key Idea

Reflecting a Point in the Coordinate Plane

- To reflect a point in the *x*-axis, use the same *x*-coordinate and take the opposite of the *y*-coordinate.
- To reflect a point in the *y*-axis, use the same *y*-coordinate and take the opposite of the *x*-coordinate.

EXAMPLE 1 Reflecting Points in One Axis

a. Reflect (−2, 4) in the *x*-axis.

Plot (−2, 4).

To reflect (−2, 4) in the *x*-axis, use the same *x*-coordinate, −2, and take the opposite of the *y*-coordinate. The opposite of 4 is −4.

So, the reflection of (−2, 4) in the *x*-axis is (−2, −4).

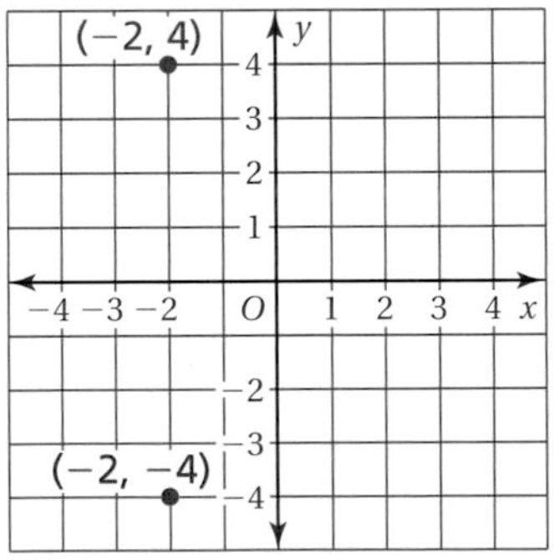

b. Reflect (−3, −1) in the *y*-axis.

Plot (−3, −1).

To reflect (−3, −1) in the *y*-axis, use the same *y*-coordinate, −1, and take the opposite of the *x*-coordinate. The opposite of −3 is 3.

So, the reflection of (−3, −1) in the *y*-axis is (3, −1).

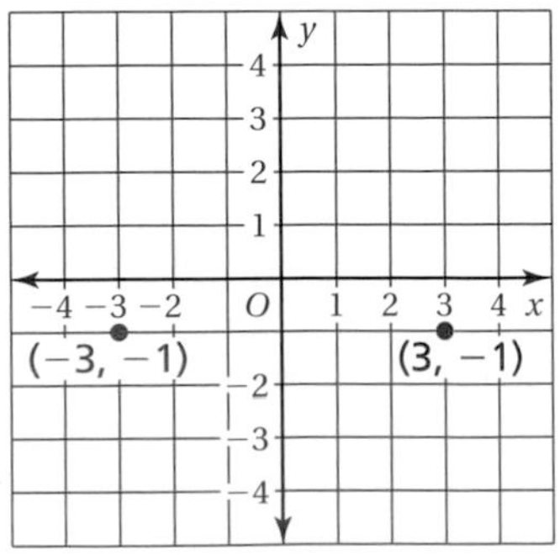

Common Core
Coordinate Plane
In this extension, you will
- understand reflections of points in the coordinate plane.

Learning Standard
6.NS.6b

EXAMPLE 2 Reflecting a Point in Both Axes

Reflect (2, 1) in the x-axis followed by the y-axis.

Step1: First, plot (2, 1).

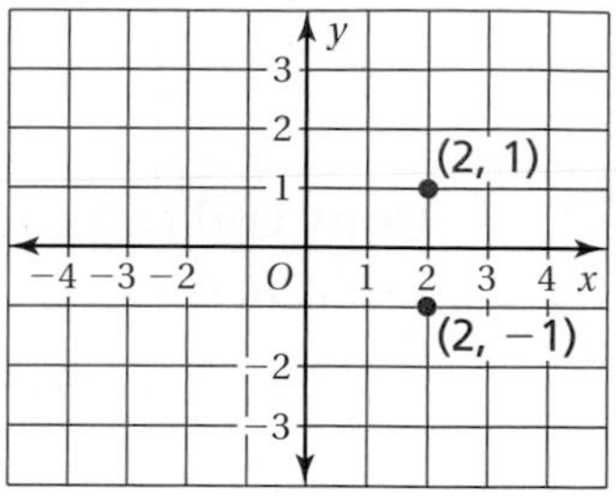

Step 2: Next, reflect (2, 1) in the x-axis. Use the same x-coordinate, 2, and take the opposite of the y-coordinate. The opposite of 1 is -1.

The point (2, 1) reflected in the x-axis is $(2, -1)$.

Common Error

When reflecting a second time, be sure to use the reflected point and not the original point.

Step 3: Finally, reflect $(2, -1)$ in the y-axis. Use the same y-coordinate, -1, and take the opposite of the x-coordinate. The opposite of 2 is -2.

The point $(2, -1)$ reflected in the y-axis is $(-2, -1)$.

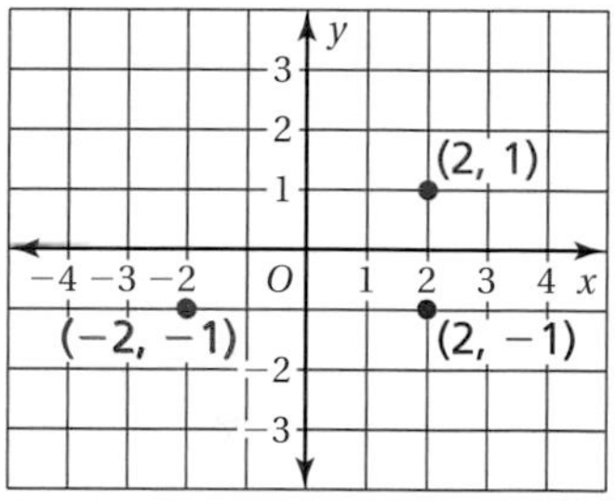

So, (2, 1) reflected in the x-axis followed by the y-axis is $(-2, -1)$.

Practice

Reflect the point in (a) the x-axis and (b) the y-axis.

1. (3, 2)
2. $(-4, 4)$
3. $(-5, -6)$
4. $(4, -7)$
5. $(0, -1)$
6. $(-8, 0)$
7. (2.5, 4.5)
8. $\left(-5\frac{1}{2}, 3\right)$

Reflect the point in the x-axis followed by the y-axis.

9. (4, 5)
10. $(-1, 7)$
11. $(-2, -2)$
12. $(6.5, -10.5)$
13. **REASONING** A point is reflected in the x-axis. The reflected point is $(3, -9)$. What is the original point? What is the distance between the points?
14. **REASONING** A point is reflected in the y-axis. The reflected point is (5.75, 0). What is the original point? What is the distance between the points?
15. a. **STRUCTURE** In Exercises 9–12, reflect the point in the y-axis followed by the x-axis. Do you get the same results? Explain.

    b. **LOGIC** Make a conjecture about how to use the coordinates of a point to find its reflection in both axes.
16. **GEOMETRY** The vertices of a triangle are $(-1, 3)$, $(-5, 3)$, and $(-5, 7)$. How would you reflect the triangle in the x-axis? in the y-axis? Give the coordinates of the reflected triangle for each case.

6.4–6.5 Quiz

Check It Out
Progress Check
BigIdeasMath.com

Find the absolute value. *(Section 6.4)*

1. $|-12|$
2. $|4|$

Copy and complete the statement using <, >, or =. *(Section 6.4)*

3. $5 \square |-9|$
4. $|-11| \square |-10|$

Write an ordered pair corresponding to the point. *(Section 6.5)*

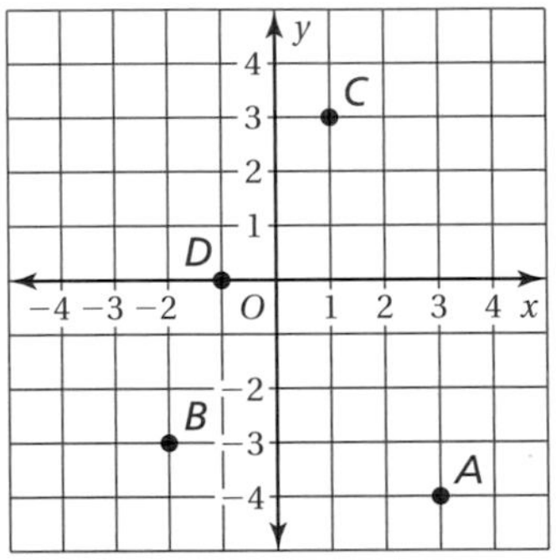

5. Point A
6. Point B
7. Point C
8. Point D

Plot the ordered pair in a coordinate plane. Describe the location of the point. *(Section 6.5)*

9. $Q(2, 5)$
10. $R(1, -4)$
11. $S(-2.5, 3.5)$
12. $T\left(0, -1\frac{1}{2}\right)$

Reflect the point in (a) the x-axis and (b) the y-axis. *(Section 6.5)*

13. $(1, 3)$
14. $(-2, 6)$

Reflect the point in the x-axis followed by the y-axis. *(Section 6.5)*

15. $(3, -2)$
16. $(-4, -5)$

17. **HIKING** The table shows the elevations of several checkpoints along a hiking trail. *(Section 6.4)*

    a. Which checkpoint is farthest from sea level?

    b. Which checkpoint is closest to sea level?

    c. Is Checkpoint 2 or Checkpoint 3 closer to sea level? Explain.

Checkpoint	Elevation (feet)
1	110
2	38
3	−24
4	12
5	−142

18. **GEOMETRY** The points $A(-4, 2)$, $B(1, 2)$, $C(1, -1)$, and $D(-4, -1)$ are the vertices of a figure. *(Section 6.5)*

    a. Draw the figure in a coordinate plane.

    b. Find the perimeter of the figure.

    c. Find the area of the figure.

6 Chapter Review

Check It Out
Vocabulary Help
BigIdeasMath.com

Review Key Vocabulary

positive numbers, *p. 250*
negative numbers, *p. 250*
opposites, *p. 250*
integers, *p. 250*
absolute value, *p. 270*
coordinate plane, *p. 276*
origin, *p. 276*
quadrants, *p. 276*

Review Examples and Exercises

6.1 Integers *(pp. 248–253)*

Write a positive or negative integer to represent losing 150 points in a pinball game.

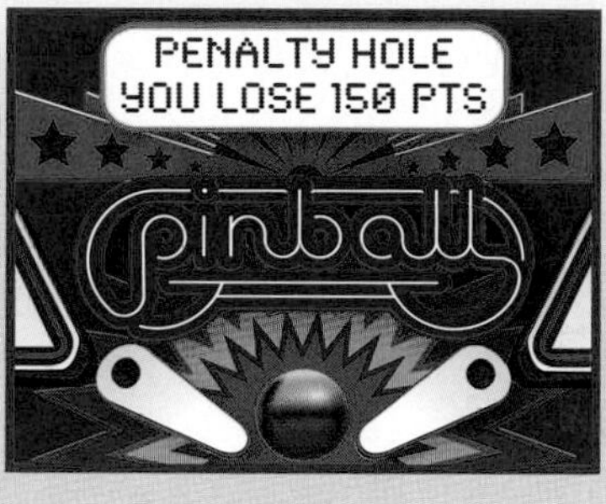

"Lose" indicates a number less than 0. So, use a negative integer.

∴ -150

Exercises

Write a positive or negative integer that represents the situation.

1. An elevator goes down 8 floors.

2. You earn $12.

Graph the integer and its opposite.

3. -7

4. 13

5. 4

6. -100

6.2 Comparing and Ordering Integers *(pp. 254–259)*

Order $-3, -4, 2, 0, -1$ from least to greatest.

Graph each integer on a number line.

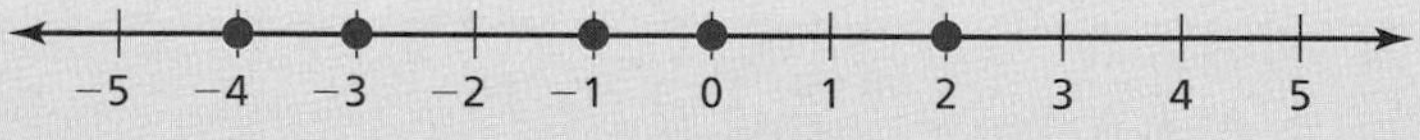

Write the integers as they appear on the number line from left to right.

∴ So, the order from least to greatest is $-4, -3, -1, 0, 2$.

Exercises

Order the integers from least to greatest.

7. $-5, 4, 2, -3, -1$

8. $5, -20, -10, 10, 15$

9. Order the temperatures $-3°C$, $8°C$, $-12°C$, $-7°C$, and $0°C$ from coldest to warmest.

6.3 Fractions and Decimals on the Number Line (pp. 260–265)

Compare $-3\frac{7}{8}$ and $-3\frac{3}{8}$.

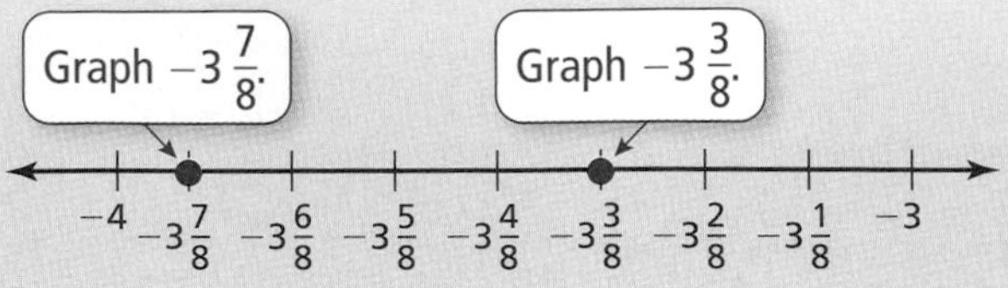

$-3\frac{7}{8}$ is to the left of $-3\frac{3}{8}$.

So, $-3\frac{7}{8} < -3\frac{3}{8}$.

Exercises

Graph the number and its opposite.

10. $-\frac{2}{5}$ **11.** $1\frac{3}{4}$ **12.** -1.6 **13.** 2.75

Copy and complete the statement using < or >.

14. $-2\frac{1}{6}$ $\square$ $-2\frac{5}{6}$ **15.** $-\frac{1}{3}$ $\square$ $-\frac{1}{8}$ **16.** -3.27 $\square$ -2.68

6.4 Absolute Value (pp. 268–273)

Find the absolute value of −3.

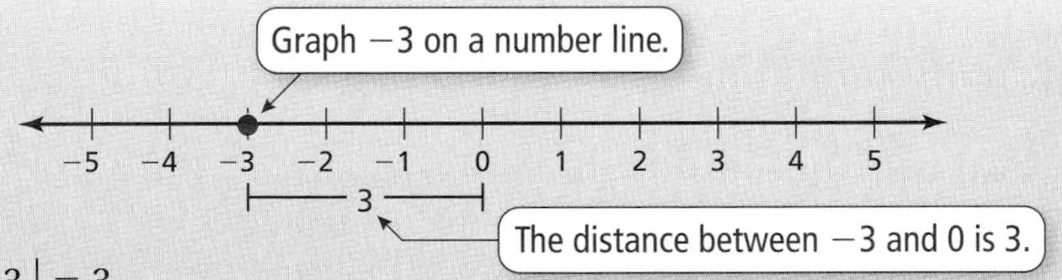

So, $|-3| = 3$.

Exercises

Find the absolute value.

17. $|-8|$ **18.** $|13|$ **19.** $\left|3\frac{6}{7}\right|$ **20.** $|-1.34|$

Copy and complete the statement using <, >, or =.

21. $|-2|$ $\square$ 2 **22.** $|4.4|$ $\square$ $|-2.8|$ **23.** $\left|\frac{1}{6}\right|$ $\square$ $\left|-\frac{2}{9}\right|$

6.5 The Coordinate Plane (pp. 274–283)

a. Plot $(-3, 0)$ and $(4, -4)$ in a coordinate plane. Describe the location of each point.

To plot $(-3, 0)$, start at the origin. Move 3 units left. Then plot the point.

To plot $(4, -4)$, start at the origin. Move 4 units right and 4 units down. Then plot the point.

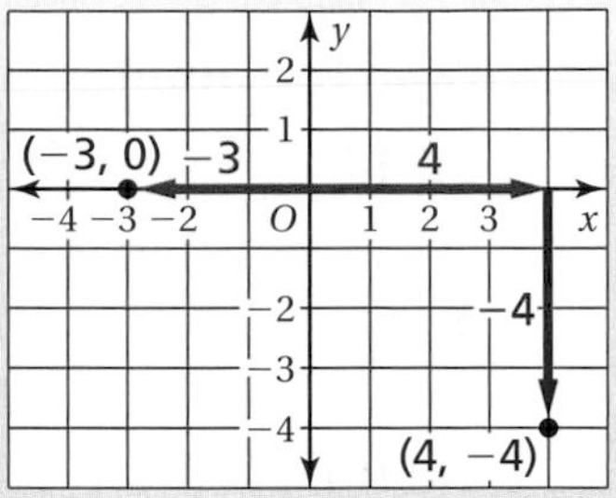

∴ The point $(-3, 0)$ is on the x-axis.
The point $(4, -4)$ is in Quadrant IV.

b. Reflect $(2, -3)$ in the x-axis.

Plot $(2, -3)$.

To reflect $(2, -3)$ in the x-axis, use the same x-coordinate, 2, and take the opposite of the y-coordinate. The opposite of -3 is 3.

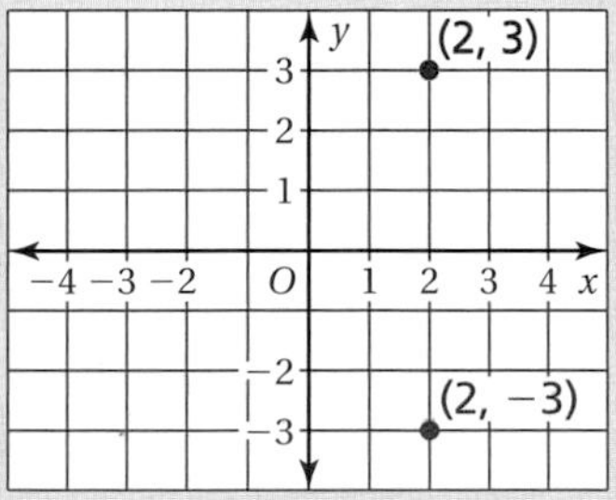

∴ So, the reflection of $(2, -3)$ in the x-axis is $(2, 3)$.

c. Reflect $(2, -3)$ in the y-axis.

Plot $(2, -3)$.

To reflect $(2, -3)$ in the y-axis, use the same y-coordinate, -3, and take the opposite of the x-coordinate. The opposite of 2 is -2.

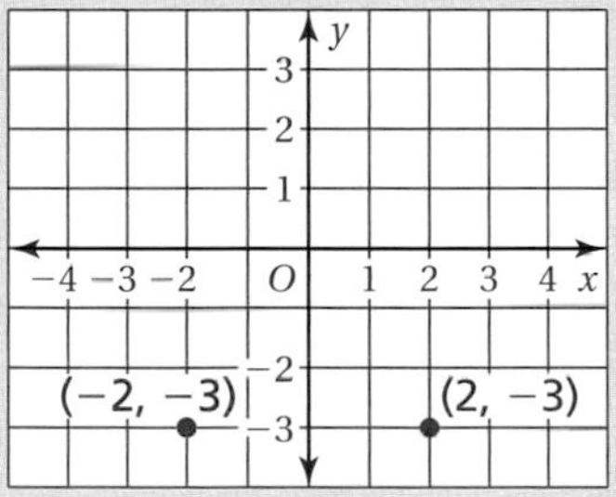

∴ So, the reflection of $(2, -3)$ in the y-axis is $(-2, -3)$.

Exercises

Plot the ordered pair in a coordinate plane. Describe the location of the point.

24. $A(1, 3)$

25. $B(0, -3)$

26. $C(-4, -2)$

27. $D(-1, 2)$

Reflect the point in (a) the x-axis and (b) the y-axis.

28. $(4, 1)$

29. $(-2, 3)$

30. $(2, -5)$

31. $(-3.5, -2.5)$

Reflect the point in the x-axis followed by the y-axis.

32. $(1, 2)$

33. $(-4, 6)$

34. $(3, -4)$

35. $(-3, -3)$

6 Chapter Test

Check It Out
Test Practice
BigIdeasMath.com

Order the integers from least to greatest.

1. 0, −2, 3, 1, −4

2. −8, −3, 5, 4, −5

Graph the number and its opposite.

3. 14

4. −40

5. $-1\frac{1}{3}$

6. 1.75

Find the absolute value.

7. $|-7|$

8. $|-11|$

Copy and complete the statement using <, >, or =.

9. $-\frac{2}{3} \ \square \ -\frac{3}{5}$

10. $1.55 \ \square \ -2.46$

11. $|-6| \ \square \ -3$

12. $-2.5 \ \square \ |2.5|$

Plot the ordered pair in a coordinate plane. Describe the location of the point.

13. $J(4, 0)$

14. $K(-3, 5)$

15. $L(1.5, -3.5)$

16. $M(-2, -3)$

Reflect the point in the x-axis followed by the y-axis.

17. (2, 4)

18. (−5, 1)

19. **POOL** A diver is on a springboard that is 3 meters above the surface of a pool. Another diver is 2 meters below the surface of the pool.

   a. Write an integer for the position of each diver relative to the surface of the pool.

   b. Find the absolute value of each integer.

   c. Who is farther from the surface of the pool?

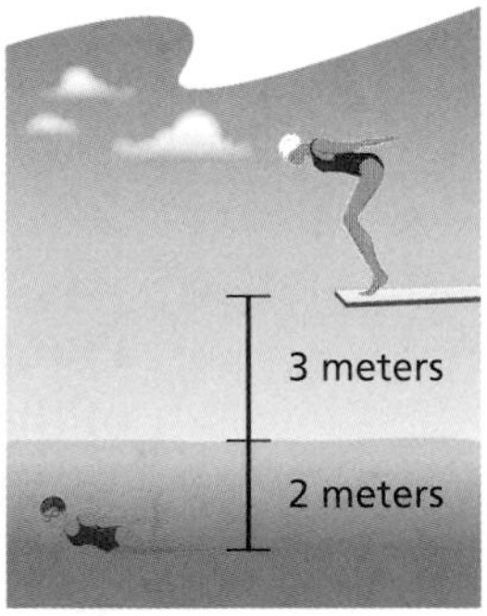

20. **OPEN-ENDED** Two vertices of a triangle are $F(1, -4)$ and $G(6, -4)$. List two possible coordinates of the third vertex so that the triangle has an area of 20 square units.

21. **MELTING POINT** The table shows the melting points (in degrees Celsius) of several elements. Compare the melting point of mercury to the melting point of each of the other elements.

Element	Mercury	Radon	Bromine	Cesium	Francium
Melting Point (°C)	−38.83	−71	−7.2	28.5	27

Test-Taking Strategy
Read All Choices Before Answering
What is the greatest number of cat treats?
A 1/2
B 3/4
C 0
D 3
D for D-licious!
"Reading all choices before answering can get you a lot more yummy treats!"

1. What is the value of the expression below when $a = 6$, $b = 5$, and $c = 4$? *(6.EE.2c)*

$$8a - 3c + 5b$$

A. 11

B. 53

C. 61

D. 107

2. Point P is plotted in the coordinate plane below.

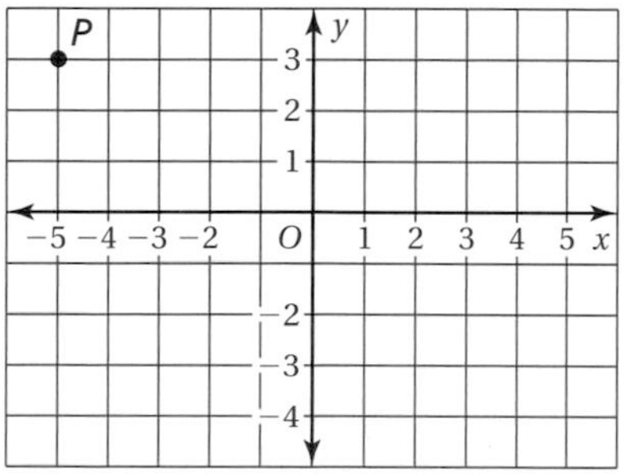

What are the coordinates of point P? *(6.NS.8)*

F. $(-5, -3)$

G. $(-5, 3)$

H. $(-3, -5)$

I. $(3, -5)$

3. What is the value of the expression below? *(6.NS.1)*

$$4\frac{1}{8} \div 5\frac{1}{2}$$

4. Which list of numbers is in order from least to greatest? *(6.NS.7)*

A. $2, |-3|, |4|, -6$

B. $-6, |4|, 2, |-3|$

C. $-6, |-3|, 2, |4|$

D. $-6, 2, |-3|, |4|$

5. Which percent is equivalent to $\frac{4}{5}$? *(6.RP.3c)*

F. 20%

G. 45%

H. 80%

I. 125%

6. Which property is illustrated by the statement below? *(6.EE.3)*

$$4 + (6 + n) = (4 + 6) + n$$

A. Associative Property of Addition

B. Commutative Property of Addition

C. Associative Property of Multiplication

D. Distributive Property

7. You bought 0.875 kilogram of mixed nuts. What was the total cost, in dollars, of the mixed nuts that you bought? *(6.NS.3)*

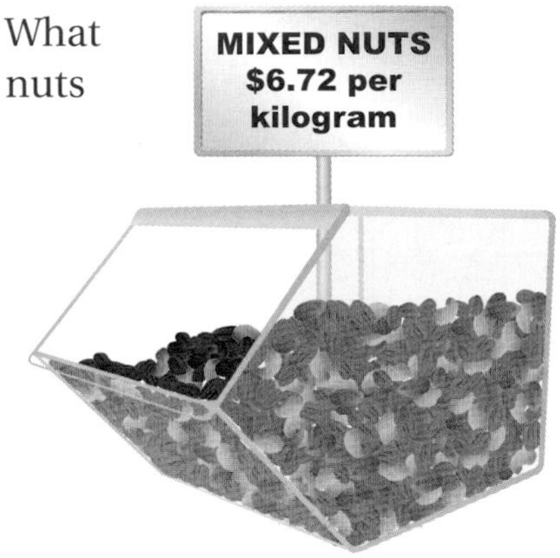

8. On Saturday, you earned $35 mowing lawns. This was x dollars more than you earned on Thursday. Which expression represents the amount, in dollars, you earned mowing lawns on Thursday? *(6.EE.6)*

F. $35x$

G. $x + 35$

H. $x - 35$

I. $35 - x$

9. Helene was finding the percent of a number in the box below.

75% of 24 is what number?

75% of $24 = 24 \div \frac{3}{4}$

$= 32$

What should Helene do to correct the error that she made? *(6.RP.3c)*

A. Divide 24 by 75.

B. Divide $\frac{3}{4}$ by 24.

C. Multiply 24 by 75.

D. Multiply 24 by $\frac{3}{4}$.

10. In the mural below, the squares that are painted red are marked with the letter R.

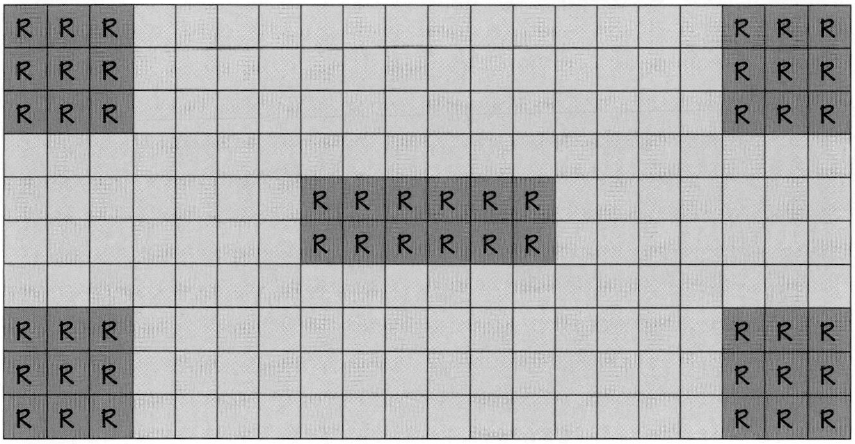

What percent of the mural is painted red? *(6.RP.3c)*

F. 24% **H.** 48%

G. 25% **I.** 50%

11. Use grid paper to complete the following. *(6.NS.8)*

Think Solve Explain

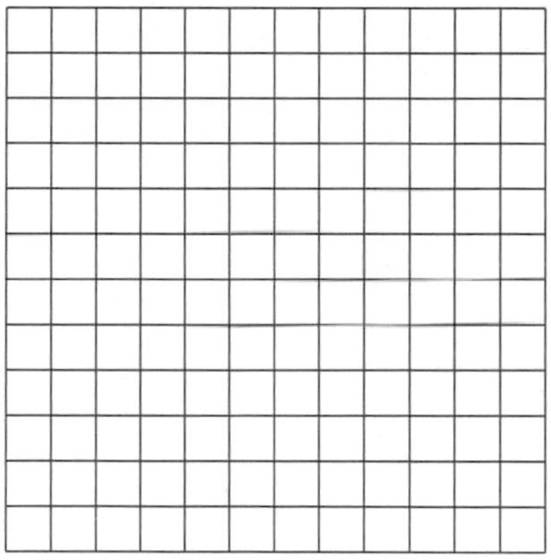

Part A Draw an x-axis and a y-axis in the coordinate plane. Then plot and label the point $(2, -3)$.

Part B Plot and label *four* points that are 3 units away from $(2, -3)$.

12. Which expression is equivalent to the expression below? *(6.NS.1)*

$$k \div 3\frac{1}{3}$$

A. $k \bullet \frac{3}{10}$ **C.** $k \div \frac{3}{10}$

B. $k \bullet \frac{10}{3}$ **D.** $k \div \frac{7}{3}$

7 Equations and Inequalities

Ten Q
"What is 7Q plus 3Q?"
"You're welcome."
"With the help of your twin brother, I think I have figured it out."
Can someone call the society for cruelty to cats?
"You weigh 36 dog biscuits."

That ought to be good for second place.
DEAR SIRS,

"Dear Sir: I have the answer to the contest question, 'How many seconds are in a year?' There are 12: January 2nd, February 2nd, ..."

What You Learned Before

Evaluating Expressions (6.EE.2c)

Example 1 **Evaluate $7x + 3y$ when $x = 2$ and $y = 4$.**

$7x + 3y = 7 \cdot 2 + 3 \cdot 4$ — Substitute 2 for x and 4 for y.

$= 14 + 12$ — Using order of operations, multiply from left to right.

$= 26$ — Add 14 and 12.

Example 2 **Evaluate $5x^2 - 2(y + 1) + 9$ when $x = 2$ and $y = 1$.**

$5x^2 - 2(y + 1) + 9 = 5(2)^2 - 2(1 + 1) + 9$ — Substitute 2 for x and 1 for y.

$= 5(2)^2 - 2 \cdot 2 + 9$ — Using order of operations, evaluate within the parentheses.

$= 5 \cdot 4 - 2 \cdot 2 + 9$ — Using order of operations, evaluate the exponent.

$= 20 - 4 + 9$ — Using order of operations, multiply from left to right.

$= 25$ — Subtract 4 from 20. Add the result to 9.

Try It Yourself

Evaluate the expression when $a = \frac{1}{2}$ and $b = 7$.

1. $6ab$
2. $16a - b$
3. $3b - 2a - 9$
4. $b^2 - 16a + 5$

Writing Expressions (6.EE.2a)

Example 3 **Write the phrase as an expression.**

a. the sum of twice a number n and five

$2n + 5$

b. twelve less than four times a number y

$4y - 12$

Try It Yourself

Write the phrase as an expression.

5. six more than three times a number w
6. the quotient of seven and a number p
7. two less than a number t
8. the product of a number x and five
9. five more than six divided by a number r
10. four less than three times a number b

7.1 Writing Equations in One Variable

Essential Question How does rewriting a word problem help you solve the word problem?

1 ACTIVITY: Rewriting a Word Problem

Work with a partner. Read the problem several times. Think about how you could rewrite the problem. Leave out information that you do not need to solve the problem.

Given Problem (63 words)

Your minivan has a flat, rectangular area in the back. When you fold down the rear seats of the van and move them forward, the width of the rectangular area in the van is increased by 2 feet, as shown in the diagram.

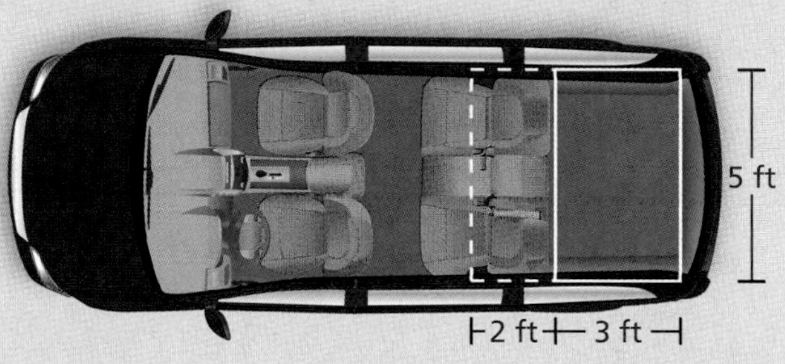

By how many square feet does the rectangular area increase when the rear seats are folded down and moved forward?

Rewritten Problem (28 words)

When you fold down the back seats of a minivan, the added area is a 5-foot by 2-foot rectangle. What is the area of this rectangle?

Can you make the problem even simpler?

Rewritten Problem (____ words)

Explain why your rewritten problem is easier to read.

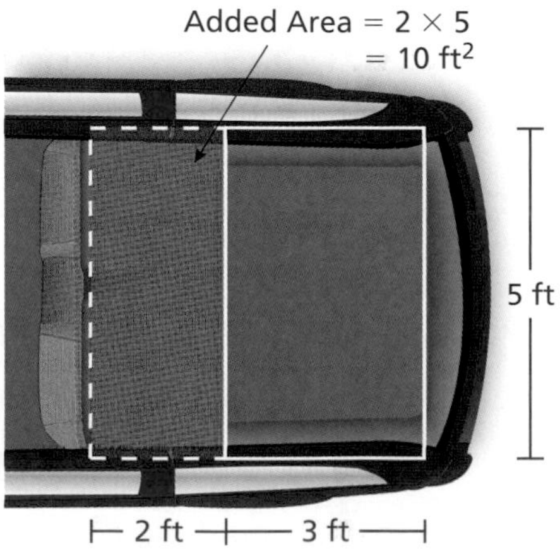

COMMON CORE

Writing Equations

In this lesson, you will

- write word sentences as equations.

Learning Standard 6.EE.6

Math Practice 1

Analyze Givens
What information do you need to solve the problem?

2 ACTIVITY: Rewriting a Word Problem

Work with a partner. Rewrite each problem using fewer words. Leave out information that you do not need to solve the problem. Then solve the problem.

a. (63 words)

A supermarket is having its grand opening on Saturday morning. Every fifth customer will receive a $10 coupon for a free turkey. Every seventh customer will receive a $3 coupon for 2 gallons of ice cream. You are the manager of the store and you expect to have 400 customers. How many of each type of coupon should you plan to give away?

b. (71 words)

You and your friend are at a football game. The stadium is 4 miles from your home. You each brought $5 to spend on refreshments. During the third quarter of the game, you say, "I read that the greatest distance that a baseball has been thrown is 445 feet 10 inches." Your friend says, "That's about one and a half times the length of the football field." Is your friend correct?

c. (90 words)

You are visiting your cousin who lives in the city. To get back home, you take a taxi. The taxi charges $2.10 for the first mile and $0.90 for each additional mile. After riding 13 miles, you decide that the fare is going to be more than the $20 you have with you. So, you tell the driver to stop and let you out. Then you call a friend and ask your friend to come and pick you up. After paying the driver, how much of your $20 is left?

What Is Your Answer?

3. **IN YOUR OWN WORDS** How does rewriting a word problem help you solve the word problem? Make up a word problem that has more than 50 words. Then show how you can rewrite the problem using at most 25 words.

"Solving a math word problem is like making maple syrup."

"You need to boil down 40 gallons of sap from a sugar maple tree to get 1 gallon of syrup."

Practice Use what you learned about writing equations to complete Exercises 4 and 5 on page 298.

7.1 Lesson

Check It Out
Lesson Tutorials
BigIdeasMath.com

Key Vocabulary
equation, *p. 296*

An **equation** is a mathematical sentence that uses an equal sign, =, to show that two expressions are equal.

Expressions	*Equations*
$4 + 8$	$4 + 8 = 12$
$x + 8$	$x + 8 = 12$

To write a word sentence as an equation, look for key words or phrases such as *is*, *the same as*, or *equals* to determine where to place the equal sign.

EXAMPLE 1 Writing Equations

Write the word sentence as an equation.

a. The sum of a number n and 7 is 15.

The sum of a number n and 7 is 15.

$n + 7 \quad = 15$ *Sum of* means *addition.*

∴ An equation is $n + 7 = 15$.

b. A number y decreased by 4 is 3.

A number y decreased by 4 is 3.

$y - 4 \quad = 3$ *Decreased by* means *subtraction.*

∴ An equation is $y - 4 = 3$.

c. 12 times a number p equals 48.

12 times a number p equals 48.

$12p \quad = \quad 48$ *Times* means *multiplication.*

∴ An equation is $12p = 48$.

On Your Own

Now You're Ready
Exercises 6–13

Write the word sentence as an equation.

1. 9 less than a number b equals 2.
2. The product of a number g and 5 is 30.
3. A number k increased by 10 is the same as 24.
4. The quotient of a number q and 4 is 12.

EXAMPLE 2 Writing an Equation

Ten servers decorate 25 tables for a wedding. Each table is decorated as shown. Let c be the total number of white and purple candles. Which equation can you use to find c?

Ⓐ $c = 25 + (4 \times 6)$ Ⓑ $c = 25(4 + 6)$

Ⓒ $c = 10(25 + 4 + 6)$ Ⓓ $c = 10(4 + 6)$

Words	The total number of candles	is	the number of tables	times	the number of candles on each table.

Variable Let c be the total number of candles.

Equation $c \quad = \quad 25 \quad \times \quad (4 + 6)$

The correct answer is Ⓑ.

EXAMPLE 3 Real-Life Application

After two rounds, 24 students are eliminated from a spelling bee. There are 96 students remaining. Write an equation you can use to find the number of students that started the spelling bee.

Reading

The word *eliminated* means *subtraction*.

Words	The number of students that started	minus	the number of students eliminated	is	the number of students remaining.

Variable Let s be the number of students that started.

Equation $s \quad - \quad 24 \quad = \quad 96$

An equation is $s - 24 = 96$.

On Your Own

5. You enter an elevator and go down 7 floors. You exit on the 10th floor. Write an equation you can use to find the floor where you entered the elevator.
6. Together you and a friend have \$52. Your friend has \$28. Write an equation you can use to find how much money you have.
7. A typical person takes about 24,000 breaths each day. Write an equation you can use to find the number of breaths a typical person takes each minute.

7.1 Exercises

Check It Out
Help with Homework
BigIdeasMath.com

Vocabulary and Concept Check

1. **VOCABULARY** How are expressions and equations different?
2. **DIFFERENT WORDS, SAME QUESTION** Which is different? Write "both" equations.

4 less than a number n is 8.	A number n is 4 less than 8.
A number n minus 4 equals 8.	4 subtracted from a number n is 8.

3. **OPEN-ENDED** Write a word sentence for the equation $28 - n = 5$.

Practice and Problem Solving

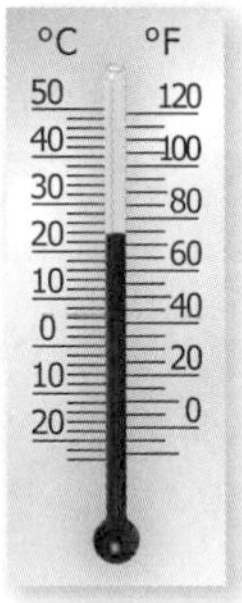

Rewrite the problem using fewer words. Leave out information that you do not need to solve the problem. Then solve the problem.

4. In a cross-country race you run at a steady rate of 7 minutes per mile. After 21 minutes, you finish in fourth place. How long is the race?
5. For a science project, you record the high temperature each day. The high temperature on Day 1 was 6° less than on Day 4 and 4° less than on Day 10. The high temperature on Day 10 was 62°F. What was the high temperature on Day 1?

Write the word sentence as an equation.

1 6. The sum of a number x and 4 equals 12.
7. A number y decreased by 9 is 8.
8. 9 times a number b is 36.
9. A number w divided by 5 equals 6.
10. 54 equals 9 more than a number t.
11. 5 is one-fourth of a number c.
12. 11 is the quotient of a number y and 6.
13. 9 less than a number n equals 27.

14. **ERROR ANALYSIS** Describe the error in writing the sentence as an equation.

✗ A number n is 5 more than 12.
$n + 5 = 12$

15. **FUNDRAISING** Students and faculty raised \$6042 for band uniforms. The faculty raised \$1780. Write an equation you can use to find the amount a raised by the students.
16. **GOLF** You hit a golf ball 90 yards. It travels three-fourths of the distance to the hole. Write an equation you can use to find the distance d from the tee to the hole.

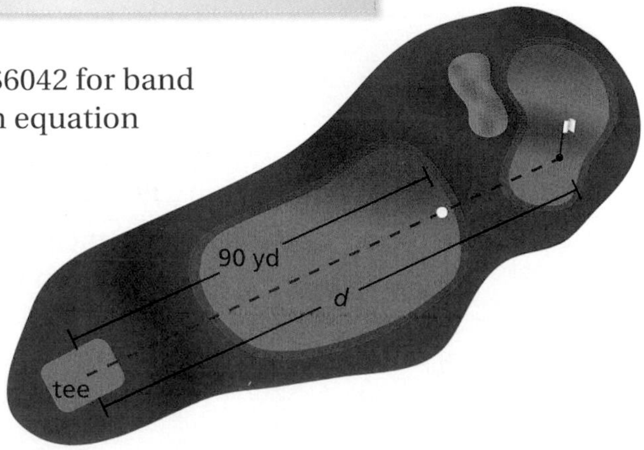

GEOMETRY Write an equation that you can use to find the value of x.

17. Perimeter of triangle: 16 in.

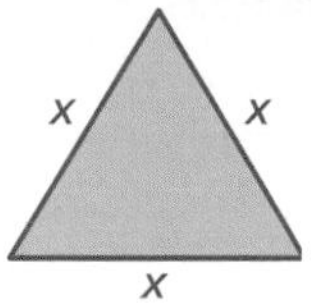

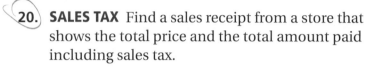

18. Perimeter of square: 30 mm

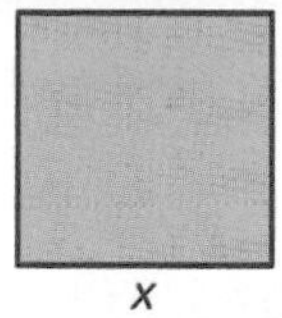

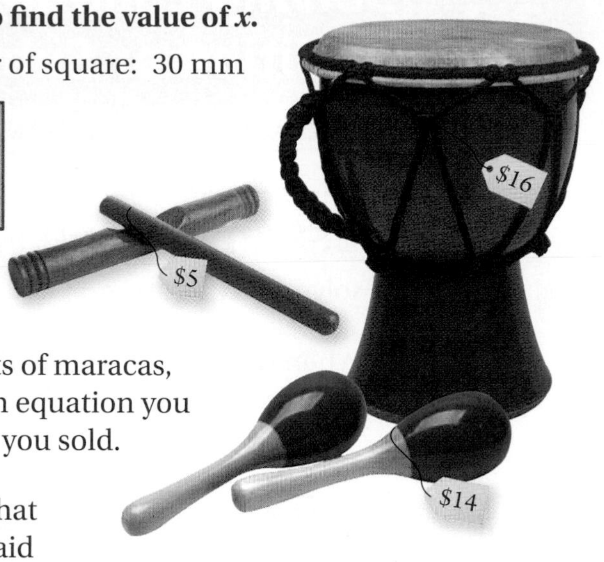

19. **MUSIC** You sell instruments at a Caribbean music festival. You earn \$326 by selling 12 sets of maracas, 6 sets of claves, and x *djembe* drums. Write an equation you can use to find the number of *djembe* drums you sold.

20. **SALES TAX** Find a sales receipt from a store that shows the total price and the total amount paid including sales tax.

a. Write an equation you can use to find the sales tax rate r.

b. Can you use r to find the *percent* for the sales tax? Explain.

21. **STRAWBERRIES** You buy a basket of 24 strawberries. You eat them as you walk to the beach. It takes the same amount of time to walk each block. When you are halfway there, half of the berries are gone. After walking 3 more blocks, you still have 5 blocks to go. You reach the beach 28 minutes after you began. One-sixth of your strawberries are left.

a. Is there enough information to find the time it takes to walk each block? Explain.

b. Is there enough information to find how many strawberries you ate while walking the last block? Explain.

22. Geometry A triangle is cut from a rectangle. The height of the triangle is half of the unknown side length s. The area of the shaded region is 84 square inches. Write an equation you can use to find the side length s.

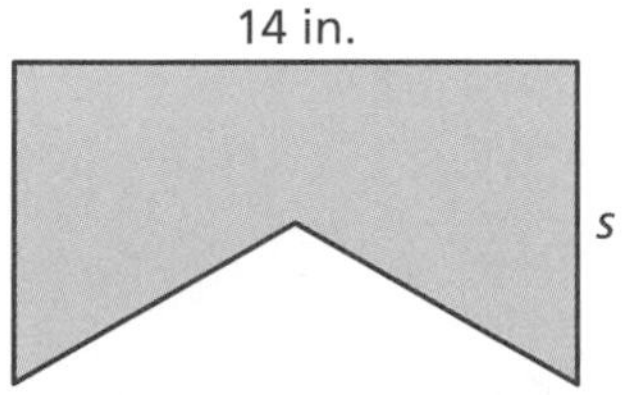

Fair Game Review What you learned in previous grades & lessons

Evaluate the expression when $a = 7$. *(Section 3.1)*

23. $6 + a$ **24.** $a - 4$ **25.** $4a$ **26.** $\frac{35}{a}$

27. **MULTIPLE CHOICE** Which expression is equivalent to $8(x + 3)$? *(Section 3.4)*

(A) $8x + 3$ (B) $8x + 24$ (C) $8x + 11$ (D) $x + 24$

7.2 Solving Equations Using Addition or Subtraction

Essential Question How can you use addition or subtraction to solve an equation?

When two sides of a scale weigh the same, the scale will balance.

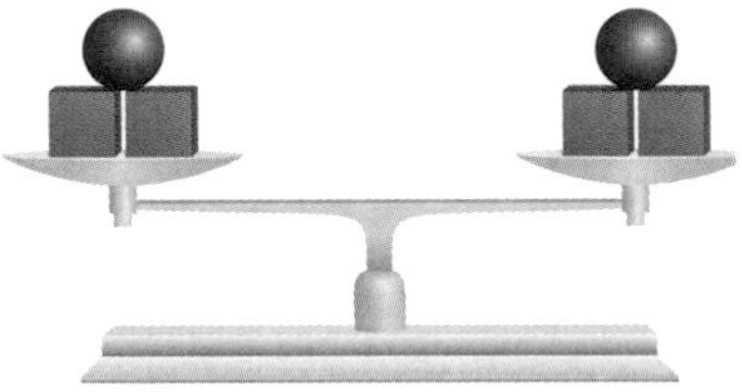

When you add or subtract the same amount on each side of the scale, it will still balance.

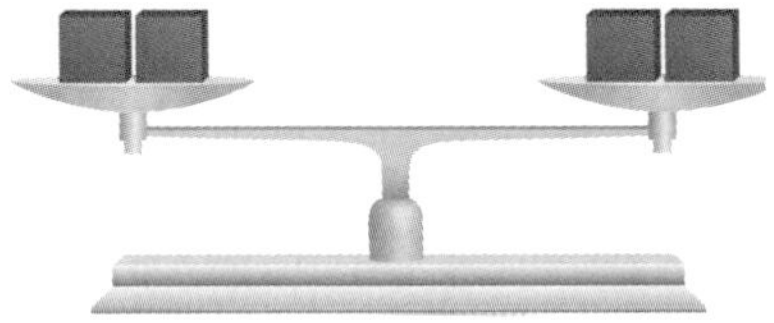

1 ACTIVITY: Solving an Equation

Work with a partner.

a. Use a model to solve $n + 3 = 7$.

- Explain how the model represents the equation $n + 3 = 7$.

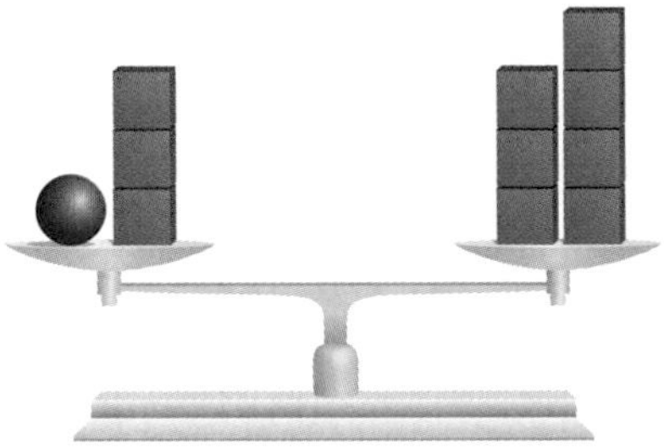

- How much does one  weigh? How do you know?

∴ The solution is $n =$ ▒.

b. Describe how you could check your answer in part (a).

c. Which model below represents the solution of $n + 1 = 9$? How do you know?

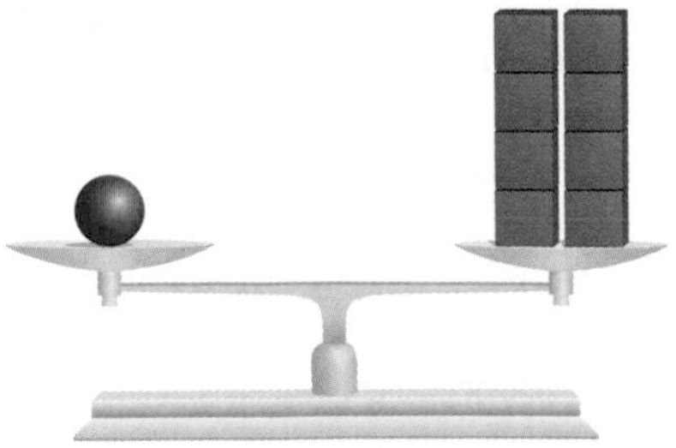

COMMON CORE

Solving Equations

In this lesson, you will

- use addition or subtraction to solve equations.
- use substitution to check answers.
- solve real-life problems.

Learning Standards
6.EE.5
6.EE.7

2 ACTIVITY: Solving Equations

Work with a partner. Solve the equation using the method in Activity 1.

Math Practice 2
Understand Quantities
What does the variable represent in the equation?

a.
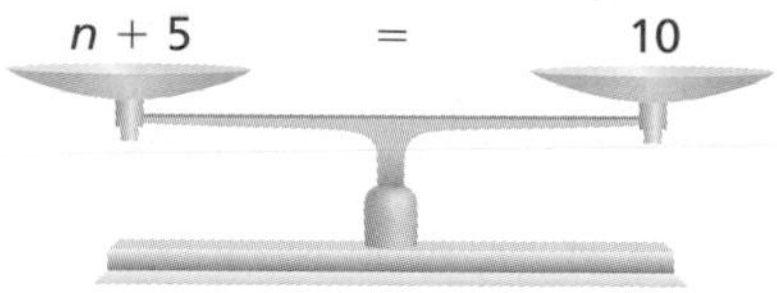

b.
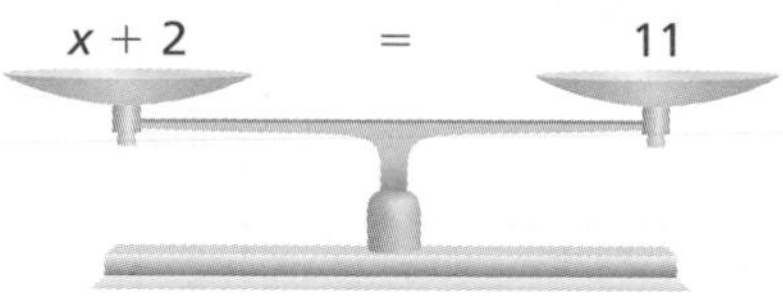

c.
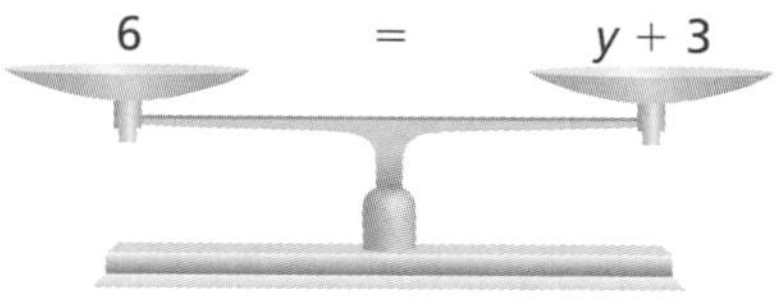

d.
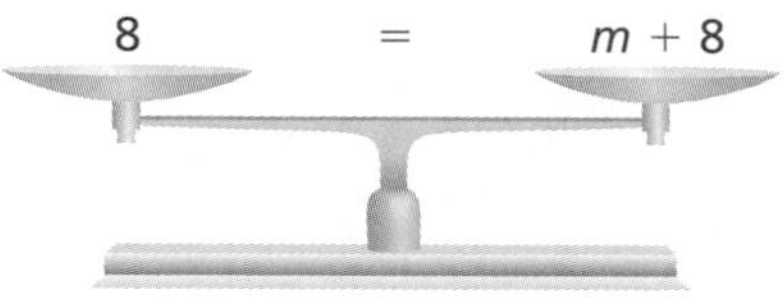

3 ACTIVITY: Solving Equations Using Mental Math

Work with a partner. Write a question that represents the equation. Use mental math to answer the question. Then check your solution.

Equation	Question	Solution	Check
a. $x + 1 = 5$			
b. $4 + m = 11$			
c. $8 = a + 3$			
d. $x - 9 = 21$			
e. $13 = p - 4$			

What Is Your Answer?

4. **REPEATED REASONING** In Activity 3, how are parts (d) and (e) different from parts (a)–(c)? Did your process to find the solution change? Explain.
5. Decide whether the statement is *true* or *false*. If false, explain your reasoning.
   a. In an equation, you can use any letter as a variable.
   b. The goal in solving an equation is to get the variable by itself.
   c. In the solution, the variable must always be on the left side of the equal sign.
   d. If you add a number to one side, you should subtract it from the other side.
6. **IN YOUR OWN WORDS** How can you use addition or subtraction to solve an equation? Give two examples to show how your procedure works.
7. Are the following equations equivalent? Explain your reasoning.

$$x - 5 = 12 \qquad \text{and} \qquad 12 = x - 5$$

Practice Use what you learned about solving equations to complete Exercises 12–17 on page 305.

7.2 Lesson

Check It Out
Lesson Tutorials
BigIdeasMath.com

Key Vocabulary
solution, *p. 302*
inverse operations, *p. 303*

Equations may be true for some values and false for others. A **solution** of an equation is a value that makes the equation true.

Value of x	$x + 3 = 7$	Are both sides equal?
3	$3 + 3 \stackrel{?}{=} 7$ $6 \neq 7$ ✗	no
4	$4 + 3 \stackrel{?}{=} 7$ $7 = 7$ ✓	yes
5	$5 + 3 \stackrel{?}{=} 7$ $8 \neq 7$ ✗	no

Reading

The symbol $\neq$ means *is not equal to.*

So, the value $x = 4$ is a solution of the equation $x + 3 = 7$.

EXAMPLE 1 Checking Solutions

Tell whether the given value is a solution of the equation.

a. $p + 10 = 38$; $p = 18$

$18 + 10 \stackrel{?}{=} 38$ — Substitute 18 for p.

$28 \neq 38$ ✗ — Sides are *not* equal.

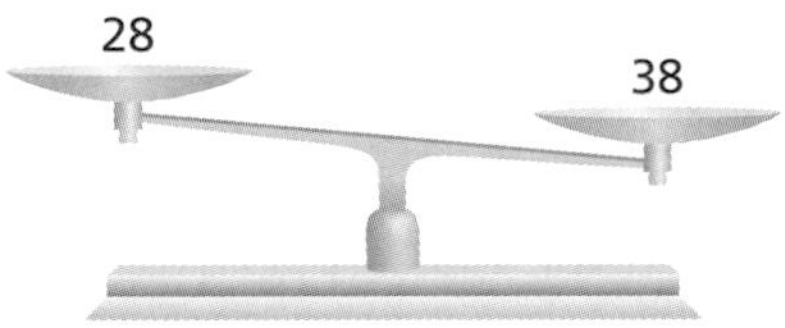

So, $p = 18$ is *not* a solution.

b. $4y = 56$; $y = 14$

$4(14) \stackrel{?}{=} 56$ — Substitute 14 for y.

$56 = 56$ ✓ — Sides are equal.

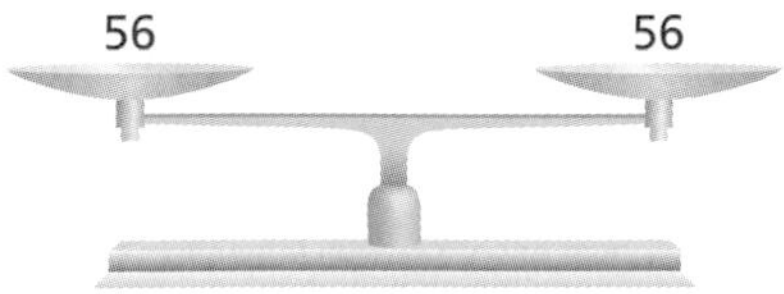

So, $y = 14$ is a solution.

On Your Own

Now You're Ready
Exercises 6–11

Tell whether the given value is a solution of the equation.

1. $a + 6 = 17$; $a = 9$
2. $9 - g = 5$; $g = 3$
3. $35 = 7n$; $n = 5$
4. $\frac{q}{2} = 28$; $q = 14$

Multi-Language Glossary at BigIdeasMath.com

You can use *inverse operations* to solve equations. **Inverse operations** "undo" each other. Addition and subtraction are inverse operations.

Key Ideas

Addition Property of Equality

Words When you add the same number to each side of an equation, the two sides remain equal.

Numbers

$$\begin{array}{rr} 8 = & 8 \\ \underline{+5} & \underline{+5} \\ 13 = & 13 \end{array}$$

Algebra

$$\begin{array}{rr} x - 4 = & 5 \\ \underline{+4} & \underline{+4} \\ x = & 9 \end{array}$$

Subtraction Property of Equality

Words When you subtract the same number from each side of an equation, the two sides remain equal.

Numbers

$$\begin{array}{rr} 8 = & 8 \\ \underline{-5} & \underline{-5} \\ 3 = & 3 \end{array}$$

Algebra

$$\begin{array}{rr} x + 4 = & 5 \\ \underline{-4} & \underline{-4} \\ x = & 1 \end{array}$$

EXAMPLE 2 Solving Equations Using Addition

a. Solve $x - 2 = 6$.

$x - 2 =$	6	Write the equation.
$\underline{+2}$	$\underline{+2}$	Addition Property of Equality
$x =$	8	Simplify.

Undo the subtraction.

The solution is $x = 8$.

Check

$x - 2 = 6$

$8 - 2 \stackrel{?}{=} 6$

$6 = 6$ ✓

Study Tip

You can check your solution by substituting it for the variable in the original equation.

b. Solve $18 = x - 7$.

$18 = x - 7$		Write the equation.
$\underline{+7}$	$\underline{+7}$	Addition Property of Equality
$25 = x$		Simplify.

The solution is $x = 25$.

Check

$18 = x - 7$

$18 \stackrel{?}{=} 25 - 7$

$18 = 18$ ✓

On Your Own

Now You're Ready
Exercises 18–20

Solve the equation. Check your solution.

5. $k - 3 = 1$ **6.** $n - 10 = 4$ **7.** $15 = r - 6$

EXAMPLE 3 Solving Equations Using Subtraction

a. Solve $x + 2 = 9$.

$$x + 2 = 9 \quad \text{Write the equation.}$$

Undo the addition. → $-2 \quad -2$ Subtraction Property of Equality

$$x = 7 \quad \text{Simplify.}$$

∴ The solution is $x = 7$.

Check

$x + 2 = 9$

$7 + 2 \stackrel{?}{=} 9$

$9 = 9$ ✓

b. Solve $26 = 11 + x$.

$$26 = 11 + x \quad \text{Write the equation.}$$

$$-11 \quad -11 \quad \text{Subtraction Property of Equality}$$

$$15 = x \quad \text{Simplify.}$$

∴ The solution is $x = 15$.

Check

$26 = 11 + x$

$26 \stackrel{?}{=} 11 + 15$

$26 = 26$ ✓

EXAMPLE 4 Real-Life Application

Your parents give you \$20 to help buy the new pair of shoes shown. After you buy the shoes, you have \$5.50 left. Write and solve an equation to find how much money you had before your parents gave you \$20.

Words The starting amount plus the amount your parents gave you minus the cost of the shoes is the amount left.

Variable Let s be the starting amount.

Equation $s \quad + \quad 20 \quad - \quad 59.95 \quad = \quad 5.50$

$s + 20 - 59.95 = 5.50$	Write the equation.
$s + 20 - 59.95 + 59.95 = 5.50 + 59.95$	Addition Property of Equality
$s + 20 = 65.45$	Simplify.
$s + 20 - 20 = 65.45 - 20$	Subtraction Property of Equality
$s = 45.45$	Simplify.

∴ You had \$45.45 before your parents gave you money.

Study Tip

In Example 4, you can solve the problem arithmetically by working backwards from \$5.50.

$5.50 + 59.95 - 20 = 45.45$

So, your answer is reasonable.

On Your Own

Now You're Ready
Exercises 21–23

Solve the equation. Check your solution.

8. $s + 8 = 17$ **9.** $9 = y + 6$ **10.** $13 + m = 20$

11. You eat 8 blueberries and your friend eats 11 blueberries from a package. There are 23 blueberries left. Write and solve an equation to find the number of blueberries in a full package.

7.2 Exercises

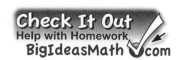

Check It Out
Help with Homework
BigIdeasMath.com

Vocabulary and Concept Check

1. **WRITING** How can you check the solution of an equation?

Name the inverse operation you can use to solve the equation.

2. $x - 8 = 12$
3. $n + 3 = 13$
4. $b + 14 = 33$

5. **WRITING** When solving $x + 5 = 16$, why do you subtract 5 from the left side of the equation? Why do you subtract 5 from the right side of the equation?

Practice and Problem Solving

Tell whether the given value is a solution of the equation.

1 6. $x + 42 = 85;\ x = 43$
7. $8b = 48;\ b = 6$
8. $19 - g = 7;\ g = 15$
9. $\frac{m}{4} = 16;\ m = 4$
10. $w + 23 = 41;\ w = 28$
11. $s - 68 = 11;\ s = 79$

Use a scale to model and solve the equation.

12. $n + 7 = 9$
13. $t + 4 = 5$
14. $c + 2 = 8$

Write a question that represents the equation. Use mental math to answer the question. Then check your solution.

15. $a + 5 = 12$
16. $v + 9 = 18$
17. $20 = d - 6$

Solve the equation. Check your solution.

2 18. $y - 7 = 3$
19. $z - 3 = 13$
20. $8 = r - 14$
3 21. $p + 5 = 8$
22. $k + 6 = 18$
23. $64 = h + 30$
24. $f - 27 = 19$
25. $25 = q + 14$
26. $\frac{3}{4} = j - \frac{1}{2}$
27. $x + \frac{2}{3} = \frac{9}{10}$
28. $1.2 = m - 2.5$
29. $a + 5.5 = 17.3$

ERROR ANALYSIS **Describe and correct the error in solving the equation.**

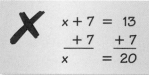

30. ✗

$$\begin{aligned} x + 7 &= 13 \\ +7 \quad & \quad +7 \\ x \quad &= 20 \end{aligned}$$

31. ✗

$$\begin{aligned} 34 &= y - 12 \\ -12 \quad & \quad +12 \\ 22 &= y \end{aligned}$$

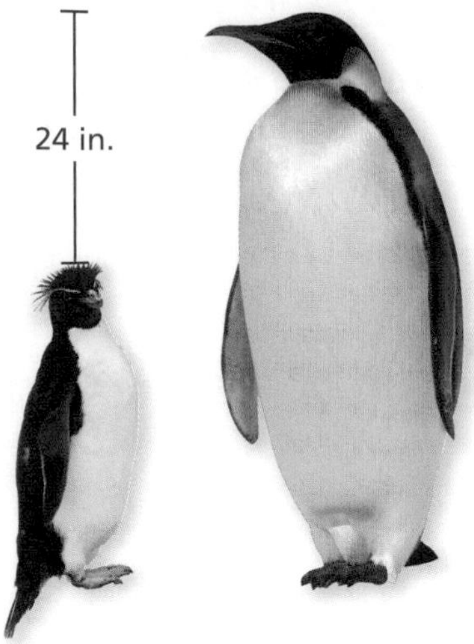

32. **PENGUINS** An emperor penguin is 45 inches tall. It is 24 inches taller than a rockhopper penguin. Write and solve an equation to find the height of a rockhopper penguin. Is your answer reasonable? Explain.

33. **ELEVATOR** You get in an elevator and go down 8 floors. You exit on the 16th floor. Write and solve an equation to find what floor you got on the elevator.

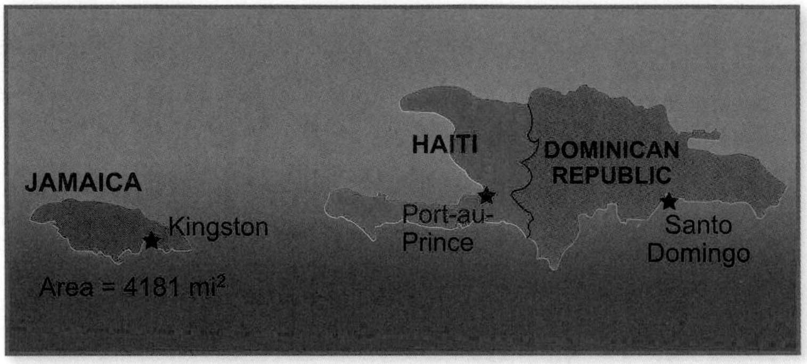

34. **AREA** The area of Jamaica is 6460 square miles less than the area of Haiti. Write and solve an equation to find the area of Haiti.

35. **REASONING** The solution of the equation $x + 3 = 12$ is shown. Explain each step. Use a property, if possible.

	Write the equation.
$x + 3 = 12$	
$x + 3 - 3 = 12 - 3$	
$x + 0 = 9$	
$x = 9$	

Write the word sentence as an equation. Then solve the equation.

36. 13 subtracted from a number w is 15.
37. A number k increased by 7 is 34.
38. 9 is the difference of a number n and 7.
39. 93 is the sum of a number g and 58.

Solve the equation. Check your solution.

40. $b + 7 + 12 = 30$
41. $y + 4 - 1 = 18$
42. $m + 18 + 23 = 71$
43. $v - 7 = 9 + 12$
44. $5 + 44 = 2 + r$
45. $22 + 15 = d - 17$

GEOMETRY Write and solve an addition equation to find x.

46. Perimeter = 48 ft

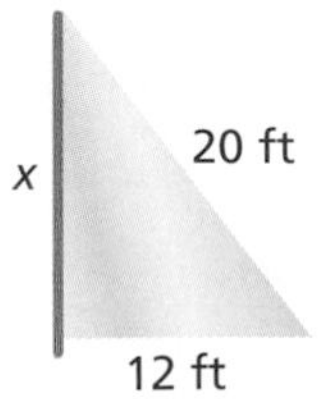

47. Perimeter = 132 in.

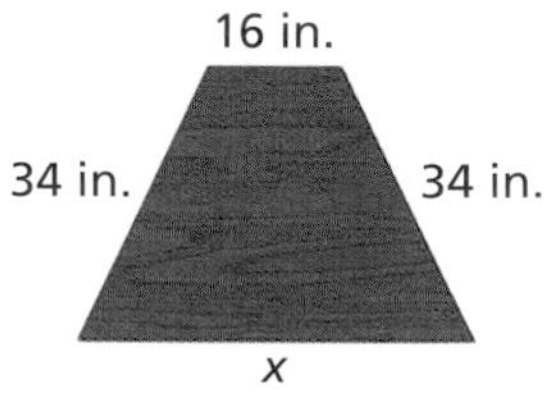

48. Perimeter = 93 ft

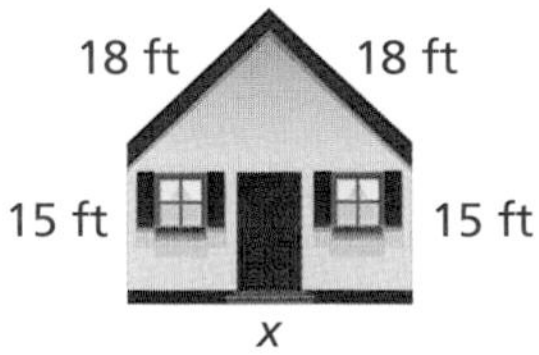

49. **REASONING** Explain why the equations $x + 4 = 13$ and $4 + x = 13$ have the same solution.

50. **REASONING** Explain why the equations $x - 13 = 4$ and $13 - x = 4$ do *not* have the same solution.

51. SIMPLIFYING AND SOLVING Compare and contrast the two problems.

Simplify the expression $2(x + 3) - 4$.

$$\begin{aligned} 2(x + 3) - 4 &= 2x + 6 - 4 \\ &= 2x + 2 \end{aligned}$$

Solve the equation $x + 3 = 4$.

$$\begin{aligned} x + 3 &= 4 \\ \underline{-3} &\quad \underline{-3} \\ x &= 1 \end{aligned}$$

52. PUZZLE In a *magic square*, the sum of the numbers in each row, column, and diagonal is the same. Write and solve equations to find the values of *a*, *b*, and *c*.

a	37	16
19	25	*b*
34	*c*	28

53. FUNDRAISER You participate in a dance-a-thon fundraiser. After your parents pledge \$15.50 and your neighbor pledges \$8.75, you have \$66.55. Write and solve an equation to find how much money you had before your parents and neighbor pledged.

54. MONEY On Saturday, you spend \$33, give \$15 to a friend, and receive \$20 for mowing your neighbor's lawn. You have \$21 left. Use two methods to find how much money you started with that day.

Bumper Cars: $1.75
Roller Coaster: $1.25 more than Ferris Wheel
Giant Slide: $0.50 less than Bumper Cars
Ferris Wheel: $1.50 more than Giant Slide

55. AMUSEMENT PARK You have \$15.

a. How much money do you have left if you ride each ride once?

b. Do you have enough money to ride each ride twice? Explain.

56. Critical Thinking Consider the equation $x + y = 15$. The value of x increases by 3. What has to happen to the value of y so that $x + y = 15$ remains true?

Fair Game Review What you learned in previous grades & lessons

Find the value of the expression. Use estimation to check your answer. *(Section 1.1)*

57. 12×8 **58.** 13×16 **59.** $75 \div 15$ **60.** $72 \div 3$

61. MULTIPLE CHOICE What is the area of the parallelogram? *(Section 4.1)*

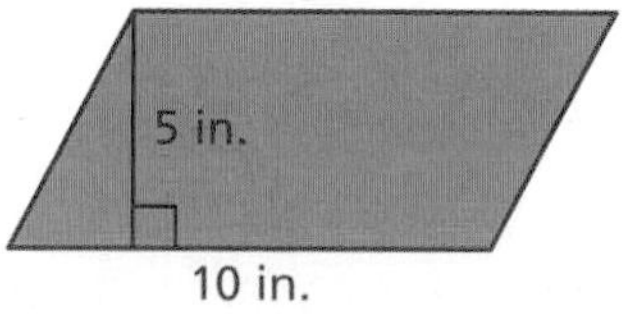
5 in.
10 in.

Ⓐ 25 in.^{2} Ⓑ 30 in.^{2}

Ⓒ 50 in.^{2} Ⓓ 100 in.^{2}

7.3 Solving Equations Using Multiplication or Division

Essential Question How can you use multiplication or division to solve an equation?

1 ACTIVITY: Finding Missing Dimensions

Work with a partner. Describe how you would find the value of *x*. Then find the value and check your result.

a. rectangle

Area = 24 square units

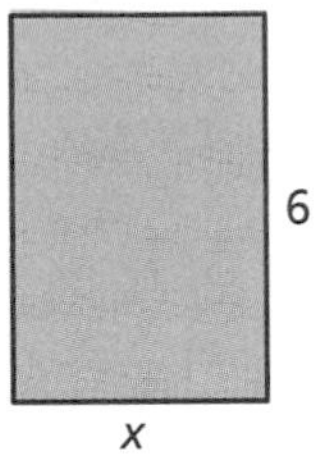

b. parallelogram

Area = 20 square units

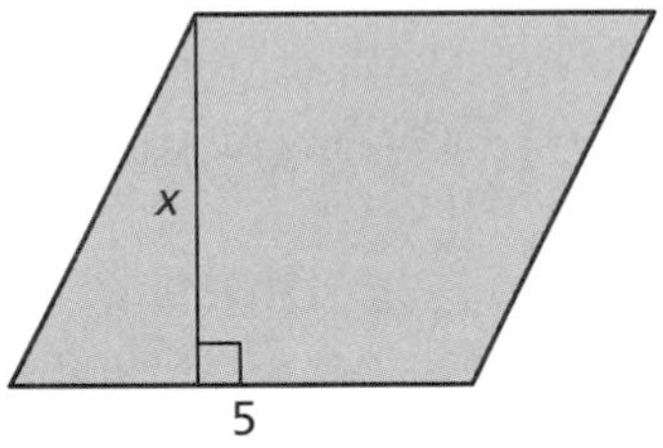

c. triangle

Area = 28 square units

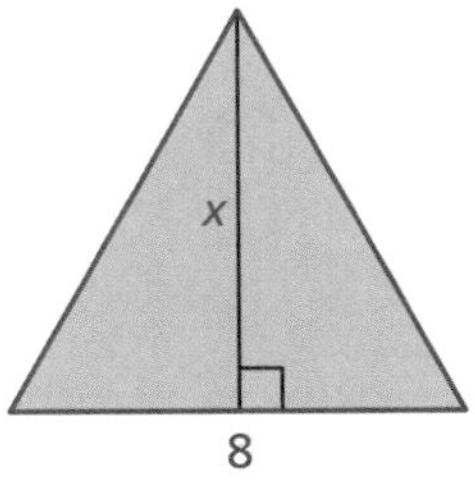

2 ACTIVITY: Using an Equation to Model a Story

Work with a partner.

a. Use a model to solve the problem.

> Three people go out to lunch. They decide to share the $12 bill evenly. How much does each person pay?

- What equation does the model represent? Explain how this represents the problem.

- How much does one weigh? How do you know?

∴ Each person pays ____.

b. Describe how you can check your answer in part (a).

COMMON CORE

Solving Equations

In this lesson, you will

- use multiplication or division to solve equations.
- use substitution to check answers.
- solve real-life problems.

Learning Standards
6.EE.5
6.EE.7

3 ACTIVITY: Using Equations to Model a Story

Work with a partner.

- **What is the unknown?**
- **Write an equation that represents each problem.**
- **What does the variable in your equation represent?**
- **Explain how you can solve the equation.**
- **Answer the question.**

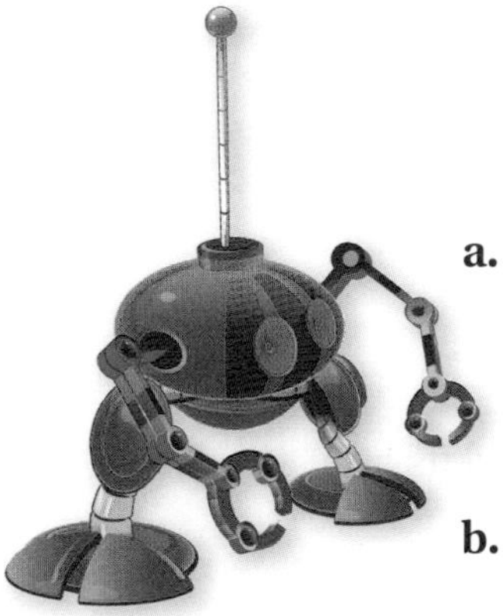

	Problem	**Equation**
a.	Three robots go out to lunch. They decide to share the $11.91 bill evenly. How much does each robot pay?	
b.	On Earth, objects weigh 6 times what they weigh on the Moon. A robot weighs 96 pounds on Earth. What does it weigh on the Moon?	
c.	At maximum speed, a robot runs 6 feet in 1 second. How many feet does the robot run in 1 minute?	
d.	Four identical robots lie on the ground head-to-toe and measure 14 feet. How tall is each robot?	

Math Practice 4

Interpret Results

What does the solution represent? Does the answer make sense?

What Is Your Answer?

4. Complete each sentence by matching.

• The inverse operation of addition	• is multiplication.
• The inverse operation of subtraction	• is subtraction.
• The inverse operation of multiplication	• is addition.
• The inverse operation of division	• is division.

5. **IN YOUR OWN WORDS** How can you use multiplication or division to solve an equation? Give two examples to show how your procedure works.

Practice Use what you learned about solving equations to complete Exercises 15–18 on page 312.

7.3 Lesson

Check It Out
Lesson Tutorials
BigIdeasMath.com

Key Ideas

Remember

Inverse operations "undo" each other. Multiplication and division are inverse operations.

Multiplication Property of Equality

Words When you multiply each side of an equation by the same nonzero number, the two sides remain equal.

Numbers $\frac{8}{4} = 2$

$\frac{8}{4} \cdot 4 = 2 \cdot 4$

$8 = 8$

Algebra $\frac{x}{4} = 2$

$\frac{x}{4} \cdot 4 = 2 \cdot 4$

$x = 8$

Multiplicative Inverse Property

Words The product of a nonzero number n and its reciprocal, $\frac{1}{n}$, is 1.

Numbers $5 \cdot \frac{1}{5} = 1$

Algebra $n \cdot \frac{1}{n} = \frac{1}{n} \cdot n = 1, n \neq 0$

EXAMPLE 1 Solving Equations Using Multiplication

a. Solve $\frac{w}{4} = 12$.

$\frac{w}{4} = 12$	Write the equation.
$\frac{w}{4} \cdot 4 = 12 \cdot 4$	Multiplication Property of Equality
$w = 48$	Simplify.

Undo the division. → $\frac{w}{4} \cdot 4 = 12 \cdot 4$

∴ The solution is $w = 48$.

Check

$\frac{w}{4} = 12$

$\frac{48}{4} \stackrel{?}{=} 12$

$12 = 12$ ✓

b. Solve $\frac{2}{7}x = 6$.

$\frac{2}{7}x = 6$	Write the equation.
$\frac{7}{2} \cdot \left(\frac{2}{7}x\right) = \frac{7}{2} \cdot 6$	Multiplication Property of Equality
$x = 21$	Simplify.

Use the Multiplicative Inverse Property. → $\frac{7}{2} \cdot \left(\frac{2}{7}x\right) = \frac{7}{2} \cdot 6$

∴ The solution is $x = 21$.

On Your Own

Now You're Ready
Exercises 7–10

Solve the equation. Check your solution.

1. $\frac{a}{8} = 6$
2. $14 = \frac{2y}{5}$
3. $3z \div 2 = 9$

Key Idea

Division Property of Equality

Words When you divide each side of an equation by the same nonzero number, the two sides remain equal.

Numbers	**Algebra**
$8 \cdot 4 = 32$	$4x = 32$
$8 \cdot 4 \div 4 = 32 \div 4$	$\frac{4x}{4} = \frac{32}{4}$
$8 = 8$	$x = 8$

EXAMPLE 2 Solving an Equation Using Division

Solve $5b = 65$.

$5b = 65$ Write the equation.

Undo the multiplication. → $\frac{5b}{5} = \frac{65}{5}$ Division Property of Equality

$b = 13$ Simplify.

∴ The solution is $b = 13$.

Check

$5b = 65$

$5(13) \stackrel{?}{=} 65$

$65 = 65$ ✓

EXAMPLE 3 Real-Life Application

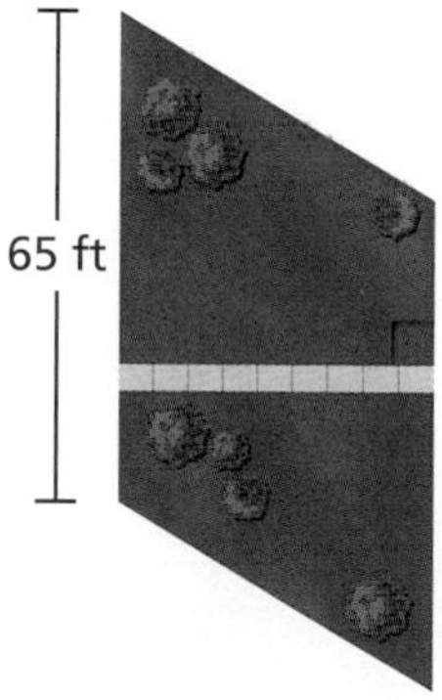

The area of the parallelogram-shaped courtyard is 2730 square feet. What is the length of the sidewalk?

The height of the parallelogram represents the length of the sidewalk.

$A = bh$ Use the formula for area of a parallelogram.

$2730 = 65h$ Substitute 2730 for A and 65 for b.

$\frac{2730}{65} = \frac{65h}{65}$ Division Property of Equality

$42 = h$ Simplify.

∴ So, the sidewalk is 42 feet long.

On Your Own

Now You're Ready
Exercises 11–14

Solve the equation. Check your solution.

4. $p \cdot 3 = 18$
5. $12q = 60$
6. $81 = 9r$
7. You and four friends buy tickets to a baseball game. The total cost is $70. Write and solve an equation to find the cost of each ticket.

7.3 Exercises

Check It Out
Help with Homework
BigIdeasMath.com

Vocabulary and Concept Check

1. **NUMBER SENSE** What number divided by 12 equals 1?

2. **WRITING** What property of equality would you use to solve $\frac{x}{6} = 7$? Explain how you would use the property.

Copy and complete the first step in the solution.

3. $4x = 24$

$\frac{4x}{\square} = \frac{24}{\square}$

4. $\frac{x}{3} = 11$

$\frac{x}{3} \cdot \square = 11 \cdot \square$

5. $8 = n \div 3$

$8 \cdot \square = (n \div 3) \cdot \square$

6. **OPEN-ENDED** Write an equation that can be solved using the Division Property of Equality.

Practice and Problem Solving

Solve the equation. Check your your solution.

1 7. $\frac{s}{10} = 7$	8. $6 = \frac{t}{5}$	9. $5x \div 6 = 20$	10. $24 = \frac{3r}{4}$
2 11. $3a = 12$	12. $5 \cdot z = 35$	13. $40 = 4y$	14. $42 = 7k$
15. $7x = 105$	16. $75 = 6 \cdot w$	17. $13 = d \div 6$	18. $9 = v \div 5$
19. $\frac{2c}{15} = 8.8$	20. $7b \div 12 = 4.2$	21. $12.5 \cdot n = 32$	22. $3.4m = 20.4$

23. **ERROR ANALYSIS** Describe and correct the error in solving the equation.

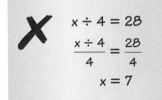

✗
$x \div 4 = 28$
$\frac{x \div 4}{4} = \frac{28}{4}$
$x = 7$

24. **ANOTHER WAY** Show how you can solve the equation $3x = 9$ by multiplying each side by the reciprocal of 3.

25. **BASKETBALL** Forty-five basketball players participate in a tournament. Write and solve an equation to find the number of 3-person teams that they can form.

26. **THEATER** A theater has 1200 seats. Each row has 20 seats. Write and solve an equation to find the number of rows in the theater.

Solve for x. Check your answer.

27. rectangle

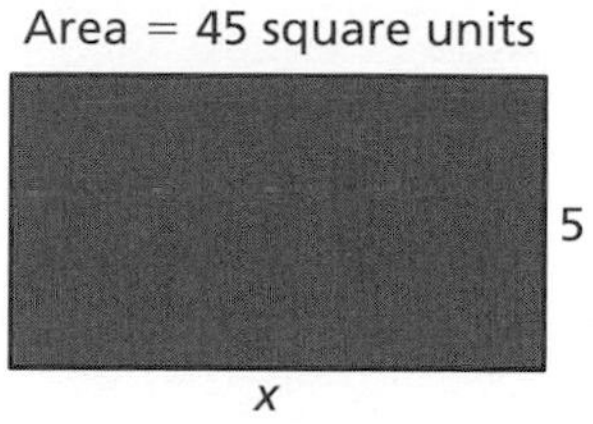

28. rectangle

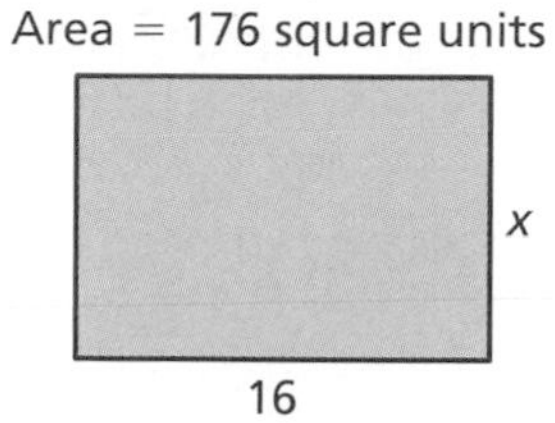

29. parallelogram

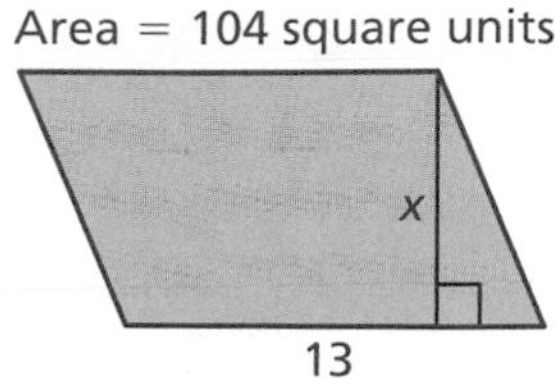

30. **TEST SCORE** On a test, you correctly answer six 5-point questions and eight 2-point questions. You earn 92% of the possible points on the test. How many points p is the test worth?

31. **CARD GAME** You use index cards to play a homemade game. The object is to be the first to get rid of all your cards. How many cards are in your friend's stack?

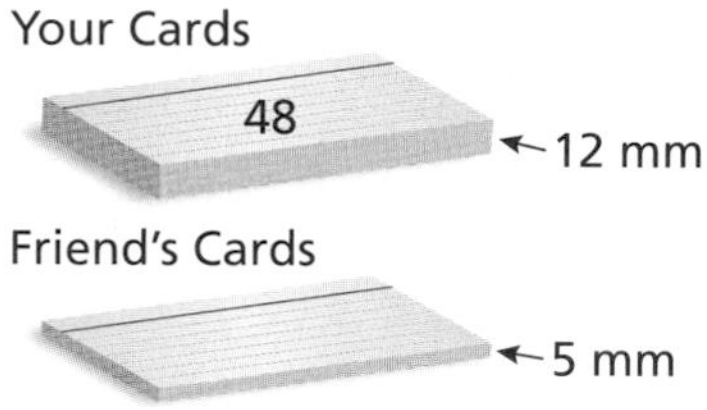

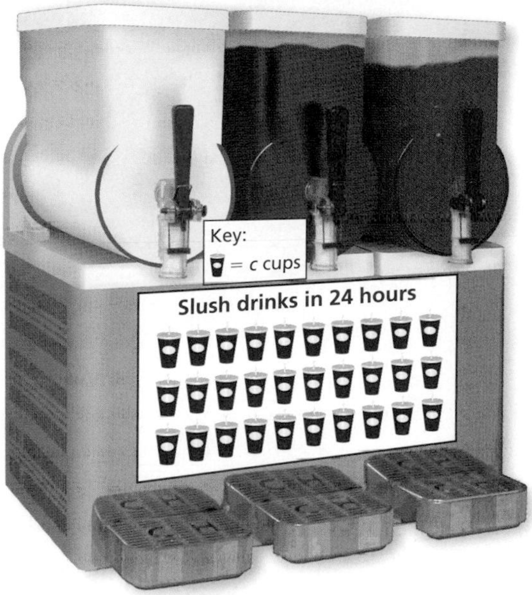

32. **SLUSH DRINKS** A slush drink machine fills 1440 cups in 24 hours.

a. Write and solve an equation to find the number c of cups each symbol represents.

b. To lower costs, you replace the cups with paper cones that hold 20% less. Write and solve an equation to find the number n of paper cones that the machine can fill in 24 hours.

STRUCTURE Solve the equation. Explain how you found your answer.

33. $5x + 3x = 5x + 18$

34. $8y + 2y = 2y + 40$

35. Number Sense The area of the picture is 100 square inches. The length is 4 times the width. Find the length and width of the picture.

Fair Game Review What you learned in previous grades & lessons

Write the word sentence as an equation. *(Section 7.1)*

36. The sum of a number b and 8 is 17.

37. A number t divided by 3 is 7.

38. **MULTIPLE CHOICE** What is the value of a^3 when $a = 4$? *(Section 3.1)*

Ⓐ 12 Ⓑ 43 Ⓒ 64 Ⓓ 81

7.4 Writing Equations in Two Variables

Essential Question How can you write an equation in two variables?

1 ACTIVITY: Writing an Equation in Two Variables

Work with a partner. You earn $8 per hour working part-time at a store.

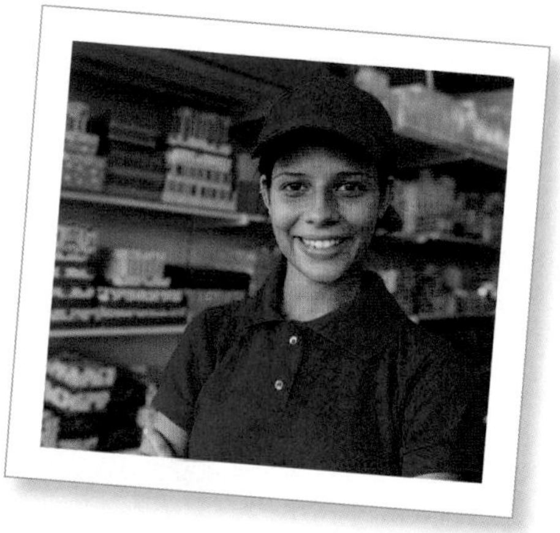

a. Complete the table.

Hours Worked	Money Earned (dollars)
1	
2	
3	
4	
5	

b. Use the values from the table to complete the graph. Then answer each question below.

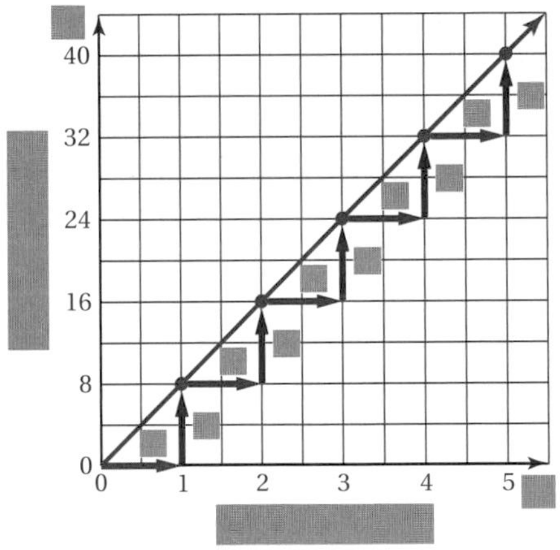

- What does the horizontal axis represent? What variable did you use to identify it?
- What does the vertical axis represent? What variable did you use to identify it?
- How are the ordered pairs in the graph related to the values in the table?
- How are the horizontal and vertical distances shown on the graph related to the values in the table?

c. How can you write an equation that shows how the two variables are related?

d. What does the green line in the graph represent?

COMMON CORE

Writing Equations

In this lesson, you will

- identify independent and dependent variables.
- write equations in two variables.
- use tables and graphs to analyze the relationship between two variables.

Learning Standard 6.EE.9

2 ACTIVITY: Describing Variables

Work with a partner. Use the equation you wrote in Activity 1.

a. How is this equation different from the equations earlier in this chapter?

b. One of the variables in this equation *depends* on the other variable. Determine which variable is which by answering the following questions:

- Does the amount of money you earn *depend* on the number of hours you work?
- Does the number of hours you work *depend* on the amount of money you earn?

What do you think is the significance of having two types of variables? How do you think you can use these types of variables in real life?

3 ACTIVITY: Describing a Formula in Two Variables

Math Practice 7
Look for Patterns
What pattern do you notice in the table for the perimeter of the square?

Work with a partner. Recall that the perimeter of a square is 4 times its side length.

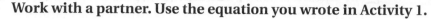

a. Write the formula for the perimeter of a square. Tell what each variable represents.

b. Describe how the perimeter of a square changes as its side length increases by 1 unit. Use a table and a graph to support your answer.

c. In your formula, which variable depends on which?

What Is Your Answer?

4. **IN YOUR OWN WORDS** How can you write an equation in two variables?

5. The equation $y = 7.75x$ shows how the number of movie tickets is related to the total amount of money spent. Describe what each part of the equation represents.

6. **CHOOSE TOOLS** In Activity 1, you want to know the amount of money you earn after working 30.5 hours during a week. Would you use the table, the graph, or the equation to find your earnings? What are your earnings? Explain your reasoning.

7. Give an example of another real-life situation that you can model by an equation in two variables.

Practice Use what you learned about equations in two variables to complete Exercises 4 and 5 on page 319.

7.4 Lesson

Check It Out
Lesson Tutorials
BigIdeasMath.com

An **equation in two variables** represents two quantities that change in relationship to one another. A **solution of an equation in two variables** is an ordered pair that makes the equation true.

Key Vocabulary
- equation in two variables, *p. 316*
- solution of an equation in two variables, *p. 316*
- independent variable, *p. 316*
- dependent variable, *p. 316*

EXAMPLE 1 Identifying Solutions of Equations in Two Variables

Tell whether the ordered pair is a solution of the equation.

a. $y = 2x$; (3, 6)

$6 \stackrel{?}{=} 2(3)$ Substitute.

$6 = 6$ ✓ Compare.

∴ So, (3, 6) is a solution.

b. $y = 4x - 3$; (4, 12)

$12 \stackrel{?}{=} 4(4) - 3$

$12 \neq 13$ ✗

∴ So, (4, 12) is *not* a solution.

You can use equations in two variables to represent situations involving two related quantities. The variable representing the quantity that can change freely is the **independent variable**. The other variable is called the **dependent variable** because its value *depends* on the independent variable.

EXAMPLE 2 Using an Equation in Two Variables

The equation $y = 128 - 8x$ gives the amount y (in fluid ounces) of milk remaining in a gallon jug after you pour x cups.

a. Identify the independent and dependent variables.

∴ Because the amount y remaining depends on the number x of cups you pour, y is the dependent variable and x is the independent variable.

b. How much milk remains in the jug after you pour 10 cups?

Use the equation to find the value of y when $x = 10$.

$y = 128 - 8x$ Write the equation.

$= 128 - 8(10)$ Substitute 10 for x.

$= 48$ Simplify.

∴ There are 48 fluid ounces remaining.

On Your Own

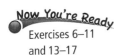

Now You're Ready
Exercises 6–11 and 13–17

Tell whether the ordered pair is a solution of the equation.

1. $y = 7x$; (2, 21)
2. $y = 5x + 1$; (3, 16)
3. The equation $y = 10x + 25$ gives the amount y (in dollars) in your savings account after x weeks.
   a. Identify the independent and dependent variables.
   b. How much is in your savings account after 8 weeks?

Multi-Language Glossary at BigIdeasMath.com

Key Idea

Study Tip

When you draw a line through the points, you graph *all* the solutions of the equation.

Tables, Graphs, and Equations

You can use tables and graphs to represent equations in two variables. The table and graph below represent the equation $y = x + 2$.

Independent Variable, x	Dependent Variable, y	Ordered Pair, (x, y)
1	3	(1, 3)
2	4	(2, 4)
3	5	(3, 5)

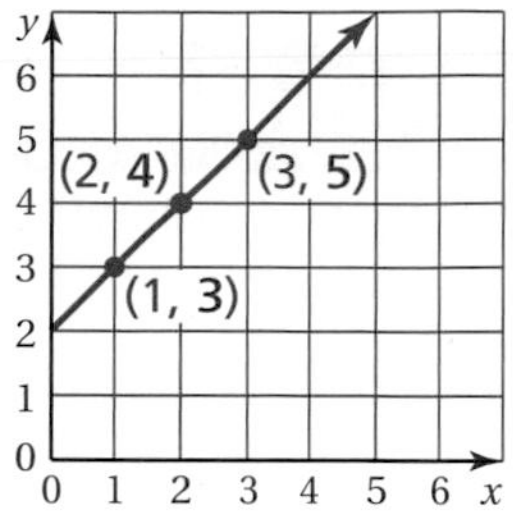

EXAMPLE 3 Writing and Graphing an Equation in Two Variables

An athlete burns 200 calories weight lifting. The athlete then works out on an elliptical trainer and burns 10 calories for every minute. Write and graph an equation in two variables that represents the total number of calories burned during the workout.

Words The total number of calories burned equals calories burned weight lifting plus calories burned per minute times the number of minutes.

Variables Let c be the total number of calories burned, and let m be the number of minutes on the elliptical trainer.

Equation $c = 200 + 10 \cdot m$

Reading

Make sure you read and understand the context of the problem. Because you cannot have a negative number of minutes, use only whole number values of *m*.

To graph the equation, first make a table. Then plot the ordered pairs and draw a line through the points.

Minutes, m	$c = 200 + 10m$	Calories, c	Ordered Pair, (m, c)
10	$c = 200 + 10(10)$	300	(10, 300)
20	$c = 200 + 10(20)$	400	(20, 400)
30	$c = 200 + 10(30)$	500	(30, 500)

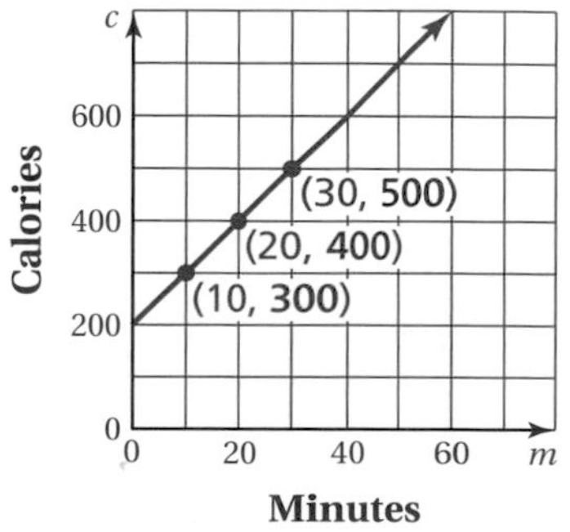

On Your Own

Now You're Ready Exercises 22 and 23

4. It costs \$25 to rent a kayak plus \$8 for each hour. Write and graph an equation in two variables that represents the total cost of renting the kayak.

You can model many rate problems by using the *distance formula* $d = rt$, where d is the distance traveled, r is the speed, and t is the time. When you are given a speed, you can use the formula to write an equation in two variables that represents the situation.

Key Idea

Remember

Speed is an example of a rate.

Distance Formula

Words To find the distance traveled d, multiply the speed r by the time t.

Algebra $d = rt$

EXAMPLE 4 Real-Life Application

A train averages 40 miles per hour between two cities. Use a graph to show the relationship between the time and the distance traveled.

Method 1: Use a ratio table.

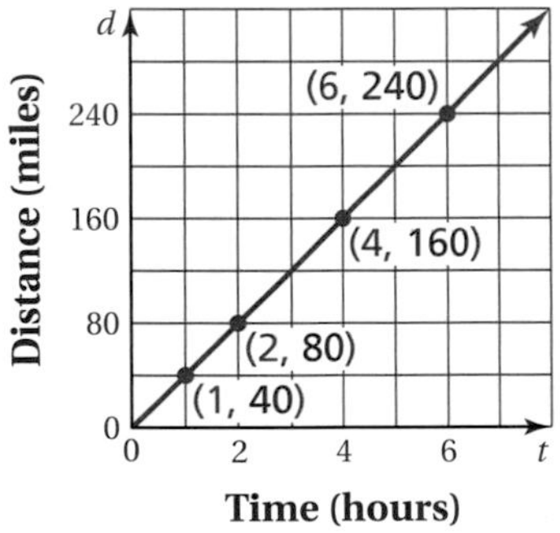

You can use a ratio table and multiplication to find equivalent rates. Then plot the ordered pairs (time, distance) from the table and draw a line through the points.

× 2 × 4 × 6

Time (hours)	1	2	4	6
Distance (miles)	40	80	160	240

× 2 × 4 × 6

Method 2: Use an equation in two variables.

Use the distance formula to write the equation $d = 40t$. Use the equation to make a table. Then plot the ordered pairs and draw a line through the points, as shown in the graph above.

Time (hours), t	$d = 40t$	Distance (miles), d	Ordered Pair, (t, d)
1	$d = 40(1)$	40	(1, 40)
2	$d = 40(2)$	80	(2, 80)
4	$d = 40(4)$	160	(4, 160)
6	$d = 40(6)$	240	(6, 240)

On Your Own

Now You're Ready Exercise 25

5. **WHAT IF?** The train averages 50 miles per hour. Use a graph to show the relationship between the time and the distance traveled.

7.4 Exercises

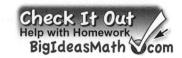

Check It Out
Help with Homework
BigIdeasMath.com

Vocabulary and Concept Check

1. **VOCABULARY** How are independent variables and dependent variables different?
2. **PRECISION** Explain how to graph an equation in two variables.
3. **WHICH ONE DOESN'T BELONG?** Which one does *not* belong with the other three? Explain your reasoning.

$y = 12x + 25$	$c = 10t - 5$	$a = 7b + 11$	$n = 4n - 6$

Practice and Problem Solving

Write a formula for the given measure. Tell what each variable represents. Identify which variable depends on which in the formula.

4. the perimeter of a rectangle with a length of 5 inches
5. the area of a trapezoid with base lengths of 7 feet and 11 feet

Tell whether the ordered pair is a solution of the equation.

1 6. $y = 4x$; (0, 4)
7. $y = 3x$; (2, 6)
8. $y = 5x - 10$; (3, 5)
9. $y = x + 7$; (1, 6)
10. $y = 7x + 2$; (2, 0)
11. $y = 2x - 3$; (4, 5)

12. **ERROR ANALYSIS** Describe and correct the error in finding a solution of the equation in two variables.

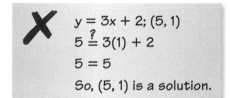

✗
$y = 3x + 2$; (5, 1)
$5 \stackrel{?}{=} 3(1) + 2$
$5 = 5$
So, (5, 1) is a solution.

Identify the independent and dependent variables.

2 13. The equation $A = 25w$ gives the area A (in square feet) of a rectangular dance floor with a width of w feet.

14. The equation $c = 0.09s$ gives the amount c (in dollars) of commission a salesperson receives for making a sale of s dollars.

15. The equation $t = 12p + 12$ gives the total cost t (in dollars) of a meal with a tip of p percent (in decimal form).

16. The equation $h = 60 - 4m$ gives the height h (in inches) of the water in a tank m minutes after it starts to drain.

17. **DRUM SET** The equation $b = 540 - 30m$ gives the balance b (in dollars) that you owe on a drum set after m monthly payments. What is the balance after 9 monthly payments?

OPEN-ENDED Complete the table by describing possible independent or dependent variables.

	Independent Variable	Dependent Variable
18.	The number of hours you study for a test	
19.	The speed you are pedaling a bike	
20.		Your monthly cell phone bill
21.		The amount of money you earn

3 **22.** **PIZZA** A cheese pizza costs \$5. Additional toppings cost \$1.50 each. Write and graph an equation in two variables that represents the total cost of a pizza.

23. **GYM MEMBERSHIP** It costs \$35 to join a gym. The monthly fee is \$25. Write and graph an equation in two variables that represents the total cost of a gym membership.

24. **TEXTING** The maximum size of a text message is 160 characters. A space counts as one character.

a. Write an equation in two variables that represents the remaining (unused) characters in a text message as you type.

b. Identify the independent and dependent variables.

c. How many characters remain in the message shown?

4 **25.** **CHOOSE TOOLS** A car averages 60 miles per hour on a road trip. Use a graph to show the relationship between the time and the distance traveled. What method did you use to create your graph?

Write and graph an equation in two variables that shows the relationship between the time and the distance traveled.

26.

Moves 2 meters every 3 hours.

27.

Rises 5 stories every 6 seconds.

28.

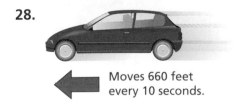

Moves 660 feet every 10 seconds.

29.

Moves 960 kilometers every 4 minutes.

Fill in the blank so that the ordered pair is a solution of the equation.

30. $y = 8x + 3$; (1, ____) **31.** $y = 12x + 2$; (____ , 14) **32.** $y = 22 - 9x$; (____ , 4)

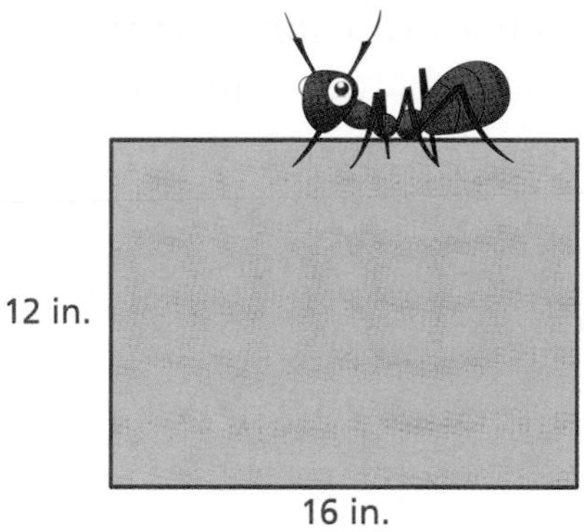

33. CRITICAL THINKING Can the dependent variable cause a change in the independent variable? Explain.

34. OPEN-ENDED Write an equation in two variables that has (3, 4) as a solution.

35. WALKING You walk 5 city blocks in 12 minutes. How many city blocks can you walk in 2 hours?

36. ANT How fast should the ant walk to go around the rectangle in 4 minutes?

37. LIGHTNING To estimate how far you are from lightning (in miles), count the number of seconds between a lightning flash and the thunder that follows. Then divide the number of seconds by 5. Use a graph to show the relationship between the time and the distance. Describe the method you used to create your graph.

38. PROBLEM SOLVING You and a friend start biking in opposite directions from the same point. You travel 108 feet every 8 seconds. Your friend travels 63 feet every 6 seconds.

a. How far apart are you and your friend after 15 minutes?

b. After 20 minutes, you take a 5-minute rest, but your friend does not. How far apart are you and your friend after 40 minutes? Explain your reasoning.

39. **Reasoning** The graph represents the cost c (in dollars) of buying n tickets to a baseball game.

a. Should the points be connected with a line to show all the solutions? Explain your reasoning.

b. Write an equation in two variables that represents the graph.

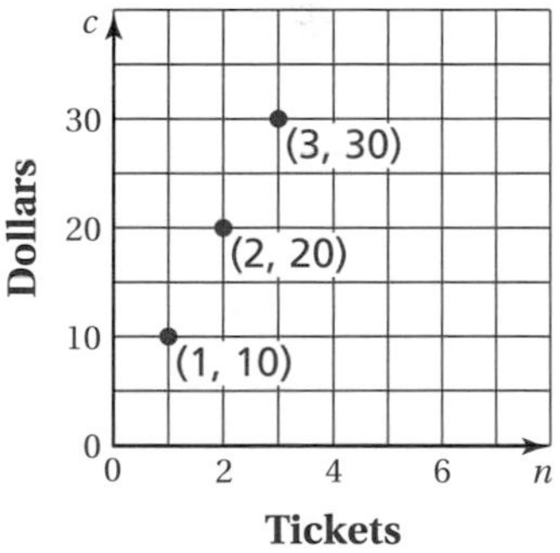

Fair Game Review What you learned in previous grades & lessons

Write the fraction as a percent. *(Section 5.5)*

40. $\frac{3}{10}$ **41.** $\frac{4}{5}$ **42.** $\frac{9}{20}$ **43.** $\frac{17}{25}$

44. MULTIPLE CHOICE What is the area of the triangle? *(Section 4.2)*

Ⓐ 36 cm^2 Ⓑ 68 cm^2

Ⓒ 72 cm^2 Ⓓ 76.5 cm^2

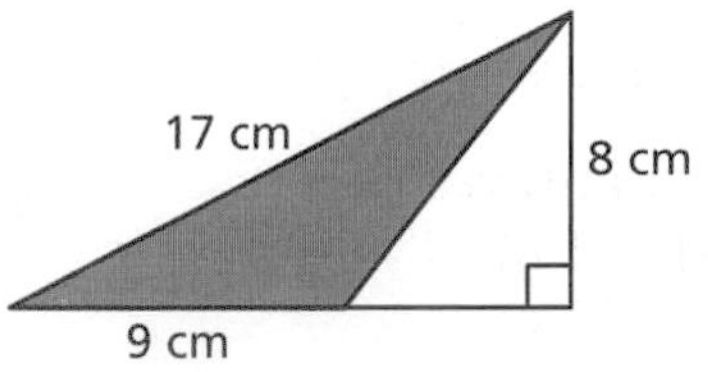

7 Study Help

Check It Out
Graphic Organizer
BigIdeasMath.com

You can use an **example and non-example chart** to list examples and non-examples of a vocabulary word or term. Here is an example and non-example chart for equations.

Equations

Examples	Non-Examples
$x = 5$	5
$2a = 16$	$2a$
$x + 4 = 19$	$x + 4$
$5 = x + 3$	$x + 3$
$12 - 7 = 5$	$12 - 7$
$\frac{3}{4}y = 6$	$\frac{3}{4}$

On Your Own

Make example and non-example charts to help you study these topics.

1. inverse operations
2. equations solved using addition or subtraction
3. equations solved using multiplication or division
4. equations in two variables

After you complete this chapter, make example and non-example charts for the following topics.

5. inequalities
6. graphs of inequalities
7. inequalities solved using addition or subtraction
8. inequalities solved using multiplication or division

"I need a good non-example of a cool animal for my example and non-example chart."

7.1–7.4 Quiz

Check It Out
Progress Check
BigIdeasMath.com

Write the word sentence as an equation. *(Section 7.1)*

1. A number x decreased by 3 is 5.

2. A number a divided by 7 equals 14.

Solve the equation. Check your solution. *(Section 7.2 and Section 7.3)*

3. $4 + k = 14$

4. $3.5 = m - 2.2$

5. $8 = \frac{4w}{3}$

6. $31 = 6.2 \cdot y$

Tell whether the ordered pair is a solution of the equation. *(Section 7.4)*

7. $y = 6x$; (3, 24)

8. $y = 3x + 4$; (4, 16)

Write and graph an equation in two variables that shows the relationship between the time and the distance traveled. *(Section 7.4)*

9.

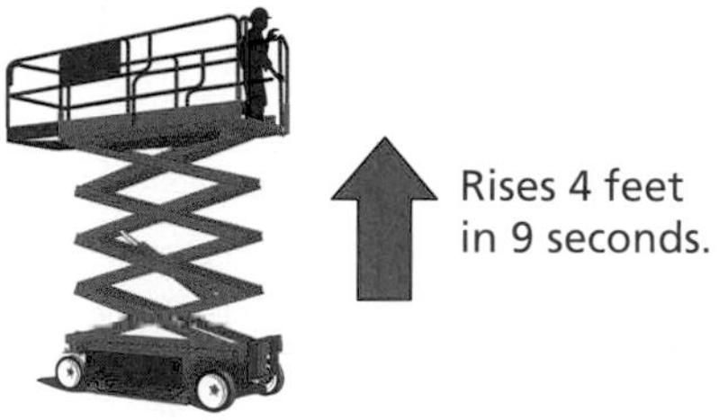

10.

11. RIBBON The length of the blue ribbon is two-thirds the length of the red ribbon. Write an equation you can use to find the length r of the red ribbon. *(Section 7.1)*

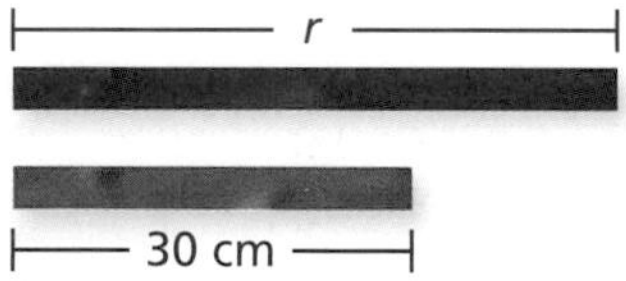

12. BRIDGES The main span of the Sunshine Skyway Bridge is 360 meters long. The Skyway's main span is 30 meters shorter than the main span of the Dames Point Bridge. Write and solve an equation to find the length ℓ of the main span of the Dames Point Bridge. *(Section 7.2)*

13. SHOPPING At a farmer's market, you buy 4 pounds of tomatoes and 2 pounds of sweet potatoes. You spend 80% of the money in your wallet. Write and solve an equation to find how much money is in your wallet before you pay. *(Section 7.3)*

14. SUNDAE A sundae costs $2. Additional toppings cost $0.50 each. Write and graph an equation in two variables that represents the total cost of a sundae. *(Section 7.4)*

7.5 Writing and Graphing Inequalities

Essential Question How can you use a number line to represent solutions of an inequality?

1 ACTIVITY: Understanding Inequality Statements

Work with a partner. Read the statement. Circle each number that makes the statement true, and then answer the questions.

a. "Your friend is more than 3 minutes late."

−3 −2 −1 0 1 2 3 4 5 6

- What do you notice about the numbers that you circled?
- Is the number 3 included? Why or why not?
- Write four other numbers that make the statement true.

b. "The temperature is at most 2 degrees."

−5 −4 −3 −2 −1 0 1 2 3 4

- What do you notice about the numbers that you circled?
- Can the temperature be exactly 2 degrees? Explain.
- Write four other numbers that make the statement true.

c. "You need at least 4 pieces of paper for your math homework."

−3 −2 −1 0 1 2 3 4 5 6

- What do you notice about the numbers that you circled?
- Can you have exactly 4 pieces of paper? Explain.
- Write four other numbers that make the statement true.

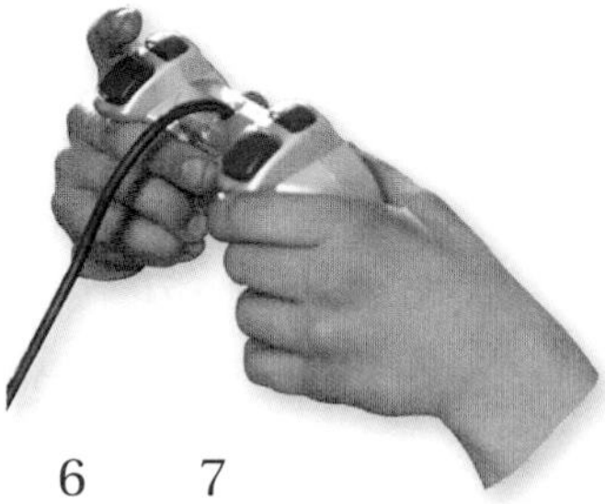

d. "After playing a video game for 20 minutes, you have fewer than 6 points."

−2 −1 0 1 2 3 4 5 6 7

- What do you notice about the numbers that you circled?
- Is the number 6 included? Why or why not?
- Write four other numbers that make the statement true.

COMMON CORE

Writing Inequalities

In this lesson, you will

- write word sentences as inequalities.
- use a number line to graph the solution set of inequalities.
- use inequalities to represent real-life situations.

Learning Standards
6.EE.5
6.EE.8

2 ACTIVITY: Understanding Inequality Symbols

Work with a partner.

a. Consider the statement "*x* is a number such that $x < 2$."

- Can the number be exactly 2? Explain.
- Circle each number that makes the statement true.

  −5 −4 −3 −2 −1 0 1 2 3 4

- Write four other numbers that make the statement true.

Math Practice 6

State the Meaning of Symbols

What do the symbols < and ≥ mean?

b. Consider the statement "*x* is a number such that $x \geq 1$."

- Can the number be exactly 1? Explain.
- Circle each number that makes the statement true.

  −5 −4 −3 −2 −1 0 1 2 3 4

- Write four other numbers that make the statement true.

3 ACTIVITY: How Close Can You Come to 0?

Work with a partner.

a. Which number line shows $x > 0$? Which number line shows $x \geq 0$? Explain your reasoning.

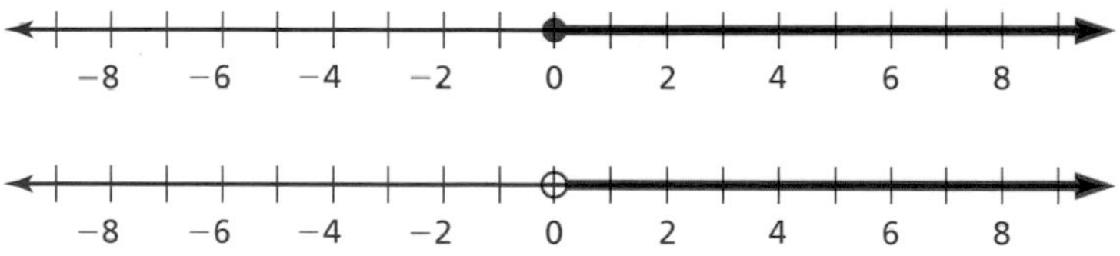

b. Write the least positive number you can think of that is still a solution of the inequality $x > 0$. Explain your reasoning.

What Is Your Answer?

4. **IN YOUR OWN WORDS** How can you use a number line to represent solutions of an inequality?
5. Write an inequality. Graph all solutions of your inequality on a number line.
6. Graph the inequalities $x > 9$ and $9 < x$ on different number lines. What do you notice?

Practice Use what you learned about graphing inequalities to complete Exercises 17–20 on page 329.

7.5 Lesson

Check It Out
Lesson Tutorials
BigIdeasMath.com

Key Vocabulary
inequality, *p. 326*
solution of an inequality, *p. 327*
solution set, *p. 327*
graph of an inequality, *p. 328*

An **inequality** is a mathematical sentence that compares expressions. It contains the symbols <, >, ≤, or ≥. To write an inequality, look for the following phrases to determine where to place the inequality symbol.

Inequality Symbols

Symbol	<	>	≤	≥
Key Phrases	• is less than • is fewer than	• is greater than • is more than	• is less than or equal to • is at most • is no more than	• is greater than or equal to • is at least • is no less than

EXAMPLE 1 Writing Inequalities

Write the word sentence as an inequality.

a. A number c is less than -4.

A number c | is less than | -4.

c | $<$ | -4

An inequality is $c < -4$.

b. A number k plus 5 is greater than or equal to 8.

A number k plus 5 | is greater than or equal to | 8.

$k + 5$ | $\geq$ | 8

An inequality is $k + 5 \geq 8$.

c. Four times a number q is at most 16.

Four times a number q | is at most | 16.

$4q$ | $\leq$ | 16

An inequality is $4q \leq 16$.

On Your Own

Now You're Ready
Exercises 5–10

Write the word sentence as an inequality.

1. A number n is greater than 1.
2. Twice a number p is fewer than 7.
3. A number w minus 3 is less than or equal to 10.
4. A number z divided by 2 is at least -6.

Multi-Language Glossary at BigIdeasMath.com

A **solution of an inequality** is a value that makes the inequality true. An inequality can have more than one solution. The set of all solutions of an inequality is called the **solution set**.

Value of x	$x + 3 \le 7$	Is the inequality true?
3	$3 + 3 \overset{?}{\le} 7$ $6 \le 7$ ✓	yes
4	$4 + 3 \overset{?}{\le} 7$ $7 \le 7$ ✓	yes
5	$5 + 3 \overset{?}{\le} 7$ $8 \nleq 7$ ✗	no

Reading

The symbol $\nleq$ means *is not less than or equal to.*

EXAMPLE 2 Checking Solutions

Tell whether the given value is a solution of the inequality.

a. $x + 1 > 7;\ x = 8$

$x + 1 > 7$ — Write the inequality.

$8 + 1 \overset{?}{>} 7$ — Substitute 8 for x.

$9 > 7$ ✓ — Add. 9 is greater than 7.

∴ So, 8 is a solution of the inequality.

b. $7y < 27;\ y = 4$

$7y < 27$ — Write the inequality.

$7(4) \overset{?}{<} 27$ — Substitute 4 for y.

$28 \nless 27$ ✗ — Multiply. 28 is *not* less than 27.

∴ So, 4 is *not* a solution of the inequality.

c. $\frac{z}{3} \ge 5;\ z = 15$

$\frac{z}{3} \ge 5$ — Write the inequality.

$\frac{15}{3} \overset{?}{\ge} 5$ — Substitute 15 for z.

$5 \ge 5$ ✓ — Divide. 5 is greater than or equal to 5.

∴ So, 15 is a solution of the inequality.

On Your Own

Now You're Ready
Exercises 11–16

Tell whether 3 is a solution of the inequality.

5. $b + 4 < 6$ **6.** $9 - n \ge 6$ **7.** $18 \div x \le 10$

The **graph of an inequality** shows all the solutions of the inequality on a number line. An open circle ○ is used when a number is *not* a solution. A closed circle ● is used when a number is a solution. An arrow to the left or right shows that the graph continues in that direction.

EXAMPLE 3 Graphing an Inequality

Reading

The inequality $g > 2$ is the same as $2 < g$.

Graph $g > 2$.

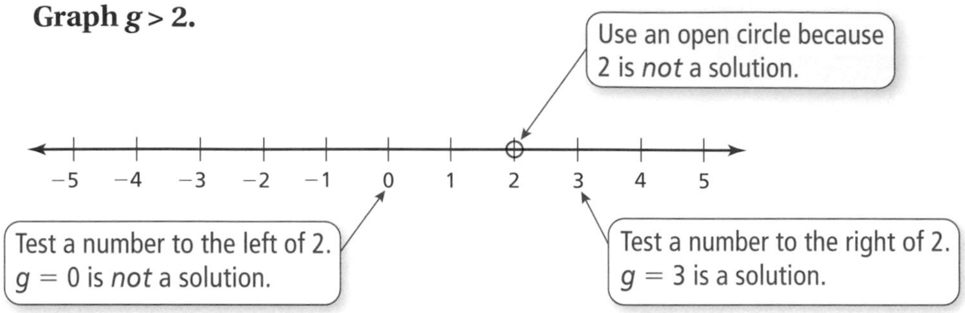
Use an open circle because 2 is *not* a solution.
−5 −4 −3 −2 −1 0 1 2 3 4 5
Test a number to the left of 2. g = 0 is *not* a solution.
Test a number to the right of 2. g = 3 is a solution.

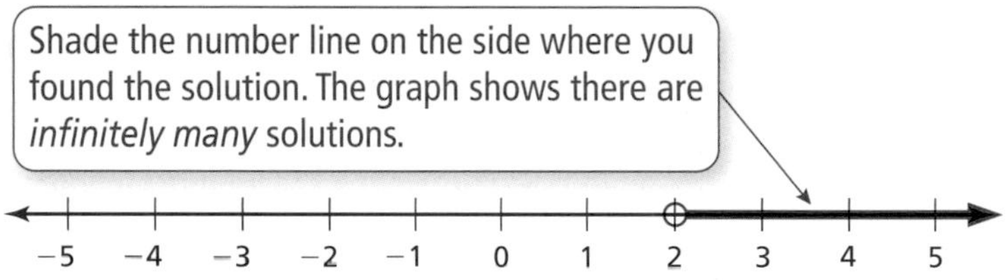
Shade the number line on the side where you found the solution. The graph shows there are *infinitely many* solutions.
−5 −4 −3 −2 −1 0 1 2 3 4 5

EXAMPLE 4 Real-Life Application

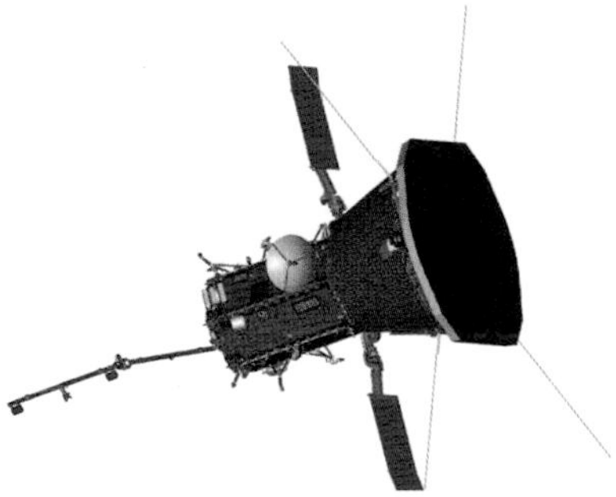

The NASA Solar Probe Plus can withstand temperatures up to and including 2600°F. Write and graph an inequality that represents the temperatures the probe can withstand.

Words	temperatures	up to and including	2600°F
Variable	Let t be the temperatures the probe can withstand.		
Inequality	t	$\leq$	2600

∴ An inequality is $t \leq 2600$.

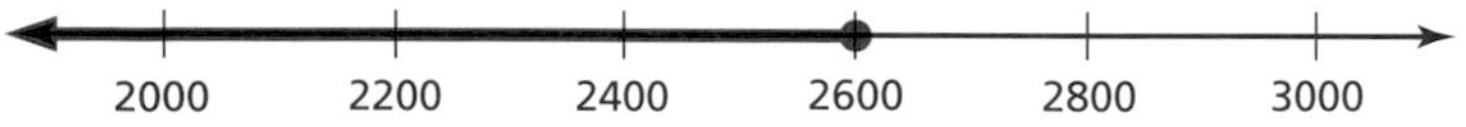
2000 2200 2400 2600 2800 3000

On Your Own

Now You're Ready
Exercises 25–36

Graph the inequality on a number line.

8. $a < 4$ **9.** $f \leq 7$ **10.** $n > 0$ **11.** $p \geq -3$

Write and graph an inequality for the situation.

12. A cruise ship can carry at most 3500 passengers.

13. A board game is designed for ages 12 and up.

7.5 Exercises

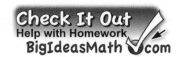

Check It Out
Help with Homework
BigIdeasMath.com

Vocabulary and Concept Check

1. **VOCABULARY** How are *greater than* and *greater than or equal to* similar? How are they different?

2. **DIFFERENT WORDS, SAME QUESTION** Which is different? Write "both" inequalities.

A number n is at most 3.	A number n is at least 3.
A number n is less than or equal to 3.	A number n is no more than 3.

3. **WRITING** Explain how the graph of $x \le 6$ is different from the graph of $x < 6$.

4. **WRITING** Are the graphs of $x \le 5$ and $5 \ge x$ the same or different? Explain.

Practice and Problem Solving

Write the word sentence as an inequality.

1 5. A number k is less than 10.

6. A number a is more than 6.

7. A number z is fewer than $\frac{3}{4}$.

8. A number b is at least -3.

9. One plus a number y is no more than -13.

10. A number x divided by 3 is at most 5.

Tell whether the given value is a solution of the inequality.

2 11. $x - 1 \le 7;\ x = 6$

12. $y + 5 < 13;\ y = 17$

13. $3z > 6;\ z = 3$

14. $\frac{b}{2} \ge 6;\ b = 10$

15. $c + 2.5 < 4.3;\ c = 1.8$

16. $a \le 0;\ a = -5$

Match the inequality with its graph.

17. $x \ge 2$

18. $x < 2$

19. $x > -2$

20. $x \le -2$

A.
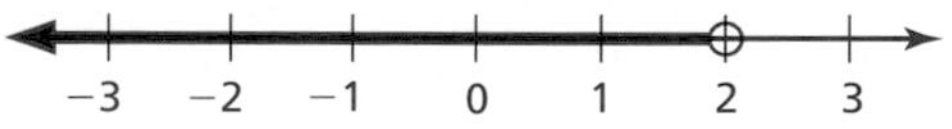

B.
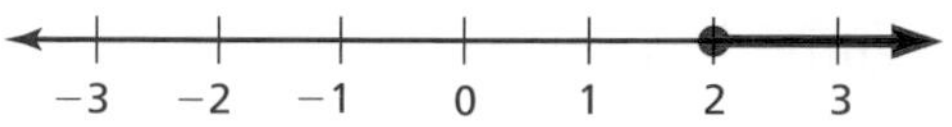

C.
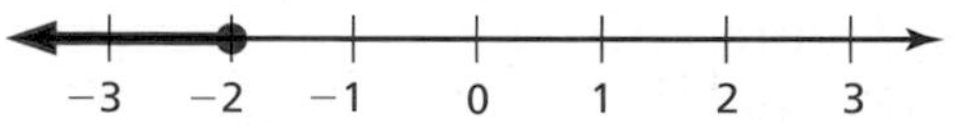

D.
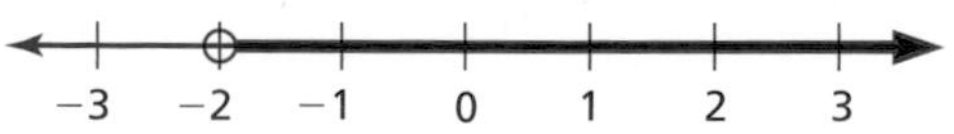

Write an inequality and a word sentence that represent the graph.

21. 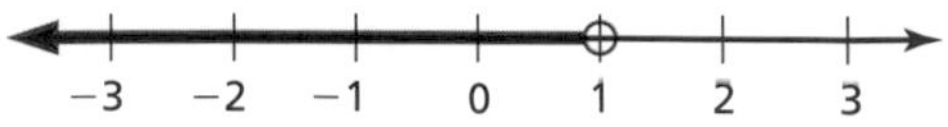

22. 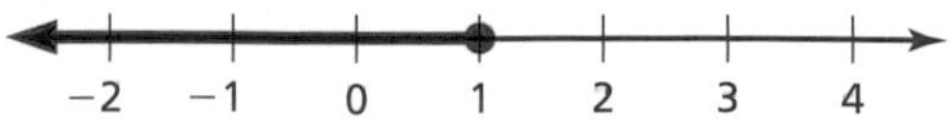

23. 

24. 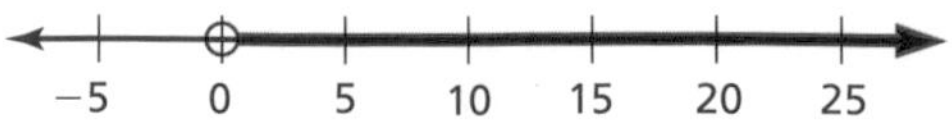

Graph the inequality on a number line.

3 **25.** $a > 4$ **26.** $n \geq 8$ **27.** $3 \geq x$ **28.** $y < \frac{1}{2}$

29. $x < \frac{2}{9}$ **30.** $-3 \geq c$ **31.** $m > -5$ **32.** $0 \leq b$

33. $1.5 > f$ **34.** $t \geq -\frac{1}{2}$ **35.** $p > -1.6$ **36.** $\frac{7}{3} \geq z$

ERROR ANALYSIS **Describe and correct the error in graphing the inequality.**

37. 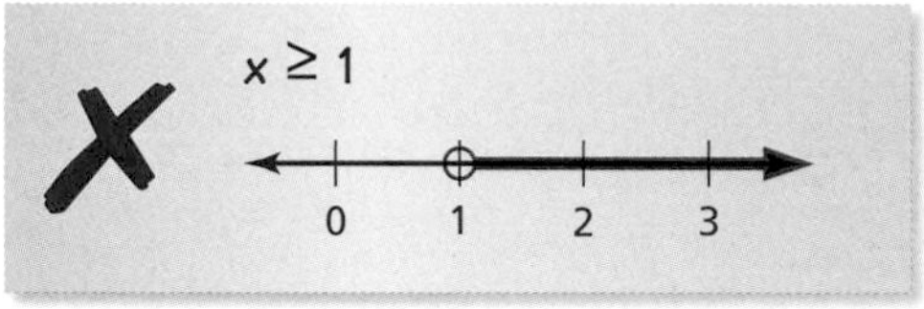

38. 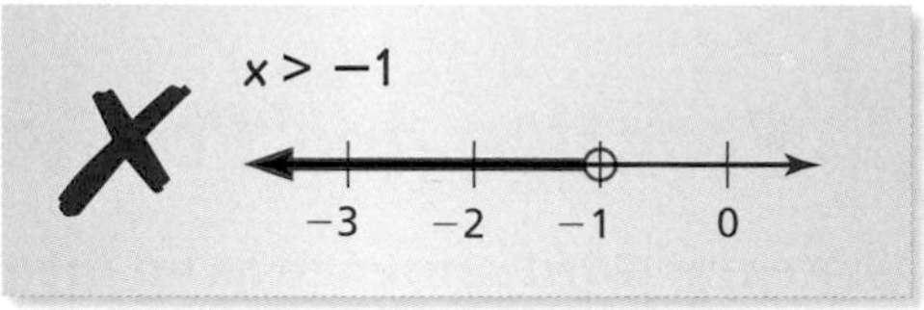

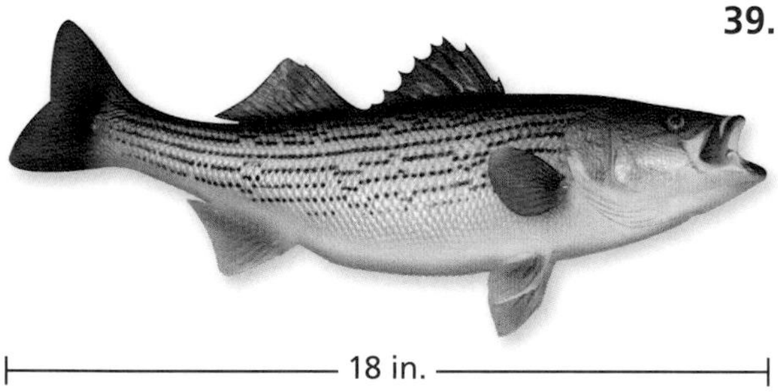

39. **FISHING** You are fishing and are allowed to catch at most 3 striped bass. Each striped bass must be no less than 18 inches long.

a. Write and graph an inequality to represent the number of striped bass you are allowed to catch.

b. Write and graph an inequality to represent the length of each striped bass you are allowed to catch.

40. **MODELING** For a food to be labeled *low sodium,* there must be no more than 140 milligrams of sodium per serving.

a. Write and graph an inequality to represent the amount of sodium in a low-sodium serving.

b. Write and graph an inequality to represent the amount of sodium in a serving that does *not* qualify as low sodium.

c. Does the food represented by the nutrition facts label qualify as a low-sodium food? Explain.

Nutrition Facts	
Serving Size ½ cup (114g)	
Servings Per Container 4	
Amount Per Serving	
Calories 90	Calories from fat 30
	% Daily Value*
Total Fat 3g	5%
Saturated Fat 0g	0%
Cholesterol 0mg	0%
Sodium 300mg	13%
Total Carbohydrate 13g	4%
Dietary Fiber 3g	12%
Sugars 3g	
Protein 3g	
Vitamin A 80% •	Vitamin C 60%
Calcium 4% •	Iron 4%

41. SHOPPING You have \$33. You want to buy a necklace and one other item from the list.

a. Write an inequality to represent the situation.

b. Can the other item be a T-shirt? Explain.

c. Can the other item be a book? Explain.

Item	Price (with tax)
T-shirt	\$ 15
Book	\$ 20
DVD	\$ 13
Necklace	\$ 16

Determine whether the statement is *sometimes*, *always*, or *never* true. Explain your reasoning.

42. A number that is a solution of the inequality $x > 5$ is also a solution of the inequality $x \geq 5$.

43. A number that is a solution of the inequality $5 \leq x$ is also a solution of the inequality $x > 5$.

44. BUS RIDE A bus ride costs \$1.50. A 30-day bus pass costs \$36. Write an inequality to represent the number of bus rides you would need to take for the bus pass to be a better deal.

45. MOVIE THEATER Fifty people are seated in a movie theater. The maximum capacity of the theater is 425 people. Write an inequality to represent the number of additional people who can still be seated.

46. Critical Thinking The map shows the elevations above sea level for an area of land.

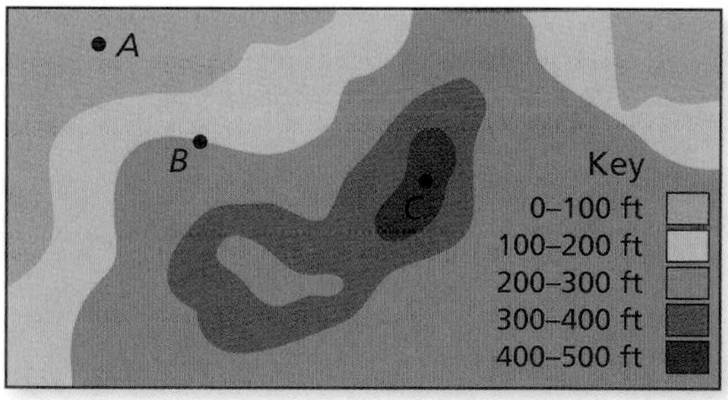

a. Graph the possible elevations of A. Write the set of elevations as two inequalities.

b. Graph the possible elevations of C. How can you write this set of elevations as a single inequality? Explain.

c. What is the elevation of B? Explain.

Fair Game Review What you learned in previous grades & lessons

Solve the equation. Check your solution. *(Section 7.2)*

47. $x + 3 = 12$ **48.** $x - 6 = 8$ **49.** $16 + x = 44$ **50.** $7.6 = x - 6.5$

51. MULTIPLE CHOICE A stack of boards is 24 inches high. Each board is $\frac{3}{8}$ of an inch thick. How many boards are in the stack? *(Section 2.2)*

Ⓐ $\frac{1}{9}$ Ⓑ $\frac{1}{6}$ Ⓒ 9 Ⓓ 64

7.6 Solving Inequalities Using Addition or Subtraction

Essential Question How can you use addition or subtraction to solve an inequality?

1 ACTIVITY: Writing an Inequality

Work with a partner. In 3 years, your friend will still not be old enough to vote.

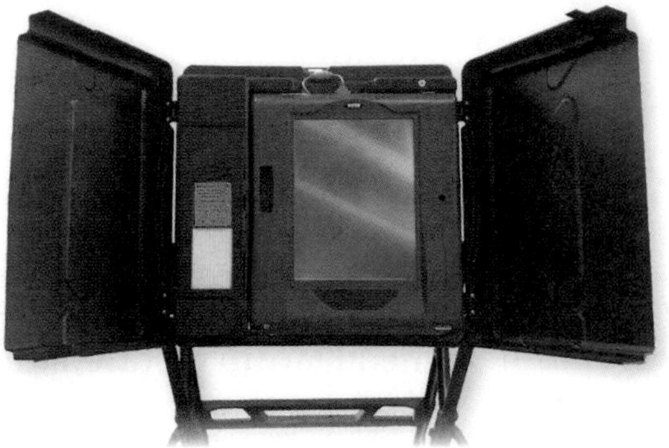

a. Which of the following represents your friend's situation? What does x represent? Explain your reasoning.

$x + 3 < 18$ $\quad$ $x + 3 \leq 18$

$x + 3 > 18$ $\quad$ $x + 3 \geq 18$

b. Graph the possible ages of your friend on a number line. Explain how you decided what to graph.

2 ACTIVITY: Writing an Inequality

Work with a partner. Baby manatees are about 4 feet long at birth. They grow to a maximum length of 13 feet.

a. Which of the following can represent a baby manatee's growth? What does x represent? Explain your reasoning.

$x + 4 < 13$ $\quad$ $x + 4 \leq 13$

$x - 4 > 13$ $\quad$ $x - 4 \geq 13$

b. Graph the solution on a number line. Explain how you decided what to graph.

COMMON CORE

Solving Inequalities

In this lesson, you will

- use addition or subtraction to solve inequalities.
- use a number line to graph the solution set of inequalities.
- solve real-life problems.

Applying Standards
6.EE.5
6.EE.8

3 ACTIVITY: Solving Inequalities

Work with a partner. Complete the following steps for Activity 1. Then repeat the steps for Activity 2.

- Use your inequality from part (a). Replace the inequality symbol with an equal sign.
- Solve the equation.
- Replace the equal sign with the original inequality symbol.
- Graph this new inequality.
- Compare the graph with your graph in part (b). What do you notice?

Math Practice 4

Interpret Results

What does the solution of the inequality represent?

4 ACTIVITY: The Triangle Inequality

Work with a partner. Draw different triangles whose sides have lengths 10 cm, 6 cm, and x cm.

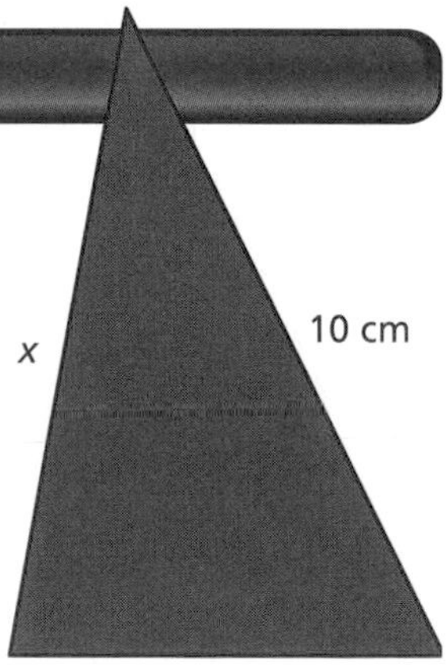

a. Which of the following describes how *small* x can be? Explain your reasoning.

$6 + x < 10$ $\quad$ $6 + x \le 10$

$6 + x > 10$ $\quad$ $6 + x \ge 10$

b. Which of the following describes how *large* x can be? Explain your reasoning.

$x - 6 < 10$ $\quad$ $x - 6 \le 10$ $\quad$ $x - 6 > 10$ $\quad$ $x - 6 \ge 10$

c. Graph the possible values of x on a number line.

What Is Your Answer?

5. **IN YOUR OWN WORDS** How can you use addition or subtraction to solve an inequality?
6. Describe a real-life situation that you can represent with an inequality. Write the inequality. Graph the solution on a number line.

Practice Use what you learned about solving inequalities to complete Exercises 5–7 on page 336.

Check It Out
Lesson Tutorials
BigIdeasMath.com

Study Tip

You can solve inequalities the same way you solve equations. Use inverse operations to get the variable by itself.

Key Ideas

Addition Property of Inequality

Words When you add the same number to each side of an inequality, the inequality remains true.

Numbers

$$\begin{array}{rcr} 3 & < & 5 \\ +2 & & +2 \\ \hline 5 & < & 7 \end{array}$$

Algebra

$$\begin{array}{rcr} x - 4 & > & 5 \\ +4 & & +4 \\ \hline x & > & 9 \end{array}$$

Graph

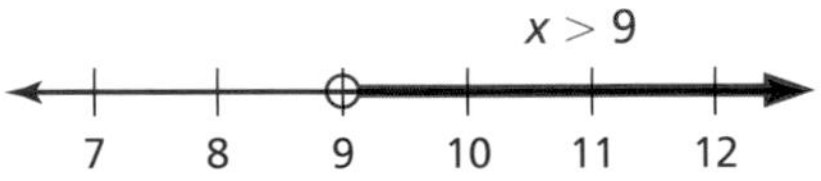

Subtraction Property of Inequality

Words When you subtract the same number from each side of an inequality, the inequality remains true.

Numbers

$$\begin{array}{rcr} 3 & < & 5 \\ -2 & & -2 \\ \hline 1 & < & 3 \end{array}$$

Algebra

$$\begin{array}{rcr} x + 4 & > & 5 \\ -4 & & -4 \\ \hline x & > & 1 \end{array}$$

Graph

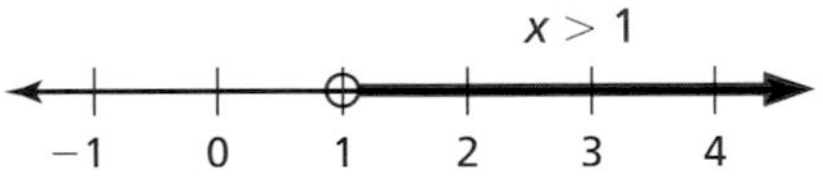

These properties are also true for ≤ and ≥.

EXAMPLE 1 Solving an Inequality Using Addition

Solve $x - 3 > 1$. Graph the solution.

$$\begin{array}{rcrl} x - 3 & > & 1 & \text{Write the inequality.} \\ +3 & & +3 & \text{Addition Property of Inequality} \\ \hline x & > & 4 & \text{Simplify.} \end{array}$$

Undo the subtraction.

∴ The solution is $x > 4$.

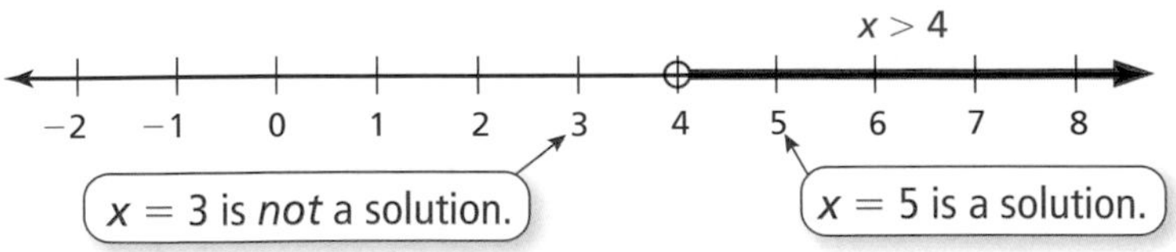

Check:

$x = 3$: $3 - 3 \overset{?}{>} 1$

$0 \not> 1$ ✗

$x = 5$: $5 - 3 \overset{?}{>} 1$

$2 > 1$ ✓

On Your Own

Solve the inequality. Graph the solution.

1. $x - 2 < 3$
2. $x - 6 \geq 4$
3. $10 \geq x - 1$

EXAMPLE 2 Solving an Inequality Using Subtraction

Solve $15 \geq 6 + x$. Graph the solution.

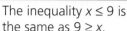

	$15 \geq 6 + x$	Write the inequality.
Undo the addition. →	$\underline{-6 \quad -6}$	Subtraction Property of Inequality
	$9 \geq x$	Simplify.

∴ The solution is $x \leq 9$.

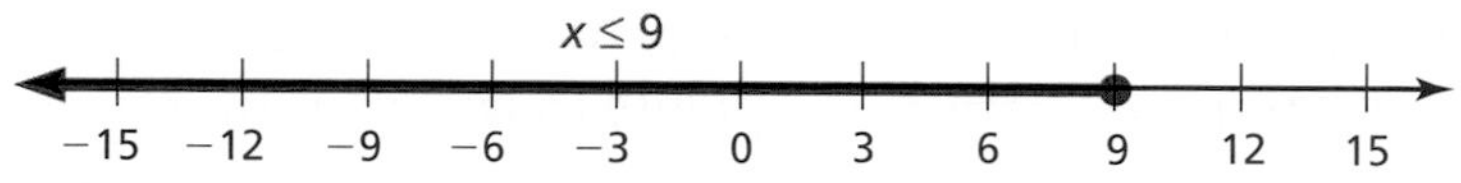

Reading

The inequality $x \leq 9$ is the same as $9 \geq x$.

On Your Own

Now You're Ready
Exercises 5–16

Solve the inequality. Graph the solution.

4. $x + 3 > 7$ **5.** $y + 2 < 17$ **6.** $16 \leq m + 9$

EXAMPLE 3 Real-Life Application

A flea market advertises that it has more than 250 vending booths. Of these, 184 are currently filled. Write and solve an inequality to represent the number of vending booths still available.

Words	The number of booths filled	plus	the number of remaining booths	is greater than	the total number of booths.
Variable	Let b be the number of remaining booths.				
Inequality	184	$+$	b	$>$	250

$184 + b > 250$	Write the inequality.
$\underline{-184 \qquad -184}$	Subtraction Property of Inequality
$b > 66$	Simplify.

∴ More than 66 vending booths are still available.

On Your Own

7. You have already spent \$24 shopping online for clothes. Write and solve an inequality to represent the additional amount you must spend to get free shipping.

7.6 Exercises

Check It Out
Help with Homework
BigIdeasMath.com

Vocabulary and Concept Check

1. **OPEN-ENDED** Write an inequality that can be solved by subtracting 7 from each side.
2. **WRITING** Explain how to solve the inequality $x - 6 > 3$.
3. **WRITING** Describe the graph of the solution of $x + 3 \leq 4$.
4. **OPEN-ENDED** Write an inequality that the graph represents. Then use the Subtraction Property of Inequality to write another inequality that the graph represents.

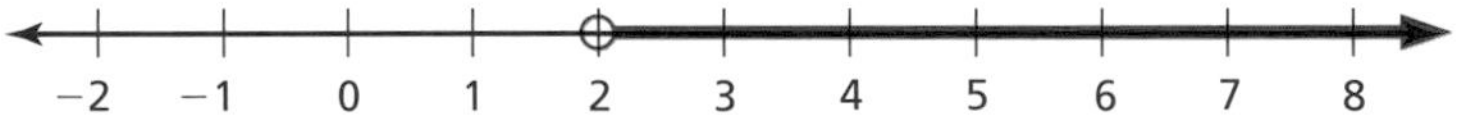

Practice and Problem Solving

Solve the inequality. Graph the solution.

5. $x - 4 < 5$	**6.** $5 + h > 7$	**7.** $3 \geq y - 2$
8. $9 \leq c + 1$	**9.** $18 > 12 + x$	**10.** $37 + z \leq 54$
11. $y - 21 < 85$	**12.** $g - 17 \geq 17$	**13.** $7.2 < x + 4.2$
14. $12.7 \geq s - 5.3$	**15.** $\frac{3}{4} \leq \frac{1}{2} + n$	**16.** $\frac{1}{3} + b > \frac{3}{4}$

17. **ERROR ANALYSIS** Describe and correct the error in solving the inequality.

✗
$$28 \geq t - 9$$
$$-9 \qquad -9$$
$$19 \geq t$$

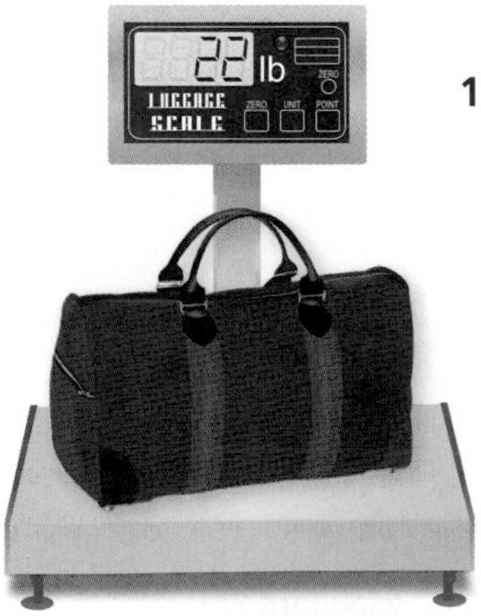

18. **AIR TRAVEL** Your carry-on bag can weigh at most 40 pounds. Write and solve an inequality to represent how much more weight you can add to the bag and still meet the requirement.

19. **SHOPPING** It costs $\$x$ for a round-trip bus ticket to the mall. You have \$24. Write and solve an inequality to represent the greatest amount of money you can spend for the bus fare and still have enough to buy the baseball cap.

Write the word sentence as an inequality. Then solve the inequality.

20. Five more than a number is less than 17.
21. Three less than a number is more than 15.

Solve the inequality. Graph the solution.

22. $x + 9 - 3 \leq 14$ **23.** $44 > 7 + s + 26$ **24.** $6.1 - 0.3 \geq c + 1$

25. VIDEO GAME The high score for a video game is 36,480. Your current score is 34,280. Each dragonfly you catch is worth 1 point. You also get a 1000-point bonus for reaching 35,000 points. Write and solve an inequality to represent the number of dragonflies you must catch to earn a new high score.

26. PICKUP TRUCKS You can register a pickup truck as a passenger vehicle if the truck is not used for commercial purposes and the weight of the truck with its contents does not exceed 8500 pounds.

a. Your pickup truck weighs 4200 pounds. Write an inequality to represent the number of pounds your truck can carry and still qualify as a passenger vehicle. Then solve the inequality.

b. A cubic yard of sand weighs about 1600 pounds. How many cubic yards of sand can you haul in your truck and still qualify as a passenger vehicle? Explain your reasoning.

27. TRIATHLON You complete two events of a triathlon. Your goal is to finish with an overall time of less than 100 minutes.

a. Write and solve an inequality to represent how many minutes you can take to finish the running event and still meet your goal.

b. The running event is 3.1 miles long. Estimate how many minutes it would take you to run 3.1 miles. Would this time allow you to reach your goal? Explain your reasoning.

Triathlon	
Event	**Your Time (minutes)**
Swimming	18.2
Biking	45.4
Running	?

28. Number Sense The possible values of x are given by $x - 3 \geq 2$. What is the least possible value of $5x$?

Fair Game Review What you learned in previous grades & lessons

Solve the equation. Check your solution. *(Section 7.3)*

29. $\frac{t}{12} = 4$ **30.** $6 = \frac{2s}{9}$ **31.** $8x = 72$ **32.** $9 = 1.5z$

33. MULTIPLE CHOICE Which brand of turkey is the best buy? *(Section 5.4)*

Ⓐ Brand A Ⓑ Brand B
Ⓒ Brand C Ⓓ Brand D

Brand	A	B	C	D
Cost (dollars)	10.38	13.47	21.45	34.93
Pounds	2	3	5	7

7.7 Solving Inequalities Using Multiplication or Division

Essential Question How can you use multiplication or division to solve an inequality?

1 ACTIVITY: Writing an Inequality

Work with a partner. A store has a clearance rack of shirts that each cost the same amount. You buy 2 shirts and have money left after paying with a $20 bill.

a. Which of the following represents your purchase? What does x represent? Explain your reasoning.

$2x < 20$	$2x \le 20$
$2x > 20$	$2x \ge 20$

b. Graph the possible values of x on a number line. Explain how you decided what to graph.

c. Can you buy a third shirt? Explain your reasoning.

2 ACTIVITY: Writing an Inequality

Work with a partner. One of your favorite stores is having a 75% off sale. You have $20. You want to buy a pair of jeans.

a. Which of the following represents your ability to buy the jeans with $20? What does x represent? Explain your reasoning.

$\frac{1}{4}x < 20$	$\frac{1}{4}x \le 20$
$\frac{1}{4}x > 20$	$\frac{1}{4}x \ge 20$

b. Graph the possible values of x on a number line. Explain how you decided what to graph.

c. Can you afford a pair of jeans that originally costs $100? Explain your reasoning.

COMMON CORE

Solving Inequalities

In this lesson, you will

- use multiplication or division to solve inequalities.
- use a number line to graph the solution set of inequalities.
- solve real-life problems.

Applying Standards
6.EE.5
6.EE.8

3 ACTIVITY: Solving Inequalities

Work with a partner. Complete the following steps for Activity 1. Then repeat the steps for Activity 2.

- Use your inequality from part (a). Replace the inequality symbol with an equal sign.
- Solve the equation.
- Replace the equal sign with the original inequality symbol.
- Graph this new inequality.
- Compare the graph with your graph in part (b). What do you notice?

4 ACTIVITY: Matching Inequalities

Work with a partner. Match the inequality with its graph. Explain your method.

a. $3x < 9$ **b.** $3x \leq 9$ **c.** $\frac{x}{2} \geq 1$

d. $6 < 2x$ **e.** $12 \leq 4x$ **f.** $\frac{x}{2} < 2$

Math Practice 1

Make a Plan

What strategy will you use to choose the correct graph?

A.
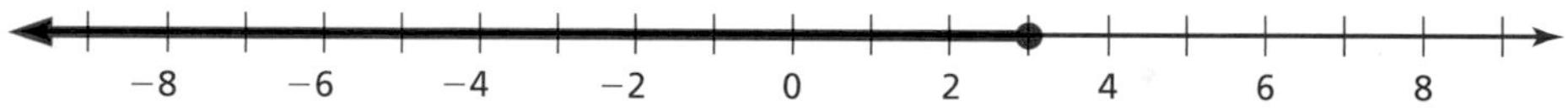

B.
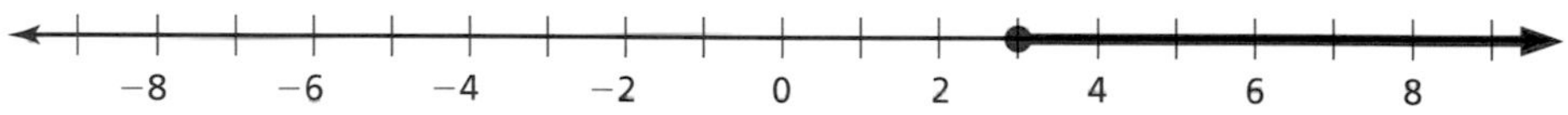

C.
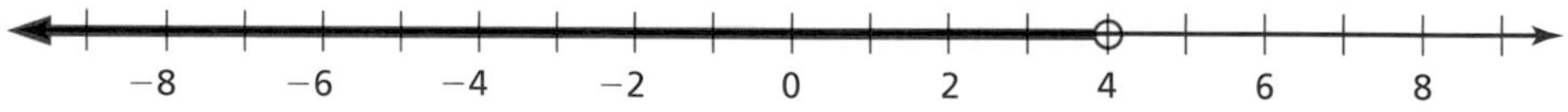

D.
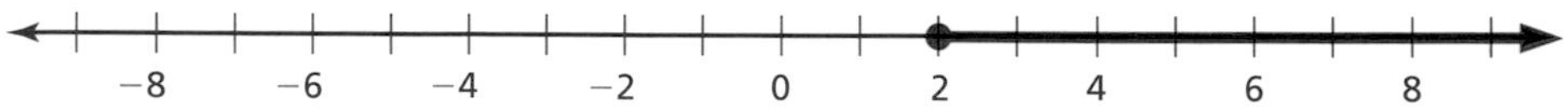

E.
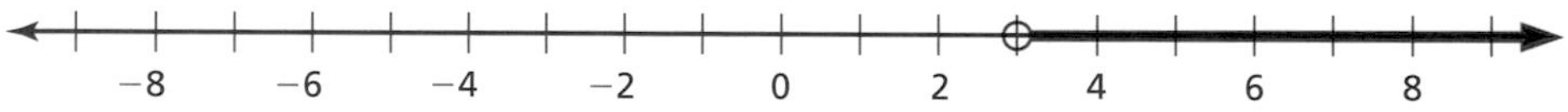

F.
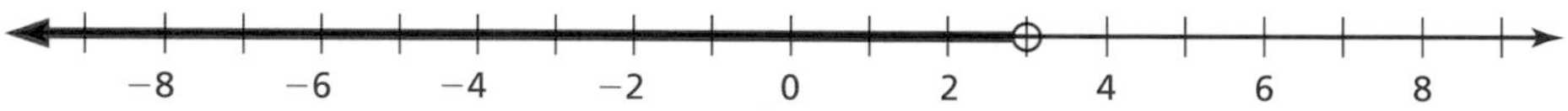

What Is Your Answer?

5. **IN YOUR OWN WORDS** How can you use multiplication or division to solve an inequality?

Practice Use what you learned about solving inequalities to complete Exercises 8–11 on page 342.

7.7 Lesson

Check It Out
Lesson Tutorials
BigIdeasMath.com

Remember

Multiplication and division are inverse operations.

Key Ideas

Multiplication Property of Inequality

Words When you multiply each side of an inequality by the same *positive* number, the inequality remains true.

Numbers

$$8 > 6$$
$$8 \times 2 > 6 \times 2$$
$$16 > 12$$

Algebra

$$\frac{x}{4} < 2$$
$$\frac{x}{4} \cdot 4 < 2 \cdot 4$$
$$x < 8$$

Division Property of Inequality

Words When you divide each side of an inequality by the same *positive* number, the inequality remains true.

Numbers

$$8 > 6$$
$$8 \div 2 > 6 \div 2$$
$$4 > 3$$

Algebra

$$4x < 8$$
$$\frac{4x}{4} < \frac{8}{4}$$
$$x < 2$$

These properties are also true for $\leq$ and $\geq$.

EXAMPLE 1 Solving an Inequality Using Multiplication

Solve $\frac{x}{5} \leq 2$. Graph the solution.

$\frac{x}{5} \leq 2$	Write the inequality.
$\frac{x}{5} \cdot 5 \leq 2 \cdot 5$	Multiplication Property of Inequality
$x \leq 10$	Simplify.

Undo the division. → (points to $\frac{x}{5} \cdot 5 \leq 2 \cdot 5$)

The solution is $x \leq 10$.

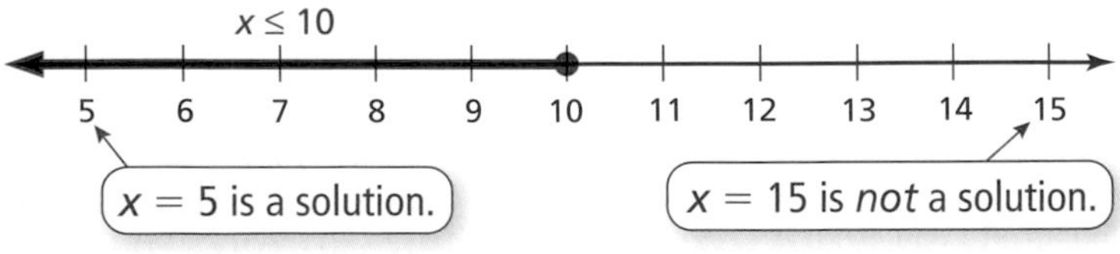

On Your Own

Now You're Ready
Exercises 6–9

Solve the inequality. Graph the solution.

1. $p \div 3 > 2$
2. $\frac{3}{5}q \leq 6$
3. $1 < \frac{s}{7}$

EXAMPLE 2 Solving an Inequality Using Division

Solve $4n > 32$. Graph the solution.

$4n > 32$ Write the inequality.

Undo the multiplication. → $\frac{4n}{4} > \frac{32}{4}$ Division Property of Inequality

$n > 8$ Simplify.

∴ The solution is $n > 8$.

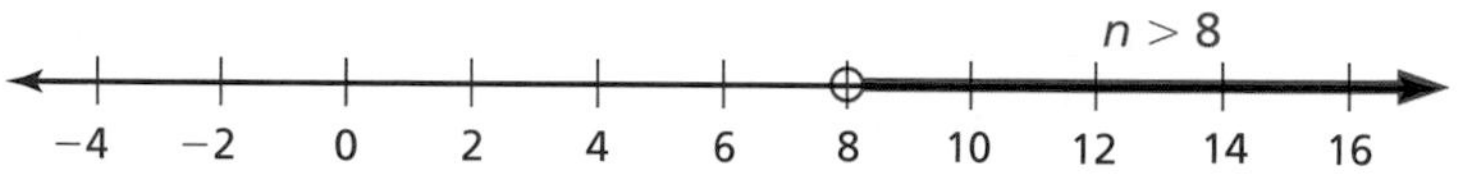

EXAMPLE 3 Real-Life Application

A one-way bus ride costs \$1.75. A 30-day bus pass costs \$42.

a. Write and solve an inequality to find the least number of one-way rides you must take for the 30-day pass to be a better deal.

b. You ride the bus an average of 20 times each month. Is the pass a better deal? Explain.

a.

Words	The price of a one-way ride	times	the number of one-way rides	is more than	\$42.
Variable	Let r be the number of one-way rides.				
Inequality	1.75	•	r	>	42

$1.75r > 42$ Write the inequality.

$\frac{1.75r}{1.75} > \frac{42}{1.75}$ Division Property of Inequality

$r > 24$ Simplify.

∴ So, you need to take more than 24 one-way rides for the pass to be a better deal.

b. No. The cost of 20 one-way rides is less than \$42. So, the pass is not a better deal.

On Your Own

Now You're Ready
Exercises 10–13

Solve the inequality. Graph the solution.

4. $11k \leq 33$

5. $5 \cdot j > 20$

6. $50 \leq 2m$

7. The sign shows the toll for driving on Alligator Alley. Write and solve an inequality to represent the number of times someone can drive on Alligator Alley with \$15.

7.7 Exercises

Check It Out
Help with Homework
BigIdeasMath.com

Vocabulary and Concept Check

1. **REASONING** How is the graph of the solution of $2x \geq 10$ different from the graph of the solution of $2x = 10$?

Name the property you should use to solve the inequality.

2. $3x \leq 27$
3. $7x > 49$
4. $\frac{x}{2} < 36$

5. **OPEN-ENDED** Write two inequalities that have the same solution set: one that you can solve using division and one that you can solve using multiplication.

Practice and Problem Solving

Solve the inequality. Graph the solution.

1 6. $\frac{m}{8} < 4$
7. $n \div 6 > 2$
8. $\frac{t}{3} \geq 15$
9. $\frac{1}{11}c \geq 9$

2 10. $12x < 96$
11. $5x \geq 25$
12. $8 \cdot w \leq 72$
13. $7p \leq 42$

14. $\frac{3}{4}b > 15$
15. $6x < 90$
16. $3s \geq 36$
17. $\frac{5}{9}v \leq 45$

18. $4t > 72$
19. $\frac{3}{4}w \leq 24$
20. $12m < 132$
21. $\frac{5x}{8} \geq 30$

22. **ERROR ANALYSIS** Describe and correct the error in solving the inequality.

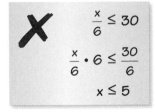

✗
$$\frac{x}{6} \leq 30$$
$$\frac{x}{6} \cdot 6 \leq \frac{30}{6}$$
$$x \leq 5$$

23. **GEOMETRY** The length of a rectangle is 8 feet, and its area is less than 168 square feet. Write and solve an inequality to represent the width of the rectangle.

24. **PLAYGROUND** Students at a playground are divided into 5 equal groups with at least 6 students in each group. Write and solve an inequality to represent the number of students at the playground.

Write the word sentence as an inequality. Then solve the inequality.

25. Eight times a number n is less than 72.
26. A number t divided by 32 is at most 4.25.
27. 225 is no less than 12 times a number w.

Graph the numbers that are solutions to both inequalities.

28. $x + 7 > 9$ and $8x \leq 64$

29. $x - 3 \leq 8$ and $6x < 72$

30. THRILL RIDE A thrill ride at an amusement park holds a maximum of 12 people per ride.

a. Write and solve an inequality to find the least number of rides needed for 15,000 people.

b. Do you think it is possible for 15,000 people to ride the thrill ride in 1 day? Explain.

Park Hours
10:00 A.M.–10:00 P.M.

31. FOOTBALL A winning football team more than doubled the offensive yards gained by its opponent. The opponent gained 272 offensive yards. The winning team had 80 offensive plays. Write and solve an inequality to find the possible number of yards per play for the winning team.

32. LOGIC Explain how you know that $7x < 7x$ has no solution.

33. OPEN-ENDED Give an example of a real-life situation in which you can list all the solutions of an inequality. Give an example of a real-life situation in which you cannot list all the solutions of an inequality.

34. FUNDRAISER You are selling items from a catalog for a school fundraiser. Write and solve two inequalities to find the range of sales that will earn you between \$40 and \$50.

Critical Thinking **Let $a > b$ and $x > y$. Tell whether the statement is *always* true. Explain your reasoning.**

35. $a + x > b + y$

36. $a - x > b - y$

37. $ax > by$

38. $\frac{a}{x} > \frac{y}{b}$

Fair Game Review What you learned in previous grades & lessons

Classify the quadrilateral. *(Skills Review Handbook)*

39.

40.

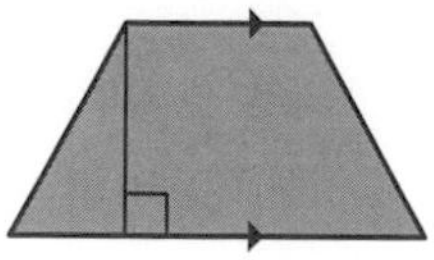

41.

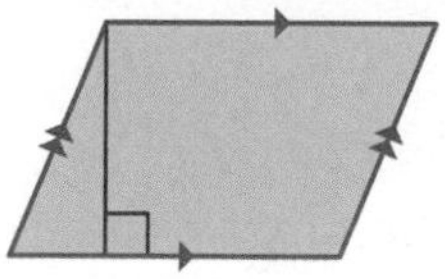

42. MULTIPLE CHOICE On a normal day, 12 airplanes arrive at an airport every 15 minutes. Which rate does not represent this situation? *(Section 5.3)*

Ⓐ 24 airplanes every 30 minutes

Ⓑ 4 airplanes every 5 minutes

Ⓒ 6 airplanes every 5 minutes

Ⓓ 48 airplanes each hour

7.5–7.7 Quiz

Check It Out
Progress Check
BigIdeasMath.com

Write the word sentence as an inequality. *(Section 7.5)*

1. A number x is greater than 0.

2. Twice a number c is at least -8.

Tell whether the given value is a solution of the inequality. *(Section 7.5)*

3. $2n > 16;\ n = 9$

4. $x - 1 \le 9;\ x = 10$

Graph the inequality on a number line. *(Section 7.5)*

5. $y > -4$

6. $m \le \frac{3}{5}$

Solve the inequality. Graph the solution. *(Section 7.6)*

7. $x + 4 \le 8$

8. $18 > 16 + g$

Write the word sentence as an inequality. Then solve the inequality. *(Section 7.6)*

9. Two less than a number is more than 15.

10. Seven more than a number is less than or equal to 27.

Solve the inequality. Graph the solution. *(Section 7.7)*

11. $\frac{3a}{2} < 24$

12. $121 \ge 11s$

Write the word sentence as an inequality. Then solve the inequality. *(Section 7.7)*

13. Three times a number x is more than 18.

14. 84 is no less than 7 times a number k.

15. WATER PARK Each visit to a water park costs \$19.95. An annual pass to the park costs \$89.95. Write an inequality to represent the number of times you would need to visit the park for the pass to be a better deal. *(Section 7.5)*

16. GARDEN You want to use a square section of your yard for a garden. You have at most 52 feet of fencing to surround the garden. Write and solve an inequality to represent the possible lengths of each side of the garden. *(Section 7.7)*

17. DELIVERY You were planning to spend \$12 on a pizza. Write and solve an inequality to represent the additional amount you must spend to get free delivery. *(Section 7.6)*

PIZZA PALACE
FREE DELIVERY
for orders of $30.00 or more
516 E 31st

7 Chapter Review

Check It Out
Vocabulary Help
BigIdeasMath.com

Review Key Vocabulary

equation, *p. 296*
solution, *p. 302*
inverse operations, *p. 303*
equation in two variables, *p. 316*
solution of an equation in two variables, *p. 316*
independent variable, *p. 316*
dependent variable, *p. 316*
inequality, *p. 326*
solution of an inequality, *p. 327*
solution set, *p. 327*
graph of an inequality, *p. 328*

Review Examples and Exercises

7.1 Writing Equations in One Variable *(pp. 294–299)*

Write the word sentence "The quotient of a number b and 6 is 9" as an equation.

The quotient of a number b and 6 is 9.

$b \div 6 = 9$ — *Quotient of* means *division*.

An equation is $b \div 6 = 9$.

Exercises

Write the word sentence as an equation.

1. The product of a number m and 2 is 8.
2. 6 less than a number t is 7.
3. A number m increased by 5 is 7.
4. 8 is the quotient of a number g and 3.

7.2 Solving Equations Using Addition or Subtraction *(pp. 300–307)*

Solve $z + 5 = 13$.

$z + 5 = 13$	Write the equation.
$\underline{-5}\quad\underline{-5}$	Subtraction Property of Equality
$z = 8$	Simplify.

Undo the addition.

The solution is $z = 8$.

Check

$z + 5 = 13$

$8 + 5 \stackrel{?}{=} 13$

$13 = 13$ ✓

Exercises

Solve the equation. Check your solution.

5. $x - 1 = 8$
6. $m + 7 = 11$
7. $21 = p - 12$

7.3 Solving Equations Using Multiplication or Division (pp. 308–313)

Solve $4c = 32$.

Undo the multiplication.

$4c = 32$	Write the equation.
$\frac{4c}{4} = \frac{32}{4}$	Division Property of Equality
$c = 8$	Simplify.

Check

$4c = 32$

$4(8) \stackrel{?}{=} 32$

$32 = 32$ ✓

Exercises

Solve the equation. Check your solution.

8. $7 \cdot q = 42$ **9.** $7k \div 3 = 21$ **10.** $\frac{5a}{7} = 25$

7.4 Writing Equations in Two Variables (pp. 314–321)

Tell whether (6, 16) is a solution of the equation $y = 3x - 4$.

$16 \stackrel{?}{=} 3(6) - 4$	Substitute.
$16 \neq 14$ ✗	Compare.

∴ So, (6, 16) is *not* a solution.

Exercises

Tell whether the ordered pair is a solution of the equation.

11. $y = 3x + 1$; (2, 7) **12.** $y = 7x - 4$; (4, 22)

13. **TAXI** A taxi ride costs \$3 plus \$2.50 per mile. Write and graph an equation in two variables that represents the total cost of a taxi ride.

7.5 Writing and Graphing Inequalities (pp. 324–331)

Write the word sentence as an inequality.

a. A number x is more than -9.

A number x is more than -9.

$x \quad > \quad -9$

∴ An inequality is $x > -9$.

b. A number r divided by 2 is at most 4.

A number r divided by 2 is at most 4.

$\frac{r}{2} \quad \leq \quad 4$

∴ An inequality is $\frac{r}{2} \leq 4$.

Exercises

Write the word sentence as an inequality.

14. A number m is less than 5. **15.** A number h is at least -12.

Graph the inequality on a number line.

16. $x < 0$ **17.** $a \geq 3$ **18.** $n \leq -1$

7.6 Solving Inequalities Using Addition or Subtraction (pp. 332–337)

Solve $1 \leq x - 4$. Graph the solution.

Undo the subtraction.

$$\begin{aligned} 1 &\leq x - 4 && \text{Write the inequality.} \\ +4 & \quad\;\; +4 && \text{Addition Property of Inequality} \\ 5 &\leq x && \text{Simplify.} \end{aligned}$$

The inequality $5 \leq x$ is the same as $x \geq 5$.

∴ The solution is $x \geq 5$.

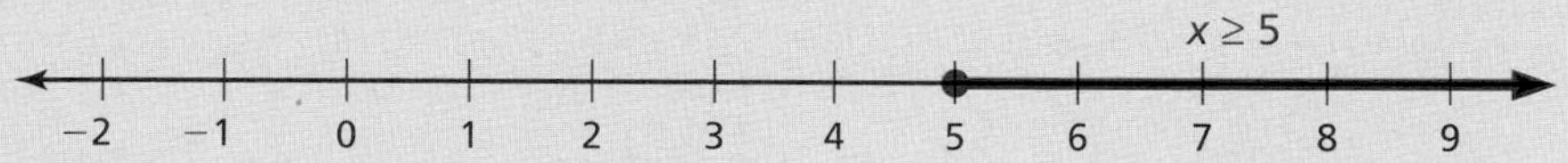

Exercises

Solve the inequality. Graph the solution.

19. $x + 1 > 3$

20. $k - 7 \leq 0$

21. $y + 8 \geq 9$

22. $24 < 11 + x$

23. $4 \leq n - 4$

24. $x - 20 > 24$

25. $b + 12 \leq 26$

26. $s - 1.5 < 2.5$

27. $\frac{1}{4} + m \leq \frac{1}{2}$

7.7 Solving Inequalities Using Multiplication or Division (pp. 338–343)

Solve $7n < 42$. Graph the solution.

Undo the multiplication.

$$\begin{aligned} 7n &< 42 && \text{Write the inequality.} \\ \frac{7n}{7} &< \frac{42}{7} && \text{Division Property of Inequality} \\ n &< 6 && \text{Simplify.} \end{aligned}$$

∴ The solution is $n < 6$.

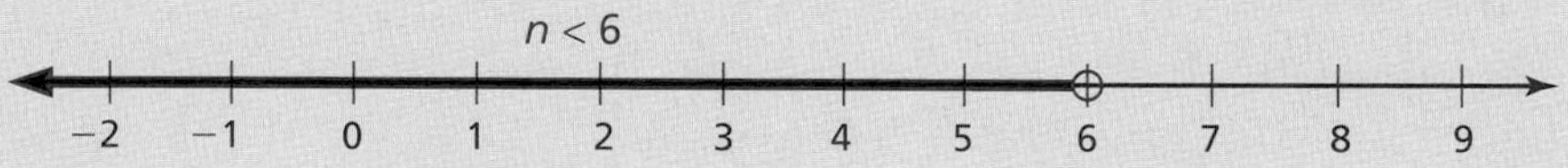

Exercises

Solve the inequality. Graph the solution.

28. $x \div 2 < 4$

29. $9n \geq 63$

30. $\frac{5}{3}x \leq 10$

31. $9 \geq 3b$

32. $10p > 40$

33. $\frac{3}{11}k < 15$

34. **TICKETS** The cost of three tickets to a movie is at least \$20. Write and solve an inequality that represents the situation.

7 Chapter Test

Check It Out
Test Practice
BigIdeasMath.com

Write the word sentence as an equation.

1. 7 times a number s is 84.

2. 13 is one-third of a number m.

Solve the equation. Check your solution.

3. $15 = 7 + b$

4. $v - 6 = 16$

5. $5x = 70$

6. $3b = 45$

7. $\frac{6m}{7} = 30$

8. $\frac{8k}{3} = 32$

Tell whether the ordered pair is a solution of the equation.

9. $y = 9x$; (3, 27)

10. $y = 4x + 2$; (8, 36)

Write an inequality for the situation.

11. An MP3 player holds up to 300 songs.

12. Riders must be at least 48 inches tall.

Graph the inequality on a number line.

13. $x \geq 5$

14. $m \leq -2$

Solve the inequality. Graph the solution.

15. $x - 3 < 7$

16. $12 \geq n + 6$

17. $\frac{4}{3}b \leq 12$

18. $72 > 12p$

19. **SCHOOL DANCE** Each ticket to a school dance is \$4. The total amount collected in ticket sales is \$332. Write and solve an equation to find the number of students attending the dance.

20. **T-SHIRTS** A soccer team will sell T-shirts for a fundraiser. The company that makes the T-shirts charges \$10 per shirt plus a \$20 shipping fee per order.

a. Write and graph an equation in two variables that represents the total cost of ordering the shirts.

b. Choose an ordered pair that lies on your graph in part (a). Interpret it in the context of the problem.

21. **HURRICANE** A hurricane has wind speeds that are greater than or equal to 74 miles per hour. Write an inequality to represent the possible wind speeds during a hurricane.

7 Standards Assessment

Test-Taking Strategy
Work Backwards

You like taking x catnaps each day, where 3x=24. How many is that?
Ⓐ 6 Ⓑ 7 Ⓒ 8 Ⓓ 9
ZZZZZZZ
"Work backwards by trying 6, 7, 8, and 9. You will see that 3(8)=24. So, C is correct."

1. What is the area of the balcony shown below? *(6.G.1)*

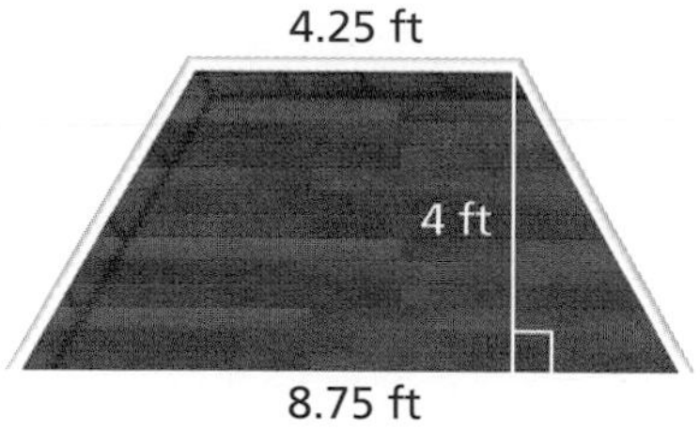
4.25 ft
4 ft
8.75 ft

A. 9 ft^2 **C.** 26 ft^2

B. 18 ft^2 **D.** 52 ft^2

2. You are making identical fruit baskets using 16 apples, 24 pears, and 32 bananas. What is the greatest number of baskets you can make using all the fruit? *(6.NS.4)*

F. 2   **H.** 8

G. 4   **I.** 16

3. Which equation represents the word sentence below? *(6.EE.7)*

The sum of 18 and 5 is equal to 9 less than a number y.

A. $18 - 5 = 9 - y$ **C.** $18 + 5 = y - 9$

B. $18 + 5 = 9 - y$ **D.** $18 - 5 = y - 9$

4. Which number line is a graph of the solution of the inequality below? *(6.EE.8)*

$$x \geq 5$$

F.
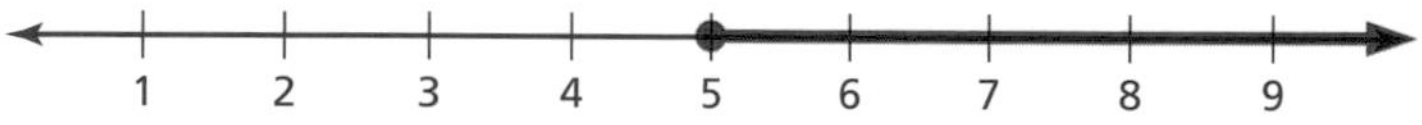

G.
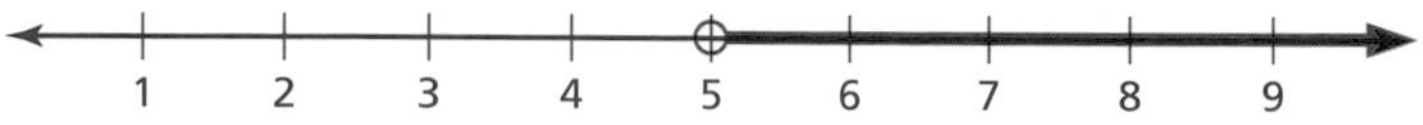

H.

I.
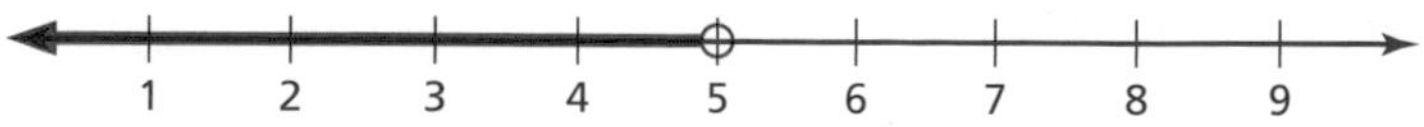

5. The steps your friend took to divide two mixed numbers are shown below.

$$3\frac{3}{5} \div 1\frac{1}{2} = \frac{18}{5} \times \frac{3}{2}$$
$$= \frac{27}{5}$$
$$= 5\frac{2}{5}$$

What should your friend change in order to divide the two mixed numbers correctly? *(6.NS.1)*

A. Find a common denominator of 5 and 2.

B. Multiply by the reciprocal of $\frac{18}{5}$.

C. Multiply by the reciprocal of $\frac{3}{2}$.

D. Rename $3\frac{3}{5}$ as $2\frac{8}{5}$.

6. An inequality is graphed on the number line below.

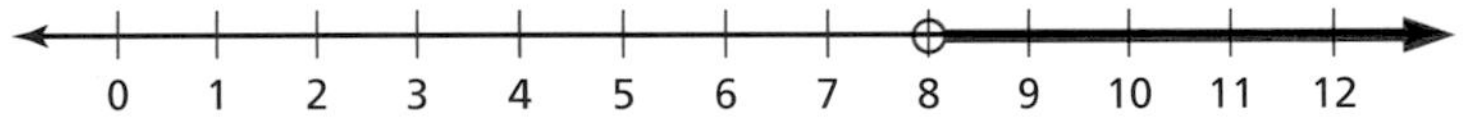

What is the least whole number value that is a solution of the inequality? *(6.EE.8)*

7. A company ordering parts receives a charge of $25 for shipping and handling plus $20 per part. Which equation represents the cost c of ordering p parts? *(6.EE.9)*

F. $c = 25 + 20p$

G. $c = 20 + 25p$

H. $p = 25 + 20c$

I. $p = 20 + 25c$

8. Which property is illustrated by the statement below? *(6.EE.3)*

$$5(3 + 6) = 5(3) + 5(6)$$

A. Associative Property of Multiplication

B. Commutative Property of Multiplication

C. Commutative Property of Addition

D. Distributive Property

9. What is the value of the expression below? *(6.NS.3)*

$$46.8 \div 0.156$$

10. In a fish tank, 75% of the fish are goldfish. How many fish are in the tank if there are 24 goldfish? *(6.RP.3c)*

F. 6 **H.** 32

G. 18 **I.** 96

11. What are the coordinates of point P in the coordinate plane below? *(6.NS.6c)*

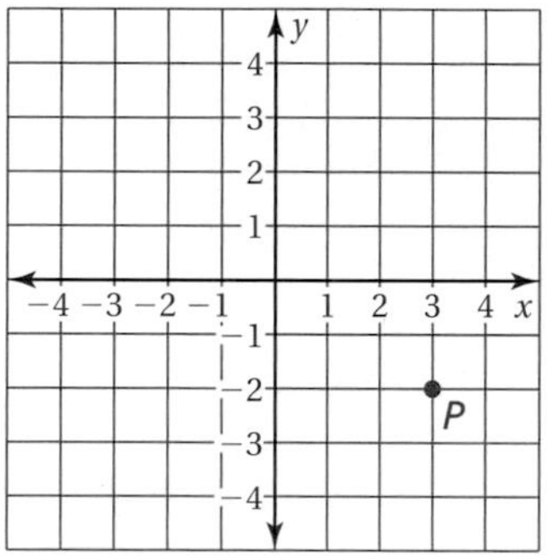

A. $(-3, -2)$ **C.** $(-2, -3)$

B. $(3, -2)$ **D.** $(-2, 3)$

12. What is the first step in evaluating the expression below? *(6.EE.1)*

$$3 \cdot (5 + 2)^2 \div 7$$

F. Multiply 3 and 5. **H.** Evaluate 5^2.

G. Add 5 and 2. **I.** Evaluate 2^2.

13. Jeff wants to save \$4000 to buy a used car. He has already saved \$850. He plans to save an additional \$150 each week. *(6.EE.7)*

Think Solve Explain

Part A Write and solve an equation to represent the number of weeks remaining until he can afford the car.

Jeff saves \$150 per week by saving $\frac{3}{4}$ of what he earns at his job each week. He works 20 hours per week.

Part B Write an equation to represent the amount per hour that Jeff must earn to save \$150 per week. Explain your reasoning.

Part C What is the amount per hour that Jeff must earn? Show your work and explain your reasoning.

8 Surface Area and Volume

"I petitioned my owner for a doghouse with greater volume."
"And this is what he built for me."

Ⓐ Yes
Ⓑ No
Ⓒ Check, please!
"I want to paint my doghouse. To make sure I buy the correct amount of paint, I want to calculate the lateral surface area."
"Then, because I want to paint the inside and the outside, I will multiply by 2. Does this seem right to you?"

What You Learned Before

Polly Prism, Prissy Pyramid, Cici Cylinder, and Connie Cone

"Name these shapes."

Classifying Figures (5.G.4)

Example 1

Identify the figure.

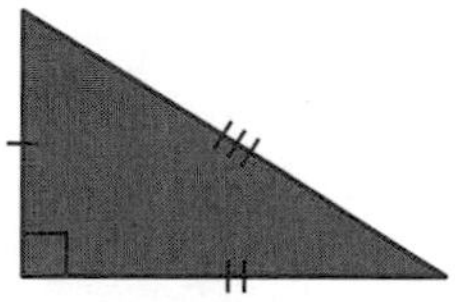

Because the figure has a right angle and three sides of different lengths, it is a right scalene triangle.

Example 2

Identify the figure.

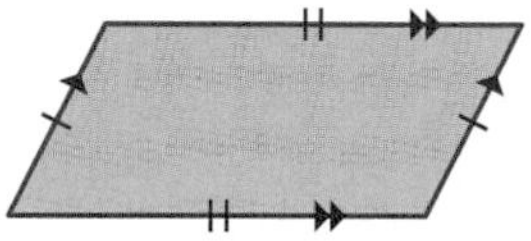

Because the figure is a quadrilateral with opposite sides that are parallel, it is a parallelogram.

Try It Yourself

Identify the figure.

1. 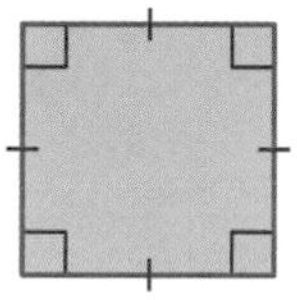

2. 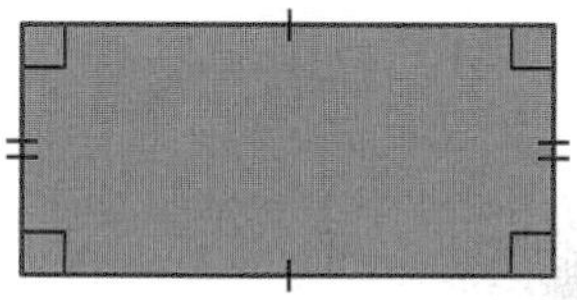

3. 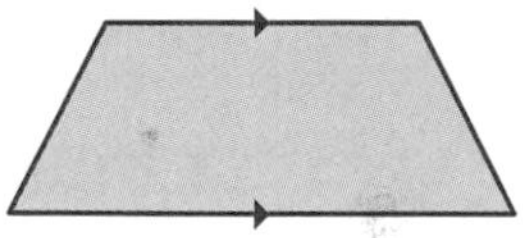

Finding Volumes of Rectangular Prisms (5.MD.5a)

Example 3

Find the volume of the rectangular prism.

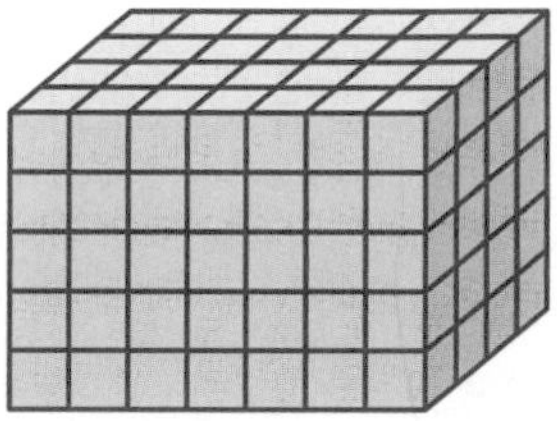

There are $4 \times 7 = 28$ unit cubes in each layer.

Because there are 5 layers, there are $5 \times 28 = 140$ unit cubes in the prism.

So, the volume is 140 cubic units.

Try It Yourself

Find the volume of the rectangular prism.

4. 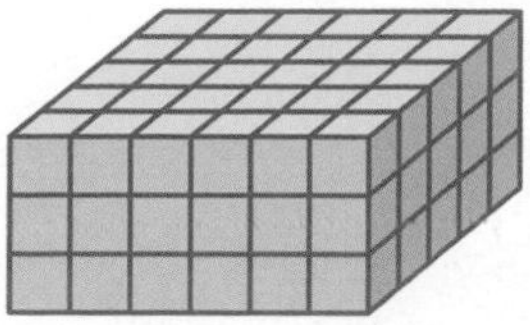

5. 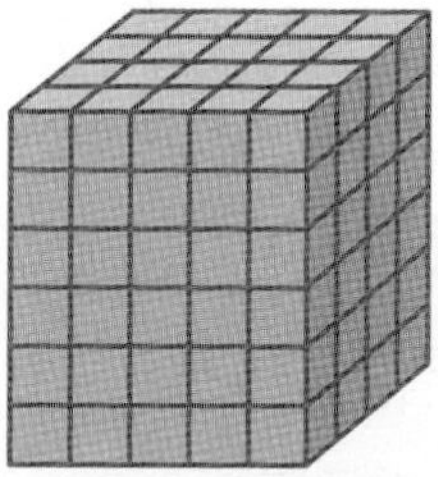

6. 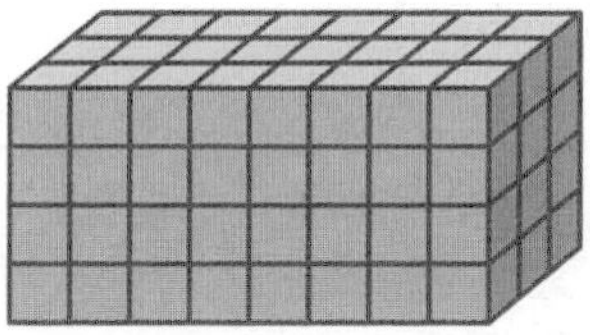

8.1 Three-Dimensional Figures

Essential Question How can you draw three-dimensional figures?

Dot paper can help you draw three-dimensional figures, or *solids.*

Square Dot Paper

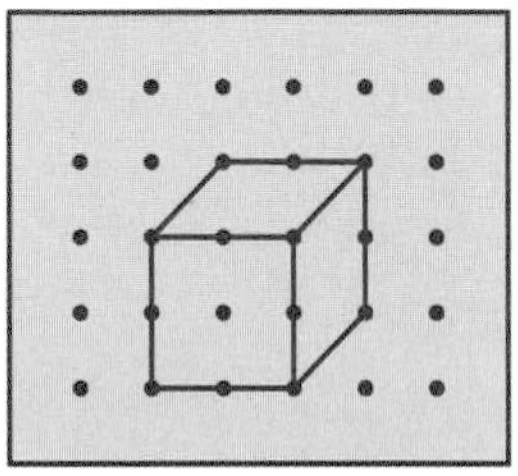

Face-On view

Isometric Dot Paper

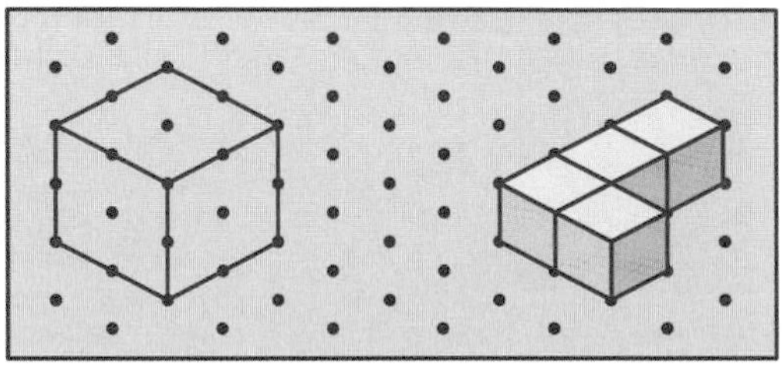

Corner view

1 ACTIVITY: Drawing Views of a Solid

Work with a partner. Draw the front, side, and top views of each stack of cubes. Then find the number of cubes in the stack.

a. Sample:

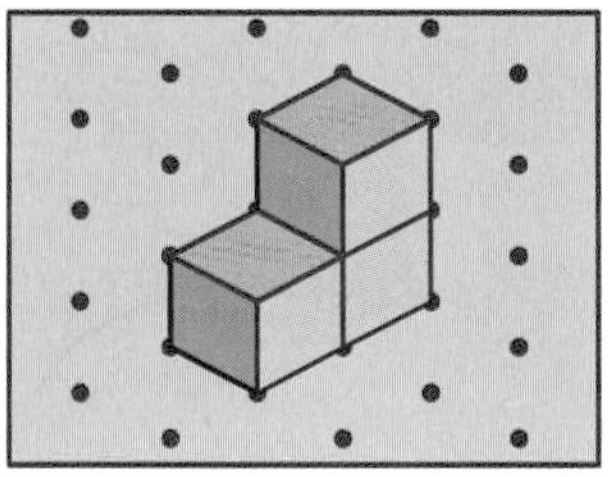

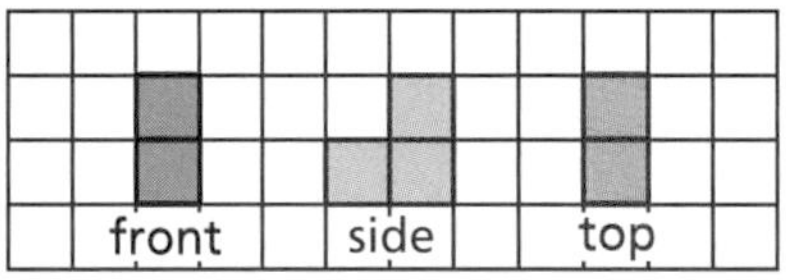

Number of cubes: 3

b.

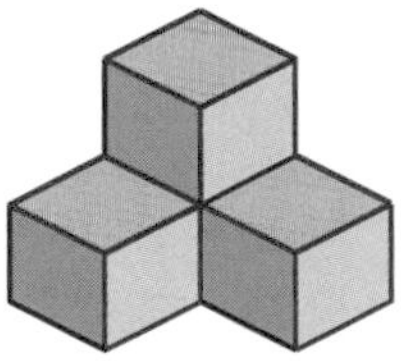

c.

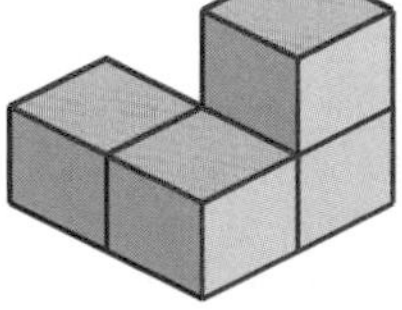

d.

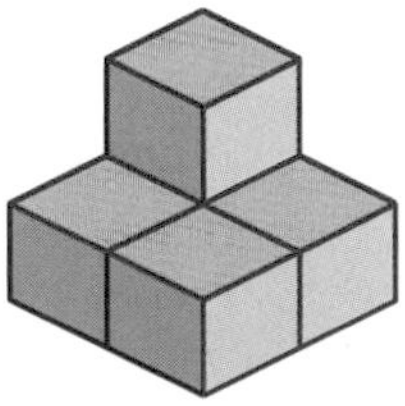

e.

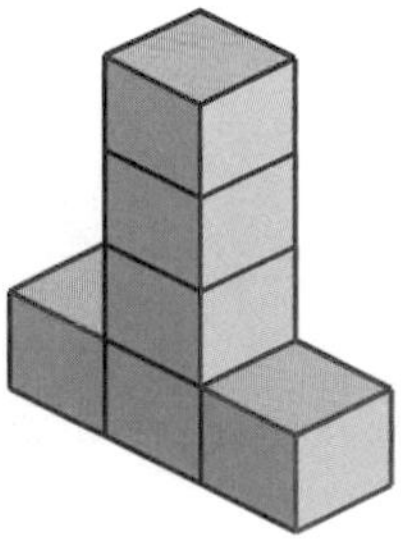

f.

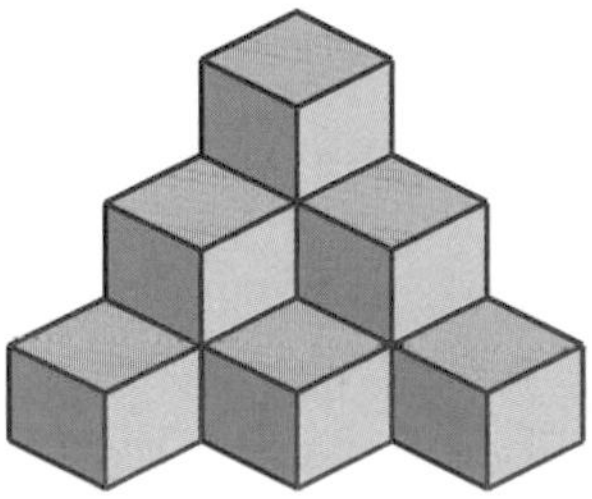

g.

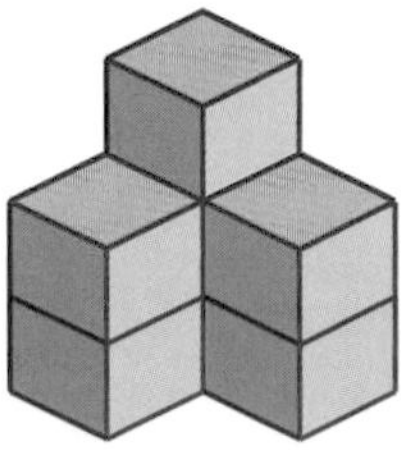

COMMON CORE

Geometry

In this lesson, you will

- draw three-dimensional figures.
- find the number of faces, edges, and vertices of solids.

Preparing for Standard 6.G.4

2 ACTIVITY: Drawing Solids

Work with a partner.

a. Use isometric dot paper to draw three different solids that use the same number of cubes as the solid at the right.

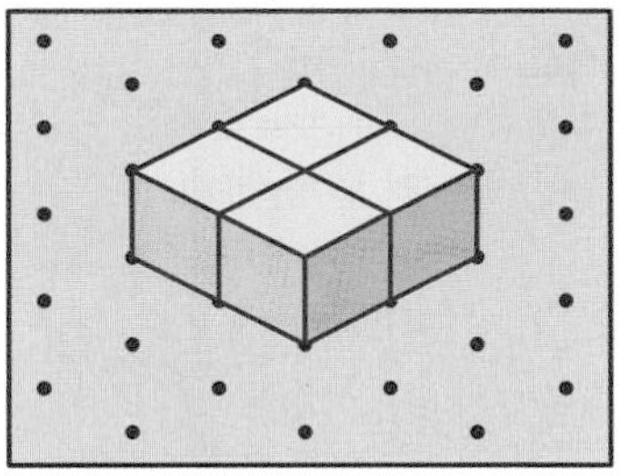

b. Use square dot paper to draw a different solid that uses the same number of *prisms* as the solid at the right.

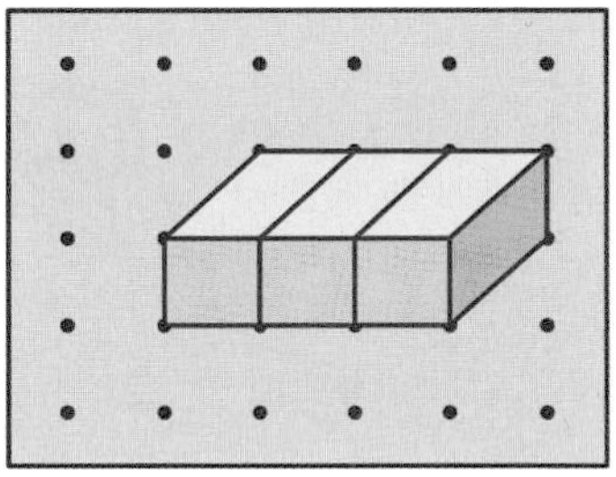

3 ACTIVITY: Exploring Faces, Edges, and Vertices

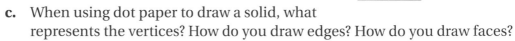

Math Practice 7

View as Components

What are the different parts of a three-dimensional object? How can dot paper help you draw the parts of the object?

Work with a partner. Use the solid shown.

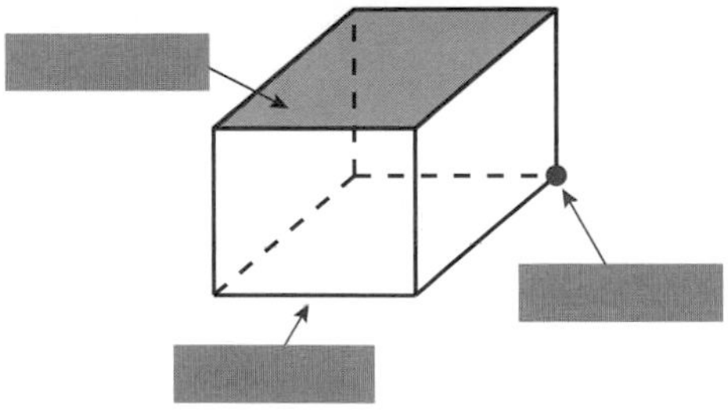

a. Match each word to the figure. Then write a definition for each word.

face *edge* *vertex*

b. Identify the number of faces, edges, and vertices in a rectangular prism.

c. When using dot paper to draw a solid, what represents the vertices? How do you draw edges? How do you draw faces?

d. What do you think it means for lines or planes to be parallel or perpendicular in three dimensions? Use drawings to identify one pair of each of the following:

- parallel faces
- parallel edges
- edge parallel to a face
- perpendicular faces
- perpendicular edges
- edge perpendicular to a face

What Is Your Answer?

4. **IN YOUR OWN WORDS** How can you draw three-dimensional figures?

Practice

Use what you learned about three-dimensional figures to complete Exercises 7–9 on page 358.

8.1 Lesson

Check It Out
Lesson Tutorials
BigIdeasMath.com

Key Vocabulary
solid, *p. 356*
polyhedron, *p. 356*
face, *p. 356*
edge, *p. 356*
vertex, *p. 356*
prism, *p. 356*
pyramid, *p. 356*

A **solid** is a three-dimensional figure that encloses a space. A **polyhedron** is a solid whose *faces* are all polygons.

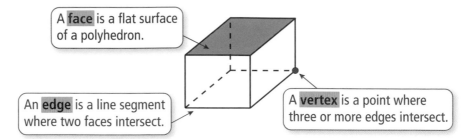

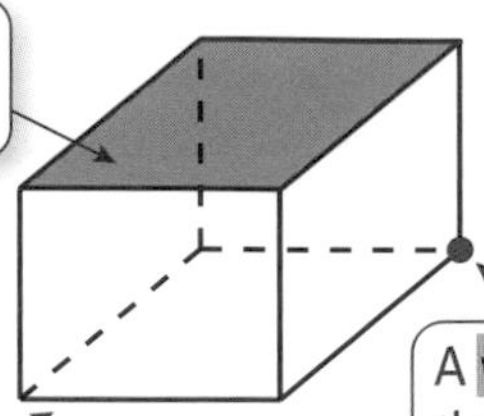

A **face** is a flat surface of a polyhedron.
An **edge** is a line segment where two faces intersect.
A **vertex** is a point where three or more edges intersect.

EXAMPLE 1 Finding the Number of Faces, Edges, and Vertices

Find the number of faces, edges, and vertices of the solid.

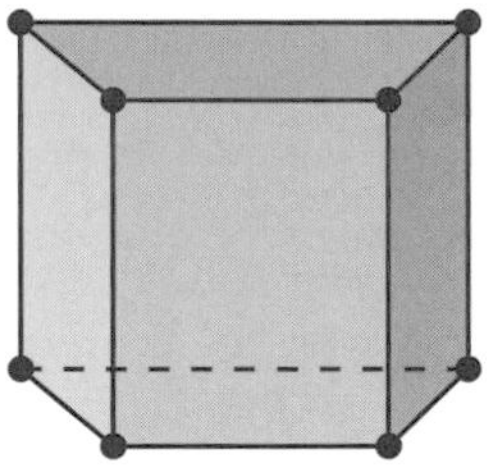

The solid has 1 face on the bottom, 1 face on the top, and 4 faces on the sides.

The faces intersect at 12 different line segments.

The edges intersect at 8 different points.

So, the solid has 6 faces, 12 edges, and 8 vertices.

On Your Own

Now You're Ready
Exercises 10–12

1. Find the number of faces, edges, and vertices of the solid.

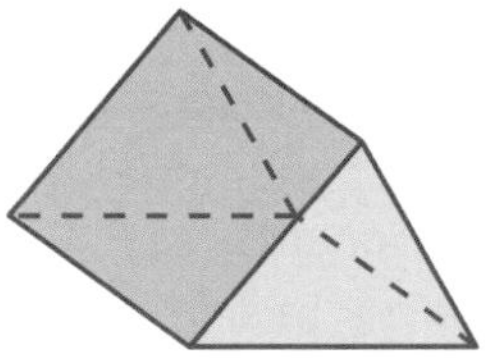

Key Ideas

Prisms

A **prism** is a polyhedron that has two parallel, identical *bases*. The *lateral faces* are parallelograms.

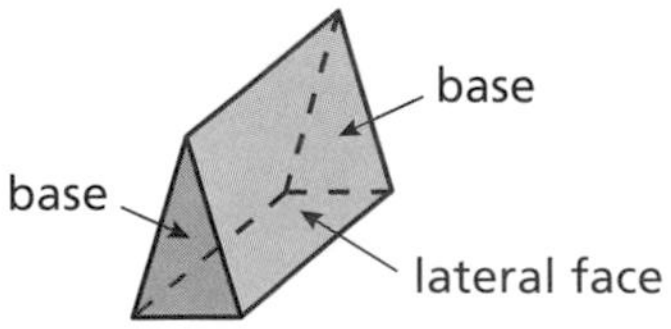

Triangular Prism

Pyramids

A **pyramid** is a polyhedron that has one base. The lateral faces are triangles.

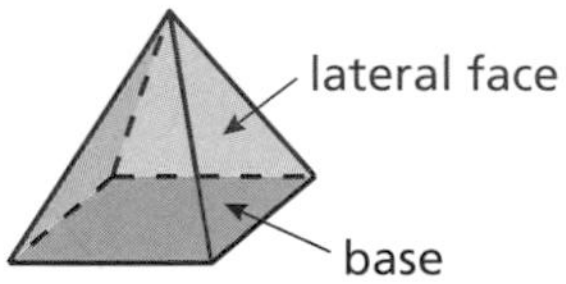

Rectangular Pyramid

The shape of the base tells the name of the prism or the pyramid.

Multi-Language Glossary at BigIdeasMath.com

EXAMPLE 2 Drawing Solids

a. Draw a rectangular prism.

Step 1:

Draw identical rectangular bases.

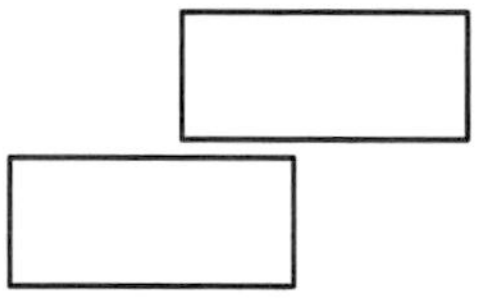

Step 2:

Connect corresponding vertices.

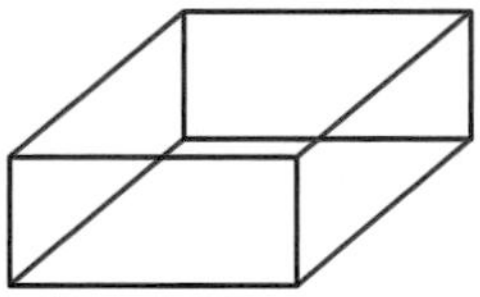

Step 3:

Change any *hidden* lines to dashed lines.

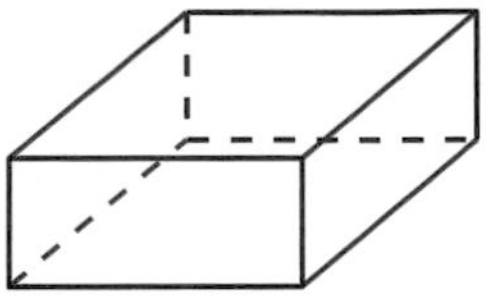

b. Draw a triangular pyramid.

Step 1:

Draw a triangular base and a point.

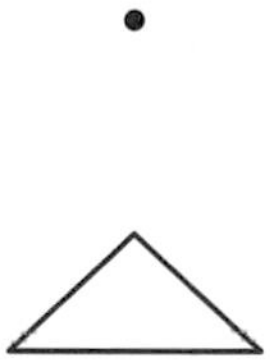

Step 2:

Connect the vertices of the triangle to the point.

Step 3:

Change any *hidden* lines to dashed lines.

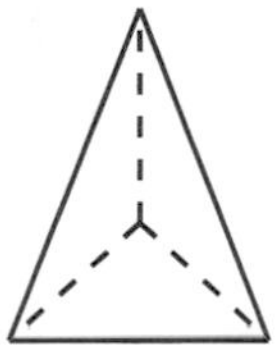

EXAMPLE 3 Drawing Views of a Solid

Draw the front, side, and top views of the eraser.

The front view is a parallelogram.

The side view is a rectangle.

The top view is a rectangle.

On Your Own

Now You're Ready
Exercises 13–22

Draw the solid.

2. square prism

3. pentagonal pyramid

Draw the front, side, and top views of the solid.

4.

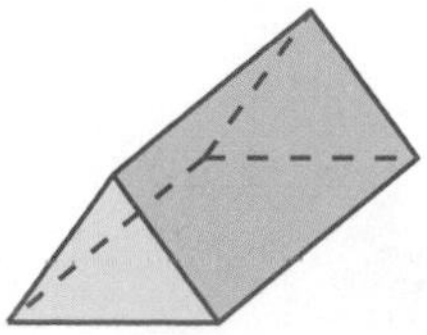

5.

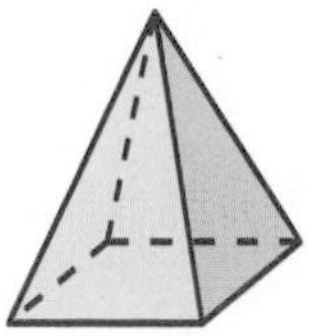

8.1 Exercises

Check It Out
Help with Homework
BigIdeasMath.com

Vocabulary and Concept Check

LOGIC Decide whether the statement is *true* or *false*. If false, explain your reasoning.

1. A triangular prism has three triangular faces.
2. A triangular prism has three rectangular faces.
3. A rectangular pyramid has one rectangular face.
4. A rectangular pyramid has three triangular faces.
5. All of the edges of a rectangular prism are parallel.
6. None of the edges of a rectangular pyramid are parallel.

Practice and Problem Solving

Draw the front, side, and top views of the stack of cubes. Then find the number of cubes in the stack.

7.

8.

9.

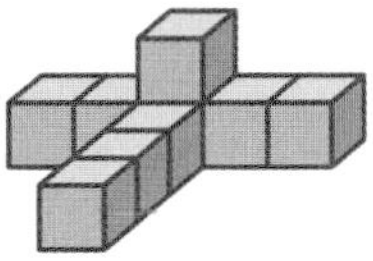

Find the number of faces, edges, and vertices of the solid.

① **10.**

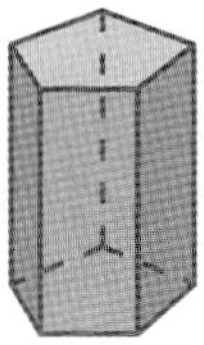

11.

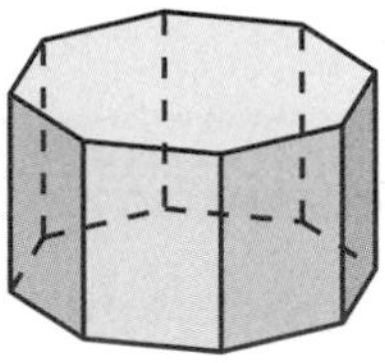

12.

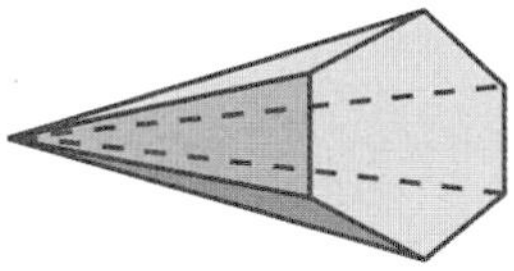

Draw the solid.

② **13.** triangular prism

14. pentagonal prism

15. rectangular pyramid

16. hexagonal pyramid

Draw the front, side, and top views of the solid.

③ **17.**

18.

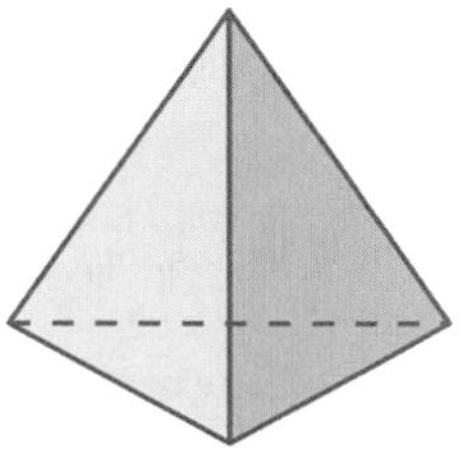

19.

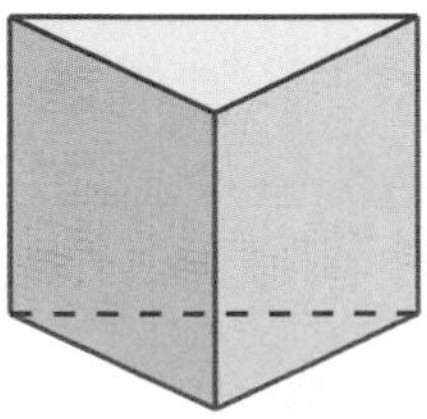

20.

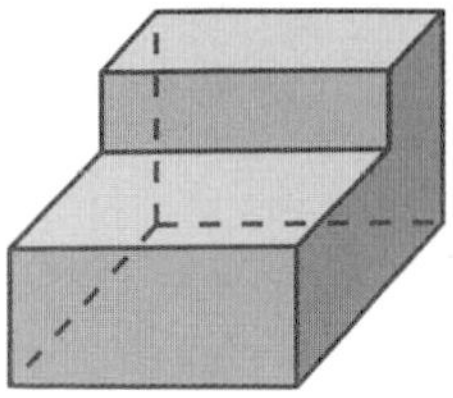

21.

22.

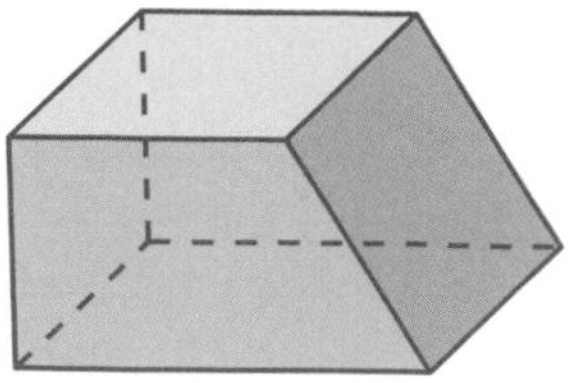

23. **PYRAMID ARENA** The Pyramid of Caius Cestius in Rome, Italy, is in the shape of a square pyramid. Draw a sketch of the pyramid.

24. **RESEARCH** Use the Internet to find a picture of the Washington Monument. Describe its shape.

Draw a solid with the following front, side, and top views.

25.

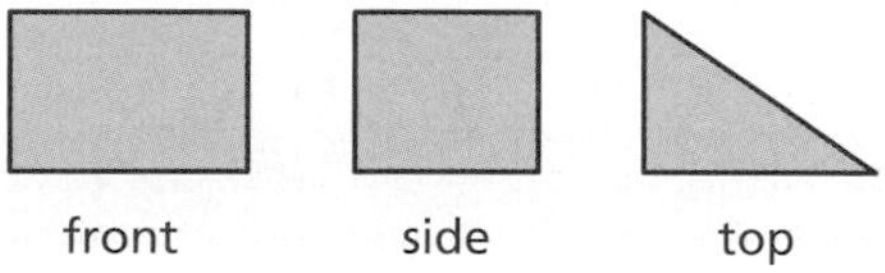

26.

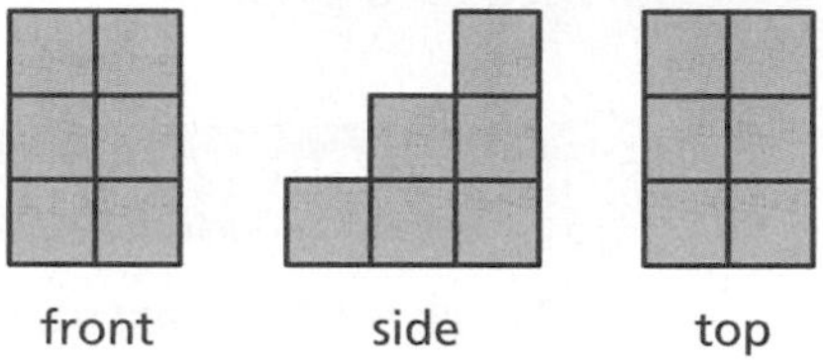

27. **PROJECT** Design and draw a house. Name the different solids that you can use to make a model of the house.

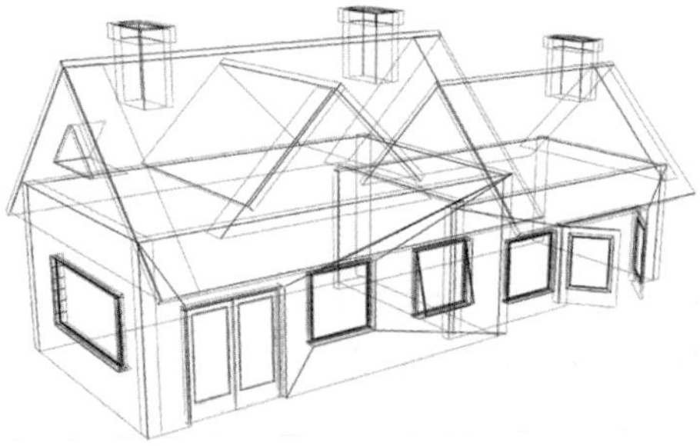

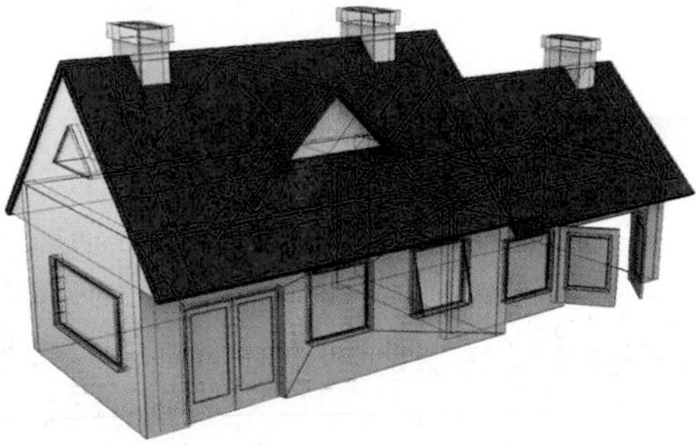

28. **REASONING** Two of the three views of a solid are shown.

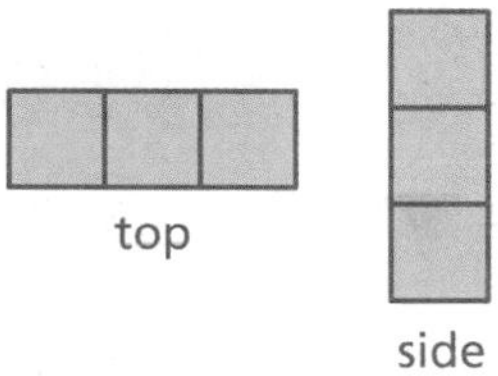

   a. What is the greatest number of unit cubes in the solid?
   b. What is the least number of unit cubes in the solid?
   c. Draw the front views of both solids in parts (a) and (b).

29. **Reasoning** Draw two different solids with five faces.

   a. Write the number of vertices and edges for each solid.
   b. Explain how knowing the numbers of edges and vertices helps you draw a three-dimensional figure.

Fair Game Review What you learned in previous grades & lessons

Find the area of the figure. *(Section 4.1, Section 4.2, and Section 4.3)*

30.

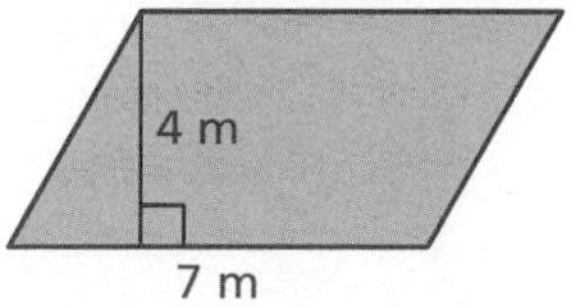

31.

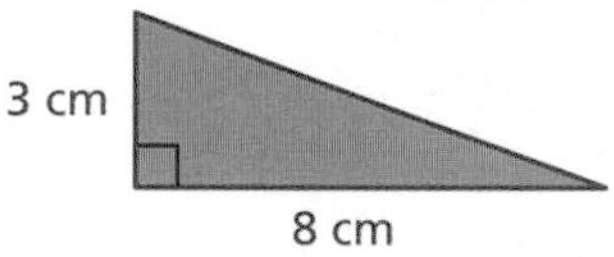

32.

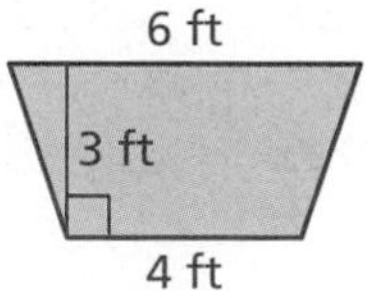

33. **MULTIPLE CHOICE** Which statement is true when $x = -2$ and $y = |-2|$? *(Section 6.4)*

   Ⓐ $x = y$ Ⓑ $y < 0$ Ⓒ $x > y$ Ⓓ $y > x$

8.2 Surface Areas of Prisms

Essential Question How can you find the area of the entire surface of a prism?

1 ACTIVITY: Identifying Prisms

Work with a partner. Label one of the faces as a "base" and the other as a "lateral face." Use the shape of the base to identify the prism.

a.

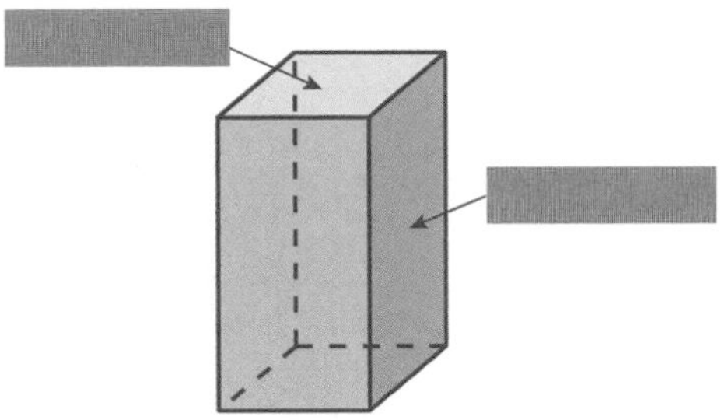

______ **Prism**

b.

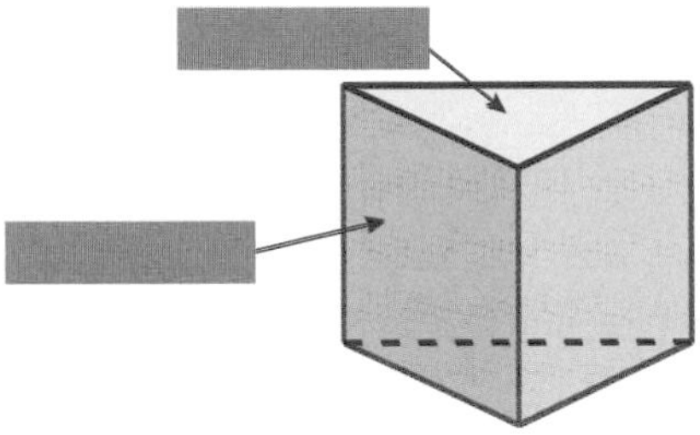

______ **Prism**

2 ACTIVITY: Using Grid Paper to Construct a Prism

Work with a partner.

a. Copy the figure shown below onto grid paper.

b. Cut out the figure and fold it to form a prism. What type of prism does it form?

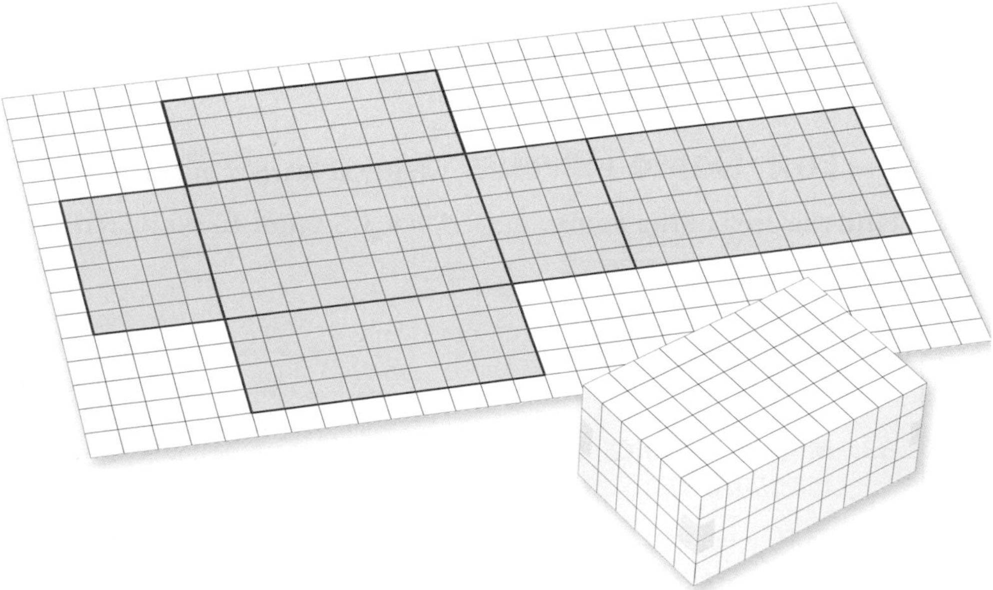

COMMON CORE

Geometry

In this lesson, you will

- use nets to represent prisms.
- find the surface area of prisms.
- solve real-life problems.

Learning Standard 6.G.4

3 ACTIVITY: Finding the Area of the Entire Surface of a Prism

Work with a partner. Label each face in the two-dimensional representation of the prism as a "base" or a "lateral face." Then find the area of the entire surface of each prism.

Math Practice 8

Repeat Calculations

When finding the areas of the faces, what calculations do you repeat?

a.

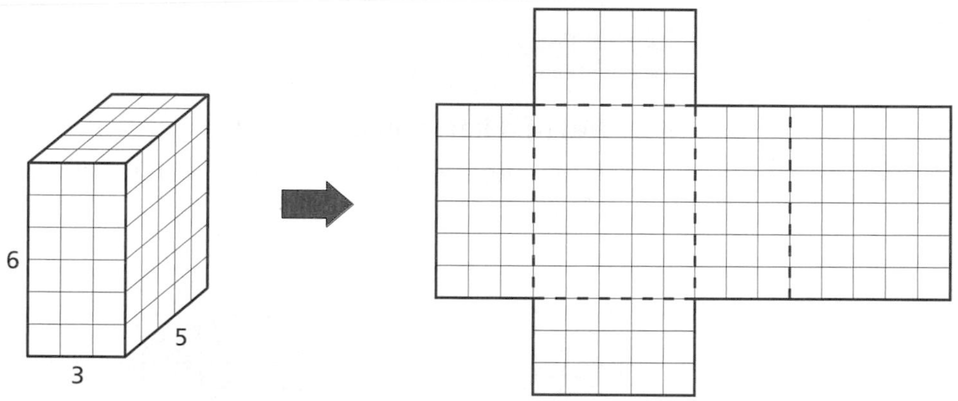

b.

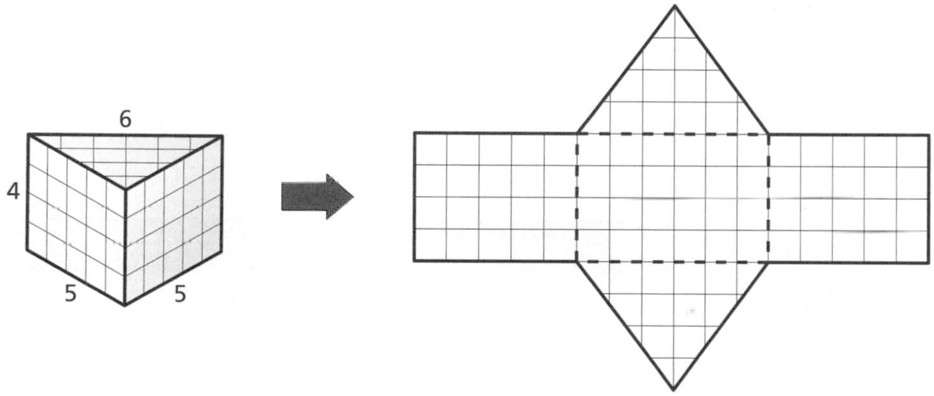

4 ACTIVITY: Drawing Two-Dimensional Representations of Prisms

Work with a partner. Draw a two-dimensional representation of each prism. Then find the area of the entire surface of each prism.

a.

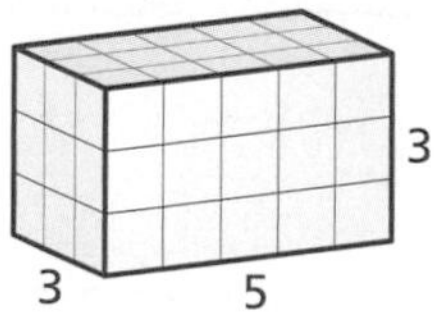

b.

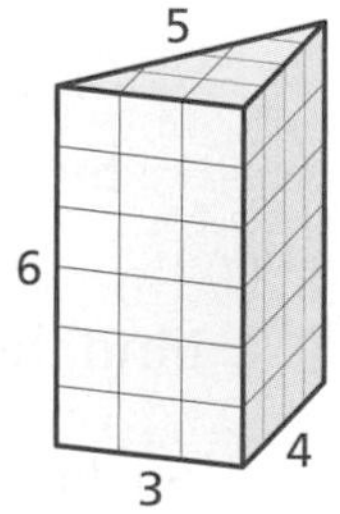

What Is Your Answer?

5. **IN YOUR OWN WORDS** How can you find the area of the entire surface of a prism?

Practice Use what you learned about the area of the entire surface of a prism to complete Exercises 3–5 on page 364.

8.2 Lesson

Check It Out
Lesson Tutorials
BigIdeasMath.com

Key Vocabulary
surface area, *p. 362*
net, *p. 362*

The **surface area** of a solid is the sum of the areas of all of its faces. You can use a two-dimensional representation of a solid, called a **net**, to find the surface area of the solid. Surface area is measured in *square units.*

Key Idea

Net of a Rectangular Prism

A *rectangular prism* is a prism with rectangular bases.

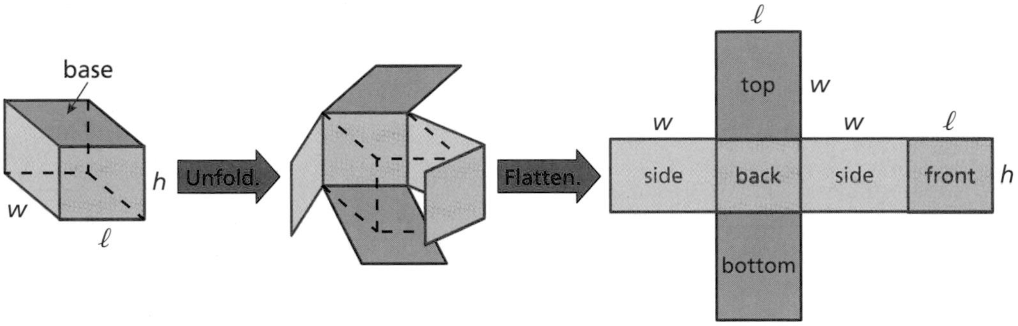

EXAMPLE 1 Finding the Surface Area of a Rectangular Prism

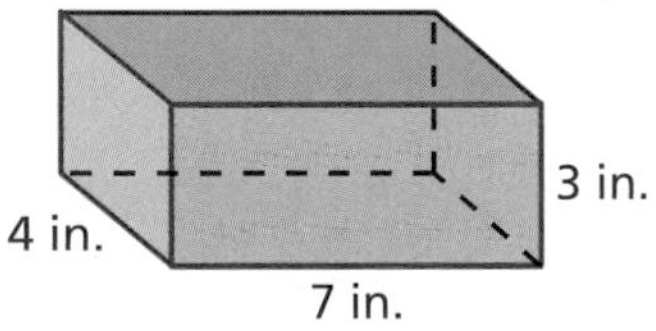

Find the surface area of the rectangular prism.

Use a net to find the area of each face.

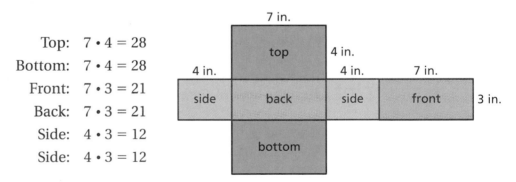

Top: $7 \cdot 4 = 28$
Bottom: $7 \cdot 4 = 28$
Front: $7 \cdot 3 = 21$
Back: $7 \cdot 3 = 21$
Side: $4 \cdot 3 = 12$
Side: $4 \cdot 3 = 12$

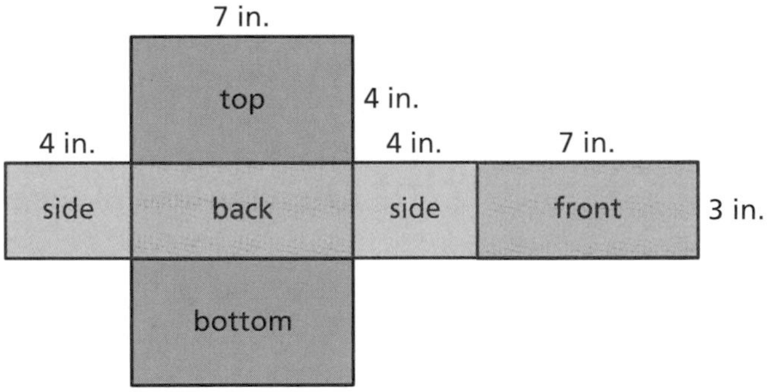

Find the sum of the areas of the faces.

$$\text{Surface Area} = \text{Area of top} + \text{Area of bottom} + \text{Area of front} + \text{Area of back} + \text{Area of a side} + \text{Area of a side}$$

$$S = 28 + 28 + 21 + 21 + 12 + 12$$

$$= 122$$

So, the surface area is 122 square inches.

Multi-Language Glossary at BigIdeasMath.com

Key Idea

Remember

The area A of a triangle with base b and height h is $A = \frac{1}{2}bh$.

Net of a Triangular Prism

A *triangular prism* is a prism with triangular bases.

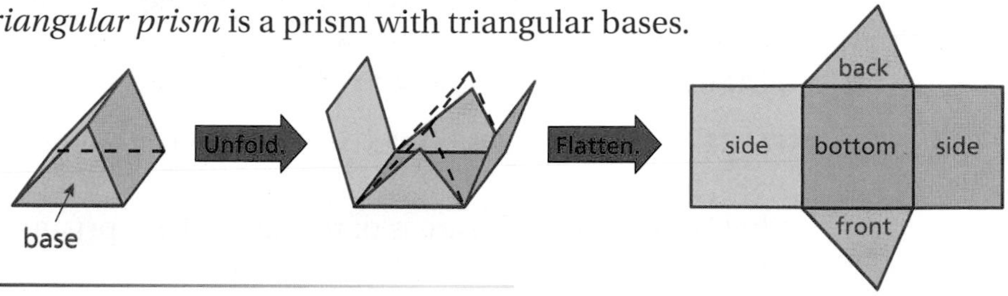

EXAMPLE 2 Finding the Surface Area of a Triangular Prism

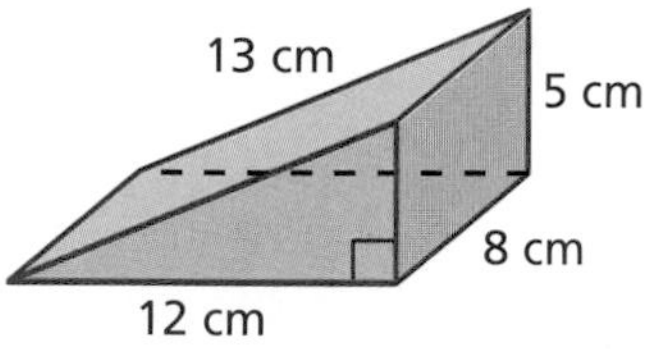

Find the surface area of the triangular prism.

Use a net to find the area of each face.

Bottom: $12 \cdot 8 = 96$

Front: $\frac{1}{2} \cdot 12 \cdot 5 = 30$

Back: $\frac{1}{2} \cdot 12 \cdot 5 = 30$

Side: $13 \cdot 8 = 104$

Side: $8 \cdot 5 = 40$

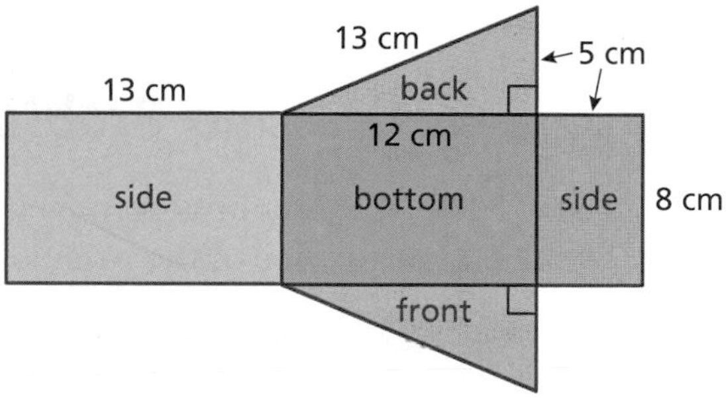

Find the sum of the areas of the faces.

Surface Area	=	Area of bottom	+	Area of front	+	Area of back	+	Area of a side	+	Area of a side	
S	=	96	+	30	+	30	+	104	+	40	= 300

So, the surface area is 300 square centimeters.

On Your Own

Now You're Ready
Exercises 6–11

Find the surface area of the rectangular prism.

1. 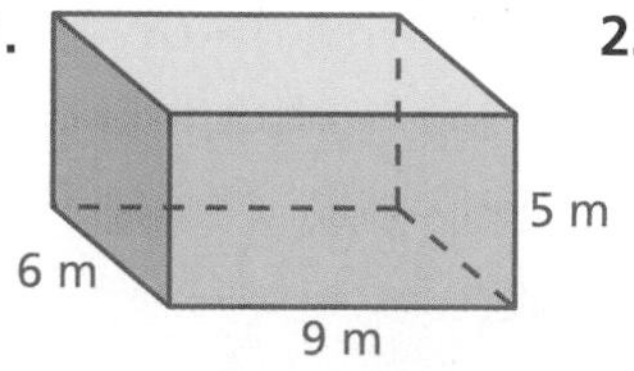

2. 

3. 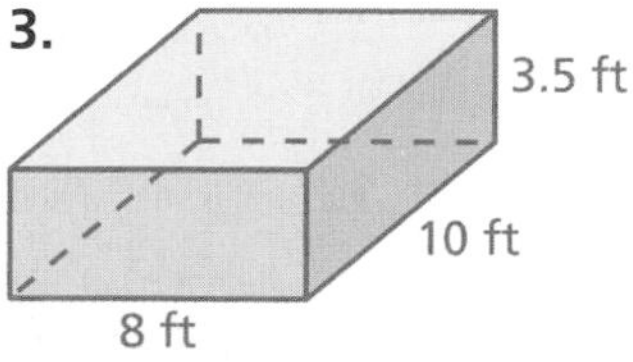

Find the surface area of the triangular prism.

4. 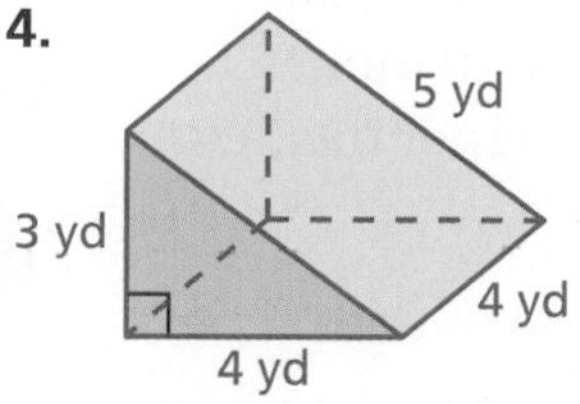

5. 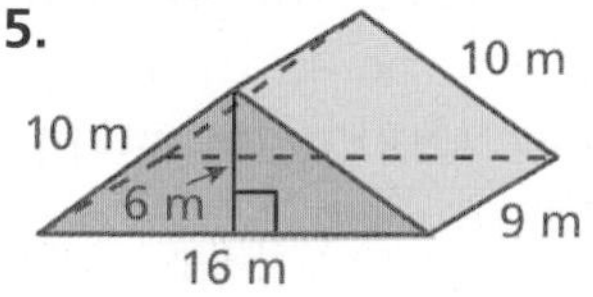

6. 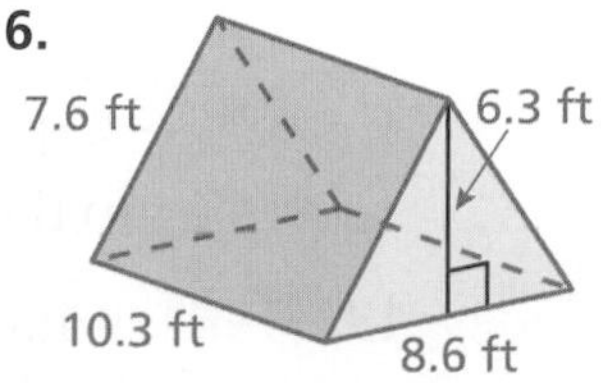

8.2 Exercises

Check It Out
Help with Homework
BigIdeasMath.com

Vocabulary and Concept Check

1. **VOCABULARY** Explain how to find the surface area of a prism.
2. **DIFFERENT WORDS, SAME QUESTION** Which is different? Find "both" answers.

What is the sum of the areas of the faces of the prism?

What is the area of the entire surface of the prism?

What is the area of the triangular faces of the prism?

What is the surface area of the prism?

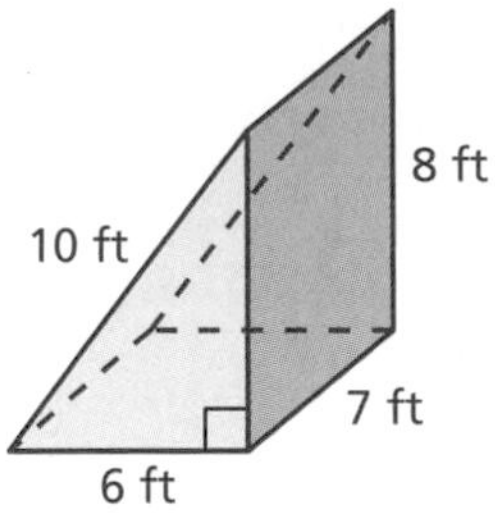

Practice and Problem Solving

Draw a two-dimensional representation of the prism. Then find the area of the entire surface of the prism.

3.
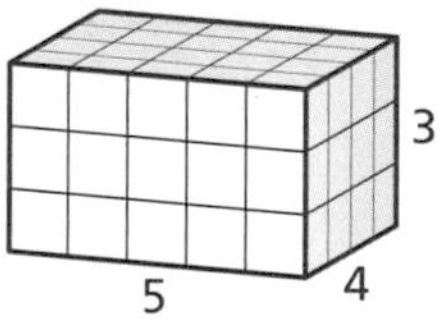

4.
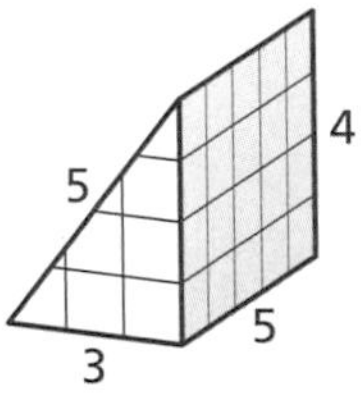

5.
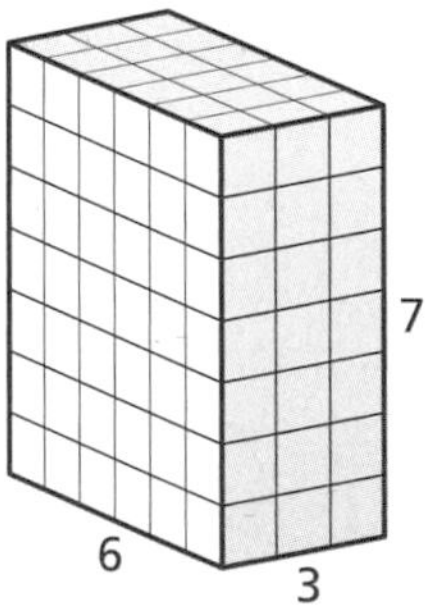

Find the surface area of the prism.

1 6.
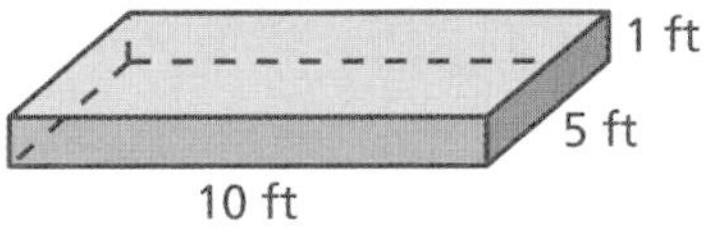

7.
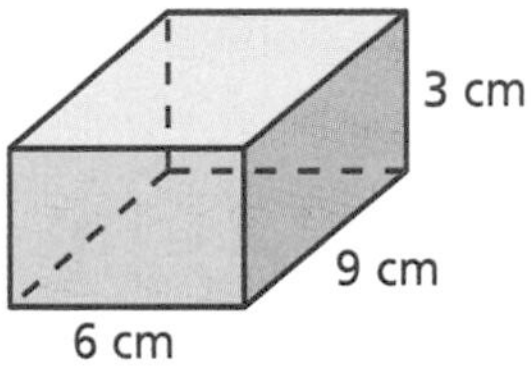

8.
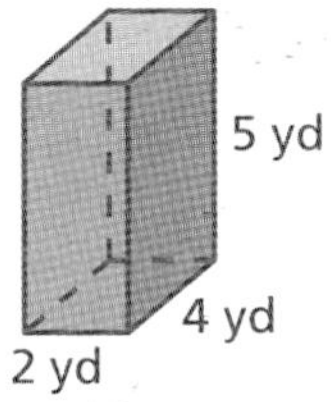

2 9.
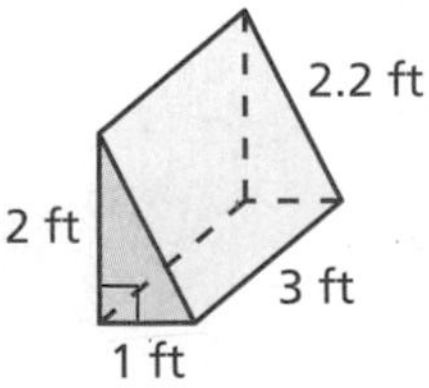

10.
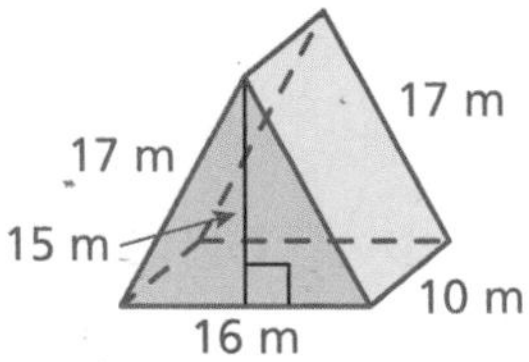

11.
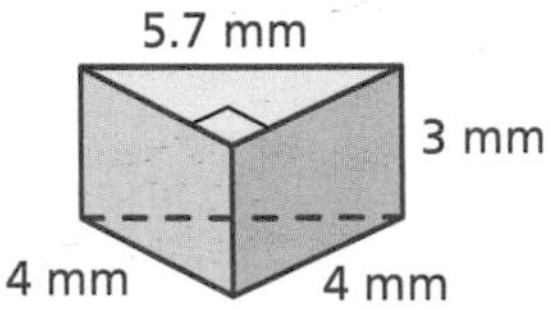

12. **GIFT BOX** A gift box in the shape of a rectangular prism measures 8 inches by 8 inches by 10 inches. What is the least amount of wrapping paper needed to wrap the gift box? Explain.
13. **TENT** What is the least amount of fabric needed to make the tent?

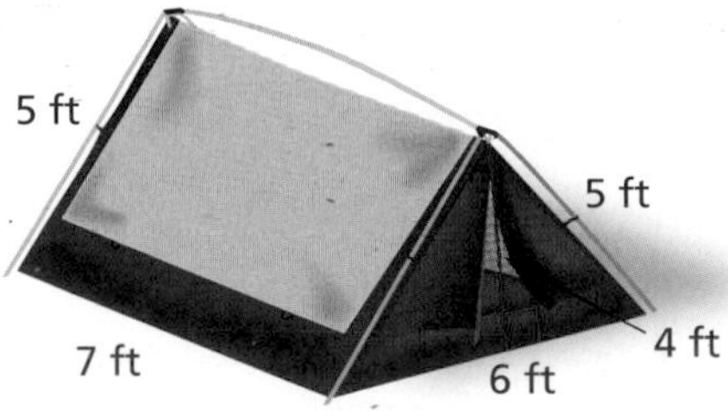

14. AQUARIUM A public library has an aquarium in the shape of a rectangular prism. The base is 6 feet by 2.5 feet. The height is 4 feet. How many square feet of glass were used to build the aquarium? (The top of the aquarium is open.)

15. STORAGE BOX The material used to make a storage box costs $1.25 per square foot. The boxes have the same volume. How much does a company save by choosing to make 50 of Box 2 instead of 50 of Box 1?

	Length	Width	Height
Box 1	20 in.	6 in.	4 in.
Box 2	15 in.	4 in.	8 in.

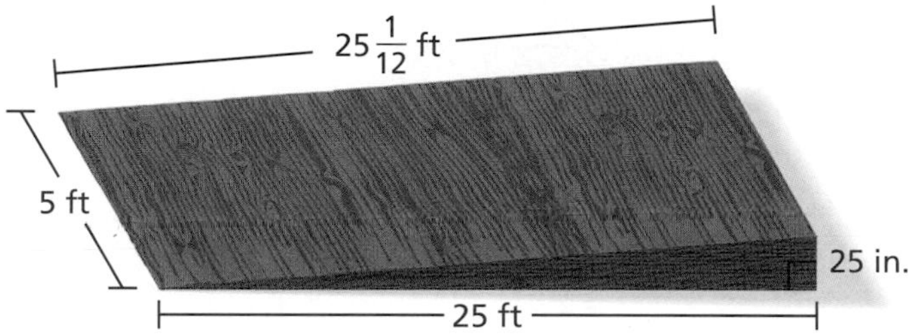

16. RAMP A quart of stain covers 100 square feet. How many quarts should you buy to stain the wheelchair ramp? (Assume you do not have to stain the bottom of the ramp.)

17. Critical Thinking A cube is removed from a rectangular prism. Find the surface area of the figure after removing the cube.

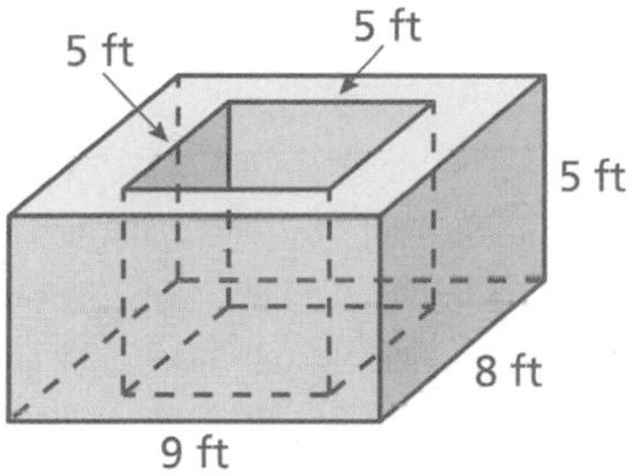

Fair Game Review What you learned in previous grades & lessons

Find the area of the triangle. *(Section 4.2)*

18.

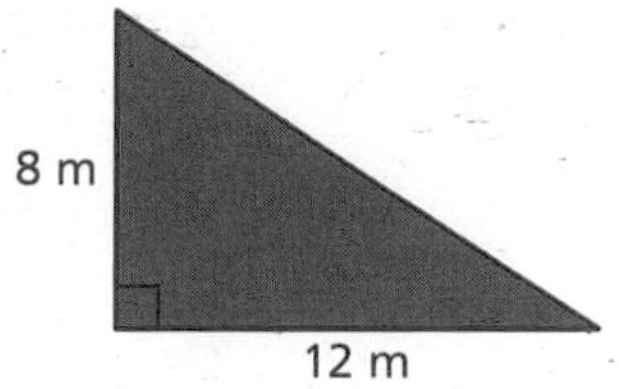

19.

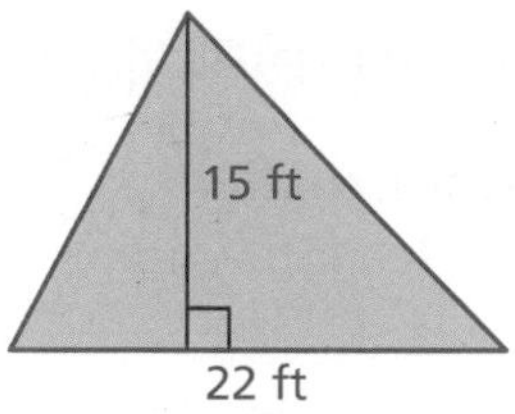

20.

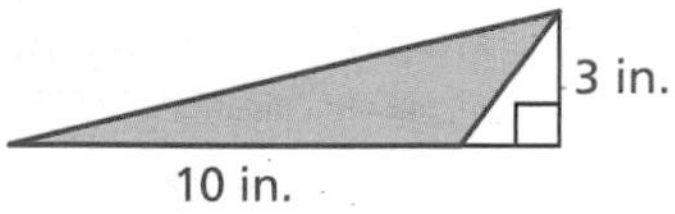

21. MULTIPLE CHOICE Which value is *not* a solution of the inequality $x - 4 \geq 2$? *(Section 7.5)*

Ⓐ $x = 10$ Ⓑ $x = 6$ Ⓒ $x = 4$ Ⓓ $x = 14$

8 Study Help

Check It Out
Graphic Organizer
BigIdeasMath.com

You can use a **process diagram** to show the steps involved in a procedure. Here is an example of a process diagram for drawing a prism.

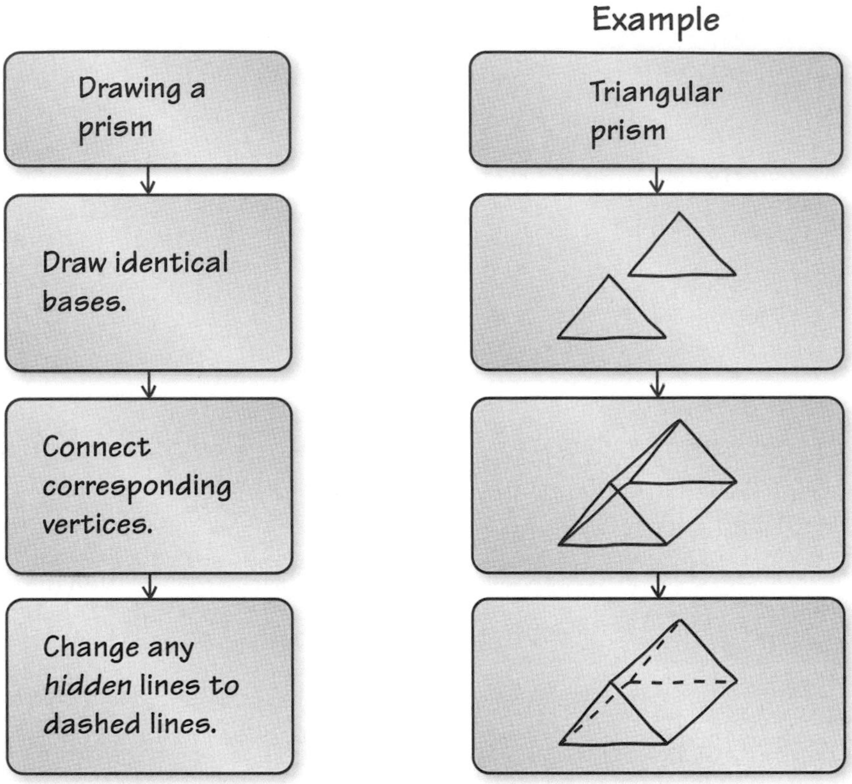

On Your Own

Make process diagrams with examples to help you study these topics.

1. drawing a pyramid
2. finding the surface area of a prism

After you complete this chapter, make process diagrams with examples for the following topics.

3. finding the surface area of a pyramid
4. finding the volume of a rectangular prism

"Descartes, you should use my process diagram when you eat your treats."

8.1–8.2 Quiz

Find the number of faces, edges, and vertices of the solid. *(Section 8.1)*

1. 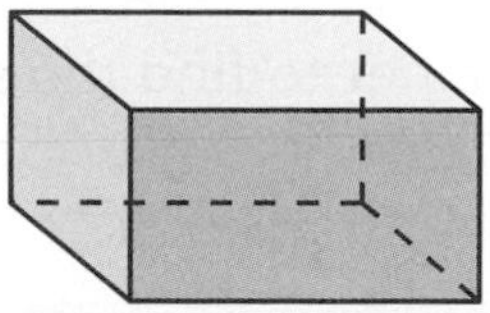

2. 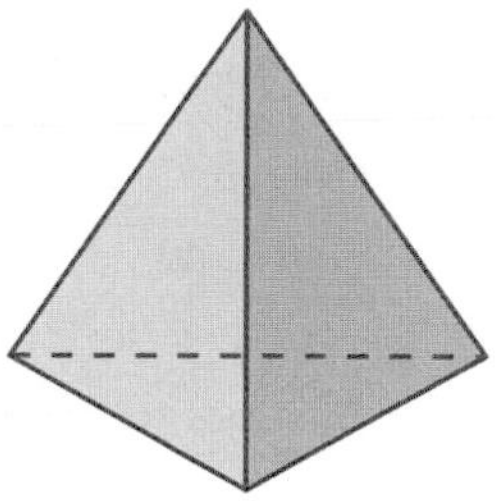

Draw the solid. *(Section 8.1)*

3. trapezoidal prism

4. octagonal pyramid

Draw the front, side, and top views of the solid. *(Section 8.1)*

5. 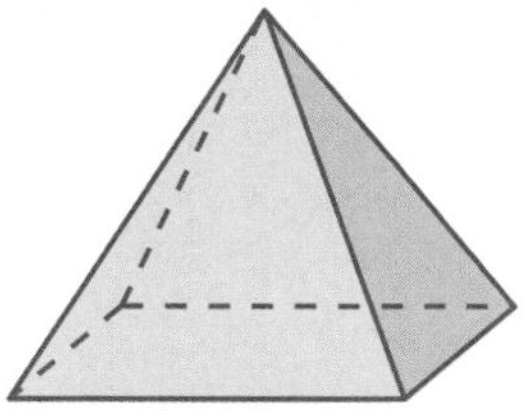

6. 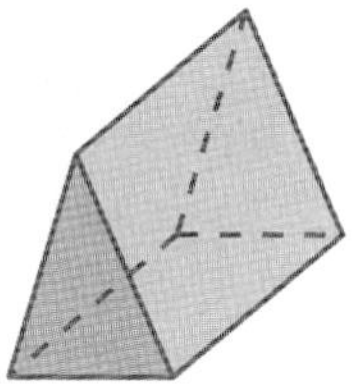

Find the surface area of the prism. *(Section 8.2)*

7. 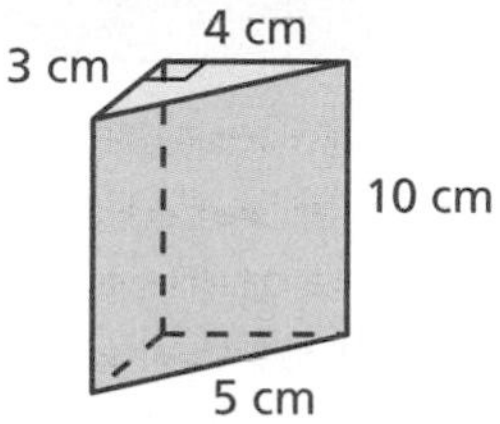

8. 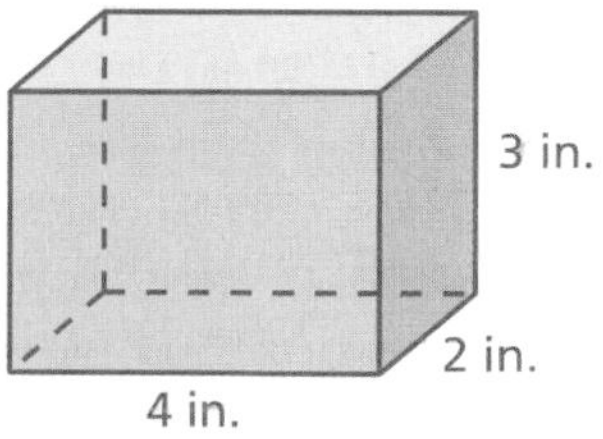

9. CEREAL A cereal box has the dimensions shown. *(Section 8.2)*

a. Find the surface area of the cereal box.

b. The manufacturer decides to decrease the size of the box by reducing each of the dimensions by 1 inch. Find the decrease in surface area.

10. GIFT BOX Find the surface area of the gift box. *(Section 8.2)*

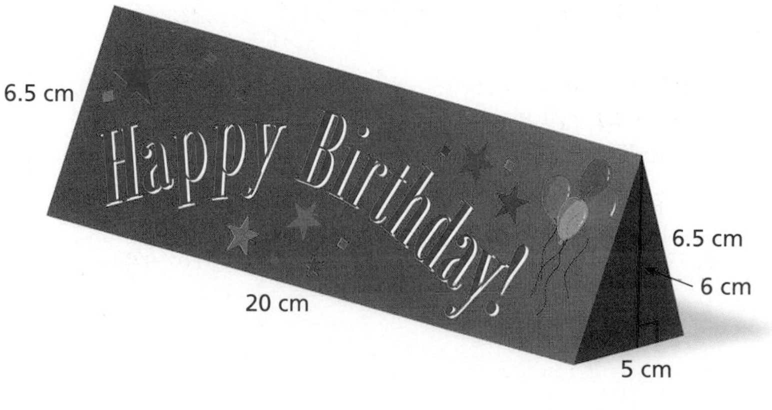

8.3 Surface Areas of Pyramids

Essential Question How can you use a net to find the surface area of a pyramid?

1 ACTIVITY: Identifying Pyramids

Work with a partner. Label one of the faces as a "base" and the other as a "lateral face." Use the shape of the base to identify the pyramid.

a.

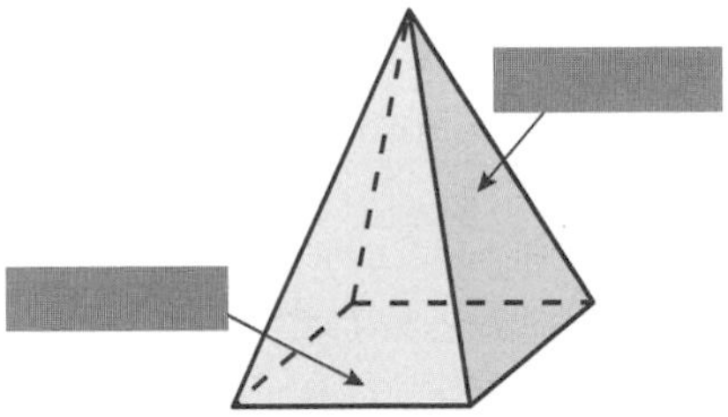

Pyramid

b.

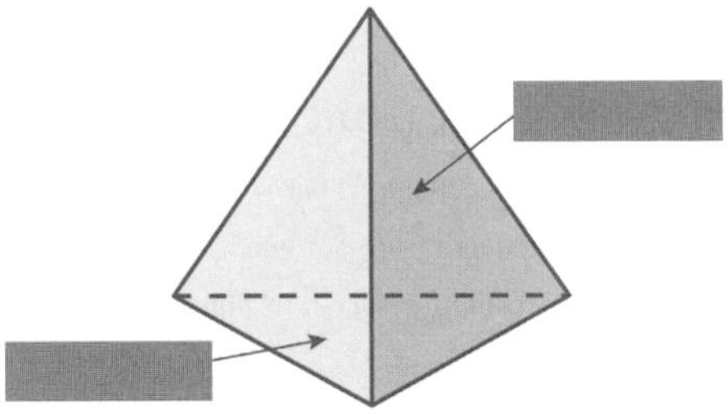

Pyramid

2 ACTIVITY: Using a Net

Work with a partner.

a. Copy the net shown below onto grid paper.

b. Cut out the net and fold it to form a pyramid. What type of rectangle is the base? Use this shape to name the pyramid.

c. Find the surface area of the pyramid.

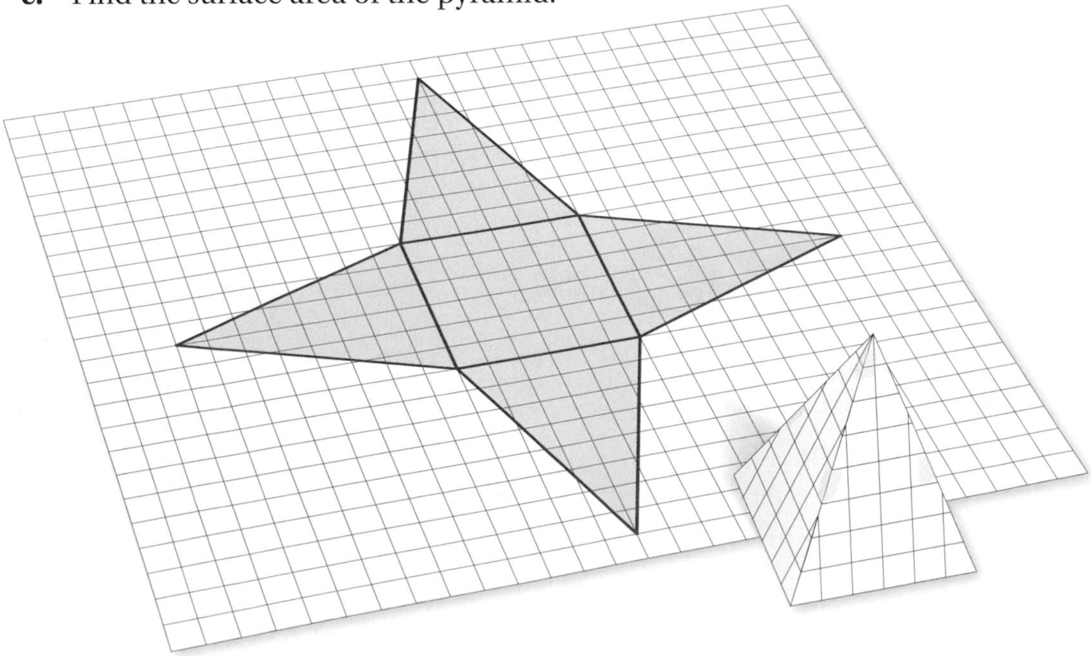

COMMON CORE

Geometry

In this lesson, you will

- use nets to represent pyramids.
- find the surface area of pyramids.
- solve real-life problems.

Learning Standard
6.G.4

3 ACTIVITY: Estimating the Surface Area of a Triangular Pyramid

Work with a partner. Label each face in the net of the triangular pyramid as a "base" or a "lateral face." Then estimate the surface area of the pyramid.

Math Practice 1

Analyze Givens

What information is given in the diagram? How does this help you estimate the surface area of the pyramid?

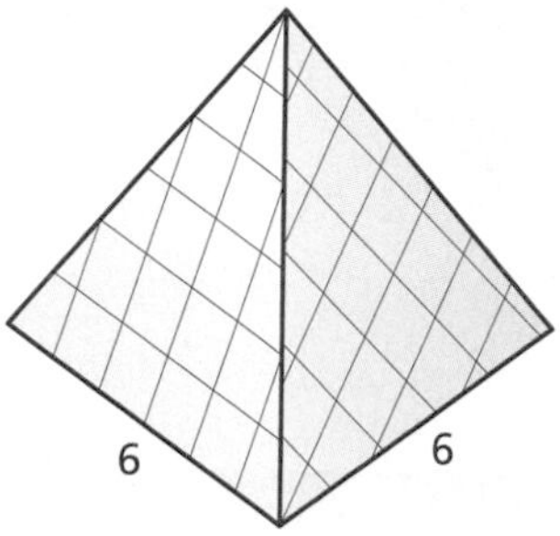

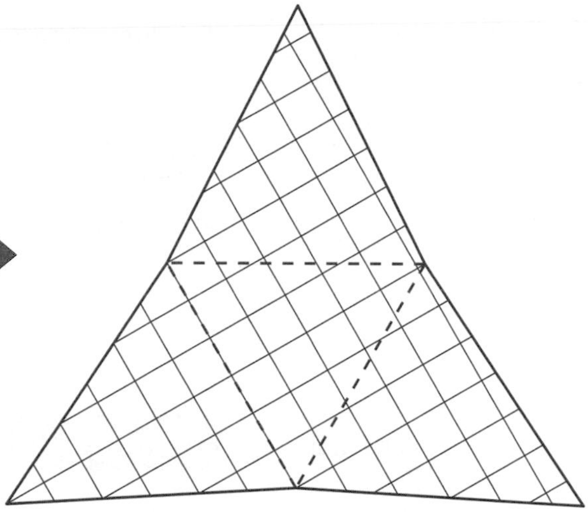

4 ACTIVITY: Finding the Surface Area of a Square Pyramid

Work with a partner. Draw a net for each square pyramid. Use the net to find the surface area of the pyramid.

a.

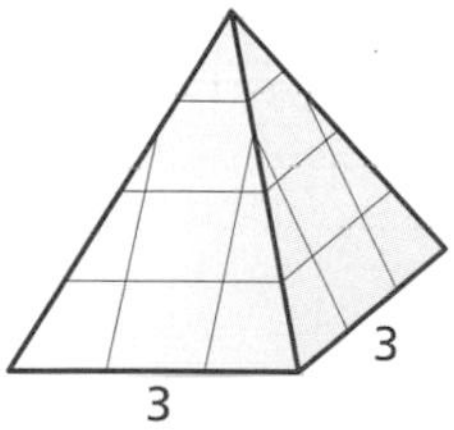

b.

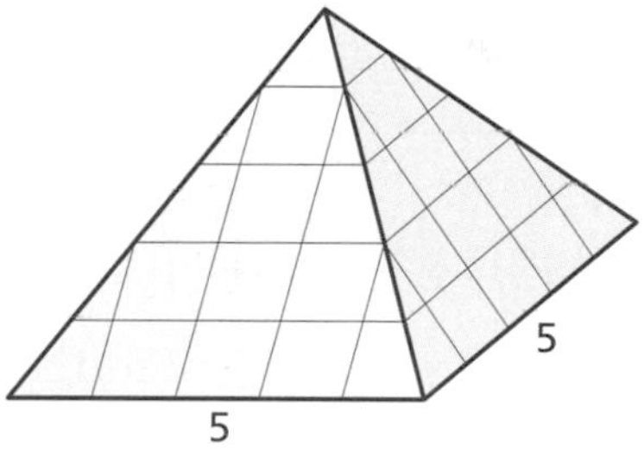

What Is Your Answer?

5. **IN YOUR OWN WORDS** How can you use a net to find the surface area of a pyramid?
6. **CONJECTURE** Make a conjecture about the lateral faces of a pyramid when the side lengths of the base have the same measure. Explain.

Practice Use what you learned about the surface area of a pyramid to complete Exercises 3–5 on page 372.

8.3 Lesson

Key Idea

Net of a Square Pyramid

A *square pyramid* is a pyramid with a square base.

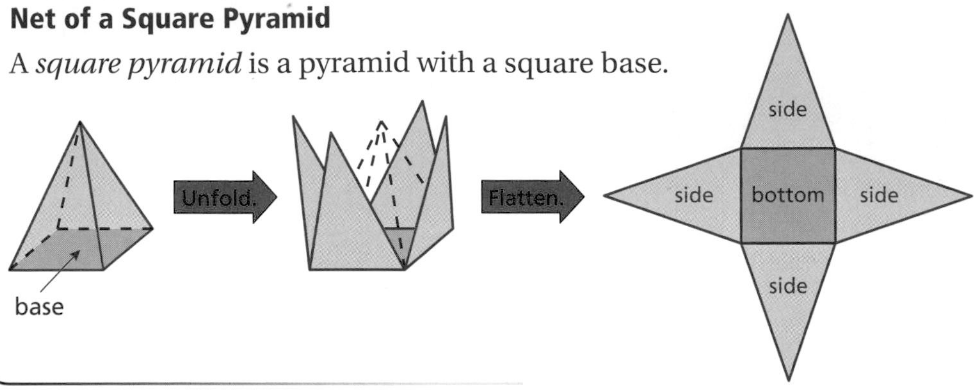

EXAMPLE 1 Finding the Surface Area of a Square Pyramid

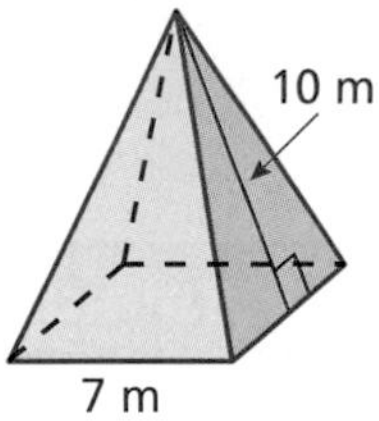

Find the surface area of the square pyramid.

Use a net to find the area of each face.

Bottom: $7 \cdot 7 = 49$

Side: $\frac{1}{2} \cdot 7 \cdot 10 = 35$

Side: $\frac{1}{2} \cdot 7 \cdot 10 = 35$

Side: $\frac{1}{2} \cdot 7 \cdot 10 = 35$

Side: $\frac{1}{2} \cdot 7 \cdot 10 = 35$

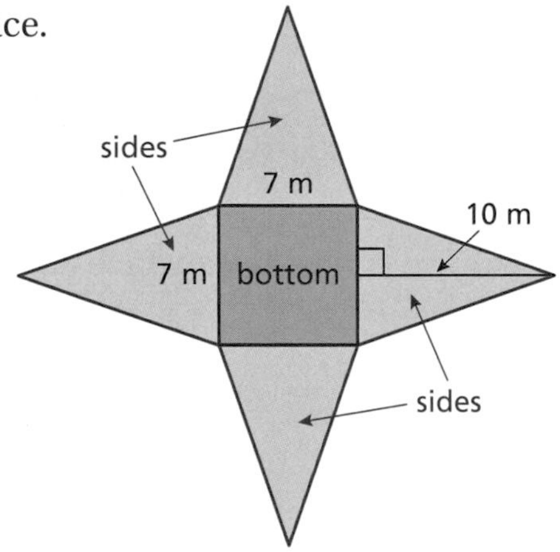

Find the sum of the areas of the faces.

Surface Area	=	Area of bottom	+	Area of a side	+	Area of a side	+	Area of a side	+	Area of a side	
S	=	49	+	35	+	35	+	35	+	35	= 189

∴ So, the surface area is 189 square meters.

On Your Own

Now You're Ready
Exercises 6–8

Find the surface area of the square pyramid.

1. 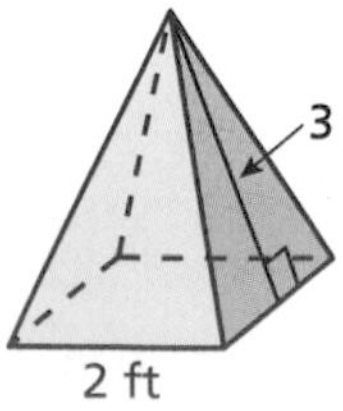

2. 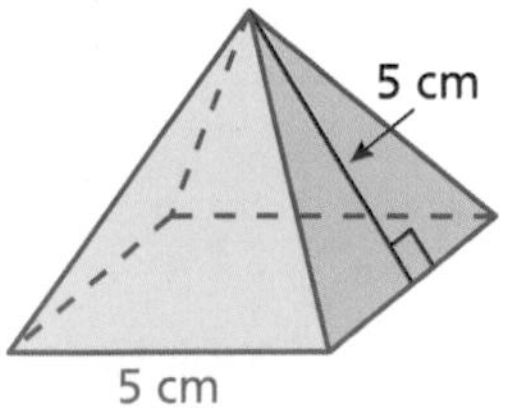

3. 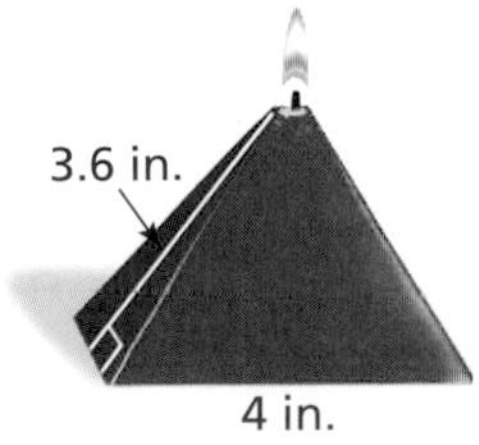

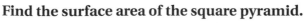

Key Idea

Net of a Triangular Pyramid

A *triangular pyramid* is a pyramid with a triangular base.

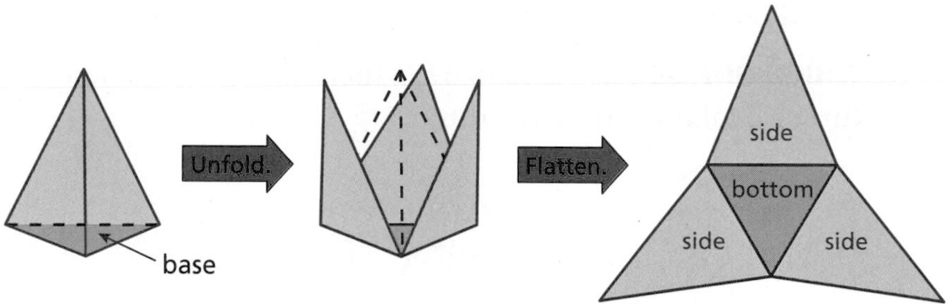

Study Tip

The base of each triangular pyramid in this section is an equilateral triangle.

EXAMPLE 2 Finding the Surface Area of a Triangular Pyramid

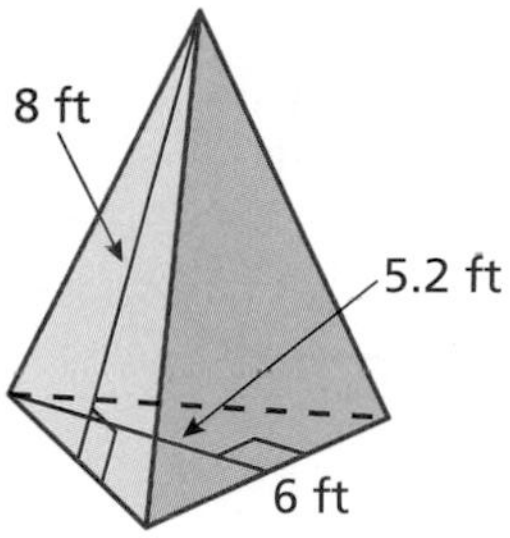

Find the surface area of the triangular pyramid.

Use a net to find the area of each face.

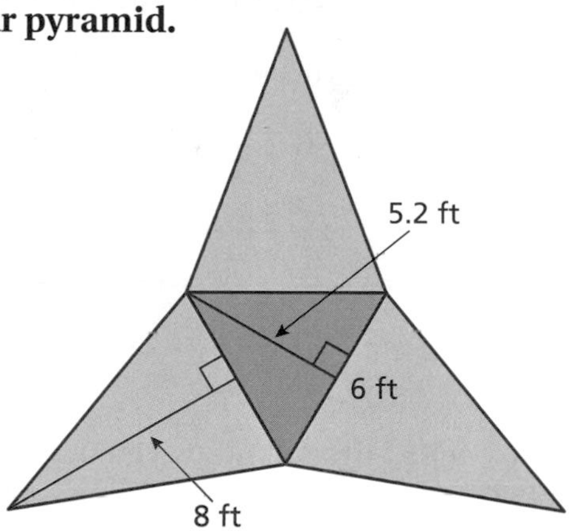

Bottom: $\frac{1}{2} \cdot 6 \cdot 5.2 = 15.6$

Side: $\frac{1}{2} \cdot 6 \cdot 8 = 24$

Side: $\frac{1}{2} \cdot 6 \cdot 8 = 24$

Side: $\frac{1}{2} \cdot 6 \cdot 8 = 24$

Find the sum of the areas of the faces.

Surface Area = Area of bottom + Area of a side + Area of a side + Area of a side

$S = 15.6 + 24 + 24 + 24$

$= 87.6$

So, the surface area is 87.6 square feet.

On Your Own

Now You're Ready Exercises 9–11

Find the surface area of the triangular pyramid.

4. 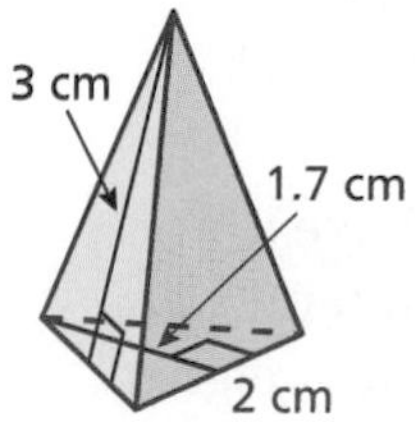

5. 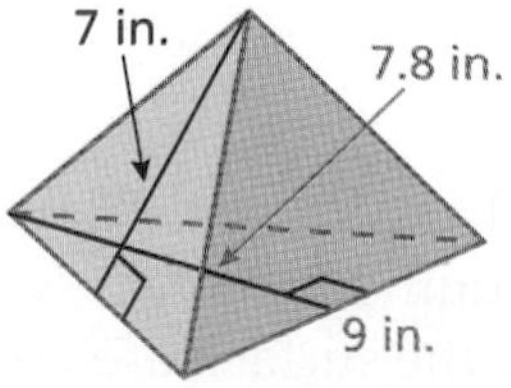

6. 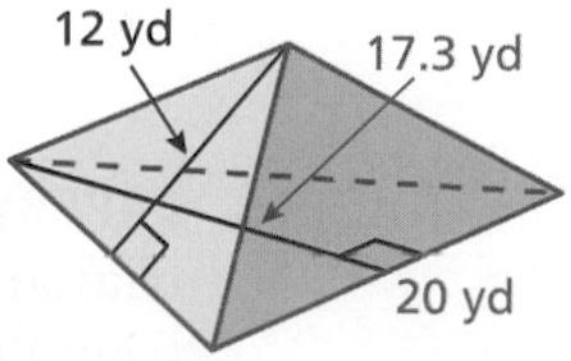

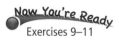

8.3 Exercises

Check It Out
Help with Homework
BigIdeasMath.com

Vocabulary and Concept Check

1. **PRECISION** Explain how to find the surface area of a pyramid.

2. **WHICH ONE DOESN'T BELONG?** Which figure does *not* belong with the other three? Explain your reasoning.

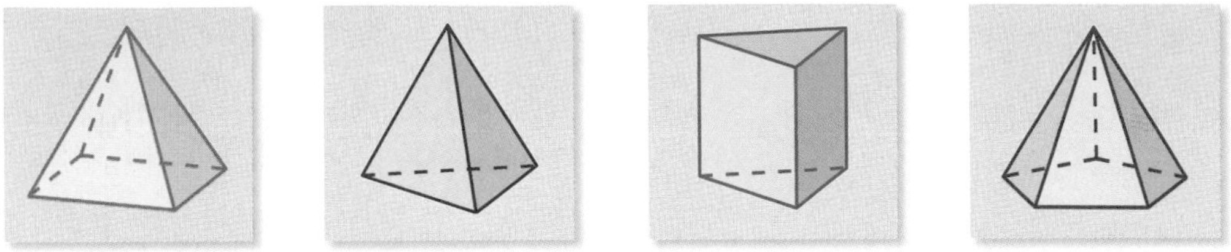

Practice and Problem Solving

Draw a net of the square pyramid. Then find the surface area of the pyramid.

3.

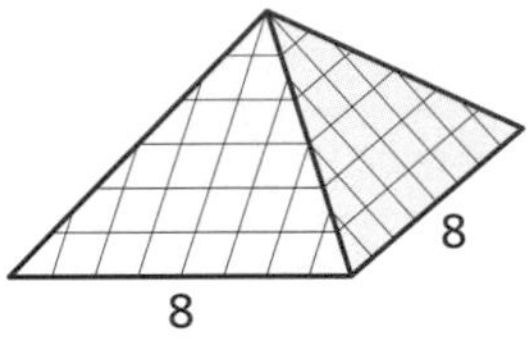

4.

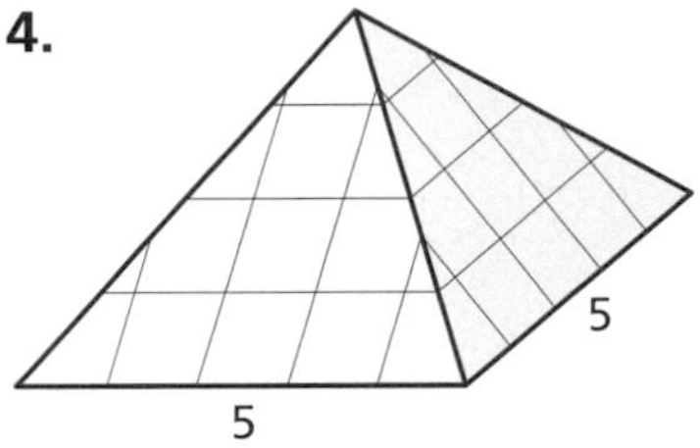

5.

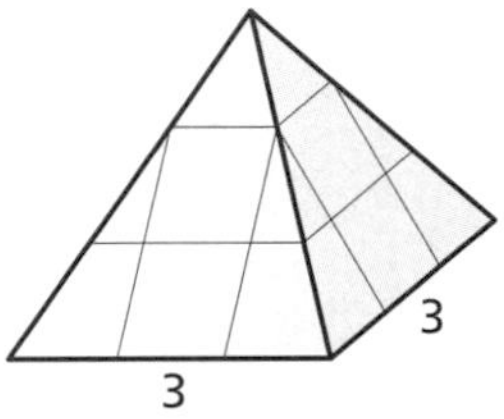

Find the surface area of the pyramid. The side lengths of the base are equal.

1 6.

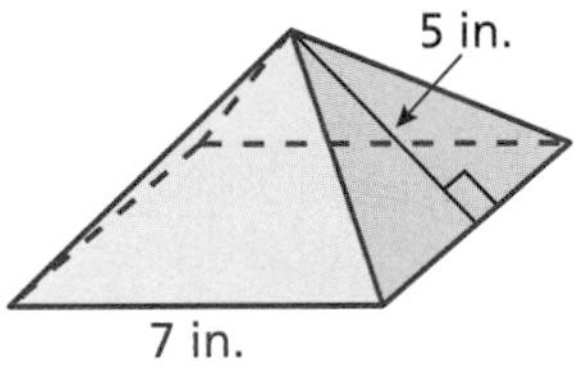

7.

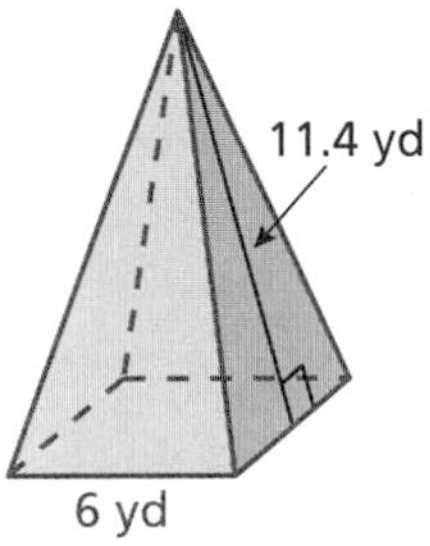

8.

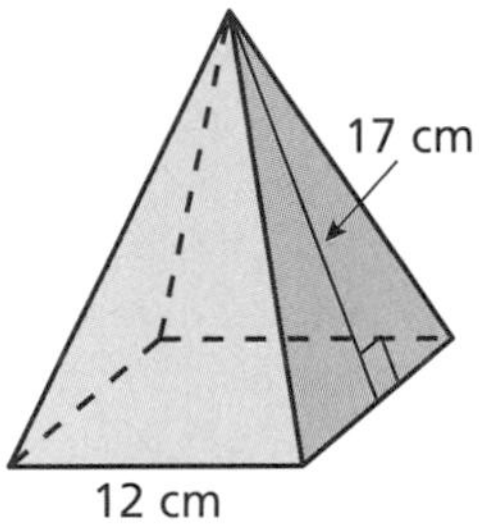

2 9.

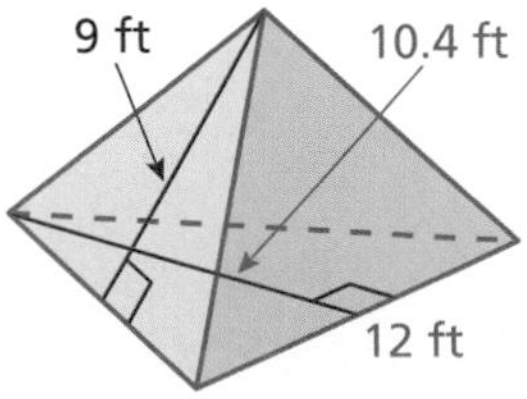

10.

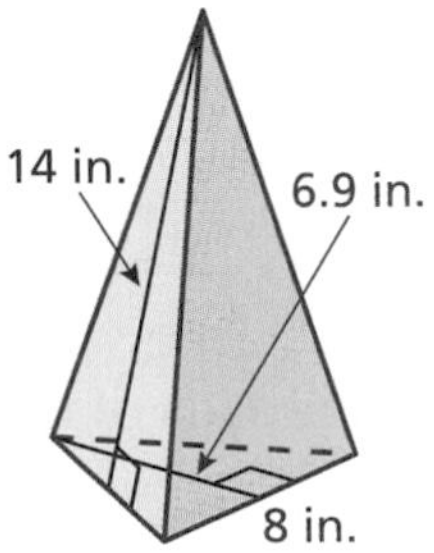

11.

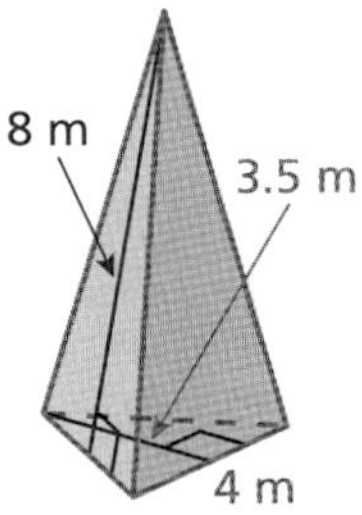

12. **PAPERWEIGHT** A paperweight is shaped like a triangular pyramid. The base is an equilateral triangle. Find the surface area of the paperweight.

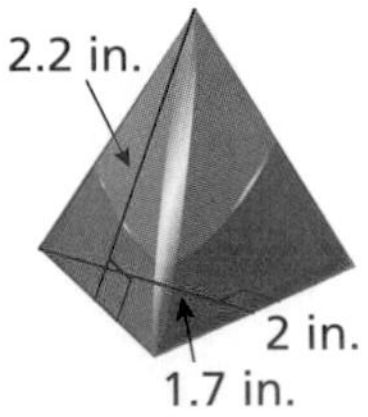

13. **LOUVRE** The entrance to the Louvre Museum in Paris, France, is a square pyramid. The side length of the base is 116 feet, and the height of one of the triangular faces is 91.7 feet. Find the surface area of the four triangular faces of the entrance to the Louvre Museum.

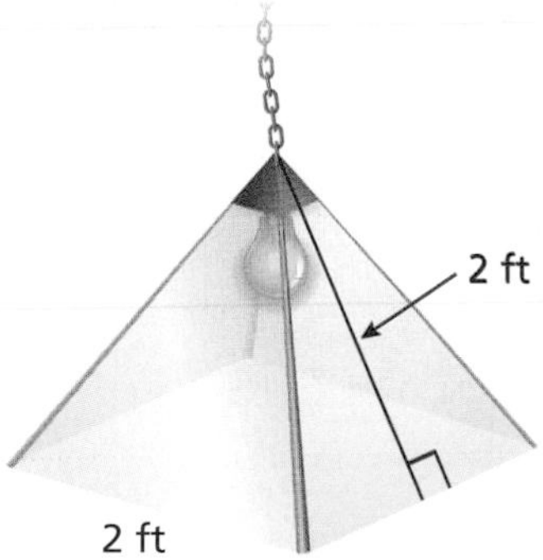

14. **LIGHT COVER** A hanging light cover made of glass is shaped like a square pyramid. The cover does not have a bottom. One square foot of the glass weighs 2.45 pounds. The chain can support 35 pounds. Will the chain support the light cover? Explain.

15. **GEOMETRY** The surface area of a square pyramid is 84 square inches. The side length of the base is 6 inches. What is the value of x?

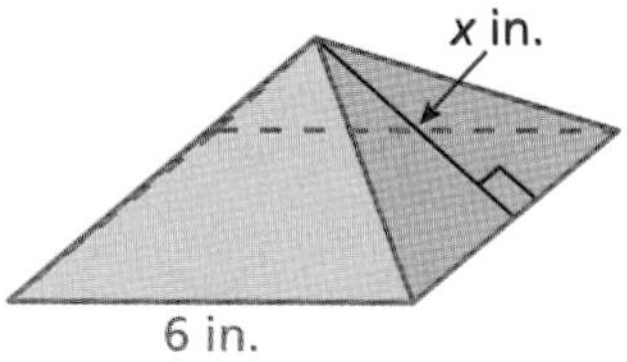

16. **STRUCTURE** In the diagram of the base of the hexagonal pyramid, all the triangles are the same. Find the surface area of the hexagonal pyramid.

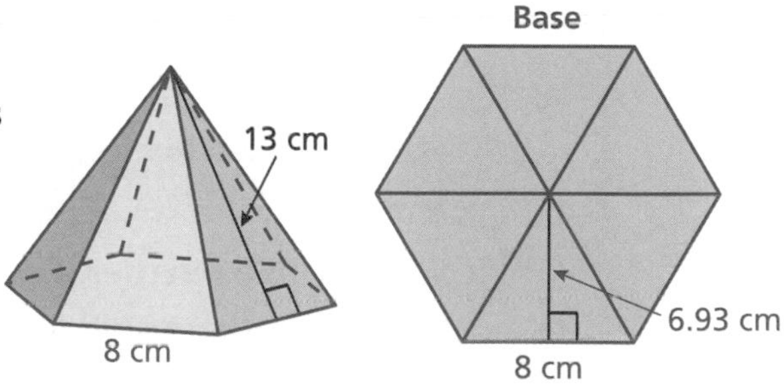

17. **Critical Thinking** Can you form a square pyramid using four of the triangles shown? Explain your reasoning.

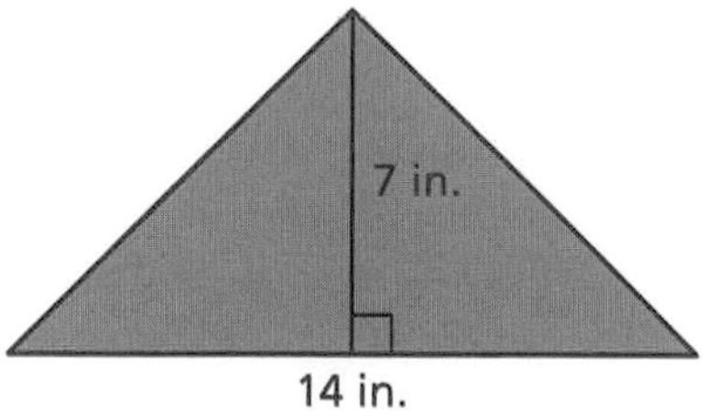

Fair Game Review What you learned in previous grades & lessons

Find the missing values in the ratio table. Then write the equivalent ratios. *(Section 5.2)*

18.

Frogs	7		28
Turtles	3	6	

19.

Apples	10	5	
Oranges	4		12

20. **MULTIPLE CHOICE** Which ordered pair is in Quadrant III? *(Section 6.5)*

Ⓐ $(5, -1)$ Ⓑ $(-2, -3)$ Ⓒ $(2, 4)$ Ⓓ $(-7, 1)$

8.4 Volumes of Rectangular Prisms

Essential Question How can you find the volume of a rectangular prism with fractional edge lengths?

Recall that the **volume** of a three-dimensional figure is a measure of the amount of space that it occupies. Volume is measured in *cubic units*.

A *unit cube* is a cube with an edge length of 1 unit.

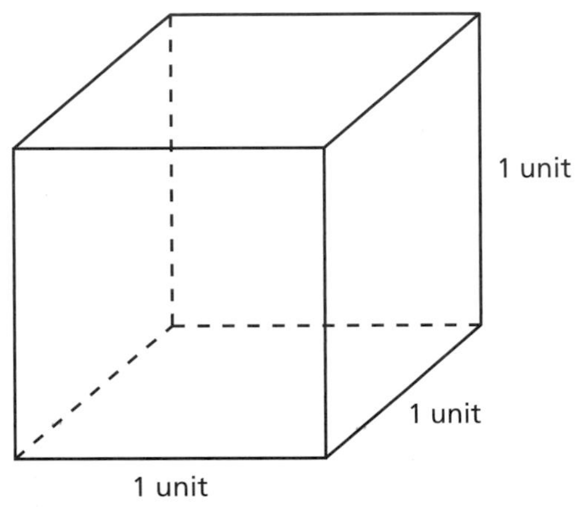

1 ACTIVITY: Using a Unit Cube

Work with a partner. The parallel edges of the unit cube have been divided into 2, 3, and 4 equal parts to create smaller rectangular prisms that are identical.

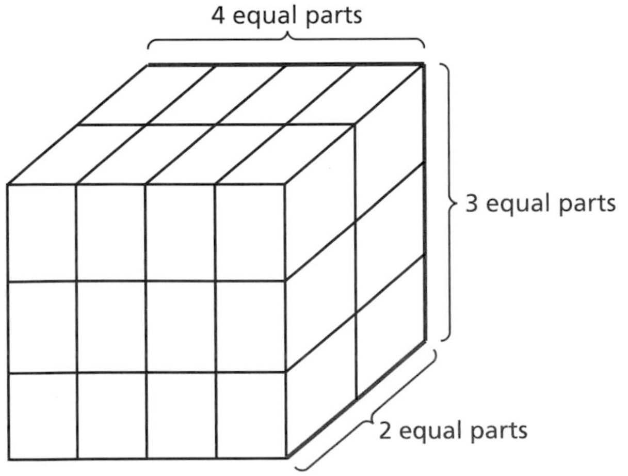

a. Draw one of these identical prisms and label its dimensions.

b. What fraction of the volume of the unit cube does one of these identical prisms represent? Use this value to find the volume of one of the identical prisms. Explain your reasoning.

COMMON CORE

Geometry

In this lesson, you will

- find the volume of prisms with fractional edge lengths by using models.
- find the volume of prisms by using formulas.

Learning Standard 6.G.2

2 ACTIVITY: Finding the Volume of a Rectangular Prism

Work with a partner.

a. How many of the identical prisms in Activity 1(a) does it take to fill the rectangular prism below? Support your answer with a drawing.

> **Math Practice 4**
> **Analyze Relationships**
> What is the relationship between the solid shown here and the solid in the previous activity?

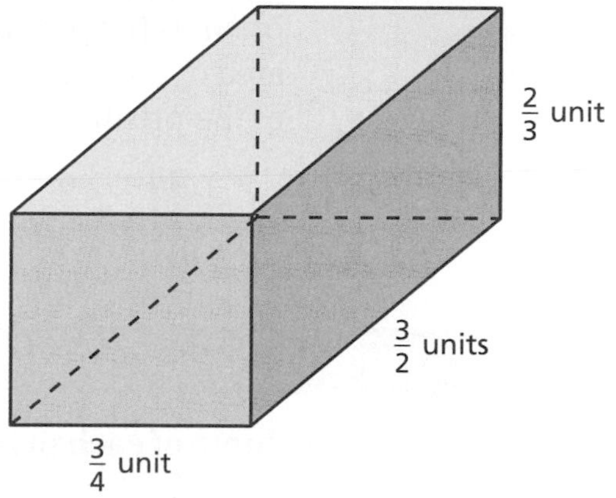

b. Use the volume of one of the identical prisms in Activity 1(a) to find the volume of the rectangular prism above. Explain your reasoning.

3 ACTIVITY: Finding the Volumes of Rectangular Prisms

Work with a partner. Explain how you can use the procedure in Activities 1 and 2 to find the volume of each rectangular prism. Then find the volume of each prism.

a.

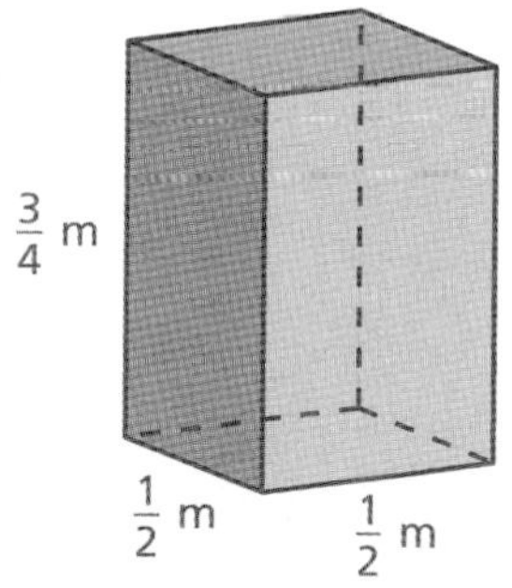

b.

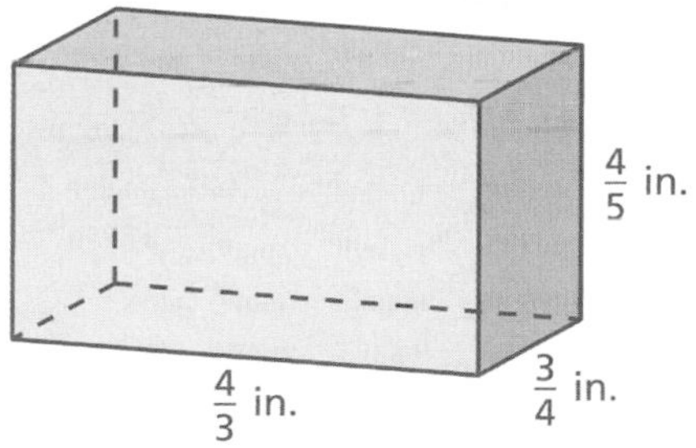

What Is Your Answer?

4. You have used the formulas $V = Bh$ and $V = \ell wh$ to find the volume V of a rectangular prism with whole number edge lengths. Do you think the formulas work for rectangular prisms with fractional edge lengths? Give examples with your answer.

5. **IN YOUR OWN WORDS** How can you find the volume of a rectangular prism with fractional edge lengths?

Practice Use what you learned about the volume of a rectangular prism to complete Exercises 4–6 on page 378.

8.4 Lesson

Check It Out
Lesson Tutorials
BigIdeasMath.com

Key Vocabulary
volume, *p. 374*

Key Idea

Volume of a Rectangular Prism

Words The volume V of a rectangular prism is the product of the area of the base and the height of the prism.

Algebra $V = Bh$ or $V = \ell wh$

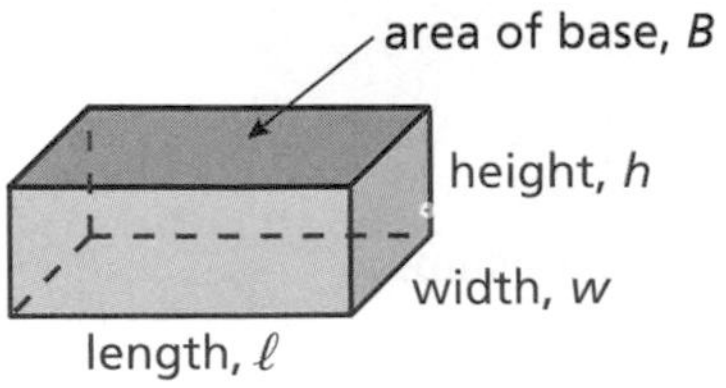

EXAMPLE 1 Finding Volumes of Rectangular Prisms

Find the volume of each prism.

a.

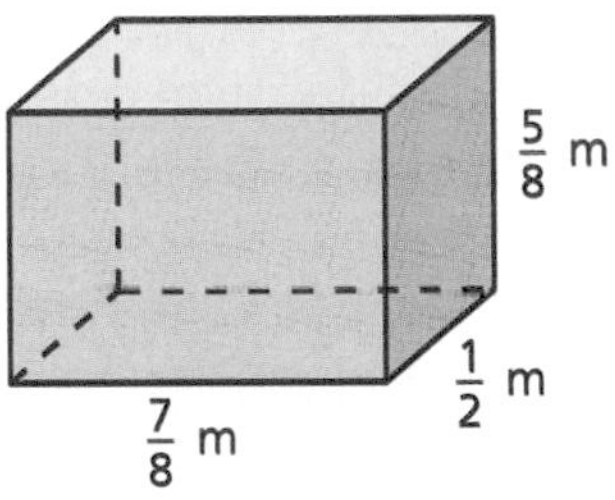

b.

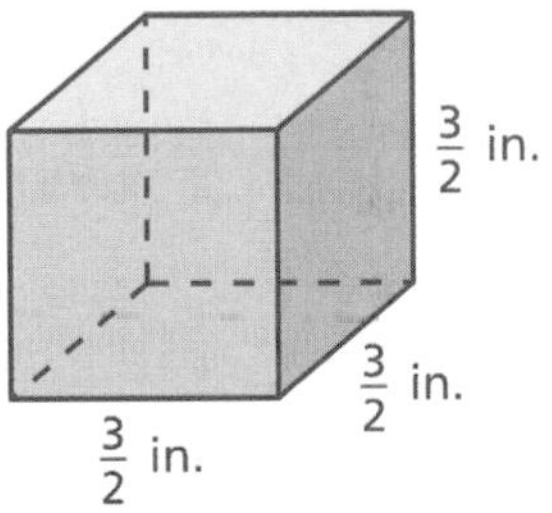

Study Tip
In Example 1(b), the rectangular prism is a cube. You can use the formula $V = s^3$ to find the volume V of a cube with an edge length of s.

a.

$V = \ell wh$ Write formula.

$= \frac{7}{8}\left(\frac{1}{2}\right)\left(\frac{5}{8}\right)$ Substitute values.

$= \frac{35}{128}$ Multiply.

So, the volume is $\frac{35}{128}$ cubic meter.

b.

$V = \ell wh$

$= \frac{3}{2}\left(\frac{3}{2}\right)\left(\frac{3}{2}\right)$

$= \frac{27}{8}$

$= 3\frac{3}{8}$

So, the volume is $3\frac{3}{8}$ cubic inches.

On Your Own

Now You're Ready
Exercises 4–9

Find the volume of the prism.

1.

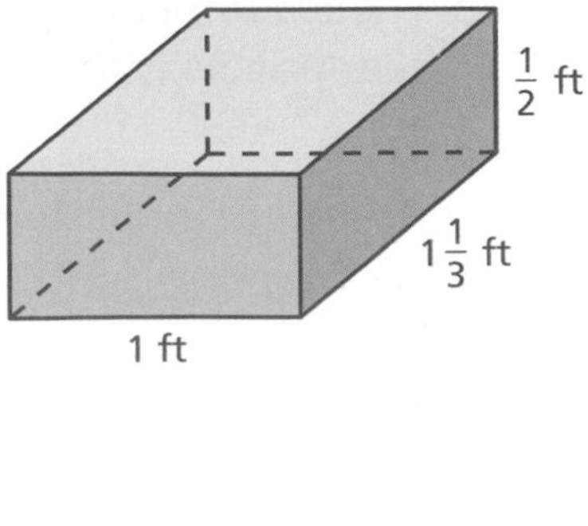

2.

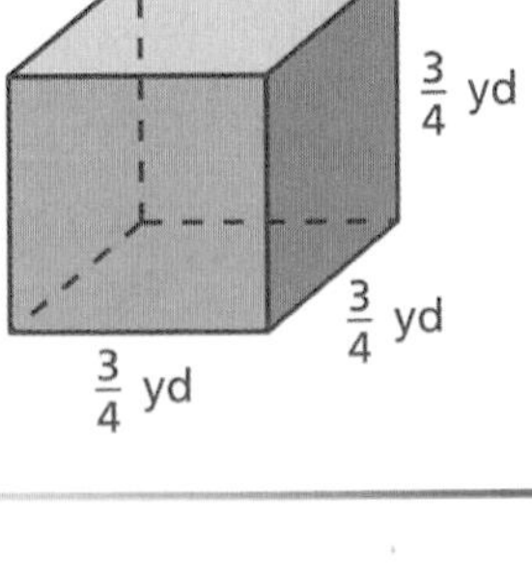

EXAMPLE 2 Using the Volume of a Rectangular Prism

One cubic foot of dirt weighs about 70 pounds. How many pounds of dirt can the dump truck haul when it is full?

Find the volume of dirt that the dump truck can haul when it is full.

$V = \ell wh$	Write formula for volume.
$= 17(8)\left(4\frac{3}{4}\right)$	Substitute values.
$= 646$	Multiply.

So, the dump truck can haul 646 cubic feet of dirt when it is full. To find the weight of the dirt, multiply by $\frac{70 \text{ lb}}{1 \text{ ft}^3}$.

$$646 \cancel{\text{ft}^3} \times \frac{70 \text{ lb}}{1 \cancel{\text{ft}^3}} = 45{,}220 \text{ lb}$$

∴ The dump truck can haul about 45,220 pounds of dirt when it is full.

EXAMPLE 3 Finding a Missing Dimension of a Rectangular Prism

Write and solve an equation to find the height of the computer tower.

Volume = 1792 in.^3

$V = \ell wh$	Write formula for volume.
$1792 = 16(7)h$	Substitute values.
$1792 = 112h$	Simplify.
$\frac{1792}{112} = \frac{112h}{112}$	Division Property of Equality
$16 = h$	Simplify.

∴ So, the height of the computer tower is 16 inches.

On Your Own

3. **WHAT IF?** In Example 2, the length of the dump truck is 20 feet. How many pounds of dirt can the dump truck haul when it is full?

Now You're Ready
Exercises 10–12

Write and solve an equation to find the missing dimension of the prism.

4. Volume = 72 in.^3

5. Volume = 1375 cm^3

8.4 Exercises

Check It Out
Help with Homework
BigIdeasMath.com

Vocabulary and Concept Check

1. **CRITICAL THINKING** Explain how volume and surface area are different.
2. **REASONING** Will the formulas for volume work for rectangular prisms with decimal edge lengths? Explain.
3. **DIFFERENT WORDS, SAME QUESTION** Which is different? Find "both" answers.

How much does it take to fill the rectangular prism?

What is the capacity of the rectangular prism?

How much does it take to cover the rectangular prism?

How much does the rectangular prism contain?

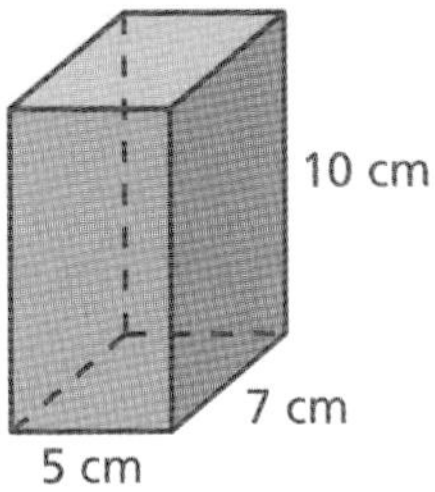

Practice and Problem Solving

Find the volume of the prism.

(1) 4.

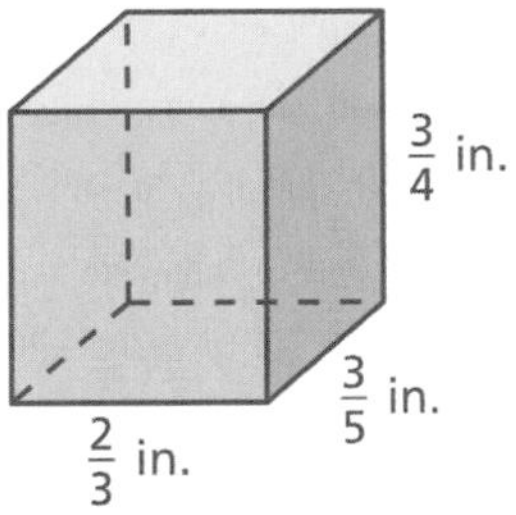

5.

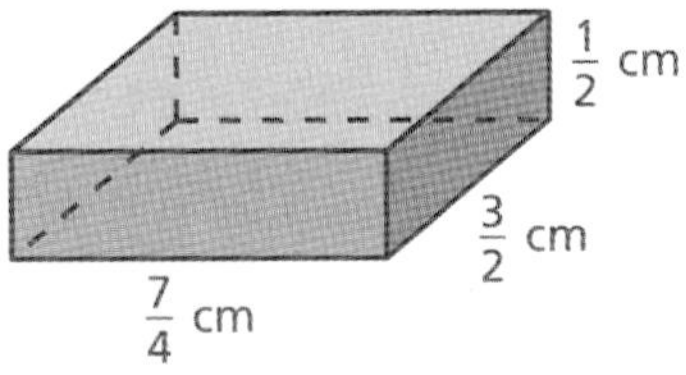

6.

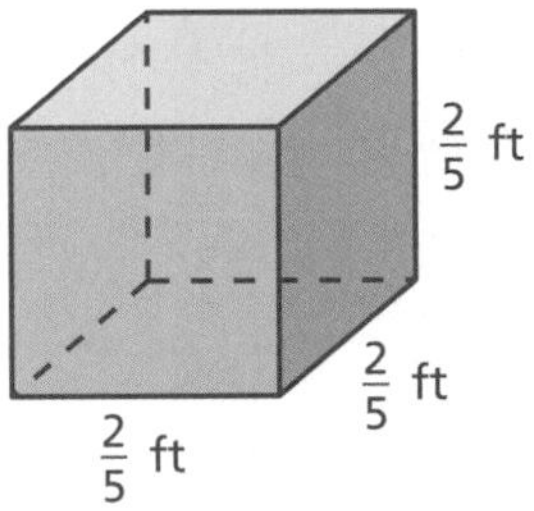

7.

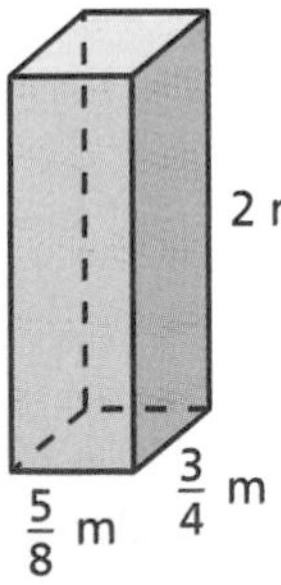

8.

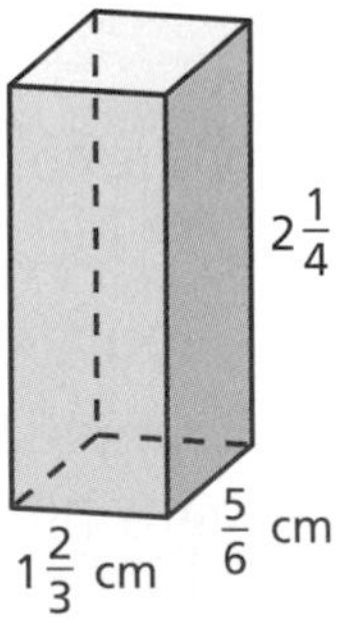

9.

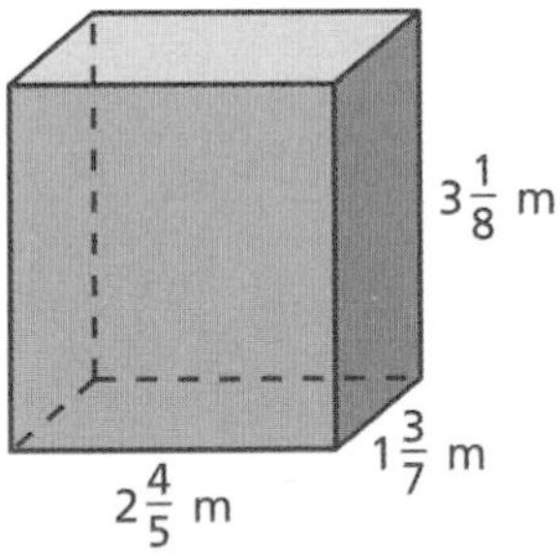

Write and solve an equation to find the missing dimension of the prism.

(3) 10. Volume = 1620 cm^3

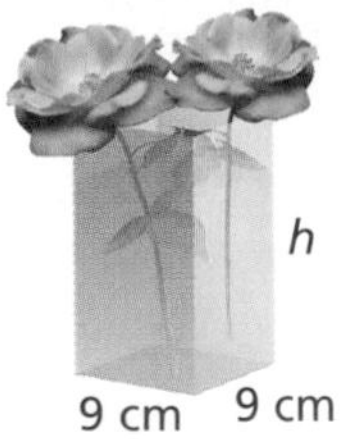

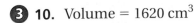

11. Volume = 220.5 cm^3

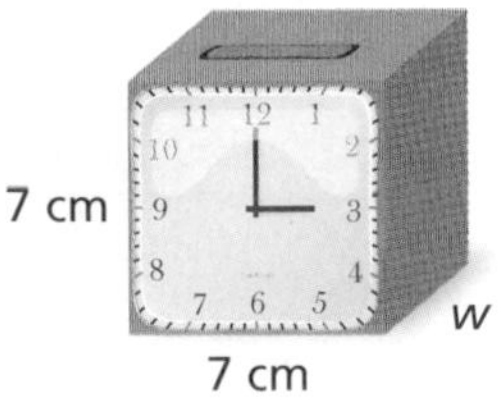

12. Volume = 532 in.^3

13. FISH TANK One cubic foot of water weighs about 62.4 pounds. How many pounds of water can the fish tank hold when it is full?

14. CUBE How many $\frac{3}{4}$-centimeter cubes do you need to create a cube with an edge length of 12 centimeters?

15. REASONING How many 1-inch cubes do you need to fill a cube that has an edge length of 1 foot? How can this result help you convert a volume from cubic inches to cubic feet? from cubic feet to cubic inches?

16. FOOD STORAGE

a. Estimate the amount of casserole left in the dish.

b. Will the casserole fit in the storage container? Explain your reasoning.

17. PROBLEM SOLVING The area of the shaded face is 96 square centimeters. What is the volume of the rectangular prism?

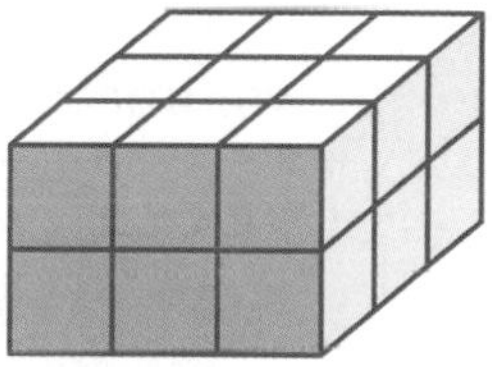

18. **Project** You have 1400 square feet of boards to use for a new tree house.

a. Design a tree house that has a volume of at least 250 cubic feet. Include sketches of your tree house.

b. Are your dimensions reasonable? Explain your reasoning.

Fair Game Review What you learned in previous grades & lessons

Tell whether the given value is a solution of the equation. *(Section 7.2)*

19. $x + 17 = 24;\ x = 7$

20. $\frac{x}{5} = 6;\ x = 35$

21. $x - 19 = 42;\ x = 21$

22. MULTIPLE CHOICE Which set of integers is ordered from least to greatest? *(Section 6.2)*

Ⓐ $-1, 3, -5, -8, 12$

Ⓑ $-1, 3, -5, -8, 12$

Ⓒ $-4, -2, 1, 7, 10$

Ⓓ $-14, -9, 6, -4, 2$

8.3–8.4 Quiz

Check It Out
Progress Check
BigIdeasMath.com

Find the surface area of the pyramid. The side lengths of the base are equal. *(Section 8.3)*

1.

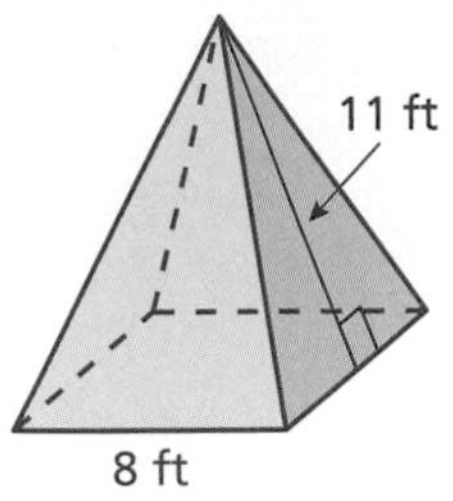

2.

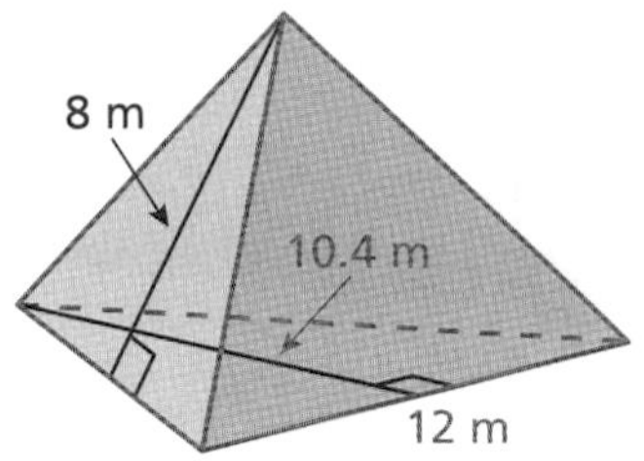

Find the volume of the prism. *(Section 8.4)*

3.

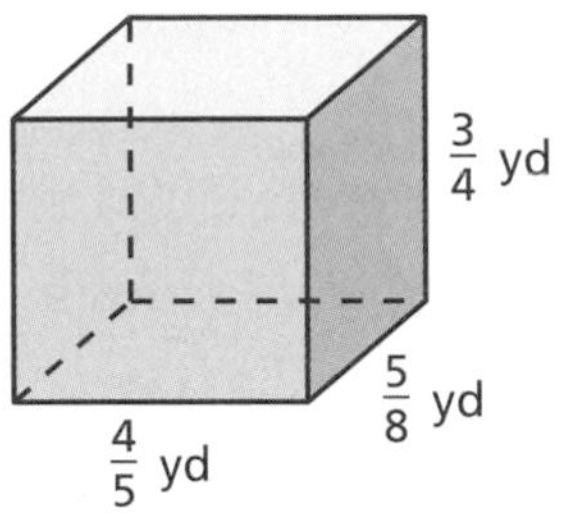

4.

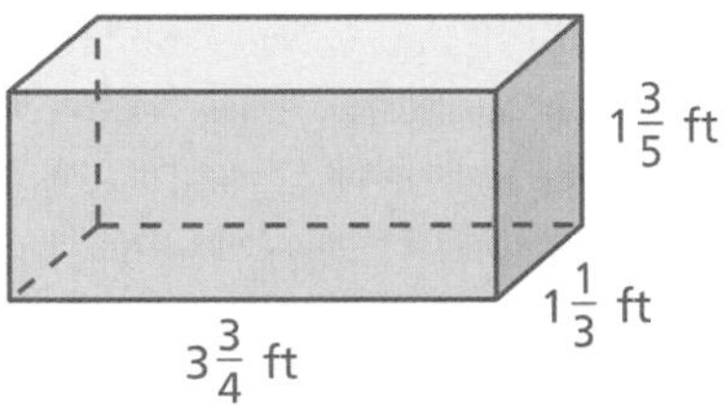

Write and solve an equation to find the missing dimension of the prism. *(Section 8.4)*

5. Volume = 1620 in.^{3}

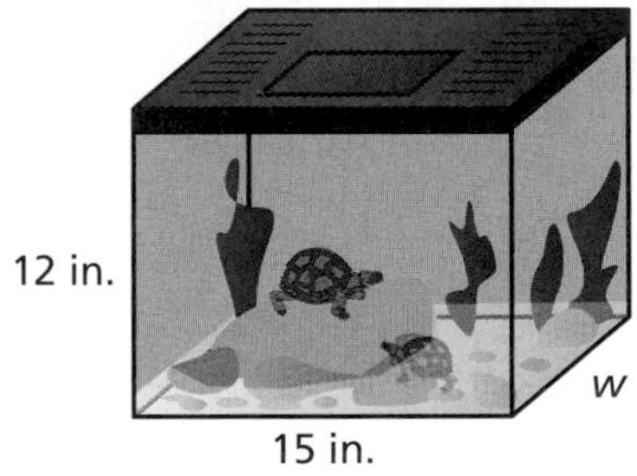

6. Volume = 154 in.^{3}

7. Volume = 4250 in.^{3}

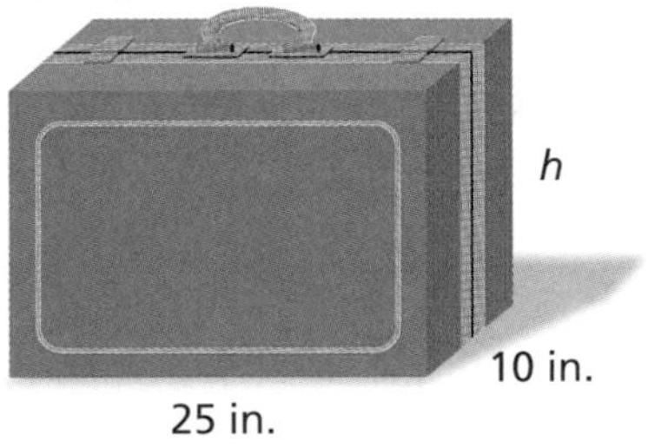

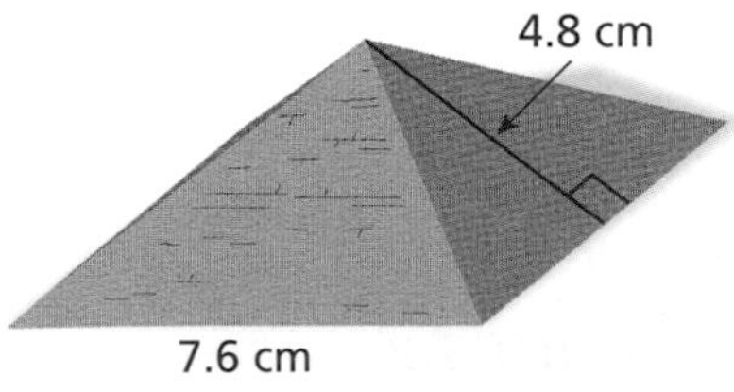

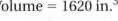

8. GREAT PYRAMID The Great Pyramid of Giza is a square pyramid. A gift shop sells miniature models of this pyramid. Find the surface area of the model shown at the left. *(Section 8.3)*

9. CUBE How many 1-inch cubes do you need to create a cube with an edge length of 7 inches? *(Section 8.4)*

10. TOY CHEST A toy company sells two different toy chests. The toy chests have different dimensions, but the same volume. What is the width w of Toy Chest 2? *(Section 8.4)*

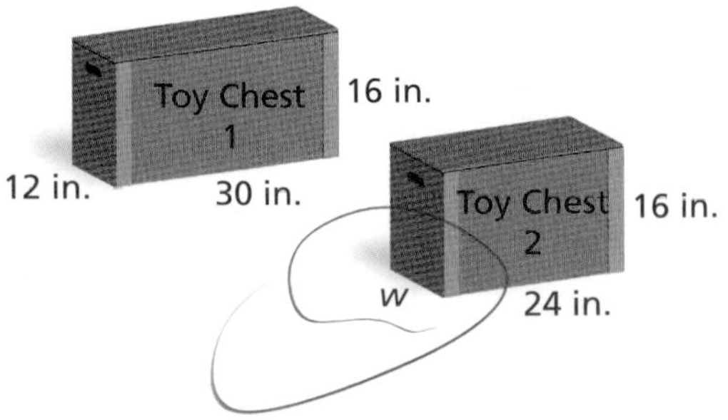

8 Chapter Review

Check It Out
Vocabulary Help
BigIdeasMath.com

Review Key Vocabulary

solid, *p. 356*
polyhedron, *p. 356*
face, *p. 356*
edge, *p. 356*
vertex, *p. 356*
prism, *p. 356*
pyramid, *p. 356*
surface area, *p. 362*
net, *p. 362*
volume, *p. 374*

Review Examples and Exercises

8.1 Three-Dimensional Figures *(pp. 354–359)*

a. Find the number of faces, edges, and vertices of the solid.

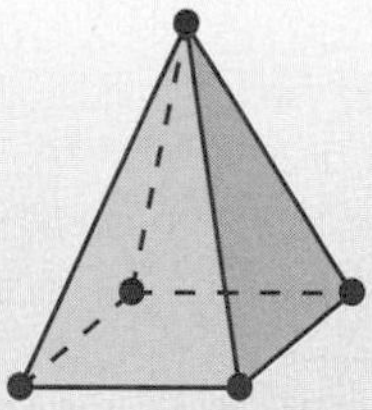

The solid has 1 face on the bottom and 4 faces on the sides.

The faces intersect at 8 different line segments.

The edges intersect at 5 different points.

So, the solid has 5 faces, 8 edges, and 5 vertices.

b. Draw a triangular prism.

Draw identical triangular bases.

Connect corresponding vertices.

Change any *hidden* lines to dashed lines.

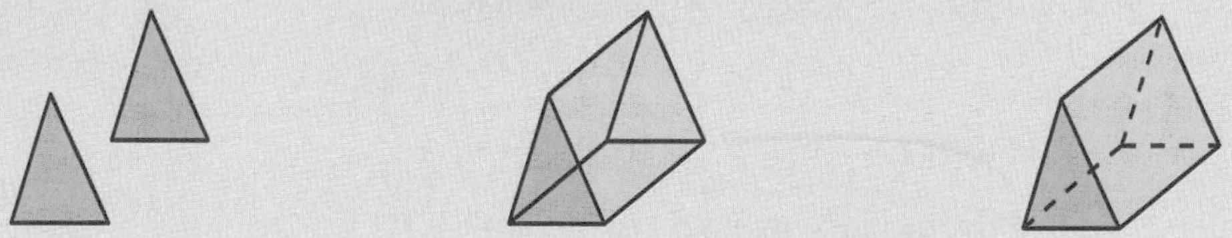

Exercises

Find the number of faces, edges, and vertices of the solid.

1. 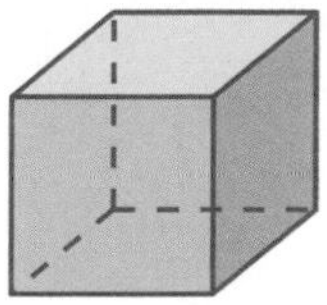

2. 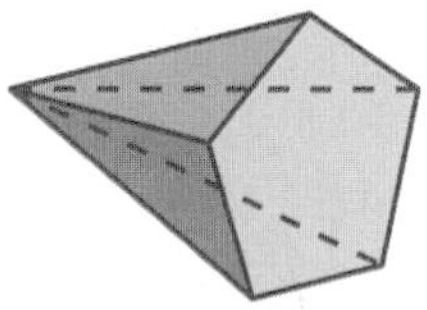

Draw the solid.

3. square pyramid

4. hexagonal prism

8.2 Surface Areas of Prisms (pp. 360–365)

Find the surface area of the rectangular prism.

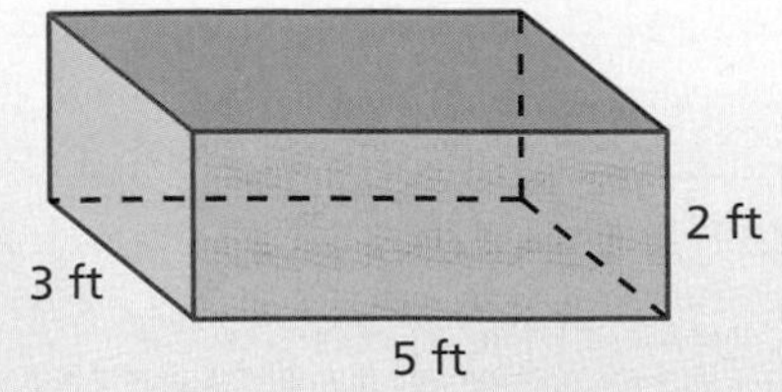

Use a net to find the area of each face.

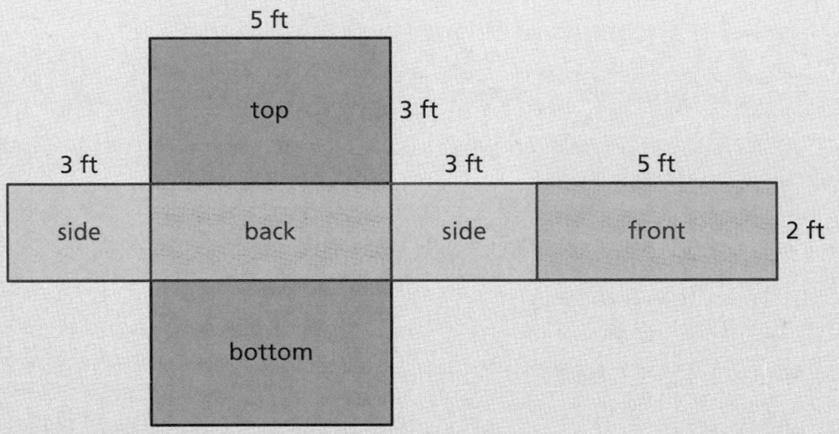

Top: $5 \cdot 3 = 15$	Front: $5 \cdot 2 = 10$	Side: $3 \cdot 2 = 6$
Bottom: $5 \cdot 3 = 15$	Back: $5 \cdot 2 = 10$	Side: $3 \cdot 2 = 6$

Find the sum of the areas of the faces.

$$S = 15 + 15 + 10 + 10 + 6 + 6$$
$$= 62$$

∴ The surface area is 62 square feet.

Exercises

Find the surface area of the prism.

5.

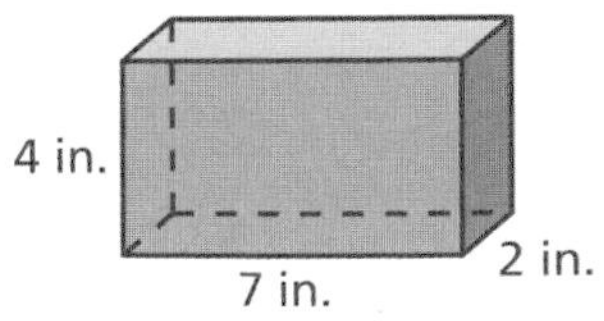

6.

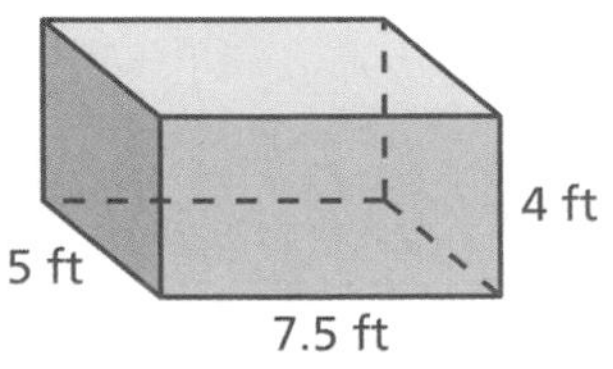

7.

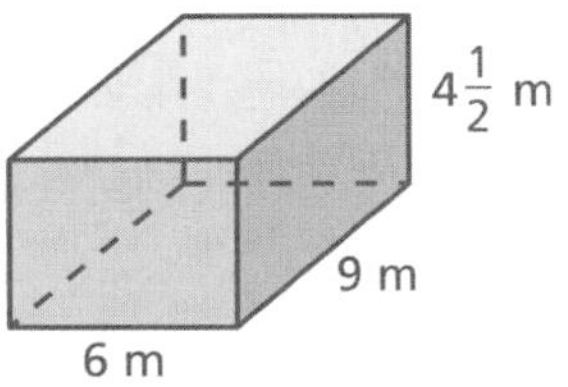

8.

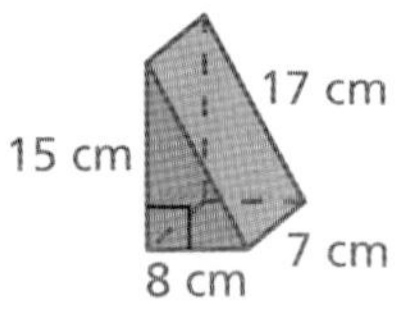

9.

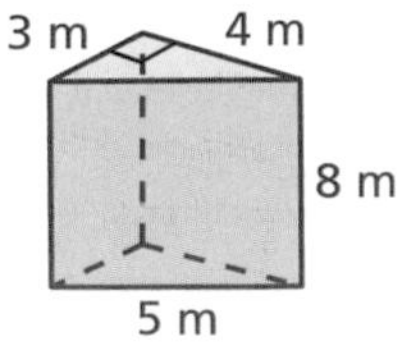

10.

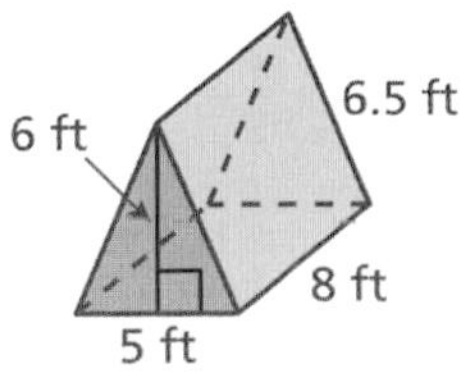

8.3 Surface Areas of Pyramids (pp. 368–373)

Find the surface area of the triangular pyramid.

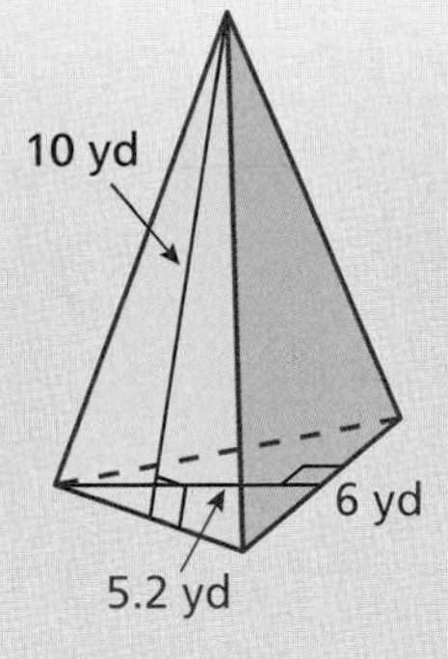

Use a net to find the area of each face.

Bottom: $\frac{1}{2} \cdot 6 \cdot 5.2 = 15.6$ Side: $\frac{1}{2} \cdot 6 \cdot 10 = 30$

Side: $\frac{1}{2} \cdot 6 \cdot 10 = 30$ Side: $\frac{1}{2} \cdot 6 \cdot 10 = 30$

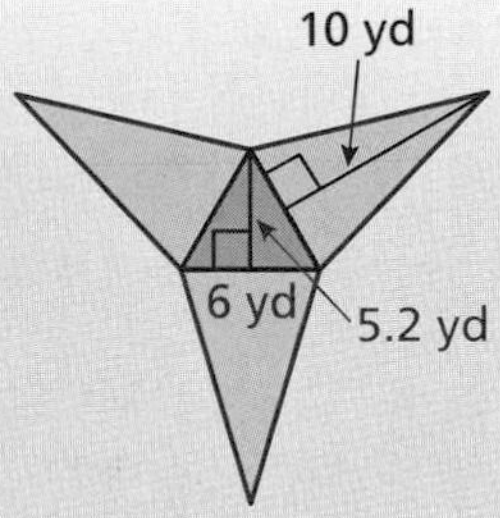

Find the sum of the areas of the faces.

$S = 15.6 + 30 + 30 + 30 = 105.6$

∴ The surface area is 105.6 square yards.

Exercises

Find the surface area of the pyramid. The side lengths of the base are equal.

11.

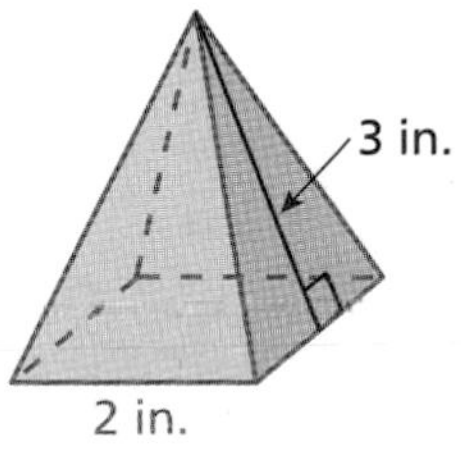

12.

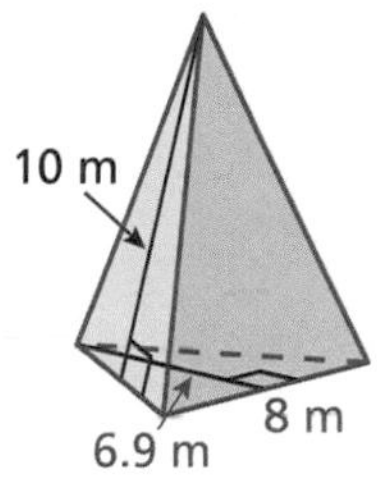

13.

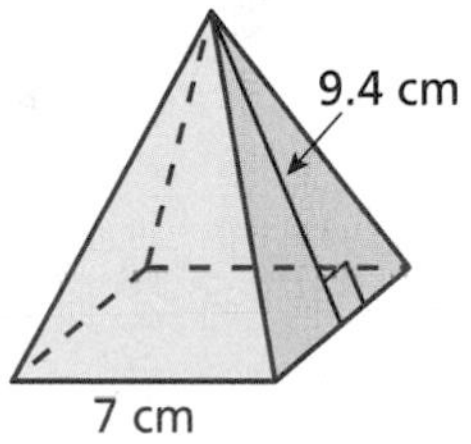

8.4 Volumes of Rectangular Prisms (pp. 374–379)

Find the volume of the prism.

$V = \ell wh$	Write formula for volume.
$= \frac{5}{6}\left(\frac{3}{4}\right)\left(\frac{4}{5}\right)$	Substitute values.
$= \frac{1}{2}$	Multiply.

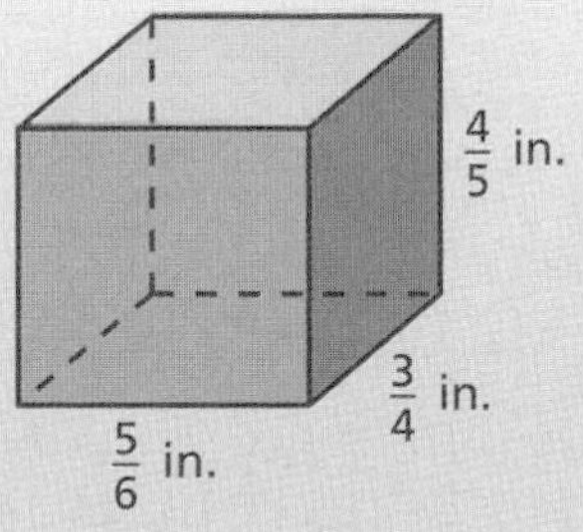

∴ The volume is $\frac{1}{2}$ cubic inch.

Exercises

Find the volume of the prism.

14.

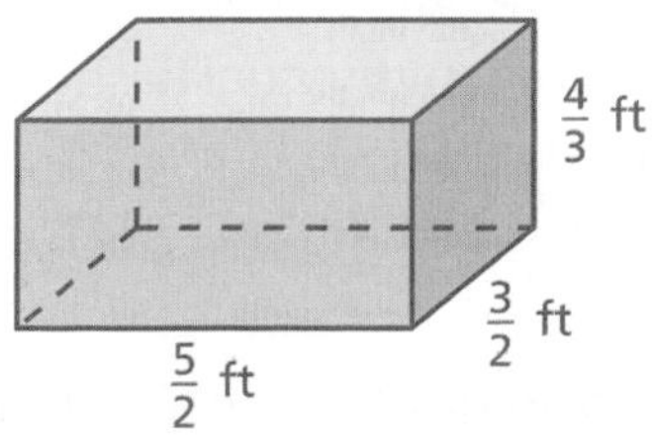

15.

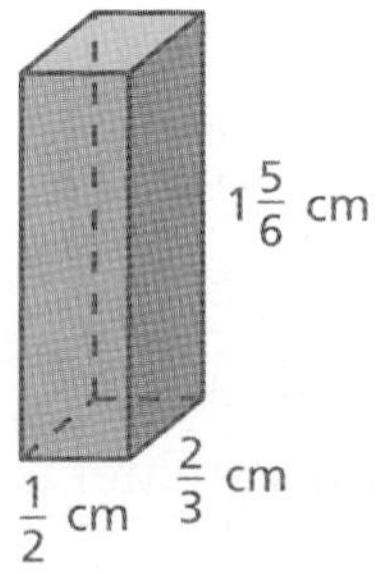

8 Chapter Test

Check It Out
Test Practice
BigIdeasMath.com

Find the number of faces, edges, and vertices of the solid.

1.

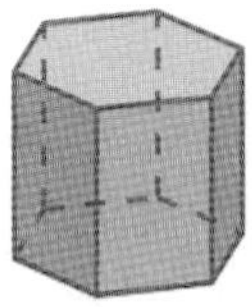

2.

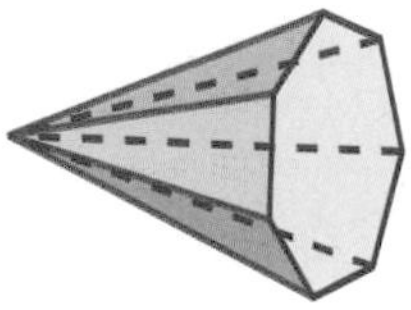

Find the surface area of the prism.

3.

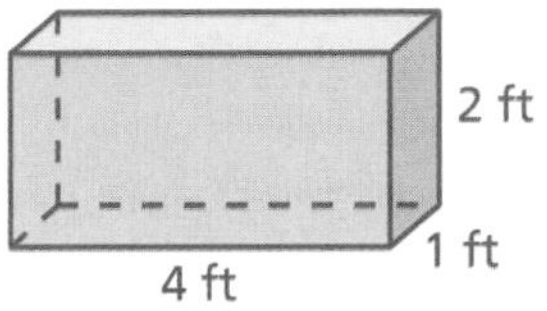

4.

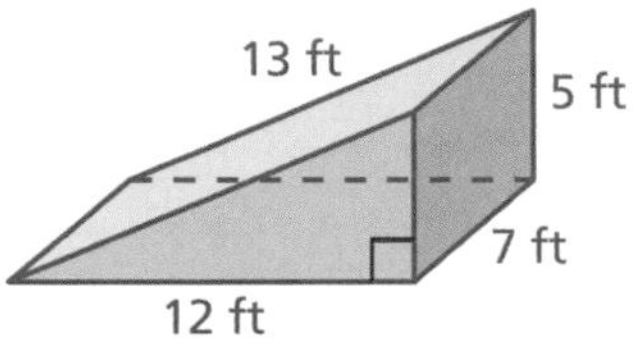

Find the surface area of the pyramid. The side lengths of the base are equal.

5.

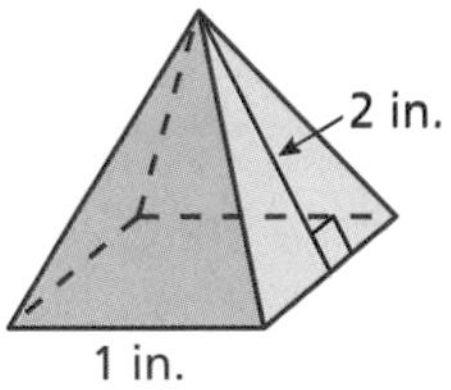

6.

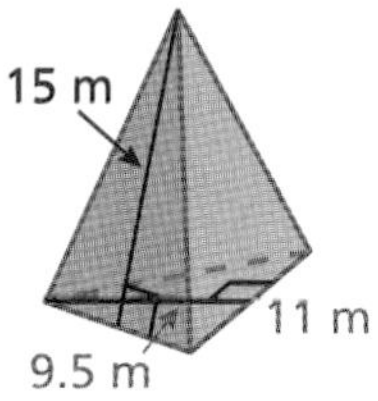

Find the volume of the prism.

7.

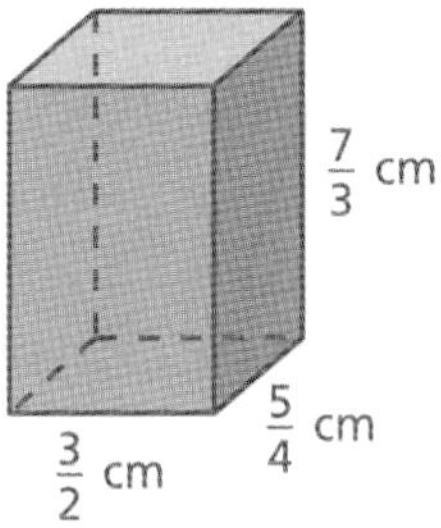

8.

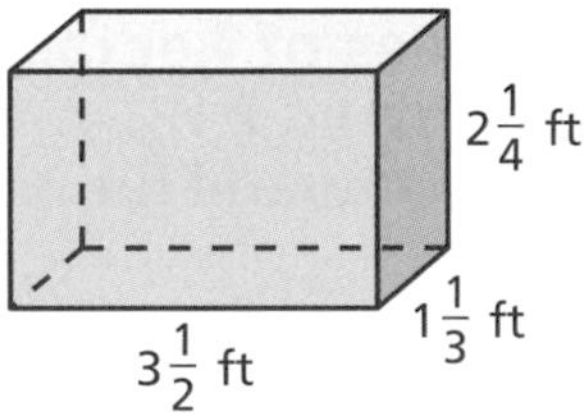

9. DRAWING A SOLID Draw an octagonal prism.

10. DVD COLLECTION You are wrapping the boxed DVD collection as a present. What is the least amount of wrapping paper needed to wrap the box?

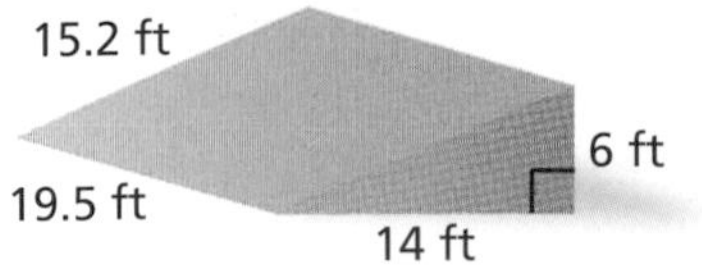

11. SKATEBOARD RAMP A quart of paint covers 80 square feet. How many quarts should you buy to paint the ramp with two coats? (Assume you will not paint the bottom of the ramp.)

12. CUBE A cube has an edge length of 4 inches. You double the side lengths. How many times greater is the volume of the new cube?

8 Standards Assessment

Test-Taking Strategy
Answer Easy Questions First

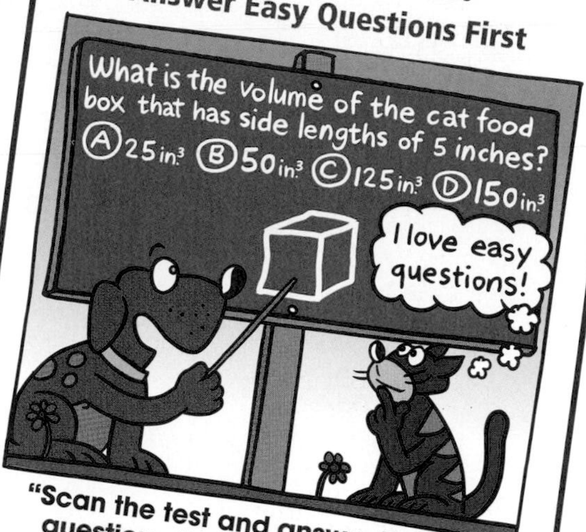
What is the volume of the cat food box that has side lengths of 5 inches?
Ⓐ 25 in.³ Ⓑ 50 in.³ Ⓒ 125 in.³ Ⓓ 150 in.³
I love easy questions!

"Scan the test and answer the easy questions first. You know that the volume of a cube is its side length cubed. So, you should choose C."

1. The temperature in a town has never been above 38 degrees Fahrenheit. Let t represent the temperature, in degrees Fahrenheit. Which inequality represents the temperature in the town? *(6.EE.8)*

A. $t < 38$ **C.** $t > 38$

B. $t \leq 38$ **D.** $t \geq 38$

2. Which number is equivalent to the expression below? *(6.EE.1)*

$$3 \cdot 4^2 + 6 \div 2$$

F. 27 **H.** 51

G. 33 **I.** 75

3. What is the volume of the package shown below? *(6.G.2)*

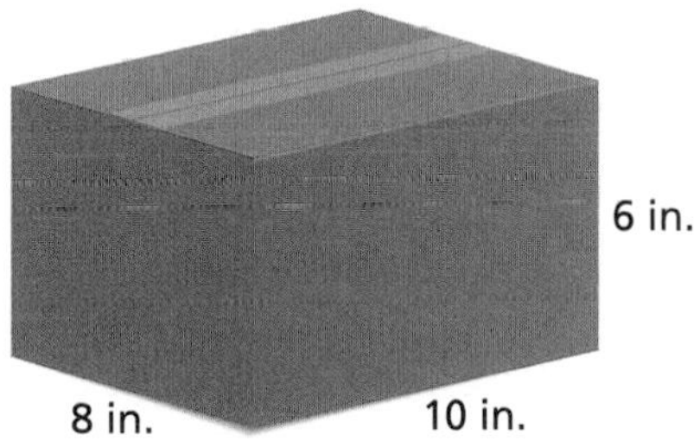
6 in.
8 in.
10 in.

A. 240 in.^3 **C.** 480 in.^3

B. 376 in.^3 **D.** 960 in.^3

4. A housing community started with 60 homes. In each of the following years, 8 more homes were built. Let y represent the number of years that have passed since the first year, and let n represent the number of homes. Which equation describes the relationship between n and y? *(6.EE.9)*

F. $n = 8y + 60$ **H.** $n = 60y + 8$

G. $n = 68y$ **I.** $n = 60 + 8 + y$

5. What is the value of m that makes the equation below true? *(6.EE.7)*

$$4m = 6$$

6. A square pyramid is shown below.

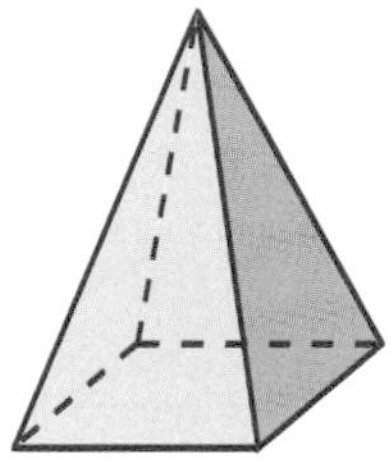

The square base and one of the triangular faces of the square pyramid are shown below with their dimensions.

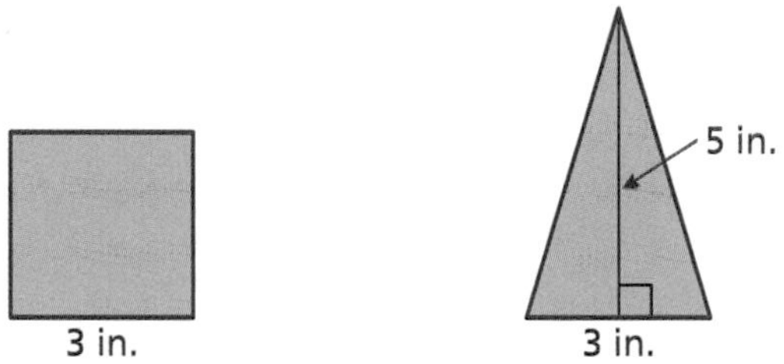

What is the total surface area of the square pyramid? *(6.G.4)*

A. 16.5 in.^2

B. 31.5 in.^2

C. 39 in.^2

D. 69 in.^2

7. A wooden box has a length of 12 inches, a width of 6 inches, and a height of 8 inches. *(6.G.2, 6.G.4)*

Think Solve Explain

Part A Draw and label a rectangular prism with the dimensions of the wooden box.

Part B What is the surface area, in square inches, of the wooden box? Show your work.

Part C You have a 2-ounce sample of wood stain that covers 900 square inches. Is this enough to give the entire box two coats of stain? Show your work and explain your reasoning.

8. A biologist measures the lengths of a crazy ant and a green anole that he has in his laboratory. His measurements are shown below.

The length of the green anole is how many times greater than the length of the crazy ant? *(6.NS.1)*

F. $\frac{9}{16}$

G. $5\frac{29}{32}$

H. 16

I. 64

9. What is the missing value in the ratio table? *(6.RP.3a)*

Castles	1	2		12
Towers	4	8	24	48

10. What is the area of the shaded figure shown below? *(6.G.1)*

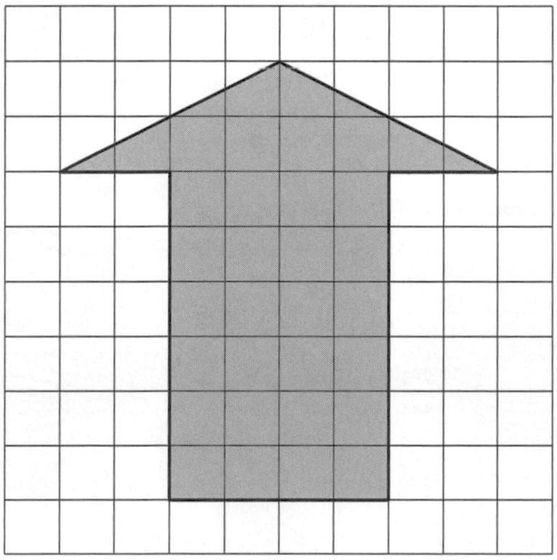

A. 32 units^2

B. 36 units^2

C. 40 units^2

D. 64 units^2

9 Statistical Measures

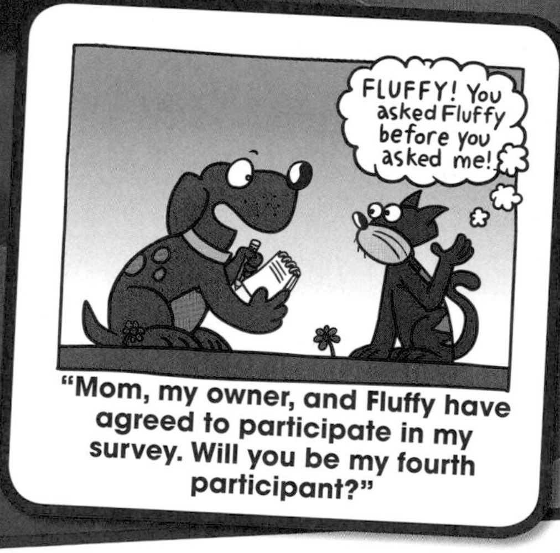

"Mom, my owner, and Fluffy have agreed to participate in my survey. Will you be my fourth participant?"

"Please hold still. I am trying to find the mean of 6, 8, and 10 by dividing their sum into three equal piles."

What You Learned Before

"Our mean weight is 18 pounds."

Ordering Decimals (4.NF.7)

Example 1 **Use a number line to order 8, 6.5, 7.25, 5.5, 4.25, and 7 from least to greatest.**

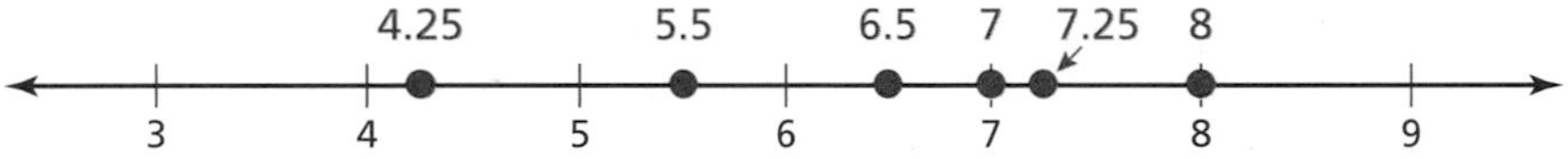

Try It Yourself

Use a number line to order the numbers from least to greatest.

1. 7.25, 4.5, 6.5, 6, 5.5, 8.75

2. 4, 2.5, 3.25, 5.5, 4.5, 6.75

3. 6.25, 3, 2.5, 3.5, 5.75, 5

4. 1.25, 5.5, 4.75, 4.5, 3.5, 2.25

Analyzing Double Bar Graphs (3.MD.3, 4.NBT.4)

Example 2 **How many more male athletes than female athletes participated in the 1992 Summer Olympics?**

$$6652 - 2704 = 3948$$

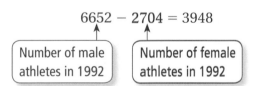

6652: Number of male athletes in 1992

2704: Number of female athletes in 1992

∴ 3948 more male athletes participated.

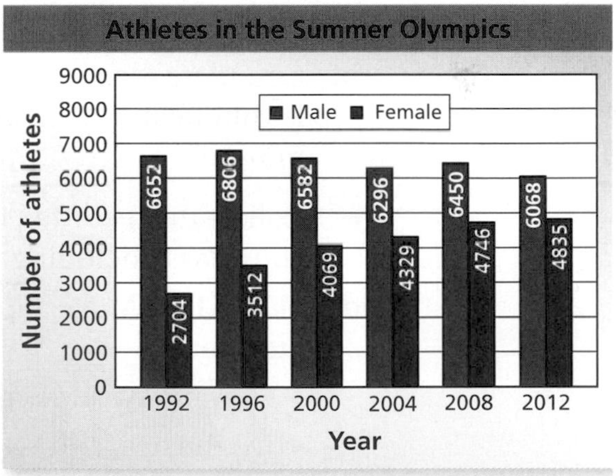

Example 3 **How many athletes participated in the 2000 Summer Olympics?**

∴ 6582 + 4069 = 10,651 athletes participated in the 2000 Summer Olympics.

Try It Yourself

5. How many more female athletes participated in the 2012 Summer Olympics than in the 1992 Summer Olympics?

6. Describe the relationship between the number of athletes in the 2000 Summer Olympics and the number of athletes in the 2004 Summer Olympics.

9.1 Introduction to Statistics

Essential Question How can you tell whether a question is a statistical question?

Your heart rate is the number of times your heart beats in a certain time period, such as 1 minute. To measure your heart rate, you can check your pulse. The illustration shows how to check your pulse by pressing lightly on your wrist.

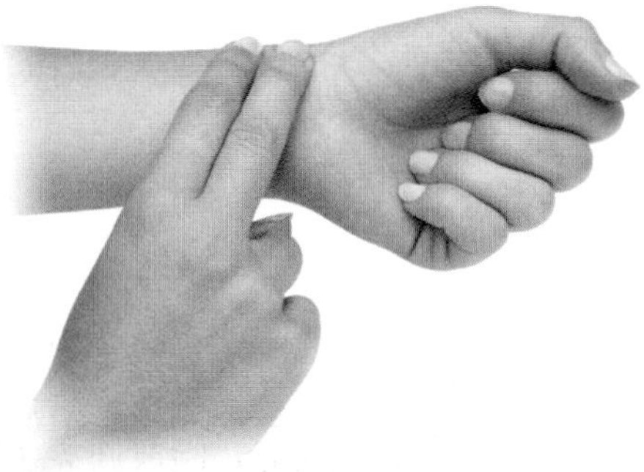

Here are other places to check your pulse:
- inside your elbow
- side of your neck
- top of your foot

1 ACTIVITY: Using Data to Answer Questions

Work with a partner.

a. Find your pulse by counting the number of beats in 10 seconds. Have your partner keep track of the time. Write a rate to describe your result.

b. Complete the ratio table. What is your heart rate in beats per minute?

Time (seconds)	10	30	60
Number of Beats			

c. Collect the recorded heart rates (in beats per minute) of the students in your class, including yourself. Compare the heart rates.

d. **MODELING** Make a *line plot* of your data. Then answer the following questions:
- How many values are in your data set?
- Do the heart rates *cluster* around a particular value or values?
- Are there any *peaks* or *gaps* in the data?
- Are there any unusual heart rates that are far removed from the other values?

e. **REASONING** How would you answer the following question by using only one value? Explain your reasoning.

"What is the heart rate of sixth grade students?"

f. **REASONING** Read and compare the following questions. How did you answer each question? Could the answer be the same for both questions? Explain.
- *What is your heart rate?*
- *What is the heart rate of sixth grade students?*

COMMON CORE

Statistics

In this lesson, you will
- recognize statistical questions.
- use dot plots to display numerical data.

Learning Standards
6.SP.1
6.SP.4

2 ACTIVITY: Identifying Types of Questions

Work with a partner.

a. Answer each question below on your own. Then compare your answers with your partner's answers. For which questions should your answers be the same? For which questions might your answers be different?

Math Practice 3

Build Arguments

How can comparing your answers help you support your conjecture?

1. What is your shoe size?
2. How many states are in the United States?
3. How many brothers and sisters do you have?
4. How many U.S. presidents have been in office?
5. What is your favorite type of movie?
6. How tall are you?

b. **CONJECTURE** Some of the questions above are considered *statistical* questions. Which ones do you think they are? Why?

3 ACTIVITY: Analyzing a Question in a Survey

Work with a partner. A student asks the following question in a survey:

"Do you prefer salty potato chips or healthy granola bars to be sold in the school's vending machines?"

a. Do you think this is a fair question to ask in a survey? Explain.

b. **LOGIC** Identify the words in the question that may influence someone's response. Then explain how you can reword the question.

c. How might the results of the survey differ when the student asks the original question and your reworded question in part (b)?

What Is Your Answer?

4. **REASONING** What do you think "statistics" means?
5. **IN YOUR OWN WORDS** How can you tell whether a question is a statistical question? Give examples to support your explanation.
6. Find the least and the greatest heart rates in your class. How can you use these two values to answer the question in Activity 1(e)?
7. Create a one-question survey. Explain why your question is a statistical question. Then conduct your survey and organize your results in a line plot. Make three observations about your data set.

Practice Use what you learned about different types of questions to complete Exercises 4–7 on page 394.

9.1 Lesson

Check It Out
Lesson Tutorials
BigIdeasMath.com

Key Vocabulary
statistics, *p. 392*
statistical question, *p. 392*

Statistics is the science of collecting, organizing, analyzing, and interpreting data. A **statistical question** is one for which you do not expect to get a single answer. Instead, you expect a variety of answers, and you are interested in the distribution and tendency of those answers.

Recall that a dot plot uses a number line to show the number of times each value in a data set occurs. Dot plots show the *spread* and the *distribution* of a data set.

EXAMPLE 1 Answering a Statistical Question

You conduct a science experiment on house mice. Your teacher asks you, "What is the weight of a mouse?"

a. Is this a statistical question? Explain.

Because you can anticipate that the weights of mice will vary, it is a statistical question.

b. You weigh some mice and record the weights (in grams) in the table. Display the data in a dot plot. Identify any clusters, peaks, or gaps in the data.

Weights (grams)			
20	19	21	20
18	20	27	21
28	23	20	19
20	21	18	27
19	22	21	20

Draw a number line that includes the least value, 18, and greatest value, 28. Then place a dot above the number line for each data value.

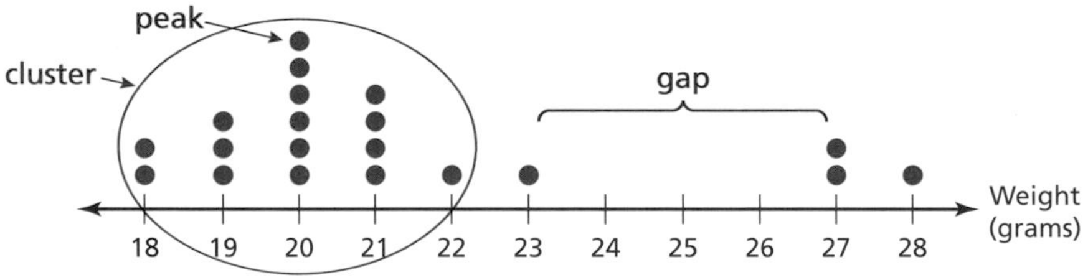

Most of the data are clustered around 20. There is a peak at 20 and a gap between 23 and 27.

c. Use the distribution of the data to answer the question.

Most mice weigh about 20 grams.

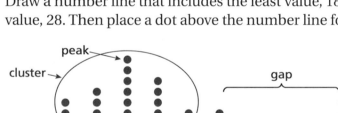

Study Tip
Dot plots are sometimes called *line plots.* It is easy to see clusters, peaks, and gaps in a dot plot.

On Your Own

Now You're Ready
Exercises 8–16

1. The table shows the ages of some people who retired early. You are asked, "How old are people who retire early?"
   a. Is this a statistical question? Explain.
   b. Display the data in a dot plot. Identify any clusters, peaks, or gaps in the data.
   c. Use the distribution of the data to answer the question.

Ages			
60	61	59	60
62	56	64	59
58	60	61	60
59	60	58	61

EXAMPLE 2 Using a Dot Plot

You record the high temperature every day while at summer camp in August. Then you create the vertical dot plot.

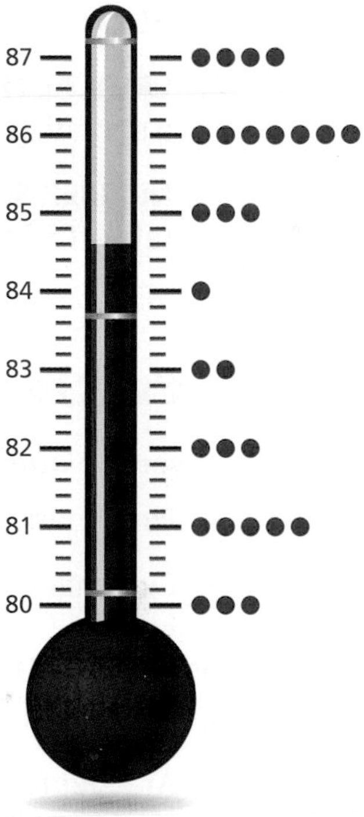

a. How many weeks were you at summer camp?

Because there are 28 data values on the dot plot, you were at camp 28 days.

$$28 \cancel{\text{days}} \cdot \frac{1 \text{ week}}{7 \cancel{\text{days}}} = 4 \text{ weeks}$$

So, you were at summer camp for 4 weeks.

b. How can you collect these data? What are the units?

You can collect these data with a thermometer. The units are degrees Fahrenheit (°F).

c. Write a statistical question that you can answer using the dot plot. Then answer the question.

One possible statistical question is:

What is the daily high temperature in August?

The high temperatures are spread out with about half of the temperatures around 81°F and half of the temperatures around 86°F.

On Your Own

Now You're Ready
Exercises 17 and 18

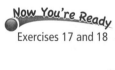

2. The dot plot shows the times of sixth grade students in a 100-meter race.

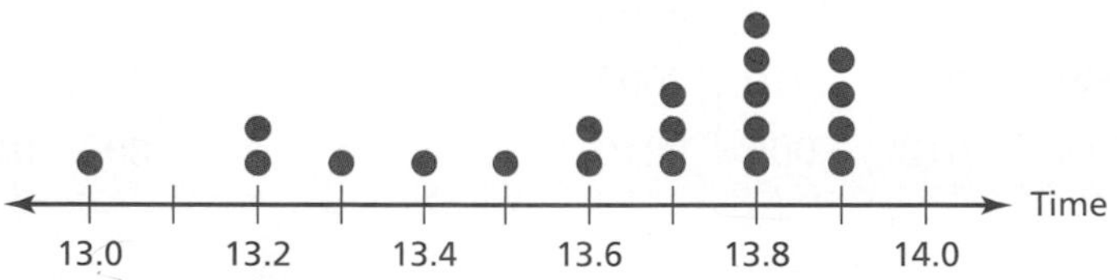

a. How many students ran in the race?

b. How can you collect these data? What are the units?

c. Write a statistical question that you can answer using the dot plot. Then answer the question.

9.1 Exercises

Check It Out
Help with Homework
BigIdeasMath.com

Vocabulary and Concept Check

1. **VOCABULARY** What is a statistical question? Give an example.
2. **CRITICAL THINKING** What process can you use to answer a statistical question?
3. **NUMBER SENSE** The results of a survey are shown in the table. Did the survey ask a statistical question? Explain.

Miles			
6	1	9	2
2	5	4	9
8	10	6	6
5	1	8	1

Practice and Problem Solving

Answer the question. Tell whether your answer would be the same as your classmates'.

4. How many inches are in 1 foot?
5. How many pets do you have?
6. On what day of the month were you born?
7. How many senators are in Congress?

Determine whether the question is a statistical question. Explain.

1 8. What is the eye color of sixth grade students?
9. At what temperature (in degrees Fahrenheit) does water freeze?
10. How many pages are in the favorite books of students your age?
11. How many hours do sixth grade students use the Internet each week?

Display the data in a dot plot. Identify any clusters, peaks, or gaps in the data.

12.

Number of Fouls					
2	1	2	0	0	2
2	1	6	1	1	0

13.

Camper Registrations				
21	25	25	22	21
23	24	26	25	16
24	26	22	25	22

14.

Years			
2011	2008	2013	2009
2009	2010	2010	2009
2010	2012	2009	2010

15.

Test Scores				
85	80	83	90	88
82	83	81	80	89
89	84	86	87	83

16. **SURVEY** You conduct a survey to answer: "How many hours does a sixth grade student spend on homework during a school night?" The table shows the results.

Hours of Homework			
2	4	3	2
1	2	2	1
2	3	5	2

a. Is this a statistical question? Explain.

b. Display the data in a dot plot. Identify any clusters, peaks, or gaps in the data.

c. Use the distribution of the data to answer the question.

2 **17. EARTHWORMS** The dot plot shows the lengths of earthworms.

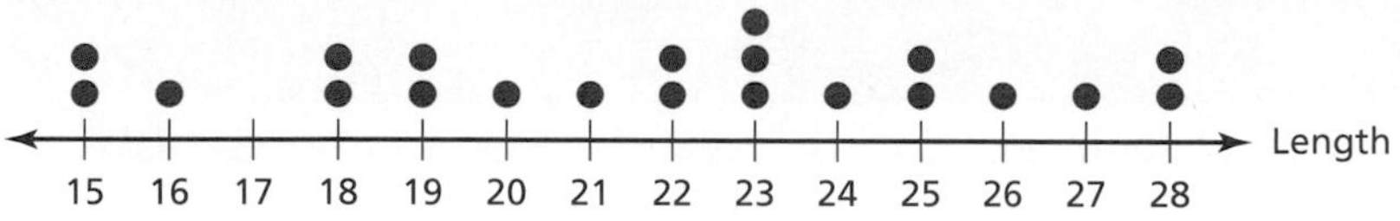

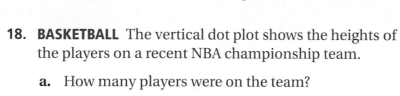

a. How many earthworms does it represent?

b. How can you collect these data? What are the units?

c. Write a statistical question that you can answer using the dot plot. Then answer the question.

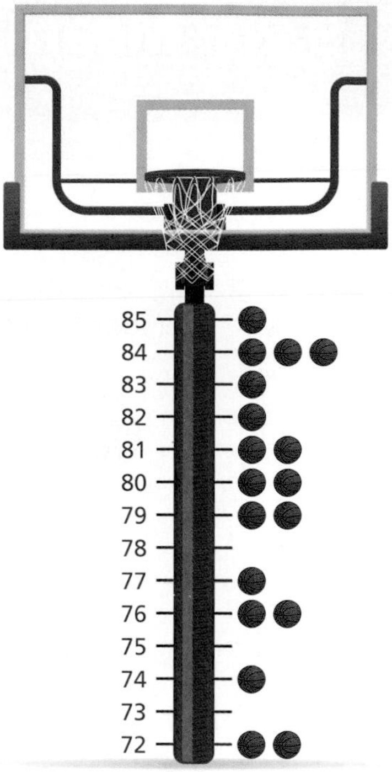

18. BASKETBALL The vertical dot plot shows the heights of the players on a recent NBA championship team.

a. How many players were on the team?

b. How can you collect these data? What are the units?

c. Write a statistical question that you can answer using the dot plot. Then answer the question.

Use the Internet to research and identify the method of measurement and the units used when collecting data about the topic.

19. wind speed

20. amount of rainfall

21. earthquake intensity

The dot plot shows the speeds of cars in a traffic study. Estimate the speed limit. Explain your reasoning.

22.

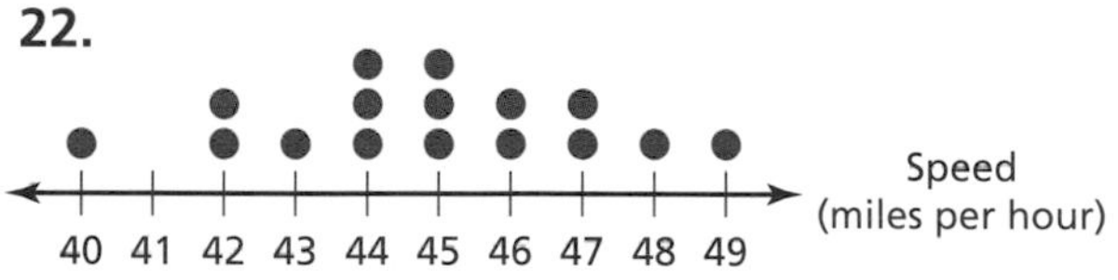

23.

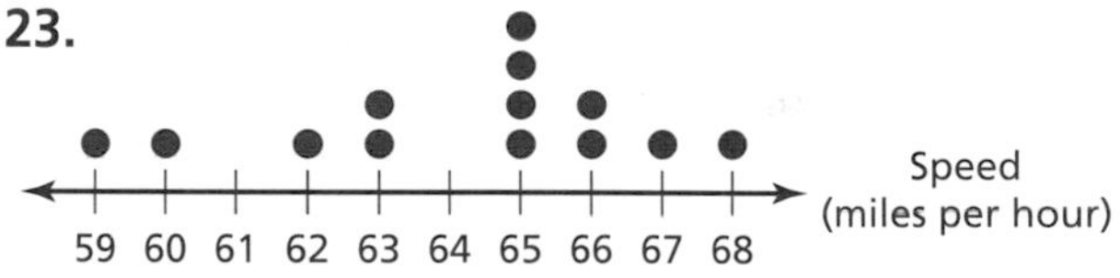

24. REASONING "How many letters are in the English alphabet?" is *not* a statistical question. Write a question about letters that is a statistical question. Explain your reasoning.

25. Reasoning A bar graph shows the favorite colors of 30 people. Does it make sense to describe the distribution of these data? Explain.

Fair Game Review What you learned in previous grades & lessons

Tell whether the ordered pair is a solution of the equation. *(Section 7.4)*

26. $y = 4x$; (2, 8)

27. $y = 3x + 5$; (3, 15)

28. $y = 6x - 15$; (4, 9)

29. MULTIPLE CHOICE A point is reflected in the x-axis. The reflected point is (4, −3). What is the original point? *(Section 6.5)*

Ⓐ (−3, 4) Ⓑ (−4, 3) Ⓒ (−4, −3) Ⓓ (4, 3)

9.2 Mean

Essential Question How can you find an average value of a data set?

1 ACTIVITY: Finding a Balance Point

Work with a partner. Discuss the distribution of the data. Where on the number line do you think the data set is *balanced*? Is this a good representation of the average? Explain.

a. number of quarters brought to a batting cage

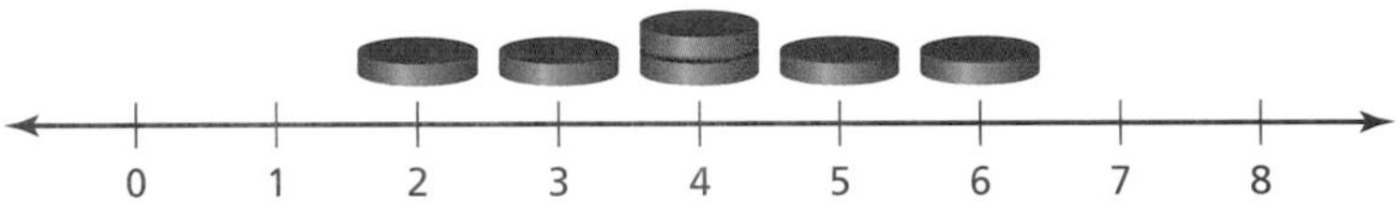

b. annual income of recent graduates (in thousands of dollars)

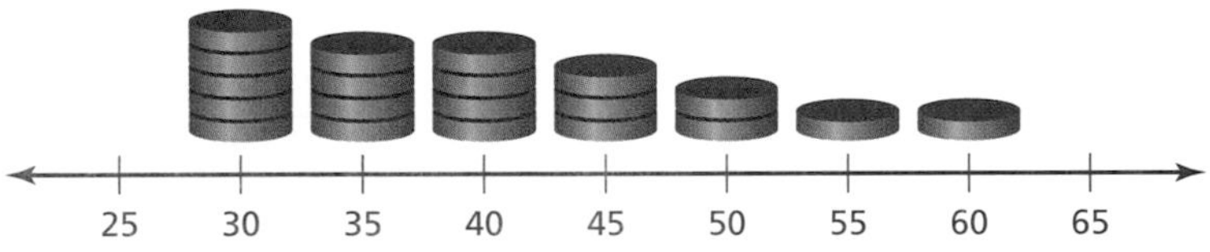

c. hybrid fuel economy (miles per gallon)

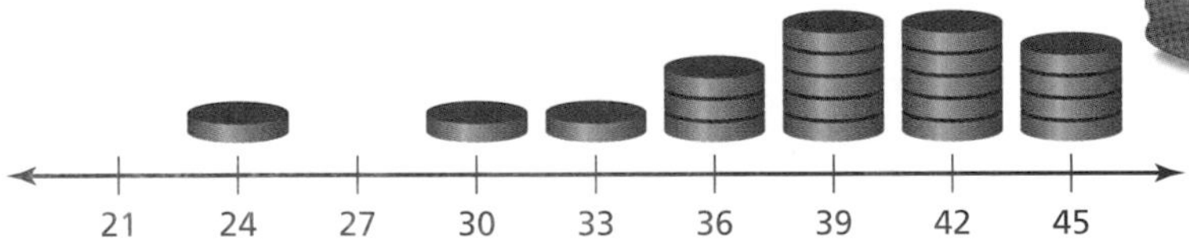

2 ACTIVITY: Finding a Fair Share

Work with a partner. It costs \$0.25 to hit 12 baseballs in a batting cage. The table shows the numbers of quarters six friends bring to the batting cage. They want to group the quarters so that everyone has the same amount.

Quarters					
John	Lisa	Miguel	Matt	Cheryl	Jean
6	3	4	5	2	4

Use counters to represent each number in the table. How can you use the counters to determine how many times each friend can use the batting cage? Explain how this procedure results in a "fair share."

COMMON CORE

Statistics

In this lesson, you will
- understand the concept of the mean of data sets.
- find the mean of data sets.
- compare and interpret the means of data sets.

Learning Standards
6.SP.2
6.SP.3
6.SP.5a
6.SP.5c

3 ACTIVITY: Finding an Average

Work with a partner. Use the information in Activity 2.

a. What is the total number of quarters the group of friends brought to the batting cage?

b. **REASONING** How can you use math to find the average number of quarters that each friend brought to the batting cage? Find the average number of quarters. Why do you think this average represents a fair share?

4 ACTIVITY: Answering a Statistical Question

Work with a partner. The table shows the numbers of quarters several people bring to a batting cage. You want to answer the question:

"How many quarters do people bring to a batting cage?"

a. Explain why this question is a statistical question.

b. **MODELING** Make a dot plot of the data. Use the distribution of the data to answer the question. Explain your reasoning.

c. **REASONING** Use an average to answer the question. Explain your reasoning.

Math Practice 6

Use Clear Definitions

What does it mean for data to have an average? How does this help you answer the question?

What Is Your Answer?

5. **IN YOUR OWN WORDS** How can you find an average value of a data set?

6. Give two real-life examples of averages.

7. Explain what it means to say the average of a data set is the point on a number line where the data set is balanced.

8. There are 5 students in the cartoon. Four of the students are 66 inches tall. One is 96 inches tall.

   a. How do you think the students decided that their average height is 6 feet?

   b. Does a height of 6 feet seem like a good representation of the average height of the 5 students? Explain why or why not.

"Yup, the average height in our class is 6 feet."

Practice Use what you learned about averages to complete Exercises 4 and 5 on page 400.

9.2 Lesson

Check It Out
Lesson Tutorials
BigIdeasMath.com

Key Vocabulary
mean, *p. 398*
outlier, *p. 399*

A mean is a type of average.

Key Idea

Mean

Words The **mean** of a data set is the sum of the data divided by the number of data values.

Numbers **Data:** 8, 5, 6, 9 **Mean:** $\frac{8+5+6+9}{4} = \frac{28}{4} = 7$

(8, 5, 6, 9 → 4 data values)

EXAMPLE 1 Finding the Mean

Text Messages Sent
Mark: 120
Laura: 95
Stacy: 101
Josh: 125
Kevin: 82
Maria: 108
Manny: 90

The table shows the number of text messages sent by a group of friends over 1 week. What is the mean number of messages sent?

Ⓐ 100 Ⓑ 102 Ⓒ 103 Ⓓ 104

$$\text{mean} = \frac{120+95+101+125+82+108+90}{7}$$

(Sum of the data; Number of values)

$$= \frac{721}{7}, \text{ or } 103$$ Simplify.

The mean number of text messages sent is 103. The correct answer is Ⓒ.

EXAMPLE 2 Comparing Means

The double bar graph shows the monthly rainfall amounts for two cities over a six-month period. Compare the mean monthly rainfalls.

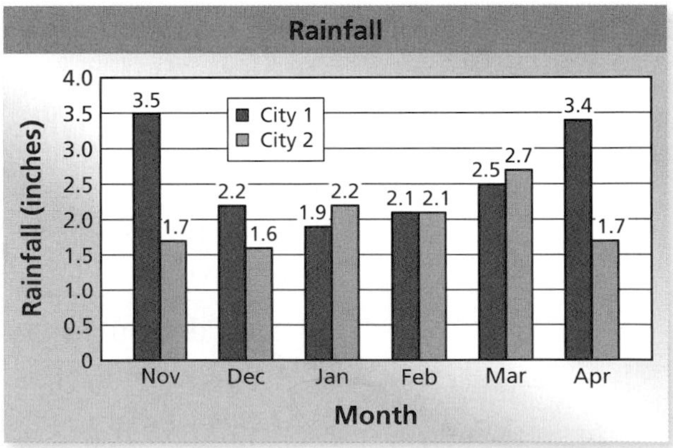

City 1 mean: $\frac{3.5+2.2+1.9+2.1+2.5+3.4}{6} = \frac{15.6}{6}$, or 2.6

City 2 mean: $\frac{1.7+1.6+2.2+2.1+2.7+1.7}{6} = \frac{12}{6}$, or 2

Because 2.6 is greater than 2, City 1 averaged more rainfall.

On Your Own

Now You're Ready
Exercises 6–9

Find the mean of the data.

1. 49, 62, 52, 54, 61, 70, 55, 53
2. 7.2, 8.5, 7.0, 8.1, 6.7

An **outlier** is a data value that is much greater or much less than the other values. When included in a data set, it can affect the mean.

EXAMPLE 3 Finding the Mean With and Without an Outlier

Shetland Pony Heights (inches)				
40	37	39	40	42
38	38	37	28	40

The table shows the heights of several Shetland ponies.

a. Identify the outlier.

b. Find the mean with and without the outlier.

c. Describe how the outlier affects the mean.

a. Display the data in a dot plot.

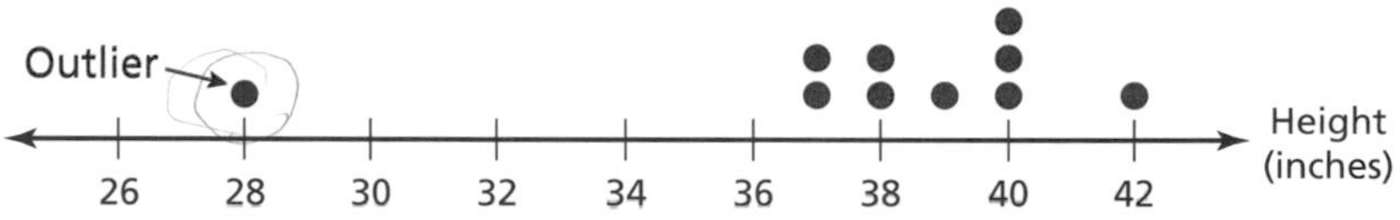

The height of 28 inches is much less than the other heights. So, it is an outlier.

b. Mean with outlier:

$$\frac{40 + 37 + 39 + 40 + 42 + 38 + 38 + 37 + 28 + 40}{10} = \frac{379}{10}, \text{ or } 37.9$$

Mean without outlier:

$$\frac{40 + 37 + 39 + 40 + 42 + 38 + 38 + 37 + 40}{9} = \frac{351}{9}, \text{ or } 39$$

c. With the outlier, the mean is less than all but three of the heights. Without the outlier, the mean better represents the heights.

On Your Own

Now You're Ready
Exercise 14

For each data set, identify the outlier. Then describe how it affects the mean.

3. Weights (in pounds) of dogs at a kennel:

   48, 50, 55, 60, 8, 37, 50

4. Prices for flights from Miami, Florida, to San Juan, Puerto Rico:

   $456, $512, $516, $900, $436, $516

9.2 Exercises

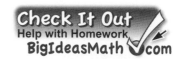

Check It Out
Help with Homework
BigIdeasMath.com

Vocabulary and Concept Check

1. **VOCABULARY** Arrange the words to explain how to find a mean.

the data values	divide by	the number of data values	add	then

2. **NUMBER SENSE** Is the mean always equal to a value in the data set? Explain.

3. **REASONING** Can you use the mean to answer a statistical question? Explain.

Practice and Problem Solving

Describe an average value of the data.

4. Ages in a class: 11, 12, 12, 12, 12, 12, 13

5. Movies seen this week: 0, 0, 0, 1, 1, 2, 3

Find the mean of the data.

1 6.

Pets Owned	
Brandon	I
Jill	III
Mark	II
Nicole	IIII
Steve	0

7.

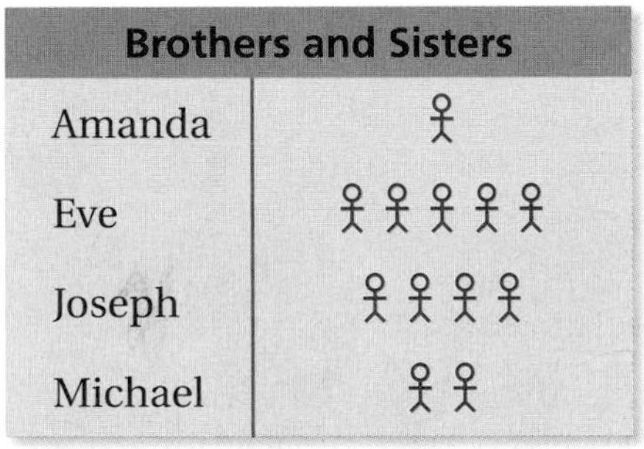
Brothers and Sisters
Amanda
Eve
Joseph
Michael

8.

Sit-ups		
108	85	94
103	112	115
98	119	126
105	82	89

9.

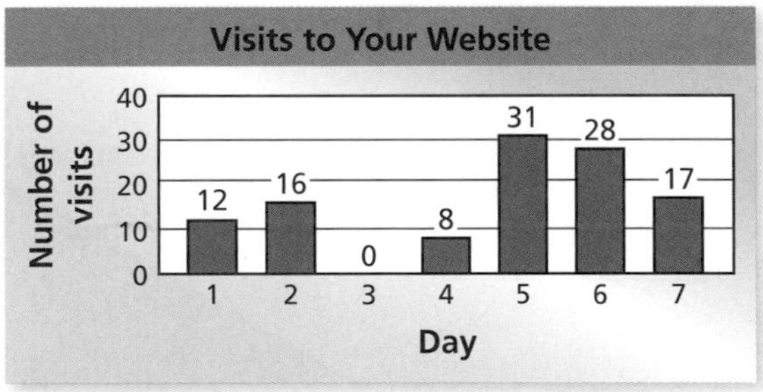

10. **GOLF** The table shows tournament finishes for a golfer.

   a. What was the golfer's mean finish?

   b. Identify two outliers for the data.

Tournament Finishes							
1	1	2	1	1	12	6	2
15	37	1	2	1	26	9	1

11. **COMMERCIALS** You and your friends are watching a television show. One of your friends asks, "How long are the commercial breaks during this show?"

   a. Is this a statistical question? Explain.

   b. Use the mean of the values in the table to answer the question.

Time (minutes)				
4.2	3.5	4.55	2.75	2.25

Month	Rainfall (inches)	Month	Rainfall (inches)
Jan	2.22	Jul	3.27
Feb	1.51	Aug	5.40
Mar	1.86	Sep	5.45
Apr	2.06	Oct	4.34
May	3.48	Nov	2.64
Jun	4.57	Dec	2.14

12. RAINFALL The table shows the monthly rainfall at a measuring station. What is the mean monthly rainfall?

13. OPEN-ENDED Create two different sets of data that have six values and a mean of 21.

3 **14. CELL PHONE** The bar graph shows your cell phone usage for five months.

a. Which data value is an outlier? Explain.

b. Find the mean with and without the outlier. Then describe how the outlier affects the mean.

c. Describe a situation that could have caused the outlier in this problem.

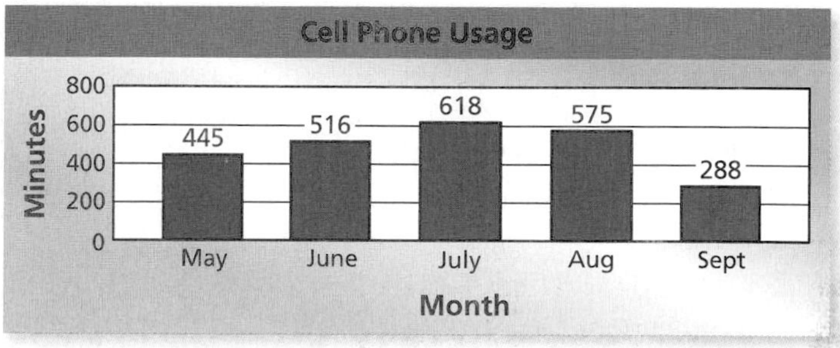

15. HEIGHT The table shows the heights of the volleyball players from two schools. What is the difference between the mean heights of the two teams? Do outliers affect either mean? Explain.

	Player Height (inches)														
Dolphins	59	65	53	56	58	61	64	68	51	56	54	57			
Tigers	63	68	66	58	54	55	61	62	53	70	64	64	62	67	69

16. REASONING Make a dot plot of the data set 11, 13, 17, 15, 12, 18, and 12. Use the dot plot to explain how the mean is the point where the data set is balanced.

17. ALLOWANCE In your class, 7 students do not receive a weekly allowance, 5 students receive \$3, 7 students receive \$5, 3 students receive \$6, and 2 students receive \$8. What is the mean weekly allowance? Explain how you found your answer.

18. **Precision** A collection of 8 backpacks has a mean weight of 14 pounds. A different collection of 12 backpacks has a mean weight of 9 pounds. What is the mean weight of the 20 backpacks? Explain how you found your answer.

Fair Game Review What you learned in previous grades & lessons

Evaluate the expression. *(Section 1.3)*

19. $\frac{8+10}{2}$ **20.** $\frac{26+34}{2}$ **21.** $\frac{18+19}{2}$ **22.** $\frac{14+17}{2}$

23. MULTIPLE CHOICE 60% of what number is 105? *(Section 5.6)*

(A) 63 (B) 175 (C) 630 (D) 1750

9.3 Measures of Center

Essential Question In what other ways can you describe an average of a data set?

1 ACTIVITY: Finding a Median

Work with a partner.

a. Write the total number of letters in the first and last names of 19 celebrities, historical figures, or people you know. Organize your data in a table. One person is already listed for you.

Person	Number of letters in first and last name
Abraham Lincoln	14

b. Order the values in your data set from least to greatest. Then write the data on a strip of grid paper with 19 boxes.

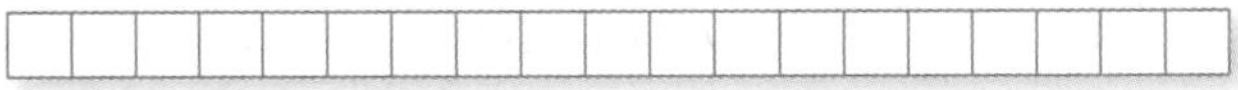

c. Place a finger on the square at each end of the strip. Move your fingers toward the center of the ordered data set until your fingers touch. On what value do your fingers touch?

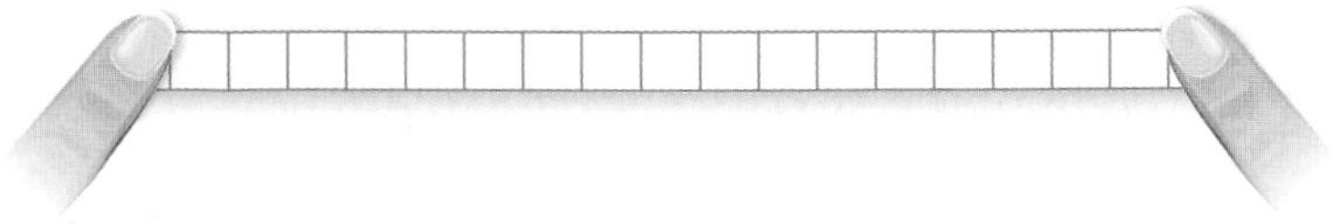

d. Now take your strip of grid paper and fold it in half. On what number is the crease? What do you notice? This value is called the *median.* How would you describe to another student what the median of a data set represents?

e. How many values are greater than the median? How many are less than the median?

f. Why do you think the median is considered an average of a data set?

COMMON CORE

Statistics

In this lesson, you will

- understand the concept of measures of center.
- find the median and mode of data sets.

Learning Standards
6.SP.2
6.SP.3
6.SP.5c

2 ACTIVITY: Adding a Value to a Data Set

Work with a partner.

a. How many total letters are in your first name and last name? Add this value to the ordered data set in Activity 1. How many values are now in your data set?

b. Write the ordered data, including your new value from part (a), on a strip of grid paper.

c. Repeat parts (c) and (d) from Activity 1. Explain your findings. How do you think you can find the median of this data set?

d. Compare the medians in Activities 1 and 2. Then answer the following questions. Explain your reasoning.

- Do you think the median always has to be a value in the data set?
- Do you think the median always has to be a whole number?

3 ACTIVITY: Finding a Mode

Work with a partner.

a. Make a dot plot for the data set in Activity 2. Describe the distribution of the data.

b. Which value occurs most often in the data set? This value is called the *mode*.

c. Do you think a data set can have no mode or more than one mode? Explain.

d. Do you think the mode always has to be a value in the data set? Explain.

e. Why do you think the mode is considered an average of a data set?

Math Practice 4

Use a Graph

How can you use the dot plot to find the mode?

What Is Your Answer?

4. **IN YOUR OWN WORDS** In what other ways can you describe an average of a data set?

5. Find the mean of your data set in Activity 2. Then compare the mean, median, and mode. Is there one measure that you think best represents your data set? Explain your reasoning.

Practice Use what you learned about the median of a data set to complete Exercises 5 and 6 on page 407.

9.3 Lesson

Check It Out
Lesson Tutorials
BigIdeasMath.com

Key Vocabulary
measure of center, *p. 404*
median, *p. 404*
mode, *p. 404*

A **measure of center** is a measure that describes the typical value of a data set. The mean is one type of measure of center. Here are two others.

Key Ideas

Median

Words Order the data. For a set with an odd number of values, the **median** is the middle value. For a set with an even number of values, the **median** is the mean of the two middle values.

Numbers **Data:** 5, 8, 9, 12, 14 — The median is 9.

Data: 2, 3, 5, 7, 10, 11 — The median is $\frac{5+7}{2}$, or 6.

Study Tip
The mode is the only measure of center that you can use to describe a set of data that is *not* made up of numbers.

Mode

Words The **mode** of a data set is the value or values that occur most often. Data can have one mode, more than one mode, or no mode. When all values occur only once, there is no mode.

Numbers **Data:** 11, 13, 15, 15, 18, 21, 24, 24

The modes are 15 and 24.

EXAMPLE 1 **Finding the Median and Mode**

Bowling Scores				
120	135	160	125	90
205	160	175	105	145

Find the median and mode of the bowling scores.

90, 105, 120, 125, 135, 145, 160, 160, 175, 205 — Order the data.

Median: $\frac{135 + 145}{2} = \frac{280}{2}$, or 140 — Add the two middle values and divide by 2.

Mode: 90, 105, 120, 125, 135, 145, 160, 160, 175, 205

The value 160 occurs most often.

∴ The median is 140. The mode is 160.

On Your Own

Now You're Ready
Exercises 7–12

Find the median and mode of the data.

1. 20, 4, 17, 8, 12, 9, 5, 20, 13
2. 100, 75, 90, 80, 110, 102

EXAMPLE 2 Finding the Mode

Favorite Types of Movies

Comedy	Drama	Horror
Horror	Drama	Horror
Comedy	Comedy	Action
Action	Comedy	Action
Horror	Drama	Comedy
Comedy	Comedy	Horror
Horror	Comedy	Action
Horror	Action	Drama

The list shows the favorite types of movies for students in a class. Organize the data in a frequency table. Then find the mode.

Type	Tally	Frequency
Action	~~IIII~~	5
Comedy	~~IIII~~ III	8
Drama	IIII	4
Horror	~~IIII~~ II	7

The number of tally marks is the frequency.

Make a tally for each vote.

Comedy received the most votes.

∴ So, the mode is comedy.

On Your Own

Now You're Ready Exercises 14–15

3. One member of the class was absent and ends up voting for horror. Does this change the mode? Explain.

EXAMPLE 3 Choosing the Best Measure of Center

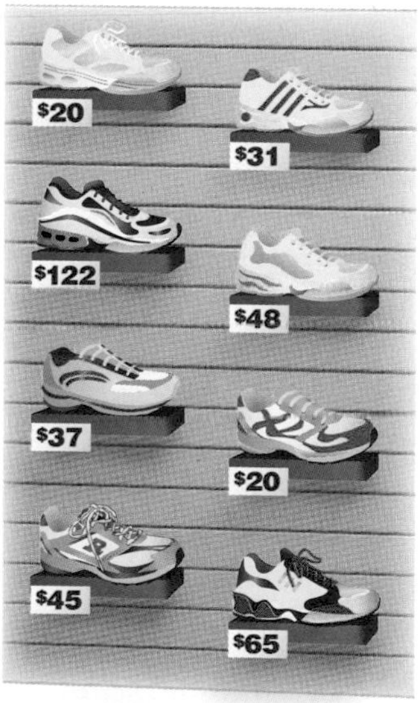

Find the mean, median, and mode of the sneaker prices. Which measure best represents the data?

Mean: $\frac{20 + 31 + 122 + 48 + 37 + 20 + 45 + 65}{8} = \frac{388}{8}$, or 48.5

Median: 20, 20, 31, 37, 45, 48, 65, 122 — Order from least to greatest.

$\frac{37 + 45}{2} = \frac{82}{2}$, or 41

Mode: 20, 20, 31, 37, 45, 48, 65, 122 — The value 20 occurs most often.

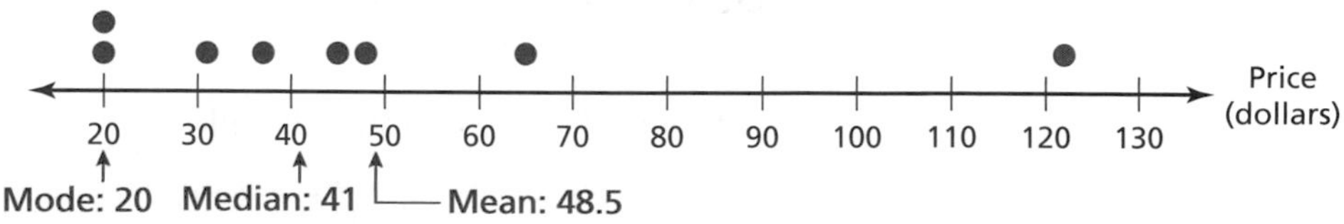

∴ The median best represents the data. The mode is less than most of the data, and the mean is greater than most of the data.

On Your Own

Now You're Ready Exercises 17–20

Find the mean, median, and mode of the data. Choose the measure that best represents the data. Explain your reasoning.

4. 1, 93, 46, 48, 34, 194, 67, 55

5. 96, 150, 102, 87, 150, 75

EXAMPLE 4 Removing an Outlier

Identify the outlier in Example 3. Find the mean, median, and mode without the outlier. Which measure does the outlier affect the most?

The price of $122 is much greater than any other price. So, it is the outlier.

	Mean	Median	Mode
With Outlier (Example 3)	48.5	41	20
Without Outlier	38	37	20

∴ The mean is affected the most by the outlier.

On Your Own

Now You're Ready
Exercises 21–22

6. The times (in minutes) it takes six students to travel to school are 8, 10, 10, 15, 20, and 45. Identify the outlier. Find the mean, median, and mode with and without the outlier. Which measure does the outlier affect the most?

EXAMPLE 5 Changing the Values of a Data Set

The prices of six video games at an online store are shown in the table. The price of each game increases by $4.98 when a shipping charge is included. How does this increase affect the mean, median, and mode?

Video Game Prices	
$53.42	$35.69
$18.99	$25.13
$27.97	$53.42

Make a new table by adding $4.98 to each price. Then find the mean, median, and mode of both data sets.

Video Game Prices with Shipping Charge	
$58.40	$40.67
$23.97	$30.11
$32.95	$58.40

	Mean	Median	Mode
Original Price	35.77	31.83	53.42
Price with Shipping Charge	40.75	36.81	58.4

Compare:

Mean: $40.75 - 35.77 = 4.98$

Median: $36.81 - 31.83 = 4.98$

Mode: $58.4 - 53.42 = 4.98$

∴ By increasing each video game price by $4.98 for shipping, the mean, median, and mode all increase by $4.98.

On Your Own

7. **WHAT IF?** The store decreases the price of each video game by $3. How does this decrease affect the mean, median, and mode?

9.3 Exercises

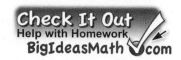

Check It Out
Help with Homework
BigIdeasMath.com

Vocabulary and Concept Check

1. **NUMBER SENSE** Give an example of a data set that has no mode.

2. **WRITING** Which is affected most by an outlier: the mean, median, or mode? Explain.

3. **WHICH ONE DOESN'T BELONG** Which word does *not* belong with the other three? Explain.

median	outlier	mode	mean

4. **NUMBER SENSE** A data set has a mean of 7, a median of 5, and a mode of 8. Which of the numbers 7, 5, and 8 *must* be in the data set? Explain.

Practice and Problem Solving

Use grid paper to find the median of the data.

5. 9, 7, 2, 4, 3, 5, 9, 6, 8, 0, 3, 8

6. 16, 24, 13, 36, 22, 26, 22, 28, 25

Find the median and mode(s) of the data.

(1) 7. 3, 5, 7, 9, 11, 3, 8

8. 14, 19, 16, 13, 16, 14

9. 93, 81, 94, 71, 89, 92, 94, 99

10. 44, 13, 36, 52, 19, 27, 33

11. 12, 33, 18, 28, 29, 12, 17, 4, 2

12. 55, 44, 40, 55, 48, 44, 58, 67

13. **ERROR ANALYSIS** Describe and correct the error in finding the median of the data.

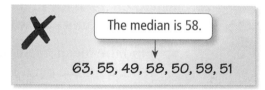

✗ The median is 58.

63, 55, 49, 58, 50, 59, 51

Find the mode(s) of the data.

(2) 14.

Shirt Color		
Black	Blue	Red
Pink	Black	Black
Gray	Green	Blue
Blue	Blue	Red
Yellow	Blue	Blue
Black	Orange	Black
Black		

15.

Talent Show Acts		
Singing	Dancing	Comedy
Singing	Singing	Dancing
Juggling	Dancing	Singing
Singing	Poetry	Dancing
Comedy	Magic	Dancing
Poetry	Singing	Singing

16. **REASONING** In Exercises 14 and 15, can you find the mean and median of the data? Explain.

Find the mean, median, and mode(s) of the data. Choose the measure that best represents the data. Explain your reasoning.

3 **17.** 48, 12, 11, 45, 48, 48, 43, 32

18. 12, 13, 40, 95, 88, 7, 95

19. 2, 8, 10, 12, 56, 9, 5, 2, 4

20. 126, 62, 144, 81, 144, 103

Find the mean, median, and mode(s) of the data with and without the outlier. Describe the effect of the outlier on the measures of center.

4 **21.** 45, 52, 17, 63, 57, 42, 54, 58

22. 85, 77, 211, 88, 91, 84, 85

Find the mean, median, and mode(s) of the data.

23. 4.7, 8.51, 6.5, 7.42, 9.64, 7.2, 9.3

24. $8\frac{1}{2}, 6\frac{5}{8}, 3\frac{1}{8}, 5\frac{3}{4}, 6\frac{5}{8}, 5\frac{1}{4}, 10\frac{5}{8}, 4\frac{1}{2}$

25. WEATHER The weather forecast for a week is shown.

	Sun	Mon	Tue	Wed	Thu	Fri	Sat

High	90° F	91° F	89° F	97° F	101° F	99° F	91° F
Low	74° F	78° F	77° F	77° F	83° F	78° F	72° F

a. Find the mean, median, and mode(s) of the high temperatures. Which measure best represents the data? Explain your reasoning.

b. Repeat part (a) for the low temperatures.

26. RESEARCH Find the unit costs of 10 different kinds of cereal. Choose one cereal whose unit cost will be an outlier.

a. Find the mean, median, and mode(s) of the data. Which measure best represents the data? Explain your reasoning.

b. Identify the outlier in the data set. Find the mean, median, and mode(s) of the data set without the outlier. Which measure does the outlier affect the most?

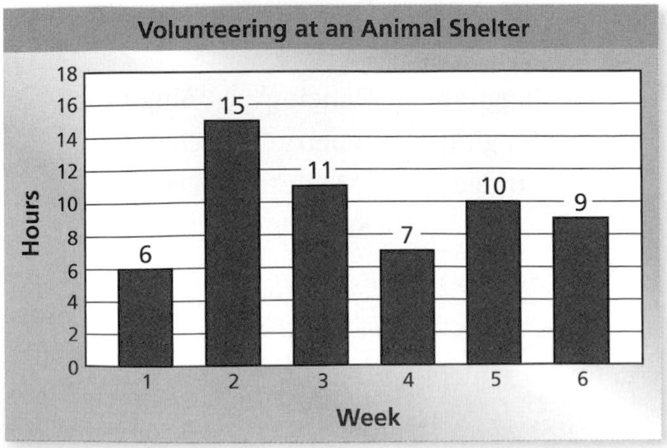

27. PROBLEM SOLVING The bar graph shows the numbers of hours you volunteered at an animal shelter. What is the minimum number of hours you need to work in the seventh week to justify that you worked an average of 10 hours for the 7 weeks? Explain your answer using measures of center.

28. REASONING Why do you think the mode is the least frequently used measure to describe a data set? Explain.

29. **MOTOCROSS** The ages of the racers in a bicycle motocross race are 14, 22, 20, 25, 26, 17, 21, 30, 27, 25, 14, and 29. The 30-year-old drops out of the race and is replaced with a 15-year-old. How are the mean, median, and mode of the ages affected?

30. **CAMERAS** The data are the prices of several digital cameras at a store.

$130 $170 $230 $130
$250 $275 $130 $185

a. Does the price shown in the advertisement represent the prices well? Explain.

b. Why might the store use this advertisement?

c. In this situation, why might a person want to know the mean? the median? the mode? Explain.

31. **SALARIES** The table shows the monthly salaries for employees at a company.

Monthly Salaries (dollars)				
1940	1660	1860	2100	1720
1540	1760	1940	1820	1600

a. Find the mean, median, and mode of the data.

b. Each employee receives a 5% raise. Find the mean, median, and mode of the data with the raise. How does this increase affect the mean, median, and mode of the data?

c. Use the original monthly salaries to calculate the annual salaries. Find the mean, median, and mode of the annual salaries. How are these values related to the mean, median, and mode of the monthly salaries?

32. **Critical Thinking** Consider the algebraic expressions $3x$, $9x$, $4x$, $23x$, $6x$, and $3x$. Assume $x > 0$.

a. Find the mean, median, and mode.

b. Is there an outlier? If so, what is it?

Fair Game Review What you learned in previous grades & lessons

Find the value of the expression. *(Section 1.1)*

33. $48 - 35$ **34.** $188 - 123$ **35.** $416 - 297$ **36.** $6249 - 3374$

37. MULTIPLE CHOICE A shelf in your room can hold at most 30 pounds. There are 12 pounds of books already on it. Which inequality represents the number of pounds you can add to the shelf? *(Section 7.6)*

(A) $x < 18$ (B) $x \geq 18$ (C) $x \leq 42$ (D) $x \leq 18$

9 Study Help

Check It Out
Graphic Organizer
BigIdeasMath.com

You can use a **concept circle** to organize information about a concept. Here is an example of a concept circle for a statistical question.

Statistical Question

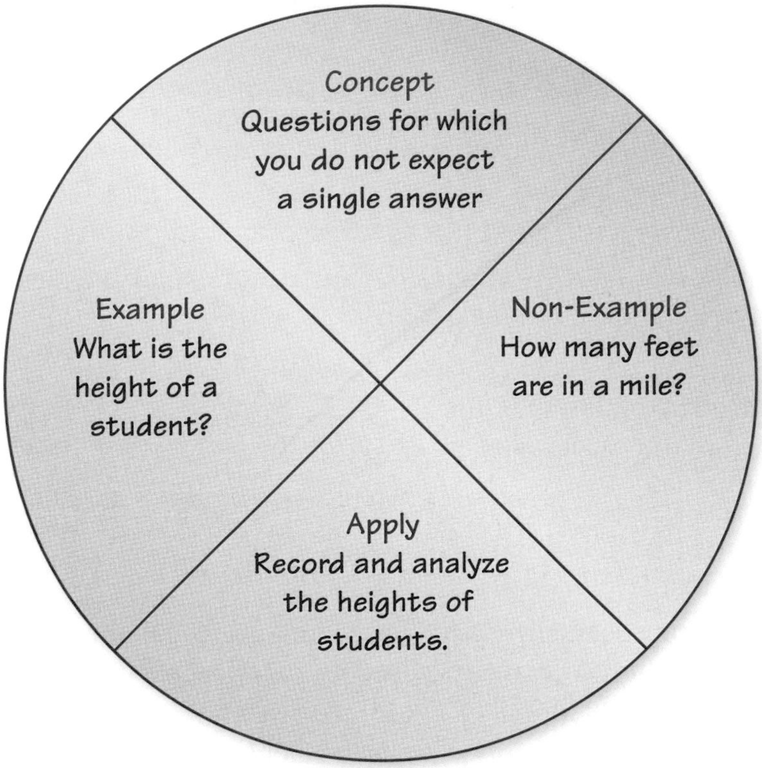

Concept
Questions for which you do not expect a single answer
Example
What is the height of a student?
Non-Example
How many feet are in a mile?
Apply
Record and analyze the heights of students.

On Your Own

Make concept circles to help you study these topics.

1. mean
2. outlier
3. measures of center
4. median
5. mode

After you complete this chapter, make concept circles for the following topics.

6. measures of variation
7. range
8. quartiles
9. interquartile range
10. mean absolute deviation

Concept
Dog biscuits as rewards
Justify
Newton is always hungry.
Example
LIVER
CHEESE
Apply
Woof! Woof!
Yes, you got your point across!

"Do you think this concept circle will help my owner understand that 'Speak' and 'Sit' need motivation?"

Display the data in a dot plot. Identify any clusters, peaks, or gaps in the data. *(Section 9.1)*

1.

Weight (grams)			
42	40	37	42
43	41	42	43
37	41	41	42

2.

Time (seconds)				
63	66	65	60	59
59	64	58	65	58
64	60	59	64	63

Find the mean of the data. *(Section 9.2)*

3.

Tour Dates	
May	III
June	卌 卌
July	卌 卌 IIII
August	卌 卌 卌 III
September	卌

4.

Scores	
Judge 1	8.9
Judge 2	9.4
Judge 3	8.6
Judge 4	9.1

Find the median and the mode(s) of the data. *(Section 9.3)*

5. 3, 5, 9, 11, 3

6. 24, 4, 37, 56, 6, 56, 45

Find the mean, median, and mode(s) of the data. Choose the measure that best represents the data. Explain your reasoning. *(Section 9.3)*

7. 47, 147, 24, 47, 38, 42

8. 34, 57, 58, 56, 21

Hours of Exercise				
5	1	5	3	5
4	5	2	5	4
3	4	6	5	6

9. EXERCISE You conduct a survey to answer: "How many hours does a sixth-grade student spend exercising during a week?" The table shows the results. *(Section 9.1)*

a. Is this a statistical question? Explain.

b. Display the data in a dot plot. Identify any clusters, peaks, or gaps in the data.

c. Use the distribution of the data to answer the question.

10. EMAILS The number of emails you received in 5 days is shown. What is the mean number of emails you received per day? *(Section 9.2)*

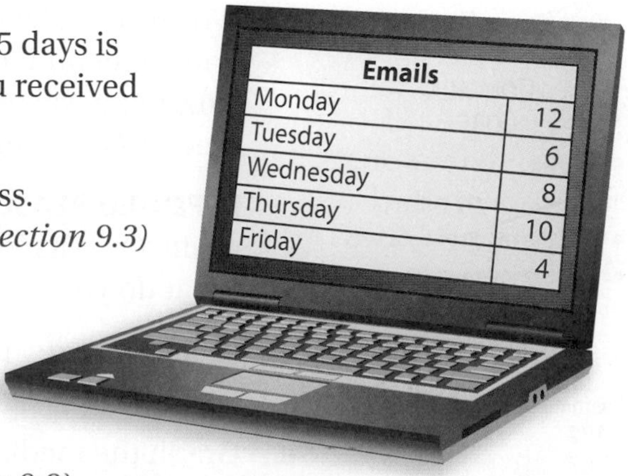

11. QUIZZES The data are your quiz scores for a class. Find the median and the mode of the data. *(Section 9.3)*

18, 19, 17, 14, 20, 20, 15, 21

12. MUSIC The data are the lengths of the songs (in minutes) on your new CD. Which measure of center best represents the data with and without the outlier? Explain. *(Section 9.3)*

2.2, 2.2, 2.4, 2.6, 2.8, 3.0, 3.2, 3.4, 14.2

9.4 Measures of Variation

Essential Question How can you describe the spread of a data set?

1 ACTIVITY: Interpreting Statements

Work with a partner. There are 24 students in your class. Your teacher makes the following statements:

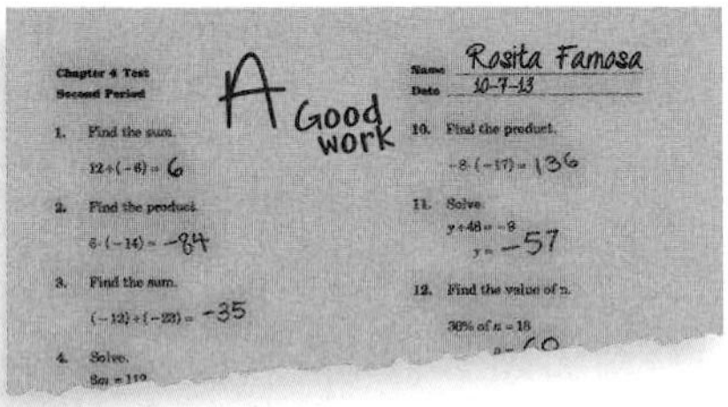

- ***"The exam scores range from 75% to 96%."***
- ***"Most of the students received high scores."***

a. What do you think the first statement means? Explain.

b. In the first statement, is your teacher describing the center of the data set? If not, what do you think your teacher is describing?

c. What do you think the scores are for most of the students in the class? Explain your reasoning.

d. Use your teacher's statements to make a dot plot that can represent the distribution of the exam scores of the class.

2 ACTIVITY: Grouping Data

Work with a partner. The numbers of U.S. states visited by each student in a sixth grade class are shown.

Number of States Visited

1	7	5	2
11	6	3	20
4	18	1	6
2	7	1	8
10	2	12	5
	3	21	

a. Between what values do the data range?

b. Write the ordered data values on a strip of grid paper and fold it to find the median. How many values are greater than the median? How many are less than the median?

c. **REPEATED REASONING** Fold the strip in half again. On what values are the two new creases? What do you think these values represent?

d. Into how many parts did you divide the data set? How many data values are in each part?

e. Graph the median and the values you found in parts (a) and (c) on a number line. Are the distances the same between these points?

f. How can you use these values to describe the spread of the data?

COMMON CORE

Statistics

In this lesson, you will

- find the range of data sets.
- find the interquartile range of data sets.
- check for outliers in data sets.

Learning Standards
6.SP.2
6.SP.3
6.SP.5c

3 ACTIVITY: Adding a Value to a Data Set

Work with a partner. A new student joins the class in Activity 2. The new student has visited 41 states.

a. Add this value to the ordered data set in Activity 2. Does your answer to part (a) change? Explain.

b. How does the distribution of the data change when this value is added? Explain your reasoning.

c. How does adding this value affect the values on your number line in part (e) of Activity 2?

4 ACTIVITY: Analyzing Data Sets

Work with a partner. Identify the data set that is the least spread out and the data set that is the most spread out. Explain your reasoning.

a.

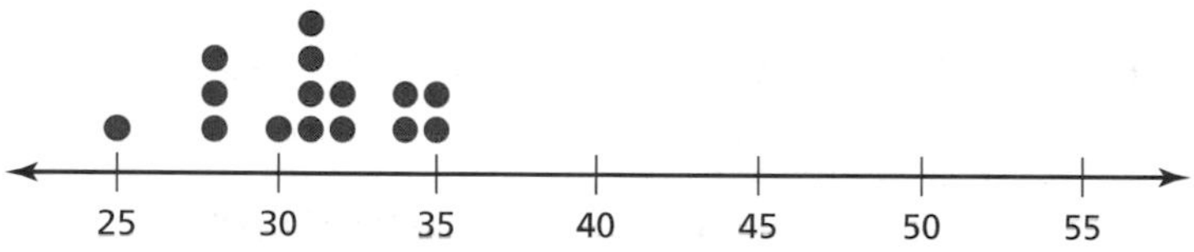

b.

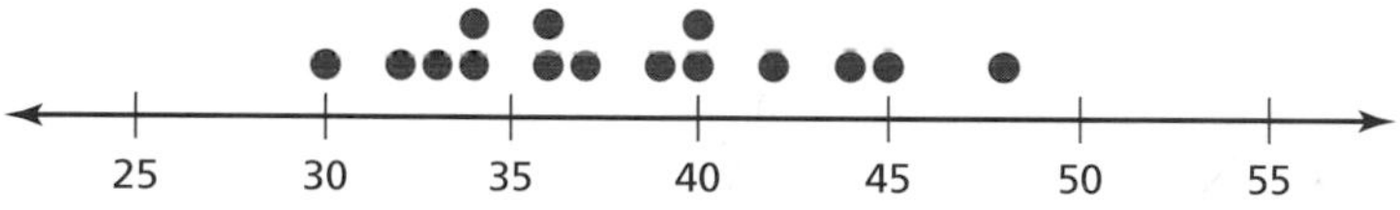

c.

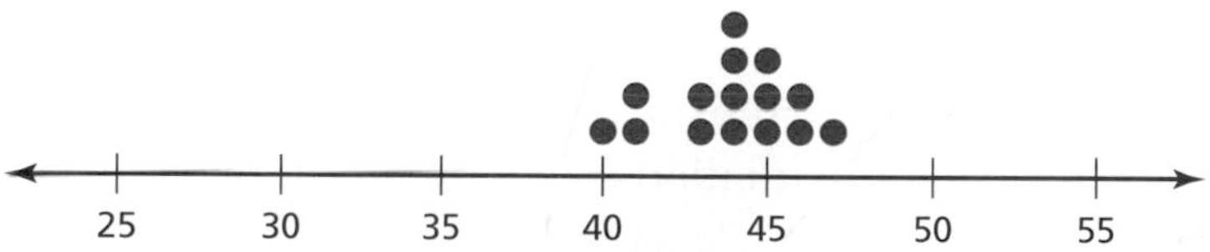

d.

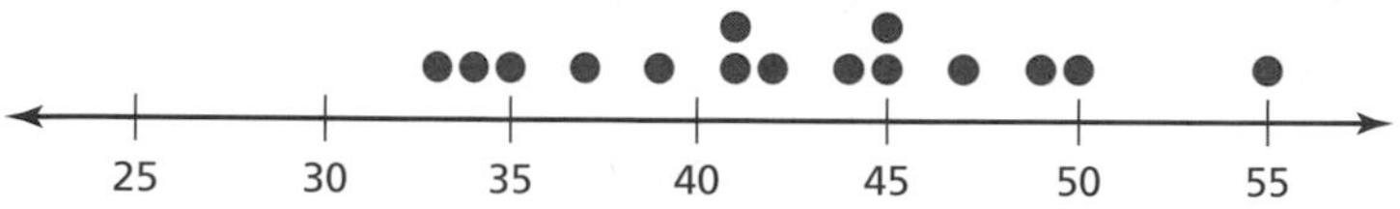

Math Practice 1

Analyze Givens

How can you use the given information to determine how spread out the data are?

What Is Your Answer?

5. **IN YOUR OWN WORDS** How can you describe the spread of a data set?

6. Make a dot plot of the data set in Activity 2. Describe any similarities between the dot plot and the number line in part (e).

Practice Use what you learned about variation to complete Exercises 4 and 5 on page 416.

9.4 Lesson

Check It Out
Lesson Tutorials
BigIdeasMath.com

A **measure of variation** is a measure that describes the distribution of a data set. A simple measure of variation to find is the *range*. The **range** of a data set is the difference between the greatest value and the least value.

Key Vocabulary
measure of variation, *p. 414*
range, *p. 414*
quartiles, *p. 414*
first quartile, *p. 414*
third quartile, *p. 414*
interquartile range, *p. 414*

EXAMPLE 1 Finding the Range

The table shows the lengths of several Burmese pythons captured for a study. Find and interpret the range of their lengths.

Lengths (feet)	
18.5	8
11	10
14	15.5
12.5	6.25
16.25	5

To find the least and the greatest values, order the lengths from least to greatest.

5, 6.25, 8, 10, 11, 12.5, 14, 15.5, 16.25, 18.5

The least value is 5. The greatest value is 18.5.

∴ So, the range of the lengths is 18.5 − 5, or 13.5 feet. This means that the lengths vary by no more than 13.5 feet.

On Your Own

Now You're Ready
Exercises 6–9

1. The ages of people in line for a roller coaster are 15, 17, 21, 32, 41, 30, 25, 52, 16, 39, 11, and 24. Find and interpret the range of their ages.

Key Ideas

Quartiles

The **quartiles** of a data set divide the data into four equal parts. Recall that the median (second quartile) divides the data set into two halves.

lower half — Median = 29 — upper half

18 21 **22** 24 28 | 30 31 **32** 36 37

The median of the lower half is the **first quartile**, Q_1.

The median of the upper half is the **third quartile**, Q_3.

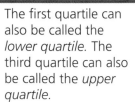

Reading
The first quartile can also be called the *lower quartile*. The third quartile can also be called the *upper quartile*.

Interquartile Range (IQR)

The difference between the third quartile and the first quartile is called the **interquartile range**. The IQR represents the range of the middle half of the data and is another measure of variation.

18 21 22 24 28 30 31 32 36 37

$$\text{IQR} = Q_3 - Q_1$$
$$= 32 - 22$$
$$= 10$$

Multi-Language Glossary at BigIdeasMath.com

EXAMPLE 2 Finding the Interquartile Range

The dot plot shows the top speeds of 12 sports cars. Find and interpret the interquartile range of the data.

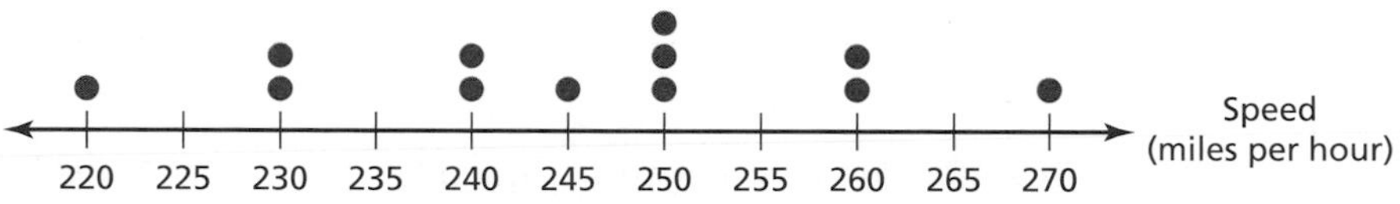

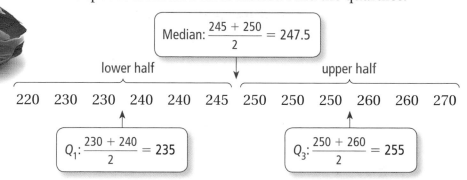

Order the speeds from slowest to fastest. Find the quartiles.

Median: $\frac{245 + 250}{2} = 247.5$

lower half: 220 230 230 240 240 245

upper half: 250 250 250 260 260 270

Q_1: $\frac{230 + 240}{2} = 235$

Q_3: $\frac{250 + 260}{2} = 255$

So, the interquartile range is $255 - 235 = 20$. This means that the middle half of the speeds vary by no more than 20 miles per hour.

You can use the quartiles and the interquartile range to check for outliers.

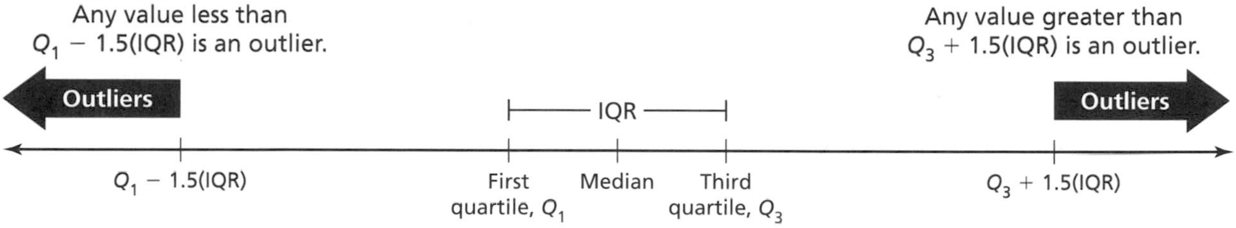

EXAMPLE 3 Checking for Outliers

Check for outliers in the data set in Example 2.

$Q_1 - 1.5(\text{IQR})$	Outlier boundaries	$Q_3 + 1.5(\text{IQR})$
$235 - 1.5(20)$	Substitute values.	$255 + 1.5(20)$
205	Simplify.	285

There are no speeds less than 205 miles per hour or greater than 285 miles per hour. So, the data set has no outliers.

On Your Own

Now You're Ready
Exercises 11–14 and 17

2. The number of pages in each of an author's novels is shown.

   356, 364, 390, 468, 400, 382, 376, 396, 350

   a. Find and interpret the interquartile range of the data.

   b. Does this data set contain any outliers? Justify your answer.

9.4 Exercises

Check It Out
Help with Homework
BigIdeasMath.com

Vocabulary and Concept Check

1. **VOCABULARY** How are measures of center different from measures of variation?
2. **VOCABULARY** How many quartiles does a data set have?
3. **DIFFERENT WORDS, SAME QUESTION** Which is different? Find "both" answers.

53, 47, 60, 45, 62, 59, 65, 50, 56, 48

What is the interquartile range of the data?	What is the range of the data?
What is the range of the middle half of the data?	What is the difference between the third quartile and the first quartile?

Practice and Problem Solving

Use grid paper to find the median of the data. Then find the median of the lower half and the median of the upper half of the data. Describe the spread of the data.

4. 5, 8, 10, 1, 7, 6, 15, 8, 6
5. 82, 62, 95, 81, 89, 51, 72, 56, 97, 98, 79, 85

Find the range of the data.

(1) 6. 26, 21, 27, 33, 24, 29
7. 52, 40, 49, 48, 62, 54, 44, 58, 39
8. 133, 117, 152, 127, 168, 146, 174
9. 4.8, 5.5, 4.2, 8.9, 3.4, 7.5, 1.6, 3.8
10. **ERROR ANALYSIS** Describe and correct the error in finding the range of the data.

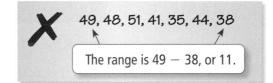

✗ 49, 48, 51, 41, 35, 44, 38

The range is 49 − 38, or 11.

Find the median, first quartile, third quartile, and interquartile range of the data.

(2) 11. 40, 33, 37, 54, 41, 34, 27, 39, 35
12. 84, 75, 90, 87, 99, 91, 85, 88, 76, 92, 94
13. 132, 127, 106, 140, 158, 135, 129, 138
14. 38, 55, 61, 56, 46, 67, 59, 75, 65, 58
15. **PAPER AIRPLANE** The table shows the distances traveled by a paper airplane. Find and interpret the range and the interquartile range of the distances.

Distances (feet)			
$13\frac{1}{2}$	$21\frac{1}{2}$	21	$16\frac{3}{4}$
$10\frac{1}{4}$	19	32	$26\frac{1}{2}$
29	$16\frac{1}{4}$	$28\frac{1}{2}$	$18\frac{1}{2}$

16. **WRITING** Consider a data set that has no mode. Which measure of variation is greater, the range or the interquartile range? Explain your reasoning.

3 17. **OUTLIERS** Use the interquartile range to identify any outliers in Exercises 11–14.

18. **REASONING** How does an outlier affect the range of a data set? Explain.

19. **BASKETBALL** The table shows the numbers of points scored by players on a basketball team.

Points Scored					
21	53	74	82	84	93
103	108	116	122	193	

a. Find the range and the interquartile range of the data.

b. Use the interquartile range to identify the outlier(s) in the data set. Find the range and the interquartile range of the data set without the outlier(s). Which measure did the outlier(s) affect more?

20. **STRUCTURE** Two data sets have the same range. Can you assume that the interquartile ranges of the two data sets are about the same? Give an example to justify your answer.

21. **SINGING** The tables show the ages of the finalists for two reality singing competitions.

a. Find the mean, median, range, and interquartile range of the ages for each show. Compare the results.

b. A 21-year-old is voted off Show A, and the 36-year-old is voted off Show B. How do these changes affect the measures in part (a)? Explain.

Ages for Show A	
18	17
15	21
22	16
18	28
24	21

Ages for Show B	
21	20
23	13
15	18
17	22
36	25

22. **Open-Ended** Create a set of data with 7 values that has a mean of 30, a median of 26, a range of 50, and an interquartile range of 36.

Fair Game Review What you learned in previous grades & lessons

Find the mean of the data. *(Section 9.2)*

23. 8, 14, 22, 7, 2, 11, 25, 7, 5, 9

24. 55, 64, 58, 43, 49, 67

25. **MULTIPLE CHOICE** What is the surface area of the rectangular prism? *(Section 8.2)*

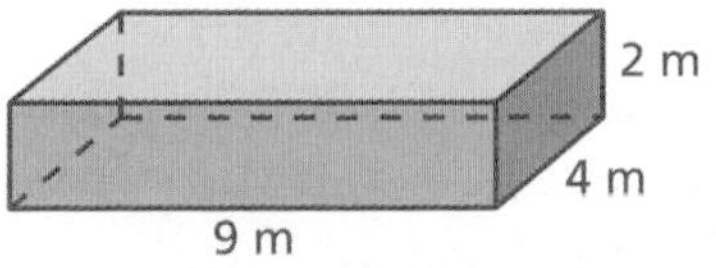

Ⓐ 62 m^2 Ⓑ 72 m^2

Ⓒ 88 m^2 Ⓓ 124 m^2

9.5 Mean Absolute Deviation

Essential Question How can you use the distances between each data value and the mean of a data set to measure the spread of a data set?

The Meaning of a Word ● Deviate

When you **deviate** from something, you stray or depart from the normal course of action.

1 ACTIVITY: Finding Distances From the Mean

Work with a partner. The table shows the exam scores of 14 students in your class.

Exam Scores			
Ben	89	Mike	95
Emma	86	Hong	96
Jeremy	80	Rob	92
Pete	80	Amy	90
Ryan	96	Sue	76
Dan	94	Kim	84
Lucy	89	Heather	85

a. What is the mean exam score?

b. Make a dot plot of the data. Place an "X" on the number line to represent the mean.

c. Is the number of exam scores that are greater than the mean equal to the number of exam scores that are less than the mean? Explain.

d. Which exam score *deviates* the most from the mean? Which exam score *deviates* the least from the mean? Explain how you found your answers.

e. Overall, do you think the exam scores are *close* to the mean or *far away* from the mean? Explain your reasoning.

COMMON CORE

Statistics

In this lesson, you will

- understand the meaning of *mean absolute deviation.*
- find the mean absolute deviation of data sets.

Learning Standards
6.SP.2
6.SP.3
6.SP.5c

2 ACTIVITY: Using Distances from the Mean

Work with a partner. Use the information in Activity 1.

a. Complete the table below. Add rows if needed. Be sure to find the sum of the values in the last column of the table.

Math Practice 2

Use Operations
What operation can you use to find the distance from the mean? Explain.

Student with Score *Less Than* the Mean	Exam Score	Distance from the Mean
	Sum:	

b. Create a table similar to the one above for students with scores *greater than* the mean.

c. **LOGIC** What do you notice about the sums you found in your tables? Why do you think this happens?

3 ACTIVITY: Interpreting Distances from the Mean

Work with a partner.

a. **LOGIC** Add the sums you found in your tables in Activity 2. Divide that amount by the total number of students. Round your result to the nearest tenth.

In your own words, what do you think this value represents?

b. **REASONING** In a data set, what do you think it means when the value you found in part (a) is close to 0? Explain.

What Is Your Answer?

4. **IN YOUR OWN WORDS** How can you use the distances between each data value and the mean of a data set to measure the spread of a data set?

5. **REASONING** Find the range and the interquartile range of the data set in Activity 1. What do you think it means when these values are close to 0? Explain.

Practice Use what you learned about distances from the mean to complete Exercises 3 and 4 on page 422.

9.5 Lesson

Check It Out
Lesson Tutorials
BigIdeasMath.com

Key Vocabulary
mean absolute deviation, *p. 420*

Another measure of variation is the *mean absolute deviation.* The **mean absolute deviation** is an average of how much data values differ from the mean.

Key Idea

Finding the Mean Absolute Deviation (MAD)

Step 1: Find the mean of the data.

Step 2: Find the distance between each data value and the mean.

Step 3: Find the sum of the distances in Step 2.

Step 4: Divide the sum in Step 3 by the total number of data values.

EXAMPLE 1 Finding the Mean Absolute Deviation

You record the numbers of raisins in 8 scoops of cereal. Find and interpret the mean absolute deviation of the data.

1, 2, 2, 2, 4, 4, 4, 5

Step 1: Mean $= \frac{1+2+2+2+4+4+4+5}{8} = \frac{24}{8} = 3$

Step 2: You can use a dot plot to organize the data. Replace each dot with its distance from the mean.

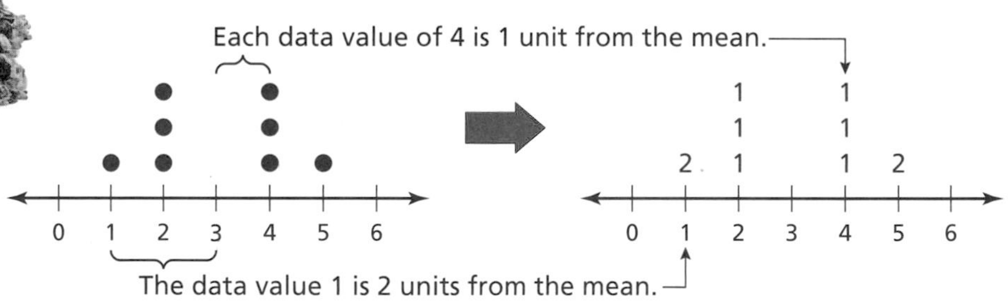

Each data value of 4 is 1 unit from the mean.
0 1 2 3 4 5 6
2 1 1 1 1 1 1 2
0 1 2 3 4 5 6
The data value 1 is 2 units from the mean.

Step 3: The sum of the distances is $2 + 1 + 1 + 1 + 1 + 1 + 1 + 2 = 10$.

Step 4: The mean absolute deviation is $\frac{10}{8} = 1.25$.

So, the data values differ from the mean by an average of 1.25 raisins.

On Your Own

Now You're Ready
Exercises 5–8

1. Find and interpret the mean absolute deviation of the data.

5, 8, 8, 10, 13, 14, 16, 22

Multi-Language Glossary at BigIdeasMath.com

EXAMPLE 2 Real-Life Application

Mendoza		
Date	Win/Loss	Runs
Aug 8	-	4
Aug 3	-	6
Jul 29	L	6
Jul 24	W	0
Jul 13	L	8
Jul 8	-	4
Jul 7	L	5
Jul 2	-	0
Jun 27	W	2
Jun 22	W	0

Rodriguez		
Date	Win/Loss	Runs
Aug 7	L	6
Aug 2	W	4
Jul 28	W	4
Jul 22	-	5
Jul 17	W	0
Jul 8	W	2
Jul 3	L	3
Jun 28	L	2
Jun 23	W	4
Jun 17	W	5

The smartphones show the numbers of runs allowed by two pitchers in their last 10 starts.

a. Find the mean, median, and mean absolute deviation of the numbers of runs allowed for each pitcher.

Order the runs allowed for Mendoza:
0, 0, 0, 2, 4, 4, 5, 6, 6, 8.

Mean $= \frac{35}{10} = 3.5$ Median $= \frac{4+4}{2} = 4$

Mean absolute deviation:

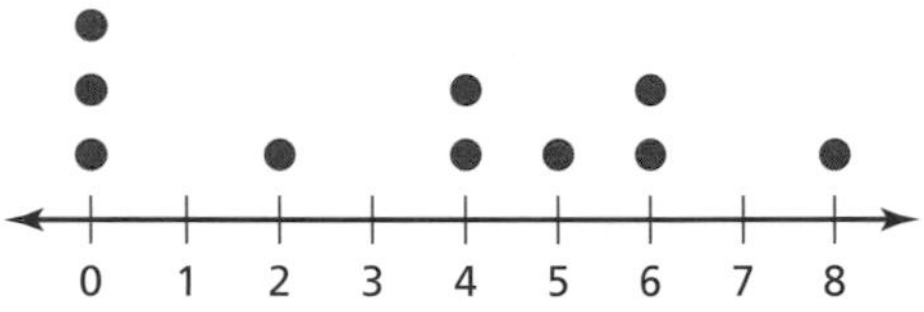

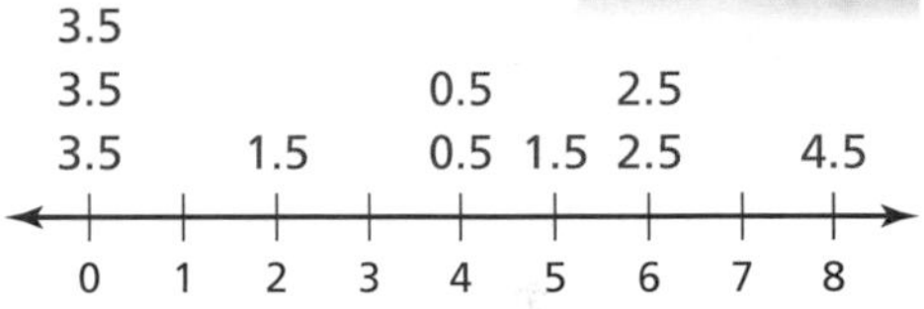

The mean absolute deviation is $\frac{24}{10} = 2.4$

Order the runs allowed for Rodriguez: 0, 2, 2, 3, 4, 4, 4, 5, 5, 6.

Mean $= \frac{35}{10} = 3.5$ Median $= \frac{4+4}{2} = 4$

Mean absolute deviation:

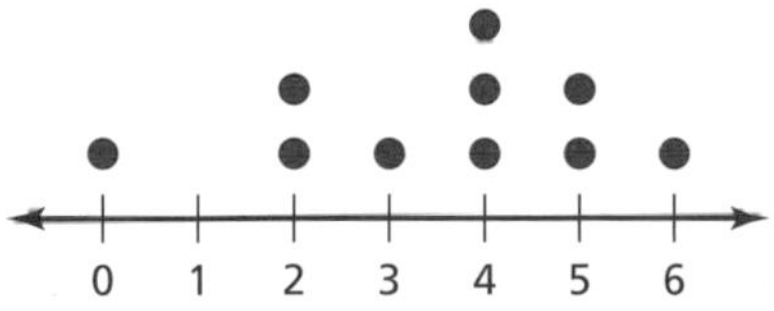

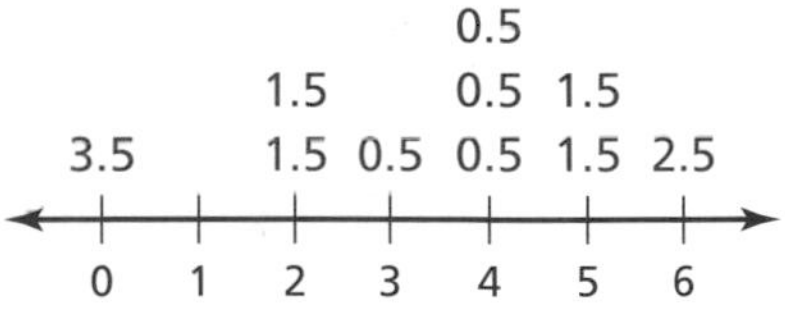

The mean absolute deviation is $\frac{14}{10} = 1.4$.

b. Which measure can you use to distinguish the data? What can you conclude about the pitchers from this measure?

You cannot use the measures of center to distinguish the data because they are the same for each data set. The measure of variation, MAD, is 2.4 for Mendoza and 1.4 for Rodriguez. This indicates that the data for Rodriguez has less variation.

∴ Using the MAD to distinguish the data, you can conclude that Rodriguez is more consistent than Mendoza.

Study Tip

The greater the mean absolute deviation, the greater the variation of the data.

On Your Own

2. **WHAT IF?** Mendoza allows 4 runs in the next game. How would you expect the mean absolute deviation to change? Explain.

9.5 Exercises

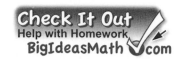

Vocabulary and Concept Check

1. **REASONING** Describe a data set that has a mean absolute deviation of 0.
2. **WHICH ONE DOESN'T BELONG?** Which one does *not* belong with the other three? Explain your reasoning.

range | interquartile range | mean | mean absolute deviation

Practice and Problem Solving

Find the average distance each data value in the set is from the mean. Round your answer to the nearest tenth, if necessary.

3. Model years of used cars on a lot: 2010, 2002, 2005, 2007, 2001
4. Prices of kites at a shop: $7, $20, $9, $35, $12, $15, $7, $10, $20, $25

Find and interpret the mean absolute deviation of the data. Round your answer to the nearest tenth, if necessary.

1 5.

Prices of Microphones (dollars)				
25	28	20	22	32
28	35	34	30	36

6.

Heights of 10-Year-Old Octuplets (inches)			
61	61	61	61
61	61	61	61

7.

Capacities of Stadiums (thousands of people)		
101.5	95.4	109.8
98.7	92.3	104.7

8.

Numbers of Visitors to a Website During a Week			
103	115	124	125
171	165	170	

9. **ERROR ANALYSIS** Describe and correct the error in finding the mean absolute deviation of the data set 35, 40, 38, 32, 42, and 41.

✗

$$\text{mean} = \frac{35 + 40 + 38 + 32 + 42 + 41}{6} = 38$$

$$\text{MAD} = \frac{3 + 2 + 6 + 4 + 3}{5} = 3.6$$

So, the values differ from the mean by an average of 3.6.

10. **MUSEUMS** The data set shows the admission prices at several museums.

$20, $20, $16, $12, $15, $25, $11

Find and interpret the range, interquartile range, and mean absolute deviation of the data.

11. **MENU** The table shows the prices of the five most-expensive and least-expensive dishes on a menu. Find the MAD of each data set. Then compare their variations.

Five Most-Expensive Dishes					Five Least-Expensive Dishes				
\$28	\$30	\$28	\$39	\$25	\$7	\$7	\$10	\$8	\$12

12. **COINS** The data sets show the years of the coins in two collections.

Derek's collection: 1950, 1952, 1908, 1902, 1955, 1954, 1901, 1910

Paul's collection: 1929, 1935, 1928, 1930, 1925, 1932, 1933, 1920

Find the measures of center and the measures of variation for each data set. Compare the measures. What can you conclude?

13. **PROBLEM SOLVING** You survey students in your class about the number of movies they watched last month. The results are shown in the table.

Movies Watched			
7	5	14	5
6	9	10	12
15	4	5	8
11	10	9	2

a. Find the measures of center and the measures of variation for the data.

b. A new student joins the class who watched 21 movies last month. Is 21 an outlier? How does including this value affect the measures of center and the measures of variation? Explain.

REASONING Which data set do you think would have the greater mean absolute deviation? Explain your reasoning.

14. guesses for number of gumballs in a jar
guesses for number of baseballs in a jar

15. monthly rainfall amounts in a city
monthly amounts of water used in a home

16. **REASONING** The MAD of a data set is considered a more reliable measure of variation than the range or the interquartile range. Why do you think this is true?

17. **Critical Thinking** Add and subtract the MAD from the mean in the original data set in Exercise 13.

a. What percent of the values are within one MAD of the mean? two MADs of the mean? Which values are more than twice the MAD from the mean?

b. What do you notice as you get more and more MADs away from the mean? Explain.

Fair Game Review What you learned in previous grades & lessons

Find the mean, median, and mode(s) of the data. *(Section 9.2 and Section 9.3)*

18. 4, 6, 7, 9, 6, 4, 5, 6, 8, 10

19. 1.2, 1.7, 1.7, 2.1, 1.4, 1.2, 1.9

20. **MULTIPLE CHOICE** What is the surface area of the square pyramid? *(Section 8.3)*

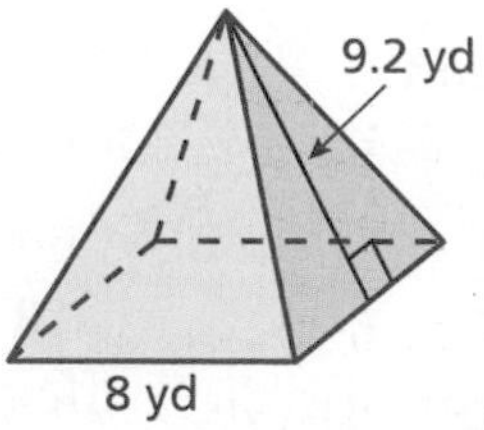

Ⓐ 100.8 yd^2 Ⓑ 147.2 yd^2

Ⓒ 211.2 yd^2 Ⓓ 368 yd^2

9.4–9.5 Quiz

Check It Out
Progress Check
BigIdeasMath.com

Find the range of the data. *(Section 9.4)*

1. 35, 76, 43, 58, 34, 67

2. 19, 21, 22, 22, 19, 25, 24, 23, 24

Find the median, first quartile, third quartile, and interquartile range of the data. *(Section 9.4)*

3. 56, 48, 72, 37, 35, 42, 48, 33, 28

4. 95, 14, 86, 62, 55, 46, 28, 37, 33, 70, 31

Find and interpret the mean absolute deviation of the data. Round your answer to the nearest tenth if necessary. *(Section 9.5)*

5.

Ages of Television Show Viewers (years)			
29	18	26	33
33	22	34	26

6.

Prices of Houses (thousands of dollars)				
80	120	95	240	140
75	135	110	90	125

7. AMUSEMENT PARKS The data set shows the admission prices at several amusement parks.

$65, $70, $40, $55, $35, $40, $60

Find and interpret the range, interquartile range, and mean absolute deviation of the data. *(Section 9.4 and Section 9.5)*

8. TEACHING EXPERIENCE The tables show the years of teaching experience of faculty members at two schools. *(Section 9.4)*

a. Find the mean, median, range, and interquartile range of the years of experience for each school. Compare the results.

b. The teacher with 11 years of experience leaves School A, and the teacher with 33 years of experience retires from School B. How does this affect the measures in part (a)? Explain.

School A: Teaching Experience (years)	
5	11
10	22
7	8
8	6
10	35

School B: Teaching Experience (years)	
4	15
6	12
10	33
12	20
8	7

9. BOOK CLUB You survey the students in your book club about the number of books they read last summer. The results are shown in the table. *(Section 9.4 and Section 9.5)*

Books Read			
8	14	15	9
6	12	9	13
11	11	7	5
12	6	10	8

a. Find the measures of center and the measures of variation for the data.

b. A new student who read 18 books last summer joins the club. Is 18 an outlier? How does adding this value to the data set affect the measures of center and variation? Explain.

9 Chapter Review

Check It Out
Vocabulary Help
BigIdeasMath.com

Review Key Vocabulary

statistics, *p. 392*
statistical question, *p. 392*
mean, *p. 398*
outlier, *p. 399*
measure of center, *p. 404*
median, *p. 404*
mode, *p. 404*
measure of variation, *p. 414*
range, *p. 414*
quartiles, *p. 414*
first quartile, *p. 414*
third quartile, *p. 414*
interquartile range, *p.414*
mean absolute deviation, *p. 420*

Review Examples and Exercises

9.1 Introduction to Statistics *(pp. 390–395)*

Display the data in a dot plot. Identify any clusters, peaks, or gaps in the data.

Heights (inches)				
70	71	70	72	69
68	69	71	64	70
64	63	72	70	67

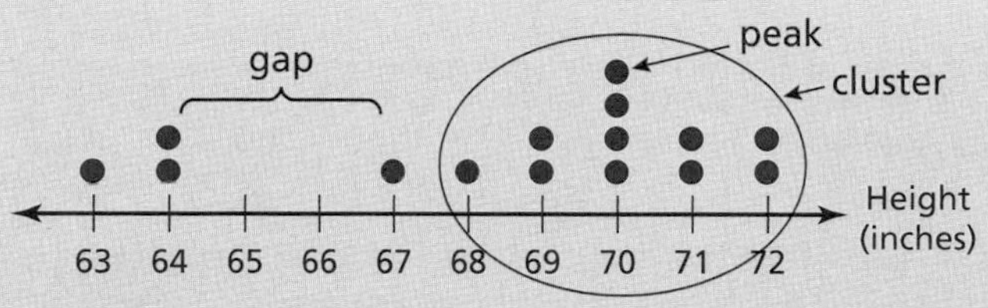

Exercises

Display the data in a dot plot. Identify any clusters, peaks, or gaps in the data.

1.

Distance (feet)			
56	55	56	57
58	54	51	55
51	56	49	56

2.

Weight (pounds)				
83	88	89	90	89
91	89	84	90	92
90	88	89	83	88

9.2 Mean *(pp. 396–401)*

Find the mean of 5, 9, 10, 6, 6, and 12.

$$\text{mean} = \frac{5 + 9 + 10 + 6 + 6 + 12}{6}$$ ← sum of the data; ← number of values

$$= \frac{48}{6}, \text{ or } 8$$ Simplify.

Exercises

Find the mean of the data.

3. 4, 5, 7, 14, 17, 12, 18

4. 15, 5, 8, 12, 5, 9, 4, 10, 2, 11

9.3 Measures of Center (pp. 402–409)

Find the median and the mode of the movie lengths in the table.

Order the data from least to greatest.

Movie Lengths (minutes)		
91	112	126
142	122	112
92	144	

Median:

91, 92, 112, 112, 122, 126, 142, 144

$$\frac{112 + 122}{2} = \frac{234}{2}, \text{ or } 117$$

Mode:

91, 92, 112, 112, 122, 126, 142, 144

The value 112 occurs most often.

The median is 117 minutes, and the mode is 112 minutes.

Exercises

Find the median and the mode(s) of the data.

5. 8, 8, 6, 8, 4, 5, 6

6. 24, 74, 61, 29, 38, 27, 68, 54

9.4 Measures of Variation (pp. 412–417)

The table shows the weights of several adult emperor penguins. (a) Find and interpret the range. (b) Find and interpret the interquartile range. (c) Check for outliers.

Weights (kilograms)	
25	27
36	23.5
33.5	31.25
30.75	32
24	29.25

a. Ordered from least to greatest, the weights are 23.5, 24, 25, 27, 29.25, 30.75, 31.25, 32, 33.5, and 36.

So, the range of the weights is 36 − 23.5, or 12.5 kilograms. The weights vary by no more than 12.5 kilograms.

b. Find the quartiles.

Median: $\frac{29.25 + 30.75}{2} = 30$

lower half: 23.5, 24, 25, 27, 29.25,

upper half: 30.75, 31.25, 32, 33.5, 36

Q_1: 25

Q_3: 32

So, the interquartile range is 32 − 25 = 7. This means that the middle half of the weights vary by no more than 7 kilograms.

c. Calculate the outlier boundaries.

$$Q_1 - 1.5(\text{IQR}) = 25 - 1.5(7) = 14.5$$

$$Q_3 + 1.5(\text{IQR}) = 32 + 1.5(7) = 42.5$$

There are no weights less than 14.5 kilograms or greater than 42.5 kilograms. So, the data set has no outliers.

Exercises

Find the range of the data.

7. 45, 76, 98, 21, 52, 39

8. 95, 63, 52, 8, 93, 16, 42, 37, 62

Find the median, first quartile, third quartile, and interquartile range of the data.

9. 28, 46, 25, 76, 18, 25, 47, 83, 44

10. 14, 25, 97, 55, 66, 28, 92, 38, 94

9.5 Mean Absolute Deviation *(pp. 418–423)*

You record the prices of 8 printers. Find and interpret the mean absolute deviation of the data.

\$120, \$150, \$90, \$110,

\$140, \$120, \$140, \$90

Step 1: Mean $= \dfrac{120 + 150 + 90 + 110 + 140 + 120 + 140 + 90}{8} = \dfrac{960}{8} = 120$

Step 2: Use a dot plot to organize the data. Replace each dot with its distance from the mean.

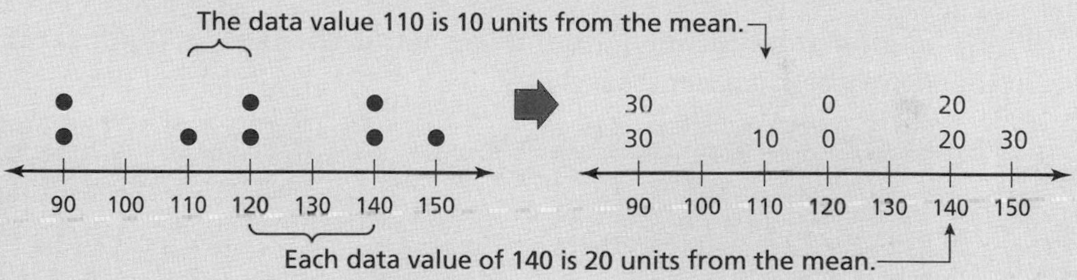

Step 3: The sum of the distances is $30 + 30 + 10 + 0 + 0 + 20 + 20 + 30 = 140$.

Step 4: The mean absolute deviation is $\dfrac{140}{8} = 17.50$.

∴ The data values differ from the mean by an average of \$17.50.

Exercises

Find and interpret the mean absolute deviation of the data. Round your answer to the nearest tenth if necessary.

11.

Shoe Sizes			
6	8.5	6	9
10	7	8	9.5

12.

Prices of Monitors (dollars)				
130	150	190	100	175
120	165	140	180	190

9 Chapter Test

Check It Out
Test Practice
BigIdeasMath.com

Display the data in a dot plot. Identify any clusters, peaks, or gaps in the data.

1.

Time (minutes)			
33	40	32	40
39	38	40	39
38	39	39	33

2.

Temperature (°F)				
81	81	80	82	81
83	76	83	76	80
75	83	82	82	81

Find the mean, median, and mode(s) of the data.

3. 2, 7, 7, 12, 4

4. 4, 5, 7, 5, 9, 9, 10, 6

Find the mean, median, and mode(s) of the data. Choose the measure that best represents the data. Explain your reasoning.

5. 5, 6, 4, 24, 18

6. 46, 27, 94, 56, 53, 65, 43

Find the range of the data.

7. 24, 56, 9, 83, 77, 14

8. 43, 12, 55, 91, 25, 86, 84, 23, 1

Find the median, first quartile, third quartile, and interquartile range of the data.

9. 32, 58, 19, 36, 44, 57, 11, 26, 74

10. 36, 24, 49, 32, 37, 28, 38, 40, 39

Find and interpret the mean absolute deviation of the data. Round your answer to the nearest tenth if necessary.

11.

Distances Driven (miles)			
312	286	196	201
158	225	206	192

12.

Prices of Sunglasses (dollars)				
15	8	19	20	18
20	22	14	10	15

13. HOTEL The table shows the numbers of guests at a hotel on different days.

Numbers of Guests					
66	58	90	57	63	55
60	62	56	54	72	

a. Find the range and the interquartile range of the data.

b. Use the interquartile range to identify the outlier(s) in the data set. Find the range and the interquartile range of the data set without the outlier(s). Which measure did the outlier(s) affect more?

14. JOBS The data sets show the numbers of hours worked each week by two friends for several weeks.

Greg's hours: 9, 18, 12, 6, 9, 21, 3, 12
Tom's hours: 12, 18, 15, 16, 14, 12, 15, 18

Find the measures of center and the measures of variation for each data set. Compare the measures. What can you conclude?

9 Standards Assessment

1. What is the value of the expression below? *(6.NS.1)*

$$8\frac{4}{9} \div 4\frac{2}{3}$$

A. $1\frac{17}{21}$

B. $2\frac{2}{3}$

C. $32\frac{8}{27}$

D. $39\frac{11}{27}$

2. What is the value of the expression below? *(6.NS.3)*

$$4.18 + 6.225 + 5.7$$

F. 15.005

G. 15.105

H. 16.005

I. 16.105

Test-Taking Strategy
Use Intelligent Guessing

What is the mean length of these hyena fangs: 4 in., 3 in., 3 in., 4 in., 5 in., 5 in.?
A 6 in.
B 1/3 ft
C 2 in.
D 5 in.
MEOW!

"The mean can't be 6 or 2 or 5 inches. So, you can use intelligent guessing to find that the answer is $\frac{1}{3}$ ft, or 4 in."

3. One number is missing from the data set in the box below.

18, 24, 22, 30, 26, _____, 25

The median of the data set is 24. What is the greatest possible value of the missing number? *(6.SP.5c)*

4. The number of hours that each of 6 students spent reading last week is shown in the bar graph below.

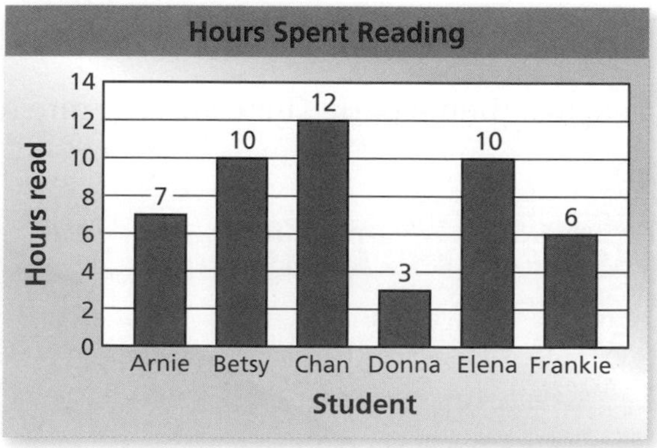

For the data in the bar graph, which measure is the *least*? *(6.SP.5c)*

A. mean

B. median

C. mode

D. range

5. You go to a beach and collect buckets of shells. Of the many shells you have collected, you notice the following.

- 9% of the seashells are auger shells.
- $\frac{1}{8}$ of the seashells are coquina shells.
- 11% of the seashells are rough scallop shells.
- 0.1 of the seashells are fighting conch shells.

Which list correctly shows the types of shells in order from least to greatest? *(6.NS.6c)*

F. auger, coquina, rough scallop, fighting conch

G. fighting conch, coquina, auger, rough scallop

H. fighting conch, auger, rough scallop, coquina

I. auger, fighting conch, rough scallop, coquina

6. What is the mean absolute deviation of the data shown in the line plot, rounded to the nearest tenth? *(6.SP.5c)*

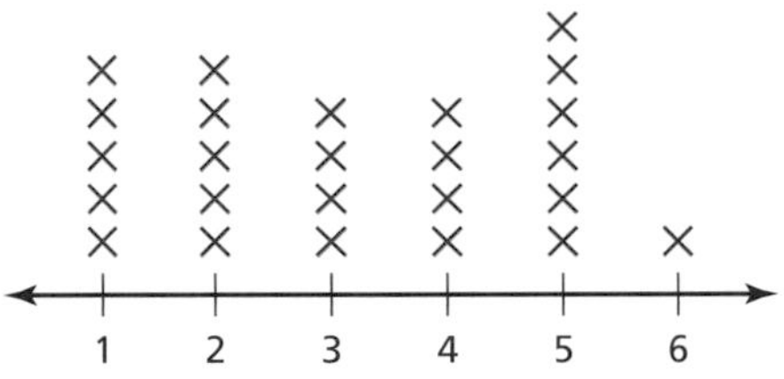

A. 1.4

B. 3

C. 3.2

D. 5

7. A family wants to buy tickets to a theme park. There are separate ticket prices for adults and children.

Which expression represents the total cost, in dollars, for a adult tickets and c child tickets? *(6.EE.2a)*

F. $600(a + c)$

G. $50(a \times c)$

H. $30a + 20c$

I. $30a \times 20c$

8. What is the value of the expression below? *(6.NS.3)*

$$52.8 \div 0.16$$

9. What is the value of the expression below when $a = 6$ and $b = 14$? *(6.EE.2c)*

$$0.8a + 0.02b$$

A. 0.4828

B. 0.8814

C. 5.08

D. 16.4

10. Which property was *not* used in the box below to simplify the expression? *(6.EE.3)*

$$\begin{aligned} 0.3 \times 53 + 53 \times 0.7 &= 53 \times 0.3 + 53 \times 0.7 \\ &= 53 \times (0.3 + 0.7) \\ &= 53 \times 1 \\ &= 53 \end{aligned}$$

F. Distributive Property

G. Associative Property of Addition

H. Identity Property of Multiplication

I. Commutative Property of Multiplication

11. Determine a data set of 5 numbers that has the following measures:

Think Solve Explain

- a mean of 7 and
- a median of 9.

Explain how you determined your data set. Then demonstrate that the mean of your data set is 7 and the median is 9. *(6.SP.5c)*

12. What is the value of the expression below? *(6.RP.3c)*

25% of 400

A. 16

B. 100

C. 1000

D. 10,000

10 Data Displays

"I took a survey of pet owners on how many times per day you should treat your dog to a biscuit."

"What do you think?"

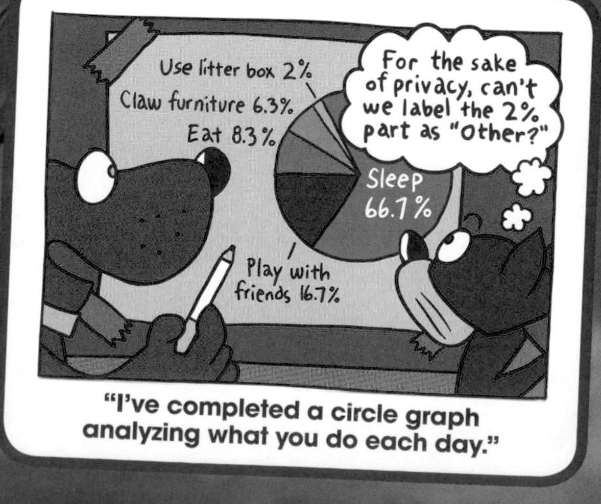

"I've completed a circle graph analyzing what you do each day."

What You Learned Before

"Okay, I have the box. But I need your help to complete my box-and-whisker plot."

Analyzing Bar Graphs (3.MD.3)

Example 1 **The bar graph shows the favorite colors of the students in a class. How many students said their favorite color is blue?**

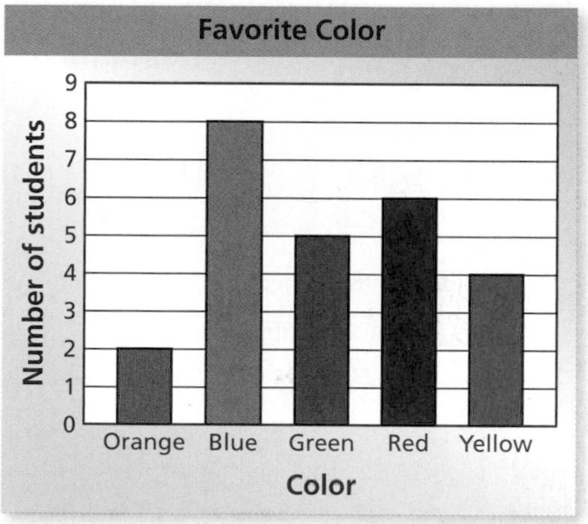

The height of the bar labeled "Blue" is 8.

∴ So, 8 students said their favorite color is blue.

Try It Yourself

1. What color was chosen the least?

2. How many students said green or red is their favorite color?

3. How many students did *not* choose yellow as their favorite color?

4. How many students are in the class?

Finding Percents (6.RP.3c)

Example 2 **The circle graph shows the favorite fruits of the students in a class. There are 20 students in the class. How many students said their favorite fruit is an orange?**

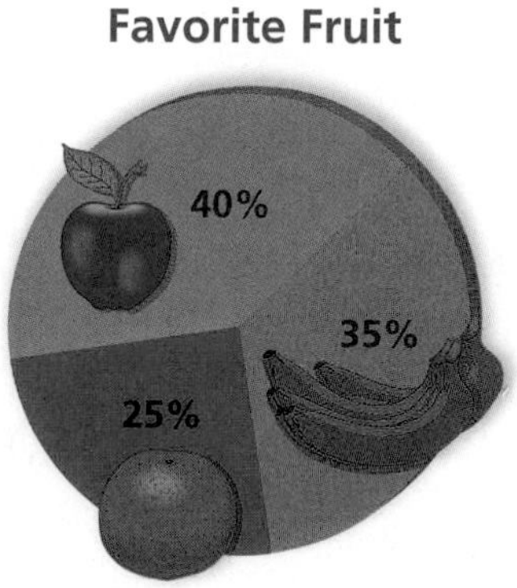

Find 25% of 20.

$$25\% \text{ of } 20 = \frac{1}{4} \cdot 20 = \frac{1 \cdot \overset{5}{\cancel{20}}}{\underset{1}{\cancel{4}}} = 5$$

∴ So, 5 students said their favorite fruit is an orange.

Try It Yourself

5. How many students said their favorite fruit is an apple?

6. How many students said their favorite fruit is a banana?

10.1 Stem-and-Leaf Plots

Essential Question How can you use place values to represent data graphically?

1 ACTIVITY: Making a Data Display

Work with a partner. The list below gives the ages of these women when they became first ladies of the United States.

THE WHITE HOUSE
WASHINGTON

Frances Cleveland – 21
Caroline Harrison – 56
Ida McKinley – 49
Edith Roosevelt – 40
Helen Taft – 48
Ellen Wilson – 52
Florence Harding – 60
Grace Coolidge – 44
Lou Hoover – 54
Eleanor Roosevelt – 48
Elizabeth Truman – 60

Mamie Eisenhower – 56
Jacqueline Kennedy – 31
Claudia Johnson – 50
Patricia Nixon – 56
Elizabeth Ford – 56
Rosalynn Carter – 49
Nancy Reagan – 59
Barbara Bush – 63
Hillary Clinton – 45
Laura Bush – 54
Michelle Obama – 45

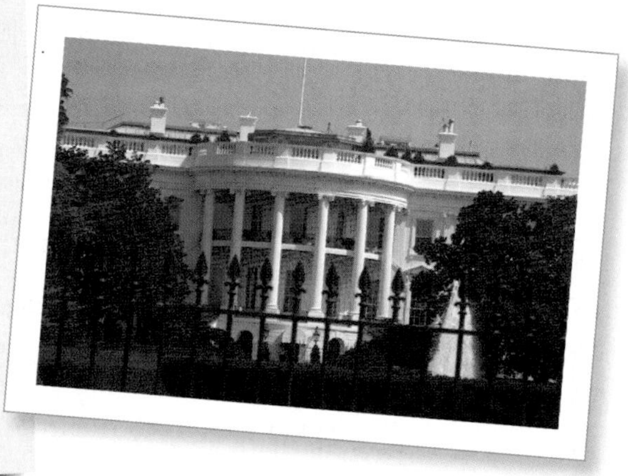

a. The incomplete data display shows the ages of the first ladies in the left column of the list above.

What do the numbers to the left of the line represent? What do the numbers to the right of the line represent?

Ages of First Ladies

2	1
3	
4	0 4 8 8 9
5	2 4 6
6	0 0

b. This data display is called a *stem-and-leaf plot.* What numbers do you think represent the *stems? leaves?* Explain your reasoning.

c. Complete the stem-and-leaf plot using the remaining ages in the right column. Order the numbers to the right of the line in numerical order.

d. REASONING Write a question about the ages of first ladies that would be easier to answer using a stem-and-leaf plot than a dot plot.

COMMON CORE
Data Displays
In this lesson, you will
- make and interpret stem-and-leaf plots.

Applying Standard 6.SP.4

2 ACTIVITY: Making a Back-to-Back Stem-and-Leaf Plot

Work with a partner. The table below shows the ages of presidents of the United States from 1885 to 2009 on their first inauguration day.

Ages of Presidents										
47	55	54	42	51	56	55	51	54	51	60
62	43	55	56	61	52	69	64	46	54	47

a. On your stem-and-leaf plot from Activity 1(c), draw a vertical line to the left of the display. Represent the ages of the presidents by including numbers to the left of the line.

b. Find the median ages of both the first ladies and the presidents of the United States.

c. Compare the distribution of each data set.

3 ACTIVITY: Conducting an Experiment

Math Practice 4

Interpret Results

How can you use the stem-and-leaf plot to interpret your results? Explain.

Work with a partner. Use two number cubes to conduct the following experiment.

- Toss the cubes and find the product of the resulting numbers.
- Repeat this process 30 times. Record your results.

a. Use a stem-and-leaf plot to organize your results.

b. Describe the distribution of the data.

What Is Your Answer?

4. **IN YOUR OWN WORDS** How can you use place values to represent data graphically?

5. How can you display data in a stem-and-leaf plot whose values range from 82 through 129?

Practice Use what you learned about stem-and-leaf plots to complete Exercises 4 and 5 on page 438.

10.1 Lesson

Check It Out
Lesson Tutorials
BigIdeasMath.com

Key Vocabulary
stem-and-leaf plot, *p. 436*
stem, *p. 436*
leaf, *p. 436*

Key Idea

Stem-and-Leaf Plots

A **stem-and-leaf plot** uses the digits of data values to organize a data set. Each data value is broken into a **stem** (digit or digits on the left) and a **leaf** (digit or digits on the right).

A stem-and-leaf plot shows how data are distributed.

Stem	Leaf
2	0 0 1 2 5 7
3	1 4 8
4	2
5	8 9

Key: 2|0 = 20

The *key* explains what the stems and leaves represent.

EXAMPLE 1 Making a Stem-and-Leaf Plot

	A	B
1	DATE	MINUTES
2	JULY 9	55
3	JULY 9	3
4	JULY 9	6
5	JULY 10	14
6	JULY 10	18
7	JULY 10	5
8	JULY 10	23
9	JULY 11	30
10	JULY 11	23
11	JULY 11	10
12	JULY 11	2
13	JULY 11	36

Make a stem-and-leaf plot of the length of the 12 cell phone calls.

Step 1: Order the data.

2, 3, 5, 6, 10, 14, 18, 23, 23, 30, 36, 55

Step 2: Choose the stems and the leaves. Because the data values range from 2 to 55, use the *tens* digits for the stems and the *ones* digits for the leaves. Be sure to include the key.

Step 3: Write the stems to the *left* of the vertical line.

Step 4: Write the leaves for each stem to the *right* of the vertical line.

Cell Phone Call Lengths

Stem	Leaf
0	2 3 5 6
1	0 4 8
2	3 3
3	0 6
4	
5	5

Key: 1|4 = 14 minutes

Order the stems vertically. The stem for data values less than 10 is 0.

Write the leaves horizontally.

Include stems without leaves.

On Your Own

Now You're Ready
Exercises 4–9

1. Make a stem-and-leaf plot of the hair lengths.

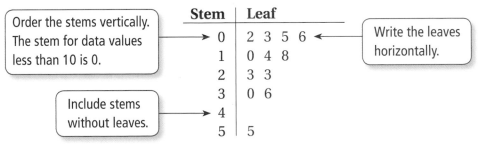

Hair Length (centimeters)									
5	1	20	12	27	2	30	5	7	38
40	47	1	2	1	32	4	44	33	23

EXAMPLE 2 Interpreting a Stem-and-Leaf Plot

Test Scores

Stem	Leaf
6	6
7	0 5 7 8
8	1 1 3 4 4 6 8 8 9
9	0 2 9
10	0

Key: 9 | 2 = 92 points

The stem-and-leaf plot shows student test scores. (a) How many students scored less than 80 points? (b) How many students scored at least 90 points? (c) How are the data distributed?

a. There are five scores less than 80 points: 66, 70, 75, 77, and 78.

∴ Five students scored less than 80 points.

b. There are four scores of at least 90 points: 90, 92, 99, and 100.

∴ Four students scored at least 90 points.

c. There are few low test scores and few high test scores. So, most of the scores are in the middle.

On Your Own

Now You're Ready
Exercises 12–15

2. Use the grading scale at the right.

a. How many students received a B on the test?

b. How many students received a C on the test?

A: 90–100
B: 80–89
C: 70–79
D: 60–69
F: 59 and below

EXAMPLE 3 Making Conclusions from a Stem-and-Leaf Plot

Which statement is *not* true?

Ⓐ Most of the plants are less than 20 inches tall.

Ⓑ The median plant height is 11 inches.

Ⓒ The range of the plant heights is 35 inches.

Ⓓ The plant height that occurs most often is 11 inches.

Plant Heights

Stem	Leaf
0	1 2 4 5 6 8 9
1	0 1 1 5 7
2	2 5
3	6

Key: 1 | 5 = 15 inches

There are 15 plant heights. So, the median is the eighth data value, 10 inches.

∴ The correct answer is Ⓑ.

On Your Own

3. You are told that three plants are taller than 20 inches. Is the statement true? Explain.

10.1 Exercises

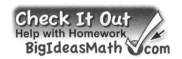

Check It Out
Help with Homework
BigIdeasMath.com

Vocabulary and Concept Check

1. **VOCABULARY** The key for a stem-and-leaf plot is $3|4 = 34$. Which number is the stem? Which number is the leaf?
2. **WRITING** Describe how to make a stem-and-leaf plot of the data values 14, 22, 9, 13, 30, 8, 25, and 29.
3. **WRITING** How does a stem-and-leaf plot show the distribution of data?

Practice and Problem Solving

Make a stem-and-leaf plot of the data.

① 4.

Books Read			
26	15	20	9
31	25	29	32
17	26	19	40

5.

Hours Online			
8	12	21	14
18	6	15	24
12	17	2	0

6.

Test Scores (%)				
87	82	95	91	69
88	68	87	65	81
97	85	80	90	62

7.

Points Scored				
58	50	42	71	75
45	51	43	38	71
42	70	56	58	43

8.

Bikes Sold			
78	112	105	99
86	96	115	100
79	81	99	108

9.

Minutes in Line			
4.0	2.6	1.9	3.1
3.6	2.2	2.7	3.8
1.6	2.0	3.1	2.9

10. **ERROR ANALYSIS** Describe and correct the error in making a stem-and-leaf plot of the data.

51, 25, 47, 42, 55, 26, 50, 44, 55

✗

Stem	Leaf
2	5 6
4	2 4 7
5	0 1 5 5

Key: $4|2 = 42$

11. **PUPPIES** The weights (in pounds) of eight puppies at a pet store are 12, 24, 17, 8, 18, 31, 24, and 15. Make a stem-and-leaf plot of the data. Describe the distribution of the data.

VOLLEYBALL The stem-and-leaf plot shows the number of *digs* for the top 15 players at a volleyball tournament.

Stem	Leaf
4	1 1 3 3 5
5	0 2 3 4
6	2 3 3 7
7	5
8	
9	7

Key: 5 | 0 = 50 digs

(2) **12.** How many players had more than 60 digs?

13. Find the mean, median, mode, range, and interquartile range of the data.

14. Describe the distribution of the data.

15. Which data value is the outlier? Describe how the outlier affects the mean.

16. REASONING Each stem-and-leaf plot below has a mean of 39. Without calculating, determine which stem-and-leaf plot has the lesser mean absolute deviation. Explain your reasoning.

Stem	Leaf
2	3 7
3	0 2 6 9
4	1 2 5 8
5	1 4

Key: 4 | 1 = 41

Stem	Leaf
2	2 4 5 8 9
3	3 8
4	5
5	3 6 7 8

Key: 5 | 3 = 53

17. TEMPERATURE The stem-and-leaf plot shows the daily high temperatures (in degrees Fahrenheit) for the first 15 days of a month.

Stem	Leaf
6	7 8
7	0 0 3 4 6 8 9
8	2 3 6 7 8 9

Key: 6 | 7 = 67°F

a. Find and interpret the mean absolute deviation of the data.

b. After you include the daily high temperatures for the rest of the month in the stem-and-leaf plot, the mean absolute deviation increases. Where do you think most of the data values for the rest of the month are located in the stem-and-leaf plot? Explain.

18. Critical Thinking The back-to-back stem-and-leaf plot shows the 9-hole golf scores for two golfers. Only one of the golfers can compete in a tournament. Use measures of center and measures of variation to give reasons why you would choose each golfer.

Rich		Will
7 5	3	
8 5 4 3 2 1	4	2 3 4 4 6 7 7 8 9
5 0	5	0

Key: 1 | 4 | 2 = 41 and 42 strokes

Fair Game Review What you learned in previous grades & lessons

Draw the solid. *(Section 8.1)*

19. square pyramid

20. hexagonal prism

21. MULTIPLE CHOICE In a bar graph, what determines the length of each bar? *(Skills Review Handbook)*

(A) frequency (B) data value (C) leaf (D) change in data

10.2 Histograms

Essential Question How can you use intervals, tables, and graphs to organize data?

1 ACTIVITY: Conducting an Experiment

Work with a partner.

a. Roll a number cube 20 times. Record your results in a tally chart.

b. Make a bar graph of the totals.

c. Go to the board and enter your totals in the class tally chart.

d. Make a second bar graph showing the class totals. Compare and contrast the two bar graphs.

	Tally Chart
1	
2	
3	
4	
5	
6	

Key: | = 1 卌 = 5

2 ACTIVITY: Using Intervals to Organize Data

Work with a partner. You are judging a paper airplane contest. A contestant flies a paper airplane 20 times. You record the following distances:

20.5 ft, 24.5 ft, 18.5 ft, 19.5 ft, 21.0 ft, 14.0 ft, 12.5 ft, 20.5 ft, 17.5 ft, 24.5 ft, 19.5 ft, 17.0 ft, 18.5 ft, 12.0 ft, 21.5 ft, 23.0 ft, 13.5 ft, 19.0 ft, 22.5 ft, 19.0 ft

a. Complete the tally chart and the bar graph of the distances.

Tally Chart		
Interval	**Tally**	**Total**
10.0–12.9		
13.0–15.9		
16.0–18.9		
19.0–21.9		
22.0–24.9		

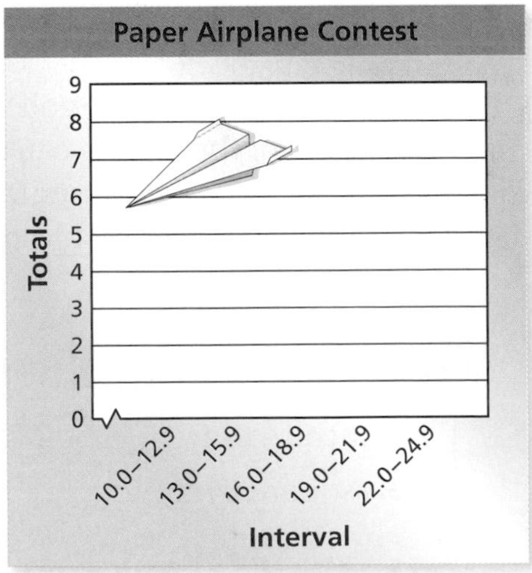

b. Make a different tally chart and bar graph of the distances. Use the following intervals:

10.0–11.9, 12.0–13.9, 14.0–15.9, 16.0–17.9, 18.0–19.9, 20.0–21.9, 22.0–23.9, 24.0–25.9

c. Which graph do you think represents the distances better? Explain.

COMMON CORE

Data Displays

In this lesson, you will

- make histograms.
- use histograms to analyze data.

Learning Standards
6.SP.2
6.SP.4

The tally chart in Activity 2 is also called a *frequency table.* A **frequency table** groups data values into intervals. The **frequency** is the number of values in an interval.

3 ACTIVITY: Developing an Experiment

Math Practice 6

Specify Units

What units will you use to measure the distance flown each time? Will the units you use affect the results in your frequency table? Explain.

Work with a partner.

a. Make the airplane shown from a single sheet of $8\frac{1}{2}$-by-11-inch paper. Then design and make your own paper airplane.

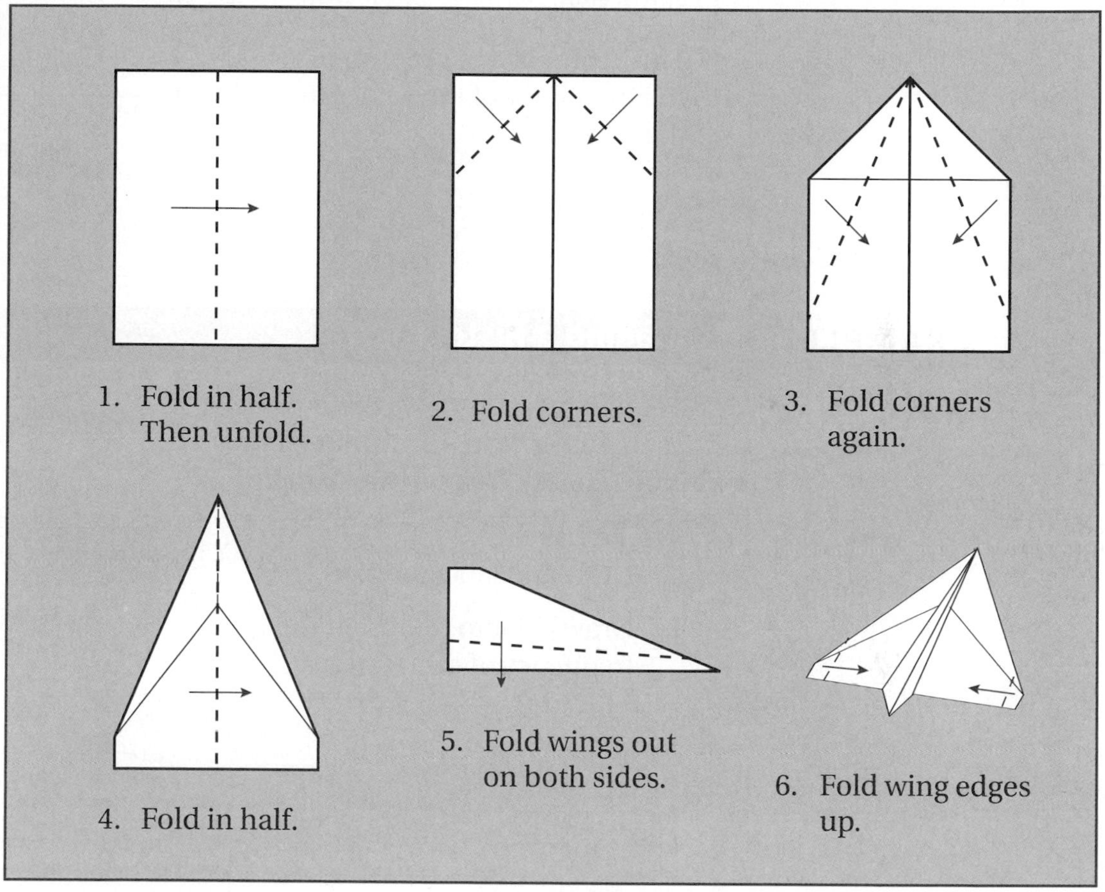

1. Fold in half. Then unfold.
2. Fold corners.
3. Fold corners again.
4. Fold in half.
5. Fold wings out on both sides.
6. Fold wing edges up.

b. **PRECISION** Fly each airplane 20 times. Keep track of the distance flown each time.

c. **MODELING** Organize the results of the flights using frequency tables and graphs. Which airplane flies farther? Explain your reasoning.

What Is Your Answer?

4. **IN YOUR OWN WORDS** How can you use intervals, tables, and graphs to organize data?

5. What intervals could you use in a graph that displays data whose values range from 40 through 59?

Practice Use what you learned about organizing data into intervals to complete Exercises 4 and 5 on page 445.

10.2 Lesson

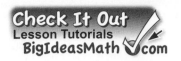

Check It Out
Lesson Tutorials
BigIdeasMath.com

Key Vocabulary
frequency table, *p. 441*
frequency, *p. 441*
histogram, *p. 442*

Key Idea

Histograms

A **histogram** is a bar graph that shows the frequency of data values in intervals of the same size.

The height of a bar represents the frequency of the values in the interval.

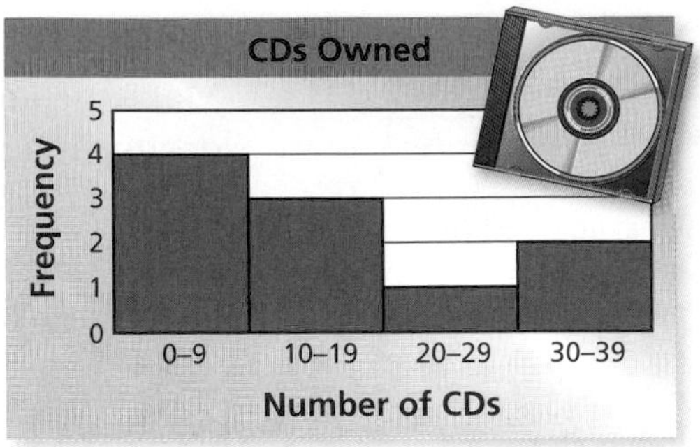

EXAMPLE 1 Making a Histogram

The frequency table shows the numbers of laps that people in a swimming class completed today. Display the data in a histogram.

Number of Laps	Frequency
1–3	11
4–6	4
7–9	0
10–12	3
13–15	6

Step 1: Draw and label the axes.

Step 2: Draw a bar to represent the frequency of each interval.

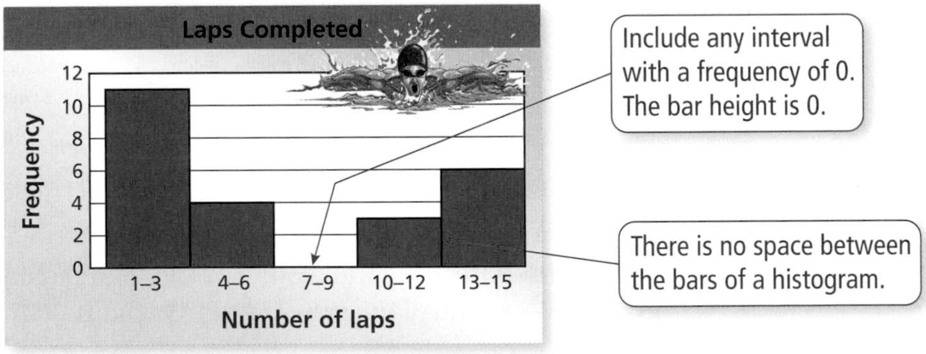

On Your Own

Now You're Ready
Exercises 6–8

1. The frequency table shows the ages of people riding a roller coaster. Display the data in a histogram.

Age	10–19	20–29	30–39	40–49	50–59
Frequency	16	11	5	2	4

Multi-Language Glossary at BigIdeasMath.com

EXAMPLE 2 Using a Histogram

The histogram shows the winning speeds at the Daytona 500. (a) Which interval contains the most data values? (b) How many of the winning speeds are less than 140 miles per hour? (c) How many of the winning speeds are at least 160 miles per hour?

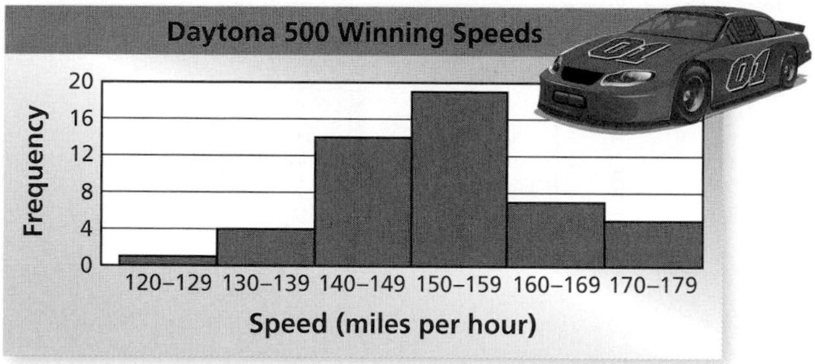

a. The interval with the tallest bar contains the most data values.

So, the 150–159 miles per hour interval contains the most data values.

b. One winning speed is in the 120–129 miles per hour interval, and four winning speeds are in the 130–139 miles per hour interval.

So, $1 + 4 = 5$ winning speeds are less than 140 miles per hour.

c. Seven winning speeds are in the 160–169 miles per hour interval, and five winning speeds are in the 170–179 miles per hour interval.

So, $7 + 5 = 12$ winning speeds are at least 160 miles per hour.

On Your Own

Now You're Ready Exercises 10–13

2. The histogram shows the numbers of hours that students in a class slept last night.

a. How many students slept at least 8 hours?

b. How many students slept less than 12 hours?

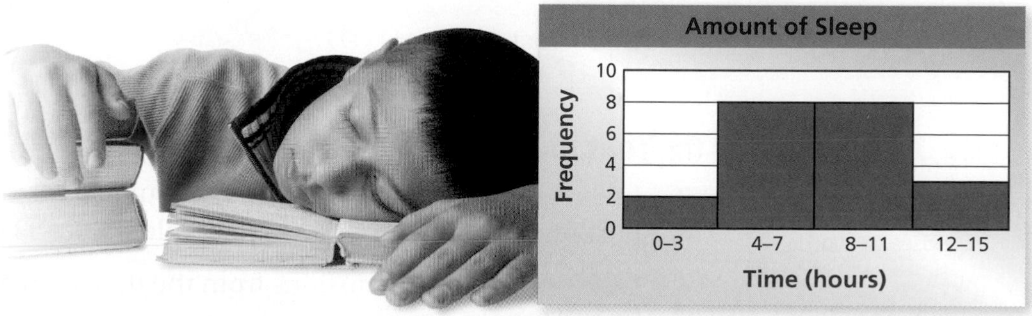

EXAMPLE 3 Comparing Data Displays

The data displays show how many push-ups students in a class completed for a physical fitness test. Which data display can you use to find how many students are in the class? Explain.

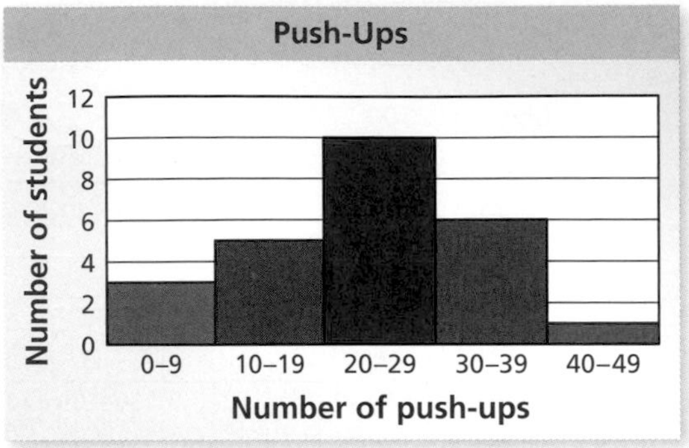

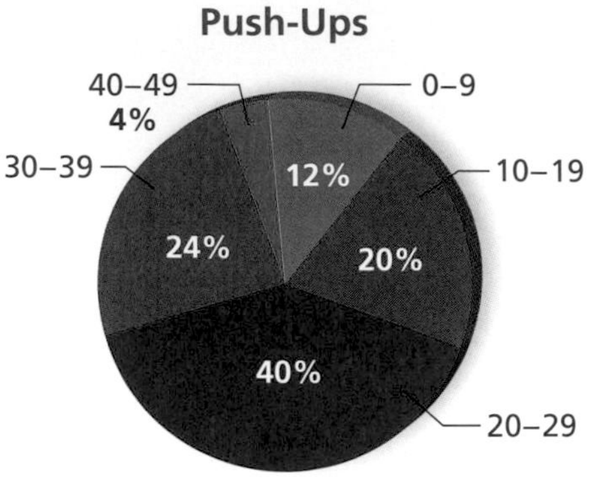

∴ You can use the histogram because it shows the number of students in each interval. The sum of these values represents the number of students in the class. You cannot use the circle graph because it does not show the number of students in each interval.

EXAMPLE 4 Making Conclusions from Data Displays

Which statement *cannot* be made using the data displays in Example 3?

(A) Twelve percent of the class completed less than 10 push-ups.

(B) Five students completed at least 10 and at most 19 push-ups.

(C) At least one student completed more than 39 push-ups.

(D) Twenty-nine percent of the class completed 30 or more push-ups.

The circle graph shows that 12% completed 0–9 push-ups. So, Statement A can be made.

In the histogram, the bar height for the 10–19 interval is 5, and the bar height for the 40–49 interval is 1. So, Statements B and C can be made.

The circle graph shows that 24% completed 30–39 push-ups, and 4% completed 40–49 push-ups. So, $24\% + 4\% = 28\%$ completed 30 or more push-ups. Statement D cannot be made.

∴ The correct answer is (D).

On Your Own

Now You're Ready
Exercises 14 and 15

3. In Example 3, which data display should you use to describe the portion of the entire class that completed 30–39 push-ups?

4. Make two more conclusions from the data displays in Example 3.

10.2 Exercises

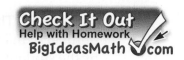

Check It Out
Help with Homework
BigIdeasMath.com

Vocabulary and Concept Check

1. **VOCABULARY** Which graph is a histogram? Explain your reasoning.

2. **REASONING** Describe the outliers in the histogram.

3. **REASONING** How can you tell when an interval of a histogram has a frequency of zero?

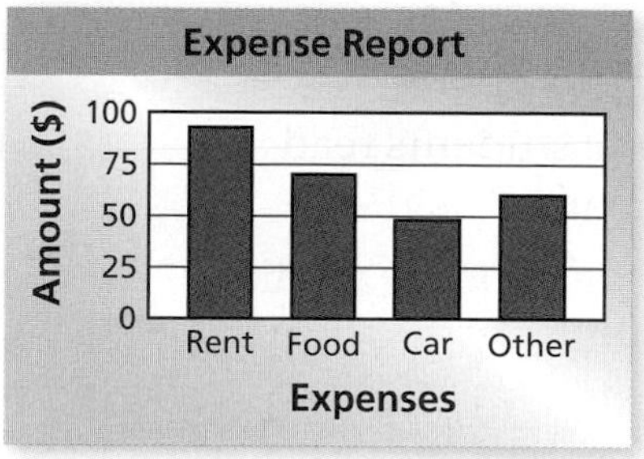

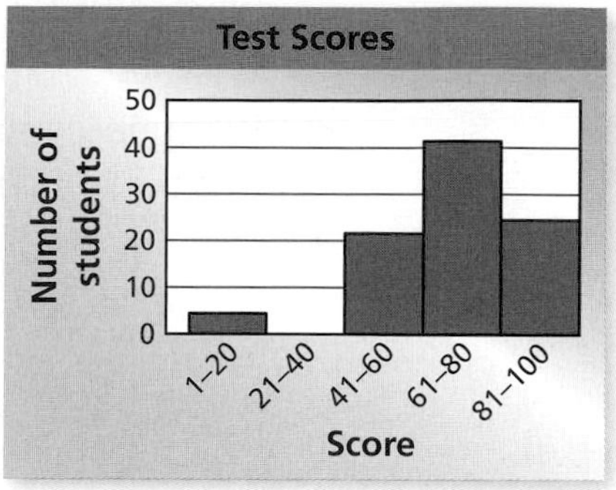

Practice and Problem Solving

Make a tally chart and a bar graph of the data.

4.

Members of Book Clubs			
6	17	13	19
13	9	18	24
11	15	21	14

5.

Points Scored				
42	45	57	39	55
38	48	36	48	46
51	29	45	54	42

Display the data in a histogram.

1 6.

States Visited	
States	**Frequency**
1–5	12
6–10	14
11–15	6
16–20	3

7.

Chess Team	
Wins	**Frequency**
10–13	3
14–17	4
18–21	4
22–25	2

8.

Movies Watched	
Movies	**Frequency**
0–1	5
2–3	11
4–5	8
6–7	1

9. **ERROR ANALYSIS** Describe and correct the error made in displaying the data in a histogram.

Confirmed Flu Cases per School	
Cases	**Frequency**
0–2	3
3–5	7
6–8	9
9–11	12

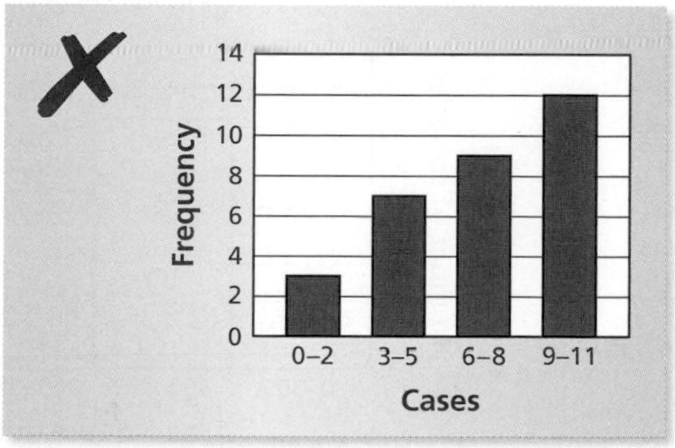

2 **10. MAGAZINES** The histogram shows the number of magazines read last month by the students in a class.

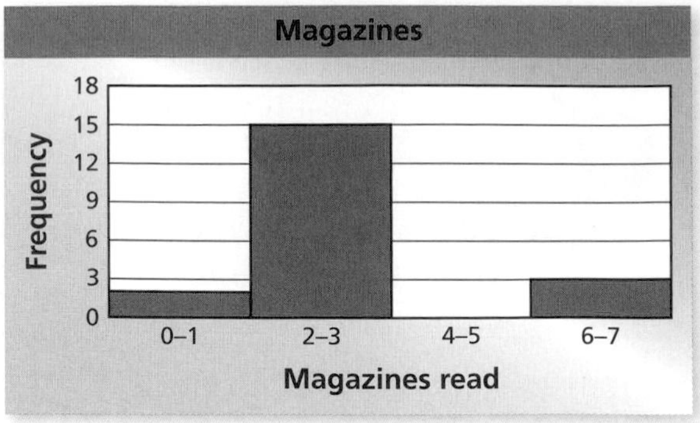

a. Which interval contains the fewest data values?

b. How many students are in the class?

c. What percent of the students read less than six magazines?

d. Can you find the mean or the median of the data? Explain.

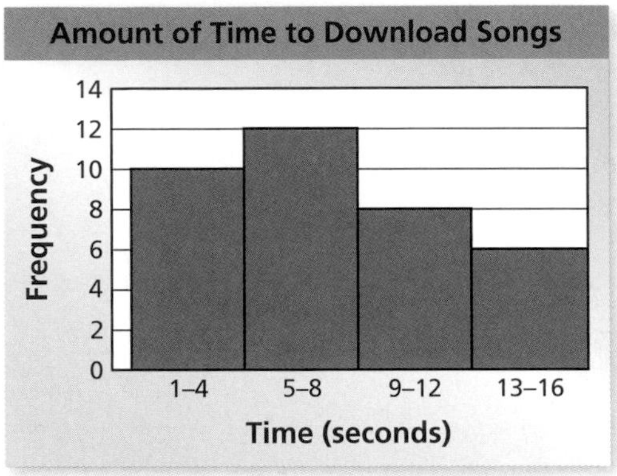

11. ERROR ANALYSIS Describe and correct the error made in reading the histogram.

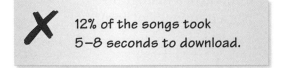

✗ 12% of the songs took 5–8 seconds to download.

12. VOTING The histogram shows the percent of the voting-age population that voted in a recent presidential election. Explain whether the graph supports each statement.

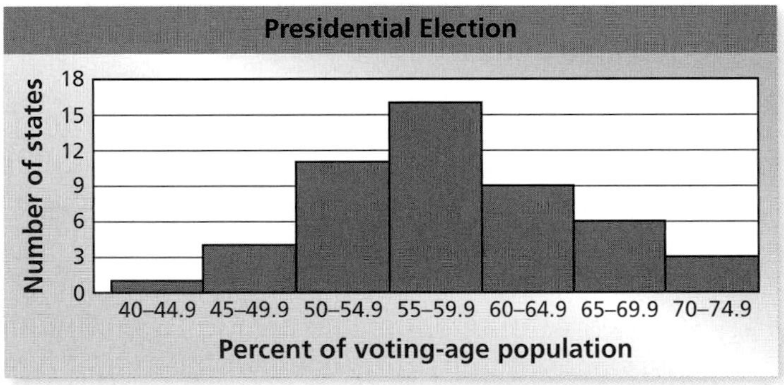

a. Only 40% of one state voted.

b. In most states, between 50% and 64.9% voted.

c. The mode of the data is between 55 and 59.9.

13. PROBLEM SOLVING The histograms show the areas of counties in Pennsylvania and Indiana. Which state do you think has the greater area? Explain.

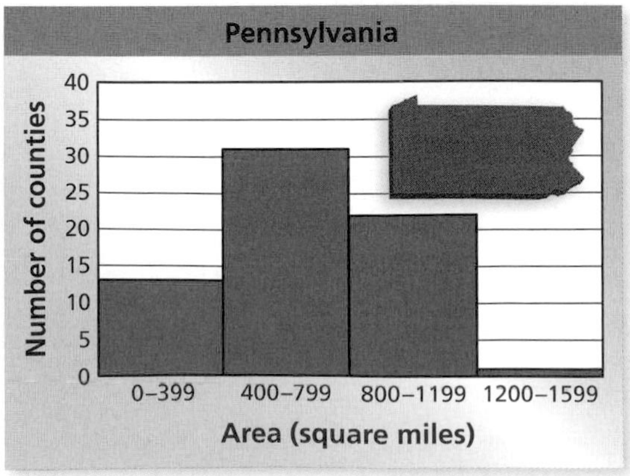

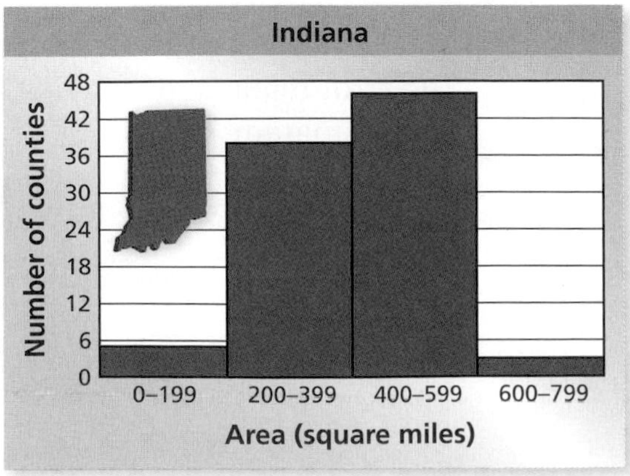

3 **14. GARBAGE** The data displays show how many pounds of garbage apartment residents produced in 1 week. Which data display can you use to find how many residents produced more than 25 pounds of garbage? Explain.

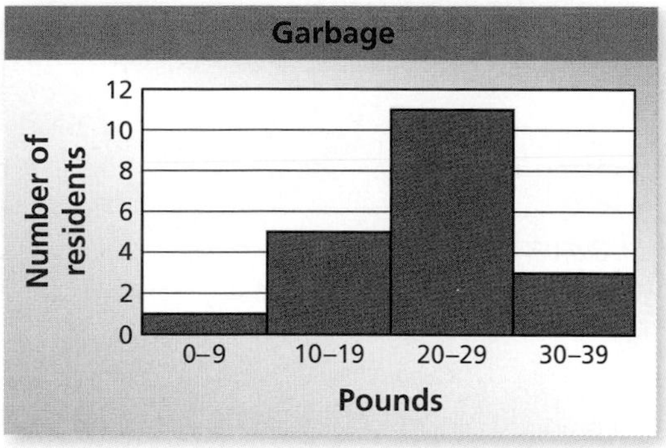

Garbage

Stem	Leaf
0	9
1	0 5 8 8 9
2	1 2 5 5 6 7 7 7 9 9 9
3	2 3 3

Key: 1 | 5 = 15 pounds

15. REASONING Determine whether you can make each statement by using the data displays in Exercise 14. Explain your reasoning.

a. One resident produced 10 pounds of garbage.

b. Twelve residents produced between 20 and 29 pounds of garbage.

16. NUMBER SENSE Can you find the range and the interquartile range of the data in Exercise 7? If so, find them. If you cannot find them, explain why not.

17. CRITICAL THINKING The table shows the weights of guide dogs enrolled in a training program.

a. Make a histogram of the data starting with the interval 51–55.

b. Make another histogram of the data using different-sized intervals.

c. Compare and contrast the two histograms.

Weights (pounds)					
81	88	57	82	70	85
71	51	82	77	79	77
83	80	54	80	81	73
59	84	75	76	68	78
83	78	55	67	85	79

18. **Logic** What are the possible values for the median in Exercise 10?

Fair Game Review What you learned in previous grades & lessons

Find the percent of the number. *(Section 5.6)*

19. 25% of 180 **20.** 30% of 90 **21.** 16% of 140 **22.** 64% of 80

23. MULTIPLE CHOICE Which is the solution of the inequality represented by "Four times a number n is at least 28"? *(Section 7.7)*

Ⓐ $n < 7$ Ⓑ $n > 7$ Ⓒ $n \le 7$ Ⓓ $n \ge 7$

Check It Out
Graphic Organizer
BigIdeasMath.com

You can use a **word magnet** to organize information associated with a vocabulary word. Here is an example of a word magnet for histogram.

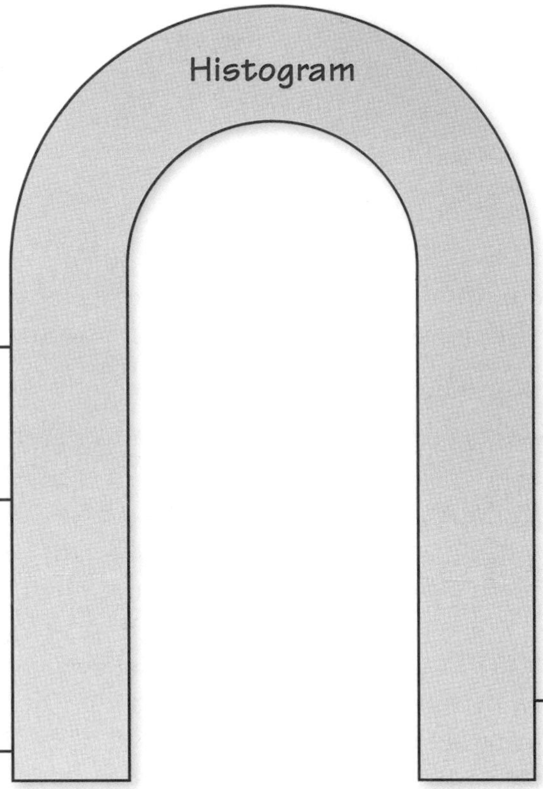

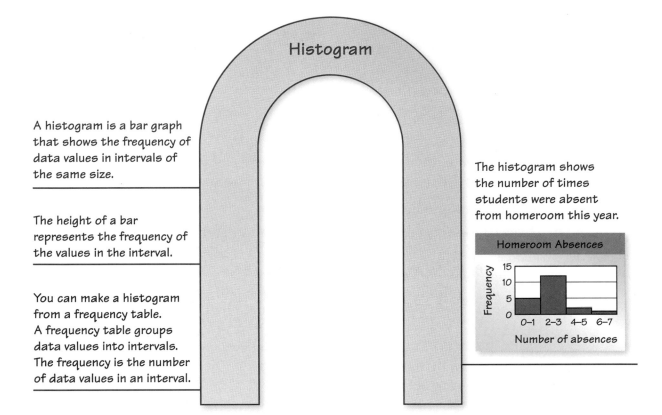

A histogram is a bar graph that shows the frequency of data values in intervals of the same size.

The height of a bar represents the frequency of the values in the interval.

You can make a histogram from a frequency table. A frequency table groups data values into intervals. The frequency is the number of data values in an interval.

The histogram shows the number of times students were absent from homeroom this year.

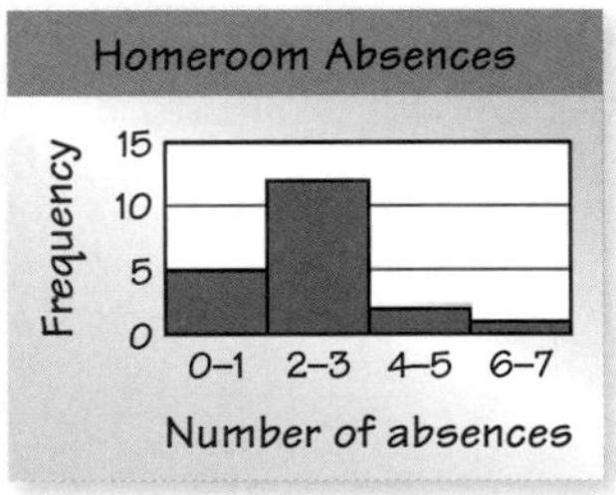

On Your Own

Make a word magnet to help you study this topic.

1. stem-and-leaf plot

After you complete this chapter, make word magnets for the following topics.

2. shapes of distributions
3. box-and-whisker plot
4. Choose three other topics that you studied earlier in this course. Make a word magnet for each topic.

"How do you like the word magnet I made for 'Beagle'?"

10.1–10.2 Quiz

Check It Out
Progress Check
BigIdeasMath.com

Make a stem-and-leaf plot of the data. *(Section 10.1)*

1.

Cans Collected Each Month			
80	90	84	92
76	83	79	59
68	55	58	61

2.

Miles Driven Each Day				
21	18	12	16	10
16	9	15	20	28
35	50	37	20	11

3.

Ages of Tortoises			
86	99	100	124
92	85	110	130
115	129	83	104

4.

Kilometers Run Each Day				
6.0	5.6	6.2	3.0	2.5
3.5	2.0	5.0	3.9	3.1
6.2	3.1	4.5	3.8	6.1

Display the data in a histogram. *(Section 10.2)*

5. Soccer Team Goals

Goals per Game	Frequency
0–1	5
2–3	4
4–5	0
6–7	1

6. Minutes Practiced

Minutes	Frequency
0–19	8
20–39	10
40–59	11
60–79	2

7. Poems Written for Class

Poems	Frequency
0–4	6
5–9	16
10–14	4
15–19	2
20–24	2

8. WEIGHTS The weights (in ounces) of nine packages are 7, 22, 16, 12, 6, 18, 15, 13, and 25. Make a stem-and-leaf plot of the data. Describe the distribution of the data. *(Section 10.1)*

9. REBOUNDS The histogram shows the number of rebounds per game for a middle school basketball player this season. *(Section 10.2)*

a. Which interval contains the most data values?

b. How many games did the player play this season?

c. What percent of the games did the player have 4 or more rebounds?

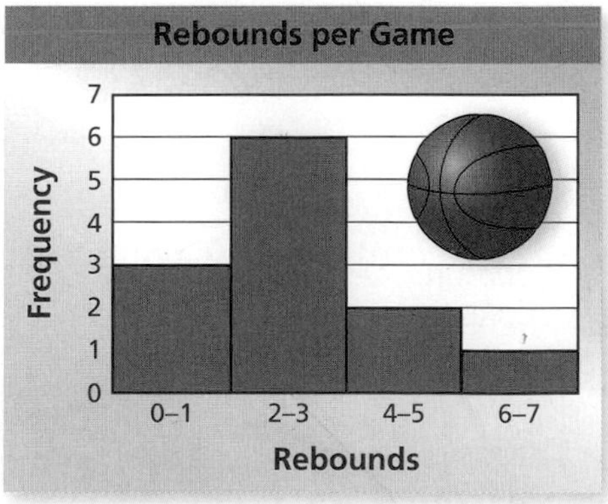

Stem	Leaf
0	6 8 8 9
1	0 1 2 3 7 8
2	0

Key: 0|9 = 9 hours

10. STAGE CREW The stem-and-leaf plot shows the number of hours 11 stage crew members spent building sets. Find the mean, median, mode, range, and interquartile range of the data. *(Section 10.1)*

10.3 Shapes of Distributions

Essential Question How can you describe the shape of the distribution of a data set?

1 ACTIVITY: Describing the Shape of a Distribution

-7253
-7290
-7200
-1192
-1142

-3500
-2531
-2079
-5897
-5341
-1392
-5406
-7875
-7335
-0494
-9018
-2184
-2367

-8678
-2063
-2911
-2103
-4328
-7826
-7957
-7246
-2119
-7845
-1109
-9154

Work with a partner. The lists at the left show the last four digits of a set of phone numbers in a phone book.

a. Create a list that represents the last digit of each phone number shown. Make a dot plot of the data.

b. In your own words, how would you describe the shape of the distribution? What single word do you think you can use to identify this type of distribution? Explain your reasoning.

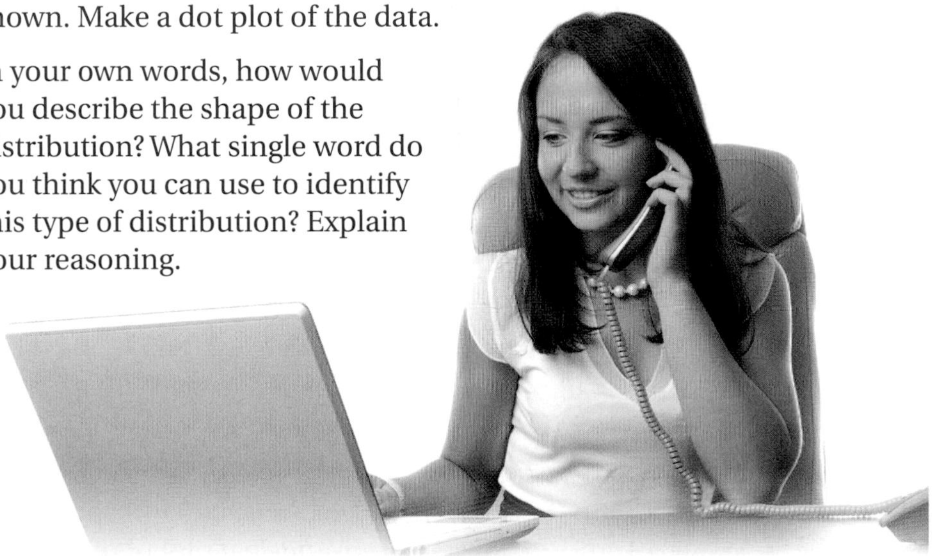

2 ACTIVITY: Describing the Shape of a Distribution

Work with a partner. The lists at the right show the first three digits of a set of phone numbers in a phone book.

a. Create a list that represents the first digit of each phone number shown. Make a dot plot of the data.

b. In your own words, how would you describe the shape of the distribution? What single word do you think you can use to identify this type of distribution? Explain your reasoning.

c. In your dot plot, draw a vertical line through the middle of the data set. What do you notice?

d. Repeat part (c) for the dot plot you constructed in Activity 1. What do you notice? Compare the distributions from Activities 1 and 2.

538-
438-
664-
761-
868-

735-
694-
599-
725-
556-
555-
456-
736-
664-
576-
664-
664-
725-

664-
664-
538-
855-
664-
538-
654-
654-
725-
538-
799-
764-

COMMON CORE

Data Displays

In this lesson, you will

- describe shapes of distributions.

Learning Standards
6.SP.2
6.SP.4

The Meaning of a Word ● Skewed

When something is **skewed**, it has a slanted direction or position.

3 ACTIVITY: Describing the Shape of a Distribution

Work with a partner. The table shows the ages of cellular phones owned by a group of students.

a. Make a dot plot of the data.

b. In your own words, how would you describe the shape of the distribution? Compare it to the distributions in Activities 1 and 2.

c. Why do you think this type of distribution is called a *skewed distribution*?

Ages of Cellular Phones (years)				
0	1	0	6	4
2	3	5	1	1
0	1	2	3	1
0	0	1	1	1
7	1	4	2	2
0	2	0	1	2

4 ACTIVITY: Finding Measures of Center

Math Practice 3

Use Prior Results

How is the distribution of the data related to the mean and the median?

Work with a partner.

a. Find the means and the medians of the data sets in Activities 1–3.

b. What do you notice about the means and the medians of the data sets and the shapes of the distributions? Explain.

c. Which measure of center do you think best describes the data set in Activity 2? in Activity 3? Explain your reasoning.

d. Using your answers to part (c), decide which measure of variation you think best describes the data set in Activity 2. Which measure of variation do you think best describes the data set in Activity 3? Explain your reasoning.

What Is Your Answer?

5. **IN YOUR OWN WORDS** How can you describe the shape of the distribution of a data set?

6. Name two other ways you can describe the distribution of a data set.

Practice Use what you learned about shapes of distributions to complete Exercises 3 and 4 on page 454.

10.3 Lesson

Check It Out
Lesson Tutorials
BigIdeasMath.com

You can use dot plots and histograms to identify shapes of distributions.

Key Ideas

Study Tip

If all the dots of a dot plot or bars of a histogram are about the same height, then the distribution is a *flat*, or *uniform*, distribution. A uniform distribution is also symmetric.

Symmetric and Skewed Distributions

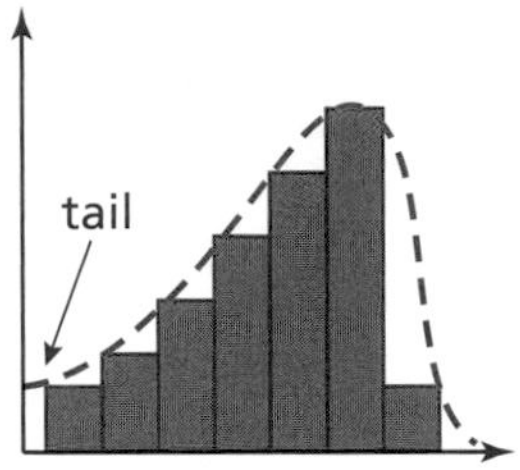

Skewed left

- The "tail" of the graph extends to the left.
- Most data are on the right.

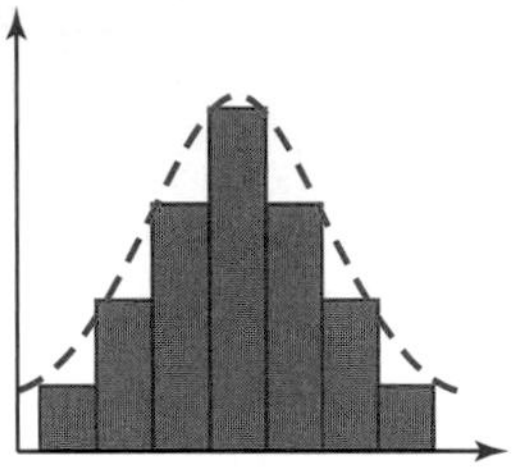

Symmetric

- The left side of the graph is a mirror image of the right side of the graph.

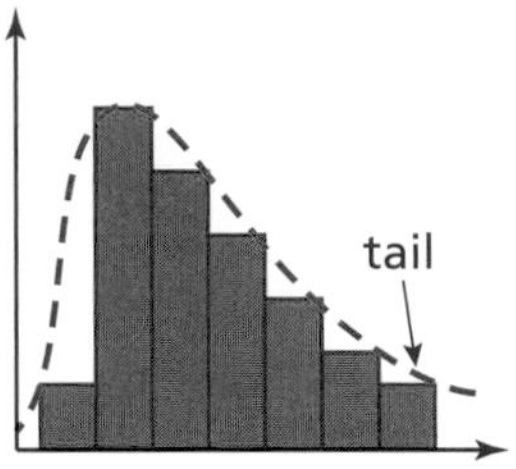

Skewed right

- The "tail" of the graph extends to the right.
- Most data are on the left.

EXAMPLE 1 Describing the Shapes of Distributions

Describe the shape of each distribution.

a.

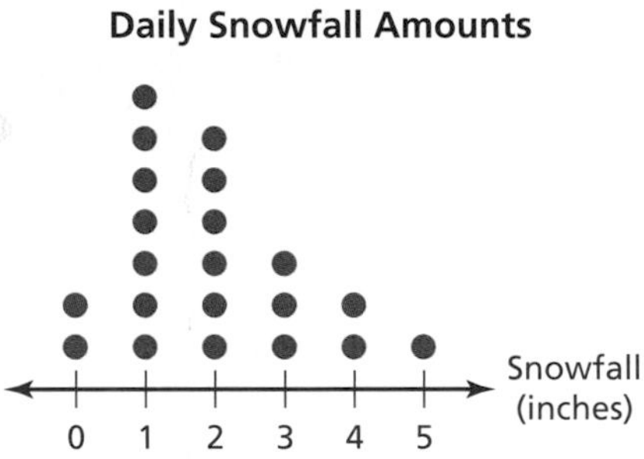

Most of the data are on the left, and the tail extends to the right.

So, the distribution is skewed right.

b.

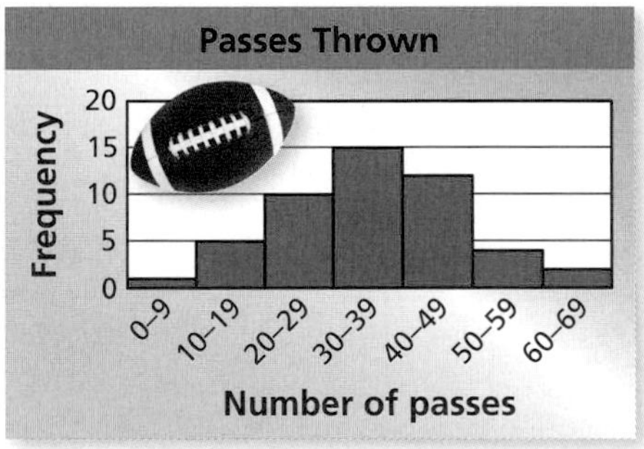

The left side of the graph is approximately a mirror image of the right side of the graph.

So, the distribution is symmetric.

On Your Own

Now You're Ready
Exercises 5–8

1. Describe the shape of the distribution.

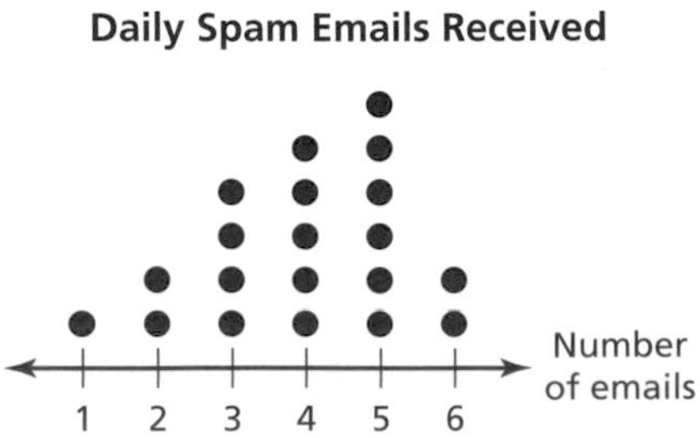

EXAMPLE 2 Describing the Shape of a Distribution

Ages	Frequency
10–13	1
14–17	3
18–21	7
22–25	12
26–29	20
30–33	18
34–37	3

The frequency table shows the ages of people watching a comedy in a theater. Display the data in a histogram. Describe the shape of the distribution.

Draw and label the axes. Then draw a bar to represent the frequency of each interval.

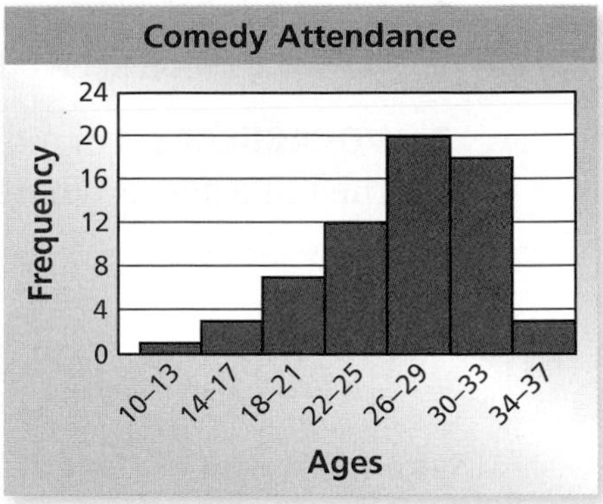

Most of the data are on the right, and the tail extends to the left.

∴ So, the distribution is skewed left.

EXAMPLE 3 Comparing Shapes of Distributions

The histogram shows the ages of people watching an animated movie in the same theater as in Example 2.

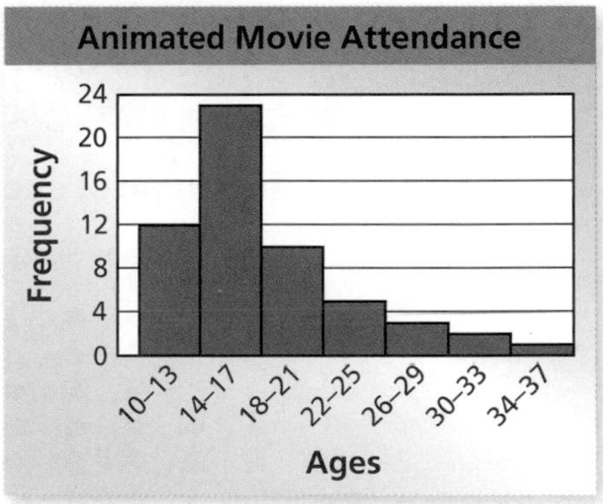

a. Describe the shape of the distribution.

Most of the data are on the left, and the tail extends to the right.

∴ So, the distribution is skewed right.

b. Which movie has an older audience?

The intervals in the histograms are the same. Most of the data for the animated movie are on the left, while most of the data for the comedy are on the right. This means that the people watching the comedy are generally older than the people watching the animated movie.

∴ So, the comedy has an older audience.

On Your Own

Now You're Ready
Exercise 9

2. The frequency table shows the ages of people watching a historical movie in a theater.

Ages	10–19	20–29	30–39	40–49	50–59	60–69
Frequency	3	18	36	40	14	5

a. Display the data in a histogram. Describe the shape of the distribution.

b. Compare the distribution of the data to the distributions in Examples 2 and 3. What can you conclude?

10.3 Exercises

Check It Out
Help with Homework
BigIdeasMath.com

Vocabulary and Concept Check

1. **VOCABULARY** How does the shape of a symmetric distribution differ from the shape of a skewed distribution?

2. **VOCABULARY** For a distribution that is skewed right, which direction does the tail extend? Where do most of the data lie?

Practice and Problem Solving

Make a dot plot of the data. In your own words, how would you describe the shape of the distribution?

3.

Miles Run per Day										
1	4	2	0	3	2	1	2	4	2	3
2	1	6	3	2	4	0	5	3	1	5

4.

Raffle Tickets Sold							
15	12	16	15	13	14	16	13
13	16	14	12	15	12	14	

Describe the shape of each distribution.

1 5.

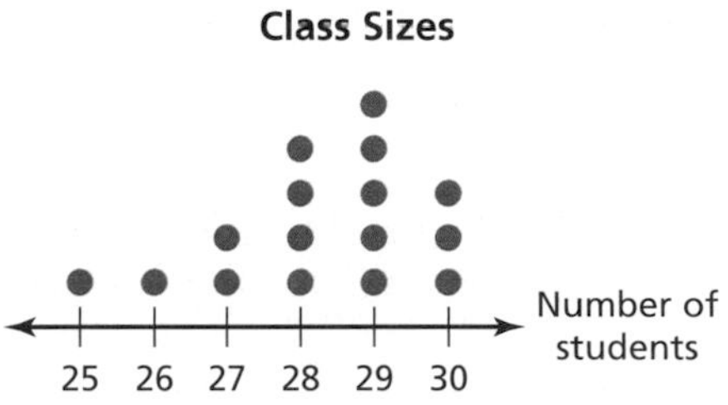

6.

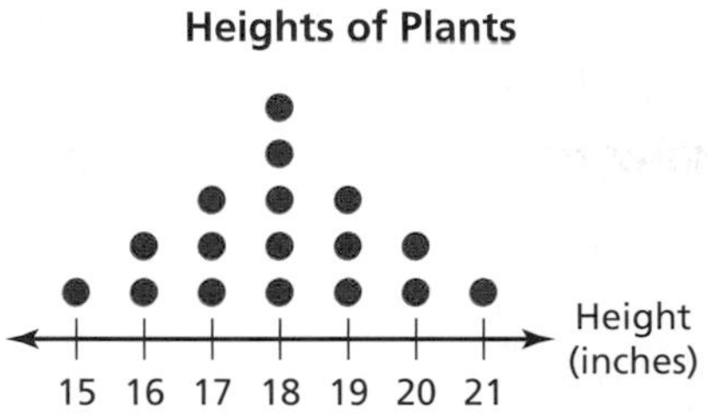

7.

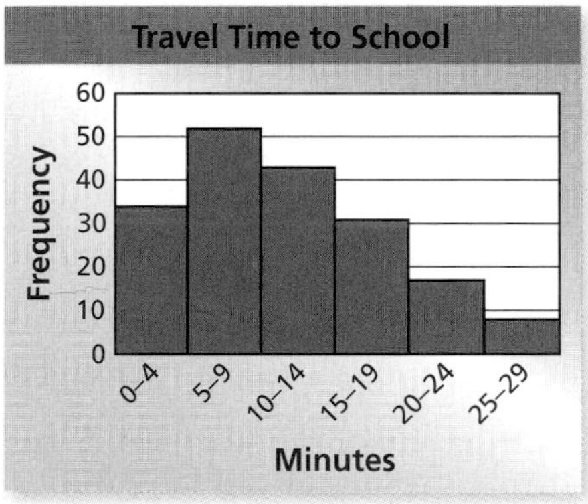

8.

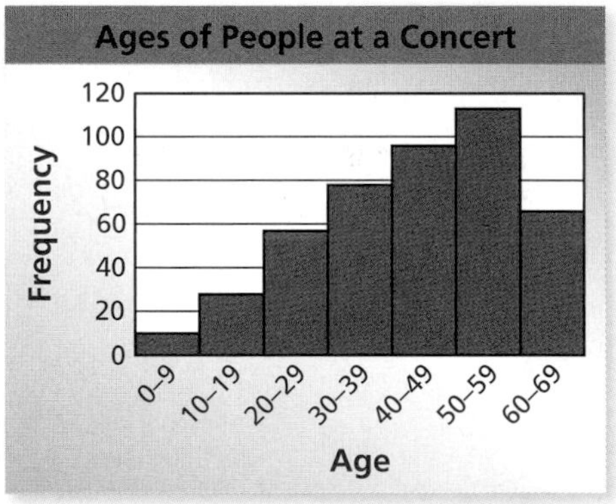

2 3 9. **POLICE** The frequency table shows the years of service for the police officers of Jones County and Pine County. Display the data for each county in a histogram. Describe the shape of each distribution. Which county's police force has less experience? Explain.

Years of Service	0–3	4–7	8–11	12–15	16–19	20–23	24–27
Frequency for Jones County	7	15	17	12	8	5	3
Frequency for Pine County	3	5	9	14	10	6	2

10. **REASONING** What is the shape of the distribution of the restaurant waiting times? Explain your reasoning.

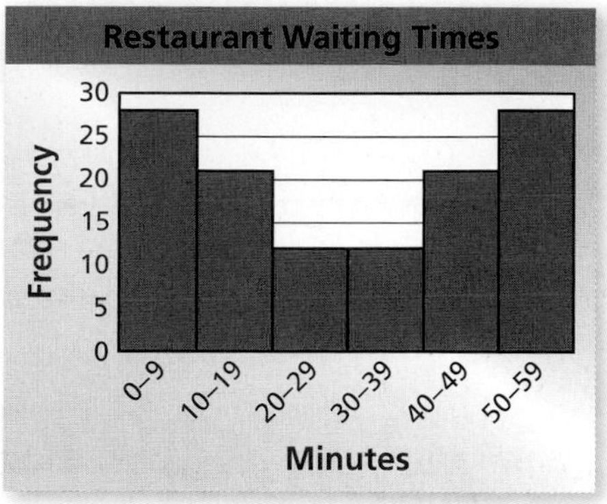

11. **LOGIC** Are all distributions either approximately symmetric or skewed? Explain. If not, give an example.

12. **REASONING** Can you use a stem-and-leaf plot to describe the shape of a distribution? Explain your reasoning.

13. **CHARITY** The table shows the donation amounts received by a charity in one day.

Donations (dollars)												
20	15	40	70	20	5	25	50	47	20	62	55	40
10	50	18	20	100	40	80	60	20	80	3	30	50
25	30	10	33	20	50	7	35	40	25	70		

a. Make a histogram of the data starting with the interval 0–14. Describe the shape of the distribution.

b. A company adds $5 to each donation. Make another histogram starting with the same first interval as in part (a). Compare the shape of this distribution with the distribution in part (a). Explain any differences in the distributions.

14. **Critical Thinking** Describe the shape of the distribution of each bar graph. Match the letters A, B, and C with the mean, the median, and the mode of the data set. Explain your reasoning.

a.

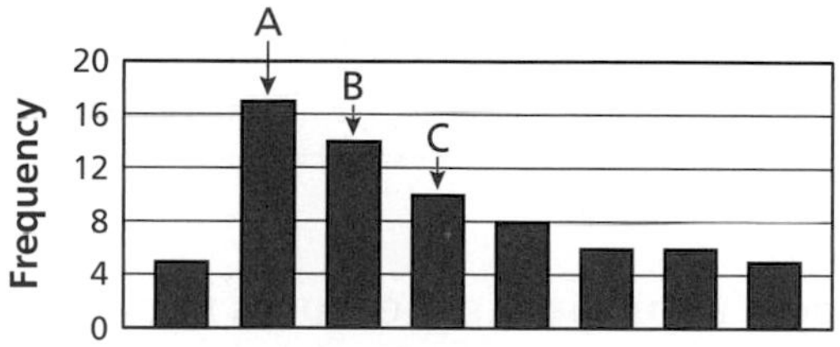

b.

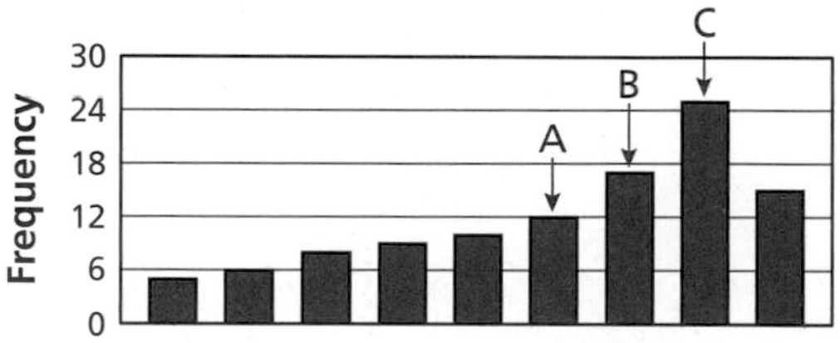

Fair Game Review What you learned in previous grades & lessons

Find the median, first quartile, third quartile, and interquartile range of the data. *(Section 9.4)*

15. 68, 74, 67, 72, 63, 70, 78, 64, 76

16. 39, 48, 33, 24, 30, 44, 36, 41, 28, 53

17. **MULTIPLE CHOICE** Sixty people participate in a trivia contest. How many four-person teams can be formed? *(Section 7.3)*

Ⓐ 15 Ⓑ 56 Ⓒ 64 Ⓓ 240

Extension 10.3 Choosing Appropriate Measures

Check It Out
Lesson Tutorials
BigIdeasMath.com

You can use a measure of center and a measure of variation to describe the distribution of a data set. The shape of the distribution can help you choose which measures are the most appropriate to use.

Key Idea

Choosing Appropriate Measures

The mean absolute deviation (MAD) uses the mean in its calculation. So, when a data distribution is *symmetric,*

- use the mean to describe the center and
- use the MAD to describe the variation.

The interquartile range (IQR) uses quartiles in its calculation. So, when a data distribution is *skewed,*

- use the median to describe the center and
- use the IQR to describe the variation.

Math Practice 2

Understand Quantities

What effect can outliers have on the mean? on the median? Explain.

EXAMPLE 1 Choosing Appropriate Measures

The dot plot shows the average number of hours students in a class sleep each night.

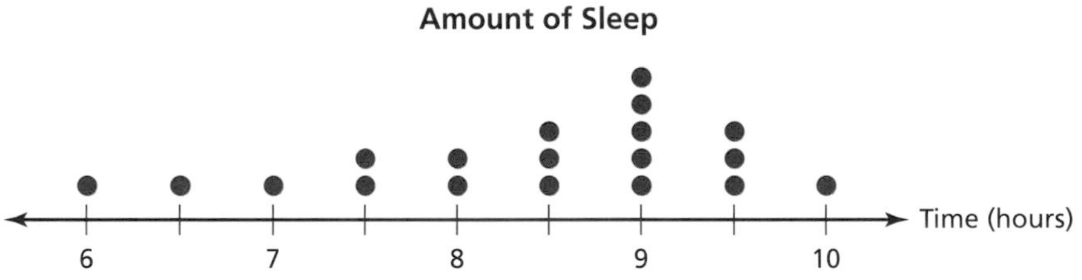

a. What are the most appropriate measures to describe the center and the variation?

Most of the data values are on the right clustered around 9, and the tail extends to the left. The distribution is skewed left.

So, the median and the interquartile range are the most appropriate measures to describe the center and the variation.

b. Describe the center and the variation of the data set.

The median is 8.5 hours. The first quartile is 7.5, and the third quartile is 9. So, the interquartile range is $9 - 7.5 = 1.5$ hours.

The data are centered around 8.5 hours. The middle half of the data varies by no more than 1.5 hours.

Common Core

Data Displays

In this extension, you will

- choose appropriate measures of center and variation to represent data sets.

Learning Standard
6.SP.5d

EXAMPLE 2 Choosing Appropriate Measures

Bordering States	Frequency
0–1	3
2–3	13
4–5	22
6–7	10
8–9	2

The frequency table shows the number of states that border each state in the United States.

a. Display the data in a histogram.

Draw and label the axes. Then draw a bar to represent the frequency of each interval.

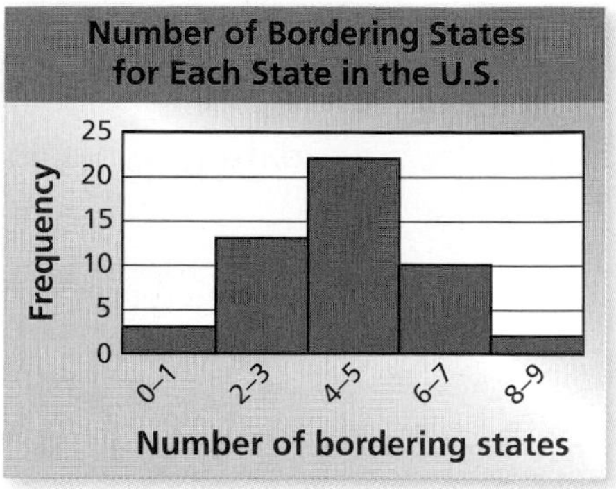

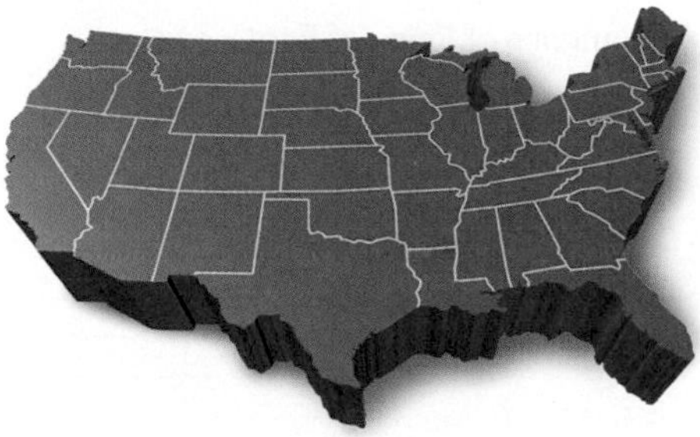

b. What are the most appropriate measures to describe the center and the variation?

The left side of the graph is approximately a mirror image of the right side of the graph. The distribution is symmetric.

So, the mean and the mean absolute deviation are the most appropriate measures to describe the center and the variation.

Practice

Choose the most appropriate measures to describe the center and the variation. Find the measures you chose.

1. Prices of Jeans

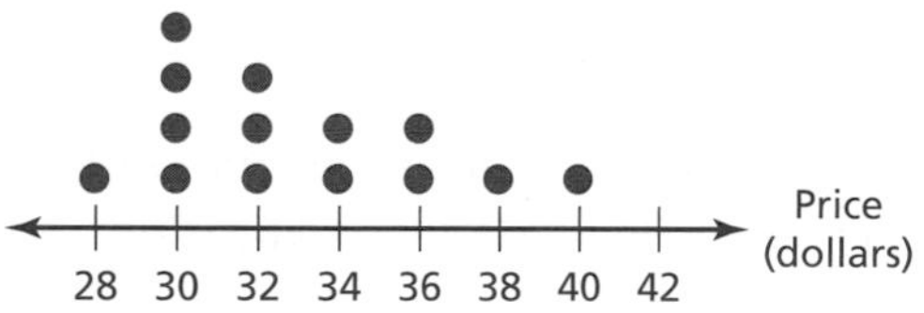

2. Weekly Biking Times

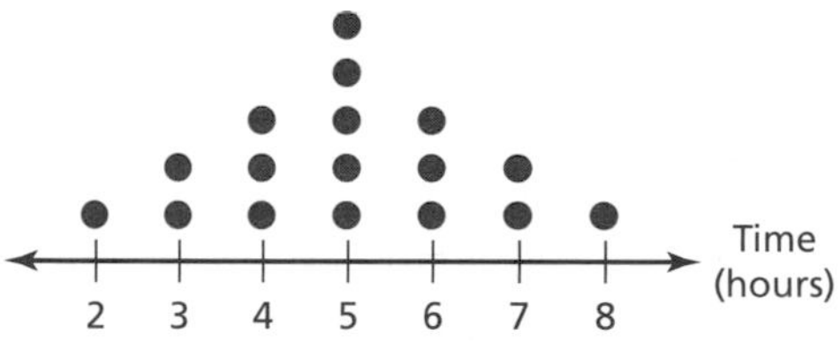

3. **REASONING** Can you find the exact values of the mean and the mean absolute deviation for the data in Example 2? Explain.

4. **GAS MILEAGE** The frequency table shows the gas mileages of several vehicles made by a company.

   a. What are the most appropriate measures to describe the center and the variation?

   b. What conclusions can you make?

Mileage (miles per gallon)	Frequency
10–14	2
15–19	1
20–24	6
25–29	8
30–34	10
35–39	3

5. **OPEN-ENDED** Construct a dot plot for which the mean is the most appropriate measure to describe the center of the distribution.

10.4 Box-and-Whisker Plots

Essential Question How can you use quartiles to represent data graphically?

1 ACTIVITY: Drawing a Box-and-Whisker Plot

Work with a partner.

The numbers of pairs of footwear owned by each student in a sixth grade class are shown.

Numbers of Pairs of Footwear

2	5	12	3
7	2	4	6
14	10	6	28
5	3	2	4
9	25	4	10
8	15	5	8

a. Order the data set from least to greatest. Then write the data on a strip of grid paper with 24 boxes.

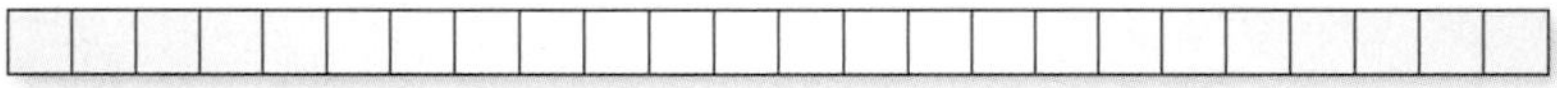

b. Use the strip of grid paper to find the median, the first quartile, and the third quartile. Identify the least value and the greatest value in the data set.

c. Graph the five numbers that you found in part (b) on the number line below.

d. The data display shown below is called a *box-and-whisker plot*. Fill in the missing labels and numbers. Explain how a box-and-whisker plot uses quartiles to represent the data.

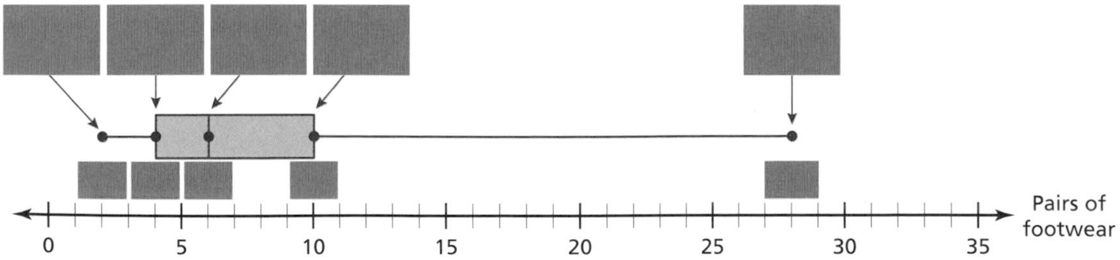

e. Using only the box-and-whisker plot, which measure(s) of center can you find for the data set? Which measure(s) of variation can you find for the data set? Explain your reasoning.

f. Why do you think this type of data display is called a box-and-whisker plot? Explain.

COMMON CORE

Data Displays

In this lesson, you will

- make and interpret box-and-whisker plots.
- compare box-and-whisker plots.

Learning Standards
6.SP.2
6.SP.4
6.SP.5c

2 ACTIVITY: Conducting a Survey

Have your class conduct a survey. Each student will write on the chalkboard the number of pairs of footwear that he or she owns.

Now, work with a partner to draw a box-and-whisker plot of the data.

3 ACTIVITY: Reading a Box-and-Whisker Plot

Math Practice 7

View as Components

What do the different components of a box-and-whisker plot represent?

Work with a partner. The box-and-whisker plots show the test score distributions of two sixth grade achievement tests. The same group of students took both tests. The students took one test in the fall and the other in the spring.

a. Compare and contrast the test results.

b. Decide which box-and-whisker plot represents the results of which test. How did you make your decision?

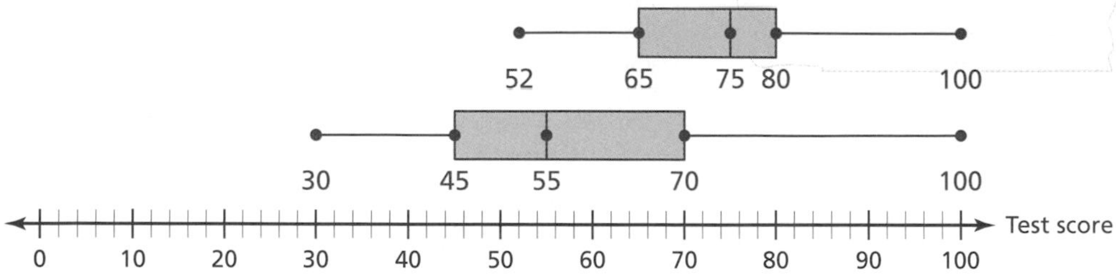

What Is Your Answer?

4. IN YOUR OWN WORDS How can you use quartiles to represent data graphically?

5. Describe who might be interested in test score distributions like those shown in Activity 3. Explain why it is important for such people to know test score distributions.

Practice Use what you learned about box-and-whisker plots to complete Exercise 4 on page 463.

Check It Out
Lesson Tutorials
BigIdeasMath.com

Key Vocabulary
box-and-whisker plot, *p. 460*
five-number summary, *p. 460*

Key Idea

Box-and-Whisker Plot

A **box-and-whisker plot** represents a data set along a number line by using the least value, the greatest value, and the quartiles of the data. A box-and-whisker plot shows the *variability* of a data set.

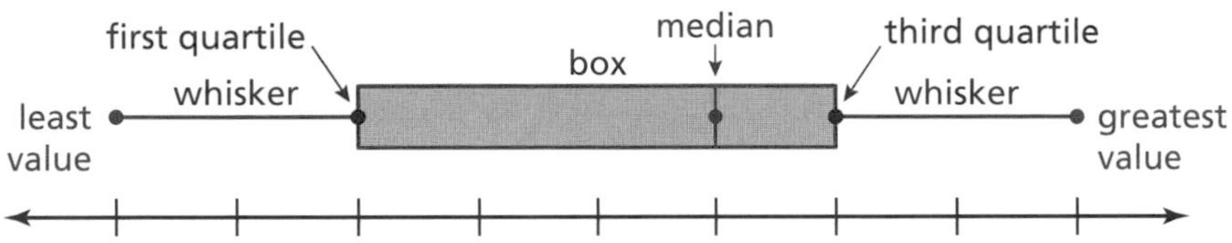

The five numbers that make up the box-and-whisker plot are called the **five-number summary** of the data set.

EXAMPLE 1 Making a Box-and-Whisker Plot

Make a box-and-whisker plot for the ages (in years) of the spider monkeys at a zoo:

15, 20, 14, 38, 30, 36, 30, 30, 27, 26, 33, 35

Step 1: Order the data. Find the median and the quartiles.

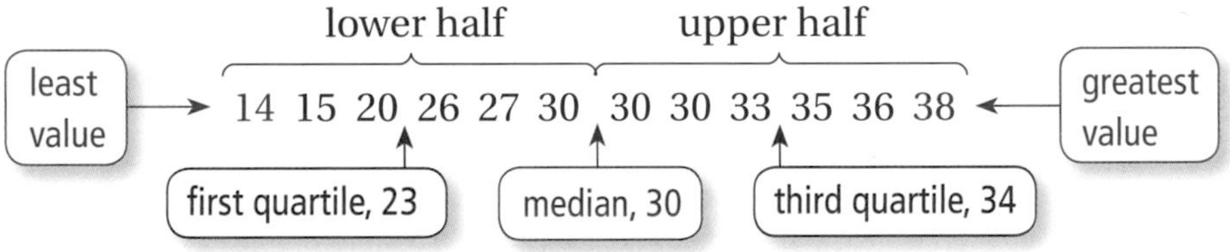

Step 2: Draw a number line that includes the least and greatest values. Graph points above the number line that represent the five-number summary.

Step 3: Draw a box using the quartiles. Draw a line through the median. Draw whiskers from the box to the least and the greatest values.

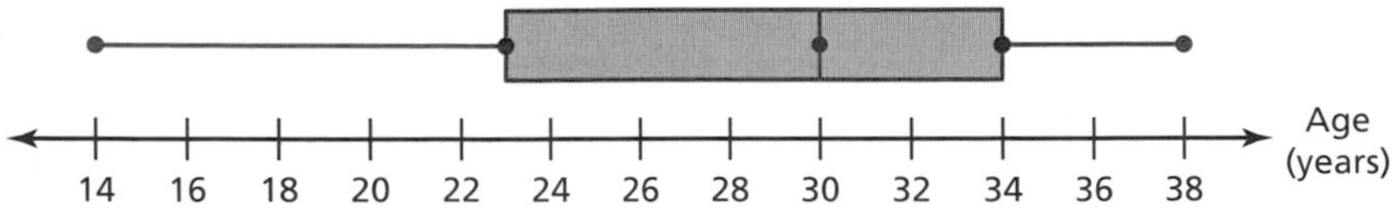

On Your Own

Now You're Ready
Exercises 5–8

1. A group of friends spent 1, 0, 2, 3, 4, 3, 6, 1, 0, 1, 2, and 2 hours online last night. Make a box-and-whisker plot for the data.

The figure shows how data are distributed in a box-and-whisker plot.

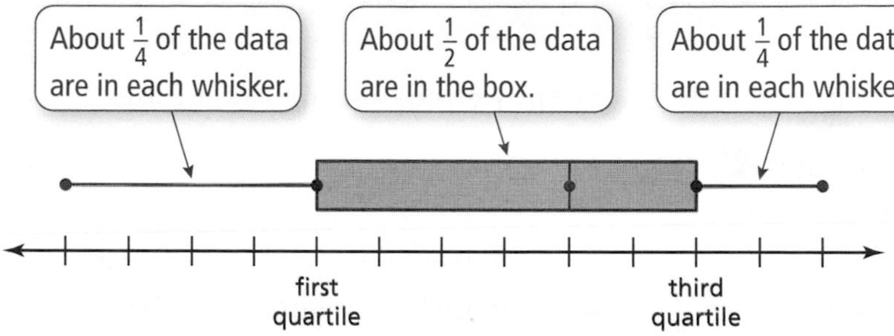

Study Tip

A long whisker or box indicates that the data are more spread out.

EXAMPLE 2 Analyzing a Box-and-Whisker Plot

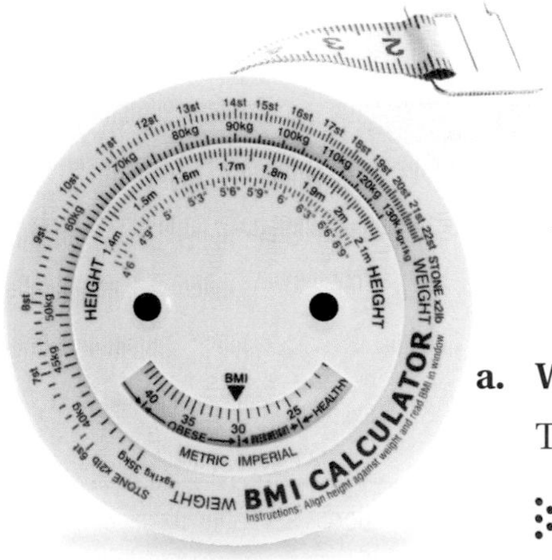

The box-and-whisker plot shows the body mass index (BMI) of a sixth grade class.

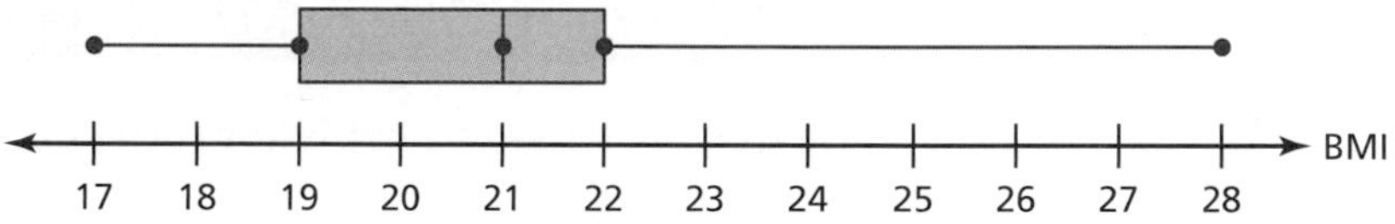

a. What fraction of the students have a BMI of at least 22?

The right whisker represents students who have a BMI of at least 22.

∴ So, about $\frac{1}{4}$ of the students have a BMI of at least 22.

b. Are the data more spread out below the first quartile or above the third quartile? Explain.

The right whisker is longer than the left whisker.

∴ So, the data are more spread out above the third quartile than below the first quartile.

c. Find and interpret the interquartile range of the data.

$$\text{interquartile range} = \text{third quartile} - \text{first quartile}$$
$$= 22 - 19 = 3$$

∴ So, the middle half of the students' BMIs varies by no more than 3.

On Your Own

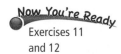

Now You're Ready
Exercises 11 and 12

2. The box-and whisker plot shows the heights of the roller coasters at an amusement park. (a) What fraction of the roller coasters are between 120 feet tall and 220 feet tall? (b) Are the data more spread out below or above the median? Explain. (c) Find and interpret the interquartile range of the data.

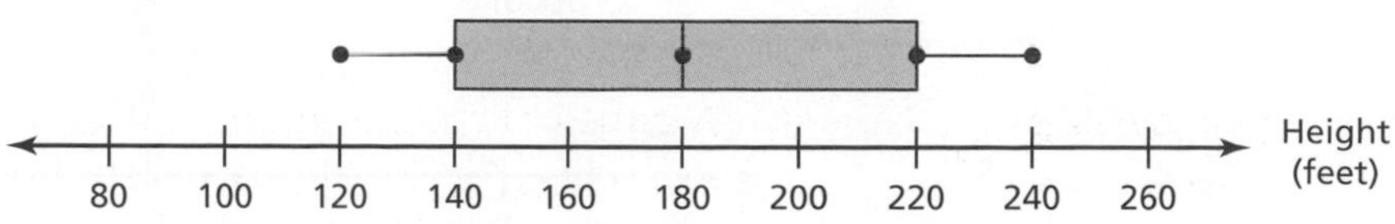

A box-and-whisker plot also shows the shape of a distribution.

Key Ideas

Study Tip

If you can draw a line through the median of a box-and-whisker plot, and each side is a mirror image of the other, then the distribution is symmetric.

Shapes of Box-and-Whisker Plots

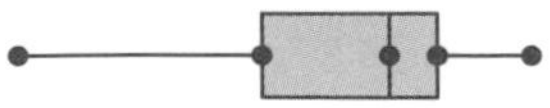

Skewed left

- Left whisker longer than right whisker
- Most data on the right

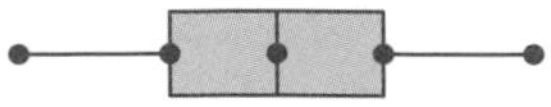

Symmetric

- Whiskers about same length
- Median in the middle of the box

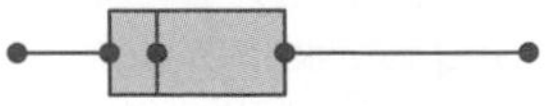

Skewed right

- Right whisker longer than left whisker
- Most data on the left

EXAMPLE 3 Comparing Box-and-Whisker Plots

The double box-and-whisker plot represents the prices of snowboards at two stores.

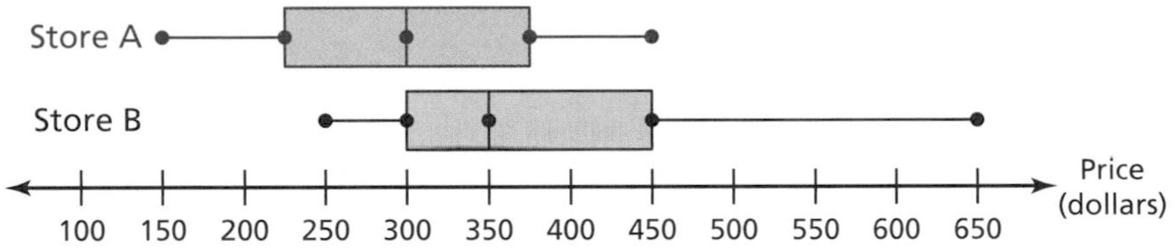

a. Identify the shape of each distribution.

For Store A, the whisker lengths are equal. The median is in the middle of the box. The data on the left are the mirror image of the data on the right. So, the distribution is symmetric.

For Store B, the right whisker is longer than the left whisker, and most of the data are on the left side of the display. So, the distribution is skewed right.

b. Which store's prices are more spread out? Explain.

Both boxes appear to be the same length. So, the interquartile range of each data set is equal. However, the range of the prices in Store B is greater than the range of the prices in Store A. So, the prices in Store B are more spread out.

On Your Own

Now You're Ready
Exercises 13–17

3. The double box-and-whisker plot represents the life spans of crocodiles and alligators at a zoo. Identify the shape of each distribution. Which reptile's life spans are more spread out? Explain.

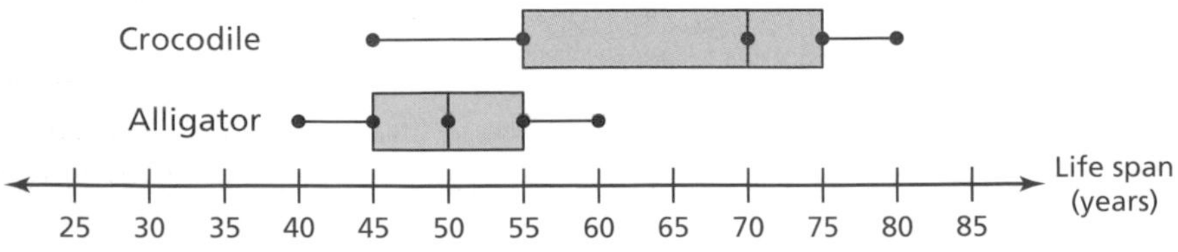

10.4 Exercises

Vocabulary and Concept Check

1. **VOCABULARY** Explain how to find the five-number summary of a data set.

2. **NUMBER SENSE** In a box-and-whisker plot, what fraction of the data is greater than the first quartile?

3. **DIFFERENT WORDS, SAME QUESTION** Which is different? Find "both" answers.

Is the distribution skewed right?

Is the left whisker longer than the right whisker?

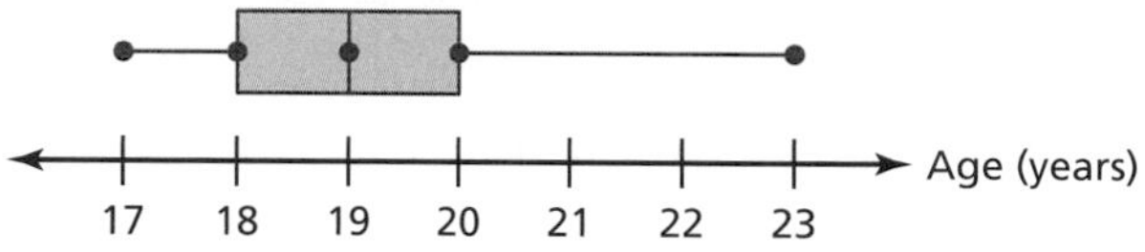

Are the data more spread out below the first quartile than above the third quartile?

Does the lower fourth of the data vary more than the upper fourth of the data?

Practice and Problem Solving

4. The box-and-whisker plots represent the daily attendance at two beaches during July. Compare and contrast the attendances for the two beaches.

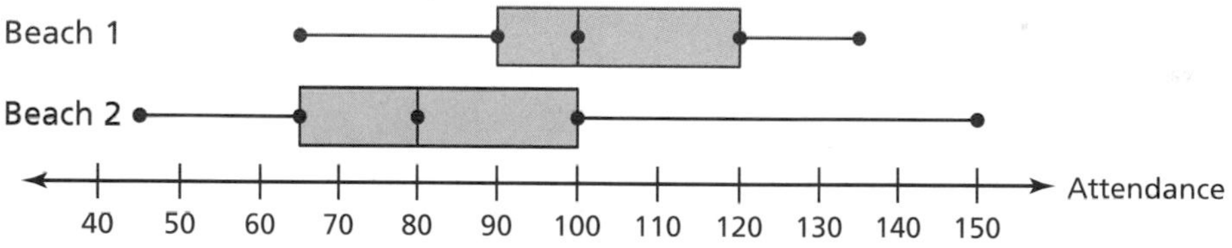

Make a box-and-whisker plot for the data.

(1) 5. Ages of teachers (in years): 30, 62, 26, 35, 45, 22, 49, 32, 28, 50, 42, 35

6. Quiz scores: 8, 12, 9, 10, 12, 8, 5, 9, 7, 10, 8, 9, 11

7. Donations (in dollars): 10, 30, 5, 15, 50, 25, 5, 20, 15, 35, 10, 30, 20

8. Ski lengths (in centimeters): 180, 175, 205, 160, 210, 175, 190, 205, 190, 160, 165, 195

9. **ERROR ANALYSIS** Describe and correct the error in making a box-and-whisker plot for the data.

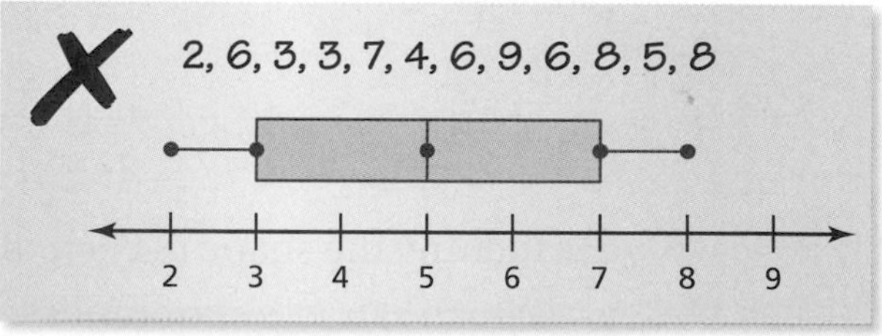

10. **CAMPING** The numbers of days 12 friends went camping during the summer are 6, 2, 0, 10, 3, 6, 6, 4, 12, 0, 6, and 2. Make a box-and-whisker plot for the data. What is the range of the data?

2 **11. DUNK TANK** The box-and-whisker plot represents the numbers of gallons of water needed to fill different types of dunk tanks offered by a company.

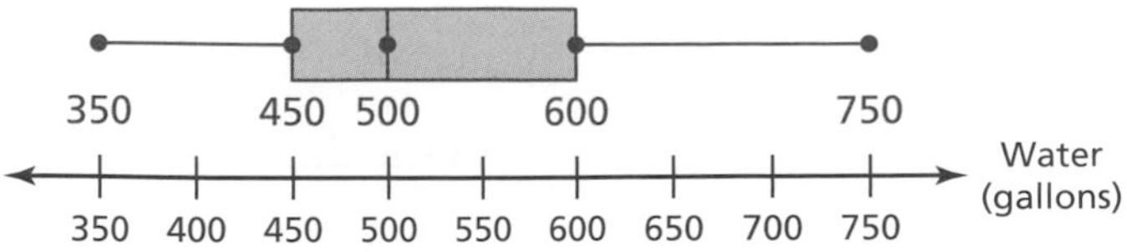

a. What fraction of the dunk tanks require at least 500 gallons of water?

b. Are the data more spread out below the first quartile or above the third quartile? Explain.

c. Find and interpret the interquartile range of the data.

12. BUILDINGS The box-and-whisker plot represents the heights (in meters) of the tallest buildings in Chicago.

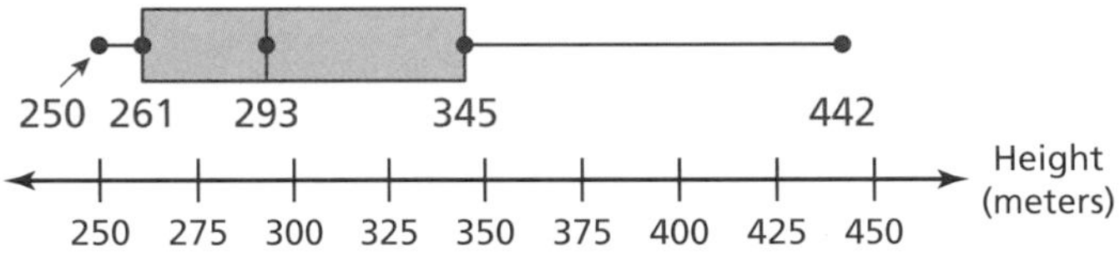

a. What percent of the buildings are no taller than 345 meters?

b. Is there more variability in the heights above 345 meters or below 261 meters? Explain.

c. Find and interpret the interquartile range of the data.

Identify the shape of the distribution. Explain.

3 **13.**

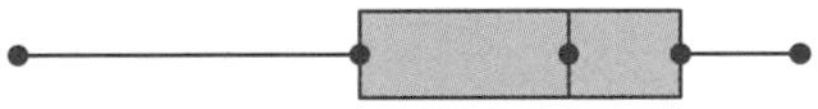

14.

15.

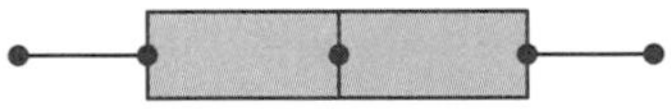

16.

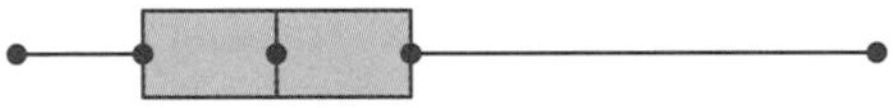

17. RECESS The double box-and-whisker plot represents the start times of recess for two schools.

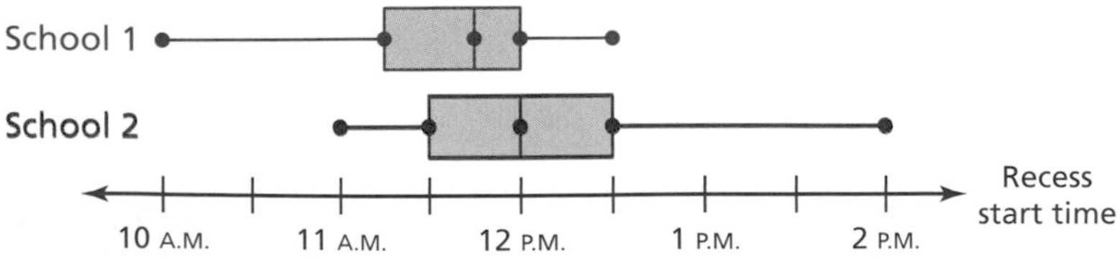

a. Identify the shape of each distribution.

b. Which school's start times for recess are more spread out? Explain.

c. Which school is more likely to have recess before lunch? Explain.

Make a box-and-whisker plot for the data.

18. Temperatures (in °C): 5, 1, 4, 0, 9, 0, −8, 5, 2, 4, −1, 10, 7, −5

19. Checking account balances (in dollars): 30, 0, −10, 50, 20, 90, −15, 40, 100, 45, −20, 70

20. **REASONING** The data set in Exercise 18 has an outlier. Describe how removing the outlier affects the box-and-whisker plot.

21. **CHOOSE TOOLS** What are the most appropriate measures to describe the center and the variation of the distribution in Exercise 12?

22. **OPEN-ENDED** Write a data set with 12 values that has a symmetric box-and-whisker plot.

23. **CRITICAL THINKING** When would a box-and-whisker plot *not* have one or both whiskers?

24. **STRUCTURE** Draw a histogram that could represent the distribution shown in Exercise 15.

25. **REASONING** The double box-and-whisker plot represents the runs scored per game by two softball teams during a 32-game season.

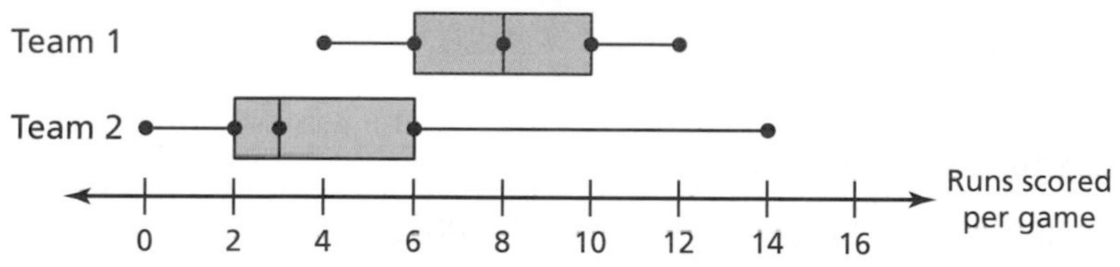

a. Which team is more consistent at scoring runs? Explain.

b. In how many games did Team 2 score 6 runs or less?

c. Team 1 played Team 2 once during the season. Which team do you think won? Explain.

d. Which team do you think has the greater mean? Explain.

26. Choose Tools A market research company wants to summarize the variability of the SAT scores of graduating seniors in the United States. Do you think the company should use a stem-and-leaf plot, a histogram, or a box-and-whisker plot? Explain.

Fair Game Review What you learned in previous grades & lessons

Copy and complete the statement using < or >. *(Section 6.3)*

27. $-\frac{2}{3}$ ▭ $-\frac{3}{4}$ **28.** $-2\frac{1}{5}$ ▭ $-2\frac{1}{6}$ **29.** -5.3 ▭ -5.5

30. **MULTIPLE CHOICE** Which of the following items is most likely represented by a rectangular prism with a volume of 1785 cubic inches? *(Section 8.4)*

Ⓐ closet Ⓑ computer tower

Ⓒ filing cabinet Ⓓ your math book

10.3–10.4 Quiz

Check It Out
Progress Check
BigIdeasMath.com

Describe the shape of each distribution. *(Section 10.3)*

1.

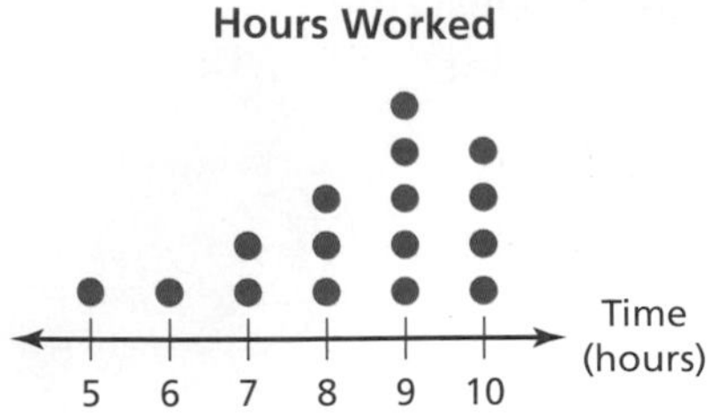

2.

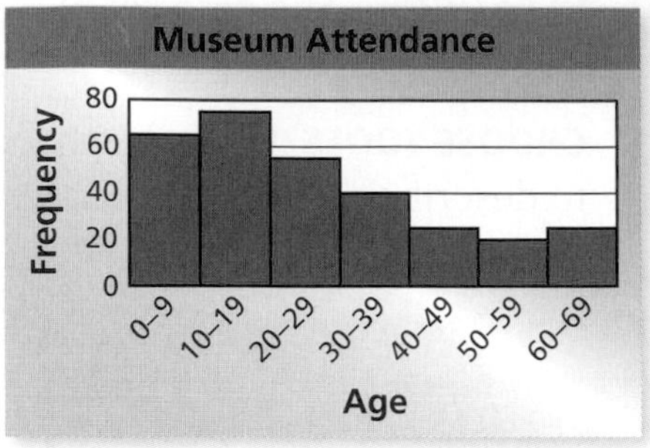

Choose the most appropriate measures to describe the center and the variation. Find the measures you chose. *(Section 10.3)*

3.

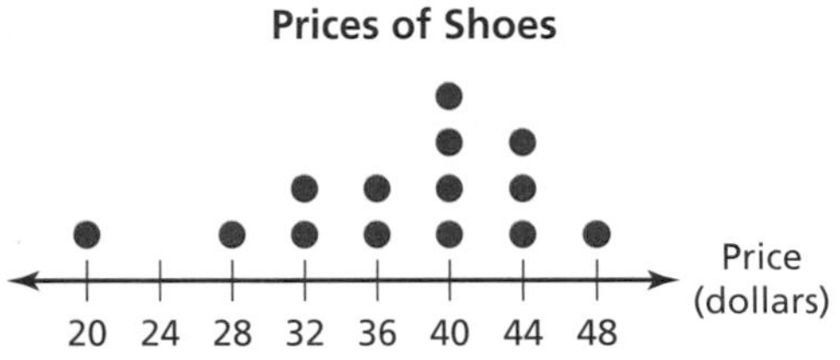

4.

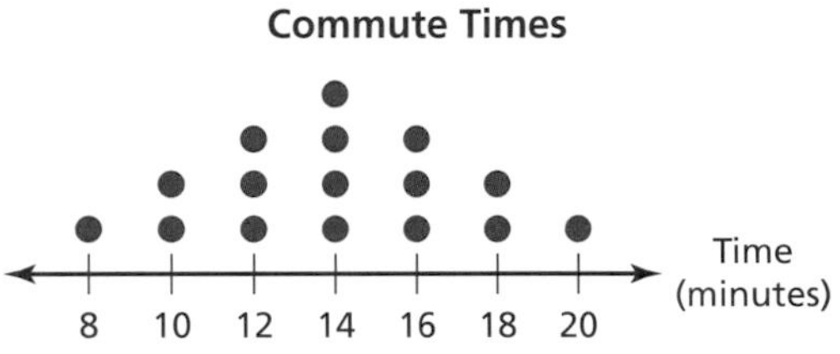

Make a box-and-whisker plot for the data. *(Section 10.4)*

5. Science test scores: 85, 76, 99, 84, 92, 95, 68, 100, 93, 88, 87, 85

6. Shoe sizes: 12, 8.5, 9, 10, 9, 11, 11.5, 9, 9, 10, 10, 10.5, 8

7. **MOVIES** The box-and-whisker plot represents the lengths (in minutes) of movies being shown at a theater. *(Section 10.4)*

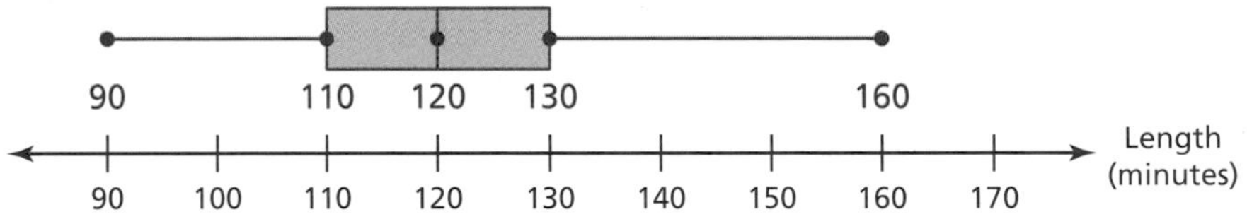

a. What percent of the movies are no longer than 120 minutes?

b. Is there more variability in the movie lengths longer than 130 minutes or shorter than 110 minutes? Explain.

c. Find and interpret the interquartile range of the data.

8. **EXPERIENCE** The frequency table shows the years of experience of employees at two branches of a company. Display the data for each branch in a histogram. Describe the shape of each distribution. Which branch has less experience? Explain. *(Section 10.3)*

Years of Experience	0–2	3–6	7–10	11–14	15–18	19–22	23–26
Frequency at Branch A	10	25	14	20	8	5	2
Frequency at Branch B	3	6	8	10	15	25	8

10 Chapter Review

Check It Out
Vocabulary Help
BigIdeasMath.com

Review Key Vocabulary

stem-and-leaf plot, *p. 436*
stem, *p. 436*
leaf, *p. 436*
frequency table, *p. 441*
frequency, *p. 441*
histogram, *p. 442*
box-and-whisker plot, *p. 460*
five-number summary, *p. 460*

Review Examples and Exercises

10.1 Stem-and-Leaf Plots *(pp. 434–439)*

Day	DVDs Rented
Sun	50
Mon	19
Tue	25
Wed	28
Thu	39
Fri	53
Sat	50

Make a stem-and-leaf plot of the number of DVDs rented each day at a store.

Step 1: Order the data. 19, 25, 28, 39, 50, 50, 53

Step 2: Choose the stems and the leaves. Because the data range from 19 to 53, use the *tens* digits for the stems and the *ones* digits for the leaves. Be sure to include the key.

Step 3: Write the stems to the *left* of the vertical line.

Step 4: Write the leaves for each stem to the *right* of the vertical line.

DVDs Rented

Stem	Leaf
1	9
2	5 8
3	9
4	
5	0 0 3

Key: 2 | 5 = 25 DVDs

Order the stems vertically. The stem for data values less than 10 is 0.

Include stems without leaves.

Write the leaves horizontally.

Exercises

Make a stem-and-leaf plot of the data.

1.

Hats Sold Each Day			
5	18	12	15
21	30	8	12
13	9	14	25

2.

Ages of Park Volunteers			
13	17	40	15
48	21	19	52
13	55	60	20

The stem-and-leaf plot shows the weights (in pounds) of yellowfin tuna caught during a fishing contest.

Stem	Leaf
7	6
8	0 2 5 7 9
9	5 6
10	2

Key: 8 | 5 = 85 pounds

3. How many tuna weigh less than 90 pounds?

4. What is the median weight of the tuna?

10.2 Histograms (pp. 440–447)

The frequency table shows the number of crafts each member of a craft club made for a fundraiser. Display the data in a histogram.

Crafts	Frequency
0–2	10
3–5	8
6–8	5
9–11	0
12–14	2

Step 1: Draw and label the axes.

Step 2: Draw a bar to represent the frequency of each interval.

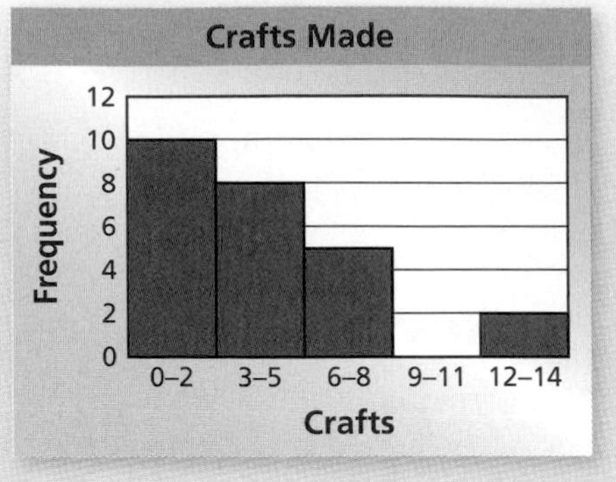

Exercises

Display the data in a histogram.

5.

Heights of Gymnasts	
Heights (in.)	**Frequency**
50–54	1
55–59	8
60–64	5
65–69	2

6.

Minutes Studied	
Minutes	**Frequency**
0–19	5
20–39	9
40–59	12
60–79	3

10.3 Shapes of Distributions (pp. 450–457)

Describe the shape of each distribution.

a.

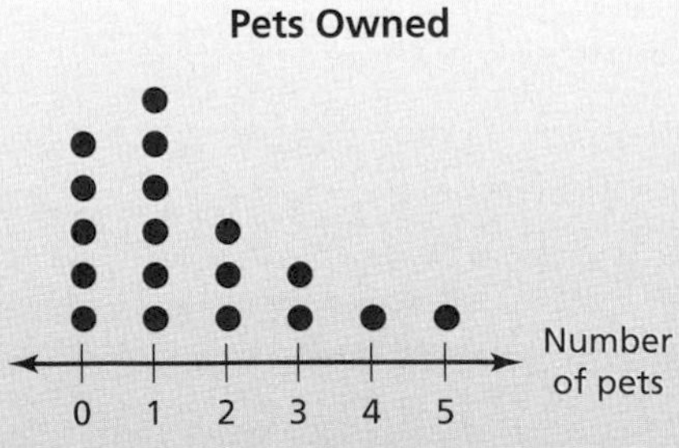

b.

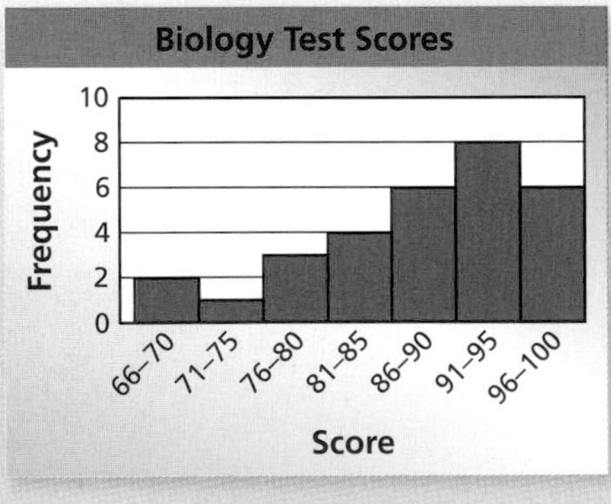

a. Most of the data are on the left, and the tail extends to the right.

∴ So, the distribution is skewed right.

b. Most of the data are on the right, and the tail extends to the left.

∴ So, the distribution is skewed left.

Exercises

7. Describe the shape of the distribution.

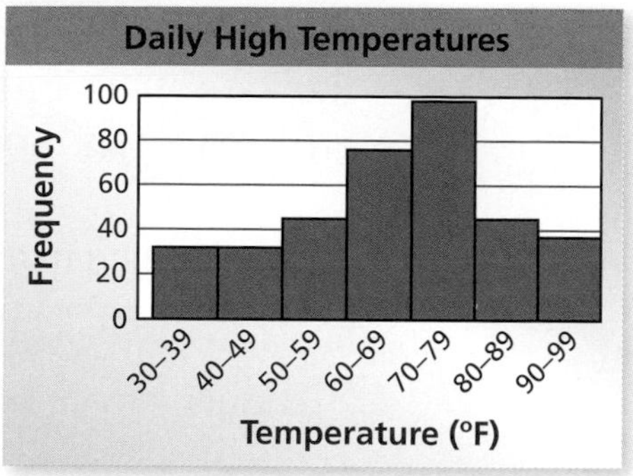

8. Choose the most appropriate measures to describe the center and the variation. Find the measures you chose.

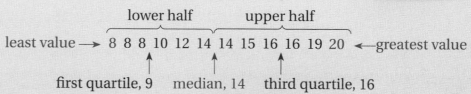

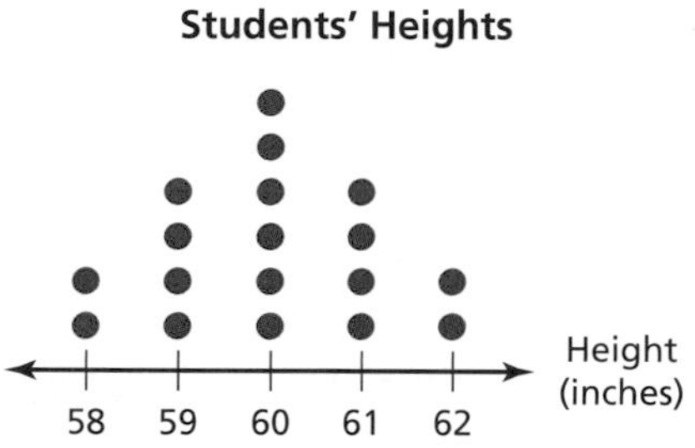

10.4 Box-and-Whisker Plots *(pp. 458–465)*

Make a box-and-whisker plot for the weights (in pounds) of pumpkins sold at a market.

16, 20, 14, 15, 12, 8, 8, 19, 14, 10, 8, 16

Step 1: Order the data. Find the median and the quartiles.

lower half / upper half

least value → 8 8 8 10 12 14 14 15 16 16 19 20 ← greatest value

first quartile, 9 — median, 14 — third quartile, 16

Step 2: Draw a number line that includes the least and the greatest values. Graph points above the number line that represent the five-number summary.

Step 3: Draw a box using the quartiles. Draw a line through the median. Draw whiskers from the box to the least and the greatest values.

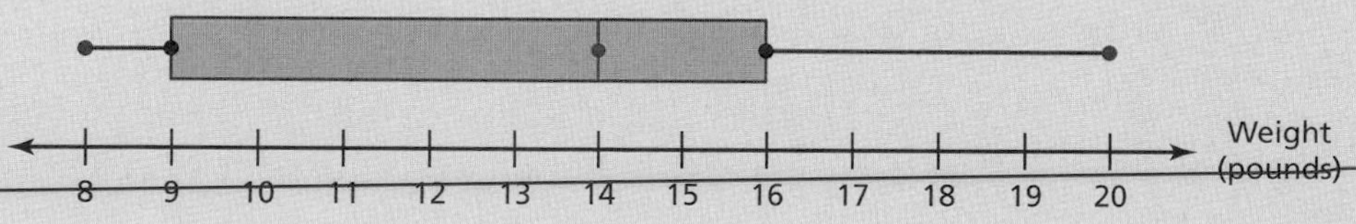

Exercises

Make a box-and-whisker plot for the data.

9. Ages of volunteers at a hospital: 14, 17, 20, 16, 17, 14, 21, 18

10. Masses (in kilograms) of lions: 120, 200, 180, 150, 200, 200, 230, 160

10 Chapter Test

Check It Out
Test Practice
BigIdeasMath.com

Make a stem-and-leaf plot of the data.

1.

Quiz Scores (%)			
96	88	80	72
80	94	92	100
76	80	68	90

2.

CDs Sold Each Day				
45	31	29	38	38
67	40	62	45	60
40	39	60	43	48

3. Find the mean, median, mode, range, and interquartile range of the data.

Cooking Time (minutes)

Stem	Leaf
3	5 8
4	0 1 8
5	0 4 4 4 5 9
6	0

Key: 4 | 1 = 41 minutes

4. Display the data in a histogram.

Television Watched Per Week	
Hours	**Frequency**
0–9	14
10–19	16
20–29	10
30–39	8

5. WATER The dot plot shows the number of glasses of water that the students in a class drink in one day.

a. Describe the shape of the distribution.

b. Choose the most appropriate measures to describe the center and the variation. Find the measures you chose.

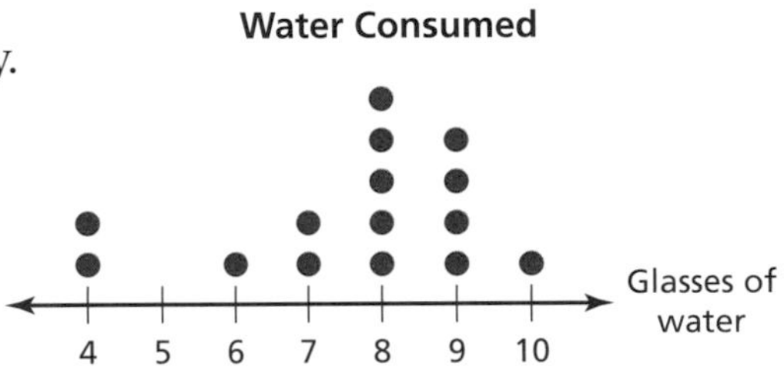

Make a box-and-whisker plot for the data.

6. Ages (in years) of dogs at a vet's office: 1, 3, 5, 11, 5, 7, 5, 9

7. Lengths (in inches) of fish in a pond: 12, 13, 7, 8, 14, 6, 13, 10

8. Hours practiced each week: 7, 6, 5, 4.5, 3.5, 7, 7.5, 2, 8, 7, 7.5, 6.5

9. CELL PHONES The double box-and-whisker plot compares the battery life (in hours) of two brands of cell phones.

a. What is the range of the upper 75% of each brand?

b. Which battery has a longer battery life? Explain.

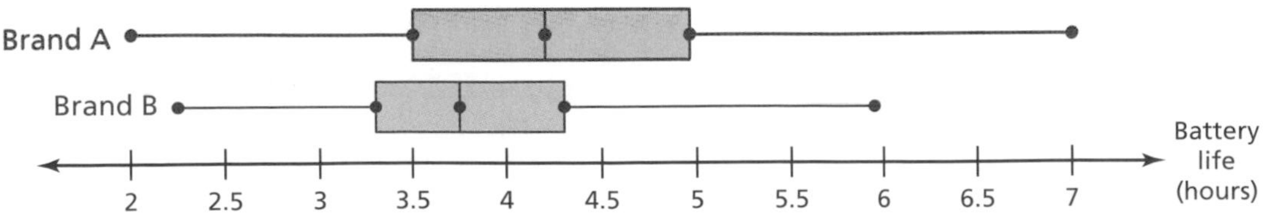

10 Standards Assessment

Test-Taking Strategy
Read Question Before Answering

Of 2048 cats, how many of them will NOT answer to "Here, kitty kitty"?
Ⓐ 100% Ⓑ 1 Ⓒ $\frac{4098}{2}$ Ⓓ 2048
Hey, it means "free food."

"Be sure to read the question before choosing your answer. You may find a word that changes the meaning."

1. Research scientists are measuring the number of days lettuce seeds take to germinate. In a study, 500 seeds were planted. Of these, 473 seeds germinated. The box-and-whisker plot summarizes the number of days it took the seeds to germinate. What can you conclude from the box-and-whisker plot? *(6.SP.4, 6.SP.5c)*

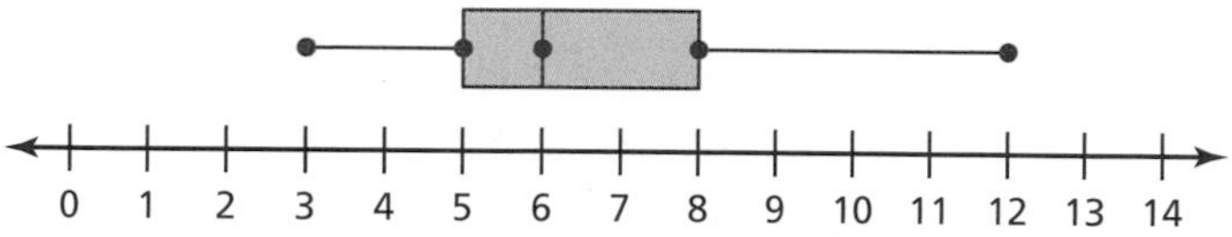

   A. The median number of days for the seeds to germinate is 12.

   B. 50% of the seeds took more than 8 days to germinate.

   C. 50% of the seeds took less than 5 days to germinate.

   D. The median number of days for the seeds to germinate was 6.

2. You are comparing the costs of buying bottles of water at the supermarket. Which of the following has the least cost per liter? *(6.RP.3b)*

   F. six 1-liter bottles for \$1.80

   G. one 2-liter bottle for \$0.65

   H. eight $\frac{1}{2}$-liter bottles for \$1.50

   I. twelve $\frac{1}{2}$-liter bottles for \$1.98

3. What number belongs in the box to make the equation true? *(6.NS.1)*

$$3\frac{1}{2} \div 5\frac{2}{3} = \frac{7}{2} \times \square$$

   A. $\frac{17}{3}$

   B. $\frac{13}{2}$

   C. $\frac{3}{17}$

   D. $\frac{3}{2}$

4. What is the mean number of seats? *(6.SP.5c)*

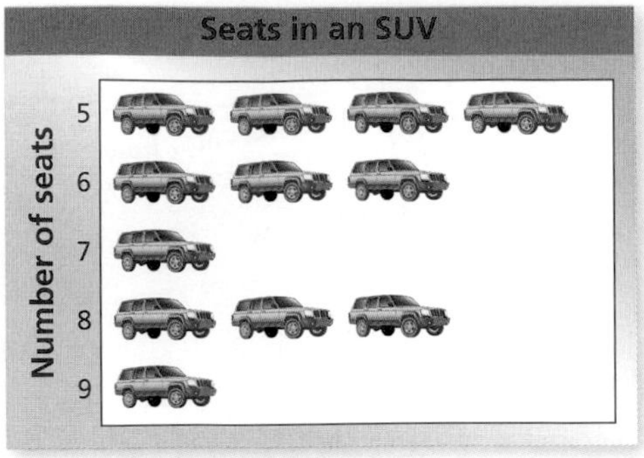

F. 2.4 seats

G. 5 seats

H. 6.5 seats

I. 7 seats

5. On Wednesday, the town of Mims received 17 millimeters of rain. This was x millimeters more rain than the town received on Tuesday. Which expression represents the amount of rain, in millimeters, the town received on Tuesday? *(6.EE.2a, 6.EE.6)*

A. $17x$

B. $17 - x$

C. $x + 17$

D. $x - 17$

6. One of the leaves is missing in the stem-and-leaf plot.

The median of the data set represented by the stem-and-leaf plot is 38. What is the value of the missing leaf? *(6.SP.4, 6.SP.5c)*

Stem	Leaf
1	3 4
2	
3	4 5 7 7 7 ? 9
4	0 1 1 4
5	0 2 3

Key: 1 | 4 = 14

7. Which property is demonstrated by the equation below? *(6.EE.3)*

$$723 + (884 + 277) = 723 + (277 + 884)$$

F. Associative Property of Addition

G. Commutative Property of Addition

H. Distributive Property

I. Identity Property of Addition

8. A student took 5 tests this marking period and had a mean score of 92. Her scores on the first 4 tests were 90, 96, 86, and 92. What was her score on the fifth test? *(6.SP.5c)*

A. 92

B. 93

C. 96

D. 98

9. At the end of the school year, your teacher counted up the number of absences for each student. The results are shown in the histogram below.

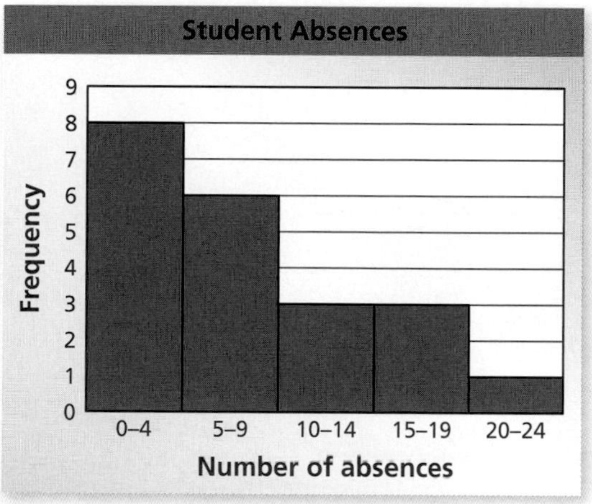

Based on the histogram, how many students had fewer than 10 absences? *(6.SP.4)*

10. The 16 members of a camera club have the ages listed below. *(6.SP.4, 6.SP.5c)*

Think Solve Explain

40, 22, 24, 58, 30, 31, 37, 25, 62, 40, 39, 37, 28, 28, 51, 44

Part A Order the ages from least to greatest.

Part B Find the median of the ages.

Part C Make a box-and-whisker plot for the ages of the camera club members.

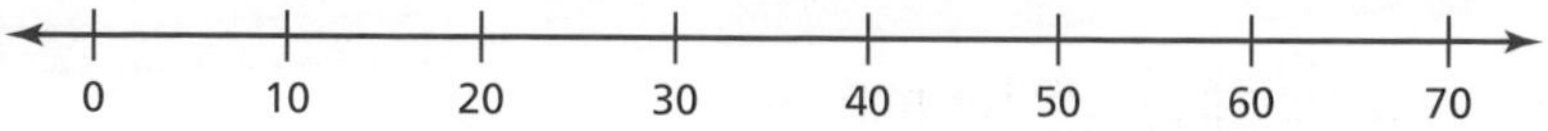

11 Integers

"Look, subtraction is not that difficult. Imagine that you have five squeaky mouse toys."
?
5 + (-1) = 4
"After your friend Fluffy comes over for a visit, you notice that one of the squeaky toys is missing."
4 - (-1) = 5
I'm going to count my toys
"Now, you go over to Fluffy's and retrieve the missing squeaky mouse toy. It's easy."

"Dear Sir: You asked me to 'find' the opposite of −1."
That's not the only thing that is missing.
"I didn't know it was missing."

What You Learned Before

"I liked it because it is the opposite of the freezing point on the Fahrenheit temperature scale."

Commutative and Associative Properties (6.EE.3)

Example 1 **a. Simplify the expression 6 + (14 + *x*).**

$6 + (14 + x) = (6 + 14) + x$ — Associative Property of Addition

$= 20 + x$ — Add 6 and 14.

b. Simplify the expression (3.1 + *x*) + 7.4.

$(3.1 + x) + 7.4 = (x + 3.1) + 7.4$ — Commutative Property of Addition

$= x + (3.1 + 7.4)$ — Associative Property of Addition

$= x + 10.5$ — Add 3.1 and 7.4.

c. Simplify the expression 5(12*y*).

$5(12y) = (5 \cdot 12)y$ — Associative Property of Multiplication

$= 60y$ — Multiply 5 and 12.

Try It Yourself

Simplify the expression. Explain each step.

1. $3 + (b + 8)$ **2.** $(d + 4) + 6$ **3.** $6(5p)$

Properties of Zero and One (6.EE.3)

Example 2 **a. Simplify the expression 6 • 0 • *q*.**

$6 \cdot 0 \cdot q = (6 \cdot 0) \cdot q$ — Associative Property of Multiplication

$= 0 \cdot q = 0$ — Multiplication Property of Zero

b. Simplify the expression 3.6 • *s* • 1.

$3.6 \cdot s \cdot 1 = 3.6 \cdot (s \cdot 1)$ — Associative Property of Multiplication

$= 3.6 \cdot s$ — Multiplication Property of One

$= 3.6s$

Try It Yourself

Simplify the expression. Explain each step.

4. $13 \cdot m \cdot 0$ **5.** $1 \cdot x \cdot 29$ **6.** $(n + 14) + 0$

11.1 Integers and Absolute Value

Essential Question How can you use integers to represent the velocity and the speed of an object?

On these two pages, you will investigate vertical motion (up or down).

- Speed tells how fast an object is moving, but it does not tell the direction.
- Velocity tells how fast an object is moving, and it also tells the direction.

  When velocity is positive, the object is moving up.

  When velocity is negative, the object is moving down.

1 ACTIVITY: Falling Parachute

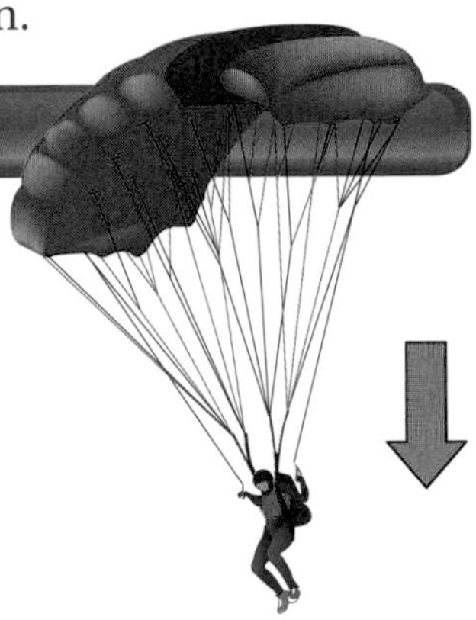

Work with a partner. You are gliding to the ground wearing a parachute. The table shows your height above the ground at different times.

Time (seconds)	0	1	2	3
Height (feet)	90	75	60	45

a. Describe the pattern in the table. How many feet do you move each second? After how many seconds will you land on the ground?

b. What integer represents your speed? Give the units.

c. Do you think your velocity should be represented by a positive or negative integer? Explain your reasoning.

d. What integer represents your velocity? Give the units.

2 ACTIVITY: Rising Balloons

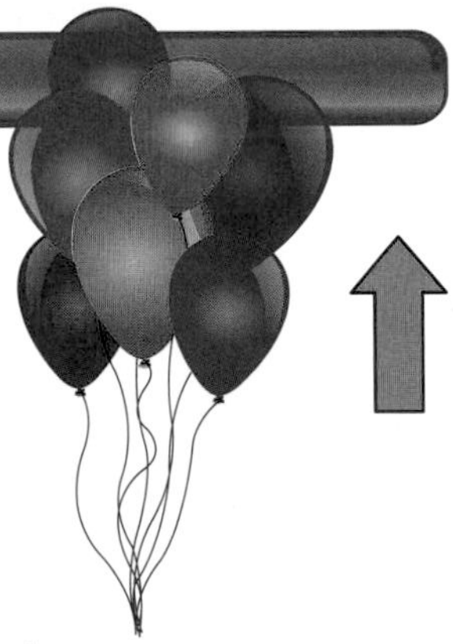

Work with a partner. You release a group of balloons. The table shows the height of the balloons above the ground at different times.

Time (seconds)	0	1	2	3
Height (feet)	8	12	16	20

a. Describe the pattern in the table. How many feet do the balloons move each second? After how many seconds will the balloons be at a height of 40 feet?

b. What integer represents the speed of the balloons? Give the units.

c. Do you think the velocity of the balloons should be represented by a positive or negative integer? Explain your reasoning.

d. What integer represents the velocity of the balloons? Give the units.

COMMON CORE

Integers

In this lesson, you will

- define the absolute value of a number.
- find absolute values of numbers.
- solve real-life problems.

Preparing for Standards
7.NS.1
7.NS.2
7.NS.3

3 ACTIVITY: Firework Parachute

Work with a partner. The table shows the height of a firework's parachute above the ground at different times.

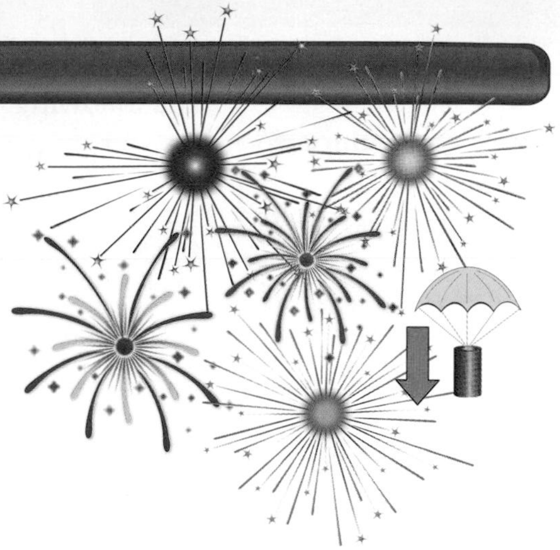

Time (seconds)	Height (feet)
0	480
1	360
2	240
3	120
4	0

Math Practice 6

Use Clear Definitions

What information can you use to support your answer?

a. Describe the pattern in the table. How many feet does the parachute move each second?

b. What integer represents the speed of the parachute? What integer represents the velocity? How are these integers similar in their relation to 0 on a number line?

Inductive Reasoning

4. Copy and complete the table.

Velocity (feet per second)	-14	20	-2	0	25	-15
Speed (feet per second)						

5. Find two different velocities for which the speed is 16 feet per second.

6. Which number is greater: -4 or 3? Use a number line to explain your reasoning.

7. One object has a velocity of -4 feet per second. Another object has a velocity of 3 feet per second. Which object has the greater speed? Explain your answer.

What Is Your Answer?

8. IN YOUR OWN WORDS How can you use integers to represent the velocity and the speed of an object?

9. LOGIC In this lesson, you will study *absolute value*. Here are some examples:

$|-16| = 16$ $|16| = 16$ $|0| = 0$ $|-2| = 2$

Which of the following is a true statement? Explain your reasoning.

$|\text{velocity}| = \text{speed}$ $|\text{speed}| = \text{velocity}$

Practice Use what you learned about absolute value to complete Exercises 4–11 on page 480.

11.1 Lesson

Check It Out
Lesson Tutorials
BigIdeasMath.com

Key Vocabulary
integer, *p. 478*
absolute value, *p. 478*

The following numbers are **integers**:

$\ldots, -3, -2, -1, 0, 1, 2, 3, \ldots$

Key Idea

Absolute Value

Words The **absolute value** of an integer is the distance between the number and 0 on a number line. The absolute value of a number a is written as $|a|$.

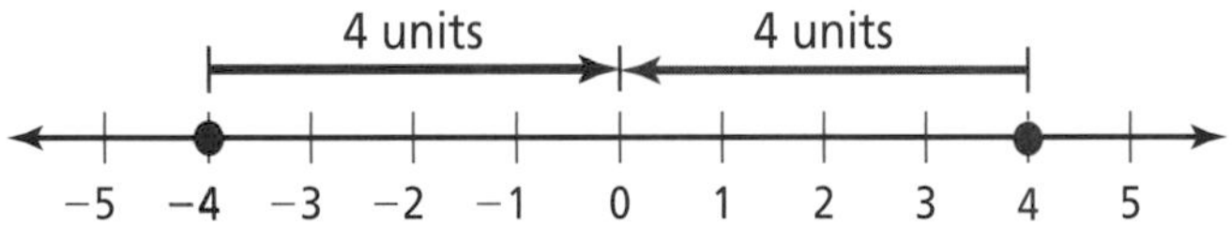

Numbers $|-4| = 4$ $|4| = 4$

EXAMPLE 1 Finding Absolute Value

Find the absolute value of 2.

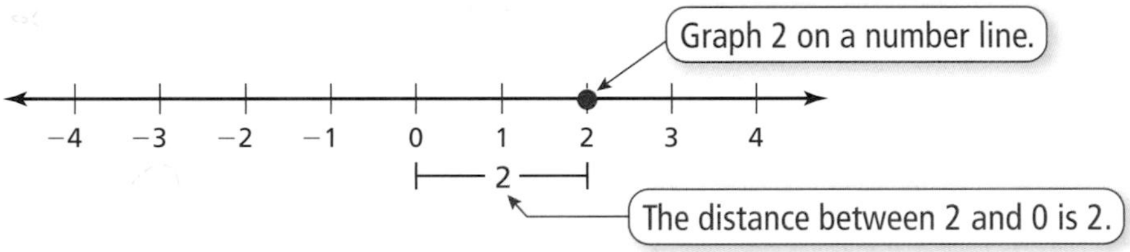

So, $|2| = 2$.

EXAMPLE 2 Finding Absolute Value

Find the absolute value of −3.

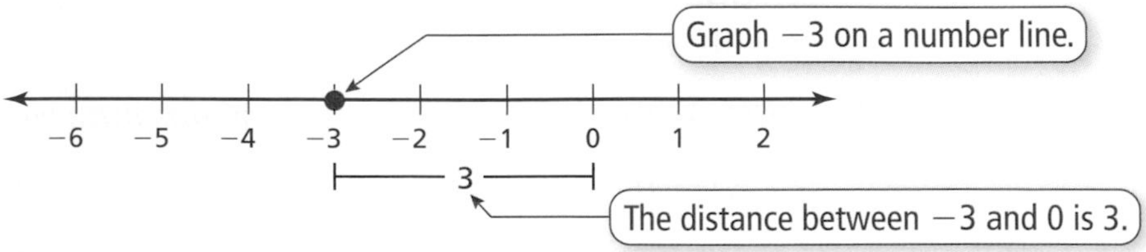

So, $|-3| = 3$.

On Your Own

Now You're Ready
Exercises 4–19

Find the absolute value.

1. $|7|$ **2.** $|-1|$ **3.** $|-5|$ **4.** $|14|$

Multi-Language Glossary at BigIdeasMath.com

EXAMPLE 3 Comparing Values

Remember

A number line can be used to compare and order integers. Numbers to the left are less than numbers to the right. Numbers to the right are greater than numbers to the left.

Compare 1 and $|-4|$.

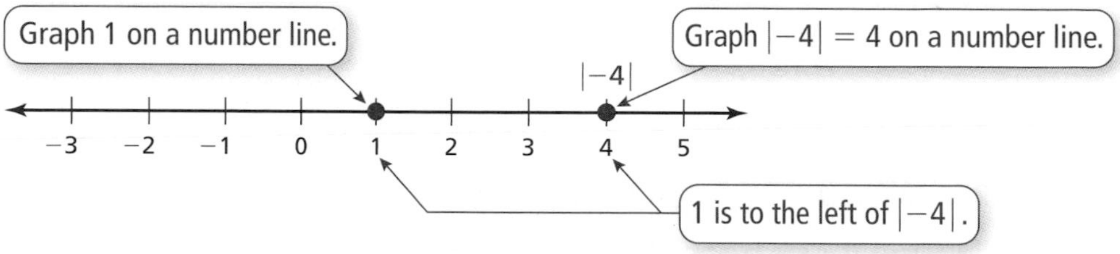

∴ So, $1 < |-4|$.

On Your Own

Now You're Ready
Exercises 20–25

Copy and complete the statement using <, >, or =.

5. $|-2|$ ▢ -1

6. -7 ▢ $|6|$

7. $|10|$ ▢ 11

8. 9 ▢ $|-9|$

EXAMPLE 4 Real-Life Application

Substance	Freezing Point (°C)
Butter	35
Airplane fuel	−53
Honey	−3
Mercury	−39
Candle wax	55

The *freezing point* is the temperature at which a liquid becomes a solid.

a. Which substance in the table has the lowest freezing point?

b. Is the freezing point of mercury or butter closer to the freezing point of water, 0°C?

a. Graph each freezing point.

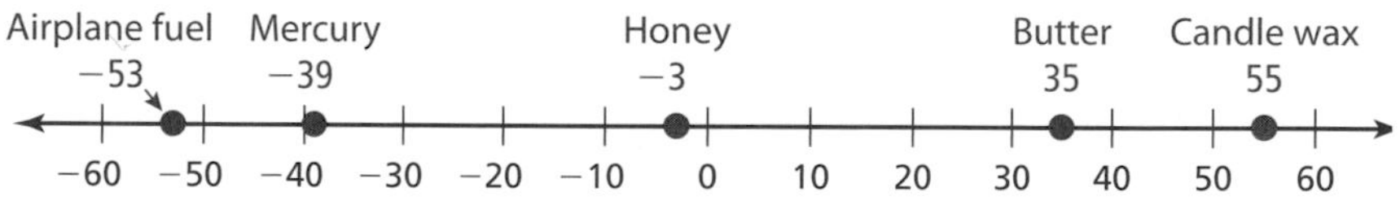

∴ Airplane fuel has the lowest freezing point, −53°C.

b. The freezing point of water is 0°C, so you can use absolute values.

Mercury: $|-39| = 39$ **Butter:** $|35| = 35$

∴ Because 35 is less than 39, the freezing point of butter is closer to the freezing point of water.

On Your Own

9. Is the freezing point of airplane fuel or candle wax closer to the freezing point of water? Explain your reasoning.

11.1 Exercises

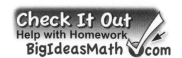

Vocabulary and Concept Check

1. **VOCABULARY** Which of the following numbers are integers?

$$9,\ 3.2,\ -1,\ \frac{1}{2},\ -0.25,\ 15$$

2. **VOCABULARY** What is the absolute value of an integer?

3. **WHICH ONE DOESN'T BELONG?** Which expression does *not* belong with the other three? Explain your reasoning.

$\lvert 6 \rvert$	6	-6	$\lvert -6 \rvert$

Practice and Problem Solving

Find the absolute value.

(1) (2)

4. $\lvert 9 \rvert$
5. $\lvert -6 \rvert$
6. $\lvert -10 \rvert$
7. $\lvert 10 \rvert$
8. $\lvert -15 \rvert$
9. $\lvert 13 \rvert$
10. $\lvert -7 \rvert$
11. $\lvert -12 \rvert$
12. $\lvert 5 \rvert$
13. $\lvert -8 \rvert$
14. $\lvert 0 \rvert$
15. $\lvert 18 \rvert$
16. $\lvert -24 \rvert$
17. $\lvert -45 \rvert$
18. $\lvert 60 \rvert$
19. $\lvert -125 \rvert$

Copy and complete the statement using <, >, or =.

(3)

20. $2 \ \square \ \lvert -5 \rvert$
21. $\lvert -4 \rvert \ \square \ 7$
22. $-5 \ \square \ \lvert -9 \rvert$
23. $\lvert -4 \rvert \ \square \ -6$
24. $\lvert -1 \rvert \ \square \ \lvert -8 \rvert$
25. $\lvert 5 \rvert \ \square \ \lvert -5 \rvert$

ERROR ANALYSIS **Describe and correct the error.**

26. ✗ $\lvert 10 \rvert = -10$

27. ✗ $\lvert -5 \rvert < 4$

28. **SAVINGS** You deposit \$50 in your savings account. One week later, you withdraw \$20. Write each amount as an integer.

29. **ELEVATOR** You go down 8 floors in an elevator. Your friend goes up 5 floors in an elevator. Write each amount as an integer.

Order the values from least to greatest.

30. $8, \lvert 3 \rvert, -5, \lvert -2 \rvert, -2$
31. $\lvert -6 \rvert, -7, 8, \lvert 5 \rvert, -6$
32. $-12, \lvert -26 \rvert, -15, \lvert -12 \rvert, \lvert 10 \rvert$
33. $\lvert -34 \rvert, 21, -17, \lvert 20 \rvert, \lvert -11 \rvert$

Simplify the expression.

34. $\lvert -30 \rvert$
35. $-\lvert 4 \rvert$
36. $-\lvert -15 \rvert$

37. **PUZZLE** Use a number line.

a. Graph and label the following points on a number line: $A = -3$, $E = 2$, $M = -6$, $T = 0$. What word do the letters spell?

b. Graph and label the absolute value of each point in part (a). What word do the letters spell now?

38. **OPEN-ENDED** Write a negative integer whose absolute value is greater than 3.

REASONING Determine whether $n \geq 0$ or $n \leq 0$.

39. $n + |-n| = 2n$

40. $n + |-n| = 0$

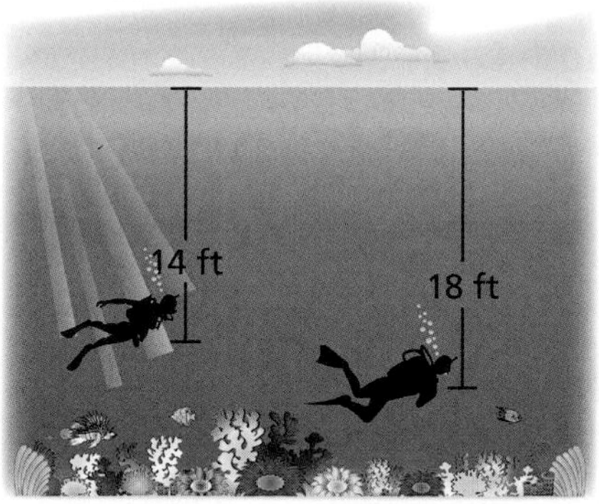

41. **CORAL REEF** The depths of two scuba divers exploring a living coral reef are shown.

a. Write an integer for the position of each diver relative to sea level.

b. Which integer in part (a) is greater?

c. Which integer in part (a) has the greater absolute value? Compare this absolute value with the depth of that diver.

42. **VOLCANOES** The *summit elevation* of a volcano is the elevation of the top of the volcano relative to sea level. The summit elevation of the volcano Kilauea in Hawaii is 1277 meters. The summit elevation of the underwater volcano Loihi in the Pacific Ocean is -969 meters. Which summit is closer to sea level?

43. **MINIATURE GOLF** The table shows golf scores, relative to *par*.

a. The player with the lowest score wins. Which player wins?

b. Which player is at par?

c. Which player is farthest from par?

Player	Score
1	+5
2	0
3	−4
4	−1
5	+2

True or False? **Determine whether the statement is *true* or *false*. Explain your reasoning.**

44. If $x < 0$, then $|x| = -x$.

45. The absolute value of every integer is positive.

Fair Game Review What you learned in previous grades & lessons

Add. *(Section 1.1)*

46. $19 + 32$

47. $50 + 94$

48. $181 + 217$

49. $1149 + 2021$

50. **MULTIPLE CHOICE** Which value is *not* a whole number? *(Skills Review Handbook)*

Ⓐ -5 Ⓑ 0 Ⓒ 4 Ⓓ 113

11.2 Adding Integers

Essential Question Is the sum of two integers *positive*, *negative*, or *zero*? How can you tell?

1 ACTIVITY: Adding Integers with the Same Sign

Work with a partner. Use integer counters to find $-4 + (-3)$.

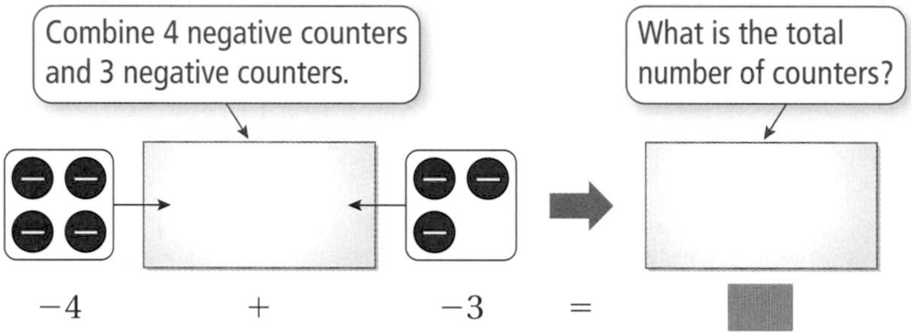

Combine 4 negative counters and 3 negative counters.
What is the total number of counters?
−4
+
−3
=

So, $-4 + (-3) =$ ____.

2 ACTIVITY: Adding Integers with Different Signs

Work with a partner. Use integer counters to find $-3 + 2$.

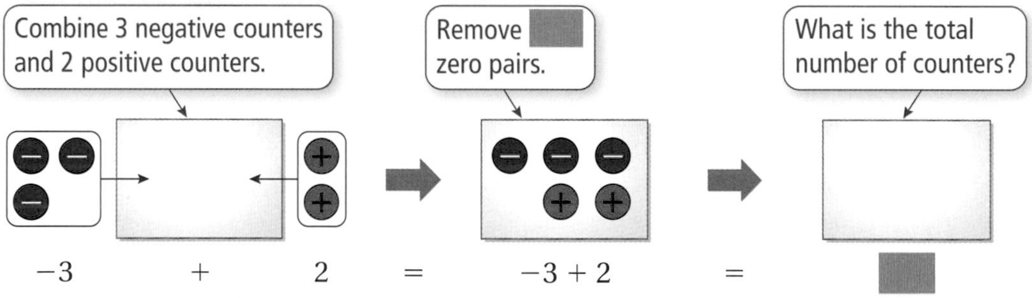

Combine 3 negative counters and 2 positive counters.
Remove ____ zero pairs.
What is the total number of counters?
−3
+
2
=
−3 + 2
=

So, $-3 + 2 =$ ____.

COMMON CORE

Integers

In this lesson, you will
- add integers.
- show that the sum of a number and its opposite is 0.
- solve real-life problems.

Learning Standards
7.NS.1a
7.NS.1b
7.NS.1d
7.NS.3

3 ACTIVITY: Adding Integers with Different Signs

Work with a partner. Use a number line to find $5 + (-3)$.

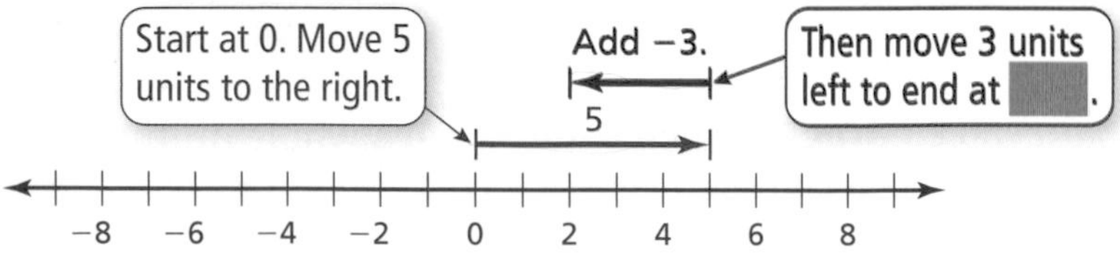

Start at 0. Move 5 units to the right.
Add −3.
5
Then move 3 units left to end at ____.
−8
−6
−4
−2
0
2
4
6
8

So, $5 + (-3) =$ ____.

4 ACTIVITY: Adding Integers with Different Signs

Math Practice 3

Make Conjectures

How can the relationship between the integers help you write a rule?

Work with a partner. Write the addition expression shown. Then find the sum. How are the integers in the expression related to 0 on a number line?

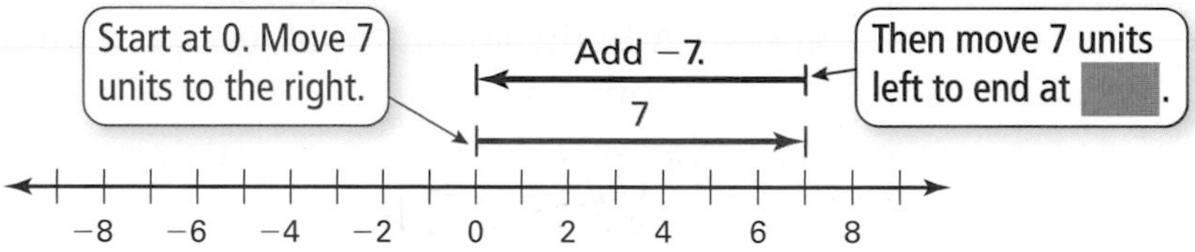

Start at 0. Move 7 units to the right.
Add −7.
7
Then move 7 units left to end at ▭.
−8
−6
−4
−2
0
2
4
6
8

Inductive Reasoning

Work with a partner. Use integer counters or a number line to complete the table.

	Exercise	Type of Sum	Sum	Sum: Positive, Negative, or Zero
1	**5.** $-4 + (-3)$	Integers with the same sign		
2	**6.** $-3 + 2$			
3	**7.** $5 + (-3)$			
4	**8.** $7 + (-7)$			
	9. $2 + 4$			
	10. $-6 + (-2)$			
	11. $-5 + 9$			
	12. $15 + (-9)$			
	13. $-10 + 10$			
	14. $-6 + (-6)$			
	15. $13 + (-13)$			

What Is Your Answer?

16. IN YOUR OWN WORDS Is the sum of two integers *positive, negative,* or *zero*? How can you tell?

17. STRUCTURE Write general rules for adding (a) two integers with the same sign, (b) two integers with different signs, and (c) two integers that vary only in sign.

Practice Use what you learned about adding integers to complete Exercises 8–15 on page 486.

11.2 Lesson

Check It Out
Lesson Tutorials
BigIdeasMath.com

Key Vocabulary
opposites, *p. 484*
additive inverse, *p. 484*

Key Idea

Adding Integers with the Same Sign

Words Add the absolute values of the integers. Then use the common sign.

Numbers $2 + 5 = 7$ $\quad -2 + (-5) = -7$

EXAMPLE 1 Adding Integers with the Same Sign

Find $-2 + (-4)$. Use a number line to check your answer.

$-2 + (-4) = -6$ Add $|-2|$ and $|-4|$.

Use the common sign.

∴ The sum is -6.

Check

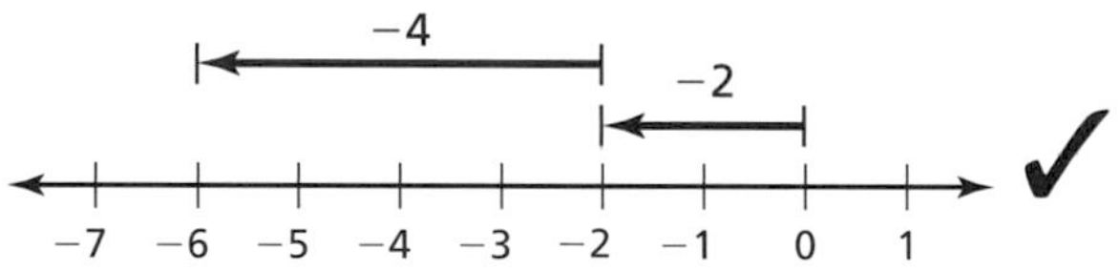

On Your Own

Add.

1. $7 + 13$
2. $-8 + (-5)$
3. $-20 + (-15)$

The Meaning of a Word

Opposite

When you walk across a street, you are moving to the **opposite** side of the street.

Two numbers that are the same distance from 0, but on opposite sides of 0, are called **opposites.** For example, -3 and 3 are opposites.

Key Ideas

Adding Integers with Different Signs

Words Subtract the lesser absolute value from the greater absolute value. Then use the sign of the integer with the greater absolute value.

Numbers $8 + (-10) = -2$ $\quad -13 + 17 = 4$

Additive Inverse Property

Words The sum of an integer and its **additive inverse**, or opposite, is 0.

Numbers $6 + (-6) = 0$ $\quad -25 + 25 = 0$ **Algebra** $a + (-a) = 0$

Multi-Language Glossary at BigIdeasMath.com

EXAMPLE 2 Adding Integers with Different Signs

a. Find $5 + (-10)$.

$5 + (-10) = -5$ — $|-10| > |5|$. So, subtract $|5|$ from $|-10|$.

Use the sign of -10.

∴ The sum is -5.

b. Find $-3 + 7$.

$-3 + 7 = 4$ — $|7| > |-3|$. So, subtract $|-3|$ from $|7|$.

Use the sign of 7.

∴ The sum is 4.

c. Find $-12 + 12$.

$-12 + 12 = 0$ — The sum is 0 by the Additive Inverse Property.

-12 and 12 are opposites.

∴ The sum is 0.

EXAMPLE 3 Adding More Than Two Integers

The list shows four bank account transactions in July. Find the change C in the account balance.

JULY TRANSACTIONS	
Withdrawal	-$40
Deposit	$50
Deposit	$75
Withdrawal	-$50

Study Tip

A deposit of $50 and a withdrawal of $50 represent opposite quantities, $+50$ and -50, which have a sum of 0.

Find the sum of the four transactions.

$C = -40 + 50 + 75 + (-50)$	Write the sum.
$= -40 + 75 + 50 + (-50)$	Commutative Property of Addition
$= -40 + 75 + [50 + (-50)]$	Associative Property of Addition
$= -40 + 75 + 0$	Additive Inverse Property
$= 35 + 0$	Add -40 and 75.
$= 35$	Addition Property of Zero

∴ Because $C = 35$, the account balance increased $35 in July.

On Your Own

Now You're Ready
Exercises 8–23 and 28–39

Add.

4. $-2 + 11$ **5.** $9 + (-10)$ **6.** $-31 + 31$

7. WHAT IF? In Example 3, the deposit amounts are $30 and $40. Find the change C in the account balance.

11.2 Exercises

Check It Out
Help with Homework
BigIdeasMath.com

Vocabulary and Concept Check

1. **WRITING** How do you find the additive inverse of an integer?
2. **NUMBER SENSE** Is $3 + (-4)$ the same as $-4 + 3$? Explain.

Tell whether the sum is *positive*, *negative*, or *zero* without adding. Explain your reasoning.

3. $-8 + 20$
4. $30 + (-30)$
5. $-10 + (-18)$

Tell whether the statement is *true* or *false*. Explain your reasoning.

6. The sum of two negative integers is always negative.
7. An integer and its absolute value are always opposites.

Practice and Problem Solving

Add.

1 2 8. $6 + 4$
9. $-4 + (-6)$
10. $-2 + (-3)$
11. $-5 + 12$
12. $5 + (-7)$
13. $8 + (-8)$
14. $9 + (-11)$
15. $-3 + 13$
16. $-4 + (-16)$
17. $-3 + (-1)$
18. $14 + (-5)$
19. $0 + (-11)$
20. $-10 + (-15)$
21. $-13 + 9$
22. $18 + (-18)$
23. $-25 + (-9)$

ERROR ANALYSIS **Describe and correct the error in finding the sum.**

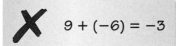

24. ✗ $9 + (-6) = -3$

25. ✗ $-10 + (-10) = 0$

26. **TEMPERATURE** The temperature is -3°F at 7:00 A.M. During the next 4 hours, the temperature increases 21°F. What is the temperature at 11:00 A.M.?

27. **BANKING** Your bank account has a balance of $-\$12$. You deposit \$60. What is your new balance?

Tell how the Commutative and Associative Properties of Addition can help you find the sum mentally. Then find the sum.

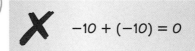

3 28. $9 + 6 + (-6)$
29. $-8 + 13 + (-13)$
30. $9 + (-17) + (-9)$
31. $7 + (-12) + (-7)$
32. $-12 + 25 + (-15)$
33. $6 + (-9) + 14$

Add.

34. $13 + (-21) + 16$
35. $22 + (-14) + (-35)$
36. $-13 + 27 + (-18)$
37. $-19 + 26 + 14$
38. $-32 + (-17) + 42$
39. $-41 + (-15) + (-29)$

40. **SCIENCE** A lithium atom has positively charged protons and negatively charged electrons. The sum of the charges represents the charge of the lithium atom. Find the charge of the atom.

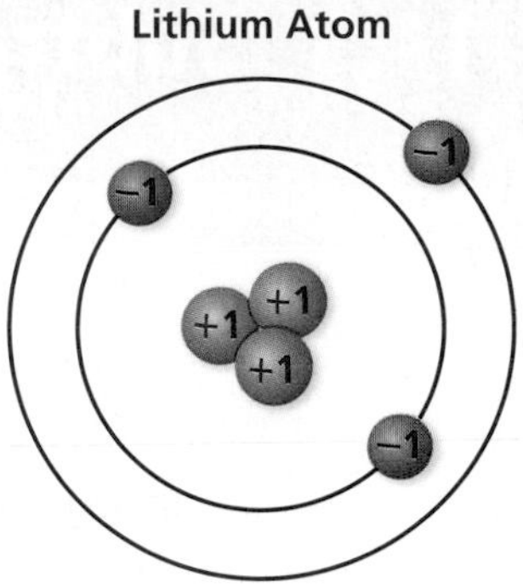

41. **OPEN-ENDED** Write two integers with different signs that have a sum of −25. Write two integers with the same sign that have a sum of −25.

ALGEBRA **Evaluate the expression when $a = 4$, $b = -5$, and $c = -8$.**

42. $a + b$

43. $-b + c$

44. $|a + b + c|$

MENTAL MATH **Use mental math to solve the equation.**

45. $d + 12 = 2$

46. $b + (-2) = 0$

47. $-8 + m = -15$

48. **PROBLEM SOLVING** Starting at point *A*, the path of a dolphin jumping out of the water is shown.

a. Is the dolphin deeper at point *C* or point *E*? Explain your reasoning.

b. Is the dolphin higher at point *B* or point *D*? Explain your reasoning.

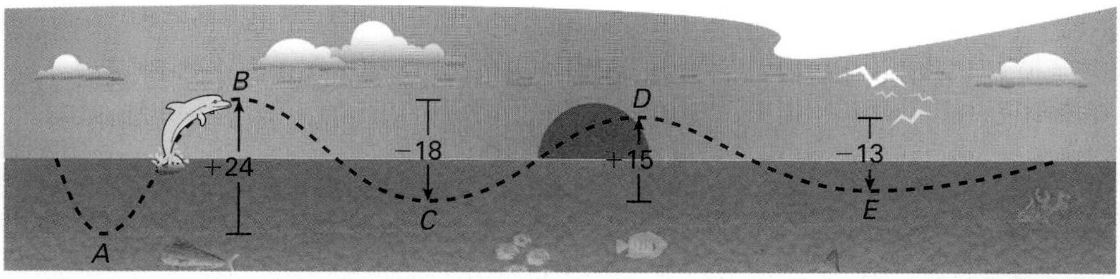

49. **Puzzle** According to a legend, the Chinese Emperor Yu-Huang saw a magic square on the back of a turtle. In a *magic square*, the numbers in each row and in each column have the same sum. This sum is called the *magic sum.*

Copy and complete the magic square so that each row and each column has a magic sum of 0. Use each integer from −4 to 4 exactly once.

Fair Game Review What you learned in previous grades & lessons

Subtract. *(Section 1.1)*

50. $69 - 38$

51. $82 - 74$

52. $177 - 63$

53. $451 - 268$

54. **MULTIPLE CHOICE** What is the range of the numbers below? *(Section 9.4)*

12, 8, 17, 12, 15, 18, 30

Ⓐ 12 Ⓑ 15 Ⓒ 18 Ⓓ 22

11.3 Subtracting Integers

Essential Question How are adding integers and subtracting integers related?

1 ACTIVITY: Subtracting Integers

Work with a partner. Use integer counters to find 4 − 2.

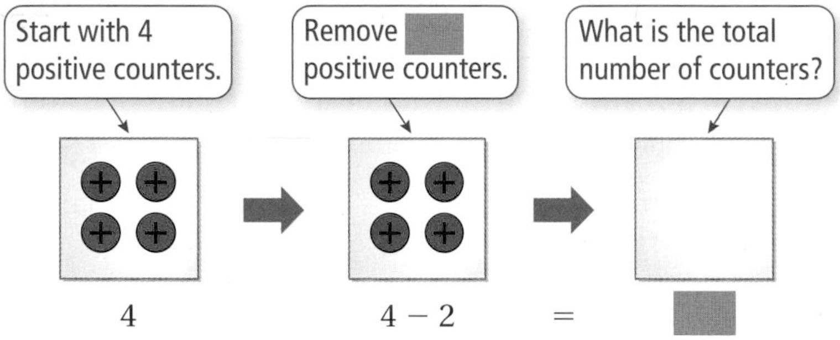

Start with 4 positive counters.
Remove ____ positive counters.
What is the total number of counters?
4
4 − 2
=

So, 4 − 2 = ____.

2 ACTIVITY: Adding Integers

Work with a partner. Use integer counters to find 4 + (−2).

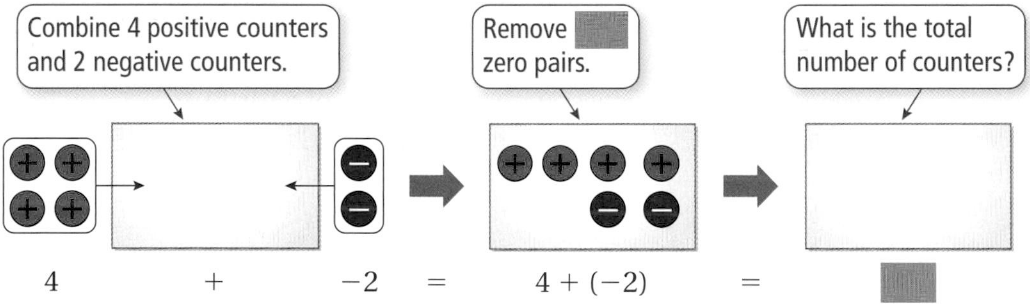

Combine 4 positive counters and 2 negative counters.
Remove ____ zero pairs.
What is the total number of counters?
4
+
−2
=
4 + (−2)
=

So, 4 + (−2) = ____.

3 ACTIVITY: Subtracting Integers

Work with a partner. Use a number line to find −3 − 1.

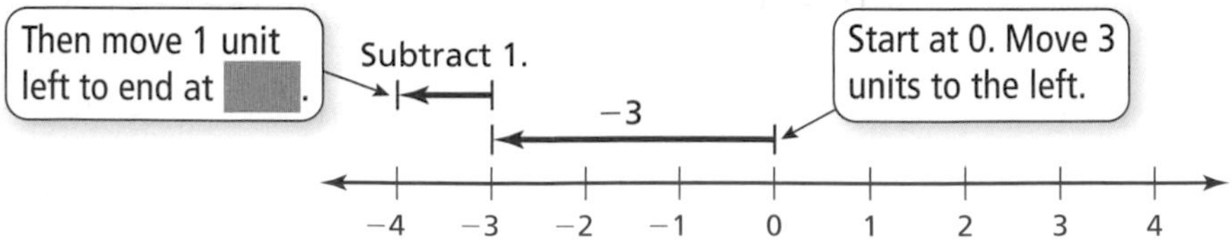

Then move 1 unit left to end at ____.
Subtract 1.
−3
Start at 0. Move 3 units to the left.
−4
−3
−2
−1
0
1
2
3
4

So, −3 − 1 = ____.

COMMON CORE

Integers

In this lesson, you will

- subtract integers.
- solve real-life problems.

Learning Standards
7.NS.1c
7.NS.1d
7.NS.3

4 ACTIVITY: Adding Integers

Math Practice 2
Make Sense of Quantities
What integers will you use in your addition expression?

Work with a partner. Write the addition expression shown. Then find the sum.

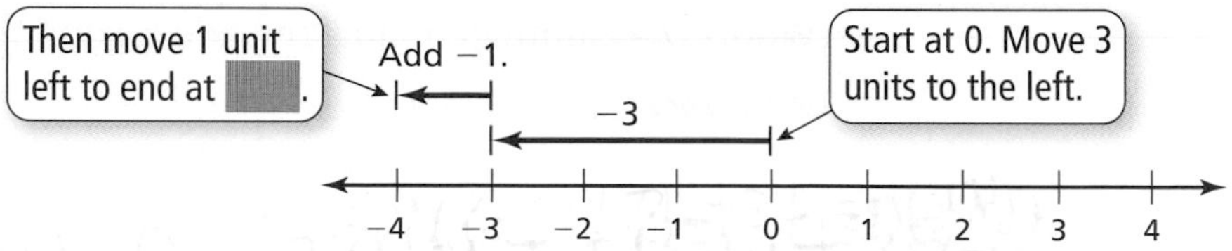

Then move 1 unit left to end at ▇.
Add −1.
−3
Start at 0. Move 3 units to the left.
−4
−3
−2
−1
0
1
2
3
4

Inductive Reasoning

Work with a partner. Use integer counters or a number line to complete the table.

	Exercise	Operation: Add or Subtract	Answer
1	**5.** $4 - 2$	Subtract 2	
2	**6.** $4 + (-2)$		
3	**7.** $-3 - 1$		
4	**8.** $-3 + (-1)$		
	9. $3 - 8$		
	10. $3 + (-8)$		
	11. $9 - 13$		
	12. $9 + (-13)$		
	13. $-6 - (-3)$		
	14. $-6 + 3$		
	15. $-5 - (-12)$		
	16. $-5 + 12$		

What Is Your Answer?

17. IN YOUR OWN WORDS How are adding integers and subtracting integers related?

18. STRUCTURE Write a general rule for subtracting integers.

19. Use a number line to find the value of the expression $-4 + 4 - 9$. What property can you use to make your calculation easier? Explain.

Practice Use what you learned about subtracting integers to complete Exercises 8–15 on page 492.

11.3 Lesson

Key Idea

Subtracting Integers

Words To subtract an integer, add its opposite.

Numbers $3 - 4 = 3 + (-4) = -1$

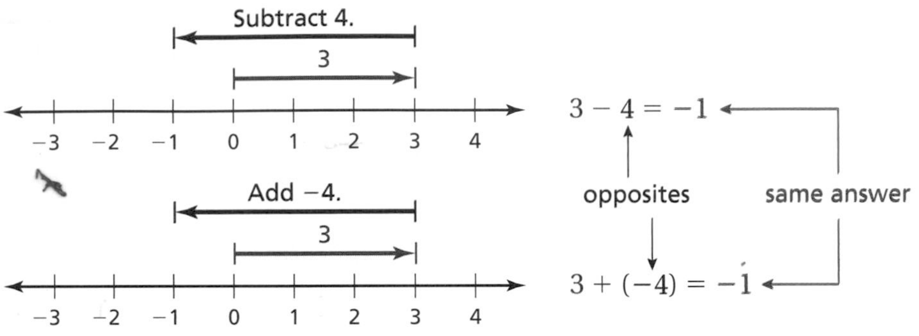

EXAMPLE 1 Subtracting Integers

a. Find 3 − 12.

$$3 - 12 = 3 + (-12) \quad \text{Add the opposite of 12.}$$
$$= -9 \quad \text{Add.}$$

∴ The difference is −9.

b. Find −8 − (−13).

$$-8 - (-13) = -8 + 13 \quad \text{Add the opposite of } -13.$$
$$= 5 \quad \text{Add.}$$

∴ The difference is 5.

c. Find 5 − (−4).

$$5 - (-4) = 5 + 4 \quad \text{Add the opposite of } -4.$$
$$= 9 \quad \text{Add.}$$

∴ The difference is 9.

On Your Own

Now You're Ready
Exercises 8–23

Subtract.

1. $8 - 3$ **2.** $9 - 17$ **3.** $-3 - 3$

4. $-14 - 9$ **5.** $9 - (-8)$ **6.** $-12 - (-12)$

EXAMPLE 2 Subtracting Integers

Evaluate $-7 - (-12) - 14$.

$-7 - (-12) - 14 = -7 + 12 - 14$	Add the opposite of -12.
$= 5 - 14$	Add -7 and 12.
$= 5 + (-14)$	Add the opposite of 14.
$= -9$	Add.

∴ So, $-7 - (-12) - 14 = -9$.

On Your Own

Now You're Ready
Exercises 27–32

Evaluate the expression.

7. $-9 - 16 - 8$

8. $-4 - 20 - 9$

9. $0 - 9 - (-5)$

10. $-8 - (-6) - 0$

11. $15 - (-20) - 20$

12. $-14 - 9 - 36$

EXAMPLE 3 Real-Life Application

Which continent has the greater range of elevations?

	North America	Africa
Highest Elevation	6198 m	5895 m
Lowest Elevation	−86 m	−155 m

To find the range of elevations for each continent, subtract the lowest elevation from the highest elevation.

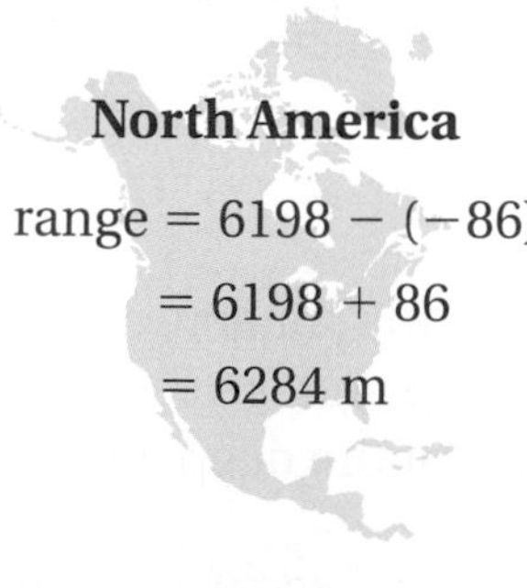

North America

range $= 6198 - (-86)$
$= 6198 + 86$
$= 6284$ m

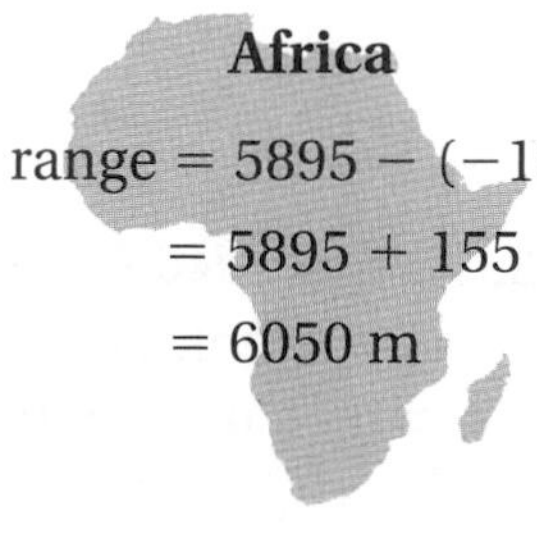

Africa

range $= 5895 - (-155)$
$= 5895 + 155$
$= 6050$ m

∴ Because 6284 is greater than 6050, North America has the greater range of elevations.

On Your Own

13. The highest elevation in Mexico is 5700 meters, on Pico de Orizaba. The lowest elevation in Mexico is -10 meters, in Laguna Salada. Find the range of elevations in Mexico.

11.3 Exercises

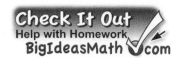

Check It Out
Help with Homework
BigIdeasMath.com

Vocabulary and Concept Check

1. **WRITING** How do you subtract one integer from another?
2. **OPEN-ENDED** Write two integers that are opposites.
3. **DIFFERENT WORDS, SAME QUESTION** Which is different? Find "both" answers.

Find the difference of 3 and -2.	What is 3 less than -2?
How much less is -2 than 3?	Subtract -2 from 3.

MATCHING Match the subtraction expression with the corresponding addition expression.

4. $9 - (-5)$
5. $-9 - 5$
6. $-9 - (-5)$
7. $9 - 5$

A. $-9 + 5$
B. $9 + (-5)$
C. $-9 + (-5)$
D. $9 + 5$

Practice and Problem Solving

Subtract.

1
8. $4 - 7$
9. $8 - (-5)$
10. $-6 - (-7)$
11. $-2 - 3$
12. $5 - 8$
13. $-4 - 6$
14. $-8 - (-3)$
15. $10 - 7$
16. $-8 - 13$
17. $15 - (-2)$
18. $-9 - (-13)$
19. $-7 - (-8)$
20. $-6 - (-6)$
21. $-10 - 12$
22. $32 - (-6)$
23. $0 - 20$

24. **ERROR ANALYSIS** Describe and correct the error in finding the difference $7 - (-12)$.

✗ $7 - (-12) = 7 + (-12) = -5$

25. **SWIMMING POOL** The floor of the shallow end of a swimming pool is at -3 feet. The floor of the deep end is 9 feet deeper. Which expression can be used to find the depth of the deep end?

$-3 + 9$ $\quad$ $-3 - 9$ $\quad$ $9 - 3$

26. **SHARKS** A shark is at -80 feet. It swims up and jumps out of the water to a height of 15 feet. Write a subtraction expression for the vertical distance the shark travels.

Evaluate the expression.

2
27. $-2 - 7 + 15$
28. $-9 + 6 - (-2)$
29. $12 - (-5) - 8$
30. $-87 - 5 - 13$
31. $-6 - (-8) + 6$
32. $-15 - 7 - (-11)$

MENTAL MATH **Use mental math to solve the equation.**

33. $m - 5 = 9$

34. $w - (-3) = 7$

35. $6 - c = -9$

ALGEBRA **Evaluate the expression when $k = -3$, $m = -6$, and $n = 9$.**

36. $4 - n$

37. $m - (-8)$

38. $-5 + k - n$

39. $|m - k|$

40. **PLATFORM DIVING** The figure shows a diver diving from a platform. The diver reaches a depth of 4 meters. What is the change in elevation of the diver?

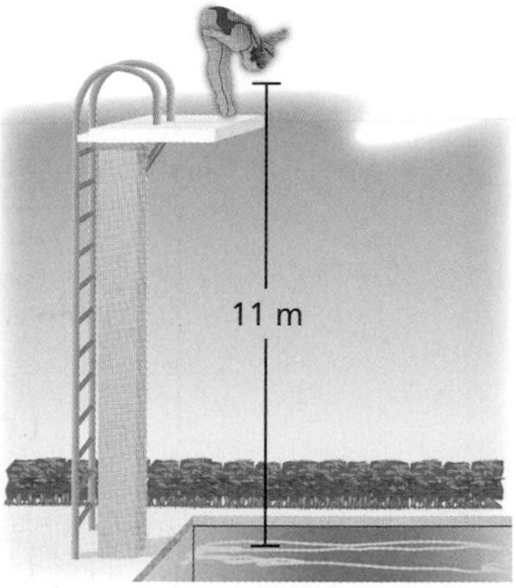

41. **OPEN-ENDED** Write two different pairs of negative integers, x and y, that make the statement $x - y = -1$ true.

42. **TEMPERATURE** The table shows the record monthly high and low temperatures for a city in Alaska.

	Jan	Feb	Mar	Apr	May	Jun	Jul	Aug	Sep	Oct	Nov	Dec
High (°F)	56	57	56	72	82	92	84	85	73	64	62	53
Low (°F)	−35	−38	−24	−15	1	29	34	31	19	−6	−21	−36

a. Find the range of temperatures for each month.

b. What are the all-time high and all-time low temperatures?

c. What is the range of the temperatures in part (b)?

REASONING **Tell whether the difference between the two integers is *always*, *sometimes*, or *never* positive. Explain your reasoning.**

43. two positive integers

44. two negative integers

45. a positive integer and a negative integer

46. a negative integer and a positive integer

Number Sense **For what values of a and b is the statement true?**

47. $|a - b| = |b - a|$

48. $|a + b| = |a| + |b|$

49. $|a - b| = |a| - |b|$

Fair Game Review What you learned in previous grades & lessons

Add. *(Section 11.2)*

50. $-5 + (-5) + (-5) + (-5)$

51. $-9 + (-9) + (-9) + (-9) + (-9)$

Multiply. *(Section 1.1)*

52. 8×5

53. 6×78

54. 36×41

55. 82×29

56. **MULTIPLE CHOICE** Which value of n makes the value of the expression $4n + 3$ a composite number? *(Skills Review Handbook)*

Ⓐ 1 Ⓑ 2 Ⓒ 3 Ⓓ 4

11 Study Help

Check It Out
Graphic Organizer
BigIdeasMath.com

You can use an **idea and examples chart** to organize information about a concept. Here is an example of an idea and examples chart for absolute value.

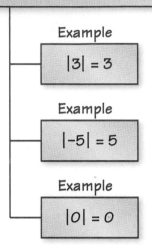

Absolute Value: the distance between a number and 0 on the number line

- Example: $|3| = 3$
- Example: $|-5| = 5$
- Example: $|0| = 0$

On Your Own

Make idea and examples charts to help you study these topics.

1. integers
2. adding integers
   a. with the same sign
   b. with different signs
3. Additive Inverse Property
4. subtracting integers

After you complete this chapter, make idea and examples charts for the following topics.

5. multiplying integers
   a. with the same sign
   b. with different signs
6. dividing integers
   a. with the same sign
   b. with different signs

"I made an idea and examples chart to give my owner ideas for my birthday next week."

11.1–11.3 Quiz

Check It Out
Progress Check
BigIdeasMath.com

Copy and complete the statement using <, >, or =. *(Section 11.1)*

1. $|-8| \;\square\; 3$

2. $7 \;\square\; |-7|$

Order the values from least to greatest. *(Section 11.1)*

3. $-4, |-5|, |-4|, 3, -6$

4. $12, -8, |-15|, -10, |-9|$

Evaluate the expression. *(Section 11.2 and Section 11.3)*

5. $-3 + (-8)$

6. $-4 + 16$

7. $3 - 9$

8. $-5 - (-5)$

Evaluate the expression when $a = -2$, $b = -8$, and $c = 5$. *(Section 11.2 and Section 11.3)*

9. $4 - a - c$

10. $|b - c|$

11. **EXPLORING** Two climbers explore a cave. *(Section 11.1)*

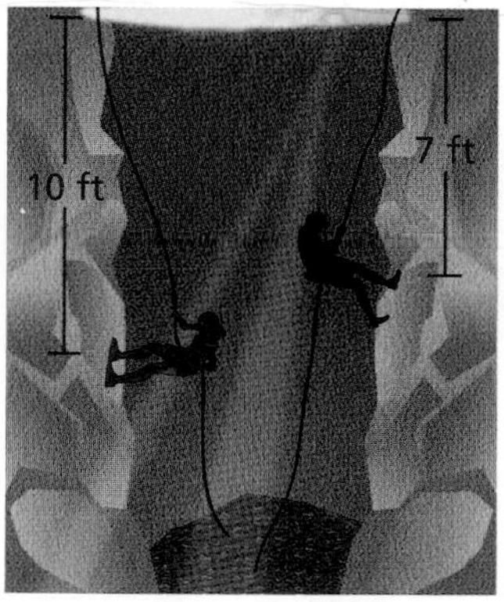

   a. Write an integer for the position of each climber relative to the surface.

   b. Which integer in part (a) is greater?

   c. Which integer in part (a) has the greater absolute value?

12. **SCHOOL CARNIVAL** The table shows the income and expenses for a school carnival. The school's goal was to raise \$1100. Did the school reach its goal? Explain. *(Section 11.2)*

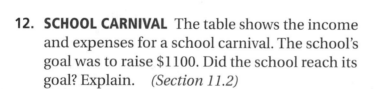

Games	Concessions	Donations	Flyers	Decorations
\$650	\$530	\$52	−\$28	−\$75

13. **TEMPERATURE** Temperatures in the Gobi Desert reach −40°F in the winter and 90°F in the summer. Find the range of the temperatures. *(Section 11.3)*

11.4 Multiplying Integers

Essential Question Is the product of two integers *positive*, *negative*, or *zero*? How can you tell?

1 ACTIVITY: Multiplying Integers with the Same Sign

Work with a partner. Use repeated addition to find 3 • 2.

Recall that multiplication is repeated addition. 3 • 2 means to add 3 groups of 2.

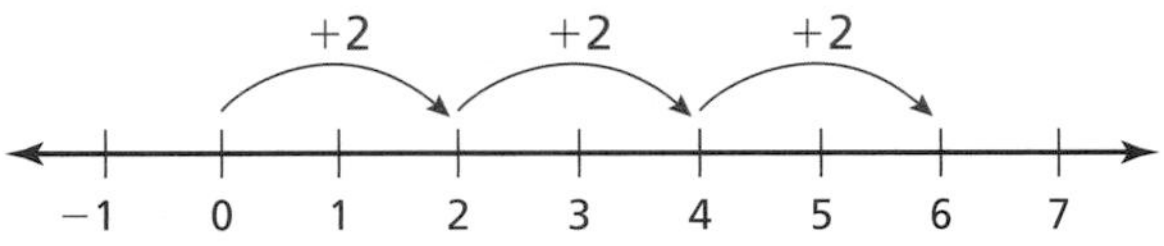

Now you can write

3 • 2 = ____ + ____ + ____

= ____.

So, 3 • 2 = ____.

2 ACTIVITY: Multiplying Integers with Different Signs

Work with a partner. Use repeated addition to find 3 • (−2).

3 • (−2) means to add 3 groups of −2.

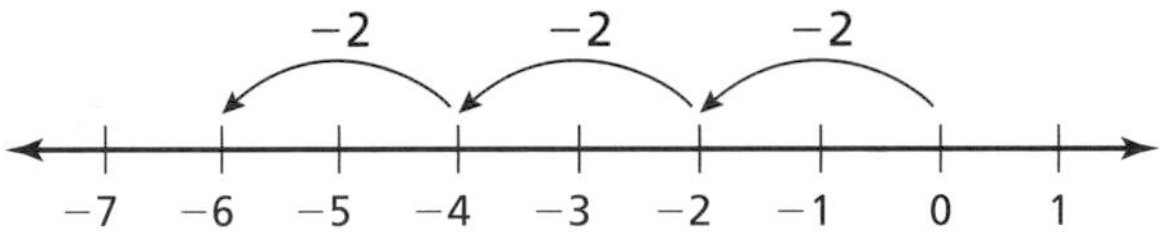

Now you can write

3 • (−2) = ____ + ____ + ____

= ____.

So, 3 • (−2) = ____.

3 ACTIVITY: Multiplying Integers with Different Signs

Work with a partner. Use a table to find −3 • 2.

Describe the pattern of the products in the table. Then complete the table.

2	•	2	=	4
1	•	2	=	2
0	•	2	=	0
−1	•	2	=	____
−2	•	2	=	____
−3	•	2	=	____

So, −3 • 2 = ____.

Common Core

Integers

In this lesson, you will
- multiply integers.
- solve real-life problems.

Learning Standards
7.NS.2a
7.NS.2c
7.NS.3

4 ACTIVITY: Multiplying Integers with the Same Sign

Work with a partner. Use a table to find $-3 \cdot (-2)$.

Math Practice 7

Look for Patterns

How can you use the pattern to complete the table?

Describe the pattern of the products in the table. Then complete the table.

-3	$\cdot$	3	$=$	-9
-3	$\cdot$	2	$=$	-6
-3	$\cdot$	1	$=$	-3
-3	$\cdot$	0	$=$	0
-3	$\cdot$	-1	$=$	3
-3	$\cdot$	-2	$=$	6

So, $-3 \cdot (-2) =$ ____.

Inductive Reasoning

Work with a partner. Complete the table.

	Exercise	Type of Product	Product	Product: Positive or Negative
1	**5.** $3 \cdot 2$	Integers with the same sign		
2	**6.** $3 \cdot (-2)$	-6		
3	**7.** $-3 \cdot 2$			
4	**8.** $-3 \cdot (-2)$			
	9. $6 \cdot 3$			
	10. $2 \cdot (-5)$			
	11. $-6 \cdot 5$			
	12. $-5 \cdot (-3)$			

What Is Your Answer?

13. Write two integers whose product is 0.

14. **IN YOUR OWN WORDS** Is the product of two integers *positive, negative,* or *zero*? How can you tell?

15. **STRUCTURE** Write general rules for multiplying (a) two integers with the same sign and (b) two integers with different signs.

Practice Use what you learned about multiplying integers to complete Exercises 8–15 on page 500.

11.4 Lesson

Check It Out
Lesson Tutorials
BigIdeasMath.com

Key Ideas

Multiplying Integers with the Same Sign

Words The product of two integers with the same sign is positive.

Numbers $2 \cdot 3 = 6$ $\quad -2 \cdot (-3) = 6$

Multiplying Integers with Different Signs

Words The product of two integers with different signs is negative.

Numbers $2 \cdot (-3) = -6$ $\quad -2 \cdot 3 = -6$

EXAMPLE 1 Multiplying Integers with the Same Sign

Find $-5 \cdot (-6)$.

The integers have the same sign.

$$-5 \cdot (-6) = 30$$

The product is positive.

∴ The product is 30.

EXAMPLE 2 Multiplying Integers with Different Signs

Multiply.

a. $3(-4)$ **b.** $-7 \cdot 4$

The integers have different signs.

$3(-4) = -12$ $\quad -7 \cdot 4 = -28$

The product is negative.

∴ The product is -12. ∴ The product is -28.

On Your Own

Now You're Ready
Exercises 8–23

Multiply.

1. $5 \cdot 5$

2. $4(11)$

3. $-1(-9)$

4. $-7 \cdot (-8)$

5. $12 \cdot (-2)$

6. $4(-6)$

7. $-10(-6)(0)$

8. $-7 \cdot (-5) \cdot (-4)$

EXAMPLE 3 Using Exponents

Study Tip

Place parentheses around a negative number to raise it to a power.

a. Evaluate $(-2)^2$.

$(-2)^2 = (-2) \cdot (-2)$ — Write $(-2)^2$ as repeated multiplication.

$= 4$ — Multiply.

b. Evaluate -5^2.

$-5^2 = -(5 \cdot 5)$ — Write 5^2 as repeated multiplication.

$= -25$ — Multiply.

c. Evaluate $(-4)^3$.

$(-4)^3 = (-4) \cdot (-4) \cdot (-4)$ — Write $(-4)^3$ as repeated multiplication.

$= 16 \cdot (-4)$ — Multiply.

$= -64$ — Multiply.

On Your Own

Now You're Ready Exercises 32–37

Evaluate the expression.

9. $(-3)^2$ **10.** $(-2)^3$ **11.** -7^2 **12.** -6^3

(-3)(-3) (-2)(-2) -7·7 = 14 -6·6·6

EXAMPLE 4 Real-Life Application

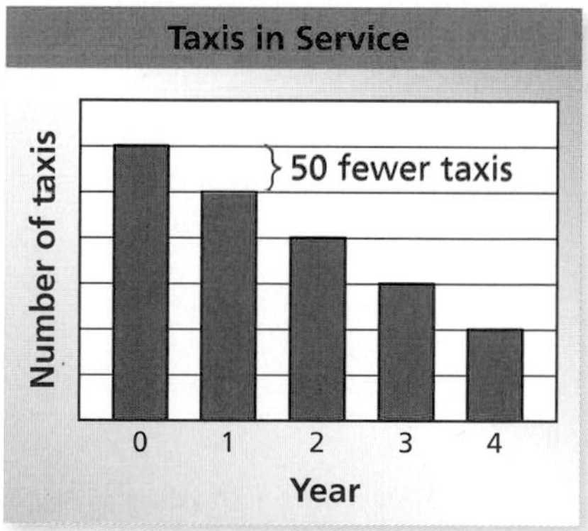

The bar graph shows the number of taxis a company has in service. The number of taxis decreases by the same amount each year for 4 years. Find the total change in the number of taxis.

The bar graph shows that the number of taxis in service decreases by 50 each year. Use a model to solve the problem.

total change = change per year • number of years

$= -50 \cdot 4$

Use -50 for the change per year because the number *decreases* each year.

$= -200$

The total change in the number of taxis is -200.

On Your Own

13. A manatee population decreases by 15 manatees each year for 3 years. Find the total change in the manatee population.

11.4 Exercises

Check It Out
Help with Homework
BigIdeasMath.com

Vocabulary and Concept Check

1. **WRITING** What can you conclude about the signs of two integers whose product is (a) positive and (b) negative?

2. **OPEN-ENDED** Write two integers whose product is negative.

Tell whether the product is *positive* or *negative* without multiplying. Explain your reasoning.

3. $4(-8)$

4. $-5(-7)$

5. $-3 \cdot 12$

Tell whether the statement is *true* or *false*. Explain your reasoning.

6. The product of three positive integers is positive.

7. The product of three negative integers is positive.

Practice and Problem Solving

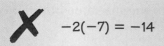

Multiply.

① ② 8. $6 \cdot 4$

9. $7(-3)$

10. $-2(8)$

11. $-3(-4)$

12. $-6 \cdot 7$

13. $3 \cdot 9$

14. $8 \cdot (-5)$

15. $-1 \cdot (-12)$

16. $-5(10)$

17. $-13(0)$

18. $-9 \cdot 9$

19. $15(-2)$

20. $-10 \cdot 11$

21. $-6 \cdot (-13)$

22. $7(-14)$

23. $-11 \cdot (-11)$

24. **JOGGING** You burn 10 calories each minute you jog. What integer represents the change in your calories after you jog for 20 minutes?

25. **WETLANDS** About 60,000 acres of wetlands are lost each year in the United States. What integer represents the change in wetlands after 4 years?

Multiply.

26. $3 \cdot (-8) \cdot (-2)$

27. $6(-9)(-1)$

28. $-3(-5)(-4)$

29. $(-5)(-7)(-20)$

30. $-6 \cdot 3 \cdot (-2)$

31. $3 \cdot (-12) \cdot 0$

Evaluate the expression.

③ 32. $(-4)^2$

33. $(-1)^3$

34. -8^2

35. -6^2

36. $-5^2 \cdot 4$

37. $-2 \cdot (-3)^3$

ERROR ANALYSIS **Describe and correct the error in evaluating the expression.**

38. ✗ $-2(-7) = -14$

39. ✗ $-10^2 = 100$

ALGEBRA **Evaluate the expression when $a = -2$, $b = 3$, and $c = -8$.**

40. ab **41.** $|a^2c|$ **42.** $-ab^3 - ac$

NUMBER SENSE **Find the next two numbers in the pattern.**

43. $-12, 60, -300, 1500, \ldots$ **44.** $7, -28, 112, -448, \ldots$

45. **GYM CLASS** You lose four points each time you attend gym class without sneakers. You forget your sneakers three times. What integer represents the change in your points?

46. **MODELING** The height of an airplane during a landing is given by $22{,}000 + (-480t)$, where t is the time in minutes.

a. Copy and complete the table.

b. Estimate how many minutes it takes the plane to land. Explain your reasoning.

Time (minutes)	5	10	15	20
Height (feet)				

47. **INLINE SKATES** In June, the price of a pair of inline skates is \$165. The price changes each of the next 3 months.

a. Copy and complete the table.

Month	Price of Skates
June	$165 = \$165$
July	$165 + (-12) = \$\_\_\_\_$
August	$165 + 2(-12) = \$\_\_\_\_$
September	$165 + 3(-12) = \$\_\_\_\_$

b. Describe the change in the price of the inline skates for each month.

c. The table at the right shows the amount of money you save each month to buy the inline skates. Do you have enough money saved to buy the inline skates in August? September? Explain your reasoning.

Amount Saved	
June	\$35
July	\$55
August	\$45
September	\$18

48. **Reasoning** Two integers, a and b, have a product of 24. What is the least possible sum of a and b?

Fair Game Review What you learned in previous grades & lessons

Divide. *(Section 1.1)*

49. $27 \div 9$ **50.** $48 \div 6$ **51.** $56 \div 4$ **52.** $153 \div 9$

53. **MULTIPLE CHOICE** What is the prime factorization of 84? *(Section 1.4)*

Ⓐ $2^2 \times 3^2$ Ⓑ $2^3 \times 7$ Ⓒ $3^3 \times 7$ Ⓓ $2^2 \times 3 \times 7$

11.5 Dividing Integers

Essential Question Is the quotient of two integers *positive*, *negative*, or *zero*? How can you tell?

1 ACTIVITY: Dividing Integers with Different Signs

Work with a partner. Use integer counters to find $-15 \div 3$.

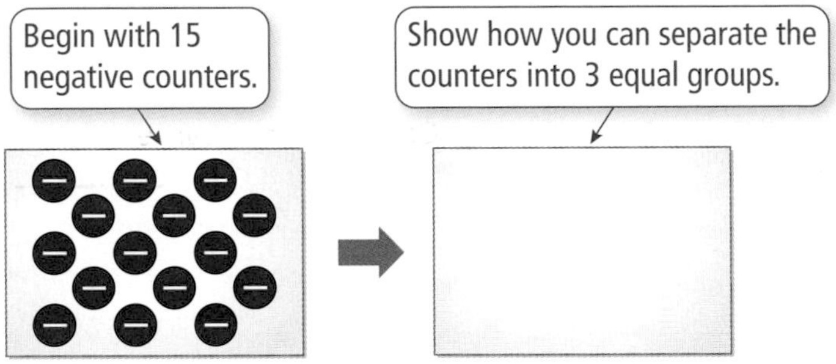

Begin with 15 negative counters.
Show how you can separate the counters into 3 equal groups.

∴ Because there are ____ negative counters in each group, $-15 \div 3 =$ ____.

2 ACTIVITY: Rewriting a Product as a Quotient

Work with a partner. Rewrite the product $3 \cdot 4 = 12$ as a quotient in two different ways.

First Way

12 is equal to 3 groups of ____.

∴ So, $12 \div 3 =$ ____.

Second Way

12 is equal to 4 groups of ____.

∴ So, $12 \div 4 =$ ____.

3 ACTIVITY: Dividing Integers with Different Signs

Work with a partner. Rewrite the product $-3 \cdot (-4) = 12$ as a quotient in two different ways. What can you conclude?

First Way

$12 \div ($ ____ $) =$ ____

Second Way

$12 \div ($ ____ $) =$ ____

∴ In each case, when you divide a ____ integer by a ____ integer, you get a ____ integer.

COMMON CORE

Integers

In this lesson, you will
- divide integers.
- solve real-life problems.

Learning Standards
7.NS.2b
7.NS.3

4 ACTIVITY: Dividing Negative Integers

Math Practice 8

Maintain Oversight

How do you know what the sign will be when you divide two integers?

Work with a partner. Rewrite the product $3 \cdot (-4) = -12$ as a quotient in two different ways. What can you conclude?

First Way

$-12 \div ($____$) = $____

Second Way

$-12 \div ($____$) = $____

When you divide a ____ integer by a ____ integer, you get a ____ integer. When you divide a ____ integer by a ____ integer, you get a ____ integer.

Inductive Reasoning

Work with a partner. Complete the table.

	Exercise	Type of Quotient	Quotient	Quotient: Positive, Negative, or Zero
1	**5.** $-15 \div 3$	Integers with different signs		
2	**6.** $12 \div 4$			
3	**7.** $12 \div (-3)$			
4	**8.** $-12 \div (-4)$			
	9. $-6 \div 2$			
	10. $-21 \div (-7)$			
	11. $10 \div (-2)$			
	12. $12 \div (-6)$			
	13. $0 \div (-15)$			
	14. $0 \div 4$			

What Is Your Answer?

15. IN YOUR OWN WORDS Is the quotient of two integers *positive, negative,* or *zero*? How can you tell?

16. STRUCTURE Write general rules for dividing (a) two integers with the same sign and (b) two integers with different signs.

Practice Use what you learned about dividing integers to complete Exercises 8–15 on page 506.

11.5 Lesson

Check It Out
Lesson Tutorials
BigIdeasMath.com

Key Ideas

Remember

Division by 0 is undefined.

Dividing Integers with the Same Sign

Words The quotient of two integers with the same sign is positive.

Numbers $8 \div 2 = 4$ $\quad -8 \div (-2) = 4$

Dividing Integers with Different Signs

Words The quotient of two integers with different signs is negative.

Numbers $8 \div (-2) = -4$ $\quad -8 \div 2 = -4$

EXAMPLE 1 Dividing Integers with the Same Sign

Find $-18 \div (-6)$.

The integers have the same sign.

$$-18 \div (-6) = 3$$

The quotient is positive.

∴ The quotient is 3.

EXAMPLE 2 Dividing Integers with Different Signs

Divide.

a. $75 \div (-25)$ **b.** $\frac{-54}{6}$

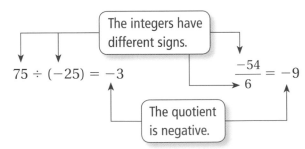

The integers have different signs.

$$75 \div (-25) = -3 \qquad \frac{-54}{6} = -9$$

The quotient is negative.

∴ The quotient is -3. ∴ The quotient is -9.

On Your Own

Now You're Ready
Exercises 8–23

Divide.

1. $14 \div 2$ **2.** $-32 \div (-4)$ **3.** $-40 \div (-8)$

4. $0 \div (-6)$ **5.** $\frac{-49}{7}$ **6.** $\frac{21}{-3}$

EXAMPLE 3 Evaluating an Expression

Remember

Use order of operations when evaluating an expression.

Evaluate $10 - x^2 \div y$ when $x = 8$ and $y = -4$.

$10 - x^2 \div y = 10 - 8^2 \div (-4)$	Substitute 8 for x and -4 for y.
$= 10 - 8 \cdot 8 \div (-4)$	Write 8^2 as repeated multiplication.
$= 10 - 64 \div (-4)$	Multiply 8 and 8.
$= 10 - (-16)$	Divide 64 by -4.
$= 26$	Subtract.

On Your Own

Now You're Ready
Exercises 28–31

Evaluate the expression when $a = -18$ and $b = -6$.

7. $a \div b$
8. $\dfrac{a + 6}{3}$
9. $\dfrac{b^2}{a} + 4$

EXAMPLE 4 Real-Life Application

You measure the height of the tide using the support beams of a pier. Your measurements are shown in the picture. What is the mean hourly change in the height?

Use a model to solve the problem.

$$\text{mean hourly change} = \frac{\text{final height} - \text{initial height}}{\text{elapsed time}}$$

$= \dfrac{8 - 59}{6}$	Substitute. The elapsed time from 2 P.M. to 8 P.M. is 6 hours.
$= \dfrac{-51}{6}$	Subtract.
$= -8.5$	Divide.

∴ The mean change in the height of the tide is -8.5 inches per hour.

On Your Own

10. The height of the tide at the Bay of Fundy in New Brunswick decreases 36 feet in 6 hours. What is the mean hourly change in the height?

11.5 Exercises

Check It Out
Help with Homework
BigIdeasMath.com

Vocabulary and Concept Check

1. **WRITING** What can you tell about two integers when their quotient is positive? negative? zero?
2. **VOCABULARY** A quotient is undefined. What does this mean?
3. **OPEN-ENDED** Write two integers whose quotient is negative.
4. **WHICH ONE DOESN'T BELONG?** Which expression does *not* belong with the other three? Explain your reasoning.

$\frac{10}{-5}$ $\frac{-10}{5}$ $\frac{-10}{-5}$ $-\left(\frac{10}{5}\right)$

Tell whether the quotient is *positive* or *negative* without dividing.

5. $-12 \div 4$
6. $\frac{-6}{-2}$
7. $15 \div (-3)$

Practice and Problem Solving

Divide, if possible.

1 2

8. $4 \div (-2)$
9. $21 \div (-7)$
10. $-20 \div 4$
11. $-18 \div (-3)$
12. $\frac{-14}{7}$
13. $\frac{0}{6}$
14. $\frac{-15}{-5}$
15. $\frac{54}{-9}$
16. $-33 \div 11$
17. $-49 \div (-7)$
18. $0 \div (-2)$
19. $60 \div (-6)$
20. $\frac{-56}{14}$
21. $\frac{18}{0}$
22. $\frac{65}{-5}$
23. $\frac{-84}{-7}$

ERROR ANALYSIS Describe and correct the error in finding the quotient.

24. ✗ $\frac{-63}{-9} = -7$
25. ✗ $0 \div (-5) = -5$

26. **ALLIGATORS** An alligator population in a nature preserve in the Everglades decreases by 60 alligators over 5 years. What is the mean yearly change in the alligator population?
27. **READING** You read 105 pages of a novel over 7 days. What is the mean number of pages you read each day?

ALGEBRA Evaluate the expression when $x = 10$, $y = -2$, and $z = -5$.

3

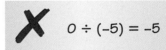

28. $x \div y$
29. $\frac{10y^2}{z}$
30. $\left|\frac{xz}{-y}\right|$
31. $\frac{-x^2 + 6z}{y}$

Find the mean of the integers.

32. 3, −10, −2, 13, 11

33. −26, 39, −10, −16, 12, 31

Evaluate the expression.

34. $-8 - 14 \div 2 + 5$

35. $24 \div (-4) + (-2) \cdot (-5)$

36. PATTERN Find the next two numbers in the pattern −128, 64, −32, 16, ... Explain your reasoning.

37. SNOWBOARDING A snowboarder descends a 1200-foot hill in 3 minutes. What is the mean change in elevation per minute?

38. GOLF The table shows a golfer's score for each round of a tournament.

a. What was the golfer's total score?

b. What was the golfer's mean score per round?

Scorecard	
Round 1	−2
Round 2	−6
Round 3	−7
Round 4	−3

39. TUNNEL The Detroit-Windsor Tunnel is an underwater highway that connects the cities of Detroit, Michigan, and Windsor, Ontario. How many times deeper is the roadway than the bottom of the ship?

40. AMUSEMENT PARK The regular admission price for an amusement park is \$72. For a group of 15 or more, the admission price is reduced by \$25. How many people need to be in a group to save \$500?

41. Number Sense Write five different integers that have a mean of −10. Explain how you found your answer.

Fair Game Review What you learned in previous grades & lessons

Graph the values on a number line. Then order the values from least to greatest. *(Section 11.1)*

42. $-6, 4, |2|, -1, |-10|$

43. $3, |0|, |-4|, -3, -8$

44. $|5|, -2, -5, |-2|, -7$

45. MULTIPLE CHOICE What is the value of $4 \cdot 3 + (12 \div 2)^2$? *(Section 1.3)*

Ⓐ 15   Ⓑ 48   Ⓒ 156   Ⓓ 324

11.4–11.5 Quiz

Check It Out
Progress Check
BigIdeasMath.com

Evaluate the expression. *(Section 11.4 and Section 11.5)*

1. $-7(6)$

2. $-1(-10)$

3. $\frac{-72}{-9}$

4. $-24 \div 3$

5. $-3 \cdot 4 \cdot (-6)$

6. $(-3)^3$

Evaluate the expression when $a = 4$, $b = -6$, and $c = -12$. *(Section 11.4 and Section 11.5)*

7. c^2

8. bc

9. $\frac{ab}{c}$

10. $\frac{|c - b|}{a}$

11. **SPEECH** In speech class, you lose 3 points for every 30 seconds you go over the time limit. Your speech is 90 seconds over the time limit. What integer represents the change in your points? *(Section 11.4)*

12. **MOUNTAIN CLIMBING** On a mountain, the temperature decreases by 18°F every 5000 feet. What integer represents the change in temperature at 20,000 feet? *(Section 11.4)*

13. **GAMING** You play a video game for 15 minutes. You lose 165 points. What is the mean change in points per minute? *(Section 11.5)*

14. **DIVING** You dive 21 feet from the surface of a lake in 7 seconds. *(Section 11.4 and Section 11.5)*

    a. What is the mean change in your position in feet per second?

    b. You continue diving. What is your position relative to the surface after 5 more seconds?

15. **HIBERNATION** A female grizzly bear weighs 500 pounds. After hibernating for 6 months, she weighs only 200 pounds. What is the mean change in weight per month? *(Section 11.5)*

11 Chapter Review

Check It Out
Vocabulary Help
BigIdeasMath.com

Review Key Vocabulary

integer, *p. 478*
absolute value, *p. 478*
opposites, *p. 484*
additive inverse, *p. 484*

Review Examples and Exercises

11.1 Integers and Absolute Value *(pp. 476–481)*

Find the absolute value of -2.

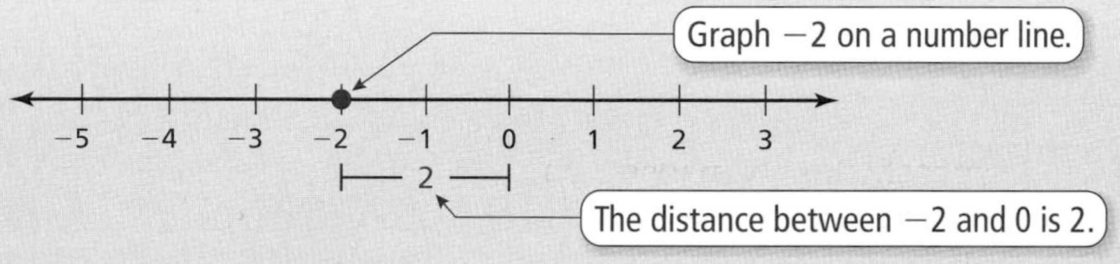

Graph −2 on a number line.
−5
−4
−3
−2
−1
0
1
2
3
2
The distance between −2 and 0 is 2.

So, $|-2| = 2$.

Exercises

Find the absolute value.

1. $|3|$
2. $|-9|$
3. $|-17|$
4. $|8|$

5. **ELEVATION** The elevation of Death Valley, California, is -282 feet. The Mississippi River in Illinois has an elevation of 279 feet. Which is closer to sea level?

11.2 Adding Integers *(pp. 482–487)*

Find $6 + (-14)$.

$6 + (-14) = -8$ — $|-14| > |6|$. So, subtract $|6|$ from $|-14|$.

Use the sign of -14.

The sum is -8.

Exercises

Add.

6. $-16 + (-11)$
-27
7. $-15 + 5$
-10
8. $100 + (-75)$
25
9. $-32 + (-2)$
-34

11.3 Subtracting Integers *(pp. 488–493)*

Subtract.

a. $7 - 19 = 7 + (-19)$ Add the opposite of 19.

$= -12$ Add.

∴ The difference is -12.

b. $-6 - (-10) = -6 + 10$ Add the opposite of -10.

$= 4$ Add.

∴ The difference is 4.

Exercises

Subtract.

10. $8 - 18$ **11.** $-16 - (-5)$ **12.** $-18 - 7$ **13.** $-12 - (-27)$

14. GAME SHOW Your score on a game show is -300. You answer the final question incorrectly, so you lose 400 points. What is your final score?

11.4 Multiplying Integers *(pp. 496–501)*

a. Find $-7 \cdot (-9)$.

The integers have the same sign.

$$-7 \cdot (-9) = 63$$

The product is positive.

∴ The product is 63.

b. Find $-6(14)$.

The integers have different signs.

$$-6(14) = -84$$

The product is negative.

∴ The product is -84.

Exercises

Multiply.

15. $-8 \cdot 6$ **16.** $10(-7)$ **17.** $-3 \cdot (-6)$ **18.** $-12(5)$

11.5 Dividing Integers *(pp. 502–507)*

a. Find $30 \div (-10)$.

The integers have different signs.

$$30 \div (-10) = -3$$

The quotient is negative.

The quotient is -3.

b. Find $\dfrac{-72}{-9}$.

The integers have the same sign.

$$\frac{-72}{-9} = 8$$

The quotient is positive.

The quotient is 8.

Exercises

Divide.

19. $-18 \div 9$ -2

20. $\dfrac{-42}{-6}$ 7

21. $\dfrac{-30}{6}$ -5

22. $84 \div (-7)$ -12

Evaluate the expression when $x = 3$, $y = -4$, and $z = -6$.

23. $z \div x$

24. $\dfrac{xy}{z}$

25. $\dfrac{z - 2x}{y}$

Find the mean of the integers.

26. $-3, -8, 12, -15, 9$

27. $-54, -32, -70, -25, -65, -42$

28. PROFITS The table shows the weekly profits of a fruit vendor. What is the mean profit for these weeks?

Week	1	2	3	4
Profit	−\$125	−\$86	\$54	−\$35

29. RETURNS You return several shirts to a store. The receipt shows that the amount placed back on your credit card is −\$30.60. Each shirt is −\$6.12. How many shirts did you return?

11 Chapter Test

Check It Out
Test Practice
BigIdeasMath.com

Find the absolute value.

1. $|-9|$ 2. $|64|$ 3. $|-22|$

Copy and complete the statement using <, >, or =.

4. $4 \;\square\; |-8|$ 5. $|-7| \;\square\; -12$ 6. $-7 \;\square\; |3|$

Evaluate the expression.

7. $-6 + (-11)$ 8. $2 - (-9)$

9. $-9 \cdot 2$ 10. $-72 \div (-3)$

Evaluate the expression when $x = 5$, $y = -3$, and $z = -2$.

11. $\dfrac{y + z}{x}$ 12. $\dfrac{x - 5z}{y}$

Find the mean of the integers.

13. 11, −7, −14, 10, −5 14. −32, −41, −39, −27, −33, −44

15. **NASCAR** A driver receives −25 points for each rule violation. What integer represents the change in points after 4 rule violations?

16. **GOLF** The table shows your scores, relative to *par*, for nine holes of golf. What is your total score for the nine holes?

Hole	1	2	3	4	5	6	7	8	9	Total
Score	+1	−2	−1	0	−1	+3	−1	−3	+1	?

17. **VISITORS** In a recent 10-year period, the change in the number of visitors to U.S. national parks was about −11,150,000 visitors.

   a. What was the mean yearly change in the number of visitors?

   b. During the seventh year, the change in the number of visitors was about 10,800,000. Explain how the change for the 10-year period can be negative.

11 Standards Assessment

Test-Taking Strategy
Solve Directly or Eliminate Choices

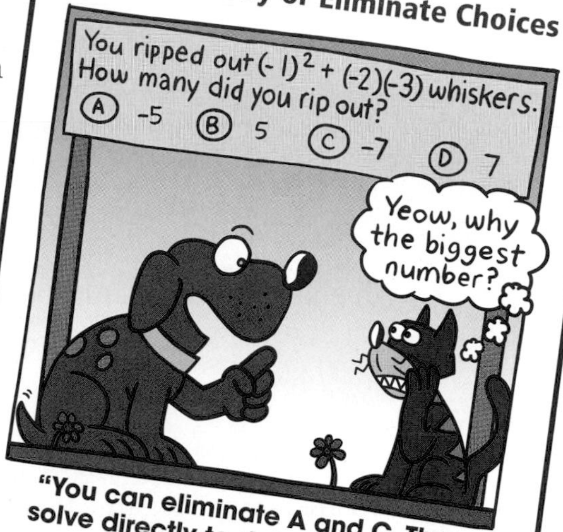

You ripped out (-1)^2 + (-2)(-3) whiskers.
How many did you rip out?
(A) -5 (B) 5 (C) -7 (D) 7
Yeow, why the biggest number?

"You can eliminate A and C. Then, solve directly to determine that the correct answer is D."

1. A football team gains 2 yards on the first play, loses 5 yards on the second play, loses 3 yards on the third play, and gains 4 yards on the fourth play. What is the team's overall gain or loss for all four plays? *(7.NS.1b)*

A. a gain of 14 yards **C.** a loss of 2 yards

B. a gain of 2 yards **D.** a loss of 14 yards

2. Which expression is *not* equal to the number 0? *(7.NS.1a)*

F. $5 - 5$ **H.** $6 - (-6)$

G. $-7 + 7$ **I.** $-8 - (-8)$

3. What is the value of the expression below when $a = -2$, $b = 3$, and $c = -5$? *(7.NS.3)*

$$\left| a^2 - 2ac + 5b \right|$$

A. -9 **C.** 1

B. -1 **D.** 9

4. What is the value of the expression below? *(7.NS.1c)*

$$17 - (-8)$$

5. Sam was evaluating an expression in the box below.

$$(-2)^3 \cdot 3 - (-5) = 8 \cdot 3 - (-5)$$
$$= 24 + 5$$
$$= 29$$

What should Sam do to correct the error that he made? *(7.NS.3)*

F. Subtract 5 from 24 instead of adding.

G. Rewrite $(-2)^3$ as -8.

H. Subtract -5 from 3 before multiplying by $(-2)^3$.

I. Multiply -2 by 3 before raising the quantity to the third power.

6. What is the value of the expression below when $x = 6$, $y = -4$, and $z = -2$? *(7.NS.3)*

$$\frac{x - 2y}{-z}$$

A. -7

B. -1

C. 1

D. 7

7. What is the missing number in the sequence below? *(7.NS.1c)*

39, 24, 9, ___, −21

8. You are playing a game using the spinner shown. You start with a score of 0 and spin the spinner four times. When you spin blue or green, you add the number to your score. When you spin red or orange, you subtract the number from your score. Which sequence of colors represents the greatest score? *(7.NS.3)*

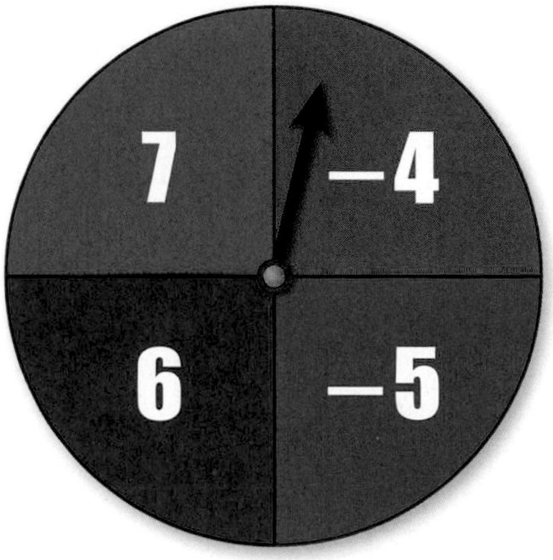

F. red, green, green, red

G. orange, orange, green, blue

H. red, blue, orange, green

I. blue, red, blue, red

9. Which expression represents a negative integer? *(7.NS.3)*

A. $5 - (-6)$

B. $(-3)^3$

C. $-12 \div (-6)$

D. $(-2)(-4)$

10. Which expression has the greatest value when $x = -2$ and $y = -3$? *(7.NS.3)*

F. $-xy$

G. xy

H. $x - y$

I. $-x - y$

11. What is the value of the expression below? *(7.NS.3)*

$$-5 \cdot (-4)^2 - (-3)$$

A. -83 **C.** 77

B. -77 **D.** 83

12. Which property does the equation below represent? *(7.NS.1d)*

$$-80 + 30 + (-30) = -80 + [30 + (-30)]$$

F. Commutative Property of Addition

G. Associative Property of Addition

H. Additive Inverse Property

I. Addition Property of Zero

13. What is the mean of the data set in the box below? *(7.NS.3)*

$-8, -6, -2, 0, -6, -8, 4, -7, -8, 1$

A. -8 **C.** -6

B. -7 **D.** -4

14. Consider the number line shown below. *(7.NS.1b, 7.NS.1c)*

Think
Solve
Explain

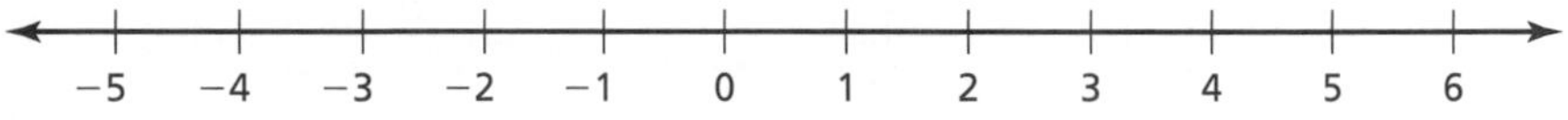

Part A Use the number line to explain how to add -2 and -3.

Part B Use the number line to explain how to subtract 5 from 2.

15. What is the value of the expression below? *(7.NS.3)*

$$\frac{-3 - 2^2}{-1}$$

F. -25 **H.** 7

G. -1 **I.** 25

12 Rational Numbers

What You Learned Before

"Let's play a game. The goal is to say a positive rational number that is less than the other pet's number... You go first."

Writing Decimals and Fractions (4.NF.6)

Example 1 Write 0.37 as a fraction.

$$0.37 = \frac{37}{100}$$

Example 2 Write $\frac{2}{5}$ as a decimal.

$$\frac{2}{5} = \frac{2 \cdot 2}{5 \cdot 2} = \frac{4}{10} = 0.4$$

Try It Yourself

Write the decimal as a fraction or the fraction as a decimal.

1. 0.51 **2.** 0.731 **3.** $\frac{3}{5}$ **4.** $\frac{7}{8}$

Adding and Subtracting Fractions (5.NF.1)

Example 3 Find $\frac{1}{3} + \frac{1}{5}$.

$$\frac{1}{3} + \frac{1}{5} = \frac{1 \cdot 5}{3 \cdot 5} + \frac{1 \cdot 3}{5 \cdot 3} = \frac{5}{15} + \frac{3}{15} = \frac{8}{15}$$

Example 4 Find $\frac{1}{4} - \frac{2}{9}$.

$$\frac{1}{4} - \frac{2}{9} = \frac{1 \cdot 9}{4 \cdot 9} - \frac{2 \cdot 4}{9 \cdot 4} = \frac{9}{36} - \frac{8}{36} = \frac{1}{36}$$

Multiplying and Dividing Fractions (5.NF.4, 6.NS.1)

Example 5 Find $\frac{5}{6} \cdot \frac{3}{4}$.

$$\frac{5}{6} \cdot \frac{3}{4} = \frac{5 \cdot \cancel{3}^{1}}{_{2}\cancel{6} \cdot 4} = \frac{5}{8}$$

Example 6 Find $\frac{2}{3} \div \frac{9}{10}$.

$$\frac{2}{3} \div \frac{9}{10} = \frac{2}{3} \cdot \frac{10}{9} = \frac{2 \cdot 10}{3 \cdot 9} = \frac{20}{27}$$

Multiply by the reciprocal of the divisor.

Try It Yourself

Evaluate the expression.

5. $\frac{1}{4} + \frac{13}{20}$ **6.** $\frac{14}{15} - \frac{1}{3}$ **7.** $\frac{3}{7} \cdot \frac{9}{10}$ **8.** $\frac{4}{5} \div \frac{16}{17}$

12.1 Rational Numbers

Essential Question How can you use a number line to order rational numbers?

The Meaning of a Word ● Rational

The word **rational** comes from the word *ratio*. Recall that you can write a ratio using fraction notation.

If you sleep for 8 hours in a day, then the ratio of your sleeping time to the total hours in a day can be written as $\dfrac{8\text{ h}}{24\text{ h}}$.

A **rational number** is a number that can be written as the ratio of two integers.

$2 = \dfrac{2}{1}$ $\qquad -3 = \dfrac{-3}{1}$ $\qquad -\dfrac{1}{2} = \dfrac{-1}{2}$ $\qquad 0.25 = \dfrac{1}{4}$

1 ACTIVITY: Ordering Rational Numbers

Work in groups of five. Order the numbers from least to greatest.

- Use masking tape and a marker to make a number line on the floor similar to the one shown.

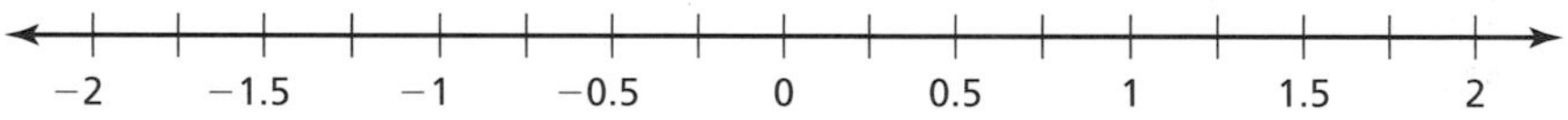

- Write the numbers on pieces of paper. Then each person should choose one.
- Stand on the location of your number on the number line.
- Use your positions to order the numbers from least to greatest.

a. $-0.5,\ 1.25,\ -\dfrac{1}{3},\ 0.5,\ -\dfrac{5}{3}$

b. $-\dfrac{7}{4},\ 1.1,\ \dfrac{1}{2},\ -\dfrac{1}{10},\ -1.3$

c. $-1.4,\ -\dfrac{3}{5},\ \dfrac{9}{2},\ \dfrac{1}{4},\ 0.9$

d. $\dfrac{5}{4},\ 0.75,\ -\dfrac{5}{4},\ -0.8,\ -1.1$

Common Core

Rational Numbers

In this lesson, you will

- understand that a rational number is an integer divided by an integer.
- convert rational numbers to decimals.

Learning Standards
7.NS.2b
7.NS.2d

2 ACTIVITY: The Game of Math Card War

Math Practice 1

Consider Similar Problems

What are some ways to determine which number is greater?

Preparation:

- Cut index cards to make 40 playing cards.
- Write each number in the table on a card.

To Play:

- Play with a partner.
- Deal 20 cards to each player facedown.
- Each player turns one card faceup. The player with the greater number wins. The winner collects both cards and places them at the bottom of his or her cards.
- Suppose there is a tie. Each player lays three cards facedown, then a new card faceup. The player with the greater of these new cards wins. The winner collects all ten cards and places them at the bottom of his or her cards.
- Continue playing until one player has all the cards. This player wins the game.

$-\frac{3}{2}$	$\frac{3}{10}$	$-\frac{3}{4}$	-0.6	1.25	-0.15	$\frac{5}{4}$	$\frac{3}{5}$	-1.6	-0.3
$\frac{3}{20}$	$\frac{8}{5}$	-1.2	$\frac{19}{10}$	0.75	-1.5	$-\frac{6}{5}$	$-\frac{3}{5}$	1.2	0.3
1.5	1.9	-0.75	-0.4	$\frac{3}{4}$	$-\frac{5}{4}$	-1.9	$\frac{2}{5}$	$-\frac{3}{20}$	$-\frac{19}{10}$
$\frac{6}{5}$	$-\frac{3}{10}$	1.6	$-\frac{2}{5}$	0.6	0.15	$\frac{3}{2}$	-1.25	0.4	$-\frac{8}{5}$

What Is Your Answer?

3. **IN YOUR OWN WORDS** How can you use a number line to order rational numbers? Give an example.

The numbers are in order from least to greatest. Fill in the blank spaces with rational numbers.

4. $-\frac{1}{2}$, ____, $\frac{1}{3}$, ____, $\frac{7}{5}$, ____

5. $-\frac{5}{2}$, ____, -1.9, ____, $-\frac{2}{3}$, ____

6. $-\frac{1}{3}$, ____, -0.1, ____, $\frac{4}{5}$, ____

7. -3.4, ____, -1.5, ____, 2.2, ____

Practice Use what you learned about ordering rational numbers to complete Exercises 28–30 on page 522.

12.1 Lesson

Check It Out
Lesson Tutorials
BigIdeasMath.com

Key Vocabulary
rational number, *p. 520*
terminating decimal, *p. 520*
repeating decimal, *p. 520*

Key Idea

Rational Numbers

A **rational number** is a number that can be written as $\frac{a}{b}$ where a and b are integers and $b \neq 0$.

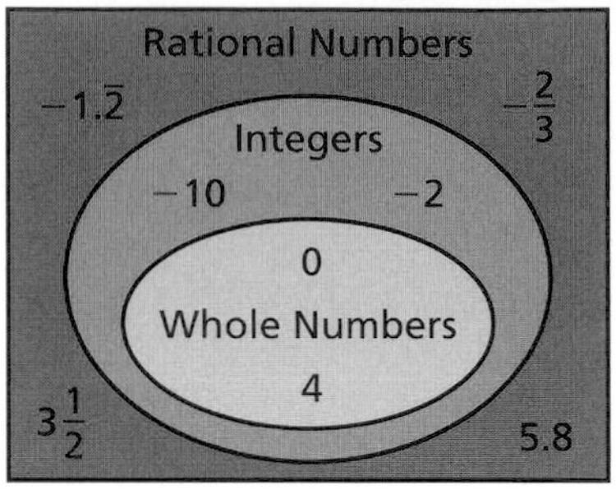

Because you can divide any integer by any nonzero integer, you can use long division to write fractions and mixed numbers as decimals. These decimals are also rational numbers and will either *terminate* or *repeat.*

A **terminating decimal** is a decimal that ends.

$1.5, -0.25, 10.625$

A **repeating decimal** is a decimal that has a pattern that repeats.

$-1.333\ldots = -1.\overline{3}$

$0.151515\ldots = 0.\overline{15}$

Use *bar notation* to show which of the digits repeat.

EXAMPLE 1 Writing Rational Numbers as Decimals

a. Write $-2\frac{1}{4}$ as a decimal.

Notice that $-2\frac{1}{4} = -\frac{9}{4}$.

Divide 9 by 4.

$$
\begin{array}{r}
2.25 \\
4\overline{)9.00} \\
-8 \\
\hline
1\,0 \\
-8 \\
\hline
20 \\
-20 \\
\hline
0
\end{array}
$$

The remainder is 0. So, it is a terminating decimal.

So, $-2\frac{1}{4} = -2.25$.

b. Write $\frac{5}{11}$ as a decimal.

Divide 5 by 11.

$$
\begin{array}{r}
0.4545 \\
11\overline{)5.0000} \\
-4\,4 \\
\hline
60 \\
-55 \\
\hline
50 \\
-44 \\
\hline
60 \\
-55 \\
\hline
5
\end{array}
$$

The remainder repeats. So, it is a repeating decimal.

So, $\frac{5}{11} = 0.\overline{45}$.

On Your Own

Now You're Ready
Exercises 11–18

Write the rational number as a decimal.

1. $-\frac{6}{5}$ **2.** $-7\frac{3}{8}$ **3.** $-\frac{3}{11}$ **4.** $1\frac{5}{27}$

Multi-Language Glossary at BigIdeasMath.com

EXAMPLE 2 Writing a Decimal as a Fraction

Write −0.26 as a fraction in simplest form.

Study Tip

If p and q are integers, then $-\frac{p}{q} = \frac{-p}{q} = \frac{p}{-q}$.

$$-0.26 = -\frac{26}{100}$$

Write the digits after the decimal point in the numerator.

The last digit is in the hundredths place. So, use 100 in the denominator.

$$= -\frac{13}{50}$$

Simplify.

On Your Own

Now You're Ready
Exercises 20–27

Write the decimal as a fraction or a mixed number in simplest form.

5. −0.7 **6.** 0.125 **7.** −3.1 **8.** −10.25

EXAMPLE 3 Ordering Rational Numbers

Creature	Elevation (kilometers)
Anglerfish	$-\frac{13}{10}$
Squid	$-2\frac{1}{5}$
Shark	$-\frac{2}{11}$
Whale	−0.8

The table shows the elevations of four sea creatures relative to sea level. Which of the sea creatures are deeper than the whale? Explain.

Write each rational number as a decimal.

$$-\frac{13}{10} = -1.3$$

$$-2\frac{1}{5} = -2.2$$

$$-\frac{2}{11} = -0.\overline{18}$$

Then graph each decimal on a number line.

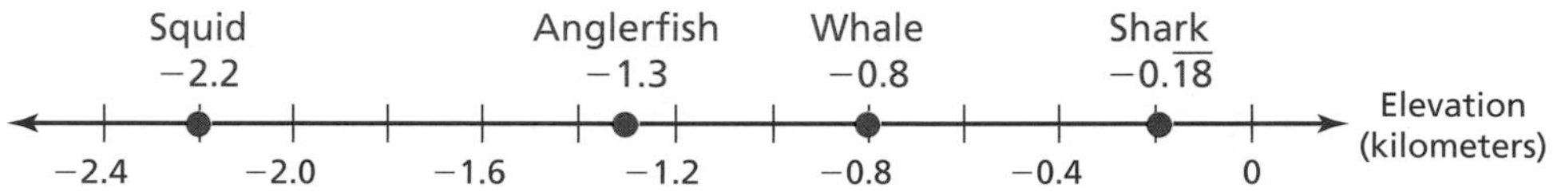

∴ Both −2.2 and −1.3 are less than −0.8. So, the squid and the anglerfish are deeper than the whale.

On Your Own

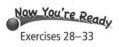

Now You're Ready
Exercises 28–33

9. WHAT IF? The elevation of a dolphin is $-\frac{1}{10}$ kilometer. Which of the sea creatures in Example 3 are deeper than the dolphin? Explain.

12.1 Exercises

Check It Out
Help with Homework
BigIdeasMath.com

Vocabulary and Concept Check

1. **VOCABULARY** Is the quotient of two integers always a rational number? Explain.
2. **WRITING** Are all terminating and repeating decimals rational numbers? Explain.

Tell whether the number belongs to each of the following number sets: *rational numbers, integers, whole numbers.*

3. -5
4. $-2.1\overline{6}$
5. 12
6. 0

Tell whether the decimal is *terminating* or *repeating*.

7. $-0.4848\ldots$
8. -0.151
9. 72.72
10. $-5.2\overline{36}$

Practice and Problem Solving

Write the rational number as a decimal.

(1) 11. $\frac{7}{8}$
12. $\frac{1}{11}$
13. $-\frac{7}{9}$
14. $-\frac{17}{40}$
15. $1\frac{5}{6}$
16. $-2\frac{17}{18}$
17. $-5\frac{7}{12}$
18. $8\frac{15}{22}$

19. **ERROR ANALYSIS** Describe and correct the error in writing the rational number as a decimal.

✗ $-\frac{7}{11} = -0.6\overline{3}$

Write the decimal as a fraction or a mixed number in simplest form.

(2) 20. -0.9
21. 0.45
22. -0.258
23. -0.312
24. -2.32
25. -1.64
26. 6.012
27. -12.405

Order the numbers from least to greatest.

(3) 28. $-\frac{3}{4}, 0.5, \frac{2}{3}, -\frac{7}{3}, 1.2$
29. $\frac{9}{5}, -2.5, -1.1, -\frac{4}{5}, 0.8$
30. $-1.4, -\frac{8}{5}, 0.6, -0.9, \frac{1}{4}$
31. $2.1, -\frac{6}{10}, -\frac{9}{4}, -0.75, \frac{5}{3}$
32. $-\frac{7}{2}, -2.8, -\frac{5}{4}, \frac{4}{3}, 1.3$
33. $-\frac{11}{5}, -2.4, 1.6, \frac{15}{10}, -2.25$

34. **COINS** You lose one quarter, two dimes, and two nickels.
    a. Write the amount as a decimal.
    b. Write the amount as a fraction in simplest form.

35. **HIBERNATION** A box turtle hibernates in sand at $-1\frac{5}{8}$ feet. A spotted turtle hibernates at $-1\frac{16}{25}$ feet. Which turtle is deeper?

Copy and complete the statement using <, >, or =.

36. -2.2 □ -2.42 **37.** -1.82 □ -1.81 **38.** $\frac{15}{8}$ □ $1\frac{7}{8}$

39. $-4\frac{6}{10}$ □ -4.65 **40.** $-5\frac{3}{11}$ □ $-5.\overline{2}$ **41.** $-2\frac{13}{16}$ □ $-2\frac{11}{14}$

42. OPEN-ENDED Find one terminating decimal and one repeating decimal between $-\frac{1}{2}$ and $-\frac{1}{3}$.

Player	Hits	At Bats
Eva	42	90
Michelle	38	80

43. SOFTBALL In softball, a batting average is the number of hits divided by the number of times at bat. Does Eva or Michelle have the higher batting average?

44. PROBLEM SOLVING You miss 3 out of 10 questions on a science quiz and 4 out of 15 questions on a math quiz. Which quiz has a higher percent of correct answers?

45. SKATING Is the half pipe deeper than the skating pool? Explain.

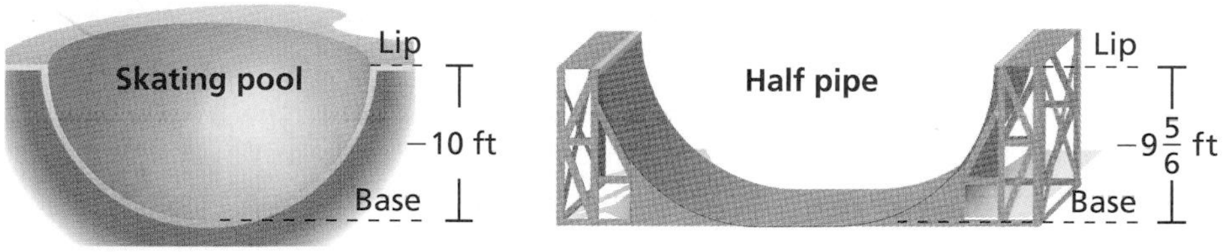

46. ENVIRONMENT The table shows the changes from the average water level of a pond over several weeks. Order the numbers from least to greatest.

Week	1	2	3	4
Change (inches)	$-\frac{7}{5}$	$-1\frac{5}{11}$	-1.45	$-1\frac{91}{200}$

47. Critical Thinking Given: a and b are integers.

a. When is $-\frac{1}{a}$ positive? **b.** When is $\frac{1}{ab}$ positive?

Fair Game Review What you learned in previous grades & lessons

Add or subtract. *(Section 1.6 and Section 2.4)*

48. $\frac{3}{5} + \frac{2}{7}$ **49.** $\frac{9}{10} - \frac{2}{3}$ **50.** $8.79 - 4.07$ **51.** $11.81 + 9.34$

52. MULTIPLE CHOICE In one year, a company has a profit of −$2 million. In the next year, the company has a profit of $7 million. How much more profit did the company make the second year? *(Section 11.3)*

Ⓐ $2 million Ⓑ $5 million Ⓒ $7 million Ⓓ $9 million

12.2 Adding Rational Numbers

Essential Question How can you use what you know about adding integers to add rational numbers?

1 ACTIVITY: Adding Rational Numbers

Work with a partner. Use a number line to find the sum.

a. $2.7 + (-3.4)$

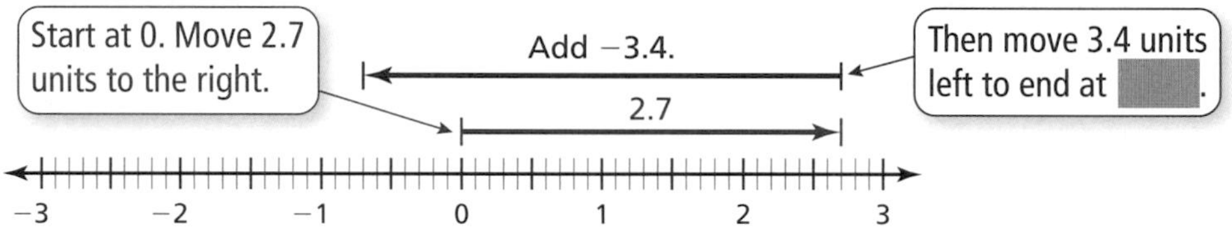

Start at 0. Move 2.7 units to the right.
Add −3.4.
2.7
Then move 3.4 units left to end at ____.
−3
−2
−1
0
1
2
3

So, $2.7 + (-3.4) =$ ____.

b. $1.3 + (-1.5)$

c. $-2.1 + 0.8$

d. $-1\frac{1}{4} + \frac{3}{4}$

e. $\frac{3}{10} + \left(-\frac{3}{10}\right)$

2 ACTIVITY: Adding Rational Numbers

Work with a partner. Use a number line to find the sum.

a. $-1\frac{2}{5} + \left(-\frac{4}{5}\right)$

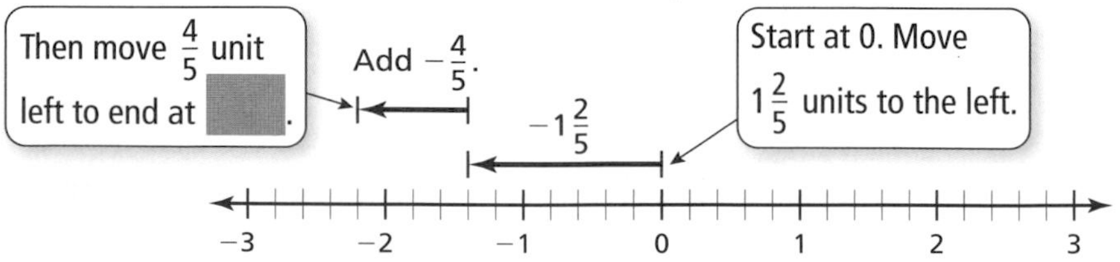

Then move $\frac{4}{5}$ unit left to end at ____.
Add $-\frac{4}{5}$.
$-1\frac{2}{5}$
Start at 0. Move $1\frac{2}{5}$ units to the left.
−3
−2
−1
0
1
2
3

So, $-1\frac{2}{5} + \left(-\frac{4}{5}\right) =$ ____.

b. $-\frac{7}{10} + \left(-1\frac{7}{10}\right)$

c. $-1\frac{2}{3} + \left(-1\frac{1}{3}\right)$

d. $-0.4 + (-1.9)$

e. $-2.3 + (-0.6)$

COMMON CORE

Rational Numbers

In this lesson, you will

- add rational numbers.
- solve real-life problems.

Learning Standards
7.NS.1a
7.NS.1b
7.NS.1d
7.NS.3

3 ACTIVITY: Writing Expressions

Work with a partner. Write the addition expression shown. Then find the sum.

Math Practice 2
Use Operations
What operation is represented in each number line? How does this help you write an expression?

a.

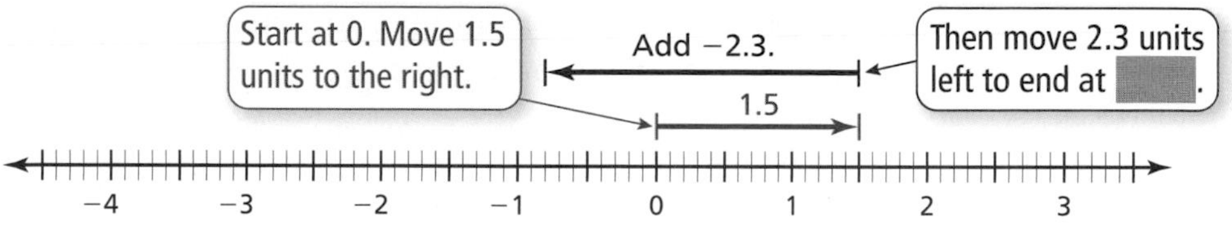
Start at 0. Move 1.5 units to the right.
Add −2.3.
Then move 2.3 units left to end at ▢.
1.5
−4 −3 −2 −1 0 1 2 3

b.

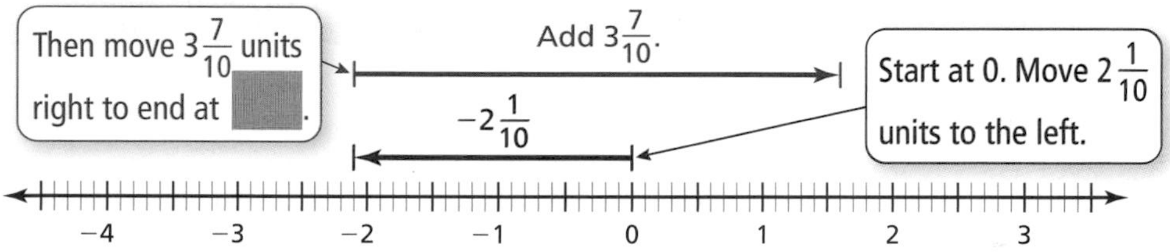
Then move $3\frac{7}{10}$ units right to end at ▢.
Add $3\frac{7}{10}$.
Start at 0. Move $2\frac{1}{10}$ units to the left.
$-2\frac{1}{10}$
−4 −3 −2 −1 0 1 2 3

c.

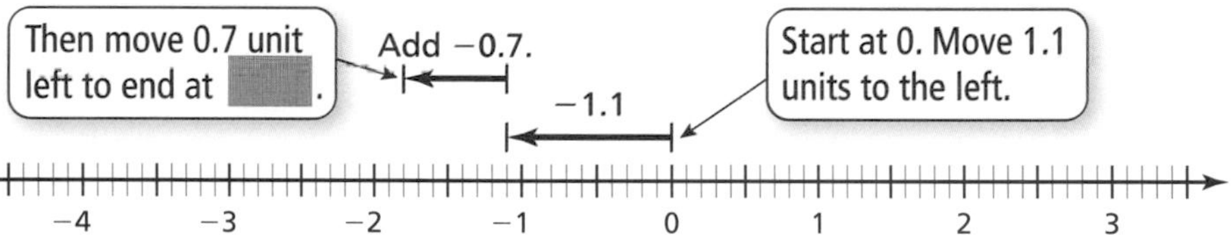
Then move 0.7 unit left to end at ▢.
Add −0.7.
Start at 0. Move 1.1 units to the left.
−1.1
−4 −3 −2 −1 0 1 2 3

What Is Your Answer?

4. **IN YOUR OWN WORDS** How can you use what you know about adding integers to add rational numbers?

PUZZLE **Find a path through the table so that the numbers add up to the sum. You can move horizontally or vertically.**

5. Sum: $\frac{3}{4}$

Start →

$\frac{1}{2}$	$\frac{2}{3}$	$-\frac{5}{7}$
$-\frac{1}{8}$	$-\frac{3}{4}$	$\frac{1}{3}$

← End

6. Sum: −0.07

Start →

2.43	1.75	−0.98
−1.09	3.47	−4.88

← End

Practice Use what you learned about adding rational numbers to complete Exercises 4–6 on page 528.

12.2 Lesson

Check It Out
Lesson Tutorials
BigIdeasMath.com

Key Idea

Adding Rational Numbers

Words To add rational numbers, use the same rules for signs as you used for integers.

Numbers $-\frac{1}{3} + \frac{1}{6} = \frac{-2}{6} + \frac{1}{6} = \frac{-2+1}{6} = \frac{-1}{6} = -\frac{1}{6}$

EXAMPLE 1 Adding Rational Numbers

Find $-\frac{8}{3} + \frac{5}{6}$. **Estimate $-3 + 1 = -2$**

Study Tip

In Example 1, notice how $-\frac{8}{3}$ is written as $-\frac{8}{3} = \frac{-8}{3} = \frac{-16}{6}$.

$-\frac{8}{3} + \frac{5}{6} = \frac{-16}{6} + \frac{5}{6}$ Rewrite using the LCD (least common denominator).

$= \frac{-16+5}{6}$ Write the sum of the numerators over the common denominator.

$= \frac{-11}{6}$ Add.

$= -1\frac{5}{6}$ Write the improper fraction as a mixed number.

∴ The sum is $-1\frac{5}{6}$. **Reasonable? $-1\frac{5}{6} \approx -2$** ✓

EXAMPLE 2 Adding Rational Numbers

Find $-4.05 + 7.62$.

$-4.05 + 7.62 = 3.57$ $|7.62| > |-4.05|$. So, subtract $|-4.05|$ from $|7.62|$.

Use the sign of 7.62.

∴ The sum is 3.57.

On Your Own

Now You're Ready
Exercises 4–12

Add.

1. $-\frac{7}{8} + \frac{1}{4}$ **2.** $-6\frac{1}{3} + \frac{20}{3}$ **3.** $2 + \left(-\frac{7}{2}\right)$

4. $-12.5 + 15.3$ **5.** $-8.15 + (-4.3)$ **6.** $0.65 + (-2.75)$

EXAMPLE 3 Evaluating Expressions

Evaluate $2x + y$ when $x = \frac{1}{4}$ and $y = -\frac{3}{2}$.

$2x + y = 2\left(\frac{1}{4}\right) + \left(-\frac{3}{2}\right)$	Substitute $\frac{1}{4}$ for x and $-\frac{3}{2}$ for y.
$= \frac{1}{2} + \left(\frac{-3}{2}\right)$	Multiply.
$= \frac{1 + (-3)}{2}$	Write the sum of the numerators over the common denominator.
$= -1$	Simplify.

EXAMPLE 4 Real-Life Application

Year	Profit (billions of dollars)
2008	−1.7
2009	−4.75
2010	1.7
2011	0.85
2012	3.6

The table shows the annual profits (in billions of dollars) of a financial company from 2008 to 2012. Positive numbers represent *gains*, and negative numbers represent *losses*. Which statement describes the profit over the five-year period?

Ⓐ gain of \$0.3 billion Ⓑ gain of \$30 million

Ⓒ loss of \$3 million Ⓓ loss of \$300 million

To determine whether there was a gain or a loss, find the sum of the profits.

five-year profit $= -1.7 + (-4.75) + 1.7 + 0.85 + 3.6$	Write the sum.
$= -1.7 + 1.7 + (-4.75) + 0.85 + 3.6$	Comm. Prop. of Add.
$= 0 + (-4.75) + 0.85 + 3.6$	Additive Inv. Prop.
$= -4.75 + 0.85 + 3.6$	Add. Prop. of Zero
$= -3.9 + 3.6$	Add -4.75 and 0.85.
$= -0.3$	Add -3.9 and 3.6.

The five-year profit is −\$0.3 billion. So, the company has a five-year loss of \$0.3 billion, or \$300 million.

∴ The correct answer is Ⓓ.

On Your Own

Now You're Ready
Exercises 15–17

Evaluate the expression when $a = \frac{1}{2}$ and $b = -\frac{5}{2}$.

7. $b + 4a$

8. $|a + b|$

9. WHAT IF? In Example 4, the 2013 profit is \$1.07 billion. State the company's gain or loss over the six-year period in millions of dollars.

12.2 Exercises

Check It Out
Help with Homework
BigIdeasMath.com

Vocabulary and Concept Check

1. **WRITING** Explain how to find the sum $-8.46 + 5.31$.

2. **OPEN-ENDED** Write an addition expression using fractions that equals $-\frac{1}{2}$.

3. **DIFFERENT WORDS, SAME QUESTION** Which is different? Find "both" answers.

Add -4.5 and 3.5.	What is the distance between -4.5 and 3.5?
What is -4.5 increased by 3.5?	Find the sum of -4.5 and 3.5.

Practice and Problem Solving

Add. Write fractions in simplest form.

(1) (2) 4. $\frac{11}{12} + \left(-\frac{7}{12}\right)$

5. $-1\frac{1}{5} + \left(-\frac{3}{5}\right)$

6. $-4.2 + 3.3$

7. $-\frac{9}{14} + \frac{2}{7}$

8. $4 + \left(-1\frac{2}{3}\right)$

9. $\frac{15}{4} + \left(-4\frac{1}{3}\right)$

10. $-3.1 + (-0.35)$

11. $12.48 + (-10.636)$

12. $20.25 + (-15.711)$

ERROR ANALYSIS Describe and correct the error in finding the sum.

13. ✗
$$\begin{array}{r} -3.7 \\ +\ (-0.25) \\ \hline -0.62 \end{array}$$

14. ✗ $-\frac{5}{8} + \frac{1}{8} = \frac{-5+1}{8} = \frac{-6}{8} = -\frac{3}{4}$

Evaluate the expression when $x = \frac{1}{3}$ and $y = -\frac{7}{4}$.

(3) 15. $x + y$

16. $3x + y$

17. $-x + |y|$

18. **BANKING** Your bank account balance is –\$20.85. You deposit \$15.50. What is your new balance?

19. **HOT DOGS** You eat $\frac{3}{10}$ of a pack of hot dogs. Your friend eats $\frac{1}{5}$ of the pack of hot dogs. What fraction of the pack of hot dogs do you and your friend eat?

Add. Write fractions in simplest form.

20. $6 + \left(-4\frac{3}{4}\right) + \left(-2\frac{1}{8}\right)$

21. $-5\frac{2}{3} + 3\frac{1}{4} + \left(-7\frac{1}{3}\right)$

22. $10.9 + (-15.6) + 2.1$

23. NUMBER SENSE When is the sum of two negative mixed numbers an integer?

24. WRITING You are adding two rational numbers with different signs. How can you tell if the sum will be *positive*, *negative*, or *zero*?

June	July	August
$-2\frac{1}{8}$	$1\frac{1}{4}$	$-\frac{9}{16}$

25. RESERVOIR The table at the left shows the water level (in inches) of a reservoir for three months compared to the yearly average. Is the water level for the three-month period greater than or less than the yearly average? Explain.

26. BREAK EVEN The table at the right shows the annual profits (in thousands of dollars) of a county fair from 2008 to 2012. What must the 2012 profit be (in hundreds of dollars) to break even over the five-year period?

Year	Profit (thousands of dollars)
2008	2.5
2009	1.75
2010	−3.3
2011	−1.4
2012	?

27. REASONING Is $|a + b| = |a| + |b|$ for all rational numbers a and b? Explain.

28. Repeated Reasoning Evaluate the expression.

$$\frac{19}{20} + \left(\frac{-18}{20}\right) + \frac{17}{20} + \left(\frac{-16}{20}\right) + \cdots + \left(\frac{-4}{20}\right) + \frac{3}{20} + \left(\frac{-2}{20}\right) + \frac{1}{20}$$

Fair Game Review What you learned in previous grades & lessons

Identify the property. Then simplify. *(Section 3.3)*

29. $8 + (-3) + 2 = 8 + 2 + (-3)$

30. $2 \cdot (4.5 \cdot 9) = (2 \cdot 4.5) \cdot 9$

31. $\frac{1}{4} + \left(\frac{3}{4} + \frac{1}{8}\right) = \left(\frac{1}{4} + \frac{3}{4}\right) + \frac{1}{8}$

32. $\frac{3}{7} \cdot \frac{4}{5} \cdot \frac{14}{27} = \frac{3}{7} \cdot \frac{14}{27} \cdot \frac{4}{5}$

33. MULTIPLE CHOICE The regular price of a photo album is \$18. You have a coupon for 15% off. How much is the discount? *(Section 5.6)*

Ⓐ \$2.70 Ⓑ \$3 Ⓒ \$15 Ⓓ \$15.30

12 Study Help

Check It Out
Graphic Organizer
BigIdeasMath.com

You can use a **process diagram** to show the steps involved in a procedure. Here is an example of a process diagram for adding rational numbers.

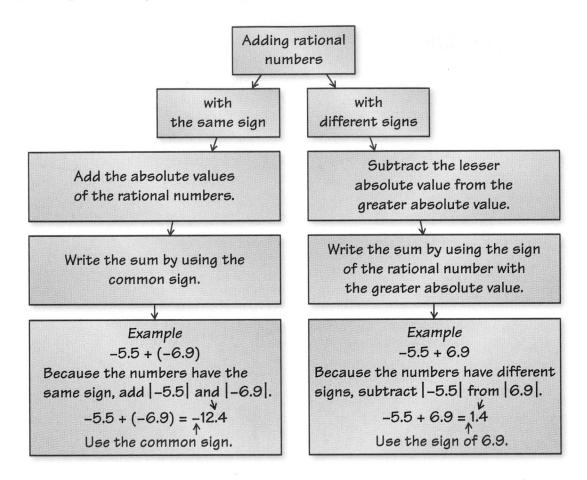

Adding rational numbers

- **with the same sign**
  - Add the absolute values of the rational numbers.
  - Write the sum by using the common sign.
  - *Example*
    $-5.5 + (-6.9)$
    Because the numbers have the same sign, add $|-5.5|$ and $|-6.9|$.
    $-5.5 + (-6.9) = -12.4$
    Use the common sign.
- **with different signs**
  - Subtract the lesser absolute value from the greater absolute value.
  - Write the sum by using the sign of the rational number with the greater absolute value.
  - *Example*
    $-5.5 + 6.9$
    Because the numbers have different signs, subtract $|-5.5|$ from $|6.9|$.
    $-5.5 + 6.9 = 1.4$
    Use the sign of 6.9.

On Your Own

Make a process diagram with examples to help you study the topic.

1. writing rational numbers as decimals

After you complete this chapter, make process diagrams with examples for the following topics.

2. subtracting rational numbers
3. multiplying rational numbers
4. dividing rational numbers

"Does this process diagram accurately show how a cat claws furniture?"

12.1–12.2 Quiz

Check It Out
Progress Check
BigIdeasMath.com

Write the rational number as a decimal. *(Section 12.1)*

1. $-\frac{3}{20}$

2. $-\frac{11}{6}$

Write the decimal as a fraction or a mixed number in simplest form. *(Section 12.1)*

3. -0.325

4. -1.28

Order the numbers from least to greatest. *(Section 12.1)*

5. $-\frac{1}{3}, -0.2, \frac{5}{3}, 0.4, 1.3$

6. $-\frac{4}{3}, -1.2, 0.3, \frac{4}{9}, -0.8$

Add. Write fractions in simplest form. *(Section 12.2)*

7. $-\frac{4}{5} + \left(-\frac{3}{8}\right)$

8. $-\frac{13}{6} + \frac{7}{12}$

9. $-5.8 + 2.6$

10. $-4.28 + (-2.56)$

Evaluate the expression when $x = \frac{3}{4}$ and $y = -\frac{1}{2}$. *(Section 12.2)*

11. $x + y$

12. $2x + y$

13. $x + |y|$

14. $|-x + y|$

15. STOCK The value of Stock A changes $-\$3.68$, and the value of Stock B changes $-\$3.72$. Which stock has the greater loss? Explain. *(Section 12.1)*

16. LEMONADE You drink $\frac{2}{7}$ of a pitcher of lemonade. Your friend drinks $\frac{3}{14}$ of the pitcher. What fraction of the pitcher do you and your friend drink? *(Section 12.2)*

17. FOOTBALL The table shows the statistics of a running back in a football game. Did he gain more than 50 yards total? Explain. *(Section 12.2)*

Quarter	1	2	3	4	Total
Yards	$-8\frac{1}{2}$	23	$42\frac{1}{2}$	$-2\frac{1}{4}$	?

12.3 Subtracting Rational Numbers

Essential Question How can you use what you know about subtracting integers to subtract rational numbers?

1 ACTIVITY: Subtracting Rational Numbers

Work with a partner. Use a number line to find the difference.

a. $-1\frac{1}{2} - \frac{1}{2}$

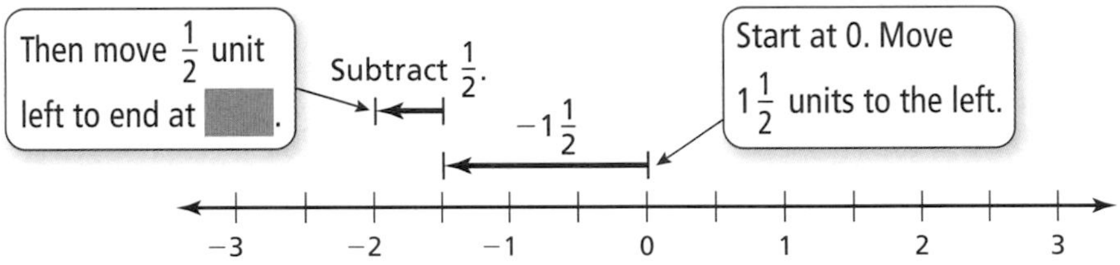

∴ So, $-1\frac{1}{2} - \frac{1}{2} =$ ____.

b. $\frac{6}{10} - 1\frac{3}{10}$

c. $-1\frac{1}{4} - 1\frac{3}{4}$

d. $-1.9 - 0.8$

e. $0.2 - 0.7$

2 ACTIVITY: Finding Distances on a Number Line

Work with a partner.

a. Plot −3 and 2 on the number line. Then find $-3 - 2$ and $2 - (-3)$. What do you notice about your results?

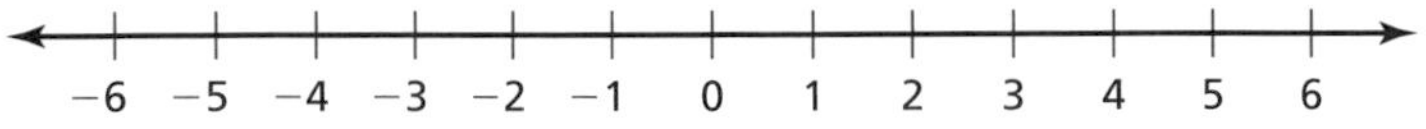

b. Plot $\frac{3}{4}$ and 1 on the number line. Then find $\frac{3}{4} - 1$ and $1 - \frac{3}{4}$. What do you notice about your results?

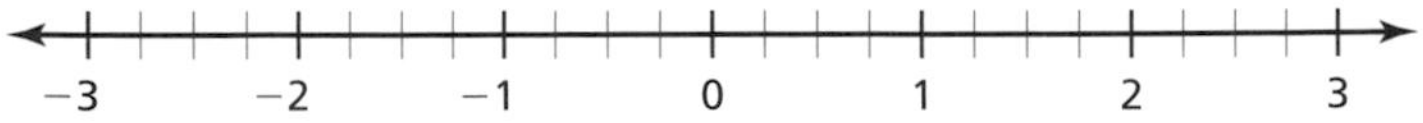

c. Choose any two points a and b on a number line. Find the values of $a - b$ and $b - a$. What do the absolute values of these differences represent? Is this true for any pair of rational numbers? Explain.

COMMON CORE

Rational Numbers

In this lesson, you will

- subtract rational numbers.
- solve real-life problems.

Learning Standards
7.NS.1c
7.NS.1d
7.NS.3

3 ACTIVITY: Financial Literacy

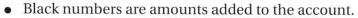

Work with a partner. The table shows the balance in a checkbook.

- Black numbers are amounts added to the account.
- **Red numbers** are amounts taken from the account.

Date	Check #	Transaction	Amount	Balance
--	--	Previous balance	--	100.00
1/02/2013	124	Groceries	**34.57**	
1/07/2013		Check deposit	875.50	
1/11/2013		ATM withdrawal	**40.00**	
1/14/2013	125	Electric company	**78.43**	
1/17/2013		Music store	**10.55**	
1/18/2013	126	Shoes	**47.21**	
1/22/2013		Check deposit	125.00	
1/24/2013		Interest	2.12	
1/25/2013	127	Cell phone	**59.99**	
1/26/2013	128	Clothes	**65.54**	
1/30/2013	129	Cable company	**75.00**	

Math Practice 4

Interpret Results

What does your answer represent? Does your answer make sense?

You can find the balance in the **second row** two different ways.

$100.00 - 34.57 = 65.43$ Subtract 34.57 from 100.00.

$100.00 + (-34.57) = 65.43$ Add −34.57 to 100.00.

a. Copy the table. Then complete the balance column.

b. How did you find the balance in the **twelfth row**?

c. Use a different way to find the balance in part (b).

What Is Your Answer?

4. **IN YOUR OWN WORDS** How can you use what you know about subtracting integers to subtract rational numbers?

5. Give two real-life examples of subtracting rational numbers that are not integers.

Practice Use what you learned about subtracting rational numbers to complete Exercises 3–5 on page 536.

12.3 Lesson

Check It Out
Lesson Tutorials
BigIdeasMath.com

Key Idea

Subtracting Rational Numbers

Words To subtract rational numbers, use the same rules for signs as you used for integers.

Numbers $\frac{2}{5} - \left(-\frac{1}{5}\right) = \frac{2}{5} + \frac{1}{5} = \frac{2+1}{5} = \frac{3}{5}$

EXAMPLE 1 Subtracting Rational Numbers

Find $-4\frac{1}{7} - \left(-\frac{6}{7}\right)$. **Estimate $-4 - (-1) = -3$**

$-4\frac{1}{7} - \left(-\frac{6}{7}\right) = -4\frac{1}{7} + \frac{6}{7}$ Add the opposite of $-\frac{6}{7}$.

$= -\frac{29}{7} + \frac{6}{7}$ Write the mixed number as an improper fraction.

$= \frac{-29 + 6}{7}$ Write the sum of the numerators over the common denominator.

$= \frac{-23}{7}$ Add.

$= -3\frac{2}{7}$ Write the improper fraction as a mixed number.

The difference is $-3\frac{2}{7}$. **Reasonable? $-3\frac{2}{7} \approx -3$** ✓

EXAMPLE 2 Subtracting Rational Numbers

Find 12.8 − 21.6.

$12.8 - 21.6 = 12.8 + (-21.6)$ Add the opposite of 21.6.

$= -8.8$ $|-21.6| > |12.8|$. So, subtract $|12.8|$ from $|-21.6|$.

Use the sign of −21.6.

The difference is −8.8.

On Your Own

Now You're Ready
Exercises 3–11

1. $\frac{1}{3} - \left(-\frac{1}{3}\right)$
2. $-3\frac{1}{3} - \frac{5}{6}$
3. $4\frac{1}{2} - 5\frac{1}{4}$
4. $-8.4 - 6.7$
5. $-20.5 - (-20.5)$
6. $0.41 - (-0.07)$

The distance between any two numbers on a number line is the absolute value of the difference of the numbers.

EXAMPLE 3 Finding Distances Between Numbers on a Number Line

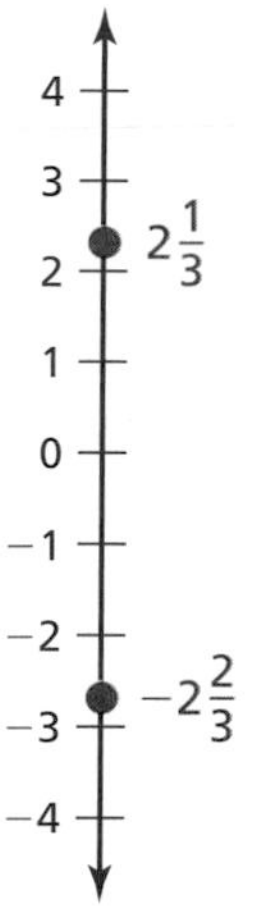

Find the distance between the two numbers on the number line.

To find the distance between the numbers, first find the difference of the numbers.

$$-2\frac{2}{3} - 2\frac{1}{3} = -2\frac{2}{3} + \left(-2\frac{1}{3}\right) \quad \text{Add the opposite of } 2\frac{1}{3}.$$

$$= -\frac{8}{3} + \left(-\frac{7}{3}\right) \quad \text{Write the mixed numbers as improper fractions.}$$

$$= \frac{-15}{3} \quad \text{Add.}$$

$$= -5 \quad \text{Simplify.}$$

∴ Because $|-5| = 5$, the distance between $-2\frac{2}{3}$ and $2\frac{1}{3}$ is 5.

EXAMPLE 4 Real-Life Application

Clearance: 11 ft 8 in.

In the water, the bottom of a boat is 2.1 feet below the surface, and the top of the boat is 8.7 feet above it. Towed on a trailer, the bottom of the boat is 1.3 feet above the ground. Can the boat and trailer pass under the bridge?

Step 1: Find the height h of the boat.

$$h = 8.7 - (-2.1) \quad \text{Subtract the lowest point from the highest point.}$$

$$= 8.7 + 2.1 \quad \text{Add the opposite of } -2.1.$$

$$= 10.8 \quad \text{Add.}$$

Step 2: Find the height t of the boat and trailer.

$$t = 10.8 + 1.3 \quad \text{Add the trailer height to the boat height.}$$

$$= 12.1 \quad \text{Add.}$$

∴ Because 12.1 feet is greater than 11 feet 8 inches, the boat and trailer cannot pass under the bridge.

On Your Own

Now You're Ready
Exercises 13–15

7. Find the distance between −7.5 and −15.3 on a number line.
8. **WHAT IF?** In Example 4, the clearance is 12 feet 1 inch. Can the boat and trailer pass under the bridge?

12.3 Exercises

Vocabulary and Concept Check

1. **WRITING** Explain how to find the difference $-\frac{4}{5} - \frac{3}{5}$.

2. **WHICH ONE DOESN'T BELONG?** Which expression does *not* belong with the other three? Explain your reasoning.

   $-\frac{5}{8} - \frac{3}{4}$ $\quad$ $-\frac{3}{4} + \frac{5}{8}$ $\quad$ $-\frac{5}{8} + \left(-\frac{3}{4}\right)$ $\quad$ $-\frac{3}{4} - \frac{5}{8}$

Practice and Problem Solving

Subtract. Write fractions in simplest form.

1 2 3. $\frac{5}{8} - \left(-\frac{7}{8}\right)$ $\quad$ 4. $-1\frac{1}{3} - 1\frac{2}{3}$ $\quad$ 5. $-1 - 2.5$

6. $-5 - \frac{5}{3}$ $\quad$ 7. $-8\frac{3}{8} - 10\frac{1}{6}$ $\quad$ 8. $-\frac{1}{2} - \left(-\frac{5}{9}\right)$

9. $5.5 - 8.1$ $\quad$ 10. $-7.34 - (-5.51)$ $\quad$ 11. $6.673 - (-8.29)$

12. **ERROR ANALYSIS** Describe and correct the error in finding the difference.

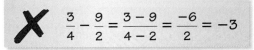

    ✗ $\frac{3}{4} - \frac{9}{2} = \frac{3-9}{4-2} = \frac{-6}{2} = -3$

Find the distance between the two numbers on a number line.

3 13. $-2\frac{1}{2}, -5\frac{3}{4}$ $\quad$ 14. $-2.2, 8.4$ $\quad$ 15. $-7, -3\frac{2}{3}$

16. **SPORTS DRINK** Your sports drink bottle is $\frac{5}{6}$ full. After practice, the bottle is $\frac{3}{8}$ full. Write the difference of the amounts after practice and before practice.

17. **SUBMARINE** The figure shows the depths of a submarine.

    a. Find the vertical distance traveled by the submarine.

    b. Find the mean hourly vertical distance traveled by the submarine.

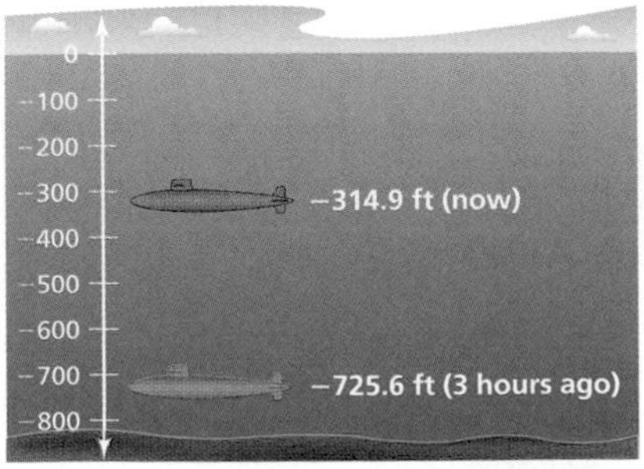

Evaluate.

18. $2\frac{1}{6} - \left(-\frac{8}{3}\right) + \left(-4\frac{7}{9}\right)$ $\quad$ 19. $6.59 + (-7.8) - (-2.41)$ $\quad$ 20. $-\frac{12}{5} + \left|-\frac{13}{6}\right| + \left(-3\frac{2}{3}\right)$

21. **REASONING** When is the difference of two decimals an integer? Explain.

22. **RECIPE** A cook has $2\frac{2}{3}$ cups of flour. A recipe calls for $2\frac{3}{4}$ cups of flour. Does the cook have enough flour? If not, how much more flour is needed?

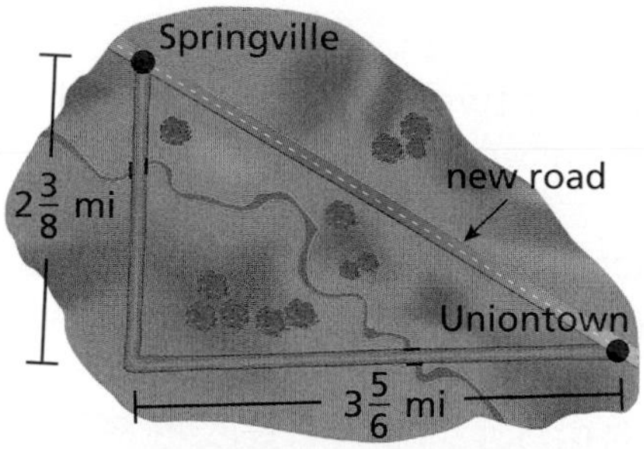

23. **ROADWAY** A new road that connects Uniontown to Springville is $4\frac{1}{3}$ miles long. What is the change in distance when using the new road instead of the dirt roads?

RAINFALL In Exercises 24–26, the bar graph shows the differences in a city's rainfall from the historical average.

24. What is the difference in rainfall between the wettest and the driest months?

25. Find the sum of the differences for the year.

26. What does the sum in Exercise 25 tell you about the rainfall for the year?

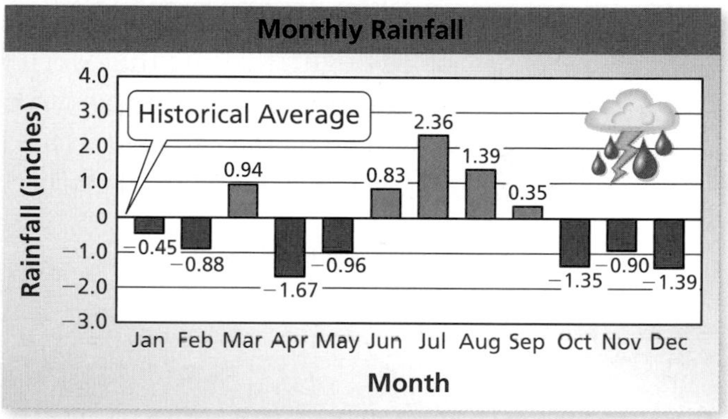

27. **OPEN-ENDED** Write two different pairs of negative decimals, x and y, that make the statement $x - y = 0.6$ true.

REASONING Tell whether the difference between the two numbers is *always, sometimes,* or *never* positive. Explain your reasoning.

28. two negative fractions

29. a positive decimal and a negative decimal

30. **Structure** Fill in the blanks to make the solution correct.

$$5.\blacksquare 4 - (\blacksquare.8\blacksquare) = -3.61$$

Fair Game Review What you learned in previous grades & lessons

Evaluate. *(Section 2.1, Section 2.3, Section 2.5, and Section 2.6)*

31. 5.2×6.9

32. $7.2 \div 2.4$

33. $2\frac{2}{3} \times 3\frac{1}{4}$

34. $9\frac{4}{5} \div 3\frac{1}{2}$

35. **MULTIPLE CHOICE** A sports store has 116 soccer balls. Over 6 months, it sells 8 soccer balls per month. How many soccer balls are in inventory at the end of the 6 months? *(Section 11.3 and Section 11.4)*

Ⓐ −48 Ⓑ 48 Ⓒ 68 Ⓓ 108

12.4 Multiplying and Dividing Rational Numbers

Essential Question Why is the product of two negative rational numbers positive?

In Section 11.4, you used a table to see that the product of two negative integers is a positive integer. In this activity, you will find that same result another way.

1 ACTIVITY: Showing $(-1)(-1) = 1$

Work with a partner. How can you show that $(-1)(-1) = 1$?

To begin, assume that $(-1)(-1) = 1$ is a true statement. From the Additive Inverse Property, you know that $1 + (-1) = 0$. So, substitute $(-1)(-1)$ for 1 to get $(-1)(-1) + (-1) = 0$. If you can show that $(-1)(-1) + (-1) = 0$ is true, then you have shown that $(-1)(-1) = 1$.

Justify each step.

$$(-1)(-1) + (-1) = (-1)(-1) + 1(-1)$$

$$= (-1)[(-1) + 1]$$

$$= (-1)0$$

$$= 0$$

So, $(-1)(-1) = 1$.

2 ACTIVITY: Multiplying by -1

Work with a partner.

a. Graph each number below on three different number lines. Then multiply each number by -1 and graph the product on the appropriate number line.

$2 \qquad 8 \qquad -1$

b. How does multiplying by -1 change the location of the points in part (a)? What is the relationship between the number and the product?

c. Graph each number below on three different number lines. Where do you think the points will be after multiplying by -1? Plot the points. Explain your reasoning.

$\frac{1}{2} \qquad 2.5 \qquad -\frac{5}{2}$

d. What is the relationship between a rational number $-a$ and the product $-1(a)$? Explain your reasoning.

COMMON CORE

Rational Numbers

In this lesson, you will

- multiply and divide rational numbers.
- solve real-life problems.

Learning Standards
7.NS.2a
7.NS.2b
7.NS.2c
7.NS.3

3 ACTIVITY: Understanding the Product of Rational Numbers

Work with a partner. Let *a* and *b* be positive rational numbers.

a. Because a and b are positive, what do you know about $-a$ and $-b$?

b. Justify each step.

$$(-a)(-b) = (-1)(a)(-1)(b)$$
$$= (-1)(-1)(a)(b)$$
$$= (1)(a)(b)$$
$$= ab$$

c. Because a and b are positive, what do you know about the product ab?

d. What does this tell you about products of rational numbers? Explain.

4 ACTIVITY: Writing a Story

Work with a partner. Write a story that uses addition, subtraction, multiplication, or division of rational numbers.

- At least one of the numbers in the story has to be negative and *not* an integer.
- Draw pictures to help illustrate what is happening in the story.
- Include the solution of the problem in the story.

Math Practice 6
Specify Units
What units are in your story?

If you are having trouble thinking of a story, here are some common uses of negative numbers:

- A profit of −\$15 is a loss of \$15.
- An elevation of −100 feet is a depth of 100 feet below sea level.
- A gain of −5 yards in football is a loss of 5 yards.
- A score of −4 in golf is 4 strokes under par.

What Is Your Answer?

5. **IN YOUR OWN WORDS** Why is the product of two negative rational numbers positive?
6. **PRECISION** Show that $(-2)(-3) = 6$.
7. How can you show that the product of a negative rational number and a positive rational number is negative?

Practice Use what you learned about multiplying rational numbers to complete Exercises 7–9 on page 542.

12.4 Lesson

Check It Out
Lesson Tutorials
BigIdeasMath.com

Key Idea

Remember

The *reciprocal* of $\frac{a}{b}$ is $\frac{b}{a}$.

Multiplying and Dividing Rational Numbers

Words To multiply or divide rational numbers, use the same rules for signs as you used for integers.

Numbers $-\frac{2}{7} \cdot \frac{1}{3} = \frac{-2 \cdot 1}{7 \cdot 3} = \frac{-2}{21} = -\frac{2}{21}$

$-\frac{1}{2} \div \frac{4}{9} = \frac{-1}{2} \cdot \frac{9}{4} = \frac{-1 \cdot 9}{2 \cdot 4} = \frac{-9}{8} = -\frac{9}{8}$

EXAMPLE 1 Dividing Rational Numbers

Find $-5\frac{1}{5} \div 2\frac{1}{3}$. **Estimate** $-5 \div 2 = -2\frac{1}{2}$

$-5\frac{1}{5} \div 2\frac{1}{3} = -\frac{26}{5} \div \frac{7}{3}$ Write mixed numbers as improper fractions.

$= \frac{-26}{5} \cdot \frac{3}{7}$ Multiply by the reciprocal of $\frac{7}{3}$.

$= \frac{-26 \cdot 3}{5 \cdot 7}$ Multiply the numerators and the denominators.

$= \frac{-78}{35}$, or $-2\frac{8}{35}$ Simplify.

∴ The quotient is $-2\frac{8}{35}$. **Reasonable?** $-2\frac{8}{35} \approx -2\frac{1}{2}$ ✓

EXAMPLE 2 Multiplying Rational Numbers

Find $-2.5 \cdot 3.6$.

$$\begin{array}{r} -2.5 \\ \times\ 3.6 \\ \hline 1\,5\,0 \\ 7\,5\,0 \\ \hline -9.0\,0 \end{array}$$

The decimals have different signs.

The product is negative.

∴ The product is -9.

EXAMPLE 3 Multiplying More Than Two Rational Numbers

Find $-\frac{1}{7} \cdot \left[\frac{4}{5} \cdot (-7)\right]$.

You can use properties of multiplication to make the product easier to find.

$$-\frac{1}{7} \cdot \left[\frac{4}{5} \cdot (-7)\right] = -\frac{1}{7} \cdot \left(-7 \cdot \frac{4}{5}\right)$$ Commutative Property of Multiplication

$$= -\frac{1}{7} \cdot (-7) \cdot \frac{4}{5}$$ Associative Property of Multiplication

$$= 1 \cdot \frac{4}{5}$$ Multiplicative Inverse Property

$$= \frac{4}{5}$$ Multiplication Property of One

The product is $\frac{4}{5}$.

On Your Own

Now You're Ready
Exercises 10–30

Multiply or divide. Write fractions in simplest form.

1. $-\frac{6}{5} \div \left(-\frac{1}{2}\right)$
2. $\frac{1}{3} \div \left(-2\frac{2}{3}\right)$
3. $1.8(-5.1)$
4. $-6.3(-0.6)$
5. $-\frac{2}{3} \cdot 7\frac{7}{8} \cdot \frac{3}{2}$
6. $-7.2 \cdot 0.1 \cdot (-100)$

EXAMPLE 4 Real-Life Application

Account Positions

Stock	Original Value	Current Value	Change
A	600.54	420.15	−180.39
B	391.10	518.38	127.28
C	380.22	99.70	−280.52

An investor owns Stocks A, B, and C. What is the mean change in the value of the stocks?

$$\text{mean} = \frac{-180.39 + 127.28 + (-280.52)}{3} = \frac{-333.63}{3} = -111.21$$

The mean change in the value of the stocks is −\$111.21.

On Your Own

7. **WHAT IF?** The change in the value of Stock D is \$568.23. What is the mean change in the value of the four stocks?

12.4 Exercises

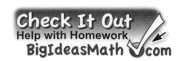

Check It Out
Help with Homework
BigIdeasMath.com

Vocabulary and Concept Check

1. **WRITING** How is multiplying and dividing rational numbers similar to multiplying and dividing integers?

2. **NUMBER SENSE** Find the reciprocal of $-\frac{2}{5}$.

Tell whether the expression is *positive* or *negative* without evaluating.

3. $-\frac{3}{10} \times \left(-\frac{8}{15}\right)$
4. $1\frac{1}{2} \div \left(-\frac{1}{4}\right)$
5. -6.2×8.18
6. $\frac{-8.16}{-2.72}$

Practice and Problem Solving

Multiply.

7. $-1\left(\frac{4}{5}\right)$
8. $-1\left(-3\frac{1}{2}\right)$
9. $-0.25(-1)$

Divide. Write fractions in simplest form.

1 10. $-\frac{7}{10} \div \frac{2}{5}$
11. $\frac{1}{4} \div \left(-\frac{3}{8}\right)$
12. $-\frac{8}{9} \div \left(-\frac{8}{9}\right)$
13. $-\frac{1}{5} \div 20$
14. $-2\frac{4}{5} \div (-7)$
15. $-10\frac{2}{7} \div \left(-4\frac{4}{11}\right)$
16. $-9 \div 7.2$
17. $8 \div 2.2$
18. $-3.45 \div (-15)$
19. $-0.18 \div 0.03$
20. $8.722 \div (-3.56)$
21. $12.42 \div (-4.8)$

Multiply. Write fractions in simplest form.

2 3 22. $-\frac{1}{4} \times \left(-\frac{4}{3}\right)$
23. $\frac{5}{6}\left(-\frac{8}{15}\right)$
24. $-2\left(-1\frac{1}{4}\right)$
25. $-3\frac{1}{3} \cdot \left(-2\frac{7}{10}\right)$
26. $0.4 \times (-0.03)$
27. $-0.05 \times (-0.5)$
28. $-8(0.09)(-0.5)$
29. $\frac{5}{6} \cdot \left(-4\frac{1}{2}\right) \cdot \left(-2\frac{1}{5}\right)$
30. $\left(-1\frac{2}{3}\right)^3$

ERROR ANALYSIS Describe and correct the error.

31. ✗ $-2.2 \times 3.7 = 8.14$

32. ✗ $-\frac{1}{4} \div \frac{3}{2} = -\frac{4}{1} \times \frac{3}{2} = -\frac{12}{2} = -6$

33. **HOUR HAND** The hour hand of a clock moves $-30°$ every hour. How many degrees does it move in $2\frac{1}{5}$ hours?

34. **SUNFLOWER SEEDS** How many 0.75-pound packages can you make with 6 pounds of sunflower seeds?

Evaluate.

35. $-4.2 + 8.1 \times (-1.9)$

36. $2.85 - 6.2 \div 2^2$

37. $-3.64 \cdot |-5.3| - 1.5^3$

38. $1\frac{5}{9} \div \left(-\frac{2}{3}\right) + \left(-2\frac{3}{5}\right)$

39. $-3\frac{3}{4} \times \frac{5}{6} - 2\frac{1}{3}$

40. $\left(-\frac{2}{3}\right)^2 - \frac{3}{4}\left(2\frac{1}{3}\right)$

41. OPEN-ENDED Write two fractions whose product is $-\frac{3}{5}$.

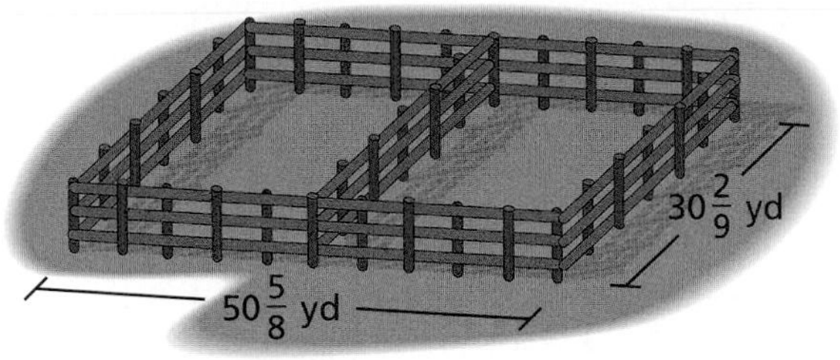

42. FENCING A farmer needs to enclose two adjacent rectangular pastures. How much fencing does the farmer need?

43. GASOLINE A 14.5-gallon gasoline tank is $\frac{3}{4}$ full. How many gallons will it take to fill the tank?

44. PRECISION A section of a boardwalk is made using 15 boards. Each board is $9\frac{1}{4}$ inches wide. The total width of the section is 144 inches. The spacing between each board is equal. What is the width of the spacing between each board?

45. RUNNING The table shows the changes in the times (in seconds) of four teammates. What is the mean change?

Teammate	Change
1	−2.43
2	−1.85
3	0.61
4	−1.45

46. Critical Thinking The daily changes in the barometric pressure for four days are −0.05, 0.09, −0.04, and −0.08 inches.

a. What is the mean change?

b. The mean change after five days is −0.01 inch. What is the change on the fifth day? Explain.

Fair Game Review What you learned in previous grades & lessons

Add or subtract. *(Section 12.2 and Section 12.3)*

47. $-6.2 + 4.7$

48. $-8.1 - (-2.7)$

49. $\frac{9}{5} - \left(-2\frac{7}{10}\right)$

50. $-4\frac{5}{6} + \left(-3\frac{4}{9}\right)$

51. MULTIPLE CHOICE What are the coordinates of the point in Quadrant IV? *(Section 6.5)*

Ⓐ (−4, 1)　Ⓑ (−3, −3)

Ⓒ (0, −2)　Ⓓ (3, −3)

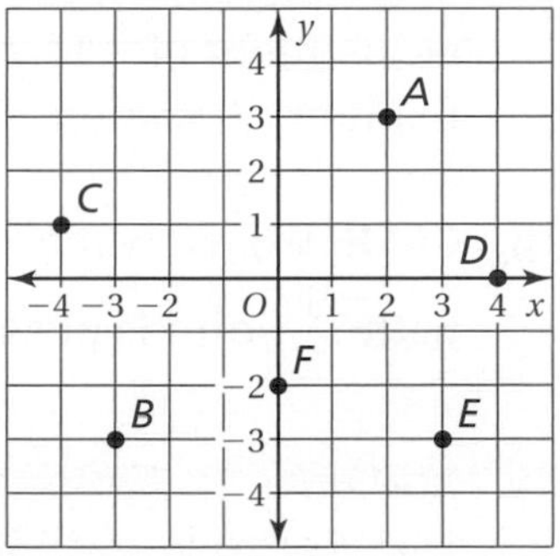

12.3–12.4 Quiz

Check It Out
Progress Check
BigIdeasMath.com

Subtract. Write fractions in simplest form. *(Section 12.3)*

1. $\frac{2}{7} - \left(\frac{6}{7}\right)$

2. $\frac{12}{7} - \left(-\frac{2}{9}\right)$

3. $9.1 - 12.9$

4. $5.647 - (-9.24)$

Find the distance between the two numbers on the number line. *(Section 12.3)*

5.

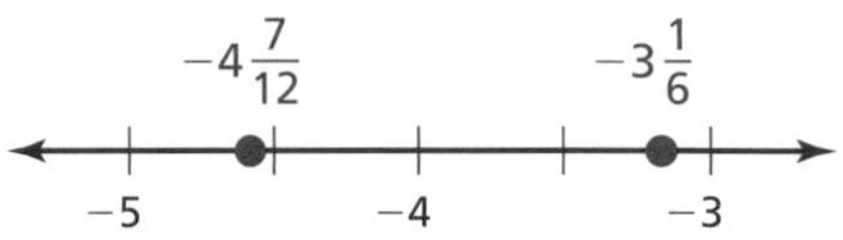

6.

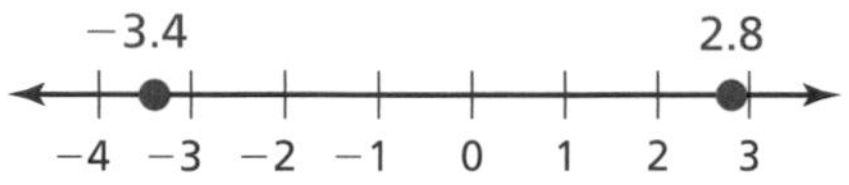

Divide. Write fractions in simplest form. *(Section 12.4)*

7. $\frac{2}{3} \div \left(-\frac{5}{6}\right)$

8. $-8\frac{5}{9} \div \left(-1\frac{4}{7}\right)$

9. $-8.4 \div 2.1$

10. $32.436 \div (-4.24)$

Multiply. Write fractions in simplest form. *(Section 12.4)*

11. $\frac{5}{8} \times \left(-\frac{4}{15}\right)$

12. $-2\frac{3}{8} \times \frac{8}{5}$

13. $-9.4 \times (-4.7)$

14. $-100(-0.6)(0.01)$

15. PARASAILING A parasail is at 200.6 feet above the water. After 5 minutes, the parasail is at 120.8 feet above the water. What is the change in height of the parasail? *(Section 12.3)*

16. TEMPERATURE Use the thermometer shown. How much did the temperature drop from 5:00 P.M. to 10:00 P.M.? *(Section 12.3)*

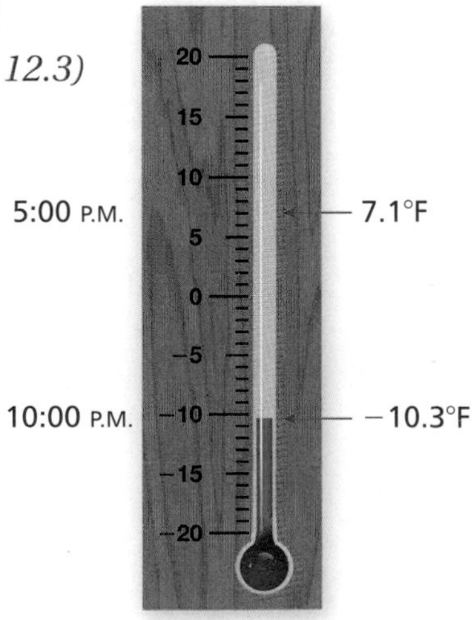

17. LATE FEES You were overcharged \$4.52 on your cell phone bill 3 months in a row. The cell phone company says that it will add −\$4.52 to your next bill for each month you were overcharged. On the next bill, you see an adjustment of −13.28. Is this amount correct? Explain. *(Section 12.4)*

18. CASHEWS How many $1\frac{1}{4}$-pound packages can you make with $7\frac{1}{2}$ pounds of cashews? *(Section 12.4)*

12 Chapter Review

Check It Out
Vocabulary Help
BigIdeasMath.com

Review Key Vocabulary

rational number, *p. 520*
terminating decimal, *p. 520*
repeating decimal, *p. 520*

Review Examples and Exercises

12.1 Rational Numbers *(pp. 518–523)*

a. Write $4\frac{3}{5}$ as a decimal.

Notice that $4\frac{3}{5} = \frac{23}{5}$.

Divide 23 by 5.

$$\begin{array}{r} 4.6 \\ 5\overline{)23.0} \\ -\,20 \\ \hline 3\,0 \\ -\,3\,0 \\ \hline 0 \end{array}$$

The remainder is 0. So, it is a terminating decimal.

∴ So, $4\frac{3}{5} = 4.6$.

b. Write −0.14 as a fraction in simplest form.

$$-0.14 = -\frac{14}{100}$$

Write the digits after the decimal point in the numerator.

The last digit is in the hundredths place. So, use 100 in the denominator.

$$= -\frac{7}{50} \qquad \text{Simplify.}$$

Exercises

Write the rational number as a decimal.

1. $-\frac{8}{15}$ **2.** $\frac{5}{8}$ **3.** $-\frac{13}{6}$ **4.** $1\frac{7}{16}$

Write the decimal as a fraction or a mixed number in simplest form.

5. −0.6 **6.** −0.35 **7.** −5.8 **8.** 24.23

12.2 Adding Rational Numbers *(pp. 524–529)*

Find $-\frac{7}{2} + \frac{5}{4}$.

$-\frac{7}{2} + \frac{5}{4} = \frac{-14}{4} + \frac{5}{4}$ Rewrite using the LCD (least common denominator).

$= \frac{-14 + 5}{4}$ Write the sum of the numerators over the common denominator.

$= \frac{-9}{4}$ Add.

$= -2\frac{1}{4}$ Write the improper fraction as a mixed number.

∴ The sum is $-2\frac{1}{4}$.

Exercises

Add. Write fractions in simplest form.

9. $\frac{9}{10} + \left(-\frac{4}{5}\right)$

10. $-4\frac{5}{9} + \frac{8}{9}$

11. $-1.6 + (-2.4)$

12.3 Subtracting Rational Numbers *(pp. 532–537)*

Find $-4\frac{2}{5} - \left(-\frac{3}{5}\right)$.

$-4\frac{2}{5} - \left(-\frac{3}{5}\right) = -4\frac{2}{5} + \frac{3}{5}$ Add the opposite of $-\frac{3}{5}$.

$= -\frac{22}{5} + \frac{3}{5}$ Write the mixed number as an improper fraction.

$= \frac{-22 + 3}{5}$ Write the sum of the numerators over the common denominator.

$= \frac{-19}{5}$, or $-3\frac{4}{5}$ Simplify.

∴ The difference is $-3\frac{4}{5}$.

Exercises

Subtract. Write fractions in simplest form.

12. $-\frac{5}{12} - \frac{3}{10}$

13. $3\frac{3}{4} - \frac{7}{8}$

14. $3.8 - (-7.45)$

15. TURTLE A turtle is $20\frac{5}{6}$ inches below the surface of a pond. It dives to a depth of $32\frac{1}{4}$ inches. What is the change in the turtle's position?

12.4 Multiplying and Dividing Rational Numbers (pp. 538–543)

a. Find $-4\frac{1}{6} \div 1\frac{1}{3}$.

$$-4\frac{1}{6} \div 1\frac{1}{3} = -\frac{25}{6} \div \frac{4}{3}$$ Write mixed numbers as improper fractions.

$$= \frac{-25}{6} \cdot \frac{3}{4}$$ Multiply by the reciprocal of $\frac{4}{3}$.

$$= \frac{-25 \cdot 3}{6 \cdot 4}$$ Multiply the numerators and the denominators.

$$= \frac{-25}{8}, \text{ or } -3\frac{1}{8}$$ Simplify.

∴ The quotient is $-3\frac{1}{8}$.

b. Find $-1.6 \cdot 2.4$.

$$\begin{array}{r} -1.6 \\ \times\ 2.4 \\ \hline 64 \\ 320 \\ \hline -3.84 \end{array}$$

The decimals have different signs.

The product is negative.

∴ The product is -3.84.

Exercises

Divide. Write fractions in simplest form.

16. $\frac{9}{10} \div \left(-\frac{6}{5}\right)$ **17.** $-\frac{4}{11} \div \frac{2}{7}$ **18.** $6.4 \div (-3.2)$ **19.** $-15.4 \div (-2.5)$

Multiply. Write fractions in simplest form.

20. $-\frac{4}{9}\left(-\frac{7}{9}\right)$ **21.** $\frac{8}{15}\left(-\frac{2}{3}\right)$ **22.** $-5.9(-9.7)$

23. $4.5(-5.26)$ **24.** $-\frac{2}{3} \cdot \left(2\frac{1}{2}\right) \cdot (-3)$ **25.** $-1.6 \cdot (0.5) \cdot (-20)$

26. SUNKEN SHIP The elevation of a sunken ship is -120 feet. Your elevation is $\frac{5}{8}$ of the ship's elevation. What is your elevation?

12 Chapter Test

Check It Out
Test Practice
BigIdeasMath.com

Write the rational number as a decimal.

1. $\frac{7}{40}$ **2.** $-\frac{1}{9}$ **3.** $-\frac{21}{16}$ **4.** $\frac{36}{5}$

Write the decimal as a fraction or a mixed number in simplest form.

5. -0.122 **6.** 0.33 **7.** -4.45 **8.** -7.09

Add or subtract. Write fractions in simplest form.

9. $-\frac{4}{9} + \left(-\frac{23}{18}\right)$ **10.** $\frac{17}{12} - \left(-\frac{1}{8}\right)$ **11.** $9.2 + (-2.8)$ **12.** $2.86 - 12.1$

Multiply or divide. Write fractions in simplest form.

13. $3\frac{9}{10} \times \left(-\frac{8}{3}\right)$

14. $-1\frac{5}{6} \div 4\frac{1}{6}$

15. $-4.4 \times (-6.02)$

16. $-5 \div 1.5$

17. $-\frac{3}{5} \cdot \left(2\frac{2}{7}\right) \cdot \left(-3\frac{3}{4}\right)$

18. $-6 \cdot (-0.05) \cdot (-0.4)$

19. ALMONDS How many 2.25-pound containers can you make with 24.75 pounds of almonds?

20. FISH The elevation of a fish is -27 feet.

a. The fish decreases its elevation by 32 feet, and then increases its elevation by 14 feet. What is its new elevation?

b. Your elevation is $\frac{2}{5}$ of the fish's new elevation. What is your elevation?

21. RAINFALL The table shows the rainfall (in inches) for three months compared to the yearly average. Is the total rainfall for the three-month period greater than or less than the yearly average? Explain.

November	December	January
-0.86	2.56	-1.24

22. BANK ACCOUNTS Bank Account A has \$750.92, and Bank Account B has \$675.44. Account A changes by –\$216.38, and Account B changes by $-$\$168.49. Which account has the greater balance? Explain.

12 Standards Assessment

Test-Taking Strategy
Estimate the Answer

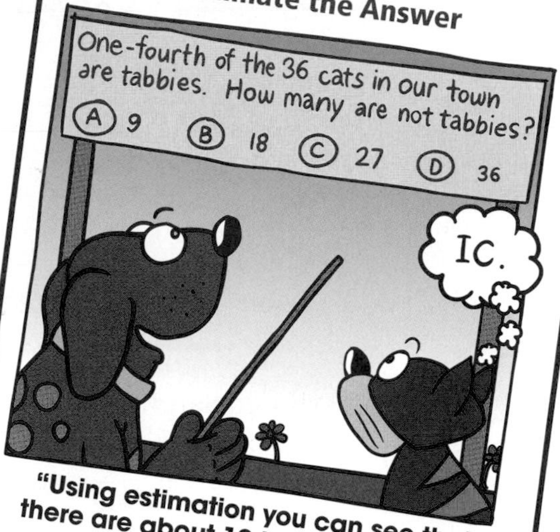

"Using estimation you can see that there are about 10 tabbies. So about 30 are not tabbies."

1. When José and Sean were each 5 years old, José was $1\frac{1}{2}$ inches taller than Sean. José grew at an average rate of $2\frac{3}{4}$ inches per year from the time that he was 5 years old until the time he was 13 years old. José was 63 inches tall when he was 13 years old. How tall was Sean when he was 5 years old? *(7.NS.3)*

A. $39\frac{1}{2}$ in.　　**C.** $44\frac{3}{4}$ in.

B. $42\frac{1}{2}$ in.　　**D.** $47\frac{3}{4}$ in.

2. Which expression represents a positive integer? *(7.NS.2a)*

F. -6^2　　**H.** $(-5)^2$

G. $(-3)^3$　　**I.** -2^3

3. What is the missing number in the sequence below? *(7.NS.2a)*

$$\frac{9}{16},\ -\frac{9}{8},\ \frac{9}{4},\ -\frac{9}{2},\ 9,\ \_\_\_\_$$

4. What is the value of the expression below? *(7.NS.1c)*

$$|-2-(-2.5)|$$

A. -4.5　　**C.** 0.5

B. -0.5　　**D.** 4.5

5. What is the distance between the two numbers on the number line? *(7.NS.1c)*

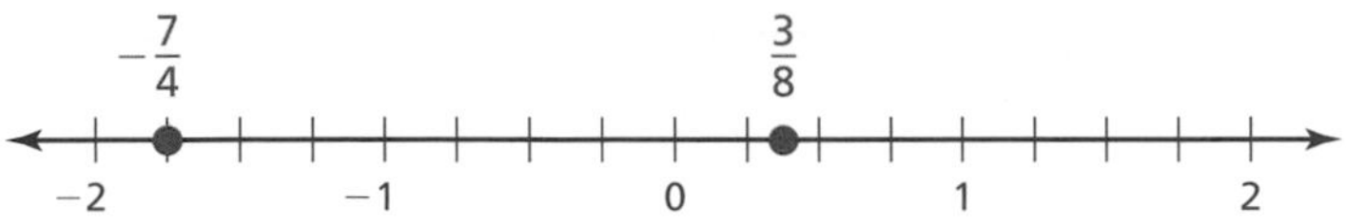

F. $-2\frac{1}{8}$　　**H.** $1\frac{3}{8}$

G. $-1\frac{3}{8}$　　**I.** $2\frac{1}{8}$

6. Sandra was evaluating an expression in the box below.

$$
\begin{aligned}
-4\frac{3}{4} \div 2\frac{1}{5} &= -\frac{19}{4} \div \frac{11}{5} \\
&= \frac{-4}{19} \bullet \frac{5}{11} \\
&= \frac{-4 \bullet 5}{19 \bullet 11} \\
&= \frac{-20}{209}
\end{aligned}
$$

What should Sandra do to correct the error that she made? *(7.NS.3)*

A. Rewrite $-\frac{19}{4}$ as $-\frac{4}{19}$ and multiply by $\frac{11}{5}$.

B. Rewrite $\frac{11}{5}$ as $\frac{5}{11}$ and multiply by $-\frac{19}{4}$.

C. Rewrite $\frac{11}{5}$ as $-\frac{5}{11}$ and multiply by $-\frac{19}{4}$.

D. Rewrite $-4\frac{3}{4}$ as $-\frac{13}{4}$ and multiply by $\frac{5}{11}$.

7. What is the value of the expression below when $q = -2$, $r = -12$, and $s = 8$? *(7.NS.3)*

$$\frac{-q^2 - r}{s}$$

F. -2

G. -1

H. 1

I. 2

8. You are stacking wooden blocks with the dimensions shown below. How many blocks do you need to stack to build a block tower that is $7\frac{1}{2}$ inches tall? *(7.NS.3)*

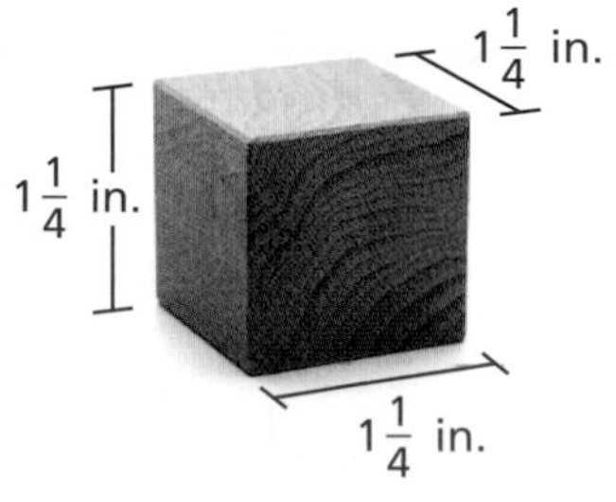

9. What is the area of a triangle with a base length of $2\frac{1}{2}$ inches and a height of 2 inches? *(7.NS.2c)*

A. $2\frac{1}{4}$ in.^{2}

B. $2\frac{1}{2}$ in.^{2}

C. $4\frac{1}{2}$ in.^{2}

D. 5 in.^{2}

10. What is the value of the expression below? *(7.NS.3)*

$$\frac{-4^2 - (-2)^3}{4}$$

F. -6

G. -2

H. 2

I. 6

11. Four points are graphed on the number line below. *(7.NS.3)*

Think Solve Explain

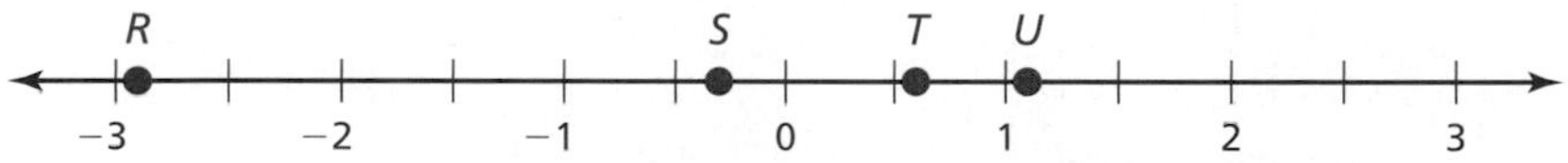

Part A Choose the two points whose values have the greatest sum. Approximate this sum. Explain your reasoning.

Part B Choose the two points whose values have the greatest difference. Approximate this difference. Explain your reasoning.

Part C Choose the two points whose values have the greatest product. Approximate this product. Explain your reasoning.

Part D Choose the two points whose values have the greatest quotient. Approximate this quotient. Explain your reasoning.

12. What number belongs in the box to make the equation true? *(7.NS.3)*

$$\frac{-0.4}{\square} + 0.8 = -1.2$$

A. -1

B. -0.2

C. 0.2

D. 1

13 Expressions and Equations

13.1 Algebraic Expressions

13.2 Adding and Subtracting Linear Expressions

13.3 Solving Equations Using Addition or Subtraction

13.4 Solving Equations Using Multiplication or Division

13.5 Solving Two-Step Equations

"I can't find my algebra tiles, so I am painting some of my dog biscuits."
I hope the paint is edible.
"Now I will be able to solve the equation 2x + (-2) = 2."

2(+ 1) = 8
2(+ 1) = 8
A three-course meal!
"Descartes, if you solve for
in the equation, what do you get?"

What You Learned Before

"Hey, Descartes ... True or False: The expressions are equivalent."

Evaluating Expressions (7.NS.3)

Example 1 **Evaluate $6x + 2y$ when $x = -3$ and $y = 5$.**

$6x + 2y = 6(-3) + 2(5)$ — Substitute −3 for *x* and 5 for *y*.

$= -18 + 10$ — Using order of operations, multiply 6 and −3, and 2 and 5.

$= -8$ — Add −18 and 10.

Example 2 **Evaluate $6x^2 - 3(y + 2) + 8$ when $x = -2$ and $y = 4$.**

$6x^2 - 3(y + 2) + 8 = 6(-2)^2 - 3(4 + 2) + 8$ — Substitute −2 for *x* and 4 for *y*.

$= 6(-2)^2 - 3(6) + 8$ — Using order of operations, evaluate within the parentheses.

$= 6(4) - 3(6) + 8$ — Using order of operations, evaluate the exponent.

$= 24 - 18 + 8$ — Using order of operations, multiply 6 and 4, and 3 and 6.

$= 14$ — Subtract 18 from 24. Add the result to 8.

Try It Yourself

Evaluate the expression when $x = -\frac{1}{4}$ and $y = 3$.

1. $2xy$ **2.** $12x - 3y$ **3.** $-4x - y + 4$ **4.** $8x - y^2 - 3$

Writing Algebraic Expressions (6.EE.2a)

Example 3 **Write the phrase as an algebraic expression.**

a. the sum of twice a number *m* and four

$2m + 4$

b. eight less than three times a number *x*

$3x - 8$

Try It Yourself

Write the phrase as an algebraic expression.

5. five more than three times a number *q*

6. nine less than a number *n*

7. the product of a number *p* and six

8. the quotient of eight and a number *h*

9. four more than three times a number *t*

10. two less than seven times a number *c*

13.1 Algebraic Expressions

Essential Question How can you simplify an algebraic expression?

1 ACTIVITY: Simplifying Algebraic Expressions

Work with a partner.

a. Evaluate each algebraic expression when $x = 0$ and when $x = 1$. Use the results to match each expression in the left table with its equivalent expression in the right table.

	Expression	Value When $x = 0$	Value When $x = 1$
A.	$3x + 2 - x + 4$		
B.	$5(x - 3) + 2$		
C.	$x + 3 - (2x + 1)$		
D.	$-4x + 2 - x + 3x$		
E.	$-(1 - x) + 3$		
F.	$2x + x - 3x + 4$		
G.	$4 - 3 + 2(x - 1)$		
H.	$2(1 - x + 4)$		
I.	$5 - (4 - x + 2x)$		
J.	$5x - (2x + 4 - x)$		

	Expression	Value When $x = 0$	Value When $x = 1$
a.	4		
b.	$-x + 1$		
c.	$4x - 4$		
d.	$2x + 6$		
e.	$5x - 13$		
f.	$-2x + 10$		
g.	$x + 2$		
h.	$2x - 1$		
i.	$-2x + 2$		
j.	$-x + 2$		

b. Compare each expression in the left table with its equivalent expression in the right table. In general, how do you think you obtain the equivalent expression in the right column?

Common Core

Algebraic Expressions

In this lesson, you will

- apply properties of operations to simplify algebraic expressions.
- solve real-life problems.

Learning Standards
7.EE.1
7.EE.2

2 ACTIVITY: Writing a Math Lesson

Math Practice 6

Communicate Precisely

What can you do to make sure that you are communicating exactly what is needed in the Key Idea?

Work with a partner. Use your results from Activity 1 to write a lesson on simplifying an algebraic expression.

Simplifying an Algebraic Expression

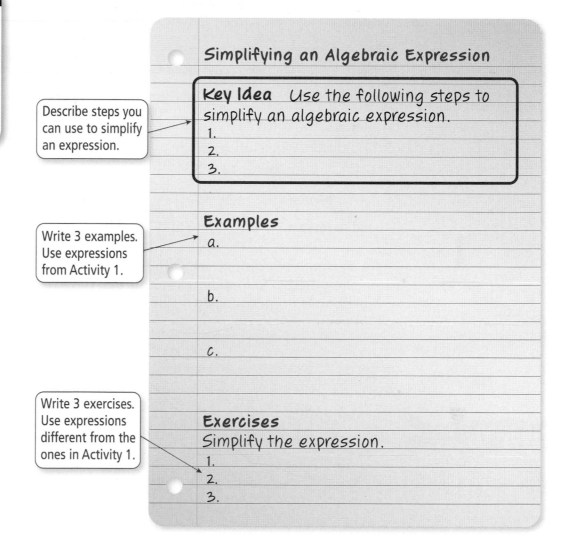

Describe steps you can use to simplify an expression.

Key Idea Use the following steps to simplify an algebraic expression.

1.
2.
3.

Write 3 examples. Use expressions from Activity 1.

Examples

a.

b.

c.

Write 3 exercises. Use expressions different from the ones in Activity 1.

Exercises

Simplify the expression.

1.
2.
3.

What Is Your Answer?

3. **IN YOUR OWN WORDS** How can you simplify an algebraic expression? Give an example that demonstrates your procedure.
4. **REASONING** Why would you want to simplify an algebraic expression? Discuss several reasons.

Practice Use what you learned about simplifying algebraic expressions to complete Exercises 12–14 on page 558.

13.1 Lesson

Check It Out
Lesson Tutorials
BigIdeasMath.com

Key Vocabulary
like terms, *p. 556*
simplest form, *p. 556*

Parts of an algebraic expression are called *terms*. **Like terms** are terms that have the same variables raised to the same exponents. Constant terms are also like terms. To identify terms and like terms in an expression, first write the expression as a sum of its terms.

EXAMPLE 1 Identifying Terms and Like Terms

Identify the terms and like terms in each expression.

a. $9x - 2 + 7 - x$

Rewrite as a sum of terms.

$$\underbrace{9x}+\underbrace{(-2)}+\underbrace{7}+\underbrace{(-x)}$$

Terms: $9x, \quad -2, \quad 7, \quad -x$

Like terms: $9x$ and $-x$, -2 and 7

b. $z^2 + 5z - 3z^2 + z$

Rewrite as a sum of terms.

$$\underbrace{z^2}+\underbrace{5z}+\underbrace{(-3z^2)}+\underbrace{z}$$

Terms: $z^2, \quad 5z, \quad -3z^2, \quad z$

Like terms: z^2 and $-3z^2$, $5z$ and z

An algebraic expression is in **simplest form** when it has no like terms and no parentheses. To *combine* like terms that have variables, use the Distributive Property to add or subtract the coefficients.

EXAMPLE 2 Simplifying an Algebraic Expression

Study Tip
To subtract a variable term, add the term with the opposite coefficient.

Simplify $\frac{3}{4}y + 12 - \frac{1}{2}y - 6$.

$$\frac{3}{4}y + 12 - \frac{1}{2}y - 6 = \frac{3}{4}y + 12 + \left(-\frac{1}{2}y\right) + (-6) \quad \text{Rewrite as a sum.}$$

$$= \frac{3}{4}y + \left(-\frac{1}{2}y\right) + 12 + (-6) \quad \text{Commutative Property of Addition}$$

$$= \left[\frac{3}{4} + \left(-\frac{1}{2}\right)\right]y + 12 + (-6) \quad \text{Distributive Property}$$

$$= \frac{1}{4}y + 6 \quad \text{Combine like terms.}$$

On Your Own

Now You're Ready
Exercises 5–10 and 12–17

Identify the terms and like terms in the expression.

1. $y + 10 - \frac{3}{2}y$
2. $2r^2 + 7r - r^2 - 9$
3. $7 + 4p - 5 + p + 2q$

Simplify the expression.

4. $14 - 3z + 8 + z$
5. $2.5x + 4.3x - 5$
6. $\frac{3}{8}b - \frac{3}{4}b$

Multi-Language Glossary at BigIdeasMath.com

EXAMPLE 3 Simplifying an Algebraic Expression

Simplify $-\frac{1}{2}(6n + 4) + 2n$.

$-\frac{1}{2}(6n + 4) + 2n = -\frac{1}{2}(6n) + \left(-\frac{1}{2}\right)(4) + 2n$	Distributive Property
$= -3n + (-2) + 2n$	Multiply.
$= -3n + 2n + (-2)$	Commutative Property of Addition
$= (-3 + 2)n + (-2)$	Distributive Property
$= -n - 2$	Simplify.

On Your Own

Now You're Ready
Exercises 18–20

Simplify the expression.

7. $3(q + 1) - 4$
8. $-2(g + 4) + 7g$
9. $7 - 4\left(\frac{3}{4}x - \frac{1}{4}\right)$

EXAMPLE 4 Real-Life Application

Each person in a group buys a ticket, a medium drink, and a large popcorn. Write an expression in simplest form that represents the amount of money the group spends at the movies. Interpret the expression.

Words Each ticket is $7.50, each medium drink is $2.75, and each large popcorn is $4.

Variable The same number of each item is purchased. So, x can represent the number of tickets, the number of medium drinks, and the number of large popcorns.

Expression $7.50x \quad + \quad 2.75x \quad + \quad 4x$

$7.50x + 2.75x + 4x = (7.50 + 2.75 + 4)x$	Distributive Property
$= 14.25x$	Add coefficients.

The expression $14.25x$ indicates that the total cost per person is $14.25.

On Your Own

10. **WHAT IF?** Each person buys a ticket, a large drink, and a small popcorn. How does the expression change? Explain.

13.1 Exercises

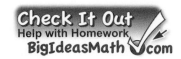

Check It Out
Help with Homework
BigIdeasMath.com

Vocabulary and Concept Check

1. **WRITING** Explain how to identify the terms of $3y - 4 - 5y$.
2. **WRITING** Describe how to combine like terms in the expression $3n + 4n - 2$.
3. **VOCABULARY** Is the expression $3x + 2x - 4$ in simplest form? Explain.
4. **REASONING** Which algebraic expression is in simplest form? Explain.

$5x - 4 + 6y$	$4x + 8 - x$
$3(7 + y)$	$12n - n$

Practice and Problem Solving

Identify the terms and like terms in the expression.

1 5. $t + 8 + 3t$

6. $3z + 4 + 2 + 4z$

7. $2n - n - 4 + 7n$

8. $-x - 9x^2 + 12x^2 + 7$

9. $1.4y + 5 - 4.2 - 5y^2 + z$

10. $\frac{1}{2}s - 4 + \frac{3}{4}s + \frac{1}{8} - s^3$

11. **ERROR ANALYSIS** Describe and correct the error in identifying the like terms in the expression.

> ✗ $3x - 5 + 2x^2 + 9x = 3x + 2x^2 + 9x - 5$
>
> Like Terms: $3x$, $2x^2$, and $9x$

Simplify the expression.

2 12. $12g + 9g$

13. $11x + 9 - 7$

14. $8s - 11s + 6$

15. $4.2v - 5 - 6.5v$

16. $8 + 4a + 6.2 - 9a$

17. $\frac{2}{5}y - 4 + 7 - \frac{9}{10}y$

3 18. $4(b - 6) + 19$

19. $4p - 5(p + 6)$

20. $-\frac{2}{3}(12c - 9) + 14c$

21. **HIKING** On a hike, each hiker carries the items shown. Write an expression in simplest form that represents the weight carried by x hikers. Interpret the expression.

22. **STRUCTURE** Evaluate the expression $-8x + 5 - 2x - 4 + 5x$ when $x = 2$ before and after simplifying. Which method do you prefer? Explain.

23. **REASONING** Are the expressions $8x^2 + 3(x^2 + y)$ and $7x^2 + 7y + 4x^2 - 4y$ equivalent? Explain your reasoning.

24. **CRITICAL THINKING** Which solution shows a correct way of simplifying $6 - 4(2 - 5x)$? Explain the errors made in the other solutions.

(A) $6 - 4(2 - 5x) = 6 - 4(-3x) = 6 + 12x$

(B) $6 - 4(2 - 5x) = 6 - 8 + 20x = -2 + 20x$

(C) $6 - 4(2 - 5x) = 2(2 - 5x) = 4 - 10x$

(D) $6 - 4(2 - 5x) = 6 - 8 - 20x = -2 - 20x$

25. **BANNER** Write an expression in simplest form that represents the area of the banner.

26. **CAR WASH** Write an expression in simplest form that represents the earnings for washing and waxing x cars and y trucks.

	Car	Truck
Wash	\$8	\$10
Wax	\$12	\$15

MODELING Draw a diagram that shows how the expression can represent the area of a figure. Then simplify the expression.

27. $5(2 + x + 3)$

28. $(4 + 1)(x + 2x)$

29. Critical Thinking You apply gold foil to a piece of red poster board to make the design shown.

   a. Write an expression in simplest form that represents the area of the gold foil.

   b. Find the area of the gold foil when $x = 3$.

   c. The pattern at the right is called "St. George's Cross." Find a country that uses this pattern as its flag.

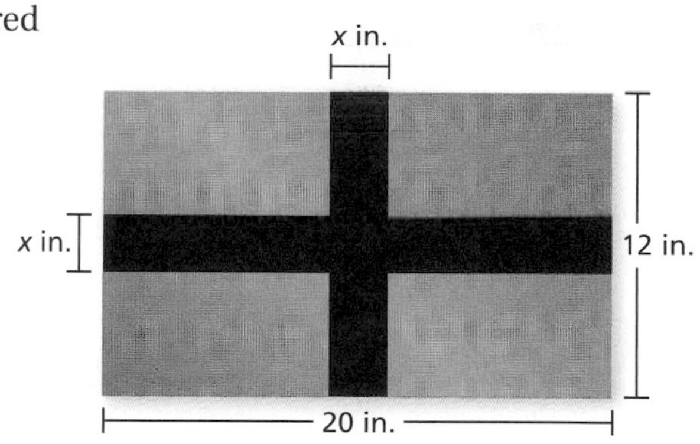

Fair Game Review What you learned in previous grades & lessons

Order the lengths from least to greatest. *(Skills Review Handbook)*

30. 15 in., 14.8 in., 15.8 in., 14.5 in., 15.3 in.

31. 0.65 m, 0.6 m, 0.52 m, 0.55 m, 0.545 m

32. **MULTIPLE CHOICE** A bird's nest is 12 feet above the ground. A mole's den is 12 inches below the ground. What is the difference in height of these two positions? *(Section 11.3)*

(A) 24 in. (B) 11 ft (C) 13 ft (D) 24 ft

13.2 Adding and Subtracting Linear Expressions

Essential Question How can you use algebra tiles to add or subtract algebraic expressions?

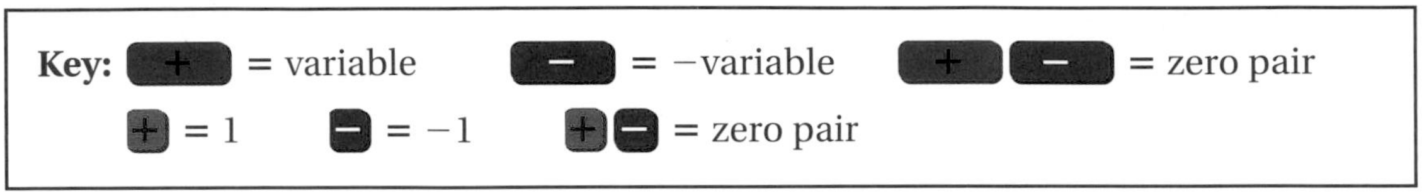

1 ACTIVITY: Writing Algebraic Expressions

Work with a partner. Write an algebraic expression shown by the algebra tiles.

a. 

b. 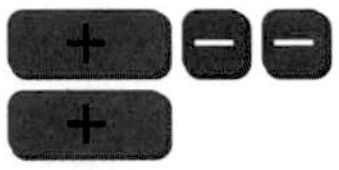

c. 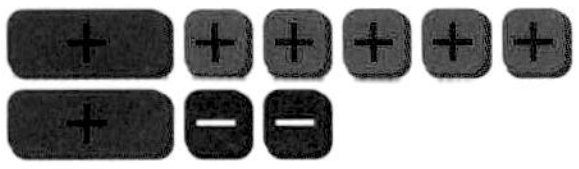

d. 

2 ACTIVITY: Adding Algebraic Expressions

Work with a partner. Write the sum of two algebraic expressions modeled by the algebra tiles. Then use the algebra tiles to simplify the expression.

a. 

b. 

c. 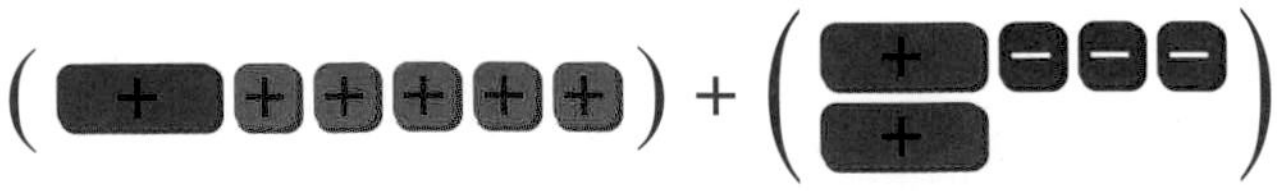

d. 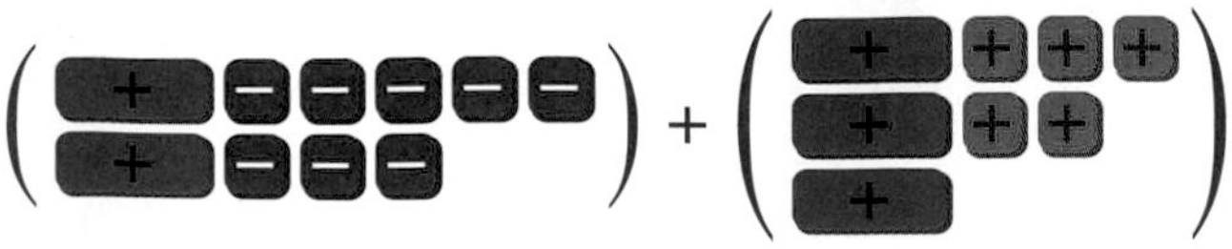

Common Core

Linear Expressions

In this lesson, you will

- apply properties of operations to add and subtract linear expressions.
- solve real-life problems.

Learning Standards
7.EE.1
7.EE.2

3 ACTIVITY: Subtracting Algebraic Expressions

Math Practice 2
Use Expressions
What do the tiles represent? How does this help you write an expression?

Work with a partner. Write the difference of two algebraic expressions modeled by the algebra tiles. Then use the algebra tiles to simplify the expression.

a. 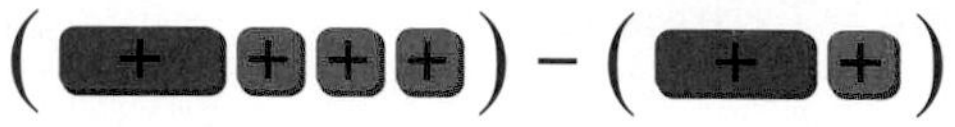

b. 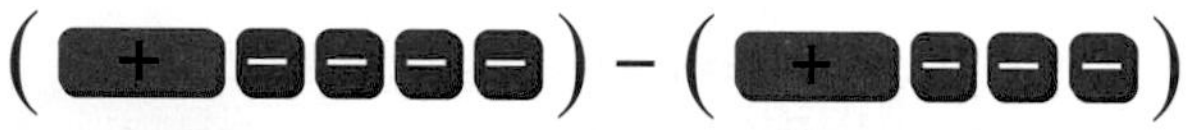

c. 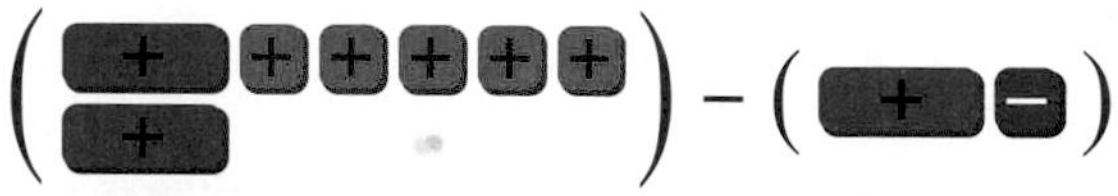

d. 

4 ACTIVITY: Adding and Subtracting Algebraic Expressions

Work with a partner. Use algebra tiles to model the sum or difference. Then use the algebra tiles to simplify the expression.

a. $(2x + 1) + (x - 1)$

b. $(2x - 6) + (3x + 2)$

c. $(2x + 4) - (x + 2)$

d. $(4x + 3) - (2x - 1)$

What Is Your Answer?

5. **IN YOUR OWN WORDS** How can you use algebra tiles to add or subtract algebraic expressions?

6. Write the difference of two algebraic expressions modeled by the algebra tiles. Then use the algebra tiles to simplify the expression.

Practice Use what you learned about adding and subtracting algebraic expressions to complete Exercises 6 and 7 on page 564.

13.2 Lesson

Check It Out
Lesson Tutorials
BigIdeasMath.com

Key Vocabulary
linear expression, p. 562

A **linear expression** is an algebraic expression in which the exponent of the variable is 1.

Linear Expressions	$-4x$	$3x + 5$	$5 - \frac{1}{6}x$
Nonlinear Expressions	x^2	$-7x^3 + x$	$x^5 + 1$

You can use a vertical or a horizontal method to add linear expressions.

EXAMPLE 1 Adding Linear Expressions

Find each sum.

a. $(x - 2) + (3x + 8)$

Vertical method: Align like terms vertically and add.

$$\begin{array}{r} x - 2 \\ +\ 3x + 8 \\ \hline 4x + 6 \end{array}$$

b. $(-4y + 3) + (11y - 5)$

Horizontal method: Use properties of operations to group like terms and simplify.

$(-4y + 3) + (11y - 5) = -4y + 3 + 11y - 5$ — Rewrite the sum.

$= -4y + 11y + 3 - 5$ — Commutative Property of Addition

$= (-4y + 11y) + (3 - 5)$ — Group like terms.

$= 7y - 2$ — Combine like terms.

EXAMPLE 2 Adding Linear Expressions

Find $2(-7.5z + 3) + (5z - 2)$.

$2(-7.5z + 3) + (5z - 2) = -15z + 6 + 5z - 2$ — Distributive Property

$= -15z + 5z + 6 - 2$ — Commutative Property of Addition

$= -10z + 4$ — Combine like terms.

On Your Own

Now You're Ready
Exercises 8–16

Find the sum.

1. $(x + 3) + (2x - 1)$
2. $(-8z + 4) + (8z - 7)$
3. $(4 - n) + 2(-5n + 3)$
4. $\frac{1}{2}(w - 6) + \frac{1}{4}(w + 12)$

Multi-Language Glossary at BigIdeasMath.com

To subtract one linear expression from another, add the opposite of each term in the expression. You can use a vertical or a horizontal method.

EXAMPLE 3 Subtracting Linear Expressions

Study Tip

To find the opposite of a linear expression, you can multiply the expression by −1.

Find each difference.

a. $(5x + 6) - (-x + 6)$ **b.** $(7y + 5) - 2(4y - 3)$

a. Vertical method: Align like terms vertically and subtract.

$$\begin{array}{r} (5x + 6) \\ -\ (-x + 6) \\ \hline \end{array} \quad \xrightarrow{\text{Add the opposite.}} \quad \begin{array}{r} 5x + 6 \\ +\quad x - 6 \\ \hline 6x \end{array}$$

b. Horizontal method: Use properties of operations to group like terms and simplify.

$(7y + 5) - 2(4y - 3) = 7y + 5 - 8y + 6$	Distributive Property
$= 7y - 8y + 5 + 6$	Commutative Property of Addition
$= (7y - 8y) + (5 + 6)$	Group like terms.
$= -y + 11$	Combine like terms.

EXAMPLE 4 Real-Life Application

The original price of a cowboy hat is d dollars. You use a coupon and buy the hat for $(d - 2)$ dollars. You decorate the hat and sell it for $(2d - 4)$ dollars. Write an expression that represents your earnings from buying and selling the hat. Interpret the expression.

earnings = selling price − purchase price	Use a model.
$= (2d - 4) - (d - 2)$	Write the difference.
$= (2d - 4) + (-d + 2)$	Add the opposite.
$= 2d - d - 4 + 2$	Group like terms.
$= d - 2$	Combine like terms.

∴ You earn $(d - 2)$ dollars. You also paid $(d - 2)$ dollars, so you doubled your money by selling the hat for twice as much as you paid for it.

On Your Own

Now You're Ready
Exercises 19–24

Find the difference.

5. $(m - 3) - (-m + 12)$ **6.** $-2(c + 2.5) - 5(1.2c + 4)$

7. WHAT IF? In Example 4, you sell the hat for $(d + 2)$ dollars. How much do you earn from buying and selling the hat?

13.2 Exercises

Check It Out
Help with Homework
BigIdeasMath.com

Vocabulary and Concept Check

VOCABULARY **Determine whether the algebraic expression is a linear expression. Explain.**

1. $x^2 + x + 1$
2. $-2x - 8$
3. $x - x^4$
4. **WRITING** Describe two methods for adding or subtracting linear expressions.
5. **DIFFERENT WORDS, SAME QUESTION** Which is different? Find "both" answers.

Subtract x from $3x - 1$.	Find $3x - 1$ decreased by x.
What is x more than $3x - 1$?	What is the difference of $3x - 1$ and x?

Practice and Problem Solving

Write the sum or difference of two algebraic expressions modeled by the algebra tiles. Then use the algebra tiles to simplify the expression.

6. 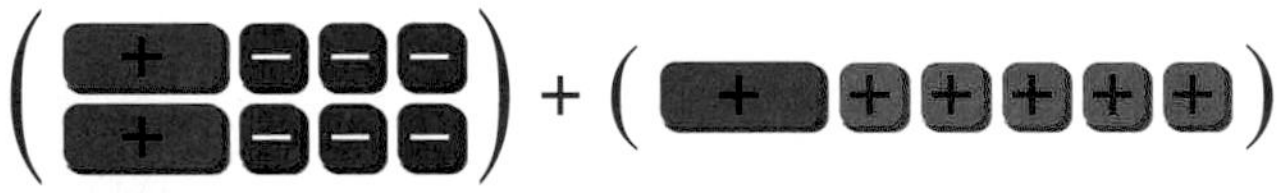

7. 

Find the sum.

1 2 8. $(n + 8) + (n - 12)$
9. $(7 - b) + (3b + 2)$
10. $(2w - 9) + (-4w - 5)$
11. $(2x - 6) + 4(x - 3)$
12. $5(-3.4k - 7) + (3k + 21)$
13. $(1 - 5q) + 2(2.5q + 8)$
14. $3(2 - 0.9h) + (-1.3h - 4)$
15. $\frac{1}{3}(9 - 6m) + \frac{1}{4}(12m - 8)$
16. $-\frac{1}{2}(7z + 4) + \frac{1}{5}(5z - 15)$
17. **BANKING** You start a new job. After w weeks, you have $(10w + 120)$ dollars in your savings account and $(45w + 25)$ dollars in your checking account. Write an expression that represents the total in both accounts.
18. **FIREFLIES** While catching fireflies, you and a friend decide to have a competition. After m minutes, you have $(3m + 13)$ fireflies and your friend has $(4m + 6)$ fireflies.
    a. Write an expression that represents the number of fireflies you and your friend caught together.
    b. The competition ends after 5 minutes. Who has more fireflies?

Find the difference.

3 19. $(-2g + 7) - (g + 11)$ 20. $(6d + 5) - (2 - 3d)$ 21. $(4 - 5y) - 2(3.5y - 8)$

22. $(2n - 9) - 5(-2.4n + 4)$ 23. $\frac{1}{8}(-8c + 16) - \frac{1}{3}(6 + 3c)$ 24. $\frac{3}{4}(3x + 6) - \frac{1}{4}(5x - 24)$

25. **ERROR ANALYSIS** Describe and correct the error in finding the difference.

✗
$$(4m + 9) - 3(2m - 5) = 4m + 9 - 6m - 15$$
$$= 4m - 6m + 9 - 15$$
$$= -2m - 6$$

26. **STRUCTURE** Refer to the expressions in Exercise 18.

a. How many fireflies are caught each minute during the competition?

b. How many fireflies are caught before the competition starts?

27. **LOGIC** Your friend says the sum of two linear expressions is always a linear expression. Is your friend correct? Explain.

28. **GEOMETRY** The expression $17n + 11$ represents the perimeter (in feet) of the triangle. Write an expression that represents the measure of the third side.

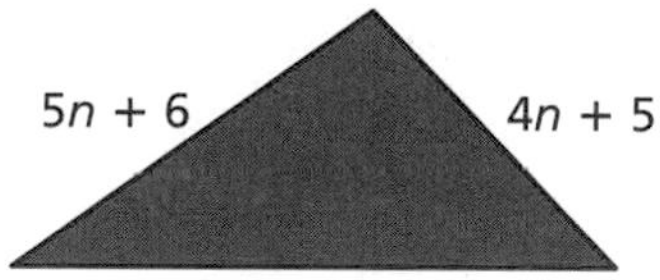

29. **TAXI** Taxi Express charges \$2.60 plus \$3.65 per mile, and Cab Cruiser charges \$2.75 plus \$3.90 per mile. Write an expression that represents how much more Cab Cruiser charges than Taxi Express.

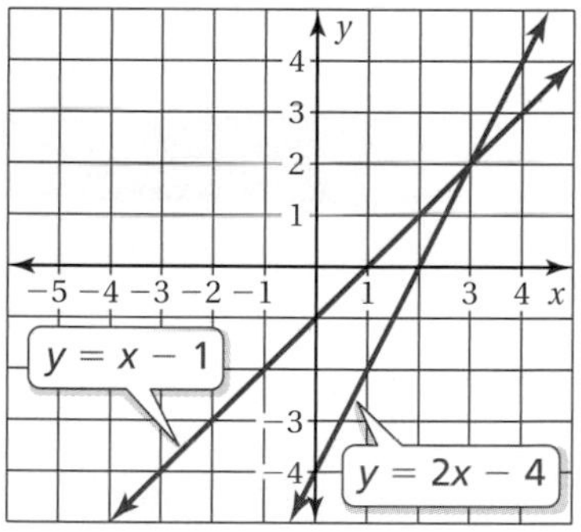

30. **MODELING** A rectangular room is 10 feet longer than it is wide. One-foot-by-one-foot tiles cover the entire floor. Write an expression that represents the number of tiles along the outside of the room.

31. **Reasoning** Write an expression in simplest form that represents the vertical distance between the two lines shown. What is the distance when $x = 3$? when $x = -3$?

Fair Game Review What you learned in previous grades & lessons

Evaluate the expression when $x = -\frac{4}{5}$ and $y = \frac{1}{3}$. *(Section 12.2)*

32. $x + y$ 33. $2x + 6y$ 34. $-x + 4y$

35. **MULTIPLE CHOICE** What is the surface area of a cube that has a side length of 5 feet? *(Section 8.2)*

Ⓐ 25 ft^2 Ⓑ 75 ft^2 Ⓒ 125 ft^2 Ⓓ 150 ft^2

Extension 13.2 Factoring Expressions

Check It Out Lesson Tutorials BigIdeasMath.com

Key Vocabulary
factoring an expression, *p. 566*

When **factoring an expression**, you write the expression as a product of factors. You can use the Distributive Property to factor expressions.

EXAMPLE 1 Factoring Out the GCF

Factor $24x - 18$ using the GCF.

Find the GCF of $24x$ and 18 by writing their prime factorizations.

$$24x = 2 \cdot 2 \cdot 2 \cdot 3 \cdot x$$
$$18 = 2 \cdot 3 \cdot 3$$

Circle the common prime factors.

So, the GCF of $24x$ and 18 is $2 \cdot 3 = 6$. Use the GCF to factor the expression.

$$24x - 18 = 6(4x) - 6(3)$$ Rewrite using GCF.

$$= 6(4x - 3)$$ Distributive Property

So, $24x - 18 = 6(4x - 3)$.

You can also use the Distributive Property to factor out any rational number from an expression.

EXAMPLE 2 Factoring Out a Fraction

Factor $\frac{1}{2}$ out of $\frac{1}{2}x + \frac{3}{2}$.

Write each term as a product of $\frac{1}{2}$ and another factor.

$$\frac{1}{2}x = \frac{1}{2} \cdot x$$ Think: $\frac{1}{2}x$ is $\frac{1}{2}$ times what?

$$\frac{3}{2} = \frac{1}{2} \cdot 3$$ Think: $\frac{3}{2}$ is $\frac{1}{2}$ times what?

Use the Distributive Property to factor out $\frac{1}{2}$.

$$\frac{1}{2}x + \frac{3}{2} = \frac{1}{2} \cdot x + \frac{1}{2} \cdot 3$$ Rewrite the expression.

$$= \frac{1}{2}(x + 3)$$ Distributive Property

So, $\frac{1}{2}x + \frac{3}{2} = \frac{1}{2}(x + 3)$.

Common Core
Linear Expressions
In this extension, you will
- factor linear expressions.

Learning Standard
7.EE.1

EXAMPLE 3 Factoring Out a Negative Number

Factor -2 out of $-4p + 10$.

Write each term as a product of -2 and another factor.

$-4p = -2 \cdot 2p$ — Think: $-4p$ is -2 times what?

$10 = -2 \cdot (-5)$ — Think: 10 is -2 times what?

Use the Distributive Property to factor out -2.

$-4p + 10 = -2 \cdot 2p + (-2) \cdot (-5)$ — Rewrite the expression.

$= -2[2p + (-5)]$ — Distributive Property

$= -2(2p - 5)$ — Simplify.

So, $-4p + 10 = -2(2p - 5)$.

Math Practice 7

View as Components

How does rewriting each term as a product help you see the common factor?

Practice

Factor the expression using the GCF.

1. $9 + 21$
2. $32 - 48$
3. $8x + 2$
4. $3y - 24$
5. $20z - 8$
6. $15w + 65$
7. $36a + 16b$
8. $21m - 49n$

Factor out the coefficient of the variable.

9. $\frac{1}{3}b - \frac{1}{3}$
10. $\frac{3}{8}d + \frac{3}{4}$
11. $2.2x + 4.4$
12. $4h - 3$

13. Factor $-\frac{1}{2}$ out of $-\frac{1}{2}x + 6$.

14. Factor $-\frac{1}{4}$ out of $-\frac{1}{2}x - \frac{5}{4}y$.

15. **WRESTLING** A square wrestling mat has a perimeter of $(12x - 32)$ feet. Write an expression that represents the side length of the mat (in feet).

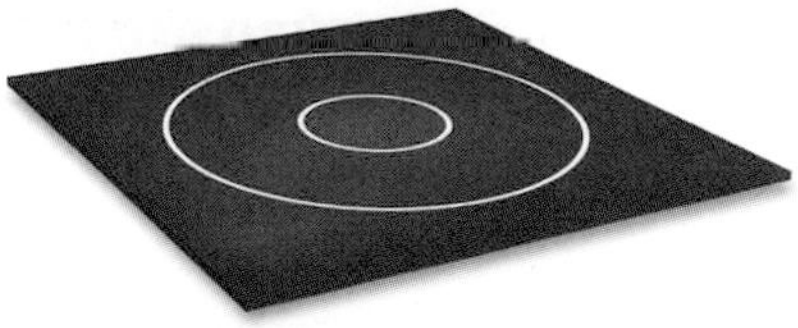

16. **MAKING A DIAGRAM** A table is 6 feet long and 3 feet wide. You extend the table by inserting two identical table *leaves*. The longest side length of each rectangular leaf is 3 feet. The extended table is rectangular with an area of $(18 + 6x)$ square feet.
    a. Make a diagram of the table and leaves.
    b. Write an expression that represents the length of the extended table. What does x represent?

17. **STRUCTURE** The area of the trapezoid is $\left(\frac{3}{4}x - \frac{1}{4}\right)$ square centimeters. Write two different pairs of expressions that represent possible lengths of the bases.

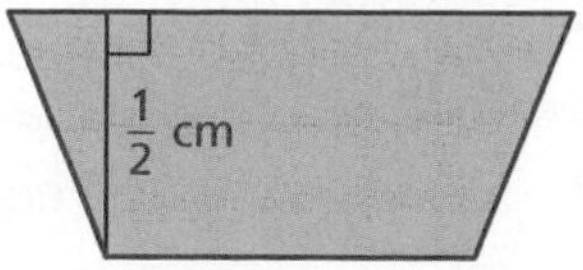

13 Study Help

Check It Out
Graphic Organizer
BigIdeasMath.com

You can use a **four square** to organize information about a topic. Each of the four squares can be a category, such as *definition, vocabulary, example, non-example, words, algebra, table, numbers, visual, graph,* or *equation*. Here is an example of a four square for like terms.

Definition Terms that have the same variables raised to the same exponents	Examples 2 and −3, 3x and −7x, x^2 and $6x^2$
Words To *combine* like terms that have variables, use the Distributive Property to add or subtract the coefficients.	Non-Examples y and 4, 3x and −4y, $6x^2$ and 2x

Like Terms

On Your Own

Make four squares to help you study these topics.

1. simplest form
2. linear expression
3. factoring expressions

After you complete this chapter, make four squares for the following topics.

4. equivalent equations
5. solving equations using addition or subtraction
6. solving equations using multiplication or division
7. solving two-step equations

"My four square shows that my new red skateboard is faster than my old blue skateboard."

13.1–13.2 Quiz

Check It Out
Progress Check
BigIdeasMath.com

Identify the terms and like terms in the expression. *(Section 13.1)*

1. $11x + 2x$

2. $9x - 5x$

3. $21x + 6 - x - 5$

4. $8x + 14 - 3x + 1$

Simplify the expression. *(Section 13.1)*

5. $2(3x + x)$

6. $-7 + 3x + 4x$

7. $2x + 4 - 3x + 2 + 3x$

8. $7x + 6 + 3x - 2 - 5x$

Find the sum or difference. *(Section 13.2)*

9. $(s + 12) + (3s - 8)$

10. $(9t + 5) + (3t - 6)$

11. $(2 - k) + 3(-4k + 2)$

12. $\frac{1}{4}(q - 12) + \frac{1}{3}(q + 9)$

13. $(n - 8) - (-2n + 2)$

14. $-3(h - 4) - 2(-6h + 5)$

Factor out the coefficient of the variable. *(Section 13.2)*

15. $5c - 15$

16. $\frac{2}{9}j + \frac{2}{3}$

17. $2.4n + 9.6$

18. $-6z + 12$

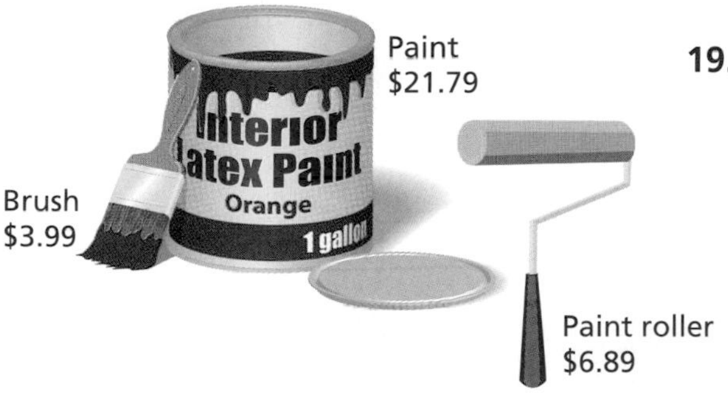

19. PAINTING You buy the same number of brushes, rollers, and paint cans. Write an expression in simplest form that represents the total amount of money you spend for painting supplies. *(Section 13.1)*

20. APPLES A basket holds n apples. You pick $2n - 3$ apples, and your friend picks $n + 4$ apples. Write an expression that represents the number of apples you and your friend picked. Interpret the expression. *(Section 13.2)*

21. EXERCISE Write an expression in simplest form for the perimeter of the exercise mat. *(Section 13.1)*

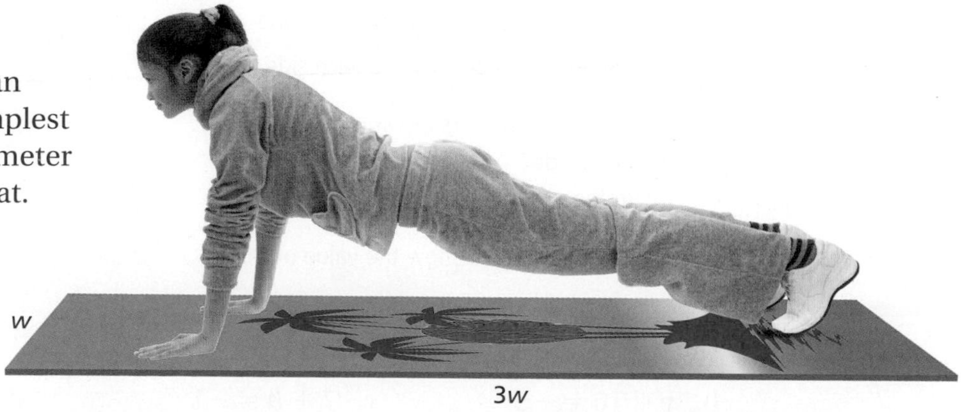

13.3 Solving Equations Using Addition or Subtraction

Essential Question How can you use algebra tiles to solve addition or subtraction equations?

1 ACTIVITY: Solving Equations

Work with a partner. Use algebra tiles to model and solve the equation.

a. $x - 3 = -4$

Model the equation $x - 3 = -4$.

To get the variable tile by itself, remove the ____ tiles on the left side by adding ____ ____ tiles to each side.

How many *zero pairs* can you remove from each side? ____
Circle them.

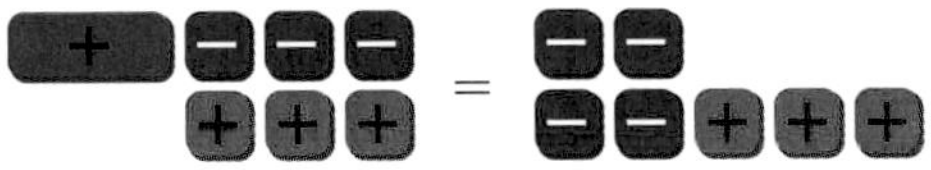

The remaining tile shows the value of x.

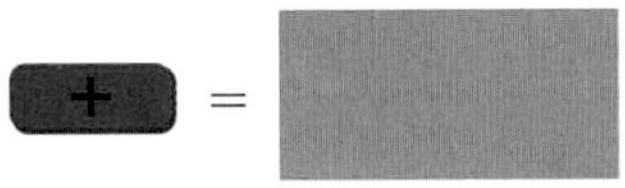

So, $x =$ ____.

b. $z - 6 = 2$ **c.** $p - 7 = -3$ **d.** $-15 = t - 5$

2 ACTIVITY: Solving Equations

Work with a partner. Use algebra tiles to model and solve the equation.

a. $-5 = n + 2$

Model the equation $-5 = n + 2$.

Remove the ____ tiles on the right side by adding ____ ____ tiles to each side.

How many *zero pairs* can you remove from the right side? ____
Circle them.

The remaining tiles show the value of n.

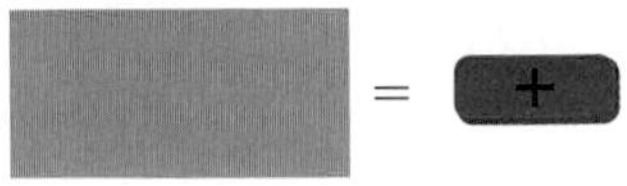

So, $n =$ ____.

b. $y + 10 = -5$ **c.** $7 + b = -1$ **d.** $8 = 12 + z$

COMMON CORE

Solving Equations

In this lesson, you will

- write simple equations.
- solve equations using addition or subtraction.
- solve real-life problems.

Learning Standard 7.EE.4a

Math Practice 4

Interpret Results

How can you add tiles to make zero pairs? Explain how this helps you solve the equation.

3 ACTIVITY: Writing and Solving Equations

Work with a partner. Write an equation shown by the algebra tiles. Then solve.

a. 

b. 

c. 

d. 

4 ACTIVITY: Using a Different Method to Find a Solution

Work with a partner. The *melting point* of a solid is the temperature at which the solid melts to become a liquid. The melting point of the element bromine is about 19°F. This is about 57°F more than the melting point of mercury.

a. Which of the following equations can you use to find the melting point of mercury? What is the melting point of mercury?

$x + 57 = 19$ $x - 57 = 19$ $x + 19 = 57$ $x + 19 = -57$

b. **CHOOSE TOOLS** How can you solve this problem without using an equation? Explain. How are these two methods related?

What Is Your Answer?

5. **IN YOUR OWN WORDS** How can you use algebra tiles to solve addition or subtraction equations? Give an example of each.

6. **STRUCTURE** Explain how you could use inverse operations to solve addition or subtraction equations without using algebra tiles.

7. What makes the cartoon funny?

"To vary or not to vary." That is the question.

"Dear Sir: Yesterday you said $x = 2$. Today you are saying $x = 3$. Please make up your mind."

8. The word *variable* comes from the word *vary*. For example, the temperature in Maine varies a lot from winter to summer.

   Write two other English sentences that use the word *vary*.

Practice Use what you learned about solving addition or subtraction equations to complete Exercises 5–8 on page 574.

13.3 Lesson

Check It Out
Lesson Tutorials
BigIdeasMath.com

Key Vocabulary
equivalent equations, *p. 572*

Two equations are **equivalent equations** if they have the same solutions. The Addition and Subtraction Properties of Equality can be used to write equivalent equations.

Key Ideas

Addition Property of Equality

Words Adding the same number to each side of an equation produces an equivalent equation.

Algebra If $a = b$, then $a + c = b + c$.

Subtraction Property of Equality

Words Subtracting the same number from each side of an equation produces an equivalent equation.

Algebra If $a = b$, then $a - c = b - c$.

Remember

Addition and subtraction are inverse operations.

EXAMPLE 1 Solving Equations

a. Solve $x - 5 = -1$.

$x - 5 = -1$ Write the equation.

Undo the subtraction. → $\underline{+5}\quad\underline{+5}$ Addition Property of Equality

$x = 4$ Simplify.

The solution is $x = 4$.

Check

$x - 5 = -1$

$4 - 5 \stackrel{?}{=} -1$

$-1 = -1$ ✓

b. Solve $z + \frac{3}{2} = \frac{1}{2}$.

$z + \frac{3}{2} = \frac{1}{2}$ Write the equation.

Undo the addition. → $\underline{-\frac{3}{2}}\quad\underline{-\frac{3}{2}}$ Subtraction Property of Equality

$z = -1$ Simplify.

The solution is $z = -1$.

Check

$z + \frac{3}{2} = \frac{1}{2}$

$-1 + \frac{3}{2} \stackrel{?}{=} \frac{1}{2}$

$\frac{1}{2} = \frac{1}{2}$ ✓

On Your Own

Now You're Ready
Exercises 5–20

Solve the equation. Check your solution.

1. $p - 5 = -2$
2. $w + 13.2 = 10.4$
3. $x - \frac{5}{6} = -\frac{1}{6}$

Multi-Language Glossary at BigIdeasMath.com

EXAMPLE 2 Writing an Equation

A company has a profit of \$750 this week. This profit is \$900 more than the profit P last week. Which equation can be used to find P?

(A) $750 = 900 - P$ (B) $750 = P + 900$

(C) $900 = P - 750$ (D) $900 = P + 750$

Words	The profit this week	is	\$900	more than	the profit last week.
Equation	750	=	P	+	900

The equation is $750 = P + 900$. The correct answer is (B).

On Your Own

Now You're Ready
Exercises 22–25

4. A company has a profit of \$120.50 today. This profit is \$145.25 less than the profit P yesterday. Write an equation that can be used to find P.

EXAMPLE 3 Real-Life Application

The line graph shows the scoring while you and your friend played a video game. Write and solve an equation to find your score after Level 4.

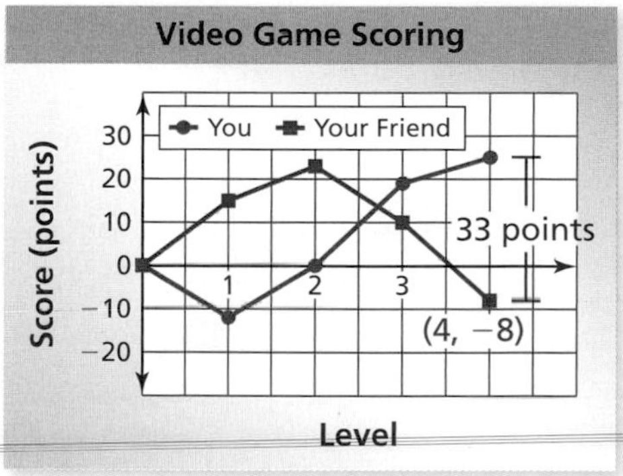

You can determine the following from the graph.

Words	Your friend's score	is	33 points	less than	your score.
Variable	Let s be your score after Level 4.				
Equation	−8	=	s	−	33

$$-8 = s - 33$$ Write equation.

$$\underline{+33} \quad \underline{+33}$$ Addition Property of Equality

$$25 = s$$ Simplify.

Your score after Level 4 is 25 points.

Reasonable? From the graph, your score after Level 4 is between 20 points and 30 points. So, 25 points is a reasonable answer.

On Your Own

5. **WHAT IF?** You have −12 points after Level 1. Your score is 27 points less than your friend's score. What is your friend's score?

13.3 Exercises

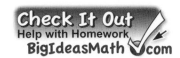

Vocabulary and Concept Check

1. **VOCABULARY** What property would you use to solve $m + 6 = -4$?
2. **VOCABULARY** Name two inverse operations.
3. **WRITING** Are the equations $m + 3 = -5$ and $m = -2$ equivalent? Explain.
4. **WHICH ONE DOESN'T BELONG?** Which equation does *not* belong with the other three? Explain your reasoning.

$x + 3 = -1$	$x + 1 = -5$	$x - 2 = -6$	$x - 9 = -13$

Practice and Problem Solving

Solve the equation. Check your solution.

① 5. $a - 6 = 13$ 6. $-3 = z - 8$ 7. $-14 = k + 6$ 8. $x + 4 = -14$

9. $c - 7.6 = -4$ 10. $-10.1 = w + 5.3$ 11. $\frac{1}{2} = q + \frac{2}{3}$ 12. $p - 3\frac{1}{6} = -2\frac{1}{2}$

13. $g - 9 = -19$ 14. $-9.3 = d - 3.4$ 15. $4.58 + y = 2.5$ 16. $x - 5.2 = -18.73$

17. $q + \frac{5}{9} = \frac{1}{6}$ 18. $-2\frac{1}{4} = r - \frac{4}{5}$ 19. $w + 3\frac{3}{8} = 1\frac{5}{6}$ 20. $4\frac{2}{5} + k = -3\frac{2}{11}$

21. **ERROR ANALYSIS** Describe and correct the error in finding the solution.

✗
$$\begin{aligned} x + 8 &= 10 \\ +8 \quad & \quad +8 \\ x &= 18 \end{aligned}$$

Write the word sentence as an equation. Then solve.

② 22. 4 less than a number n is -15. 23. 10 more than a number c is 3.

24. The sum of a number y and -3 is -8.

25. The difference between a number p and 6 is -14.

In Exercises 26–28, write an equation. Then solve.

26. **DRY ICE** The temperature of dry ice is -109.3°F. This is 184.9°F less than the outside temperature. What is the outside temperature?

27. **PROFIT** A company makes a profit of \$1.38 million. This is \$2.54 million more than last year. What was the profit last year?

28. **HELICOPTER** The difference in elevation of a helicopter and a submarine is $18\frac{1}{2}$ meters. The elevation of the submarine is $-7\frac{3}{4}$ meters. What is the elevation of the helicopter?

GEOMETRY Write and solve an equation to find the unknown side length.

29. Perimeter $= 12$ cm

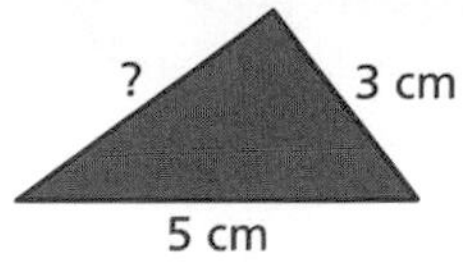

30. Perimeter $= 24.2$ in.

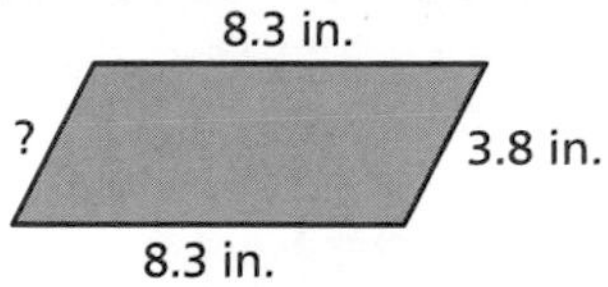

31. Perimeter $= 34.6$ ft

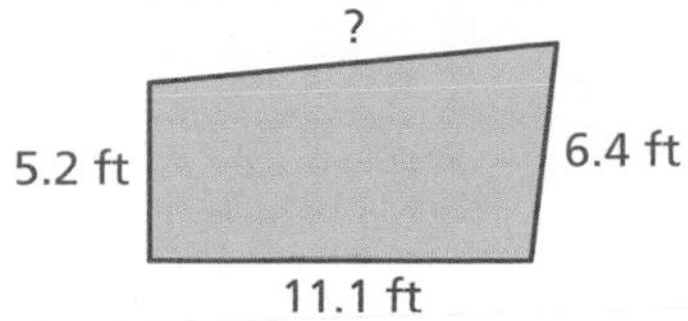

In Exercises 32–36, write an equation. Then solve.

32. STATUE OF LIBERTY The total height of the Statue of Liberty and its pedestal is 153 feet more than the height of the statue. What is the height of the statue?

33. BUNGEE JUMPING Your first jump is $50\frac{1}{6}$ feet higher than your second jump. Your first jump reaches $-200\frac{2}{5}$ feet. What is the height of your second jump?

34. TRAVEL Boatesville is $65\frac{3}{5}$ kilometers from Stanton. A bus traveling from Stanton is $24\frac{1}{3}$ kilometers from Boatesville. How far has the bus traveled?

35. GEOMETRY The sum of the measures of the angles of a triangle equals 180°. What is the measure of the missing angle?

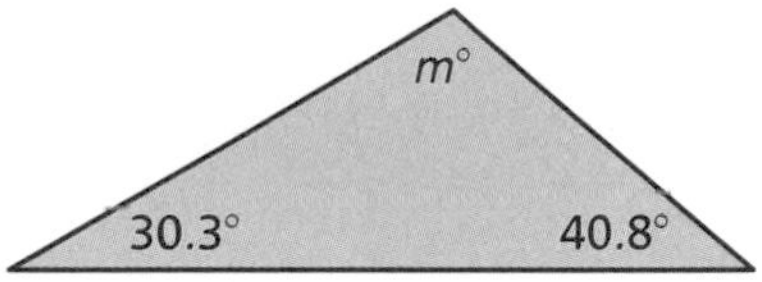

36. SKATEBOARDING The table shows your scores in a skateboarding competition. The leader has 311.62 points. What score do you need in the fourth round to win?

Round	1	2	3	4
Points	63.43	87.15	81.96	?

37. CRITICAL THINKING Find the value of $2x - 1$ when $x + 6 = 2$.

Critical Thinking **Find the values of *x*.**

38. $|x| = 2$

39. $|x| - 2 = 4$

40. $|x| + 5 = 18$

Fair Game Review What you learned in previous grades & lessons

Multiply or divide. *(Section 11.4 and Section 11.5)*

41. -7×8

42. $6 \times (-12)$

43. $18 \div (-2)$

44. $-26 \div 4$

45. MULTIPLE CHOICE A class of 144 students voted for a class president. Three-fourths of the students voted for you. Of the students who voted for you, $\frac{5}{9}$ are female. How many female students voted for you? *(Section 12.4)*

Ⓐ 50 Ⓑ 60 Ⓒ 80 Ⓓ 108

13.4 Solving Equations Using Multiplication or Division

Essential Question How can you use multiplication or division to solve equations?

1 ACTIVITY: Using Division to Solve Equations

Work with a partner. Use algebra tiles to model and solve the equation.

a. $3x = -12$

Model the equation $3x = -12$.

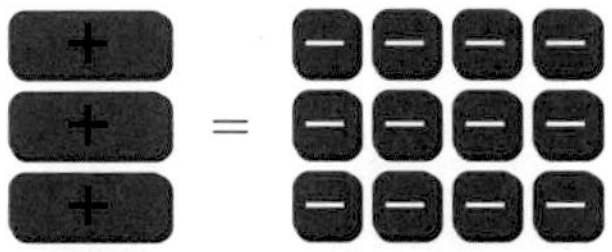

Your goal is to get one variable tile by itself. Because there are ____ variable tiles, divide the ____ tiles into ____ equal groups. Circle the groups.

Keep one of the groups. This shows the value of x.

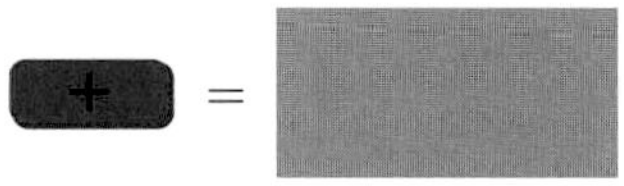

So, $x =$ ____.

b. $2k = -8$

c. $-15 = 3t$

d. $-20 = 5m$

e. $4h = -16$

2 ACTIVITY: Writing and Solving Equations

Work with a partner. Write an equation shown by the algebra tiles. Then solve.

a.

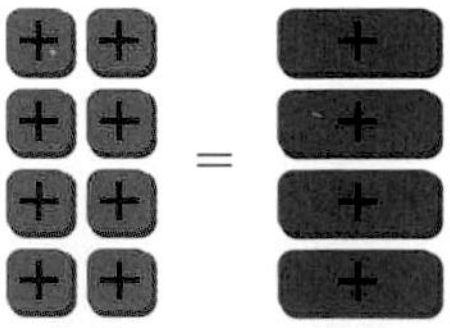

b.

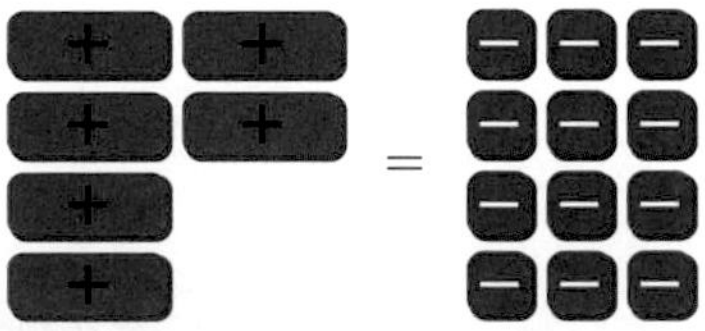

c.

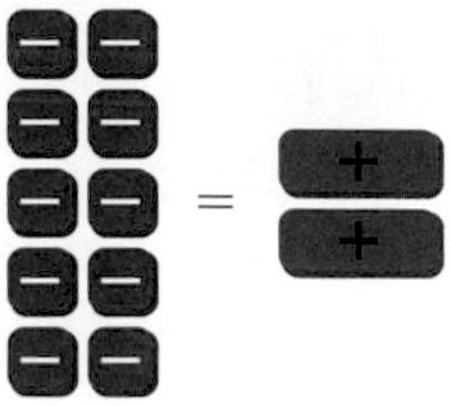

d.

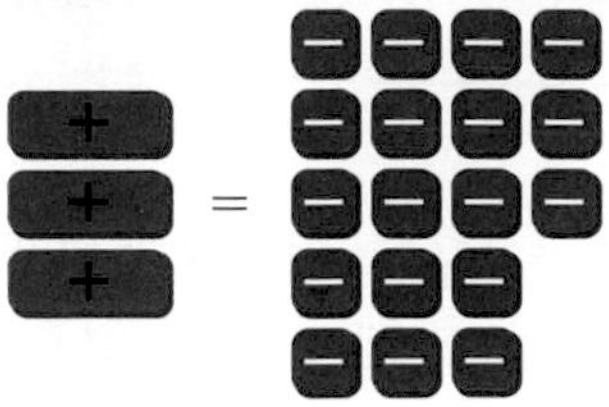

COMMON CORE

Solving Equations

In this lesson, you will

- solve equations using multiplication or division.
- solve real-life problems.

Learning Standard 7.EE.4a

3 ACTIVITY: Using a Different Method to Find a Solution

Math Practice 1

Analyze Givens

How can you use the given information to decide which equation represents the situation?

Work with a partner. Choose the equation you can use to solve each problem. Solve the equation. Then explain how to solve the problem without using an equation. How are the two methods related?

a. For the final part of a race, a handcyclist travels 32 feet each second across a distance of 400 feet. How many seconds does it take for the handcyclist to travel the last 400 feet of the race?

$32x = 400$	$400x = 32$
$\frac{x}{32} = 400$	$\frac{x}{400} = 32$

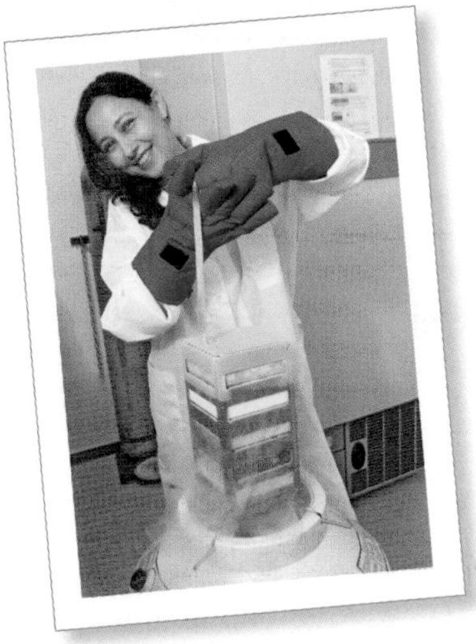

b. The melting point of the element radon is about -96°F. The melting point of nitrogen is about 3.6 times the melting point of radon. What is the melting point of nitrogen?

$3.6x = -96$	$x + 96 = 3.6$
$\frac{x}{3.6} = -96$	$-96x = 3.6$

c. This year, a hardware store has a profit of $-\$6.0$ million. This profit is $\frac{3}{4}$ of last year's profit. What is last year's profit?

$\frac{x}{-6} = \frac{3}{4}$	$-6x = \frac{3}{4}$
$\frac{3}{4} + x = -6$	$\frac{3}{4}x = -6$

What Is Your Answer?

4. **IN YOUR OWN WORDS** How can you use multiplication or division to solve equations? Give an example of each.

Practice Use what you learned about solving equations to complete Exercises 7–10 on page 580.

13.4 Lesson

Check It Out
Lesson Tutorials
BigIdeasMath.com

Key Ideas

Multiplication Property of Equality

Words Multiplying each side of an equation by the same number produces an equivalent equation.

Algebra If $a = b$, then $a \cdot c = b \cdot c$.

Division Property of Equality

Words Dividing each side of an equation by the same number produces an equivalent equation.

Algebra If $a = b$, then $a \div c = b \div c, c \neq 0$.

Remember

Multiplication and division are inverse operations.

EXAMPLE 1 Solving Equations

a. Solve $\frac{x}{3} = -6$.

$\frac{x}{3} = -6$	Write the equation.
Undo the division. → $3 \cdot \frac{x}{3} = 3 \cdot (-6)$	Multiplication Property of Equality
$x = -18$	Simplify.

The solution is $x = -18$.

Check

$\frac{x}{3} = -6$

$\frac{-18}{3} \stackrel{?}{=} -6$

$-6 = -6$ ✓

b. Solve $18 = -4y$.

$18 = -4y$	Write the equation.
Undo the multiplication. → $\frac{18}{-4} = \frac{-4y}{-4}$	Division Property of Equality
$-4.5 = y$	Simplify.

The solution is $y = -4.5$.

Check

$18 = -4y$

$18 \stackrel{?}{=} -4(-4.5)$

$18 = 18$ ✓

On Your Own

Now You're Ready
Exercises 7–18

Solve the equation. Check your solution.

1. $\frac{x}{5} = -2$ **2.** $-a = -24$ **3.** $3 = -1.5n$

EXAMPLE 2 Solving an Equation Using a Reciprocal

Solve $-\frac{4}{5}x = -8$.

$$-\frac{4}{5}x = -8$$ Write the equation.

Multiply each side by $-\frac{5}{4}$, the reciprocal of $-\frac{4}{5}$.

$$-\frac{5}{4} \cdot \left(-\frac{4}{5}x\right) = -\frac{5}{4} \cdot (-8)$$ Multiplicative Inverse Property

$$x = 10$$ Simplify.

∴ The solution is $x = 10$.

On Your Own

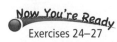

Now You're Ready
Exercises 19–22

Solve the equation. Check your solution.

4. $-14 = \frac{2}{3}x$ **5.** $-\frac{8}{5}b = 5$ **6.** $\frac{3}{8}h = -9$

EXAMPLE 3 Real-Life Application

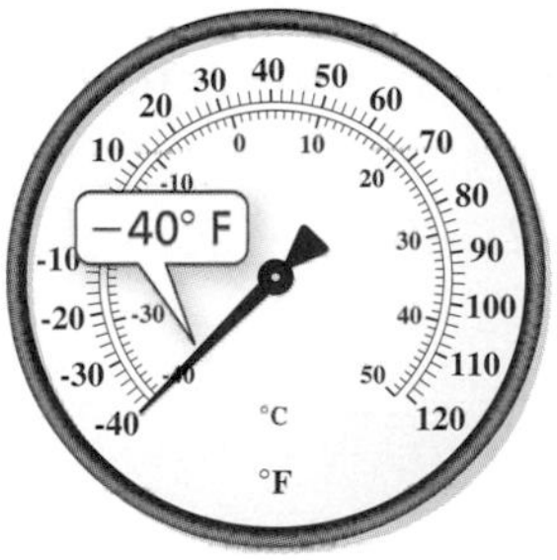

Record low temperature in Arizona

The record low temperature in Arizona is 1.6 times the record low temperature in Rhode Island. What is the record low temperature in Rhode Island?

Words The record low in Arizona is 1.6 times the record low in Rhode Island.

Variable Let t be the record low in Rhode Island.

Equation $-40 = 1.6 \times t$

$$-40 = 1.6t$$ Write equation.

$$\frac{-40}{1.6} = \frac{1.6t}{1.6}$$ Division Property of Equality

$$-25 = t$$ Simplify.

∴ The record low temperature in Rhode Island is −25°F.

On Your Own

Now You're Ready
Exercises 24–27

7. The record low temperature in Hawaii is –0.15 times the record low temperature in Alaska. The record low temperature in Hawaii is 12°F. What is the record low temperature in Alaska?

13.4 Exercises

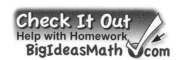

Check It Out
Help with Homework
BigIdeasMath.com

Vocabulary and Concept Check

1. **WRITING** Explain why you can use multiplication to solve equations involving division.

2. **OPEN-ENDED** Turning a light on and then turning the light off are considered to be inverse operations. Describe two other real-life situations that can be thought of as inverse operations.

Describe the inverse operation that will undo the given operation.

3. multiplying by 5 4. subtracting 12 5. dividing by -8 6. adding -6

Practice and Problem Solving

Solve the equation. Check your solution.

1 7. $3h = 15$ 8. $-5t = -45$ 9. $\frac{n}{2} = -7$ 10. $\frac{k}{-3} = 9$

11. $5m = -10$ 12. $8t = -32$ 13. $-0.2x = 1.6$ 14. $-10 = -\frac{b}{4}$

15. $-6p = 48$ 16. $-72 = 8d$ 17. $\frac{n}{1.6} = 5$ 18. $-14.4 = -0.6p$

2 19. $\frac{3}{4}g = -12$ 20. $8 = -\frac{2}{5}c$ 21. $-\frac{4}{9}f = -3$ 22. $26 = -\frac{8}{5}y$

23. **ERROR ANALYSIS** Describe and correct the error in finding the solution.

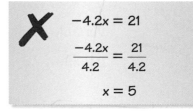

✗
$$-4.2x = 21$$
$$\frac{-4.2x}{4.2} = \frac{21}{4.2}$$
$$x = 5$$

Write the word sentence as an equation. Then solve.

3 24. A number divided by -9 is -16. 25. A number multiplied by $\frac{2}{5}$ is $\frac{3}{20}$.

26. The product of 15 and a number is -75.

27. The quotient of a number and -1.5 is 21.

In Exercises 28 and 29, write an equation. Then solve.

28. **NEWSPAPERS** You make \$0.75 for every newspaper you sell. How many newspapers do you have to sell to buy the soccer cleats?

29. **ROCK CLIMBING** A rock climber averages $12\frac{3}{5}$ feet per minute. How many feet does the rock climber climb in 30 minutes?

OPEN-ENDED (a) Write a multiplication equation that has the given solution. (b) Write a division equation that has the same solution.

30. -3 **31.** -2.2 **32.** $-\frac{1}{2}$ **33.** $-1\frac{1}{4}$

34. REASONING Which of the methods can you use to solve $-\frac{2}{3}c = 16$?

Multiply each side by $-\frac{2}{3}$.	Multiply each side by $-\frac{3}{2}$.
Divide each side by $-\frac{2}{3}$.	Multiply each side by 3, then divide each side by -2.

35. STOCK A stock has a return of $-\$1.26$ per day. Write and solve an equation to find the number of days until the total return is $-\$10.08$.

36. ELECTION In a school election, $\frac{3}{4}$ of the students vote. There are 1464 ballots. Write and solve an equation to find the number of students.

37. OCEANOGRAPHY Aquarius is an underwater ocean laboratory located in the Florida Keys National Marine Sanctuary. Solve the equation $\frac{31}{25}x = -62$ to find the value of x.

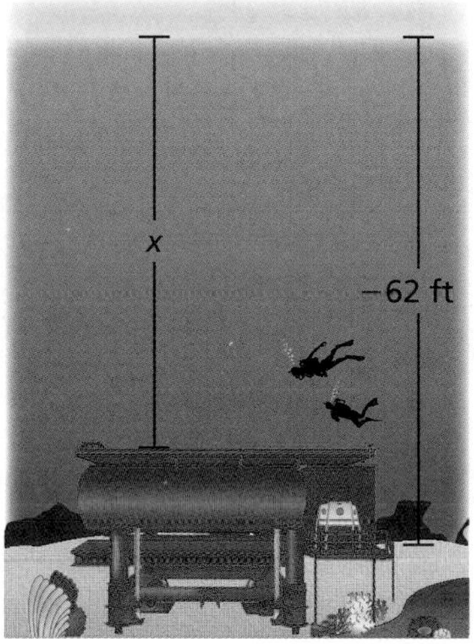

38. SHOPPING The price of a bike at Store A is $\frac{5}{6}$ the price at Store B. The price at Store A is \$150.60. Write and solve an equation to find how much you save by buying the bike at Store A.

39. CRITICAL THINKING Solve $-2|m| = -10$.

40. Number Sense In four days, your family drives $\frac{5}{7}$ of a trip. Your rate of travel is the same throughout the trip. The total trip is 1250 miles. In how many more days will you reach your destination?

Fair Game Review What you learned in previous grades & lessons

Subtract. *(Section 11.3)*

41. $5 - 12$ **42.** $-7 - 2$ **43.** $4 - (-8)$ **44.** $-14 - (-5)$

45. MULTIPLE CHOICE Of the 120 apartments in a building, 75 have been scheduled to receive new carpet. What fraction of the apartments have not been scheduled to receive new carpet? *(Skills Review Handbook)*

Ⓐ $\frac{1}{4}$ Ⓑ $\frac{3}{8}$ Ⓒ $\frac{5}{8}$ Ⓓ $\frac{3}{4}$

13.5 Solving Two-Step Equations

Essential Question How can you use algebra tiles to solve a two-step equation?

1 ACTIVITY: Solving a Two-Step Equation

Work with a partner. Use algebra tiles to model and solve $2x - 3 = -5$.

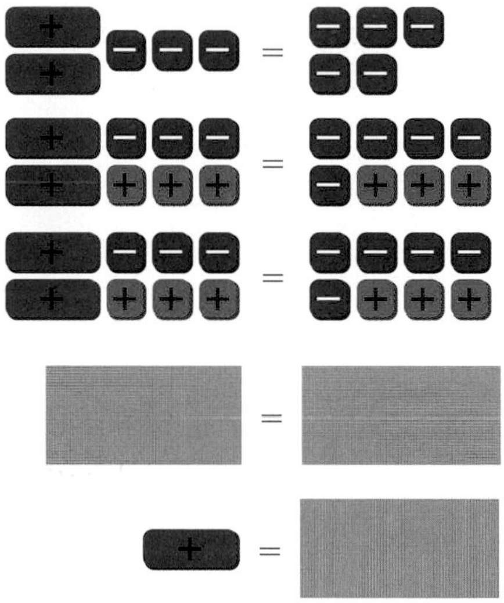

Model the equation $2x - 3 = -5$.

Remove the ____ red tiles on the left side by adding ____ yellow tiles to each side.

How many *zero pairs* can you remove from each side? ____ Circle them.

Because there are ____ green tiles, divide the red tiles into ____ equal groups. Circle the groups.

Keep one of the groups. This shows the value of x.

So, $x =$ ____.

2 ACTIVITY: The Math behind the Tiles

Work with a partner. Solve $2x - 3 = -5$ without using algebra tiles. Complete each step. Then answer the questions.

Use the steps in Activity 1 as a guide.

$2x - 3 = -5$	Write the equation.
$2x - 3 + \square = -5 + \square$	Add ____ to each side.
$2x = \square$	Simplify.
$\dfrac{2x}{\square} = \dfrac{\square}{\square}$	Divide each side by ____.
$x = \square$	Simplify.

So, $x =$ ____.

a. Which step is first, adding 3 to each side or dividing each side by 2?

b. How are the above steps related to the steps in Activity 1?

COMMON CORE

Solving Equations

In this lesson, you will

- solve two-step equations.
- solve real-life problems.

Learning Standard 7.EE.4a

3 ACTIVITY: Solving Equations Using Algebra Tiles

Work with a partner.

- **Write an equation shown by the algebra tiles.**
- **Use algebra tiles to model and solve the equation.**
- **Check your answer by solving the equation without using algebra tiles.**

a. 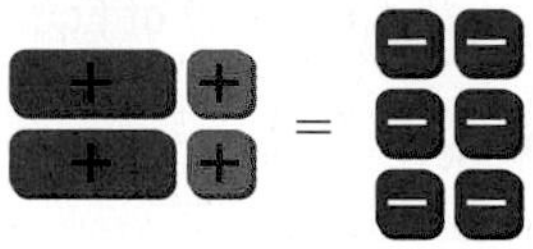

b. 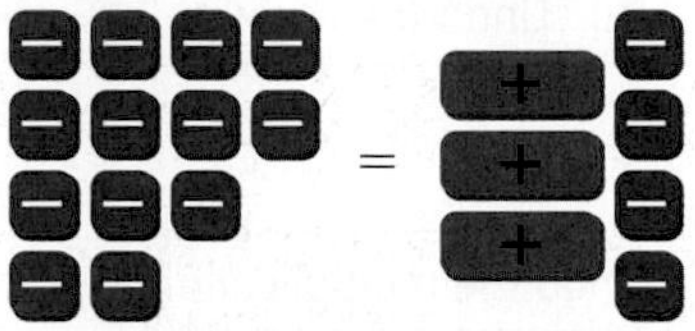

4 ACTIVITY: Working Backwards

Math Practice 8

Maintain Oversight

How does working backwards help you decide which operation to do first? Explain.

Work with a partner.

a. Sample: Your friend pauses a video game to get a drink. You continue the game. You double the score by saving a princess. Then you lose 75 points because you do not collect the treasure. You finish the game with −25 points. How many points did you have when you started?

One way to solve the problem is to work backwards. To do this, start with the end result and retrace the events.

You have −25 points at the end of the game.	-25
You lost 75 points for not collecting the treasure, so add 75 to −25.	$-25 + 75 = 50$
You doubled your score for saving the princess, so find half of 50.	$50 \div 2 = 25$

So, you started the game with 25 points.

b. You triple your account balance by making a deposit. Then you withdraw $127.32 to buy groceries. Your account is now overdrawn by $10.56. By working backwards, find your account balance before you made the deposit.

What Is Your Answer?

5. **IN YOUR OWN WORDS** How can you use algebra tiles to solve a two-step equation?
6. When solving the equation $4x + 1 = -11$, what is the first step?
7. **REPEATED REASONING** Solve the equation $2x - 75 = -25$. How do your steps compare with the strategy of working backwards in Activity 4?

Practice Use what you learned about solving two-step equations to complete Exercises 6–11 on page 586.

13.5 Lesson

Check It Out
Lesson Tutorials
BigIdeasMath.com

EXAMPLE 1 Solving a Two-Step Equation

Solve $-3x + 5 = 2$. Check your solution.

	$-3x + 5 = 2$	Write the equation.
Undo the addition. →	$-5 \quad -5$	Subtraction Property of Equality
	$-3x = -3$	Simplify.
Undo the multiplication. →	$\frac{-3x}{-3} = \frac{-3}{-3}$	Division Property of Equality
	$x = 1$	Simplify.

Check

$-3x + 5 = 2$

$-3(1) + 5 \stackrel{?}{=} 2$

$-3 + 5 \stackrel{?}{=} 2$

$2 = 2$ ✓

The solution is $x = 1$.

On Your Own

Now You're Ready
Exercises 6–17

Solve the equation. Check your solution.

1. $2x + 12 = 4$
2. $-5c + 9 = -16$
3. $3(x - 4) = 9$

EXAMPLE 2 Solving a Two-Step Equation

Solve $\frac{x}{8} - \frac{1}{2} = -\frac{7}{2}$. Check your solution.

$\frac{x}{8} - \frac{1}{2} = -\frac{7}{2}$	Write the equation.
$+\frac{1}{2} \quad +\frac{1}{2}$	Addition Property of Equality
$\frac{x}{8} = -3$	Simplify.
$8 \cdot \frac{x}{8} = 8 \cdot (-3)$	Multiplication Property of Equality
$x = -24$	Simplify.

Check

$\frac{x}{8} - \frac{1}{2} = -\frac{7}{2}$

$\frac{-24}{8} - \frac{1}{2} \stackrel{?}{=} -\frac{7}{2}$

$-3 - \frac{1}{2} \stackrel{?}{=} -\frac{7}{2}$

$-\frac{7}{2} = -\frac{7}{2}$ ✓

The solution is $x = -24$.

Study Tip

You can simplify the equation in Example 2 before solving. Multiply each side by the LCD of the fractions, 8.

$\frac{x}{8} - \frac{1}{2} = -\frac{7}{2}$

$x - 4 = -28$

$x = -24$

On Your Own

Now You're Ready
Exercises 20–25

Solve the equation. Check your solution.

4. $\frac{m}{2} + 6 = 10$
5. $-\frac{z}{3} + 5 = 9$
6. $\frac{2}{5} + 4a = -\frac{6}{5}$

EXAMPLE 3 Combining Like Terms Before Solving

Solve $3y - 8y = 25$.

$3y - 8y = 25$	Write the equation.
$-5y = 25$	Combine like terms.
$y = -5$	Divide each side by -5.

∴ The solution is $y = -5$.

EXAMPLE 4 Real-Life Application

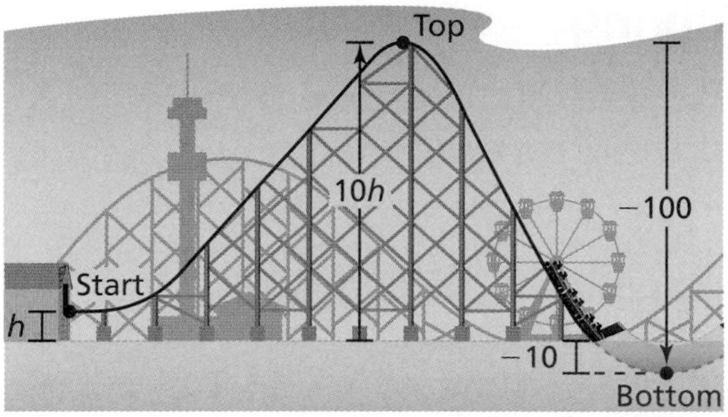

The height at the top of a roller coaster hill is 10 times the height h of the starting point. The height decreases 100 feet from the top to the bottom of the hill. The height at the bottom of the hill is -10 feet. Find h.

Location	Verbal Description	Expression
Start	The height at the start is h.	h
Top of hill	The height at the top of the hill is 10 times the starting height h.	$10h$
Bottom of hill	The height decreases by 100 feet. So, subtract 100.	$10h - 100$

The height at the bottom of the hill is -10 feet. Solve $10h - 100 = -10$ to find h.

$10h - 100 = -10$	Write equation.
$10h = 90$	Add 100 to each side.
$h = 9$	Divide each side by 10.

∴ So, the height at the start is 9 feet.

On Your Own

Now You're Ready
Exercises 29–34

Solve the equation. Check your solution.

7. $4 - 2y + 3 = -9$ **8.** $7x - 10x = 15$ **9.** $-8 = 1.3m - 2.1m$

10. **WHAT IF?** In Example 4, the height at the bottom of the hill is -5 feet. Find the height h.

13.5 Exercises

Check It Out
Help with Homework
BigIdeasMath.com

Vocabulary and Concept Check

1. **WRITING** How do you solve two-step equations?

Match the equation with the first step to solve it.

2. $4 + 4n = -12$ 3. $4n = -12$ 4. $\frac{n}{4} = -12$ 5. $\frac{n}{4} - 4 = -12$

A. Add 4. B. Subtract 4. C. Multiply by 4. D. Divide by 4.

Practice and Problem Solving

Solve the equation. Check your solution.

1
6. $2v + 7 = 3$
7. $4b + 3 = -9$
8. $17 = 5k - 2$
9. $-6t - 7 = 17$
10. $8n + 16.2 = 1.6$
11. $-5g + 2.3 = -18.8$
12. $2t - 5 = -10$
13. $-4p + 9 = -5$
14. $11 = -5x - 2$
15. $4 + 2.2h = -3.7$
16. $-4.8f + 6.4 = -8.48$
17. $7.3y - 5.18 = -51.9$

ERROR ANALYSIS Describe and correct the error in finding the solution.

18. ✗

$$-6 + 2x = -10$$
$$-6 + \frac{2x}{2} = -\frac{10}{2}$$
$$-6 + x = -5$$
$$x = 1$$

19. ✗

$$-3x + 2 = -7$$
$$-3x = -9$$
$$-\frac{3x}{3} = \frac{-9}{3}$$
$$x = -3$$

Solve the equation. Check your solution.

2
20. $\frac{3}{5}g - \frac{1}{3} = -\frac{10}{3}$
21. $\frac{a}{4} - \frac{5}{6} = -\frac{1}{2}$
22. $-\frac{1}{3} + 2z = -\frac{5}{6}$
23. $2 - \frac{b}{3} = -\frac{5}{2}$
24. $-\frac{2}{3}x + \frac{3}{7} = \frac{1}{2}$
25. $-\frac{9}{4}v + \frac{4}{5} = \frac{7}{8}$

In Exercises 26–28, write an equation. Then solve.

26. **WEATHER** Starting at 1:00 P.M., the temperature changes -4 degrees per hour. How long will it take to reach $-1°$?

Temperature at 1:00 P.M.

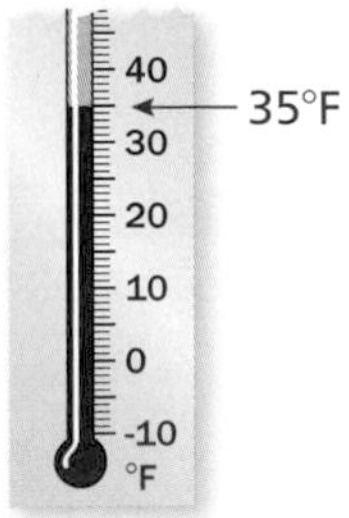

27. **BOWLING** It costs \$2.50 to rent bowling shoes. Each game costs \$2.25. You have \$9.25. How many games can you bowl?

28. **CELL PHONES** A cell phone company charges a monthly fee plus \$0.25 for each text message. The monthly fee is \$30.00 and you owe \$59.50. How many text messages did you have?

Solve the equation. Check your solution.

3 **29.** $3v - 9v = 30$ **30.** $12t - 8t = -52$ **31.** $-8d - 5d + 7d = 72$

32. $6(x - 2) = -18$ **33.** $-4(m + 3) = 24$ **34.** $-8(y + 9) = -40$

35. WRITING Write a real-world problem that can be modeled by $\frac{1}{2}x - 2 = 8$. Then solve the equation.

36. GEOMETRY The perimeter of the parallelogram is 102 feet. Find m.

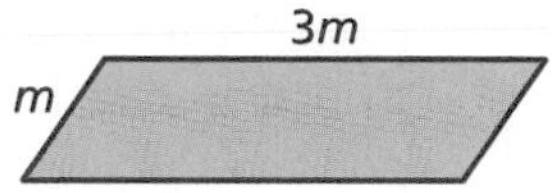

REASONING Exercises 37 and 38 are missing information. Tell what information you need to solve the problem.

37. TAXI A taxi service charges an initial fee plus \$1.80 per mile. How far can you travel for \$12?

38. EARTH The coldest surface temperature on the Moon is 57 degrees colder than twice the coldest surface temperature on Earth. What is the coldest surface temperature on Earth?

39. PROBLEM SOLVING On Saturday, you catch insects for your science class. Five of the insects escape. The remaining insects are divided into three groups to share in class. Each group has nine insects. How many insects did you catch on Saturday?

a. Solve the problem by working backwards.

b. Solve the equation $\frac{x - 5}{3} = 9$. How does the answer compare with the answer to part (a)?

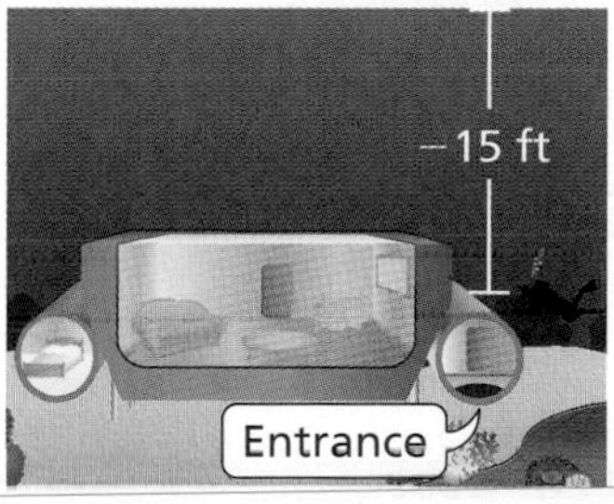

40. UNDERWATER HOTEL You must scuba dive to the entrance of your room at Jules' Undersea Lodge in Key Largo, Florida. The diver is 1 foot deeper than $\frac{2}{3}$ of the elevation of the entrance. What is the elevation of the entrance?

41. Geometry How much should you change the length of the rectangle so that the perimeter is 54 centimeters? Write an equation that shows how you found your answer.

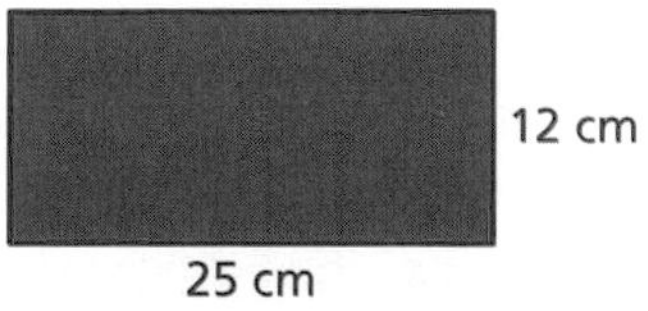

Fair Game Review What you learned in previous grades & lessons

Multiply or divide. *(Section 12.4)*

42. -6.2×5.6 **43.** $\frac{8}{3} \times \left(-2\frac{1}{2}\right)$ **44.** $\frac{5}{2} \div \left(-\frac{4}{5}\right)$ **45.** $-18.6 \div (-3)$

46. MULTIPLE CHOICE Which fraction is *not* equivalent to 0.75? *(Skills Review Handbook)*

(A) $\frac{15}{20}$ (B) $\frac{9}{12}$ (C) $\frac{6}{9}$ (D) $\frac{3}{4}$

13.3–13.5 Quiz

Check It Out
Progress Check
BigIdeasMath.com

Solve the equation. Check your solution. *(Section 13.3, Section 13.4, and Section 13.5)*

1. $-6.5 + x = -4.12$

2. $4\frac{1}{2} + p = -5\frac{3}{4}$

3. $-\frac{b}{7} = 4$

4. $-2w + 3.7 = -0.5$

Write the word sentence as an equation. Then solve. *(Section 13.3 and Section 13.4)*

5. The difference between a number b and 7.4 is -6.8.

6. $5\frac{2}{5}$ more than a number a is $7\frac{1}{2}$.

7. A number x multiplied by $\frac{3}{8}$ is $-\frac{15}{32}$.

8. The quotient of two times a number k and -2.6 is 12.

Write and solve an equation to find the value of *x*. *(Section 13.3 and Section 13.5)*

9. Perimeter = 26

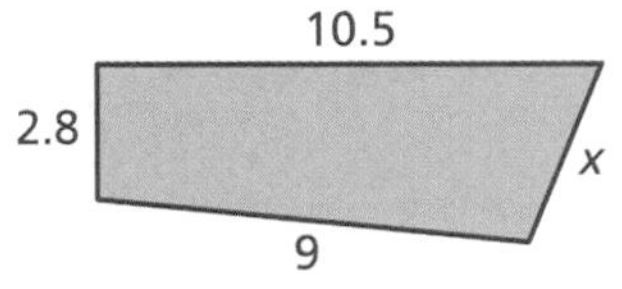

10. Perimeter = 23.59

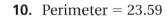

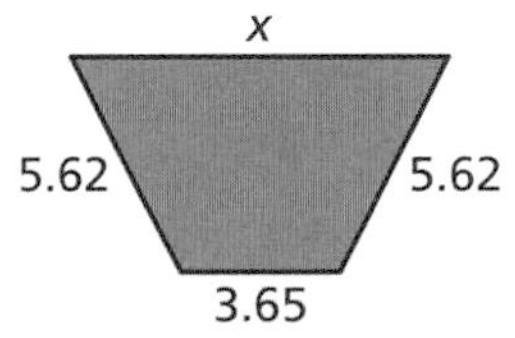

11. Perimeter = 33

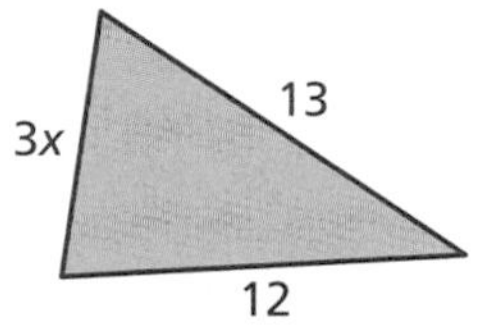

12. BANKING You withdraw $29.79 from your bank account. Now your balance is −$20.51. Write and solve an equation to find the amount of money in your bank account before you withdrew the money. *(Section 13.3)*

13. WATER LEVEL During a drought, the water level of a lake changes $-3\frac{1}{5}$ feet per day. Write and solve an equation to find how long it takes for the water level to change -16 feet. *(Section 13.4)*

14. BASKETBALL A basketball game has four quarters. The length of a game is 32 minutes. You play the entire game except for $4\frac{1}{2}$ minutes. Write and solve an equation to find the mean time you play per quarter. *(Section 13.5)*

15. SCRAPBOOKING The mat needs to be cut to have a 0.5-inch border on all four sides. *(Section 13.5)*

a. How much should you cut from the left and right sides?

b. How much should you cut from the top and bottom?

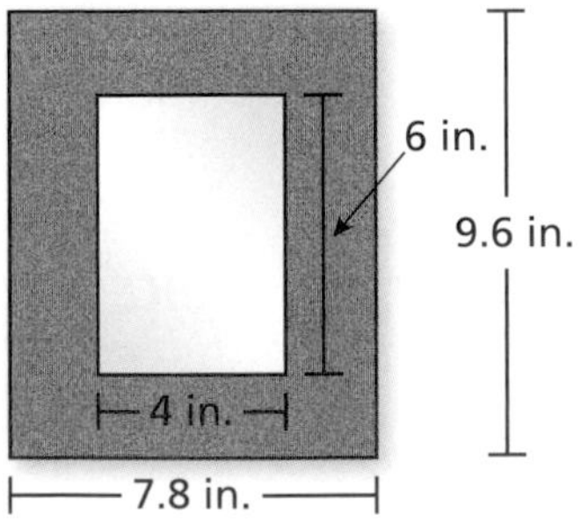

13 Chapter Review

Check It Out
Vocabulary Help
BigIdeasMath.com

Review Key Vocabulary

like terms, *p. 556*
simplest form, *p. 556*
linear expression, *p. 562*
factoring an expression, *p. 566*
equivalent equations, *p. 572*

Review Examples and Exercises

13.1 Algebraic Expressions *(pp. 554–559)*

a. Identify the terms and like terms in the expression $6y + 9 + 3y - 7$.

Rewrite as a sum of terms.

$$6y + 9 + 3y + (-7)$$

Terms: $6y,\ 9,\ 3y,\ -7$

Like terms: $6y$ and $3y$, 9 and -7

b. Simplify $\frac{2}{3}y + 14 - \frac{1}{6}y - 8$.

$\frac{2}{3}y + 14 - \frac{1}{6}y - 8 = \frac{2}{3}y + 14 + \left(-\frac{1}{6}y\right) + (-8)$ Rewrite as a sum.

$= \frac{2}{3}y + \left(-\frac{1}{6}y\right) + 14 + (-8)$ Commutative Property of Addition

$= \left[\frac{2}{3} + \left(-\frac{1}{6}\right)\right]y + 14 + (-8)$ Distributive Property

$= \frac{1}{2}y + 6$ Combine like terms.

Exercises

Identify the terms and like terms in the expression.

1. $z + 8 - 4z$
2. $3n + 7 - n - 3$
3. $10x^2 - y + 12 - 3x^2$

Simplify the expression.

4. $4h - 8h$
5. $6.4r - 7 - 2.9r$
6. $\frac{3}{5}x + 19 - \frac{3}{20}x - 7$
7. $3(2 + q) + 15$
8. $\frac{1}{8}(16m - 8) - 17$
9. $-1.5(4 - n) + 2.8$

13.2 Adding and Subtracting Linear Expressions (pp. 560–567)

a. Find $(5z + 4) + (3z - 6)$.

$$\begin{array}{r} 5z + 4 \\ + \; 3z - 6 \\ \hline 8z - 2 \end{array}$$

Align like terms vertically and add.

b. Factor $\frac{1}{4}$ out of $\frac{1}{4}x - \frac{3}{4}$.

Write each term as a product of $\frac{1}{4}$ and another factor.

$$\frac{1}{4}x = \frac{1}{4} \cdot x \qquad -\frac{3}{4} = \frac{1}{4} \cdot (-3)$$

Use the Distributive Property to factor out $\frac{1}{4}$.

$$\frac{1}{4}x - \frac{3}{4} = \frac{1}{4} \cdot x + \frac{1}{4} \cdot (-3) = \frac{1}{4}(x - 3)$$

So, $\frac{1}{4}x - \frac{3}{4} = \frac{1}{4}(x - 3)$.

Exercises

Find the sum or difference.

10. $(c - 4) + (3c + 9)$

11. $\frac{2}{5}(d - 10) - \frac{2}{3}(d + 6)$

Factor out the coefficient of the variable.

12. $2b + 8$

13. $\frac{1}{4}y + \frac{3}{8}$

14. $1.7j - 3.4$

15. $-5p + 20$

13.3 Solving Equations Using Addition or Subtraction (pp. 570–575)

Solve $x - 9 = -6$.

	$x - 9 = -6$	Write the equation.
Undo the subtraction. →	$+9 \quad +9$	Addition Property of Equality
	$x = 3$	Simplify.

Check

$x - 9 = -6$

$3 - 9 \stackrel{?}{=} -6$

$-6 = -6$ ✓

Exercises

Solve the equation. Check your solution.

16. $p - 3 = -4$

17. $6 + q = 1$

18. $-2 + j = -22$

19. $b - 19 = -11$

20. $n + \frac{3}{4} = \frac{1}{4}$

21. $v - \frac{5}{6} = -\frac{7}{8}$

22. $t - 3.7 = 1.2$

23. $\ell + 15.2 = -4.5$

13.4 Solving Equations Using Multiplication or Division (pp. 576–581)

Solve $\frac{x}{5} = -7$.

	$\frac{x}{5} = -7$	Write the equation.
Undo the division. →	$5 \cdot \frac{x}{5} = 5 \cdot (-7)$	Multiplication Property of Equality
	$x = -35$	Simplify.

Check

$\frac{x}{5} = -7$

$\frac{-35}{5} \stackrel{?}{=} -7$

$-7 = -7$ ✓

Exercises

Solve the equation. Check your solution.

24. $\frac{x}{3} = -8$ **25.** $-7 = \frac{y}{7}$ **26.** $-\frac{z}{4} = -\frac{3}{4}$ **27.** $-\frac{w}{20} = -2.5$

28. $4x = -8$ **29.** $-10 = 2y$ **30.** $-5.4z = -32.4$ **31.** $-6.8w = 3.4$

32. TEMPERATURE The mean temperature change is −3.2°F per day for 5 days. What is the total change over the 5-day period?

13.5 Solving Two-Step Equations (pp. 582–587)

Solve $-6y + 7 = -5$. Check your solution.

$-6y + 7 = -5$	Write the equation.
$-7 \quad -7$	Subtraction Property of Equality
$-6y = -12$	Simplify.
$\frac{-6y}{-6} = \frac{-12}{-6}$	Division Property of Equality
$y = 2$	Simplify.

∴ The solution is $y = 2$.

Check

$-6y + 7 = -5$

$-6(2) + 7 \stackrel{?}{=} -5$

$-12 + 7 \stackrel{?}{=} -5$

$-5 = -5$ ✓

Exercises

Solve the equation. Check your solution.

33. $-2c + 6 = -8$ **34.** $3(3w - 4) = -20$

35. $\frac{w}{6} + \frac{5}{8} = -1\frac{3}{8}$ **36.** $-3x - 4.6 = 5.9$

37. EROSION The floor of a canyon has an elevation of −14.5 feet. Erosion causes the elevation to change by −1.5 feet per year. How many years will it take for the canyon floor to have an elevation of −31 feet?

13 Chapter Test

Check It Out
Test Practice
BigIdeasMath.com

Simplify the expression.

1. $8x - 5 + 2x$
2. $2.5w - 3y + 4w$
3. $3(5 - 2n) + 9n$
4. $\frac{5}{7}x + 15 - \frac{9}{14}x - 9$

Find the sum or difference.

5. $(3j + 11) + (8j - 7)$
6. $\frac{3}{4}(8p + 12) + \frac{3}{8}(16p - 8)$
7. $(2r - 13) - (-6r + 4)$
8. $-2.5(2s - 5) - 3(4.5s - 5.2)$

Factor out the coefficient of the variable.

9. $3n - 24$
10. $\frac{1}{2}q + \frac{5}{2}$

Solve the equation. Check your solution.

11. $7x = -3$
12. $2(x + 1) = -2$
13. $\frac{2}{9}g = -8$
14. $z + 14.5 = 5.4$
15. $-14 = 6c$
16. $\frac{2}{7}k - \frac{3}{8} = -\frac{19}{8}$

17. **HAIR SALON** Write an expression in simplest form that represents the income from w women and m men getting a haircut and a shampoo.

	Women	Men
Haircut	\$45	\$15
Shampoo	\$12	\$7

18. **RECORD** A runner is compared with the world record holder during a race. A negative number means the runner is ahead of the time of the world record holder. A positive number means that the runner is behind the time of the world record holder. The table shows the time difference between the runner and the world record holder for each lap. What time difference does the runner need for the fourth lap to match the world record?

Lap	Time Difference
1	−1.23
2	0.45
3	0.18
4	?

19. **GYMNASTICS** You lose 0.3 point for stepping out of bounds during a floor routine. Your final score is 9.124. Write and solve an equation to find your score before the penalty.

20. **PERIMETER** The perimeter of the triangle is 45. Find the value of x.

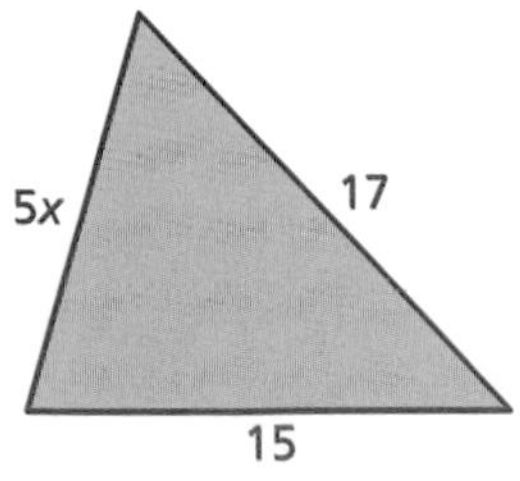

13 Standards Assessment

Test-Taking Strategy
After Answering Easy Questions, Relax

"After answering the easy questions, relax and try the harder ones. For this, 2x = 12, so x = 6 hyenas."

1. Which equation represents the word sentence shown below? *(7.EE.4a)*

> The quotient of a number b and 0.3 equals negative 10.

A. $0.3b = 10$

B. $\dfrac{b}{0.3} = -10$

C. $\dfrac{0.3}{b} = -10$

D. $\dfrac{b}{0.3} = 10$

2. What is the value of the expression below when $c = 0$ and $d = -6$? *(7.NS.2c)*

$$\frac{cd - d^2}{4}$$

3. What is the value of the expression below? *(7.NS.1c)*

$$-38 - (-14)$$

F. -52

G. -24

H. 24

I. 52

4. The daily low temperatures last week are shown below.

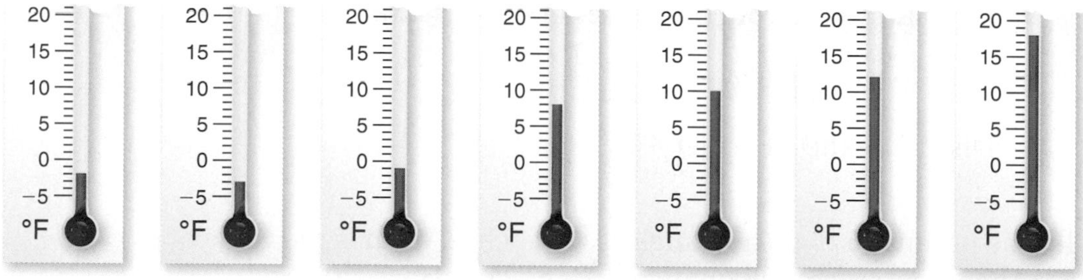

What is the mean low temperature of last week? *(7.NS.3)*

A. -2°F

B. 6°F

C. 8°F

D. 10°F

5. Which equation is equivalent to the equation shown below? *(7.EE.4a)*

$$-\frac{3}{4}x + \frac{1}{8} = -\frac{3}{8}$$

F. $-\frac{3}{4}x = -\frac{3}{8} - \frac{1}{8}$

G. $-\frac{3}{4}x = -\frac{3}{8} + \frac{1}{8}$

H. $x + \frac{1}{8} = -\frac{3}{8} \cdot \left(-\frac{4}{3}\right)$

I. $x + \frac{1}{8} = -\frac{3}{8} \cdot \left(-\frac{3}{4}\right)$

6. What is the value of the expression below? *(7.NS.2c)*

$$-0.28 \div (-0.07)$$

7. Karina was solving the equation in the box below.

$$-96 = -6(x - 15)$$
$$-96 = -6x - 90$$
$$-96 + 90 = -6x - 90 + 90$$
$$-6 = -6x$$
$$\frac{-6}{-6} = \frac{-6x}{-6}$$
$$1 = x$$

What should Karina do to correct the error that she made? *(7.EE.4a)*

A. First add 6 to both sides of the equation.

B. First subtract x from both sides of the equation.

C. Distribute the -6 to get $6x - 90$.

D. Distribute the -6 to get $-6x + 90$.

8. The perimeter of the rectangle is 400 inches. What is the value of j? (All measurements are in inches.) *(7.EE.4a)*

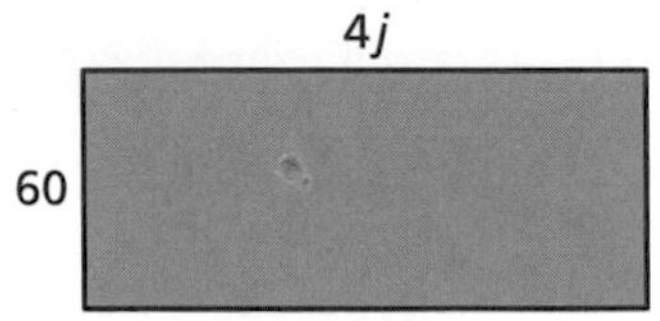

F. 35

G. 85

H. 140

I. 200

9. Jacob was evaluating the expression below when $x = -2$ and $y = 4$.

$$3 + x^2 \div y$$

His work is in the box below.

$$3 + x^2 \div y = 3 + (-2^2) \div 4$$
$$= 3 - 4 \div 4$$
$$= 3 - 1$$
$$= 3$$

What should Jacob do to correct the error that he made? *(7.NS.3)*

A. Divide 3 by 4 before subtracting.

B. Square -2, then divide.

C. Square then divide.

D. Subtract 4 from 3 before dividing.

10. Which number is equivalent to the expression shown below? *(7.NS.3)*

$$-2\frac{1}{4} - \left(-8\frac{3}{8}\right)$$

F. $-10\frac{5}{8}$

G. $-10\frac{1}{3}$

H. $6\frac{1}{8}$

I. $6\frac{1}{2}$

11. You want to buy the bicycle. You already have \$43.50 saved and plan to save an additional \$7.25 every week. *(7.EE.4a)*

Think
Solve
Explain

Part A Write and solve an equation to find the number of weeks you need to save before you can purchase the bicycle.

Part B How much sooner could you purchase the bicycle if you had a coupon for \$20 off and saved \$8.75 every week? Explain your reasoning.

14 Ratios and Proportions

"I am doing an experiment with slope. I want you to run up and down the board 10 times."

"Now with 2 more dog biscuits, do it again and we'll compare your rates."

"Dear Sir: I counted the number of bacon, cheese, and chicken dog biscuits in the box I bought."

"There were 16 bacon, 12 cheese, and only 8 chicken. That's a ratio of 4:3:2. Please go back to the original ratio of 1:1:1."

What You Learned Before

...or possibly proportional to our stupidity!

"I wonder if our rate is proportional to the slope of the hill."

Simplifying Fractions (4.NF.1)

Example 1 Simplify $\frac{4}{8}$.

$$\frac{4 \div 4}{8 \div 4} = \frac{1}{2}$$

Example 2 Simplify $\frac{10}{15}$.

$$\frac{10 \div 5}{15 \div 5} = \frac{2}{3}$$

Identifying Equivalent Fractions (4.NF.1)

Example 3 Is $\frac{1}{4}$ equivalent to $\frac{13}{52}$?

$$\frac{13 \div 13}{52 \div 13} = \frac{1}{4}$$

∴ $\frac{1}{4}$ is equivalent to $\frac{13}{52}$.

Example 4 Is $\frac{30}{54}$ equivalent to $\frac{5}{8}$?

$$\frac{30 \div 6}{54 \div 6} = \frac{5}{9}$$

∴ $\frac{30}{54}$ is *not* equivalent to $\frac{5}{8}$.

Solving Equations (6.EE.7)

Example 5 Solve $12x = 168$.

$12x = 168$	Write the equation.
$\frac{12x}{12} = \frac{168}{12}$	Division Property of Equality
$x = 14$	Simplify.

Check

$12x = 168$

$12(14) \stackrel{?}{=} 168$

$168 = 168$ ✓

Try It Yourself

Simplify.

1. $\frac{12}{144}$
2. $\frac{15}{45}$
3. $\frac{75}{100}$
4. $\frac{16}{24}$

Are the fractions equivalent? Explain.

5. $\frac{15}{60} \stackrel{?}{=} \frac{3}{4}$
6. $\frac{2}{5} \stackrel{?}{=} \frac{24}{144}$
7. $\frac{15}{20} \stackrel{?}{=} \frac{3}{5}$
8. $\frac{2}{8} \stackrel{?}{=} \frac{16}{64}$

Solve the equation. Check your solution.

9. $\frac{y}{-5} = 3$
10. $0.6 = 0.2a$
11. $-2w = -9$
12. $\frac{1}{7}n = -4$

14.1 Ratios and Rates

Essential Question How do rates help you describe real-life problems?

The Meaning of a Word ● Rate

When you rent snorkel gear at the beach, you should pay attention to the rental **rate**. The rental rate is in dollars per hour.

1 ACTIVITY: Finding Reasonable Rates

Work with a partner.

a. Match each description with a verbal rate.

b. Match each verbal rate with a numerical rate.

c. Give a reasonable numerical rate for each description. Then give an unreasonable rate.

Description	*Verbal Rate*	*Numerical Rate*
Your running rate in a 100-meter dash	Dollars per year	$\frac{\square \text{ in.}}{\text{yr}}$
The fertilization rate for an apple orchard	Inches per year	$\frac{\square \text{ lb}}{\text{acre}}$
The average pay rate for a professional athlete	Meters per second	$\frac{\$\square}{\text{yr}}$
The average rainfall rate in a rain forest	Pounds per acre	$\frac{\square \text{ m}}{\text{sec}}$

2 ACTIVITY: Simplifying Expressions That Contain Fractions

Work with a partner. Describe a situation where the given expression may apply. Show how you can rewrite each expression as a division problem. Then simplify and interpret your result.

a. $\dfrac{\frac{1}{2}\text{ c}}{4\text{ fl oz}}$ **b.** $\dfrac{2\text{ in.}}{\frac{3}{4}\text{ sec}}$ **c.** $\dfrac{\frac{3}{8}\text{ c sugar}}{\frac{3}{5}\text{ c flour}}$ **d.** $\dfrac{\frac{5}{6}\text{ gal}}{\frac{2}{3}\text{ sec}}$

COMMON CORE

Ratios and Rates

In this lesson, you will
- find ratios, rates, and unit rates.
- find ratios and rates involving ratios of fractions.

Learning Standards
7.RP.1
7.RP.3

3 ACTIVITY: Using Ratio Tables to Find Equivalent Rates

Work with a partner. A communications satellite in orbit travels about 18 miles every 4 seconds.

a. Identify the rate in this problem.

b. Recall that you can use *ratio tables* to find and organize equivalent ratios and rates. Complete the ratio table below.

Time (seconds)	4	8	12	16	20
Distance (miles)					

c. How can you use a ratio table to find the speed of the satellite in miles per minute? miles per hour?

d. How far does the satellite travel in 1 second? Solve this problem (1) by using a ratio table and (2) by evaluating a quotient.

e. How far does the satellite travel in $\frac{1}{2}$ second? Explain your steps.

4 ACTIVITY: Unit Analysis

Math Practice 7

View as Components

What is the product of the numbers? What is the product of the units? Explain.

Work with a partner. Describe a situation where the product may apply. Then find each product and list the units.

a. $10 \text{ gal} \times \frac{22 \text{ mi}}{\text{gal}}$ **b.** $\frac{7}{2} \text{ lb} \times \frac{\$3}{\frac{1}{2} \text{ lb}}$ **c.** $\frac{1}{2} \text{ sec} \times \frac{30 \text{ ft}^2}{\text{sec}}$

What Is Your Answer?

5. **IN YOUR OWN WORDS** How do rates help you describe real-life problems? Give two examples.

6. To estimate the annual salary for a given hourly pay rate, multiply by 2 and insert "000" at the end.

Sample: $10 per hour is about $20,000 per year.

a. Explain why this works. Assume the person is working 40 hours a week.

b. Estimate the annual salary for an hourly pay rate of $8 per hour.

c. You earn $1 million per month. What is your annual salary?

d. Why is the cartoon funny?

"We had someone apply for the job. He says he would like $1 million a month, but will settle for $8 an hour."

Practice Use what you discovered about ratios and rates to complete Exercises 7–10 on page 603.

14.1 Lesson

Check It Out
Lesson Tutorials
BigIdeasMath.com

Key Vocabulary
ratio, *p. 600*
rate, *p. 600*
unit rate, *p. 600*
complex fraction, *p. 601*

A **ratio** is a comparison of two quantities using division.

$\frac{3}{4}$, 3 to 4, 3 : 4

A **rate** is a ratio of two quantities with different units.

$\frac{60 \text{ miles}}{2 \text{ hours}}$

A rate with a denominator of 1 is called a **unit rate**.

$\frac{30 \text{ miles}}{1 \text{ hour}}$

EXAMPLE 1 Finding Ratios and Rates

There are 45 males and 60 females in a subway car. The subway car travels 2.5 miles in 5 minutes.

a. Find the ratio of males to females.

$$\frac{\text{males}}{\text{females}} = \frac{45}{60} = \frac{3}{4}$$

The ratio of males to females is $\frac{3}{4}$.

b. Find the speed of the subway car.

$$\text{2.5 miles in 5 minutes} = \frac{2.5 \text{ mi}}{5 \text{ min}} = \frac{2.5 \text{ mi} \div 5}{5 \text{ min} \div 5} = \frac{0.5 \text{ mi}}{1 \text{ min}}$$

The speed is 0.5 mile per minute.

EXAMPLE 2 Finding a Rate from a Ratio Table

The ratio table shows the costs for different amounts of artificial turf. Find the unit rate in dollars per square foot.

(× 4 between consecutive columns in each row)

Amount (square feet)	25	100	400	1600
Cost (dollars)	100	400	1600	6400

Use a ratio from the table to find the unit rate.

$$\frac{\text{cost}}{\text{amount}} = \frac{\$100}{25 \text{ ft}^2}$$ Use the first ratio in the table.

$$= \frac{\$4}{1 \text{ ft}^2}$$ Simplify.

So, the unit rate is $4 per square foot.

Remember
The abbreviation ft^2 means *square feet.*

On Your Own

Now You're Ready
Exercises 11–24

1. In Example 1, find the ratio of females to males.
2. In Example 1, find the ratio of females to total passengers.
3. The ratio table shows the distance that the *International Space Station* travels while orbiting Earth. Find the speed in miles per second.

Time (seconds)	3	6	9	12
Distance (miles)	14.4	28.8	43.2	57.6

A **complex fraction** has at least one fraction in the numerator, denominator, or both. You may need to simplify complex fractions when finding ratios and rates.

EXAMPLE 3 Finding a Rate from a Graph

The graph shows the speed of a subway car. Find the speed in miles per minute. Compare the speed to the speed of the subway car in Example 1.

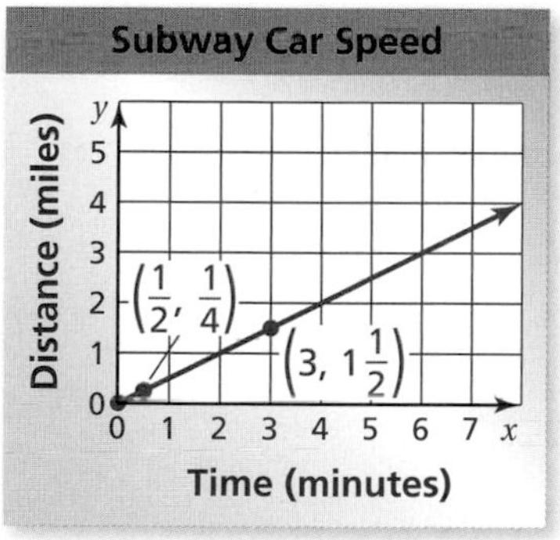

Step 1: Choose and interpret a point on the line.

The point $\left(\frac{1}{2}, \frac{1}{4}\right)$ indicates that the subway car travels $\frac{1}{4}$ mile in $\frac{1}{2}$ minute.

Step 2: Find the speed.

$$\frac{\text{distance traveled}}{\text{elapsed time}} = \frac{\frac{1}{4}}{\frac{1}{2}}$$ ← miles (numerator), ← minutes (denominator)

$$= \frac{1}{4} \div \frac{1}{2}$$ Rewrite the quotient.

$$= \frac{1}{4} \cdot 2 = \frac{1}{2}$$ Simplify.

∴ The speed of the subway car is $\frac{1}{2}$ mile per minute.

Because $\frac{1}{2}$ mile per minute = 0.5 mile per minute, the speeds of the two subway cars are the same.

On Your Own

Now You're Ready
Exercise 28

4. You use the point $\left(3, 1\frac{1}{2}\right)$ to find the speed of the subway car. Does your answer change? Explain your reasoning.

EXAMPLE 4 Solving a Ratio Problem

You mix $\frac{1}{2}$ cup of yellow paint for every $\frac{3}{4}$ cup of blue paint to make 15 cups of green paint. How much yellow paint and blue paint do you use?

Math Practice 1

Analyze Givens

What information is given in the problem? How does this help you know that the ratio table needs a "total" column? Explain.

Method 1: The ratio of yellow paint to blue paint is $\frac{1}{2}$ to $\frac{3}{4}$. Use a ratio table to find an equivalent ratio in which the total amount of yellow paint and blue paint is 15 cups.

Yellow (cups)	Blue (cups)	Total (cups)
$\frac{1}{2}$	$\frac{3}{4}$	$\frac{1}{2}+\frac{3}{4}=\frac{5}{4}$
2	3	5
6	9	15

(×4 from first row to second row; ×3 from second row to third row)

So, you use 6 cups of yellow paint and 9 cups of blue paint.

Method 2: Use the fraction of the green paint that is made from yellow paint and the fraction of the green paint that is made from blue paint. You use $\frac{1}{2}$ cup of yellow paint for every $\frac{3}{4}$ cup of blue paint, so the fraction of the green paint that is made from yellow paint is

$$\frac{\frac{1}{2}}{\frac{1}{2}+\frac{3}{4}} = \frac{\frac{1}{2}}{\frac{5}{4}} = \frac{1}{2} \cdot \frac{4}{5} = \frac{2}{5}.$$

(yellow → numerator; green → denominator)

Similarly, the fraction of the green paint that is made from blue paint is

$$\frac{\frac{3}{4}}{\frac{1}{2}+\frac{3}{4}} = \frac{\frac{3}{4}}{\frac{5}{4}} = \frac{3}{4} \cdot \frac{4}{5} = \frac{3}{5}.$$

(blue → numerator; green → denominator)

So, you use $\frac{2}{5} \cdot 15 = 6$ cups of yellow paint and $\frac{3}{5} \cdot 15 = 9$ cups of blue paint.

On Your Own

Now You're Ready
Exercises 33 and 34

5. How much yellow paint and blue paint do you use to make 20 cups of green paint?

14.1 Exercises

Check It Out
Help with Homework
BigIdeasMath.com

Vocabulary and Concept Check

1. **VOCABULARY** How can you tell when a rate is a unit rate?
2. **WRITING** Why do you think rates are usually written as unit rates?
3. **OPEN-ENDED** Write a real-life rate that applies to you.

Estimate the unit rate.

4. \$74.75

5. \$1.19

6. \$2.35

Practice and Problem Solving

Find the product. List the units.

7. $8\text{ h} \times \dfrac{\$9}{\text{h}}$
8. $8\text{ lb} \times \dfrac{\$3.50}{\text{lb}}$
9. $\dfrac{29}{2}\text{ sec} \times \dfrac{60\text{ MB}}{\text{sec}}$
10. $\dfrac{3}{4}\text{ h} \times \dfrac{19\text{ mi}}{\frac{1}{4}\text{ h}}$

Write the ratio as a fraction in simplest form.

1 11. 25 to 45
12. 63 : 28
13. 35 girls : 15 boys
14. 51 correct : 9 incorrect
15. 16 dogs to 12 cats
16. $2\frac{1}{3}$ feet : $4\frac{1}{2}$ feet

Find the unit rate.

17. 180 miles in 3 hours
18. 256 miles per 8 gallons
19. \$9.60 for 4 pounds
20. \$4.80 for 6 cans
21. 297 words in 5.5 minutes
22. $21\frac{3}{4}$ meters in $2\frac{1}{2}$ hours

Use the ratio table to find the unit rate with the specified units.

2 23. servings per package

Packages	3	6	9	12
Servings	13.5	27	40.5	54

24. feet per year

Years	2	6	10	14
Feet	7.2	21.6	36	50.4

25. **DOWNLOAD** At 1:00 P.M., you have 24 megabytes of a movie. At 1:15 P.M., you have 96 megabytes. What is the download rate in megabytes per minute?

26. **POPULATION** In 2007, the U.S. population was 302 million people. In 2012, it was 314 million. What was the rate of population change per year?

27. **PAINTING** A painter can paint 350 square feet in 1.25 hours. What is the painting rate in square feet per hour?

3 28. **TICKETS** The graph shows the cost of buying tickets to a concert.

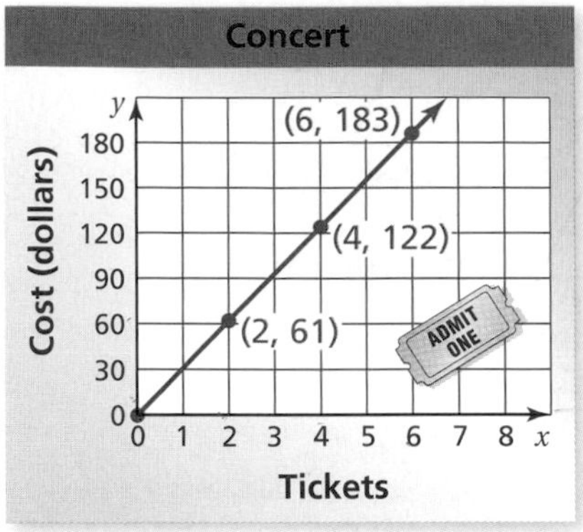

a. What does the point (4, 122) represent?

b. What is the unit rate?

c. What is the cost of buying 10 tickets?

29. **CRITICAL THINKING** Are the two statements equivalent? Explain your reasoning.

- The ratio of boys to girls is 2 to 3.
- The ratio of girls to boys is 3 to 2.

30. **TENNIS** A sports store sells three different packs of tennis balls. Which pack is the best buy? Explain.

31. **FLOORING** It costs $68 for 16 square feet of flooring. How much does it cost for 12 square feet of flooring?

32. **OIL SPILL** An oil spill spreads 25 square meters every $\frac{1}{6}$ hour. How much area does the oil spill cover after 2 hours?

4 33. **JUICE** You mix $\frac{1}{4}$ cup of juice concentrate for every 2 cups of water to make 18 cups of juice. How much juice concentrate and water do you use?

34. **LANDSCAPING** A supplier sells $2\frac{1}{4}$ pounds of mulch for every $1\frac{1}{3}$ pounds of gravel. The supplier sells 172 pounds of mulch and gravel combined. How many pounds of each item does the supplier sell?

35. **HEART RATE** Your friend's heart beats 18 times in 15 seconds when at rest. While running, your friend's heart beats 25 times in 10 seconds.

a. Find the heart rate in beats per minute at rest and while running.

b. How many more times does your friend's heart beat in 3 minutes while running than while at rest?

36. PRECISION The table shows nutritional information for three beverages.

a. Which has the most calories per fluid ounce?

b. Which has the least sodium per fluid ounce?

Beverage	Serving Size	Calories	Sodium
Whole milk	1 c	146	98 mg
Orange juice	1 pt	210	10 mg
Apple juice	24 fl oz	351	21 mg

37. RESEARCH Fire hydrants are painted one of four different colors to indicate the rate at which water comes from the hydrant.

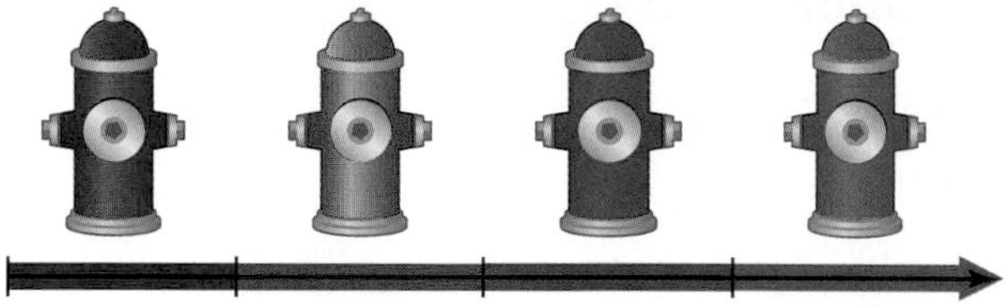

a. Use the Internet to find the ranges of the rates for each color.

b. Research why a firefighter needs to know the rate at which water comes out of a hydrant.

38. PAINT You mix $\frac{2}{5}$ cup of red paint for every $\frac{1}{4}$ cup of blue paint to make $1\frac{5}{8}$ gallons of purple paint.

a. How much red paint and blue paint do you use?

b. You decide that you want to make a lighter purple paint. You make the new mixture by adding $\frac{1}{10}$ cup of white paint for every $\frac{2}{5}$ cup of red paint and $\frac{1}{4}$ cup of blue paint. How much red paint, blue paint, and white paint do you use to make $\frac{3}{8}$ gallon of lighter purple paint?

39. Critical Thinking You and a friend start hiking toward each other from opposite ends of a 17.5-mile hiking trail. You hike $\frac{2}{3}$ mile every $\frac{1}{4}$ hour. Your friend hikes $2\frac{1}{3}$ miles per hour.

a. Who hikes faster? How much faster?

b. After how many hours do you meet?

c. When you meet, who hiked farther? How much farther?

Fair Game Review What you learned in previous grades & lessons

Copy and complete the statement using <, >, or =. *(Section 12.1)*

40. $\frac{9}{2} \;\square\; \frac{8}{3}$

41. $-\frac{8}{15} \;\square\; \frac{10}{18}$

42. $\frac{-6}{24} \;\square\; \frac{-2}{8}$

43. MULTIPLE CHOICE Which fraction is greater than $-\frac{2}{3}$ and less than $-\frac{1}{2}$? *(Section 12.1)*

Ⓐ $-\frac{3}{4}$ Ⓑ $-\frac{7}{12}$ Ⓒ $-\frac{5}{12}$ Ⓓ $-\frac{3}{8}$

14.2 Proportions

Essential Question How can proportions help you decide when things are "fair"?

The Meaning of a Word ● Proportional

When you work toward a goal, your success is usually **proportional** to the amount of work you put in.

An equation stating that two ratios are equal is a **proportion**.

1 ACTIVITY: Determining Proportions

Work with a partner. Tell whether the two ratios are equivalent. If they are not equivalent, change the next day to make the ratios equivalent. Explain your reasoning.

a. On the first day, you pay \$5 for 2 boxes of popcorn. The next day, you pay \$7.50 for 3 boxes.

First Day / Next Day

$$\frac{\$5.00}{2 \text{ boxes}} \stackrel{?}{=} \frac{\$7.50}{3 \text{ boxes}}$$

b. On the first day, it takes you $3\frac{1}{2}$ hours to drive 175 miles. The next day, it takes you 5 hours to drive 200 miles.

First Day / Next Day

$$\frac{3\frac{1}{2} \text{ h}}{175 \text{ mi}} \stackrel{?}{=} \frac{5 \text{ h}}{200 \text{ mi}}$$

c. On the first day, you walk 4 miles and burn 300 calories. The next day, you walk $3\frac{1}{3}$ miles and burn 250 calories.

First Day / Next Day

$$\frac{4 \text{ mi}}{300 \text{ cal}} \stackrel{?}{=} \frac{3\frac{1}{3} \text{ mi}}{250 \text{ cal}}$$

d. On the first day, you paint 150 square feet in $2\frac{1}{2}$ hours. The next day, you paint 200 square feet in 4 hours.

First Day / Next Day

$$\frac{150 \text{ ft}^2}{2\frac{1}{2} \text{ h}} \stackrel{?}{=} \frac{200 \text{ ft}^2}{4 \text{ h}}$$

COMMON CORE

Proportions

In this lesson, you will

- use equivalent ratios to determine whether two ratios form a proportion.
- use the Cross Products Property to determine whether two ratios form a proportion.

Learning Standard 7.RP.2a

2 ACTIVITY: Checking a Proportion

Work with a partner.

a. It is said that "one year in a dog's life is equivalent to seven years in a human's life." Explain why Newton thinks he has a score of 105 points. Did he solve the proportion correctly?

$$\frac{1 \text{ year}}{7 \text{ years}} \stackrel{?}{=} \frac{15 \text{ points}}{105 \text{ points}}$$

b. If Newton thinks his score is 98 points, how many points does he actually have? Explain your reasoning.

"I got 15 on my online test. That's 105 in dog points! Isn't that an A+?"

3 ACTIVITY: Determining Fairness

Math Practice 3
Justify Conclusions
What information can you use to justify your conclusion?

Work with a partner. Write a ratio for each sentence. Compare the ratios. If they are equal, then the answer is "It is fair." If they are not equal, then the answer is "It is not fair." Explain your reasoning.

a. You pay \$184 for 2 tickets to a concert. & I pay \$266 for 3 tickets to the same concert. → **Is this fair?**

b. You get 75 points for answering 15 questions correctly. & I get 70 points for answering 14 questions correctly. → **Is this fair?**

c. You trade 24 football cards for 15 baseball cards. & I trade 20 football cards for 32 baseball cards. → **Is this fair?**

What Is Your Answer?

4. Find a recipe for something you like to eat. Then show how two of the ingredient amounts are proportional when you double or triple the recipe.
5. **IN YOUR OWN WORDS** How can proportions help you decide when things are "fair"? Give an example.

Practice Use what you discovered about proportions to complete Exercises 15–20 on page 610.

14.2 Lesson

Check It Out
Lesson Tutorials
BigIdeasMath.com

Key Vocabulary
proportion, *p. 608*
proportional, *p. 608*
cross products, *p. 609*

Key Idea

Proportions

Words A **proportion** is an equation stating that two ratios are equivalent. Two quantities that form a proportion are **proportional**.

Numbers $\frac{2}{3} = \frac{4}{6}$ The proportion is read "2 is to 3 as 4 is to 6."

EXAMPLE 1 Determining Whether Ratios Form a Proportion

Tell whether $\frac{6}{4}$ and $\frac{8}{12}$ form a proportion.

Compare the ratios in simplest form.

$$\frac{6}{4} = \frac{6 \div 2}{4 \div 2} = \frac{3}{2}$$

$$\frac{8}{12} = \frac{8 \div 4}{12 \div 4} = \frac{2}{3}$$

The ratios are *not* equivalent.

So, $\frac{6}{4}$ and $\frac{8}{12}$ do *not* form a proportion.

EXAMPLE 2 Determining Whether Two Quantities Are Proportional

Tell whether x and y are proportional.

x	y
$\frac{1}{2}$	3
1	6
$\frac{3}{2}$	9
2	12

Compare each ratio x to y in simplest form.

$$\frac{\frac{1}{2}}{3} = \frac{1}{6} \qquad \frac{1}{6} \qquad \frac{\frac{3}{2}}{9} = \frac{1}{6} \qquad \frac{2}{12} = \frac{1}{6}$$

The ratios are equivalent.

So, x and y are proportional.

Reading
Two quantities that are proportional are in a *proportional relationship.*

On Your Own

Now You're Ready
Exercises 5–14

Tell whether the ratios form a proportion.

1. $\frac{1}{2}, \frac{5}{10}$
2. $\frac{4}{6}, \frac{18}{24}$
3. $\frac{10}{3}, \frac{5}{6}$
4. $\frac{25}{20}, \frac{15}{12}$
5. Tell whether x and y are proportional.

Birdhouses Built, x	1	2	4	6
Nails Used, y	12	24	48	72

Multi-Language Glossary at BigIdeasMath.com

Key Ideas

Cross Products

In the proportion $\frac{a}{b} = \frac{c}{d}$, the products $a \cdot d$ and $b \cdot c$ are called **cross products**.

Cross Products Property

Words The cross products of a proportion are equal.

Numbers

$$\frac{2}{3} = \frac{4}{6}$$

$2 \cdot 6 = 3 \cdot 4$

Algebra

$$\frac{a}{b} = \frac{c}{d}$$

$ad = bc$,
where $b \neq 0$ and $d \neq 0$

Study Tip

You can use the Multiplication Property of Equality to show that the cross products are equal.

$$\frac{a}{b} = \frac{c}{d}$$

$$bd \cdot \frac{a}{b} = bd \cdot \frac{c}{d}$$

$$ad = bc$$

EXAMPLE 3 Identifying Proportional Relationships

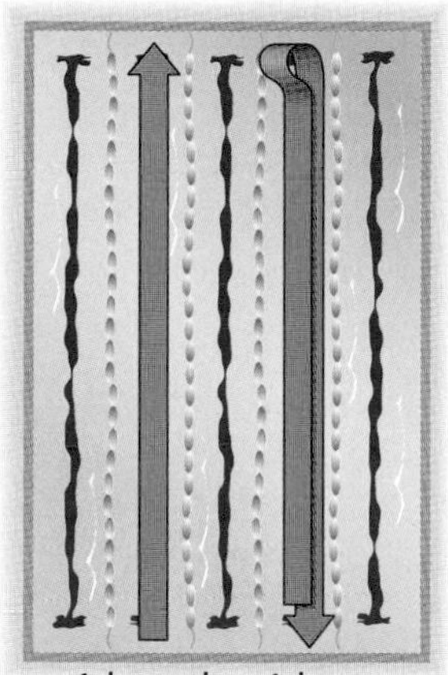

1 length 1 lap

You swim your first 4 laps in 2.4 minutes. You complete 16 laps in 12 minutes. Is the number of laps proportional to your time?

Method 1: Compare unit rates.

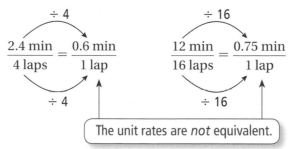

$$\frac{2.4 \text{ min}}{4 \text{ laps}} = \frac{0.6 \text{ min}}{1 \text{ lap}} \quad (\div 4)$$

$$\frac{12 \text{ min}}{16 \text{ laps}} = \frac{0.75 \text{ min}}{1 \text{ lap}} \quad (\div 16)$$

The unit rates are *not* equivalent.

So, the number of laps is *not* proportional to the time.

Method 2: Use the Cross Products Property.

$\frac{2.4 \text{ min}}{4 \text{ laps}} \stackrel{?}{=} \frac{12 \text{ min}}{16 \text{ laps}}$	Test to see if the rates are equivalent.
$2.4 \cdot 16 \stackrel{?}{=} 4 \cdot 12$	Find the cross products.
$38.4 \neq 48$	The cross products are *not* equal.

So, the number of laps is *not* proportional to the time.

On Your Own

Now You're Ready
Exercises 15–20

6. You read the first 20 pages of a book in 25 minutes. You read 36 pages in 45 minutes. Is the number of pages read proportional to your time?

14.2 Exercises

Check It Out
Help with Homework
BigIdeasMath.com

Vocabulary and Concept Check

1. **VOCABULARY** What does it mean for two ratios to form a proportion?
2. **VOCABULARY** What are two ways you can tell that two ratios form a proportion?
3. **OPEN-ENDED** Write two ratios that are equivalent to $\frac{3}{5}$.
4. **WHICH ONE DOESN'T BELONG?** Which ratio does *not* belong with the other three? Explain your reasoning.

$\frac{4}{10}$ $\frac{2}{5}$ $\frac{3}{5}$ $\frac{6}{15}$

Practice and Problem Solving

Tell whether the ratios form a proportion.

1 5. $\frac{1}{3}, \frac{7}{21}$
6. $\frac{1}{5}, \frac{6}{30}$
7. $\frac{3}{4}, \frac{24}{18}$
8. $\frac{2}{5}, \frac{40}{16}$
9. $\frac{48}{9}, \frac{16}{3}$
10. $\frac{18}{27}, \frac{33}{44}$
11. $\frac{7}{2}, \frac{16}{6}$
12. $\frac{12}{10}, \frac{14}{12}$

Tell whether *x* and *y* are proportional.

2 13.

x	1	2	3	4
y	7	8	9	10

14.

x	2	4	6	8
y	5	10	15	20

Tell whether the two rates form a proportion.

3 15. 7 inches in 9 hours; 42 inches in 54 hours
16. 12 players from 21 teams; 15 players from 24 teams
17. 440 calories in 4 servings; 300 calories in 3 servings
18. 120 units made in 5 days; 88 units made in 4 days
19. 66 wins in 82 games; 99 wins in 123 games
20. 68 hits in 172 at bats; 43 hits in 123 at bats

21. **FITNESS** You can do 90 sit-ups in 2 minutes. Your friend can do 135 sit-ups in 3 minutes. Do these rates form a proportion? Explain.
22. **HEART RATES** Find the heart rates of you and your friend. Do these rates form a proportion? Explain.

	Heartbeats	Seconds
You	22	20
Friend	18	15

Tell whether the ratios form a proportion.

23. $\frac{2.5}{4}, \frac{7}{11.2}$

24. 2 to 4, 11 to $\frac{11}{2}$

25. $2 : \frac{4}{5}, \frac{3}{4} : \frac{3}{10}$

26. **PAY RATE** You earn \$56 walking your neighbor's dog for 8 hours. Your friend earns \$36 painting your neighbor's fence for 4 hours.

- **a.** What is your pay rate?
- **b.** What is your friend's pay rate?
- **c.** Are the pay rates equivalent? Explain.

27. **GEOMETRY** Are the heights and bases of the two triangles proportional? Explain.

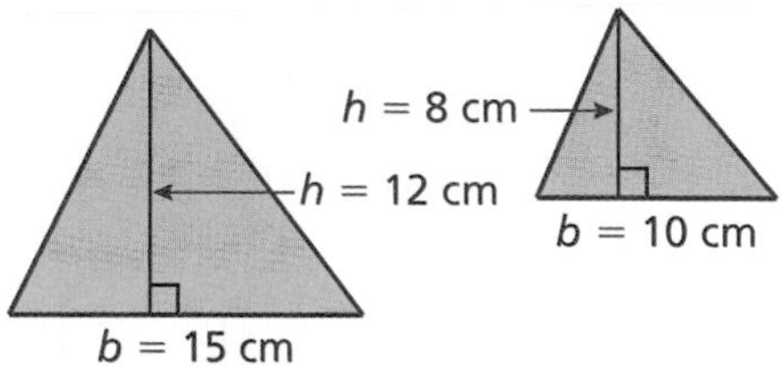

28. **BASEBALL** A pitcher coming back from an injury limits the number of pitches thrown in bull pen sessions as shown.

- **a.** Which quantities are proportional?
- **b.** How many pitches that are not curveballs do you think the pitcher will throw in Session 5?

Session Number, x	Pitches, y	Curveballs, z
1	10	4
2	20	8
3	30	12
4	40	16

29. **NAIL POLISH** A specific shade of red nail polish requires 7 parts red to 2 parts yellow. A mixture contains 35 quarts of red and 8 quarts of yellow. How can you fix the mixture to make the correct shade of red?

30. **COIN COLLECTION** The ratio of quarters to dimes in a coin collection is 5 : 3. You add the same number of new quarters as dimes to the collection.

- **a.** Is the ratio of quarters to dimes still 5 : 3?
- **b.** If so, illustrate your answer with an example. If not, show why with a "counterexample."

31. **AGE** You are 13 years old, and your cousin is 19 years old. As you grow older, is your age proportional to your cousin's age? Explain your reasoning.

32. Critical Thinking Ratio *A* is equivalent to Ratio *B*. Ratio *B* is equivalent to Ratio *C*. Is Ratio *A* equivalent to Ratio *C*? Explain.

Fair Game Review What you learned in previous grades & lessons

Add or subtract. *(Section 11.2 and Section 11.3)*

33. $-28 + 15$

34. $-6 + (-11)$

35. $-10 - 8$

36. $-17 - (-14)$

37. **MULTIPLE CHOICE** Which fraction is not equivalent to $\frac{2}{6}$? *(Skills Review Handbook)*

Ⓐ $\frac{1}{3}$ Ⓑ $\frac{12}{36}$ Ⓒ $\frac{4}{12}$ Ⓓ $\frac{6}{9}$

Extension 14.2 Graphing Proportional Relationships

Check It Out
Lesson Tutorials
BigIdeasMath.com

Recall that you can graph the values from a ratio table.

	Time, x (seconds)	Height, y (meters)	
	3	2	
+ 3	6	4	+ 2
+ 3	9	6	+ 2
+ 3	12	8	+ 2

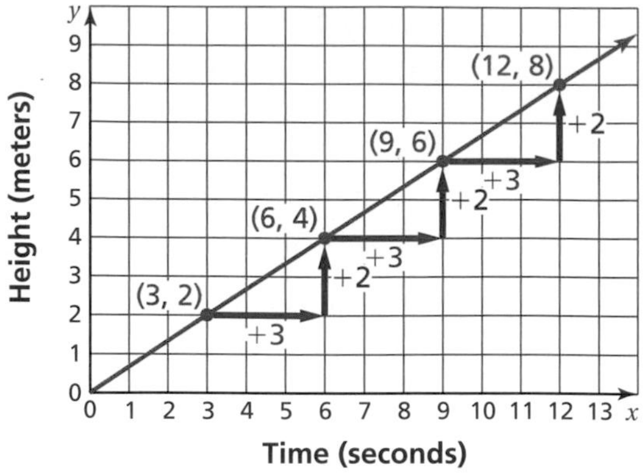

The structure in the ratio table shows why the graph has a constant *rate of change*. You can use the constant rate of change to show that the graph passes through the origin. The graph of every proportional relationship is a line through the origin.

EXAMPLE 1 Determining Whether Two Quantities Are Proportional

Use a graph to tell whether x and y are in a proportional relationship.

a.

x	2	4	6
y	6	8	10

Plot (2, 6), (4, 8), and (6, 10).
Draw a line through the points.

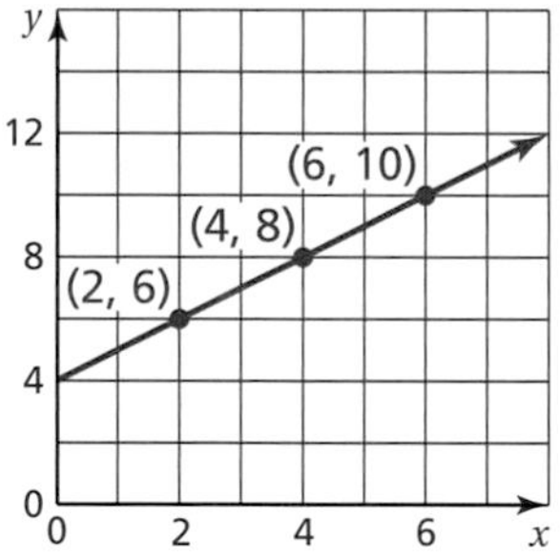

The graph is a line that does not pass through the origin.

∴ So, x and y are not in a proportional relationship.

b.

x	1	2	3
y	2	4	6

Plot (1, 2), (2, 4), and (3, 6).
Draw a line through the points.

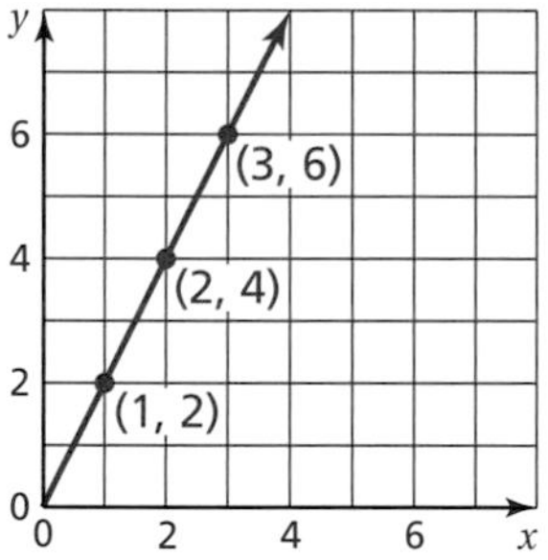

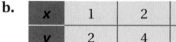

The graph is a line that passes through the origin.

∴ So, x and y are in a proportional relationship.

Common Core
Proportions
In this extension, you will
- use graphs to determine whether two ratios form a proportion.
- interpret graphs of proportional relationships.

Learning Standards
7.RP.2a
7.RP.2b
7.RP.2d

Practice

Use a graph to tell whether x and y are in a proportional relationship.

1.

x	1	2	3	4
y	3	4	5	6

2.

x	1	3	5	7
y	0.5	1.5	2.5	3.5

EXAMPLE 2 Interpreting the Graph of a Proportional Relationship

The graph shows that the distance traveled by the Mars rover *Curiosity* is proportional to the time traveled. Interpret each plotted point in the graph.

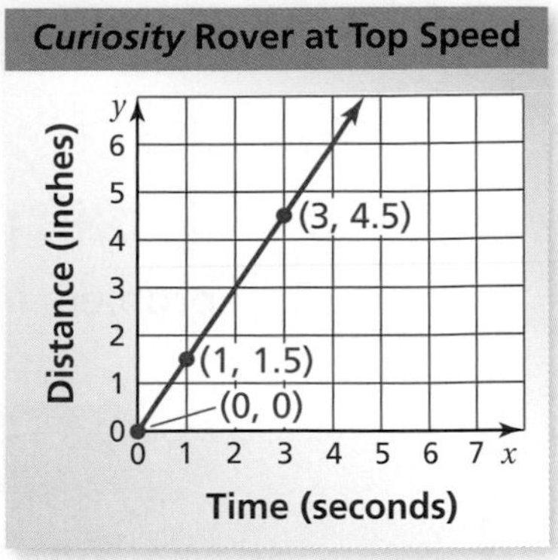

(0, 0): The rover travels 0 inches in 0 seconds.

(1, 1.5): The rover travels 1.5 inches in 1 second. So, the unit rate is 1.5 inches per second.

(3, 4.5): The rover travels 4.5 inches in 3 seconds. Because the relationship is proportional, you can also use this point to find the unit rate.

$$\frac{4.5 \text{ in.}}{3 \text{ sec}} = \frac{1.5 \text{ in.}}{1 \text{ sec}}, \text{ or } 1.5 \text{ inches per second}$$

Study Tip

In the graph of a proportional relationship, you can find the unit rate r from the point $(1, r)$.

Practice

Interpret each plotted point in the graph of the proportional relationship.

3. 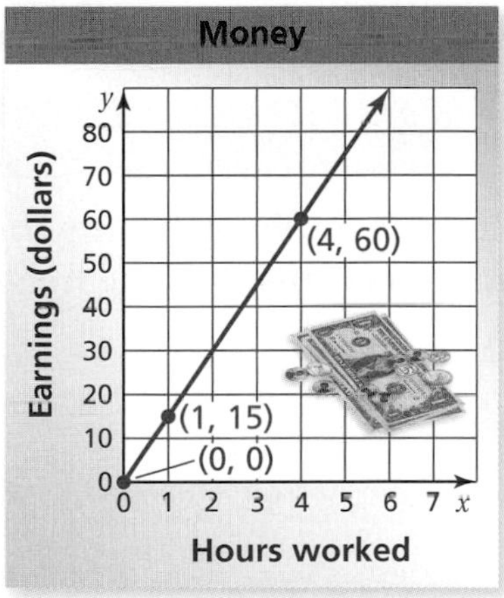

4. 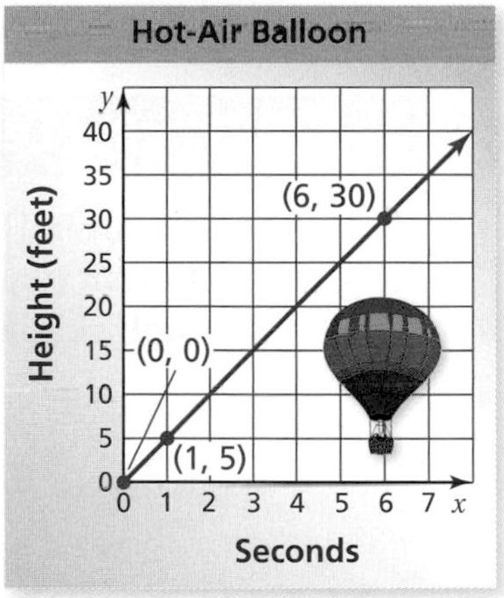

Tell whether x and y are in a proportional relationship. If so, find the unit rate.

5.

x (hours)	1	4	7	10
y (feet)	5	20	35	50

6. Let y be the temperature x hours after midnight. The temperature is 60°F at midnight and decreases 2°F every $\frac{1}{2}$ hour.

7. **REASONING** The graph of a proportional relationship passes through (12, 16) and $(1, y)$. Find y.

8. **MOVIE RENTAL** You pay \$1 to rent a movie plus an additional \$0.50 per day until you return the movie. Your friend pays \$1.25 per day to rent a movie.

   a. Make tables showing the costs to rent a movie up to 5 days.

   b. Which person pays an amount proportional to the number of days rented?

14.3 Writing Proportions

Essential Question How can you write a proportion that solves a problem in real life?

1 ACTIVITY: Writing Proportions

Work with a partner. A rough rule for finding the correct bat length is "the bat length should be half of the batter's height." So, a 62-inch-tall batter uses a bat that is 31 inches long. Write a proportion to find the bat length for each given batter height.

a. 58 inches

b. 60 inches

c. 64 inches

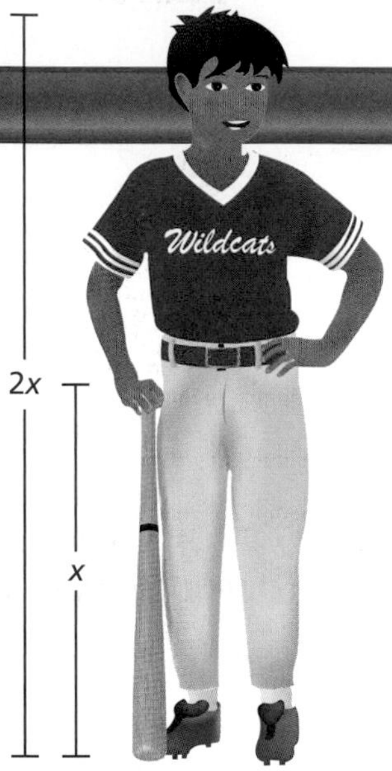

2 ACTIVITY: Bat Lengths

Work with a partner. Here is a more accurate table for determining the bat length for a batter. Find all the batter heights and corresponding weights for which the rough rule in Activity 1 is exact.

Weight of Batter (pounds)	Height of Batter (inches)							
	45–48	49–52	53–56	57–60	61–64	65–68	69–72	Over 72
Under 61	28	29	29					
61–70	28	29	30	30				
71–80	28	29	30	30	31			
81–90	29	29	30	30	31	32		
91–100	29	30	30	31	31	32		
101–110	29	30	30	31	31	32		
111–120	29	30	30	31	31	32		
121–130	29	30	30	31	32	33	33	
131–140	30	30	31	31	32	33	33	
141–150	30	30	31	31	32	33	33	
151–160	30	31	31	32	32	33	33	33
161–170		31	31	32	32	33	33	34
171–180				32	33	33	34	34
Over 180					33	33	34	34

COMMON CORE

Proportions

In this lesson, you will
- write proportions.
- solve proportions using mental math.

Learning Standards
7.RP.2c
7.RP.3

3 ACTIVITY: Writing Proportions

Math Practice 8

Evaluate Results

How do you know if your results are reasonable? Explain.

Work with a partner. The batting average of a baseball player is the number of "hits" divided by the number of "at bats."

$$\text{batting average} = \frac{\text{hits } (H)}{\text{at bats } (A)}$$

A player whose batting average is 0.250 is said to be "batting 250."

$$\frac{20 \text{ hits}}{80 \text{ at bats}} = 0.250 = \frac{250 \text{ hits}}{1000 \text{ at bats}}$$

Actual hits

Actual at bats

Batting average

Batting 250 out of 1000

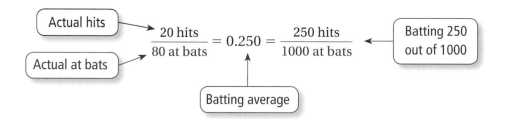

Write a proportion to find how many hits H a player needs to achieve the given batting average. Then solve the proportion.

a. 50 times at bat; batting average is 0.200.

b. 84 times at bat; batting average is 0.250.

c. 80 times at bat; batting average is 0.350.

d. 1 time at bat; batting average is 1.000.

What Is Your Answer?

4. **IN YOUR OWN WORDS** How can you write a proportion that solves a problem in real life?

5. Two players have the same batting average.

	At Bats	Hits	Batting Average
Player 1	132	45	
Player 2	132	45	

Player 1 gets four hits in the next five at bats. Player 2 gets three hits in the next three at bats.

a. Who has the higher batting average?

b. Does this seem fair? Explain your reasoning.

Practice Use what you discovered about proportions to complete Exercises 4–7 on page 618.

14.3 Lesson

Check It Out
Lesson Tutorials
BigIdeasMath.com

One way to write a proportion is to use a table.

	Last Month	This Month
Purchase	2 ringtones	3 ringtones
Total Cost	6 dollars	x dollars

Use the columns or the rows to write a proportion.

Use columns:

$$\frac{2 \text{ ringtones}}{6 \text{ dollars}} = \frac{3 \text{ ringtones}}{x \text{ dollars}}$$

Numerators have the same units.

Denominators have the same units.

Use rows:

$$\frac{2 \text{ ringtones}}{3 \text{ ringtones}} = \frac{6 \text{ dollars}}{x \text{ dollars}}$$

The units are the same on each side of the proportion.

EXAMPLE 1 Writing a Proportion

Black Bean Soup

1.5 cups black beans
0.5 cup salsa
2 cups water
1 tomato
2 teaspoons seasoning

A chef increases the amounts of ingredients in a recipe to make a proportional recipe. The new recipe has 6 cups of black beans. Write a proportion that gives the number x of tomatoes in the new recipe.

Organize the information in a table.

	Original Recipe	New Recipe
Black Beans	1.5 cups	6 cups
Tomatoes	1 tomato	x tomatoes

∴ One proportion is $\frac{1.5 \text{ cups beans}}{1 \text{ tomato}} = \frac{6 \text{ cups beans}}{x \text{ tomatoes}}$.

On Your Own

Now You're Ready
Exercises 8–11

1. Write a different proportion that gives the number x of tomatoes in the new recipe.
2. Write a proportion that gives the amount y of water in the new recipe.

EXAMPLE 2 Solving Proportions Using Mental Math

Solve $\frac{3}{2} = \frac{x}{8}$.

Step 1: Think: The product of 2 and what number is 8?

$$\frac{3}{2} = \frac{x}{8}$$

$2 \times ? = 8$

Step 2: Because the product of 2 and 4 is 8, multiply the numerator by 4 to find x.

$3 \times 4 = 12$

$$\frac{3}{2} = \frac{x}{8}$$

$2 \times 4 = 8$

∴ The solution is $x = 12$.

EXAMPLE 3 Solving Proportions Using Mental Math

In Example 1, how many tomatoes are in the new recipe?

Solve the proportion $\frac{1.5}{1} = \frac{6}{x}$.

(6 → cups black beans; x → tomatoes)

Step 1: Think: The product of 1.5 and what number is 6?

$1.5 \times ? = 6$

$$\frac{1.5}{1} = \frac{6}{x}$$

Step 2: Because the product of 1.5 and 4 is 6, multiply the denominator by 4 to find x.

$1.5 \times 4 = 6$

$$\frac{1.5}{1} = \frac{6}{x}$$

$1 \times 4 = 4$

∴ So, there are 4 tomatoes in the new recipe.

On Your Own

Now You're Ready
Exercises 16–21

Solve the proportion.

3. $\frac{5}{8} = \frac{20}{d}$

4. $\frac{7}{z} = \frac{14}{10}$

5. $\frac{21}{24} = \frac{x}{8}$

6. A school has 950 students. The ratio of female students to all students is $\frac{48}{95}$. Write and solve a proportion to find the number f of students who are female.

14.3 Exercises

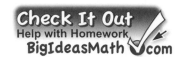

Check It Out
Help with Homework
BigIdeasMath.com

Vocabulary and Concept Check

1. **WRITING** Describe two ways you can use a table to write a proportion.
2. **WRITING** What is your first step when solving $\frac{x}{15} = \frac{3}{5}$? Explain.
3. **OPEN-ENDED** Write a proportion using an unknown value x and the ratio 5 : 6. Then solve it.

Practice and Problem Solving

Write a proportion to find how many points a student needs to score on the test to get the given score.

4. test worth 50 points; test score of 40%
5. test worth 50 points; test score of 78%
6. test worth 80 points; test score of 80%
7. test worth 150 points; test score of 96%

Use the table to write a proportion.

(1) 8.

	Game 1	Game 2
Points	12	18
Shots	14	w

9.

	May	June
Winners	n	34
Entries	85	170

10.

	Today	Yesterday
Miles	15	m
Hours	2.5	4

11.

	Race 1	Race 2
Meters	100	200
Seconds	x	22.4

12. **ERROR ANALYSIS** Describe and correct the error in writing the proportion.

✗

	Monday	Tuesday
Dollars	2.08	d
Ounces	8	16

$$\frac{2.08}{16} = \frac{d}{8}$$

13. **T-SHIRTS** You can buy 3 T-shirts for \$24. Write a proportion that gives the cost c of buying 7 T-shirts.
14. **COMPUTERS** A school requires 2 computers for every 5 students. Write a proportion that gives the number c of computers needed for 145 students.
15. **SWIM TEAM** The school team has 80 swimmers. The ratio of seventh-grade swimmers to all swimmers is 5 : 16. Write a proportion that gives the number s of seventh-grade swimmers.

Solve the proportion.

2 3 **16.** $\frac{1}{4} = \frac{z}{20}$ **17.** $\frac{3}{4} = \frac{12}{y}$ **18.** $\frac{35}{k} = \frac{7}{3}$

19. $\frac{15}{8} = \frac{45}{c}$ **20.** $\frac{b}{36} = \frac{5}{9}$ **21.** $\frac{1.4}{2.5} = \frac{g}{25}$

22. ORCHESTRA In an orchestra, the ratio of trombones to violas is 1 to 3.

a. There are 9 violas. Write a proportion that gives the number t of trombones in the orchestra.

b. How many trombones are in the orchestra?

23. ATLANTIS Your science teacher has a 1 : 200 scale model of the space shuttle *Atlantis*. Which of the proportions can you use to find the actual length x of *Atlantis*? Explain.

$\frac{1}{200} = \frac{19.5}{x}$ $\frac{1}{200} = \frac{x}{19.5}$ $\frac{200}{19.5} = \frac{x}{1}$ $\frac{x}{200} = \frac{1}{19.5}$

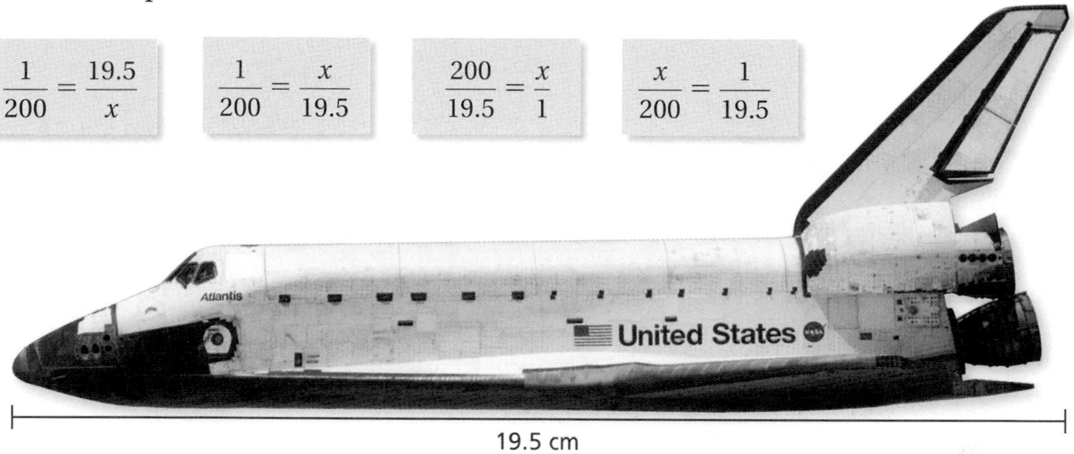

24. YOU BE THE TEACHER Your friend says "$48x = 6 \cdot 12$." Is your friend right? Explain.

Solve $\frac{6}{x} = \frac{12}{48}$.

25. Reasoning There are 180 white lockers in the school. There are 3 white lockers for every 5 blue lockers. How many lockers are in the school?

Fair Game Review What you learned in previous grades & lessons

Solve the equation. *(Section 13.4)*

26. $\frac{x}{6} = 25$ **27.** $8x = 72$ **28.** $150 = 2x$ **29.** $35 = \frac{x}{4}$

30. MULTIPLE CHOICE What is the value of $-\frac{9}{4} + \left|-\frac{8}{5}\right| - 2\frac{1}{2}$? *(Section 12.3)*

Ⓐ $-6\frac{7}{20}$ Ⓑ $-5\frac{7}{20}$ Ⓒ $-3\frac{3}{20}$ Ⓓ $-2\frac{3}{20}$

14 Study Help

Check It Out
Graphic Organizer
BigIdeasMath.com

You can use an **information wheel** to organize information about a concept. Here is an example of an information wheel for ratio.

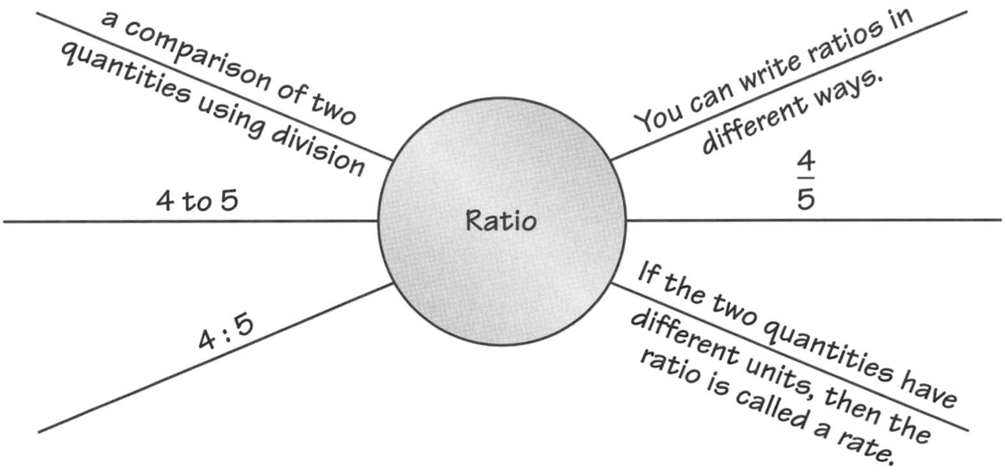

Ratio
a comparison of two quantities using division
You can write ratios in different ways.
4 to 5
$\frac{4}{5}$
4 : 5
If the two quantities have different units, then the ratio is called a rate.

On Your Own

Make information wheels to help you study these topics.

1. rate
2. unit rate
3. proportion
4. cross products
5. graphing proportional relationships

After you complete this chapter, make information wheels for the following topics.

6. solving proportions
7. slope
8. direct variation

"My information wheel summarizes how cats act when they get baths."

14.1–14.3 Quiz

Check It Out
Progress Check
BigIdeasMath.com

Write the ratio as a fraction in simplest form. *(Section 14.1)*

1. 18 red buttons : 12 blue buttons

2. $\frac{5}{4}$ inches to $\frac{2}{3}$ inch

Use the ratio table to find the unit rate with the specified units. *(Section 14.1)*

3. cost per song

Songs	0	2	4	6
Cost	\$0	\$1.98	\$3.96	\$5.94

4. gallons per hour

Hours	3	6	9	12
Gallons	10.5	21	31.5	42

Tell whether the ratios form a proportion. *(Section 14.2)*

5. $\frac{1}{8}, \frac{4}{32}$

6. $\frac{2}{3}, \frac{10}{30}$

7. $\frac{7}{4}, \frac{28}{16}$

Tell whether the two rates form a proportion. *(Section 14.2)*

8. 75 miles in 3 hours; 140 miles in 4 hours

9. 12 gallons in 4 minutes; 21 gallons in 7 minutes

10. 150 steps in 50 feet; 72 steps in 24 feet

11. 3 rotations in 675 days; 2 rotations in 730 days

Use the table to write a proportion. *(Section 14.3)*

12.

	Monday	Tuesday
Dollars	42	56
Hours	6	h

13.

	Series 1	Series 2
Games	g	6
Wins	4	3

14. MUSIC DOWNLOAD The amount of time needed to download music is shown in the table. Find the unit rate in megabytes per second. *(Section 14.1)*

Seconds	6	12	18	24
Megabytes	2	4	6	8

15. SOUND The graph shows the distance that sound travels through steel. Interpret each plotted point in the graph of the proportional relationship. *(Section 14.2)*

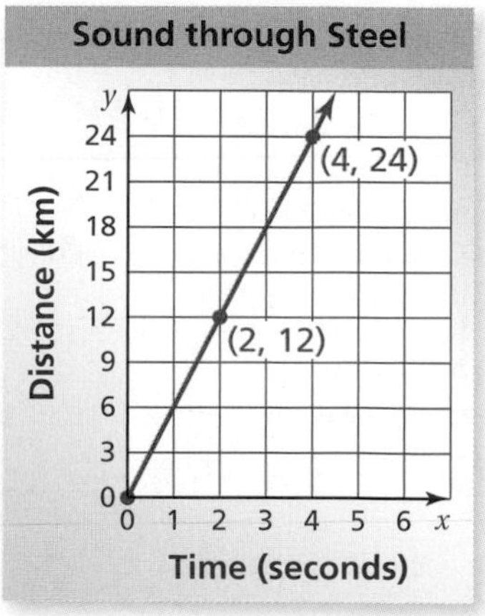

16. GAMING You advance 3 levels in 15 minutes. Your friend advances 5 levels in 20 minutes. Do these rates form a proportion? Explain. *(Section 14.2)*

17. CLASS TIME You spend 150 minutes in 3 classes. Write and solve a proportion to find how many minutes you spend in 5 classes. *(Section 14.3)*

14.4 Solving Proportions

Essential Question How can you use ratio tables and cross products to solve proportions?

1 ACTIVITY: Solving a Proportion in Science

Work with a partner. You can use ratio tables to determine the amount of a compound (like salt) that is dissolved in a solution. Determine the unknown quantity. Explain your procedure.

a. Salt Water

Salt Water	1 L	3 L
Salt	250 g	x g

1 liter 3 liters

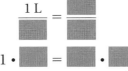

$\frac{1\text{ L}}{\square} = \frac{\square}{\square}$ Write proportion.

$1 \cdot \square = \square \cdot \square$ Set cross products equal.

$\square = \square$ Simplify.

There are $\square$ grams of salt in the 3-liter solution.

b. White Glue Solution

Water	$\frac{1}{2}$ cup	1 cup
White Glue	$\frac{1}{2}$ cup	x cups

c. Borax Solution

Borax	1 tsp	2 tsp
Water	1 cup	x cups

d. Slime (See recipe.)

Borax Solution	$\frac{1}{2}$ cup	1 cup
White Glue Solution	y cups	x cups

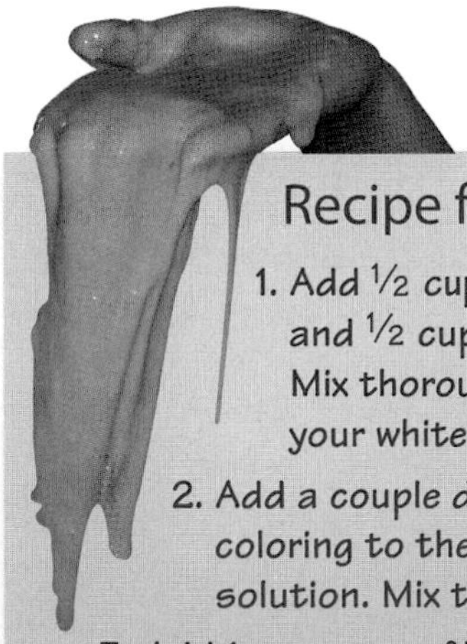

Recipe for SLIME

1. Add $\frac{1}{2}$ cup of water and $\frac{1}{2}$ cup white glue. Mix thoroughly. This is your white glue solution.
2. Add a couple drops of food coloring to the white glue solution. Mix thoroughly.
3. Add 1 teaspoon of borax to 1 cup of water. Mix thoroughly. This is your borax solution (about 1 cup).
4. Pour the borax solution and the glue solution into a separate bowl.
5. Place the slime that forms into a plastic bag. Squeeze the mixture repeatedly to mix it up.

COMMON CORE

Proportions

In this lesson, you will

- solve proportions using multiplication or the Cross Products Property.
- use a point on a graph to write and solve proportions.

Learning Standards
7.RP.2b
7.RP.2c

2 ACTIVITY: The Game of Criss Cross

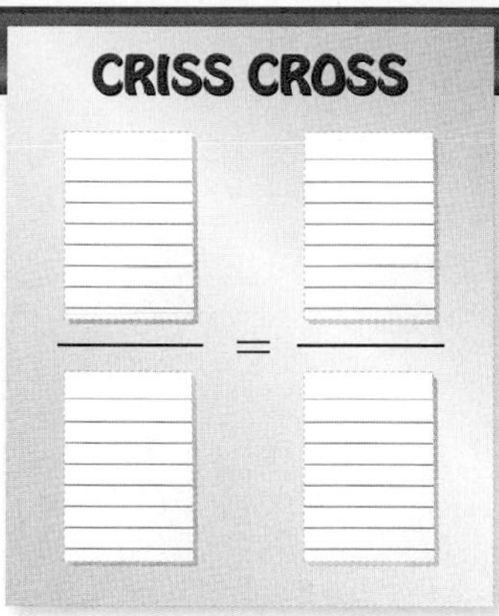

Math Practice 2

Use Operations

How can you use the name of the game to determine which operation to use?

Preparation:

- Cut index cards to make 48 playing cards.
- Write each number on a card.

  1, 1, 1, 2, 2, 2, 3, 3, 3, 4, 4, 4, 5, 5, 5, 6, 6, 6, 7, 7, 7, 8, 8, 8, 9, 9, 9, 10, 10, 10, 12, 12, 12, 13, 13, 13, 14, 14, 14, 15, 15, 15, 16, 16, 16, 18, 20, 25
- Make a copy of the game board.

To Play:

- Play with a partner.
- Deal eight cards to each player.
- Begin by drawing a card from the remaining cards. Use four of your cards to try to form a proportion.
- Lay the four cards on the game board. If you form a proportion, then say "Criss Cross." You earn 4 points. Place the four cards in a discard pile. Now it is your partner's turn.
- If you cannot form a proportion, then it is your partner's turn.
- When the original pile of cards is empty, shuffle the cards in the discard pile. Start again.
- The first player to reach 20 points wins.

What Is Your Answer?

3. **IN YOUR OWN WORDS** How can you use ratio tables and cross products to solve proportions? Give an example.
4. **PUZZLE** Use each number once to form three proportions.

1	2	10	4	12	20
15	5	16	6	8	3

Practice Use what you discovered about solving proportions to complete Exercises 10–13 on page 626.

14.4 Lesson

Check It Out
Lesson Tutorials
BigIdeasMath.com

Key Idea

Solving Proportions

Method 1 Use mental math. *(Section 14.3)*

Method 2 Use the Multiplication Property of Equality. *(Section 14.4)*

Method 3 Use the Cross Products Property. *(Section 14.4)*

EXAMPLE 1 Solving Proportions Using Multiplication

Solve $\frac{5}{7} = \frac{x}{21}$.

$\frac{5}{7} = \frac{x}{21}$	Write the proportion.
$21 \cdot \frac{5}{7} = 21 \cdot \frac{x}{21}$	Multiplication Property of Equality
$15 = x$	Simplify.

∴ The solution is 15.

On Your Own

Now You're Ready
Exercises 4–9

Use multiplication to solve the proportion.

1. $\frac{w}{6} = \frac{6}{9}$ **2.** $\frac{12}{10} = \frac{a}{15}$ **3.** $\frac{y}{6} = \frac{2}{4}$

EXAMPLE 2 Solving Proportions Using the Cross Products Property

Solve each proportion.

a. $\frac{x}{8} = \frac{7}{10}$		**b.** $\frac{9}{y} = \frac{3}{17}$
$x \cdot 10 = 8 \cdot 7$	Cross Products Property	$9 \cdot 17 = y \cdot 3$
$10x = 56$	Multiply.	$153 = 3y$
$x = 5.6$	Divide.	$51 = y$

∴ The solution is 5.6.

∴ The solution is 51.

On Your Own

Now You're Ready
Exercises 10–21

Use the Cross Products Property to solve the proportion.

4. $\dfrac{2}{7} = \dfrac{x}{28}$ **5.** $\dfrac{12}{5} = \dfrac{6}{y}$ **6.** $\dfrac{40}{z+1} = \dfrac{15}{6}$

EXAMPLE 3 Real-Life Application

The graph shows the toll y due on a turnpike for driving x miles. Your toll is \$7.50. How many *kilometers* did you drive?

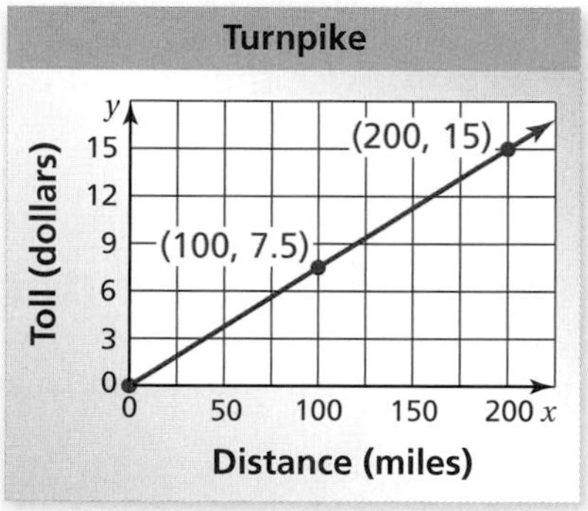

The point (100, 7.5) on the graph shows that the toll is \$7.50 for driving 100 miles. Convert 100 miles to kilometers.

Method 1: Convert using a ratio.

1 mi ≈ 1.61 km

$$100 \cancel{\text{mi}} \times \frac{1.61 \text{ km}}{1 \cancel{\text{mi}}} = 161 \text{ km}$$

∴ So, you drove about 161 kilometers.

Method 2: Convert using a proportion.

Let x be the number of kilometers equivalent to 100 miles.

kilometers → $\dfrac{1.61}{1} = \dfrac{x}{100}$ ← kilometers
miles → ← miles

Write a proportion. Use 1.61 km ≈ 1 mi.

$1.61 \cdot 100 = 1 \cdot x$ Cross Products Property

$161 = x$ Simplify.

∴ So, you drove about 161 kilometers.

On Your Own

Now You're Ready
Exercises 28–30

Write and solve a proportion to complete the statement. Round to the nearest hundredth, if necessary.

7. 7.5 in. ≈ ____ cm **8.** 100 g ≈ ____ oz

9. 2 L ≈ ____ qt **10.** 4 m ≈ ____ ft

14.4 Exercises

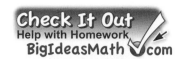

Check It Out
Help with Homework
BigIdeasMath.com

Vocabulary and Concept Check

1. **WRITING** What are three ways you can solve a proportion?

2. **OPEN-ENDED** Which way would you choose to solve $\frac{3}{x} = \frac{6}{14}$? Explain your reasoning.

3. **NUMBER SENSE** Does $\frac{x}{4} = \frac{15}{3}$ have the same solution as $\frac{x}{15} = \frac{4}{3}$? Use the Cross Products Property to explain your answer.

Practice and Problem Solving

Use multiplication to solve the proportion.

(1) 4. $\frac{9}{5} = \frac{z}{20}$
5. $\frac{h}{15} = \frac{16}{3}$
6. $\frac{w}{4} = \frac{42}{24}$
7. $\frac{35}{28} = \frac{n}{12}$
8. $\frac{7}{16} = \frac{x}{4}$
9. $\frac{y}{9} = \frac{44}{54}$

Use the Cross Products Property to solve the proportion.

(2) 10. $\frac{a}{6} = \frac{15}{2}$
11. $\frac{10}{7} = \frac{8}{k}$
12. $\frac{3}{4} = \frac{v}{14}$
13. $\frac{5}{n} = \frac{16}{32}$
14. $\frac{36}{42} = \frac{24}{r}$
15. $\frac{9}{10} = \frac{d}{6.4}$
16. $\frac{x}{8} = \frac{3}{12}$
17. $\frac{8}{m} = \frac{6}{15}$
18. $\frac{4}{24} = \frac{c}{36}$
19. $\frac{20}{16} = \frac{d}{12}$
20. $\frac{30}{20} = \frac{w}{14}$
21. $\frac{2.4}{1.8} = \frac{7.2}{k}$

22. **ERROR ANALYSIS** Describe and correct the error in solving the proportion $\frac{m}{8} = \frac{15}{24}$.

✗

$$\frac{m}{8} = \frac{15}{24}$$
$$8 \cdot m = 24 \cdot 15$$
$$m = 45$$

23. **PENS** Forty-eight pens are packaged in 4 boxes. How many pens are packaged in 9 boxes?

24. **PIZZA PARTY** How much does it cost to buy 10 medium pizzas?

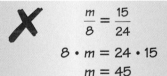

Solve the proportion.

25. $\frac{2x}{5} = \frac{9}{15}$
26. $\frac{5}{2} = \frac{d-2}{4}$
27. $\frac{4}{k+3} = \frac{8}{14}$

Write and solve a proportion to complete the statement. Round to the nearest hundredth if necessary.

3 **28.** 6 km ≈ ▒ mi **29.** 2.5 L ≈ ▒ gal **30.** 90 lb ≈ ▒ kg

31. TRUE OR FALSE? Tell whether the statement is *true* or *false*. Explain.

$$\text{If } \frac{a}{b} = \frac{2}{3}, \text{ then } \frac{3}{2} = \frac{b}{a}.$$

32. CLASS TRIP It costs \$95 for 20 students to visit an aquarium. How much does it cost for 162 students?

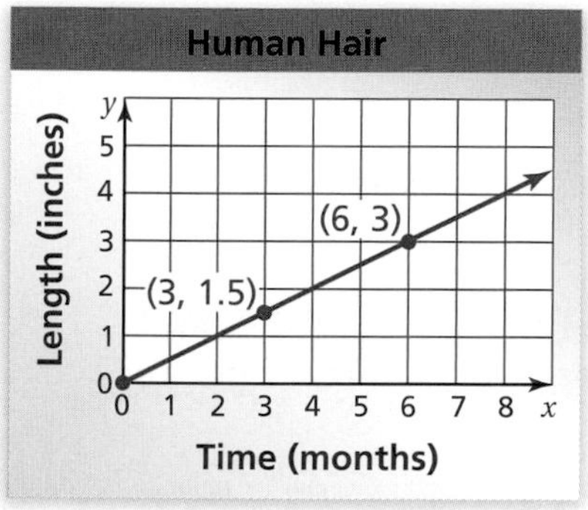

33. GRAVITY A person who weighs 120 pounds on Earth weighs 20 pounds on the Moon. How much does a 93-pound person weigh on the Moon?

34. HAIR The length of human hair is proportional to the number of months it has grown.

a. What is the hair length in *centimeters* after 6 months?

b. How long does it take hair to grow 8 inches?

c. Use a different method than the one in part (b) to find how long it takes hair to grow 20 inches.

35. SWING SET It takes 6 hours for 2 people to build a swing set. Can you use the proportion $\frac{2}{6} = \frac{5}{h}$ to determine the number of hours h it will take 5 people to build the swing set? Explain.

36. REASONING There are 144 people in an audience. The ratio of adults to children is 5 to 3. How many are adults?

37. PROBLEM SOLVING Three pounds of lawn seed covers 1800 square feet. How many bags are needed to cover 8400 square feet?

38. Critical Thinking Consider the proportions $\frac{m}{n} = \frac{1}{2}$ and $\frac{n}{k} = \frac{2}{5}$. What is the ratio $\frac{m}{k}$? Explain your reasoning.

Fair Game Review What you learned in previous grades & lessons

Plot the ordered pair in a coordinate plane. *(Section 6.5)*

39. $A(-5, -2)$ **40.** $B(-3, 0)$ **41.** $C(-1, 2)$ **42.** $D(1, 4)$

43. MULTIPLE CHOICE What is the value of $(3w - 8) - 4(2w + 3)$? *(Section 13.2)*

Ⓐ $11w + 4$ Ⓑ $-5w - 5$ Ⓒ $-5w + 4$ Ⓓ $-5w - 20$

14.5 Slope

Essential Question How can you compare two rates graphically?

1 ACTIVITY: Comparing Unit Rates

Work with a partner. The table shows the maximum speeds of several animals.

a. Find the missing speeds. Round your answers to the nearest tenth.

b. Which animal is fastest? Which animal is slowest?

c. Explain how you convert between the two units of speed.

Animal	Speed (miles per hour)	Speed (feet per second)
Antelope	61.0	
Black mamba snake		29.3
Cheetah		102.6
Chicken		13.2
Coyote	43.0	
Domestic pig		16.0
Elephant		36.6
Elk		66.0
Giant tortoise	0.2	
Giraffe	32.0	
Gray fox		61.6
Greyhound	39.4	
Grizzly bear		44.0
Human		41.0
Hyena	40.0	
Jackal	35.0	
Lion		73.3
Peregrine falcon	200.0	
Quarter horse	47.5	
Spider		1.76
Squirrel	12.0	
Thomson's gazelle	50.0	
Three-toed sloth		0.2
Tuna	47.0	

COMMON CORE

Slope

In this lesson, you will

- find the slopes of lines.
- interpret the slopes of lines as rates.

Learning Standard 7.RP.2b

2 ACTIVITY: Comparing Two Rates Graphically

Math Practice 4

Apply Mathematics

How can you use the graph to determine which animal has the greater speed?

Work with a partner. A cheetah and a Thomson's gazelle run at maximum speed.

a. Use the table in Activity 1 to calculate the missing distances.

	Cheetah	Gazelle
Time (seconds)	Distance (feet)	Distance (feet)
0		
1		
2		
3		
4		
5		
6		
7		

b. Use the table to write ordered pairs. Then plot the ordered pairs and connect the points for each animal. What do you notice about the graphs?

c. Which graph is steeper? The speed of which animal is greater?

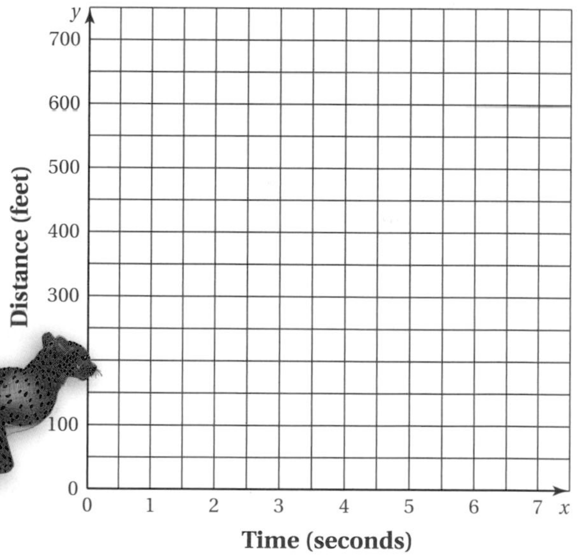

What Is Your Answer?

3. **IN YOUR OWN WORDS** How can you compare two rates graphically? Explain your reasoning. Give some examples with your answer.

4. **REPEATED REASONING** Choose 10 animals from Activity 1.

   a. Make a table for each animal similar to the table in Activity 2.

   b. Sketch a graph of the distances for each animal.

   c. Compare the steepness of the 10 graphs. What can you conclude?

14.5 Lesson

Check It Out
Lesson Tutorials
BigIdeasMath.com

Key Vocabulary
slope, *p. 630*

Key Idea

Slope

Slope is the rate of change between any two points on a line. It is a measure of the *steepness* of a line.

To find the slope of a line, find the ratio of the change in y (vertical change) to the change in x (horizontal change).

$$\text{slope} = \frac{\text{change in } y}{\text{change in } x}$$

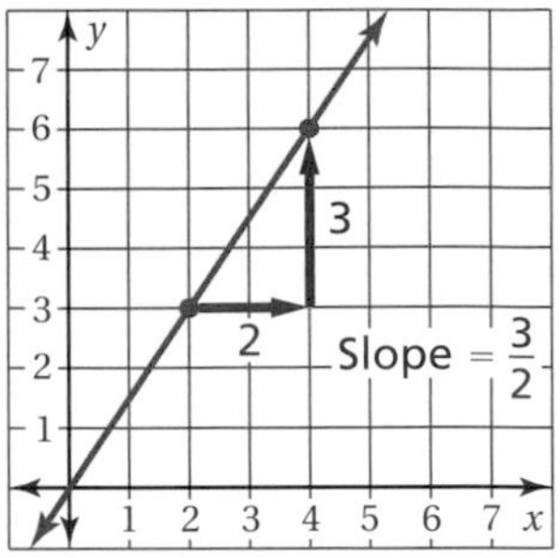

Study Tip

The slope of a line is the same between any two points on the line because lines have a *constant* rate of change.

EXAMPLE 1 Finding Slopes

Find the slope of each line.

a.

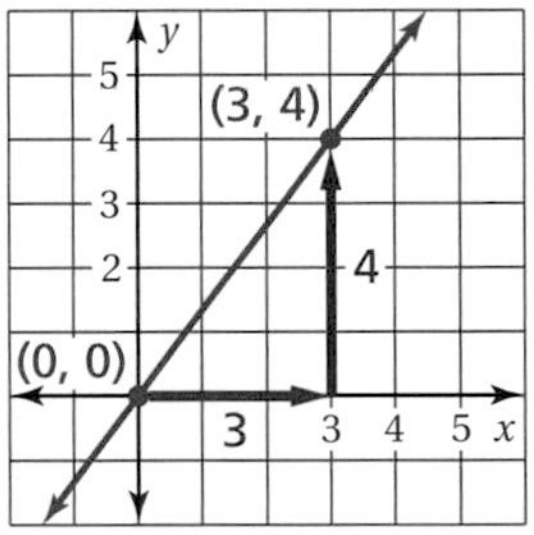

$$\text{slope} = \frac{\text{change in } y}{\text{change in } x}$$

$$= \frac{4}{3}$$

∴ The slope of the line is $\frac{4}{3}$.

b.

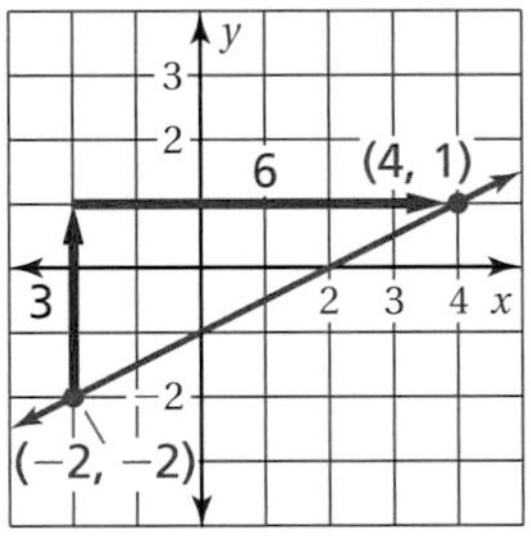

$$\text{slope} = \frac{\text{change in } y}{\text{change in } x}$$

$$= \frac{3}{6} = \frac{1}{2}$$

∴ The slope of the line is $\frac{1}{2}$.

On Your Own

Now You're Ready
Exercises 4–9

Find the slope of the line.

1.

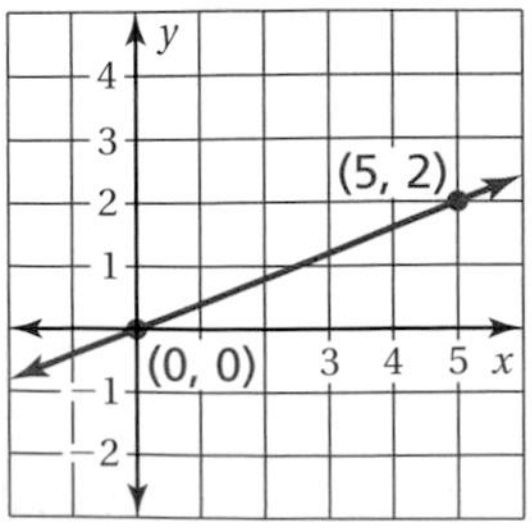

2.

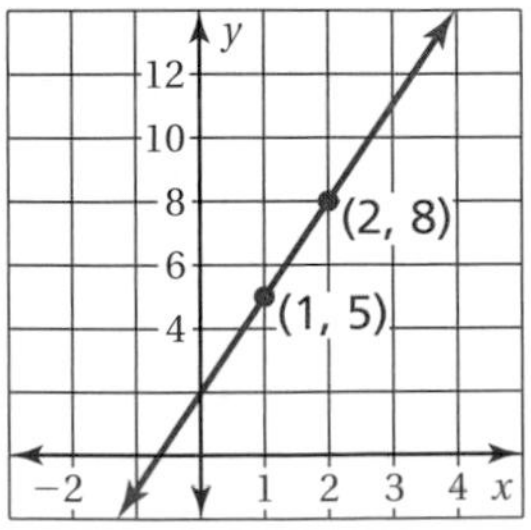

Multi-Language Glossary at BigIdeasMath.com

EXAMPLE 2 Interpreting a Slope

The table shows your earnings for babysitting.

a. Graph the data.

b. Find and interpret the slope of the line through the points.

Hours, x	0	2	4	6	8	10
Earnings, y (dollars)	0	10	20	30	40	50

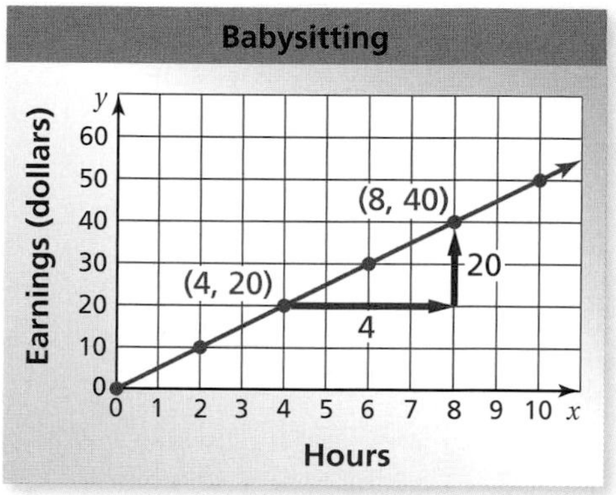

a. Graph the data. Draw a line through the points.

b. Choose any two points to find the slope of the line.

$$\text{slope} = \frac{\text{change in } y}{\text{change in } x}$$

$$= \frac{20}{4} \quad \begin{matrix} \leftarrow \text{dollars} \\ \leftarrow \text{hours} \end{matrix}$$

$$= 5$$

∴ The slope of the line represents the unit rate. The slope is 5. So, you earn \$5 per hour babysitting.

On Your Own

Now You're Ready
Exercises 10 and 11

3. In Example 2, use two other points to find the slope. Does the slope change?

4. The graph shows the amounts you and your friend earn babysitting.

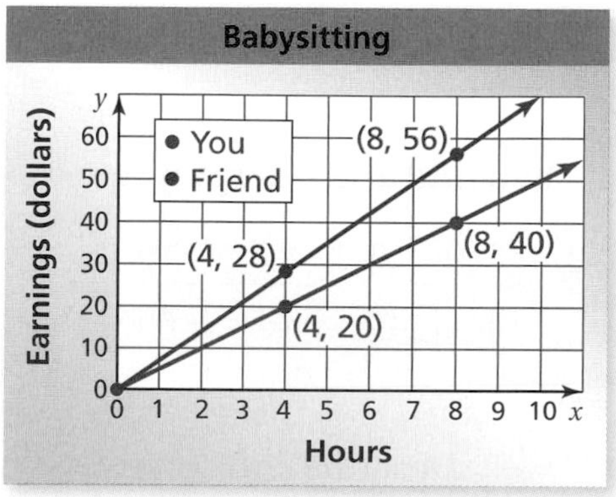

a. Compare the steepness of the lines. What does this mean in the context of the problem?

b. Find and interpret the slope of the blue line.

14.5 Exercises

Check It Out
Help with Homework
BigIdeasMath.com

Vocabulary and Concept Check

1. **VOCABULARY** Is there a connection between rate and slope? Explain.
2. **REASONING** Which line has the greatest slope?
3. **REASONING** Is it more difficult to run up a ramp with a slope of $\frac{1}{5}$ or a ramp with a slope of 5? Explain.

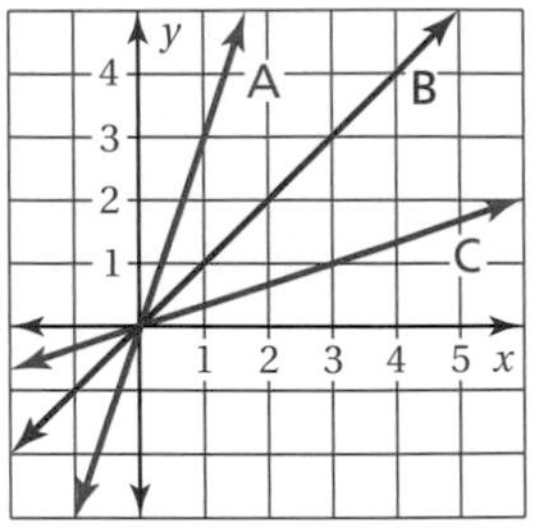

Practice and Problem Solving

Find the slope of the line.

1 4.
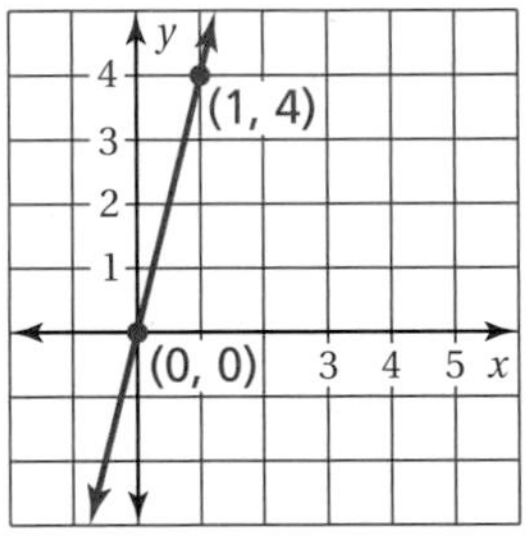

5.
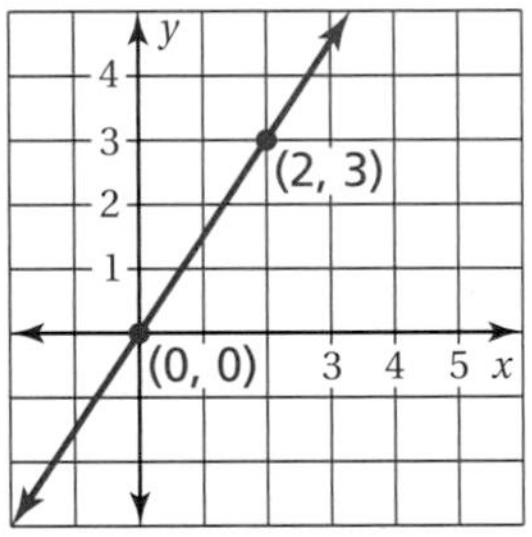

6.
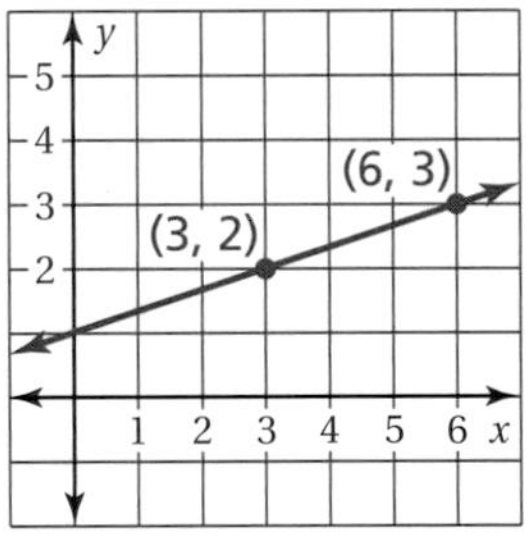

7.
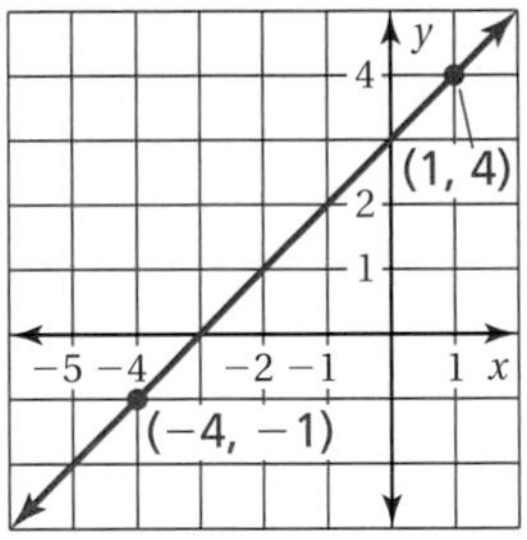

8.
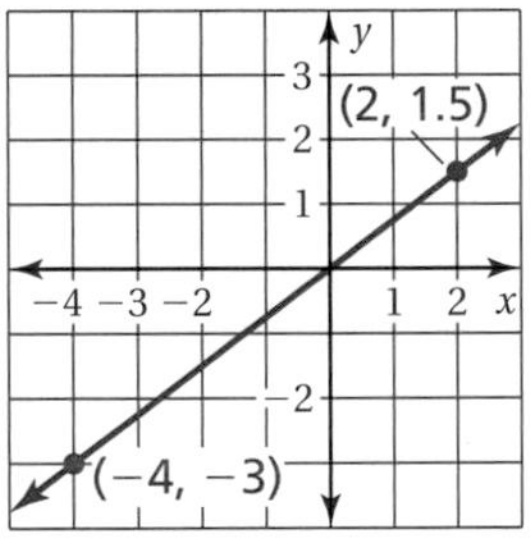

9.
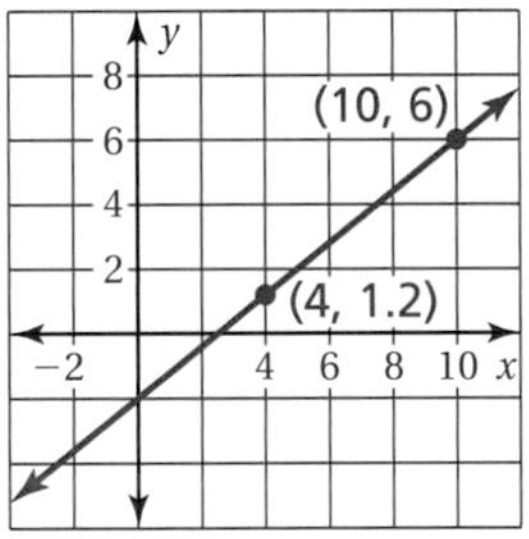

Graph the data. Then find and interpret the slope of the line through the points.

2 10.

Minutes, x	3	5	7	9
Words, y	135	225	315	405

11.

Gallons, x	5	10	15	20
Miles, y	162.5	325	487.5	650

12. **ERROR ANALYSIS** Describe and correct the error in finding the slope of the line passing through (0, 0) and (4, 5).

✗

slope $= \frac{4}{5}$

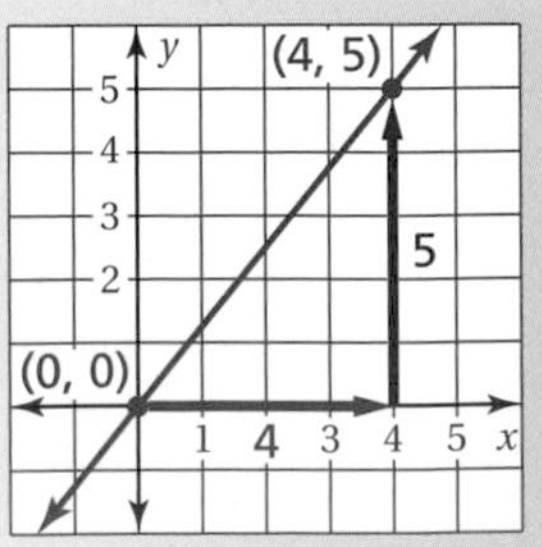

Graph the line that passes through the two points. Then find the slope of the line.

13. $(0, 0), \left(\frac{1}{3}, \frac{7}{3}\right)$

14. $\left(-\frac{3}{2}, -\frac{3}{2}\right), \left(\frac{3}{2}, \frac{3}{2}\right)$

15. $\left(1, \frac{5}{2}\right), \left(-\frac{1}{2}, -\frac{1}{4}\right)$

16. CAMPING The graph shows the amount of money you and a friend are saving for a camping trip.

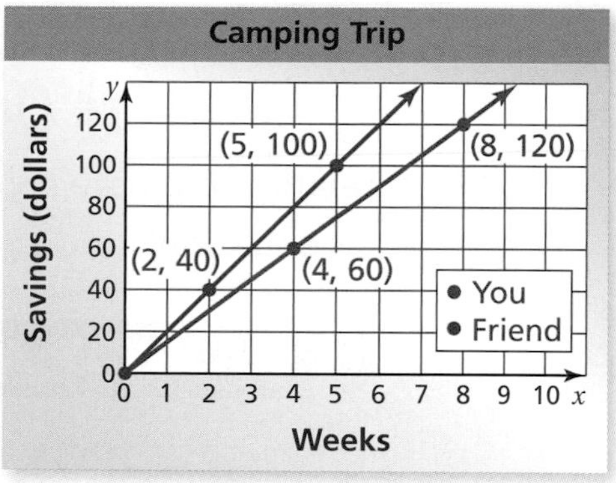

a. Compare the steepness of the lines. What does this mean in the context of the problem?

b. Find the slope of each line.

c. How much more money does your friend save each week than you?

d. The camping trip costs $165. How long will it take you to save enough money?

17. MAPS An atlas contains a map of Ohio. The table shows data from the key on the map.

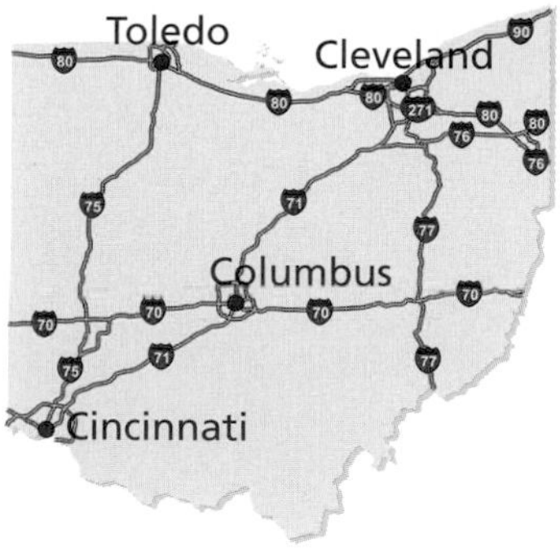

Distance on Map (mm), x	10	20	30	40
Actual Distance (mi), y	25	50	75	100

a. Graph the data.

b. Find the slope of the line. What does this mean in the context of the problem?

c. The map distance between Toledo and Columbus is 48 millimeters. What is the actual distance?

d. Cincinnati is about 225 miles from Cleveland. What is the distance between these cities on the map?

18. CRITICAL THINKING What is the slope of a line that passes through the points (2, 0) and (5, 0)? Explain.

19. Number Sense A line has a slope of 2. It passes through the points (1, 2) and $(3, y)$. What is the value of y?

Fair Game Review What you learned in previous grades & lessons

Multiply. *(Section 12.4)*

20. $-\frac{3}{5} \times \frac{8}{6}$

21. $1\frac{1}{2} \times \left(-\frac{6}{15}\right)$

22. $-2\frac{1}{4} \times \left(-1\frac{1}{3}\right)$

23. MULTIPLE CHOICE You have 18 stamps from Mexico in your stamp collection. These stamps represent $\frac{3}{8}$ of your collection. The rest of the stamps are from the United States. How many stamps are from the United States? *(Section 13.4)*

Ⓐ 12 Ⓑ 24 Ⓒ 30 Ⓓ 48

14.6 Direct Variation

Essential Question How can you use a graph to show the relationship between two quantities that vary directly? How can you use an equation?

1 ACTIVITY: Math in Literature

Gulliver's Travels was written by Jonathan Swift and published in 1726. Gulliver was shipwrecked on the island Lilliput, where the people were only 6 inches tall. When the Lilliputians decided to make a shirt for Gulliver, a Lilliputian tailor stated that he could determine Gulliver's measurements by simply measuring the distance around Gulliver's thumb. He said "Twice around the thumb equals once around the wrist. Twice around the wrist is once around the neck. Twice around the neck is once around the waist."

Work with a partner. Use the tailor's statement to complete the table.

Thumb, t	Wrist, w	Neck, n	Waist, x
0 in.			
1 in.			
	4 in.		
		12 in.	
			32 in.
	10 in.		

COMMON CORE

Direct Variation

In this lesson, you will
- identify direct variation from graphs or equations.
- use direct variation models to solve problems.

Learning Standards
7.RP.2a
7.RP.2b
7.RP.2c
7.RP.2d

2 ACTIVITY: Drawing a Graph

Work with a partner. Use the information from Activity 1.

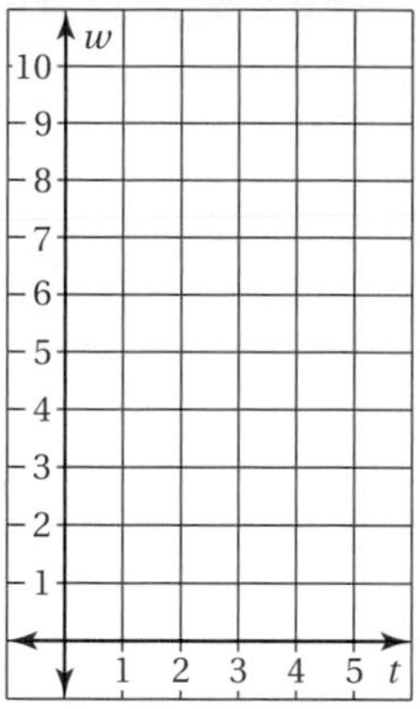

a. In your own words, describe the relationship between *t* and *w*.

b. Use the table to write the ordered pairs (*t*, *w*). Then plot the ordered pairs.

c. What do you notice about the graph of the ordered pairs?

d. Choose two points and find the slope of the line between them.

e. The quantities *t* and *w* are said to *vary directly*. An equation that describes the relationship is

$$w = \blacksquare\, t.$$

3 ACTIVITY: Drawing a Graph and Writing an Equation

Math Practice 6

Label Axes

How do you know which labels to use for the axes? Explain.

Work with a partner. Use the information from Activity 1 to draw a graph of the relationship. Write an equation that describes the relationship between the two quantities.

a. Thumb *t* and neck *n* ($n = \blacksquare\, t$)

b. Wrist *w* and waist *x* ($x = \blacksquare\, w$)

c. Wrist *w* and thumb *t* ($t = \blacksquare\, w$)

d. Waist *x* and wrist *w* ($w = \blacksquare\, x$)

What Is Your Answer?

4. **IN YOUR OWN WORDS** How can you use a graph to show the relationship between two quantities that vary directly? How can you use an equation?
5. **STRUCTURE** How are all the graphs in Activity 3 alike?
6. Give a real-life example of two variables that vary directly.
7. Work with a partner. Use string to find the distance around your thumb, wrist, and neck. Do your measurements agree with the tailor's statement in *Gulliver's Travels*? Explain your reasoning.

Practice Use what you learned about quantities that vary directly to complete Exercises 4 and 5 on page 638.

14.6 Lesson

Key Vocabulary
direct variation, *p. 636*
constant of proportionality, *p. 636*

Key Idea

Direct Variation

Words Two quantities x and y show **direct variation** when $y = kx$, where k is a number and $k \neq 0$. The number k is called the **constant of proportionality**.

Graph The graph of $y = kx$ is a line with a slope of k that passes through the origin. So, two quantities that show direct variation are in a proportional relationship.

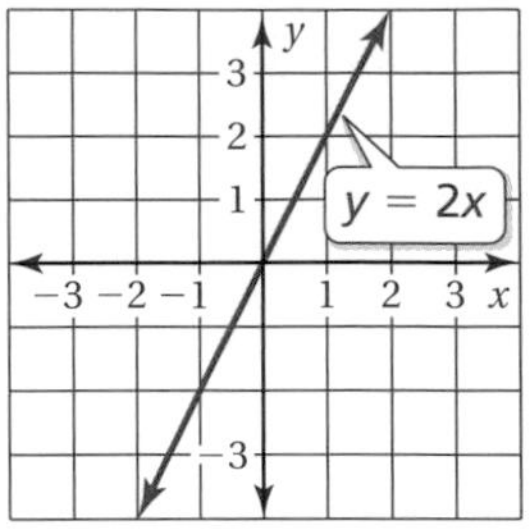

EXAMPLE 1 Identifying Direct Variation

Tell whether x and y show direct variation. Explain your reasoning.

a.

x	1	2	3	4
y	−2	0	2	4

Plot the points. Draw a line through the points.

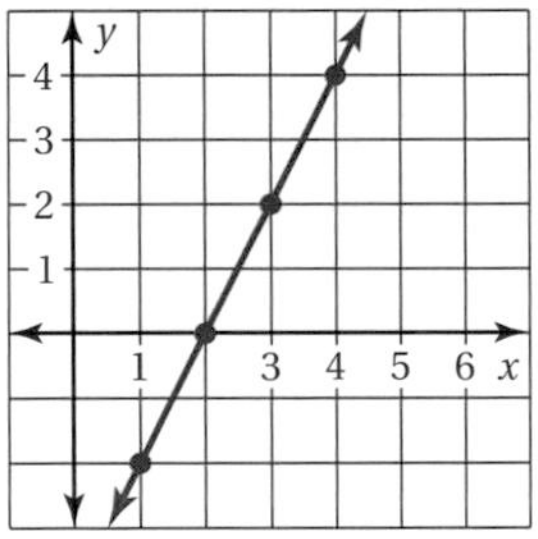

∴ The line does *not* pass through the origin. So, x and y do *not* show direct variation.

b.

x	0	2	4	6
y	0	2	4	6

Plot the points. Draw a line through the points.

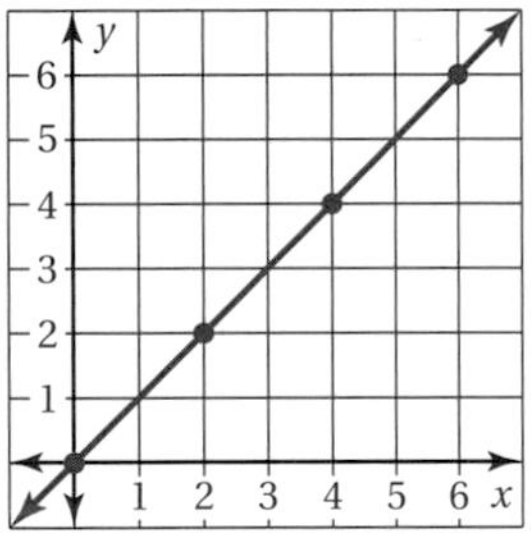

∴ The line passes through the origin. So, x and y show direct variation.

Study Tip

Other ways to say that x and y show direct variation are "y varies directly with x" and "x and y are directly proportional."

EXAMPLE 2 Identifying Direct Variation

Tell whether x and y show direct variation. Explain your reasoning.

a. $y + 1 = 2x$

$y = 2x - 1$ Solve for y.

∴ The equation *cannot* be written as $y = kx$. So, x and y do *not* show direct variation.

b. $\frac{1}{2}y = x$

$y = 2x$ Solve for y.

∴ The equation can be written as $y = kx$. So, x and y show direct variation.

On Your Own

Now You're Ready
Exercises 6–17

Tell whether x and y show direct variation. Explain your reasoning.

1.

x	y
0	−2
1	1
2	4
3	7

2.

x	y
1	4
2	8
3	12
4	16

3.

x	y
−2	4
−1	2
0	0
1	2

4. $xy = 3$

5. $x = \frac{1}{3}y$

6. $y + 1 = x$

EXAMPLE 3 Real-Life Application

x	y
$\frac{1}{2}$	8
1	16
$\frac{3}{2}$	24
2	32

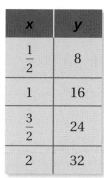

The table shows the area y (in square feet) that a robotic vacuum cleans in x minutes.

a. Graph the data. Tell whether x and y are directly proportional.

Graph the data. Draw a line through the points.

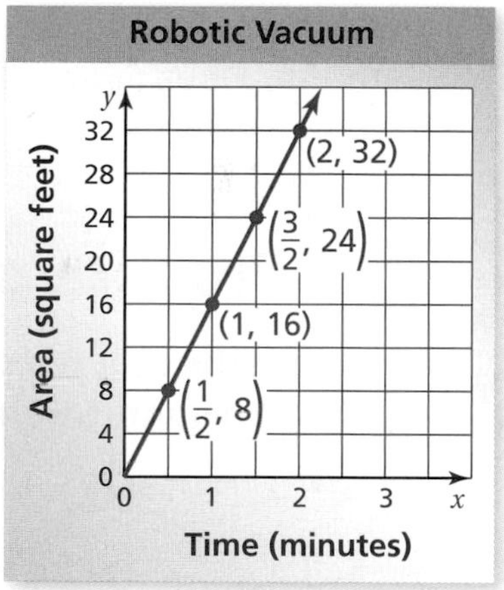

∴ The graph is a line through the origin. So, x and y are directly proportional.

b. Write an equation that represents the line.

Choose any two points to find the slope of the line.

$$\text{slope} = \frac{\text{change in } y}{\text{change in } x} = \frac{16}{1} = 16$$

∴ The slope of the line is the constant of proportionality, k. So, an equation of the line is $y = 16x$.

c. Use the equation to find the area cleaned in 10 minutes.

$y = 16x$	Write the equation.
$= 16(10)$	Substitute 10 for x.
$= 160$	Multiply.

∴ So, the vacuum cleans 160 square feet in 10 minutes.

On Your Own

Now You're Ready
Exercise 19

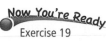

7. **WHAT IF?** The battery weakens and the robot begins cleaning less and less area each minute. Do x and y show direct variation? Explain.

14.6 Exercises

Check It Out
Help with Homework
BigIdeasMath.com

Vocabulary and Concept Check

1. **VOCABULARY** What does it mean for x and y to vary directly?
2. **WRITING** What point is on the graph of every direct variation equation?
3. **DIFFERENT WORDS, SAME QUESTION** Which is different? Find "both" answers.

Do x and y show direct variation?

Are x and y in a proportional relationship?

Is the graph of the relationship a line?

Does y vary directly with x?

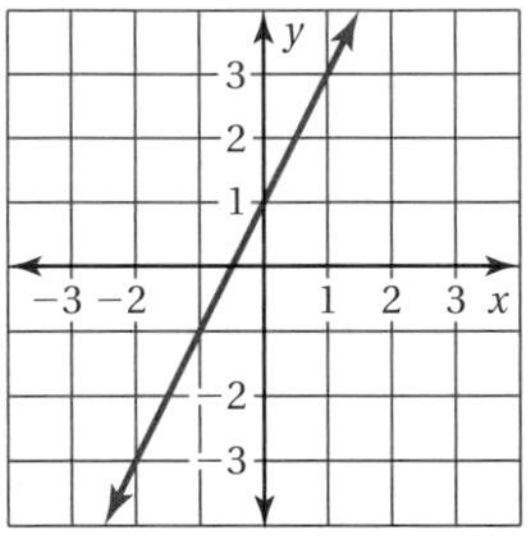

Practice and Problem Solving

Graph the ordered pairs in a coordinate plane. Do you think that graph shows that the quantities vary directly? Explain your reasoning.

4. $(-1, -1), (0, 0), (1, 1), (2, 2)$
5. $(-4, -2), (-2, 0), (0, 2), (2, 4)$

Tell whether x and y show direct variation. Explain your reasoning. If so, find k.

① 6.

x	1	2	3	4
y	2	4	6	8

7.

x	−2	−1	0	1
y	0	2	4	6

8.

x	−1	0	1	2
y	−2	−1	0	1

9.

x	3	6	9	12
y	2	4	6	8

② 10. $y - x = 4$

11. $x = \frac{2}{5}y$

12. $y + 3 = x + 6$

13. $y - 5 = 2x$

14. $x - y = 0$

15. $\frac{x}{y} = 2$

16. $8 = xy$

17. $x^2 = y$

18. **ERROR ANALYSIS** Describe and correct the error in telling whether x and y show direct variation.

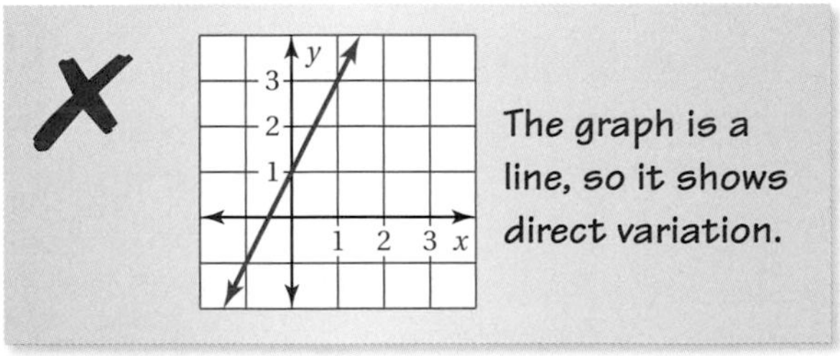

③ 19. **RECYCLING** The table shows the profit y for recycling x pounds of aluminum. Graph the data. Tell whether x and y show direct variation. If so, write an equation that represents the line.

Aluminum (lb), x	10	20	30	40
Profit, y	\$4.50	\$9.00	\$13.50	\$18.00

The variables x and y vary directly. Use the values to find the constant of proportionality. Then write an equation that relates x and y.

20. $y = 72; x = 3$ **21.** $y = 20; x = 12$ **22.** $y = 45; x = 40$

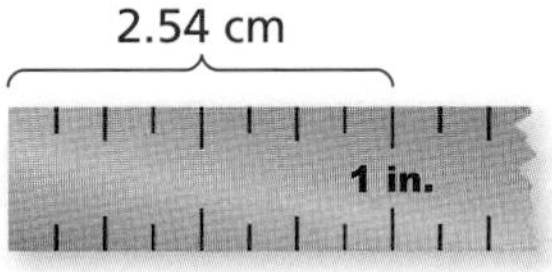

23. MEASUREMENT Write a direct variation equation that relates x inches to y centimeters.

24. MODELING Design a waterskiing ramp. Show how you can use direct variation to plan the heights of the vertical supports.

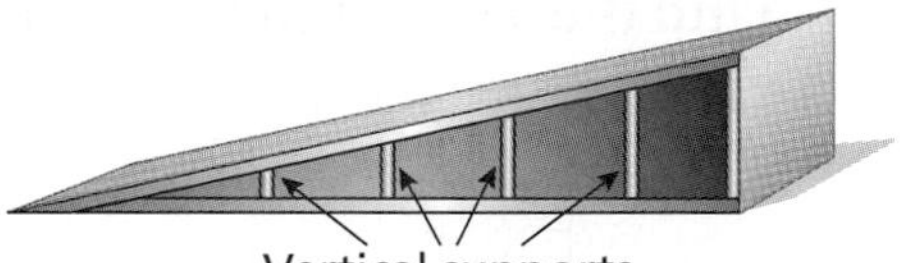

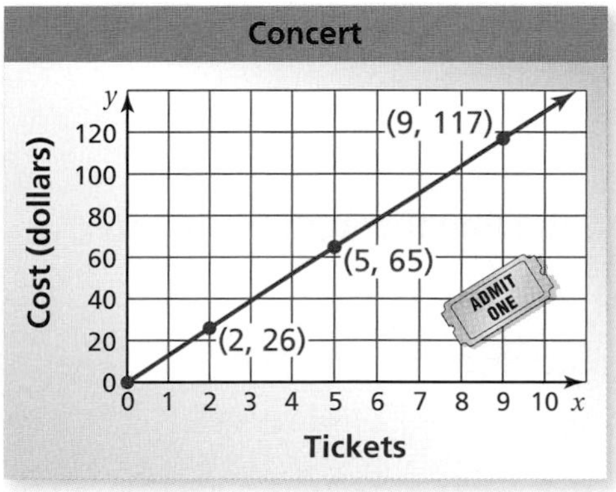

25. REASONING Use $y = kx$ to show why the graph of a proportional relationship always passes through the origin.

26. TICKETS The graph shows the cost of buying concert tickets. Tell whether x and y show direct variation. If so, find and interpret the constant of proportionality. Then write an equation and find the cost of 14 tickets.

27. CELL PHONE PLANS Tell whether x and y show direct variation. If so, write an equation of direct variation.

Minutes, x	500	700	900	1200
Cost, y	\$40	\$50	\$60	\$75

28. CHLORINE The amount of chlorine in a swimming pool varies directly with the volume of water. The pool has 2.5 milligrams of chlorine per liter of water. How much chlorine is in the pool?

29. Critical Thinking Is the graph of every direct variation equation a line? Does the graph of every line represent a direct variation equation? Explain your reasoning.

Fair Game Review What you learned in previous grades & lessons

Write the fraction as a decimal. *(Section 12.1)*

30. $\frac{13}{20}$ **31.** $\frac{9}{16}$ **32.** $\frac{21}{40}$ **33.** $\frac{24}{25}$

34. MULTIPLE CHOICE Which rate is *not* equivalent to 180 feet per 8 seconds? *(Section 14.1)*

Ⓐ $\frac{225 \text{ ft}}{10 \text{ sec}}$ Ⓑ $\frac{45 \text{ ft}}{2 \text{ sec}}$ Ⓒ $\frac{135 \text{ ft}}{6 \text{ sec}}$ Ⓓ $\frac{180 \text{ ft}}{1 \text{ sec}}$

14.4–14.6 Quiz

Check It Out
Progress Check
BigIdeasMath.com

Solve the proportion. *(Section 14.4)*

1. $\frac{7}{n} = \frac{42}{48}$

2. $\frac{x}{2} = \frac{40}{16}$

3. $\frac{3}{11} = \frac{27}{z}$

Find the slope of the line. *(Section 14.5)*

4.

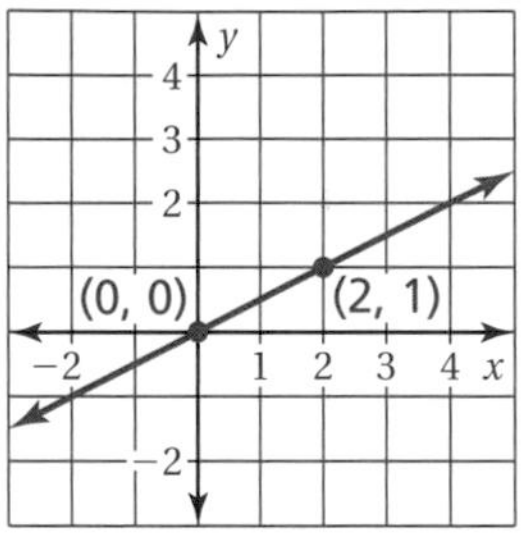

5.

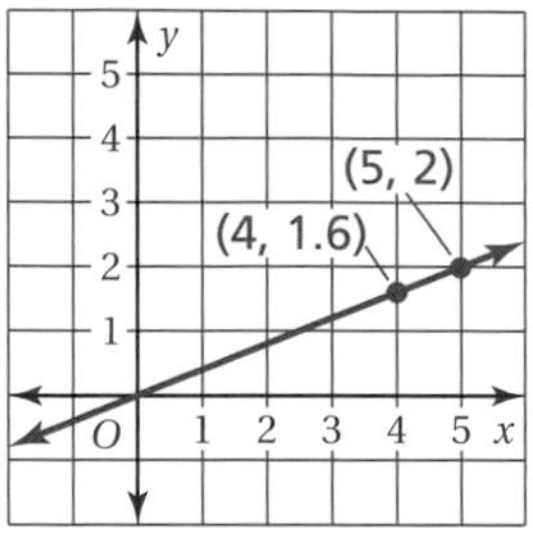

Graph the data. Then find and interpret the slope of the line through the points. *(Section 14.5)*

6.

Hours, *x*	2	4	6	8
Miles, *y*	10	20	30	40

7.

Packages, *x*	6	10	14	18
Servings, *y*	9	15	21	27

Tell whether *x* and *y* show direct variation. Explain your reasoning. *(Section 14.6)*

8. $y - 9 = 6 + x$

9. $x = \frac{5}{8}y$

10. CONCERT A benefit concert with three performers lasts 8 hours. At this rate, how many hours is a concert with four performers? *(Section 14.4)*

11. LAWN MOWING The graph shows how much you and your friend each earn mowing lawns. *(Section 14.5)*

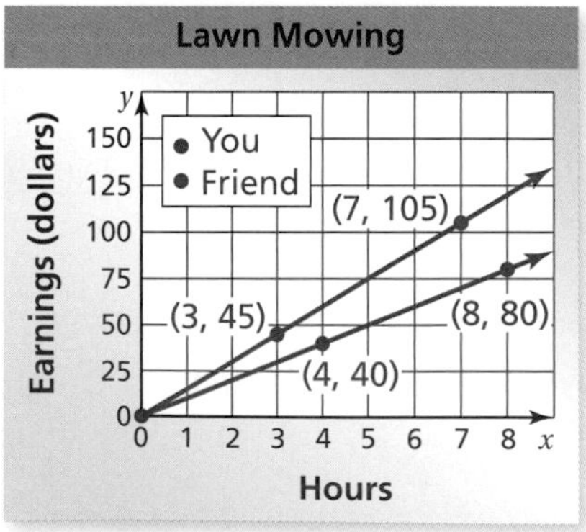

a. Compare the steepness of the lines. What does this mean in the context of the problem?

b. Find and interpret the slope of each line.

c. How much more money do you earn per hour than your friend?

12. PIE SALE The table shows the profits of a pie sale. Tell whether *x* and *y* show direct variation. If so, write the equation of direct variation. *(Section 14.6)*

Pies Sold, *x*	10	12	14	16
Profit, *y*	\$79.50	\$95.40	\$111.30	\$127.20

14 Chapter Review

Check It Out
Vocabulary Help
BigIdeasMath.com

Review Key Vocabulary

ratio, *p. 600*
rate, *p. 600*
unit rate, *p. 600*
complex fraction, *p. 601*
proportion, *p. 608*
proportional, *p. 608*
cross products, *p. 609*
slope, *p. 630*
direct variation, *p. 636*
constant of proportionality, *p. 636*

Review Examples and Exercises

14.1 Ratios and Rates *(pp. 598–605)*

There are 15 orangutans and 25 gorillas in a nature preserve. One of the orangutans swings 75 feet in 15 seconds on a rope.

a. Find the ratio of orangutans to gorillas.

b. How fast is the orangutan swinging?

a. $\dfrac{\text{orangutans}}{\text{gorillas}} = \dfrac{15}{25} = \dfrac{3}{5}$

The ratio of orangutans to gorillas is $\dfrac{3}{5}$.

b. 75 feet in 15 seconds $= \dfrac{75 \text{ ft}}{15 \text{ sec}}$

$= \dfrac{75 \text{ ft} \div 15}{15 \text{ sec} \div 15}$

$= \dfrac{5 \text{ ft}}{1 \text{ sec}}$

The orangutan is swinging 5 feet per second.

Exercises

Find the unit rate.

1. 289 miles on 10 gallons

2. $6\frac{2}{5}$ revolutions in $2\frac{2}{3}$ seconds

3. calories per serving

Servings	2	4	6	8
Calories	240	480	720	960

14.2 Proportions *(pp. 606–613)*

Tell whether the ratios $\dfrac{9}{12}$ and $\dfrac{6}{8}$ form a proportion.

$\dfrac{9}{12} = \dfrac{9 \div 3}{12 \div 3} = \dfrac{3}{4}$

$\dfrac{6}{8} = \dfrac{6 \div 2}{8 \div 2} = \dfrac{3}{4}$

The ratios are equivalent.

So, $\dfrac{9}{12}$ and $\dfrac{6}{8}$ form a proportion.

Exercises

Tell whether the ratios form a proportion.

4. $\frac{4}{9}, \frac{2}{3}$

5. $\frac{12}{22}, \frac{18}{33}$

6. $\frac{8}{50}, \frac{4}{10}$

7. $\frac{32}{40}, \frac{12}{15}$

8. Use a graph to determine whether x and y are in a proportional relationship.

x	1	3	6	8
y	4	12	24	32

14.3 Writing Proportions (pp. 614–619)

Write a proportion that gives the number r of returns on Saturday.

	Friday	Saturday
Sales	40	85
Returns	32	r

∴ One proportion is $\frac{40 \text{ sales}}{32 \text{ returns}} = \frac{85 \text{ sales}}{r \text{ returns}}$.

Exercises

Use the table to write a proportion.

9.

	Game 1	Game 2
Penalties	6	8
Minutes	16	m

10.

	Concert 1	Concert 2
Songs	15	18
Hours	2.5	h

14.4 Solving Proportions (pp. 622–627)

Solve $\frac{15}{2} = \frac{30}{y}$.

$15 \cdot y = 2 \cdot 30$ — Cross Products Property

$15y = 60$ — Multiply.

$y = 4$ — Divide.

∴ The solution is 4.

Exercises

Solve the proportion.

11. $\frac{x}{4} = \frac{2}{5}$

12. $\frac{5}{12} = \frac{y}{15}$

13. $\frac{8}{20} = \frac{6}{w}$

14. $\frac{s+1}{4} = \frac{4}{8}$

14.5 Slope (pp. 628–633)

The graph shows the number of visits your website received over the past 6 months. Find and interpret the slope.

Choose any two points to find the slope of the line.

$$\text{slope} = \frac{\text{change in } y}{\text{change in } x}$$

$$= \frac{50}{1} \quad \begin{matrix}\leftarrow \text{visits} \\ \leftarrow \text{months}\end{matrix}$$

$$= 50$$

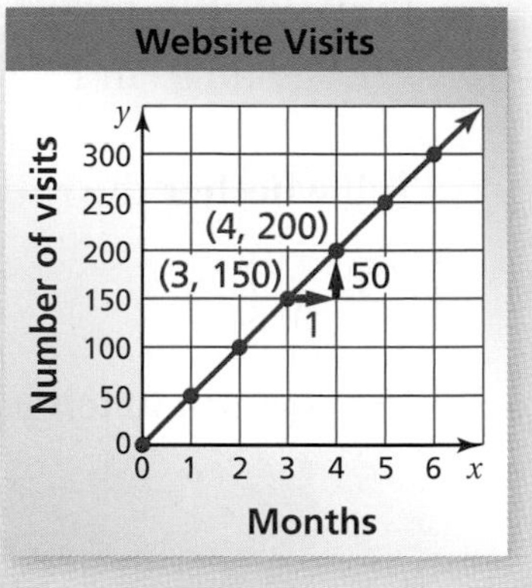

∴ The slope of the line represents the unit rate. The slope is 50. So, the number of visits increased by 50 each month.

Exercises

Find the slope of the line.

15.

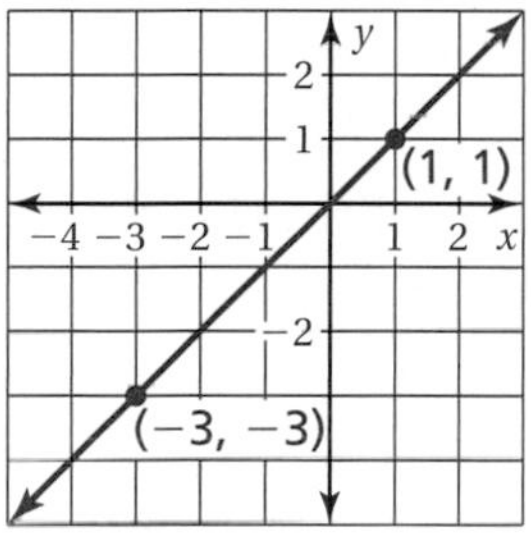

16.

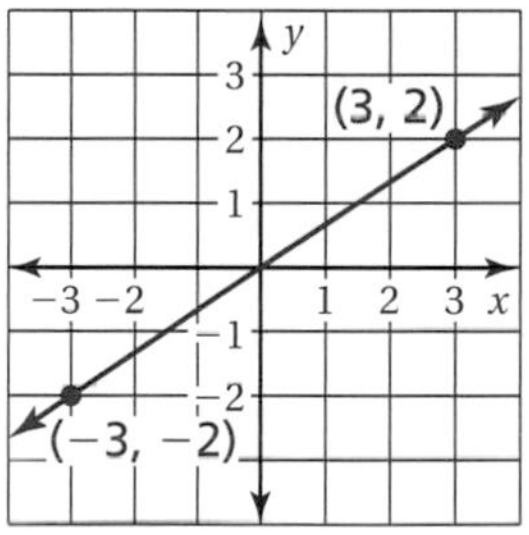

17.

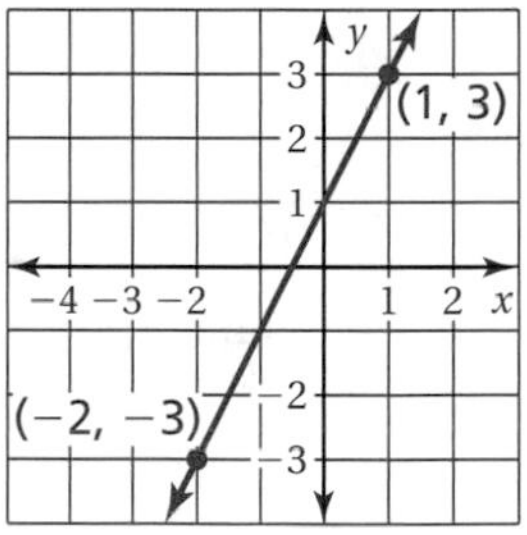

14.6 Direct Variation (pp. 634–639)

Tell whether *x* and *y* show direct variation. Explain your reasoning.

a. $x + y - 1 = 3$

$y = 4 - x$ Solve for *y*.

∴ The equation *cannot* be written as $y = kx$. So, *x* and *y* do *not* show direct variation.

b. $x = 8y$

$\frac{1}{8}x = y$ Solve for *y*.

∴ The equation can be written as $y = kx$. So, *x* and *y* show direct variation.

Exercises

Tell whether *x* and *y* show direct variation. Explain your reasoning.

18. $x + y = 6$ **19.** $y - x = 0$ **20.** $\frac{x}{y} = 20$ **21.** $x = y + 2$

14 Chapter Test

Check It Out
Test Practice
BigIdeasMath.com

Find the unit rate.

1. 84 miles in 12 days

2. $2\frac{2}{5}$ kilometers in $3\frac{3}{4}$ minutes

Tell whether the ratios form a proportion.

3. $\frac{1}{9}, \frac{6}{54}$

4. $\frac{9}{12}, \frac{8}{72}$

Use a graph to tell whether x and y are in a proportional relationship.

5.

x	2	4	6	8
y	10	20	30	40

6.

x	1	3	5	7
y	3	7	11	15

Use the table to write a proportion.

7.

	Monday	Tuesday
Gallons	6	8
Miles	180	m

8.

	Thursday	Friday
Classes	6	c
Hours	8	4

Solve the proportion.

9. $\frac{x}{8} = \frac{9}{4}$

10. $\frac{17}{3} = \frac{y}{6}$

Graph the line that passes through the two points. Then find the slope of the line.

11. $(15, 9), (-5, -3)$

12. $(2, 9), (4, 18)$

Tell whether x and y show direct variation. Explain your reasoning.

13. $xy - 11 = 5$

14. $x = \frac{3}{y}$

15. $\frac{y}{x} = 8$

16. **MOVIE TICKETS** Five movie tickets cost $36.25. What is the cost of 8 movie tickets?

17. **CROSSWALK** The graph shows the number of cycles of a crosswalk signal during the day and during the night.

    a. Compare the steepness of the lines. What does this mean in the context of the problem?

    b. Find and interpret the slope of each line.

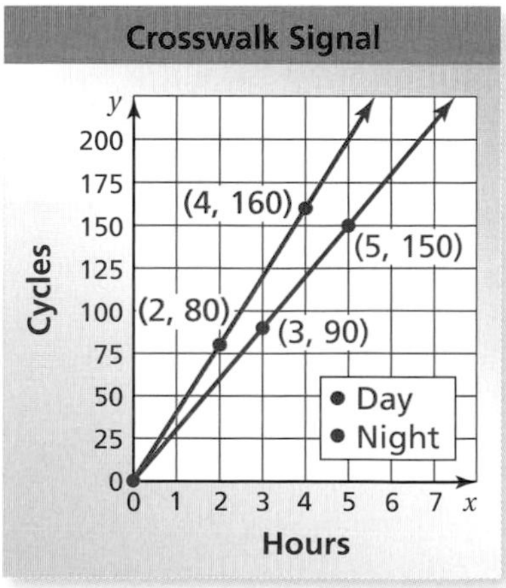

18. **GLAZE** A specific shade of green glaze requires 5 parts blue to 3 parts yellow. A glaze mixture contains 25 quarts of blue and 9 quarts of yellow. How can you fix the mixture to make the specific shade of green glaze?

14 Standards Assessment

Test-Taking Strategy
Read Question Before Answering

"Be sure to read the question before choosing your answer. You may find a word that changes the meaning."

1. The school store sells 4 pencils for \$0.80. What is the unit cost of a pencil? *(7.RP.1)*

A. \$0.20 **C.** \$3.20

B. \$0.80 **D.** \$5.00

2. Which expressions do *not* have a value of 3? *(7.NS.3)*

I. $2 + (-1)$ II. $2 - (-1)$

III. $-3 \times (-1)$ IV. $-3 \div (-1)$

F. I only **H.** II only

G. III and IV **I.** I, III, and IV

3. What is the value of the expression below? *(7.NS.3)*

$$-4 \times (-6) - (-5)$$

4. What is the slope of the line shown? *(7.RP.2b)*

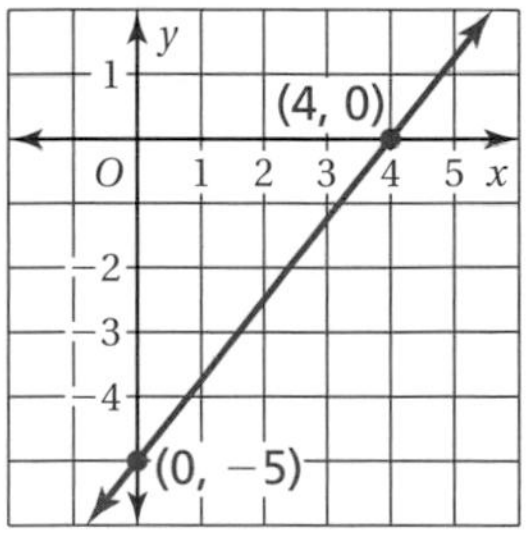

A. $\frac{4}{5}$ **C.** 4

B. $\frac{5}{4}$ **D.** 5

5. The graph below represents which inequality? *(7.EE.4b)*

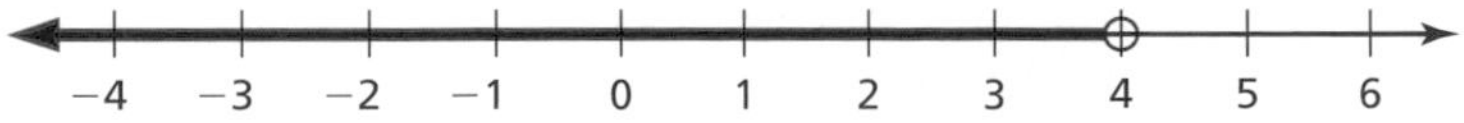

F. $-3 - 6x < -27$ **H.** $5 - 3x > -7$

G. $2x + 6 \geq 14$ **I.** $2x + 3 \leq 11$

6. The quantities x and y are proportional. What is the missing value in the table? *(7.RP.2a)*

x	y
$\frac{2}{3}$	6
$\frac{4}{3}$	12
$\frac{8}{3}$	24
5	

A. 30

B. 36

C. 45

D. 48

7. You are selling tomatoes. You have already earned $16 today. How many additional pounds of tomatoes do you need to sell to earn a total of $60? *(7.EE.4a)*

F. 4

G. 11

H. 15

I. 19

8. The distance traveled by the a high-speed train is proportional to the number of hours traveled. Which of the following is *not* a valid interpretation of the graph below? *(7.RP.2d)*

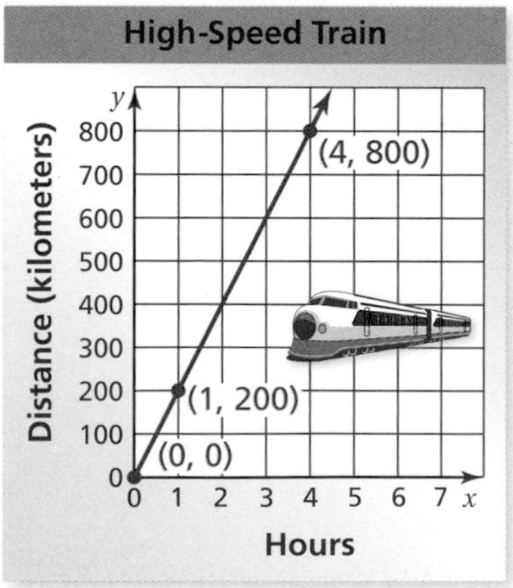

A. The train travels 0 kilometers in 0 hours.

B. The unit rate is 200 kilometers per hour.

C. After 4 hours, the train is traveling 800 kilometers per hour.

D. The train travels 800 kilometers in 4 hours.

9. Regina was evaluating the expression below. What should Regina do to correct the error she made? *(7.NS.3)*

$$-\frac{3}{2} \div \left(-\frac{8}{7}\right) = -\frac{2}{3} \times \left(-\frac{7}{8}\right)$$

$$= \frac{2 \times 7}{3 \times 8}$$

$$= \frac{14}{24}$$

$$= \frac{7}{12}$$

F. Rewrite $-\frac{3}{2} \div \left(-\frac{8}{7}\right)$ as $-\frac{2}{3} \times \left(-\frac{8}{7}\right)$.

G. Rewrite $-\frac{3}{2} \div \left(-\frac{8}{7}\right)$ as $-\frac{3}{2} \times \left(-\frac{7}{8}\right)$.

H. Rewrite $-\frac{3}{2} \div \left(-\frac{8}{7}\right)$ as $-\frac{3}{7} \times \left(-\frac{8}{2}\right)$.

I. Rewrite $-\frac{2}{3} \times \left(-\frac{7}{8}\right)$ as $-\frac{2 \times 7}{3 \times 8}$.

10. What is the least value of t for which the inequality is true? *(7.EE.4b)*

$$3 - 6t \leq -15$$

11. You can mow 800 square feet of lawn in 15 minutes. At this rate, how many minutes will you take to mow a lawn that measures 6000 square feet? *(7.RP.2c)*

Part A Write a proportion to represent the problem. Use m to represent the number of minutes. Explain your reasoning.

Part B Solve the proportion you wrote in Part A. Then use it to answer the problem. Show your work.

12. What value of p makes the equation below true? *(7.EE.4a)*

$$6 - 2p = -48$$

A. -27

B. -21

C. 21

D. 27

15 Percents

15.1 Percents and Decimals

15.2 Comparing and Ordering Fractions, Decimals, and Percents

15.3 The Percent Proportion

15.4 The Percent Equation

15.5 Percents of Increase and Decrease

15.6 Discounts and Markups

15.7 Simple Interest

75% off $1 $0.25
"Here's my sales strategy. I buy each dog bone for $0.05."
I guess P.T. Barnum was right.
"Then I mark each one up to $1. Then, I have a 75% off sale. Cool, huh?"

"At 4 a day, I have chewed 17,536 dog biscuits. At only 99.9% pure, that means that..."
I hope the same logic doesn't apply for mice!
"I have swallowed seventeen and a half contaminated dog biscuits during the past twelve years."

What You Learned Before

"The fact that these two percents do not total 100 is a sad commentary on humans."

Writing Percents as Fractions
(6.RP.3C)

Example 1 **Write 45% as a fraction in simplest form.**

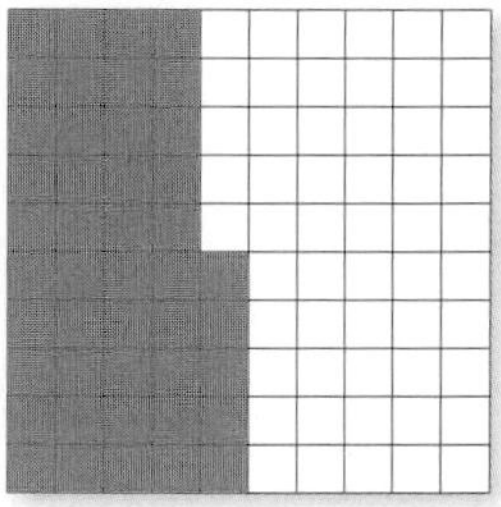

$45\% = \frac{45}{100}$ — Write as a fraction with a denominator of 100.

$= \frac{9}{20}$ — Simplify.

So, $45\% = \frac{9}{20}$.

Try It Yourself

Write the percent as a fraction or mixed number in simplest form.

1. 16% **2.** 40% **3.** 68% **4.** 85%

5. 148% **6.** 150% **7.** 105% **8.** 276%

Writing Fractions as Percents (6.RP.3C)

Example 2 **Write $\frac{3}{25}$ as a percent.**

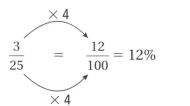

$$\frac{3}{25} \xrightarrow{\times 4} = \frac{12}{100} = 12\%$$

(numerator × 4, denominator × 4)

Because 25 × 4 = 100, multiply the numerator and denominator by 4. Write the numerator with a percent symbol.

Try It Yourself

Write the fraction or mixed number as a percent.

9. $\frac{9}{25}$ **10.** $\frac{43}{50}$ **11.** $\frac{11}{20}$ **12.** $\frac{3}{5}$

13. $1\frac{1}{4}$ **14.** $1\frac{12}{25}$ **15.** $1\frac{4}{5}$ **16.** $2\frac{3}{10}$

15.1 Percents and Decimals

Essential Question How does the decimal point move when you rewrite a percent as a decimal and when you rewrite a decimal as a percent?

1 ACTIVITY: Writing Percents as Decimals

Work with a partner. Write the percent shown by the model. Write the percent as a decimal.

a.

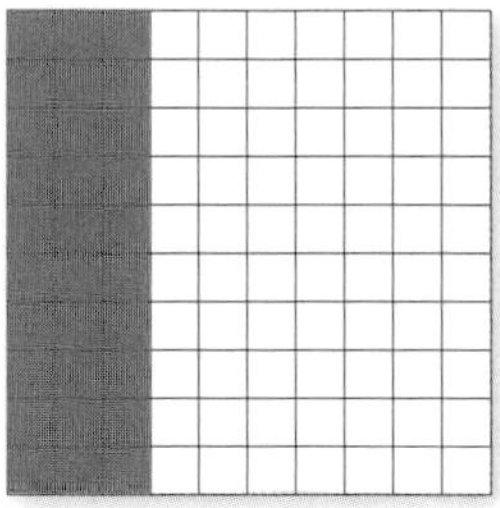

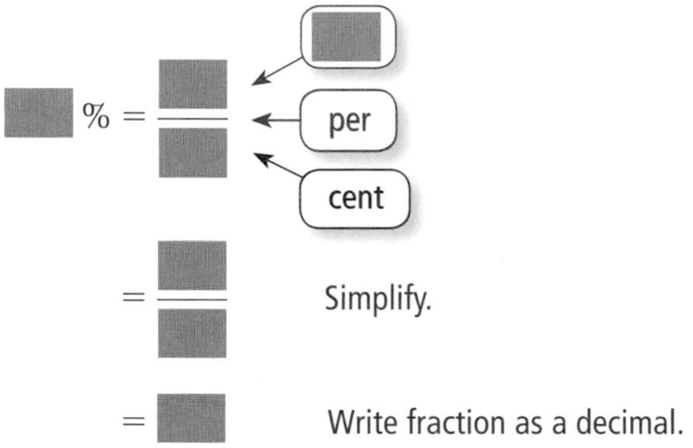

b.

c.

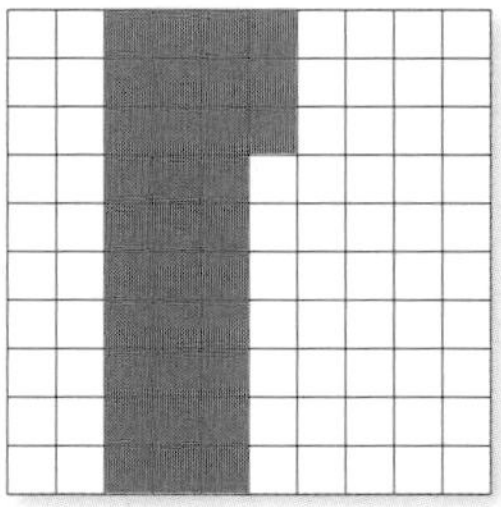

d.

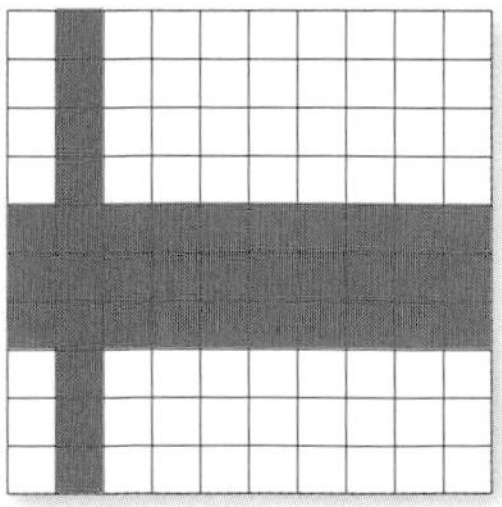

e.

f.

g.

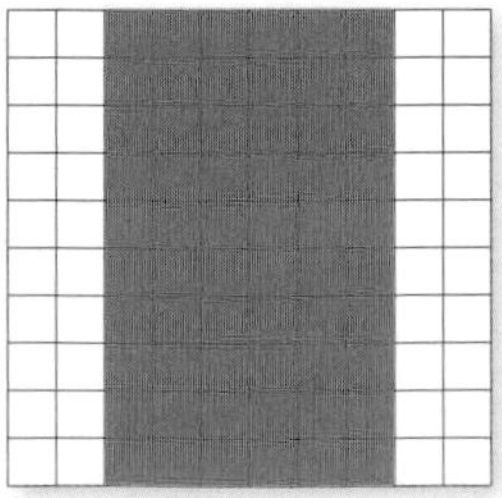

COMMON CORE

Percents and Decimals

In this lesson, you will

- write percents as decimals.
- write decimals as percents.
- solve real-life problems.

Learning Standard
7.EE.3

2 ACTIVITY: Writing Percents as Decimals

Math Practice 6

Communicate Precisely

How can reading the fraction aloud help you write it as a decimal?

Work with a partner. Write the percent as a decimal.

a. 13.5%

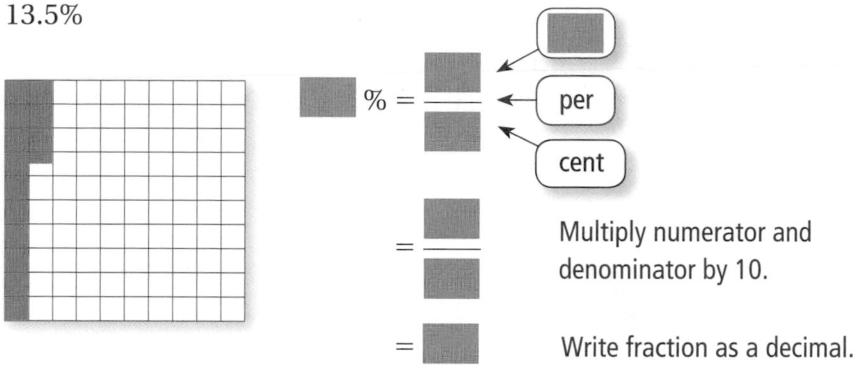

b. 12.5% **c.** 3.8% **d.** 0.5%

3 ACTIVITY: Writing Decimals as Percents

Work with a partner. Draw a model to represent the decimal. Write the decimal as a percent.

a. 0.1

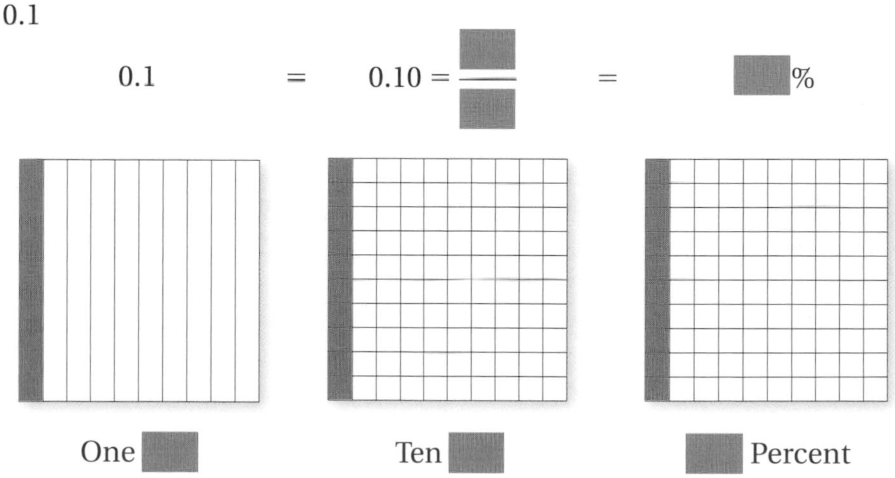

b. 0.24 **c.** 0.58 **d.** 0.05

What Is Your Answer?

4. **IN YOUR OWN WORDS** How does the decimal point move when you rewrite a percent as a decimal and when you rewrite a decimal as a percent?
5. Explain why the decimal point moves when you rewrite a percent as a decimal and when you rewrite a decimal as a percent.

Practice Use what you learned about percents and decimals to complete Exercises 7–12 and 19–24 on page 654.

15.1 Lesson

Check It Out
Lesson Tutorials
BigIdeasMath.com

Key Idea

Writing Percents as Decimals

Words Remove the percent symbol. Then divide by 100, or just move the decimal point two places to the left.

Numbers 23% = 23.% = 0.23

EXAMPLE 1 Writing Percents as Decimals

a. Write 52% as a decimal.

52% = 52.% = 0.52

Check

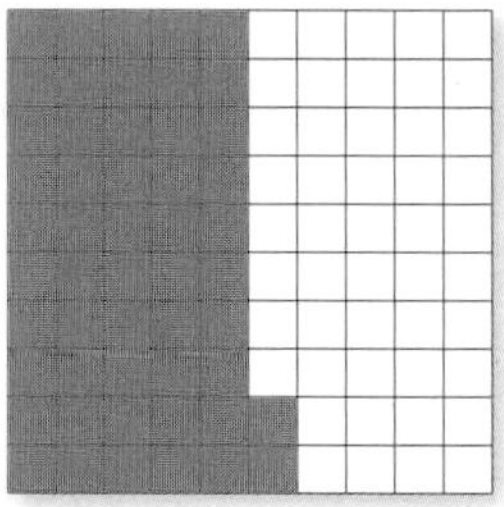

b. Write 7% as a decimal.

7% = 07.% = 0.07

Check

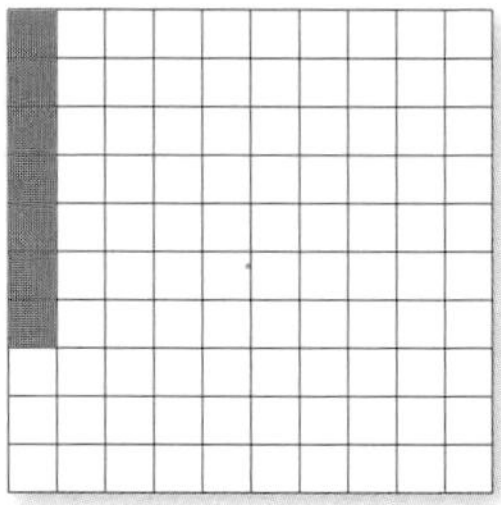

Study Tip

When moving the decimal point, you may need to place one or more zeros in the number.

On Your Own

Now You're Ready
Exercises 7–18

Write the percent as a decimal. Use a model to check your answer.

1. 24% **2.** 3% **3.** 107% **4.** 92.7%

Key Idea

Writing Decimals as Percents

Words Multiply by 100, or just move the decimal point two places to the right. Then add a percent symbol.

Numbers 0.36 = 0.36 = 36%

EXAMPLE 2 Writing Decimals as Percents

a. Write 0.47 as a percent.

0.47 = 0.47 = 47%

b. Write 0.663 as a percent.

0.663 = 0.663 = 66.3%

c. Write 1.8 as a percent.

1.8 = 1.80 = 180%

d. Write 0.009 as a percent.

0.009 = 0.009 = 0.9%

On Your Own

Now You're Ready
Exercises 19–30

Write the decimal as a percent. Use a model to check your answer.

5. 0.94 **6.** 1.2 **7.** 0.316 **8.** 0.005

EXAMPLE 3 Writing a Fraction as a Percent and a Decimal

On a math test, you get 92 out of a possible 100 points. Which of the following is *not* another way of expressing 92 out of 100?

Ⓐ $\frac{23}{25}$ Ⓑ 92% Ⓒ $\frac{17}{20}$ Ⓓ 0.92

$$92 \text{ out of } 100 = \frac{92}{100} \begin{cases} = 92\% & \text{Eliminate Choice B.} \\ = \frac{23}{25} & \text{Eliminate Choice A.} \\ = 0.92 & \text{Eliminate Choice D.} \end{cases}$$

∴ So, the correct answer is Ⓒ.

EXAMPLE 4 Real-Life Application

The figure shows the portions of ultraviolet (UV) rays reflected by four different surfaces. How many times more UV rays are reflected by water than by sea foam?

Write 25% and $\frac{21}{25}$ as decimals.

Sea foam: $25\% = 25.\% = 0.25$ **Water:** $\frac{21}{25} = \frac{84}{100} = 0.84$

Divide 0.84 by 0.25: $0.25\overline{)0.84} \longrightarrow 25\overline{)84.00}$ = 3.36

∴ So, water reflects about 3.4 times more UV rays than sea foam.

On Your Own

9. Write "18 out of 100" as a percent, a fraction, and a decimal.

10. In Example 4, how many times more UV rays are reflected by water than by sand?

15.1 Exercises

Check It Out
Help with Homework
BigIdeasMath.com

Vocabulary and Concept Check

MATCHING **Match the decimal with its equivalent percent.**

1. 0.42 **2.** 4.02 **3.** 0.042 **4.** 0.0402

A. 4.02% **B.** 42% **C.** 4.2% **D.** 402%

5. **OPEN-ENDED** Write three different decimals that are between 10% and 20%.

6. **WHICH ONE DOESN'T BELONG?** Which one does *not* belong with the other three? Explain your reasoning.

70% 0.7 $\frac{7}{10}$ 0.07

Practice and Problem Solving

Write the percent as a decimal.

1 **7.** 78% **8.** 55% **9.** 18.5%

10. 57.4% **11.** 33% **12.** 9%

13. 47.63% **14.** 91.25% **15.** 166%

16. 217% **17.** 0.06% **18.** 0.034%

Write the decimal as a percent.

2 **19.** 0.74 **20.** 0.52 **21.** 0.89

22. 0.768 **23.** 0.99 **24.** 0.49

25. 0.487 **26.** 0.128 **27.** 3.68

28. 5.12 **29.** 0.0371 **30.** 0.0046

31. **ERROR ANALYSIS** Describe and correct the error in writing 0.86 as a percent.

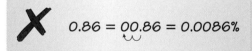

✗ $0.86 = 00.86 = 0.0086\%$

32. **MUSIC** Thirty-six percent of the songs on your MP3 player are pop songs. Write this percent as a decimal.

33. **CAT** About 0.34 of the length of a cat is its tail. Write this decimal as a percent.

34. **COMPUTER** Write the percent of free space on the computer as a decimal.

Volume	Capacity	Free Space	% Free Space
(C:)	149 GB	133 GB	89 %

Write the percent as a fraction in simplest form and as a decimal.

35. 36% **36.** 23.5% **37.** 16.24%

38. **SCHOOL** The percents of students who travel to school by car, bus, and bicycle are shown for a school of 825 students.

a. Write the percents as decimals.

b. Write the percents as fractions.

c. What percent of students use another method to travel to school?

d. **RESEARCH** Make a bar graph that represents how the students in your class travel to school.

39. **ELECTIONS** In an election, the winning candidate receives 60% of the votes. What percent of the votes does the other candidate receive?

40. **COLORS** Students in a class were asked to tell their favorite color.

a. What percent said red, blue, or yellow?

b. How many times more students said red than yellow?

c. Use two methods to find the percent of students who said green. Which method do you prefer?

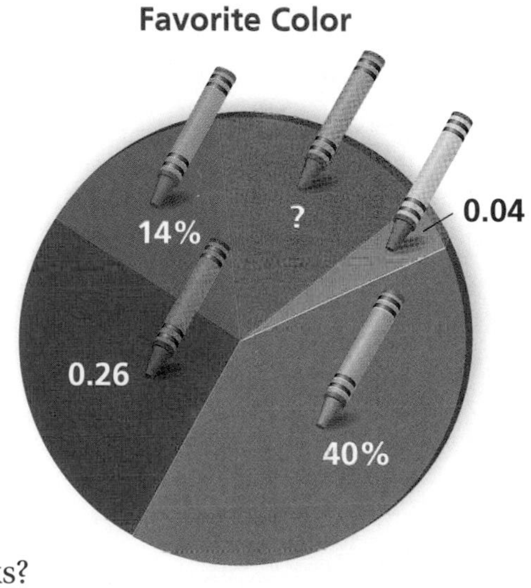

41. Problem Solving In the first 42 Super Bowls, $0.1\overline{6}$ of the MVPs (most valuable players) were running backs.

a. What percent of the MVPs were running backs?

b. What fraction of the MVPs were *not* running backs?

Fair Game Review What you learned in previous grades & lessons

Write the decimal as a fraction or mixed number in simplest form. *(Skills Review Handbook)*

42. 0.46 **43.** 0.31 **44.** 2.2 **45.** 4.32

Simplify the expression. *(Section 13.1)*

46. $4x + 3 - 9x$

47. $5 + 3.2n - 6 - 4.8n$

48. $2y - 5(y - 3)$

49. $-\frac{1}{2}(8b + 3) + 3b$

50. **MULTIPLE CHOICE** Ham costs \$4.48 per pound. Cheese costs \$6.36 per pound. You buy 1.5 pounds of ham and 0.75 pound of cheese. How much more do you pay for the ham? *(Section 2.4 and Section 2.5)*

Ⓐ \$1.41 Ⓑ \$1.95 Ⓒ \$4.77 Ⓓ \$6.18

15.2 Comparing and Ordering Fractions, Decimals, and Percents

Essential Question How can you order numbers that are written as fractions, decimals, and percents?

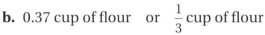

1 ACTIVITY: Using Fractions, Decimals, and Percents

Work with a partner. Decide which number form (fraction, decimal, or percent) is more common. Then find which is greater.

a. 7% sales tax or $\frac{1}{20}$ sales tax

b. 0.37 cup of flour or $\frac{1}{3}$ cup of flour

c. $\frac{5}{8}$-inch wrench or 0.375-inch wrench

d. $12\frac{3}{5}$ dollars or 12.56 dollars

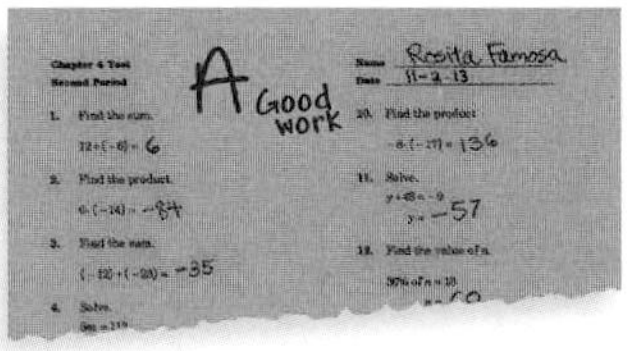

e. 93% test score or $\frac{7}{8}$ test score

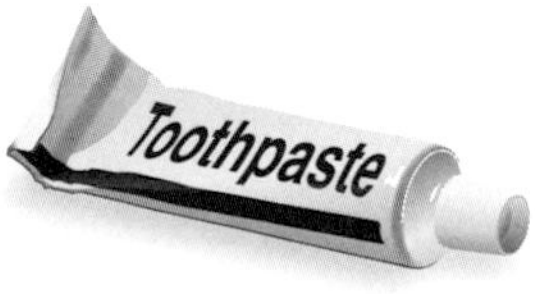

f. $5\frac{5}{6}$ fluid ounces or 5.6 fluid ounces

COMMON CORE

Fractions, Decimals, and Percents

In this lesson, you will
- compare and order fractions, decimals, and percents.
- solve real-life problems.

Learning Standard 7.EE.3

2 ACTIVITY: Ordering Numbers

Work with a partner to order the following numbers.

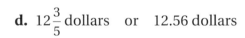

$\frac{1}{8}$ 11% $\frac{3}{20}$ 0.172 0.32 43% 7% 0.7 $\frac{5}{6}$

a. Decide on a strategy for ordering the numbers. Will you write them all as fractions, decimals, or percents?

b. Use your strategy and a number line to order the numbers from least to greatest. (Note: Label the number line appropriately.)

3 ACTIVITY: The Game of Math Card War

Math Practice 2

Make Sense of Quantities

What strategies can you use to determine which number is greater?

Preparation:

- Cut index cards to make 40 playing cards.
- Write each number in the table onto a card.

To Play:

- Play with a partner.
- Deal 20 cards facedown to each player.
- Each player turns one card faceup. The player with the greater number wins. The winner collects both cards and places them at the bottom of his or her cards.
- Suppose there is a tie. Each player lays three cards facedown, then a new card faceup. The player with the greater of these new cards wins. The winner collects all 10 cards and places them at the bottom of his or her cards.
- Continue playing until one player has all the cards. This player wins the game.

75%	$\frac{3}{4}$	$\frac{1}{3}$	$\frac{3}{10}$	0.3	25%	0.4	0.25	100%	0.27
0.75	$66\frac{2}{3}\%$	12.5%	40%	$\frac{1}{4}$	4%	0.5%	0.04	$\frac{1}{100}$	$\frac{2}{3}$
0	30%	5%	$\frac{27}{100}$	0.05	$33\frac{1}{3}\%$	$\frac{2}{5}$	0.333. . .	27%	1%
1	0.01	$\frac{1}{20}$	$\frac{1}{8}$	0.125	$\frac{1}{25}$	$\frac{1}{200}$	0.005	0.666. . .	0%

What Is Your Answer?

4. **IN YOUR OWN WORDS** How can you order numbers that are written as fractions, decimals, and percents? Give an example with your answer.

5. All but one of the U.S. coins shown has a name that is related to its value. Which one is it? How are the names of the others related to their values?

Practice Use what you learned about ordering numbers to complete Exercises 4–7, 16, and 17 on page 660.

15.2 Lesson

Check It Out
Lesson Tutorials
BigIdeasMath.com

When comparing and ordering fractions, decimals, and percents, write the numbers as all fractions, all decimals, or all percents.

EXAMPLE 1 Comparing Fractions, Decimals, and Percents

Study Tip

It is usually easier to order decimals or percents than to order fractions.

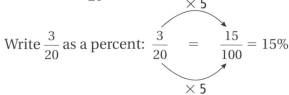

a. Which is greater, $\frac{3}{20}$ or 16%?

Write $\frac{3}{20}$ as a percent: $\frac{3}{20} = \frac{15}{100} = 15\%$ (× 5 on numerator and denominator)

∴ 15% is less than 16%. So, 16% is the greater number.

b. Which is greater, 79% or 0.08?

Write 79% as a decimal: 79% = 79.% = 0.79

∴ 0.79 is greater than 0.08. So, 79% is the greater number.

On Your Own

Now You're Ready
Exercises 4–15

1. Which is greater, 25% or $\frac{7}{25}$?
2. Which is greater, 0.49 or 94%?

EXAMPLE 2 Real-Life Application

Remember

To order numbers from least to greatest, write them as they appear on a number line from left to right.

You, your sister, and a friend each take the same number of shots at a soccer goal. You make 72% of your shots, your sister makes $\frac{19}{25}$ of her shots, and your friend makes 0.67 of his shots. Who made the fewest shots?

Write 72% and $\frac{19}{25}$ as decimals.

You: 72% = 72.% = 0.72

Sister: $\frac{19}{25} = \frac{76}{100} = 0.76$ (× 4 on numerator and denominator)

Graph the decimals on a number line.

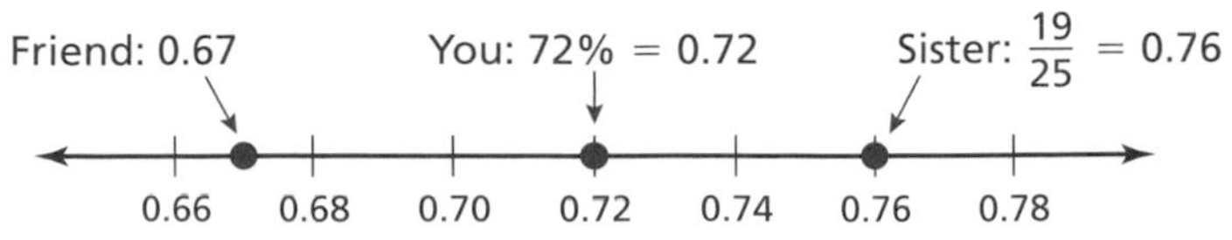

∴ 0.67 is the least number. So, your friend made the fewest shots.

On Your Own

Now You're Ready
Exercises 16–21

3. You make 75% of your shots, your sister makes $\frac{13}{20}$ of her shots, and your friend makes 0.7 of his shots. Who made the most shots?

EXAMPLE 3 Real-Life Application

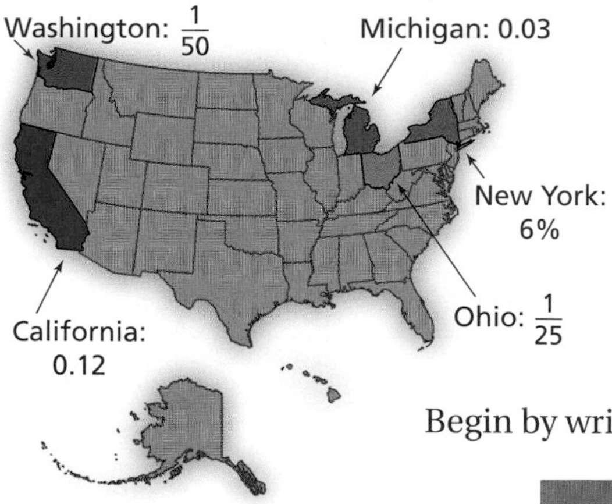

The map shows the portions of the U.S. population that live in five states.

List the five states in order by population from least to greatest.

Begin by writing each portion as a fraction, a decimal, and a percent.

State	Fraction	Decimal	Percent
Michigan	$\frac{3}{100}$	0.03	3%
New York	$\frac{6}{100}$	0.06	6%
Washington	$\frac{1}{50}$	0.02	2%
California	$\frac{12}{100}$	0.12	12%
Ohio	$\frac{1}{25}$	0.04	4%

Graph the percent for each state on a number line.

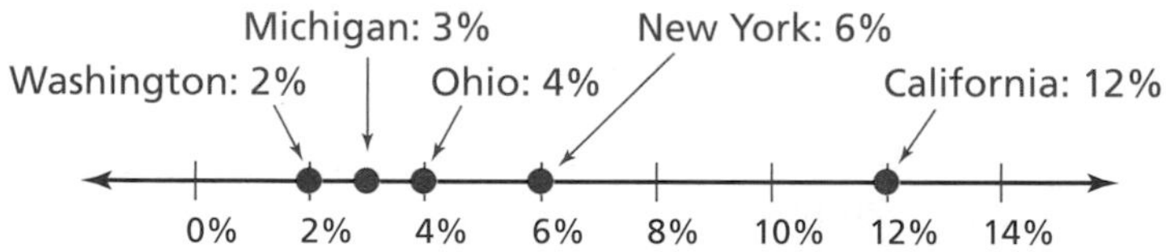

∴ The states in order by population from least to greatest are Washington, Michigan, Ohio, New York, and California.

On Your Own

4. The portion of the U.S. population that lives in Texas is $\frac{2}{25}$. The portion that lives in Illinois is 0.042. Reorder the states in Example 3 including Texas and Illinois.

15.2 Exercises

Check It Out
Help with Homework
BigIdeasMath.com

Vocabulary and Concept Check

1. **NUMBER SENSE** Copy and complete the table.

Fraction	Decimal	Percent
$\frac{18}{25}$	0.72	
$\frac{17}{20}$		85%
$\frac{13}{50}$		
	0.62	
		45%

2. **NUMBER SENSE** How would you decide whether $\frac{3}{5}$ or 59% is greater? Explain.

3. **WHICH ONE DOESN'T BELONG?** Which one does *not* belong with the other three? Explain your reasoning.

40%	$\frac{2}{5}$
0.4	0.04

Practice and Problem Solving

Tell which number is greater.

1 4. 0.9, 95%
5. 20%, 0.02
6. $\frac{37}{50}$, 37%
7. 50%, $\frac{13}{25}$
8. 0.086, 86%
9. 76%, 0.67
10. 60%, $\frac{5}{8}$
11. 0.12, 1.2%
12. 17%, $\frac{4}{25}$
13. 140%, 0.14
14. $\frac{1}{3}$, 30%
15. 80%, $\frac{7}{9}$

Use a number line to order the numbers from least to greatest.

2 16. 38%, $\frac{8}{25}$, 0.41
17. 68%, 0.63, $\frac{13}{20}$
18. $\frac{43}{50}$, 0.91, $\frac{7}{8}$, 84%
19. 0.15%, $\frac{3}{20}$, 0.015
20. 2.62, $2\frac{2}{5}$, 26.8%, 2.26, 271%
21. $\frac{87}{200}$, 0.44, 43.7%, $\frac{21}{50}$

22. **TEST** You answered 21 out of 25 questions correctly on a test. Did you reach your goal of getting at least 80%?

23. **POPULATION** The table shows the portions of the world population that live in four countries. Order the countries by population from least to greatest.

Country	Brazil	India	Russia	United States
Portion of World Population	2.8%	$\frac{7}{40}$	$\frac{1}{50}$	0.044

PRECISION **Order the numbers from least to greatest.**

24. 66.1%, 0.66, $\frac{2}{3}$, 0.667

25. $\frac{2}{9}$, 21%, $0.2\overline{1}$, $\frac{11}{50}$

Tell which letter shows the graph of the number.

26. $\frac{2}{5}$

27. 45.2%

28. 0.435

29. $\frac{4}{9}$

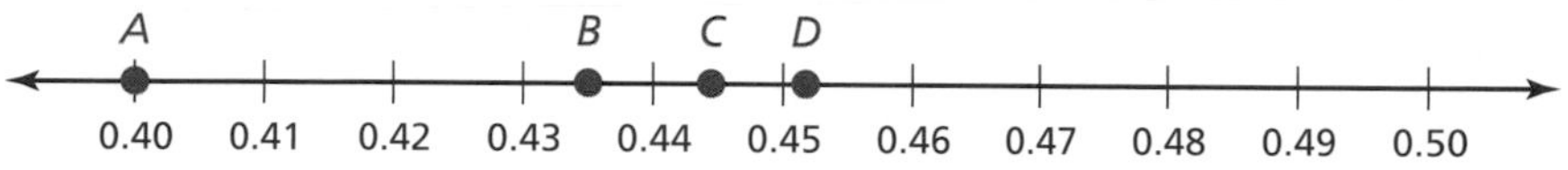

30. **TOUR DE FRANCE** The Tour de France is a bicycle road race. The whole race is made up of 21 small races called *stages*. The table shows how several stages compare to the whole Tour de France in a recent year. Order the stages from shortest to longest.

Stage	1	7	8	17	21
Portion of Total Distance	$\frac{11}{200}$	0.044	$\frac{6}{125}$	0.06	4%

31. **SLEEP** The table shows the portions of the day that several animals sleep.

a. Order the animals by sleep time from least to greatest.

b. Estimate the portion of the day that you sleep.

c. Where do you fit on the ordered list?

Animal	Portion of Day Sleeping
Dolphin	0.433
Lion	56.3%
Rabbit	$\frac{19}{40}$
Squirrel	$\frac{31}{50}$
Tiger	65.8%

32. Number Sense Tell what whole number you can substitute for a in each list so the numbers are ordered from least to greatest. If there is none, explain why.

a. $\frac{2}{a}$, $\frac{a}{22}$, 33%

b. $\frac{1}{a}$, $\frac{a}{8}$, 33%

Fair Game Review What you learned in previous grades & lessons

Tell whether the ratios form a proportion. *(Section 14.2)*

33. $\frac{6}{10}$, $\frac{9}{15}$

34. $\frac{7}{16}$, $\frac{28}{80}$

35. $\frac{20}{12}$, $\frac{35}{21}$

36. **MULTIPLE CHOICE** What is the solution of $2n - 4 > -12$? *(Section 7.6 and Section 7.7)*

Ⓐ $n < -10$　Ⓑ $n < -4$　Ⓒ $n > -2$　Ⓓ $n > -4$

15.3 The Percent Proportion

Essential Question How can you use models to estimate percent questions?

The statement "25% of 12 is 3" has three numbers. In real-life problems, any one of these numbers can be unknown.

Question	Which number is missing?	Type of Question
What is 25% of 12?	3	Find a part of a number.
3 is what percent of 12?	25%	Find a percent.
3 is 25% of what?	12	Find the whole.

1 ACTIVITY: Estimating a Part

Work with a partner. Use a model to estimate the answer to each question.

a. What number is 50% of 30?

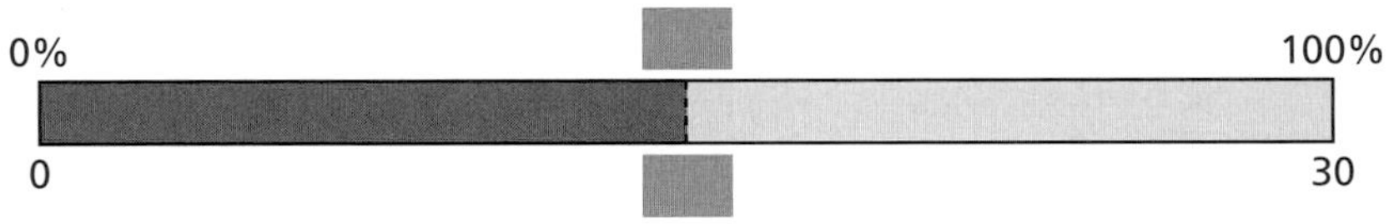

So, from the model, ___ is 50% of 30.

b. What number is 75% of 30?

c. What number is 40% of 30?

d. What number is 6% of 30?

e. What number is 65% of 30?

2 ACTIVITY: Estimating a Percent

Work with a partner. Use a model to estimate the answer to each question.

a. 15 is what percent of 75?

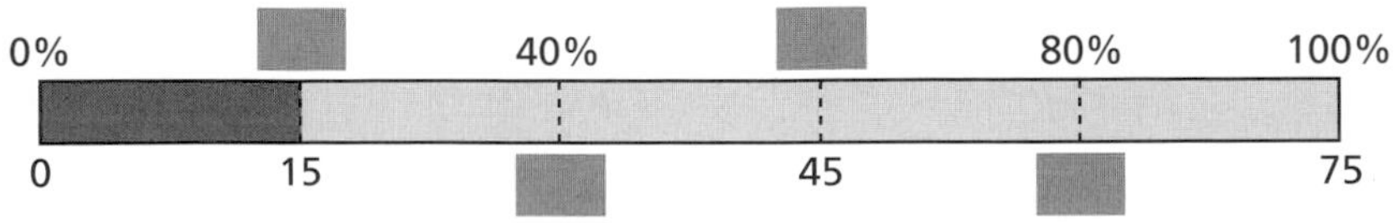

So, from the model, 15 is ___ of 75.

b. 5 is what percent of 20?

c. 18 is what percent of 40?

d. 50 is what percent of 80?

e. 75 is what percent of 50?

Common Core

Percent Proportion

In this lesson, you will

- use the percent proportion to find parts, wholes, and percents.

Learning Standard 7.RP.3

3 ACTIVITY: Estimating a Whole

Math Practice 4

Use a Model

What quantities are given? How can you use the model to find the unknown quantity?

Work with a partner. Use a model to estimate the answer to each question.

a. 24 is $33\frac{1}{3}\%$ of what number?

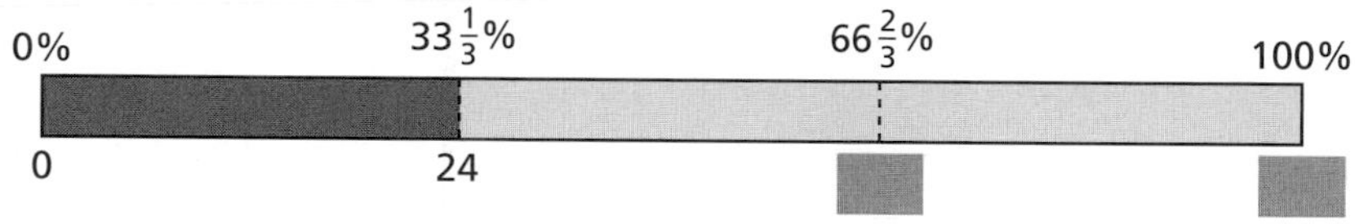

∴ So, from the model, 24 is $33\frac{1}{3}\%$ of ▭.

b. 13 is 25% of what number?

c. 110 is 20% of what number?

d. 75 is 75% of what number?

e. 81 is 45% of what number?

4 ACTIVITY: Using Ratio Tables

Work with a partner. Use a ratio table to answer each question. Then compare your answer to the estimate you found using the model.

(1d) **a.** What number is 6% of 30?

Part	6		
Whole	100		30

(1e) **b.** What number is 65% of 30?

Part	65		
Whole	100		30

(2c) **c.** 18 is what percent of 40?

Part	18		
Whole	40		100

(3e) **d.** 81 is 45% of what number?

Part	45		81
Whole	100		

What Is Your Answer?

5. **IN YOUR OWN WORDS** How can you use models to estimate percent questions? Give examples to support your answer.

6. Complete the proportion below using the given labels.

percent
whole
100
part

$\frac{\square}{\square} = \frac{\square}{\square}$

Practice Use what you learned about estimating percent questions to complete Exercises 5–10 on page 666.

15.3 Lesson

Check It Out
Lesson Tutorials
BigIdeasMath.com

Key Idea

The Percent Proportion

Words You can represent "a is p percent of w" with the proportion

$$\frac{a}{w} = \frac{p}{100}$$

where a is part of the whole w, and $p\%$, or $\frac{p}{100}$, is the percent.

Numbers 3 out of 4 is 75%.

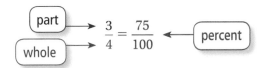

$$\frac{3}{4} = \frac{75}{100}$$

part → 3; whole → 4; percent → 75/100

Study Tip

In percent problems, the word *of* is usually followed by the whole.

EXAMPLE 1 Finding a Percent

What percent of 15 is 12?

$$\frac{a}{w} = \frac{p}{100}$$ Write the percent proportion.

$$\frac{12}{15} = \frac{p}{100}$$ Substitute 12 for a and 15 for w.

$$100 \cdot \frac{12}{15} = 100 \cdot \frac{p}{100}$$ Multiplication Property of Equality

$$80 = p$$ Simplify.

So, 80% of 15 is 12.

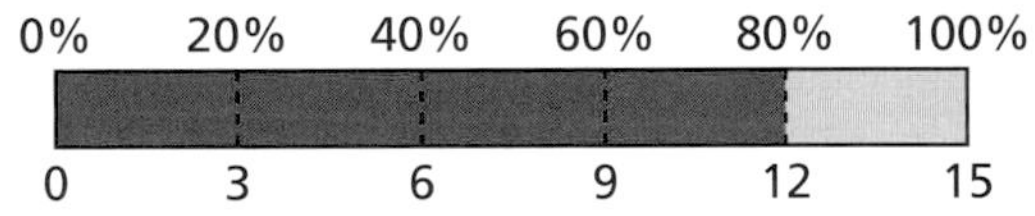

EXAMPLE 2 Finding a Part

What number is 36% of 50?

$$\frac{a}{w} = \frac{p}{100}$$ Write the percent proportion.

$$\frac{a}{50} = \frac{36}{100}$$ Substitute 50 for w and 36 for p.

$$50 \cdot \frac{a}{50} = 50 \cdot \frac{36}{100}$$ Multiplication Property of Equality

$$a = 18$$ Simplify.

So, 18 is 36% of 50.

EXAMPLE 3 Finding a Whole

150% of what number is 24?

$$\frac{a}{w} = \frac{p}{100}$$ Write the percent proportion.

$$\frac{24}{w} = \frac{150}{100}$$ Substitute 24 for *a* and 150 for *p*.

$$24 \cdot 100 = w \cdot 150$$ Cross Products Property

$$2400 = 150w$$ Multiply.

$$16 = w$$ Divide each side by 150.

So, 150% of 16 is 24.

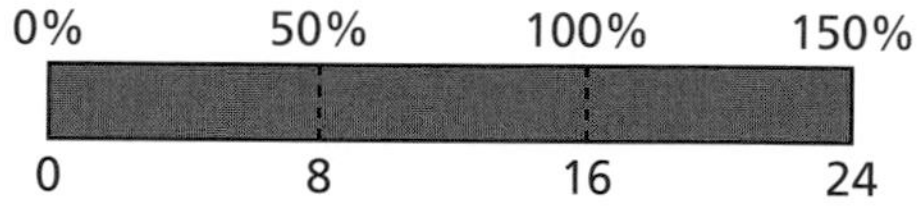

On Your Own

Now You're Ready
Exercises 11–18

Write and solve a proportion to answer the question.

1. What percent of 5 is 3?
2. 25 is what percent of 20?
3. What number is 80% of 60?
4. 10% of 40.5 is what number?
5. 0.1% of what number is 4?
6. $\frac{1}{2}$ is 25% of what number?

EXAMPLE 4 Real-Life Application

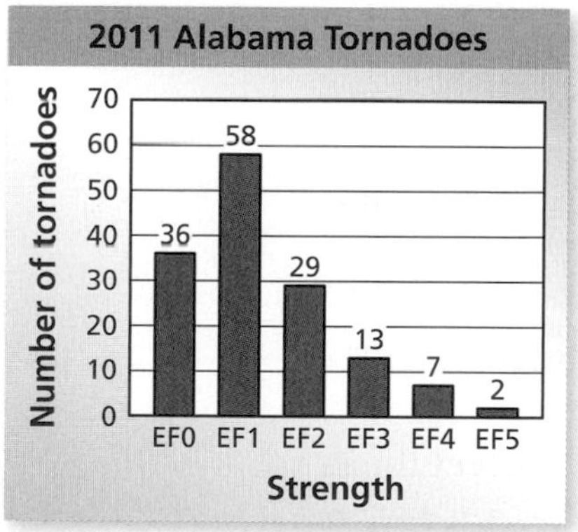

The bar graph shows the strengths of tornadoes that occurred in Alabama in 2011. What percent of the tornadoes were EF1s?

The total number of tornadoes, 145, is the *whole*, and the number of EF1 tornadoes, 58, is the *part*.

$$\frac{a}{w} = \frac{p}{100}$$ Write the percent proportion.

$$\frac{58}{145} = \frac{p}{100}$$ Substitute 58 for *a* and 145 for *w*.

$$100 \cdot \frac{58}{145} = 100 \cdot \frac{p}{100}$$ Multiplication Property of Equality

$$40 = p$$ Simplify.

So, 40% of the tornadoes were EF1s.

On Your Own

7. Twenty percent of the tornadoes occurred in central Alabama on April 27. How many tornadoes does this represent?

15.3 Exercises

Check It Out
Help with Homework
BigIdeasMath.com

Vocabulary and Concept Check

1. **VOCABULARY** Write the percent proportion in words.
2. **WRITING** Explain how to use a proportion to find 30% of a number.
3. **NUMBER SENSE** Write and solve the percent proportion represented by the model.

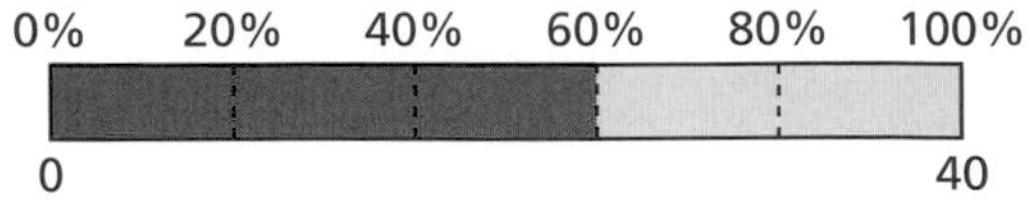

4. **WHICH ONE DOESN'T BELONG?** Which proportion does *not* belong with the other three? Explain your reasoning.

$\frac{15}{w} = \frac{50}{100}$ $\frac{12}{15} = \frac{40}{n}$ $\frac{15}{25} = \frac{p}{100}$ $\frac{a}{20} = \frac{35}{100}$

Practice and Problem Solving

Use a model to estimate the answer to the question. Use a ratio table to check your answer.

5. What number is 24% of 80?
6. 15 is what percent of 40?
7. 15 is 30% of what number?
8. What number is 120% of 70?
9. 20 is what percent of 52?
10. 48 is 75% of what number?

Write and solve a proportion to answer the question.

1 11. What percent of 25 is 12?
12. 14 is what percent of 56?
2 13. 25% of what number is 9?
14. 36 is 0.9% of what number?
3 15. 75% of 124 is what number?
16. 110% of 90 is what number?
17. What number is 0.4% of 40?
18. 72 is what percent of 45?

✗

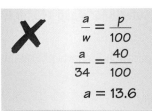

$\frac{a}{w} = \frac{p}{100}$

$\frac{a}{34} = \frac{40}{100}$

$a = 13.6$

19. **ERROR ANALYSIS** Describe and correct the error in using the percent proportion to answer the question below.

"40% of what number is 34?"

20. **FITNESS** Of 140 seventh-grade students, 15% earn the Presidential Physical Fitness Award. How many students earn the award?
21. **COMMISSION** A salesperson receives a 3% commission on sales. The salesperson receives $180 in commission. What is the amount of sales?

Write and solve a proportion to answer the question.

22. 0.5 is what percent of 20?

23. 14.2 is 35.5% of what number?

24. $\frac{3}{4}$ is 60% of what number?

25. What number is 25% of $\frac{7}{8}$?

26. **HOMEWORK** You are assigned 32 math exercises for homework. You complete 87.5% of these before dinner. How many do you have left to do after dinner?

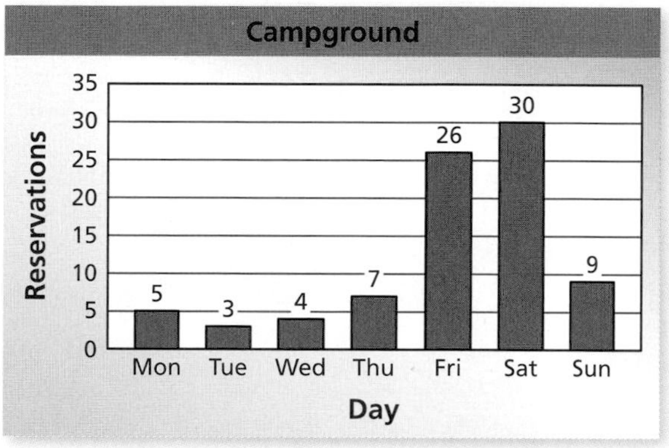

27. **HOURLY WAGE** Your friend earns \$10.50 per hour. This is 125% of her hourly wage last year. How much did your friend earn per hour last year?

28. **CAMPSITE** The bar graph shows the numbers of reserved campsites at a campground for one week. What percent of the reservations were for Friday or Saturday?

29. **PROBLEM SOLVING** A classmate displays the results of a class president election in the bar graph shown.

a. What is missing from the bar graph?

b. What percent of the votes does the last-place candidate receive? Explain your reasoning.

c. There are 124 votes total. How many votes does Chloe receive?

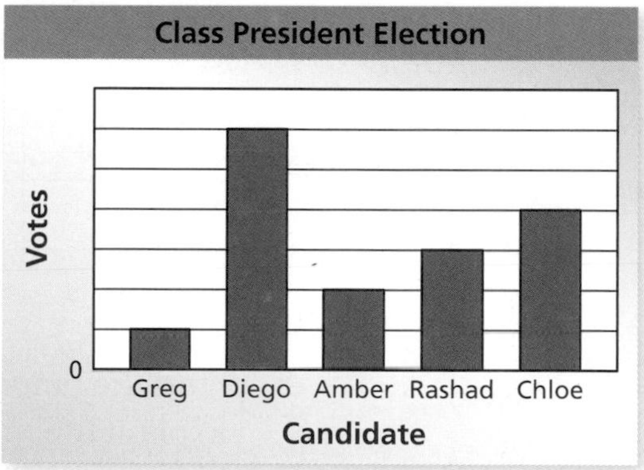

30. **REASONING** 20% of a number is x. What is 100% of the number? Assume $x > 0$.

31. **Structure** Answer each question. Assume $x > 0$.

a. What percent of $8x$ is $5x$?

b. What is 65% of $80x$?

Fair Game Review What you learned in previous grades & lessons

Evaluate the expression when $a = -15$ and $b = -5$. *(Section 11.5)*

32. $a \div b$

33. $\frac{b + 14}{a}$

34. $\frac{b^2}{a + 5}$

35. **MULTIPLE CHOICE** What is the solution of $9x = -1.8$? *(Section 13.4)*

Ⓐ $x = -5$ Ⓑ $x = -0.2$ Ⓒ $x = 0.2$ Ⓓ $x = 5$

15.4 The Percent Equation

Essential Question How can you use an equivalent form of the percent proportion to solve a percent problem?

1 ACTIVITY: Solving Percent Problems Using Different Methods

Work with a partner. The circle graph shows the number of votes received by each candidate during a school election. So far, only half the students have voted.

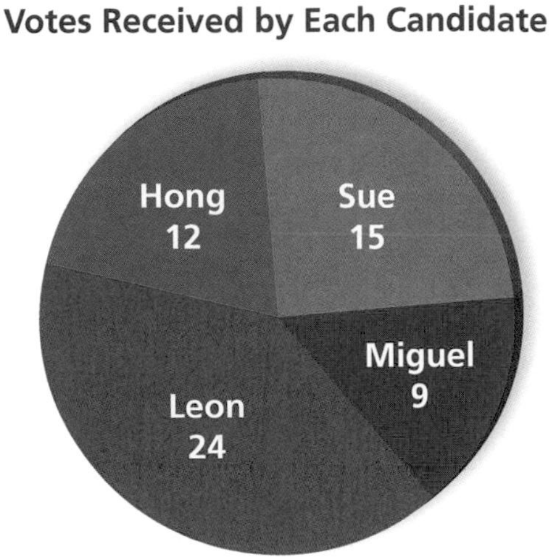

a. Complete the table.

Candidate	$\frac{\text{Number of votes received}}{\text{Total number of votes}}$
Sue	
Miguel	
Leon	
Hong	

b. Find the percent of students who voted for each candidate. Explain the method you used to find your answers.

c. Compare the method you used in part (b) with the methods used by other students in your class. Which method do you prefer? Explain.

2 ACTIVITY: Finding Parts Using Different Methods

Work with a partner. The circle graph shows the final results of the election.

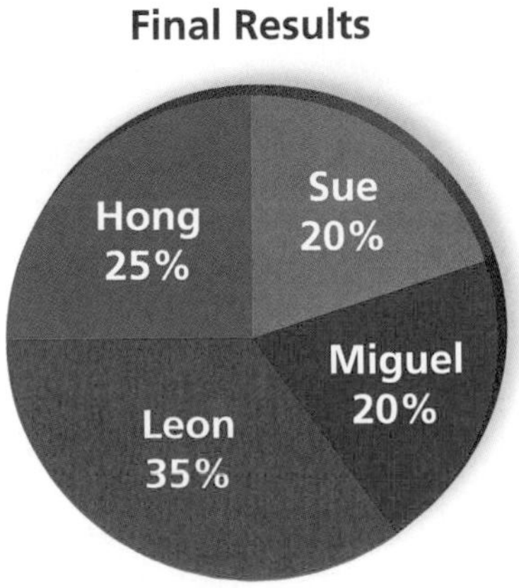

a. Find the number of students who voted for each candidate. Explain the method you used to find your answers.

b. Compare the method you used in part (a) with the methods used by other students in your class. Which method do you prefer? Explain.

COMMON CORE

Percent Equation

In this lesson, you will

- use the percent equation to find parts, wholes, and percents.
- solve real-life problems.

Learning Standards
7.RP.3
7.EE.3

3 ACTIVITY: Deriving the Percent Equation

Work with a partner. In Section 15.3, you used the percent proportion to find the missing percent, part, or whole. You can also use the *percent equation* to find these missing values.

a. Complete the steps below to find the percent equation.

$\frac{\text{part}}{\text{whole}} = \text{percent}$	Definition of percent
$\frac{\text{part}}{\text{whole}} \cdot \square = \square \cdot \square$	Multiply each side by the $\square$.
$\text{part} = \square \cdot \square$	Divide out common factors. This is the percent equation.

b. Use the percent equation to find the number of students who voted for each candidate in Activity 2. How does this method compare to the percent proportion?

4 ACTIVITY: Identifying Different Equations

Work with a partner. Without doing any calculations, choose the equation that you cannot use to answer each question.

Math Practice 3

Justify Conclusions

How can you justify the equations that you chose?

a. What number is 55% of 80?

$a = 0.55 \cdot 80$ $\qquad$ $a = \frac{11}{20} \cdot 80$ $\qquad$ $80a = 0.55$ $\qquad$ $\frac{a}{80} = \frac{55}{100}$

b. 24 is 60% of what number?

$\frac{24}{w} = \frac{60}{100}$ $\qquad$ $24 = 0.6 \cdot w$ $\qquad$ $\frac{24}{60} = w$ $\qquad$ $24 = \frac{3}{5} \cdot w$

What Is Your Answer?

5. **IN YOUR OWN WORDS** How can you use an equivalent form of the percent proportion to solve a percent problem?

6. Write a percent proportion and a percent equation that you can use to answer the question below.

16 is what percent of 250?

Practice Use what you learned about solving percent problems to complete Exercises 4–9 on page 672.

15.4 Lesson

Check It Out
Lesson Tutorials
BigIdeasMath.com

Key Idea

The Percent Equation

Words To represent "*a* is *p* percent of *w*," use an equation.

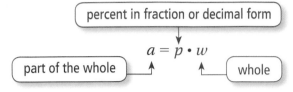

percent in fraction or decimal form

$$a = p \cdot w$$

part of the whole → a; whole → w

Numbers $15 = 0.5 \cdot 30$

EXAMPLE 1 Finding a Part of a Number

What number is 24% of 50? **Estimate**

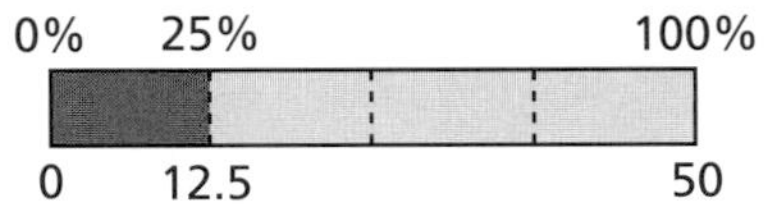

Common Error

Remember to convert a percent to a fraction or a decimal before using the percent equation. For Example 1, write 24% as $\frac{24}{100}$.

$a = p \cdot w$	Write percent equation.
$= \frac{24}{100} \cdot 50$	Substitute $\frac{24}{100}$ for p and 50 for w.
$= 12$	Simplify.

So, 12 is 24% of 50. **Reasonable? $12 \approx 12.5$ ✓**

EXAMPLE 2 Finding a Percent

9.5 is what percent of 25? **Estimate**

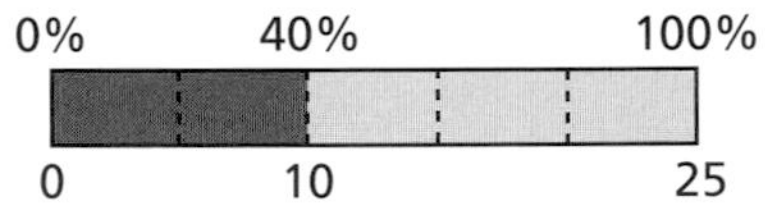

$a = p \cdot w$	Write percent equation.
$9.5 = p \cdot 25$	Substitute 9.5 for a and 25 for w.
$\frac{9.5}{25} = \frac{p \cdot 25}{25}$	Division Property of Equality
$0.38 = p$	Simplify.

Because 0.38 equals 38%, 9.5 is 38% of 25. **Reasonable? $38\% \approx 40\%$ ✓**

EXAMPLE 3 Finding a Whole

39 is 52% of what number? **Estimate**

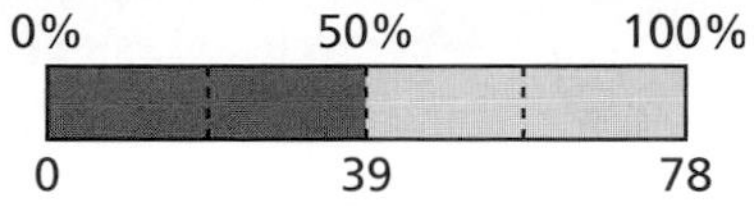

$a = p \cdot w$	Write percent equation.
$39 = 0.52 \cdot w$	Substitute 39 for *a* and 0.52 for *p*.
$75 = w$	Divide each side by 0.52.

So, 39 is 52% of 75. **Reasonable? $75 \approx 78$** ✓

On Your Own

Now You're Ready
Exercises 10–17

Write and solve an equation to answer the question.

1. What number is 10% of 20?
2. What number is 150% of 40?
3. 3 is what percent of 600?
4. 18 is what percent of 20?
5. 8 is 80% of what number?
6. 90 is 18% of what number?

EXAMPLE 4 Real-Life Application

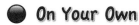

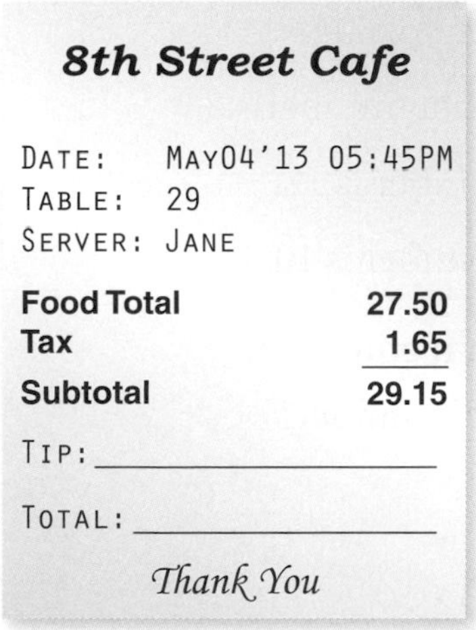
8th Street Cafe

DATE: MAY04'13 05:45PM
TABLE: 29
SERVER: JANE

Food Total	**27.50**
Tax	**1.65**
Subtotal	**29.15**

TIP: ______

TOTAL: ______

Thank You

a. Find the percent of sales tax on the food total.

Answer the question: $1.65 is what percent of $27.50?

$a = p \cdot w$	Write percent equation.
$1.65 = p \cdot 27.50$	Substitute 1.65 for *a* and 27.50 for *w*.
$0.06 = p$	Divide each side by 27.50.

Because 0.06 equals 6%, the percent of sales tax is 6%.

b. Find the amount of a 16% tip on the food total.

Answer the question: What tip amount is 16% of $27.50?

$a = p \cdot w$	Write percent equation.
$= 0.16 \cdot 27.50$	Substitute 0.16 for *p* and 27.50 for *w*.
$= 4.40$	Multiply.

So, the amount of the tip is $4.40.

On Your Own

7. **WHAT IF?** Find the amount of a 20% tip on the food total.

15.4 Exercises

Check It Out
Help with Homework
BigIdeasMath.com

Vocabulary and Concept Check

1. **VOCABULARY** Write the percent equation in words.

2. **REASONING** A number n is 150% of number m. Is n *greater than, less than,* or *equal to* m? Explain your reasoning.

3. **DIFFERENT WORDS, SAME QUESTION** Which is different? Find "both" answers.

What number is 20% of 55?	55 is 20% of what number?
20% of 55 is what number?	0.2 • 55 is what number?

Practice and Problem Solving

Answer the question. Explain the method you chose.

4. What number is 24% of 80?
5. 15 is what percent of 40?
6. 15 is 30% of what number?
7. What number is 120% of 70?
8. 20 is what percent of 52?
9. 48 is 75% of what number?

Write and solve an equation to answer the question.

1 10. 20% of 150 is what number?
11. 45 is what percent of 60?
2 12. 35% of what number is 35?
13. 0.8% of 150 is what number?
3 14. 29 is what percent of 20?
15. 0.5% of what number is 12?
16. What percent of 300 is 51?
17. 120% of what number is 102?

ERROR ANALYSIS Describe and correct the error in using the percent equation.

18. What number is 35% of 20?

✗
$$a = p \cdot w$$
$$= 35 \cdot 20$$
$$= 700$$

19. 30 is 60% of what number?

✗
$$a = p \cdot w$$
$$= 0.6 \cdot 30$$
$$= 18$$

20. **COMMISSION** A salesperson receives a 2.5% commission on sales. What commission does the salesperson receive for $8000 in sales?

21. **FUNDRAISING** Your school raised 125% of its fundraising goal. The school raised $6750. What was the goal?

22. **SURFBOARD** The sales tax on a surfboard is $12. What is the percent of sales tax?

PUZZLE **There were w signers of the Declaration of Independence. The youngest was Edward Rutledge, who was x years old. The oldest was Benjamin Franklin, who was y years old.**

23. x is 25% of 104. What was Rutledge's age?

24. 7 is 10% of y. What was Franklin's age?

25. w is 80% of y. How many signers were there?

26. y is what percent of $(w + y - x)$?

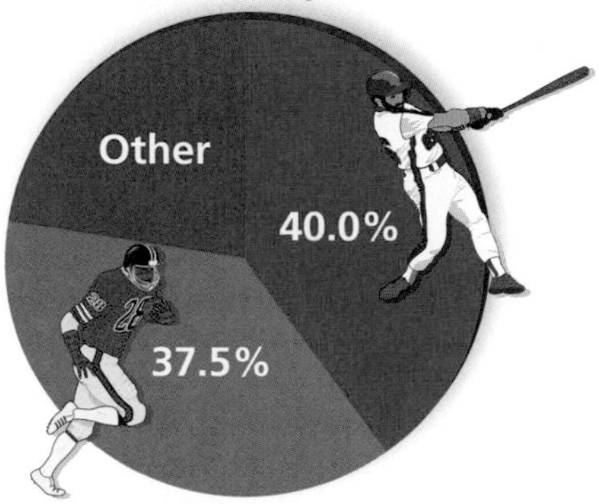

27. **LOGIC** How can you tell whether the percent of a number will be *greater than, less than,* or *equal to* the number? Give examples to support your answer.

28. **SURVEY** In a survey, a group of students were asked their favorite sport. Eighteen students chose "other" sports.
    a. How many students participated?
    b. How many chose football?

29. **WATER TANK** Water tank A has a capacity of 550 gallons and is 66% full. Water tank B is 53% full. The ratio of the capacity of Tank A to Tank B is 11 : 15.
    a. How much water is in Tank A?
    b. What is the capacity of Tank B?
    c. How much water is in Tank B?

30. **TRUE OR FALSE?** Tell whether the statement is *true* or *false*. Explain your reasoning.

    If W is 25% of Z, then $Z : W$ is 75 : 25.

31. **Reasoning** The table shows your test results for math class. What test score do you need on the last exam to earn 90% of the total points?

Test Score	Point Value
83%	100
91.6%	250
88%	150
?	300

Fair Game Review What you learned in previous grades & lessons

Simplify. Write the answer as a decimal. *(Skills Review Handbook)*

32. $\frac{10 - 4}{10}$

33. $\frac{25 - 3}{25}$

34. $\frac{105 - 84}{84}$

35. $\frac{170 - 125}{125}$

36. **MULTIPLE CHOICE** There are 160 people in a grade. The ratio of boys to girls is 3 to 5. Which proportion can you use to find the number x of boys? *(Section 14.3)*

Ⓐ $\frac{3}{8} = \frac{x}{160}$ Ⓑ $\frac{3}{5} = \frac{x}{160}$ Ⓒ $\frac{5}{8} = \frac{x}{160}$ Ⓓ $\frac{3}{5} = \frac{160}{x}$

15 Study Help

Check It Out
Graphic Organizer
BigIdeasMath.com

You can use a **summary triangle** to explain a concept. Here is an example of a summary triangle for writing a percent as a decimal.

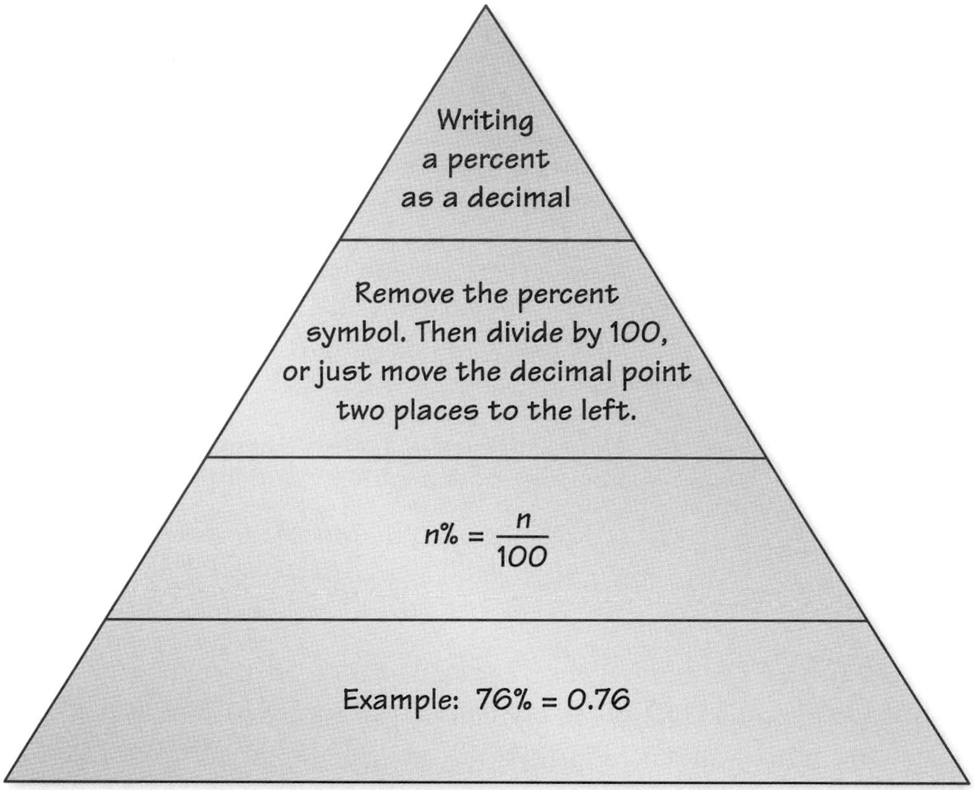

Writing a percent as a decimal
Remove the percent symbol. Then divide by 100, or just move the decimal point two places to the left.
$n\% = \frac{n}{100}$
Example: 76% = 0.76

On Your Own

Make summary triangles to help you study these topics.

1. writing a decimal as a percent
2. comparing and ordering fractions, decimals, and percents
3. the percent proportion
4. the percent equation

After you complete this chapter, make summary triangles for the following topics.

5. percent of change
6. discount
7. markup
8. simple interest

I'm writing to Classy Calicos about this one.
Lonely?
Buy a pet.
Dogs are cool!
Example: Beagles

"I found this great summary triangle in my *Beautiful Beagle Magazine*."

15.1–15.4 Quiz

Check It Out
Progress Check
BigIdeasMath.com

Write the percent as a decimal. *(Section 15.1)*

1. 34% **2.** 0.12% **3.** 62.5%

Write the decimal as a percent. *(Section 15.1)*

4. 0.67 **5.** 5.35 **6.** 0.685

Tell which number is greater. *(Section 15.2)*

7. $\frac{11}{15}$, 74% **8.** 3%, 0.3

Use a number line to order the numbers from least to greatest. *(Section 15.2)*

9. 125%, $\frac{6}{5}$, 1.22 **10.** 42%, 0.43, $\frac{17}{40}$

Write and solve a proportion to answer the question. *(Section 15.3)*

11. What percent of 15 is 6? **12.** 35 is what percent of 25?

13. What number is 40% of 50? **14.** 0.5% of what number is 5?

Write and solve an equation to answer the question. *(Section 15.4)*

15. What number is 28% of 75? **16.** 42 is 21% of what number?

17. FISHING On a fishing trip, 38% of the fish that you catch are perch. Write this percent as a decimal. *(Section 15.1)*

18. SCAVENGER HUNT The table shows the results of 8 teams competing in a scavenger hunt. Which team collected the most items? Which team collected the fewest items? *(Section 15.2)*

Team	1	2	3	4	5	6	7	8
Portion Collected	$\frac{3}{4}$	0.8	77.5%	0.825	$\frac{29}{40}$	76.25%	$\frac{63}{80}$	81.25%

19. COMPLETIONS A quarterback completed 68% of his passes in a game. He threw 25 passes. How many passes did the quarterback complete? *(Section 15.3)*

20. TEXT MESSAGES You have 44 text messages in your inbox. How many messages can your cell phone hold? *(Section 15.4)*

15.5 Percents of Increase and Decrease

Essential Question What is a percent of decrease? What is a percent of increase?

1 ACTIVITY: Percent of Decrease

Work with a partner.

Each year in the Columbia River Basin, adult salmon swim upriver to streams to lay eggs and hatch their young.

To go up the river, the adult salmon use fish ladders. But to go down the river, the young salmon must pass through several dams.

At one time, there were electric turbines at each of the eight dams on the main stem of the Columbia and Snake Rivers. About 88% of the young salmon passed through these turbines unharmed.

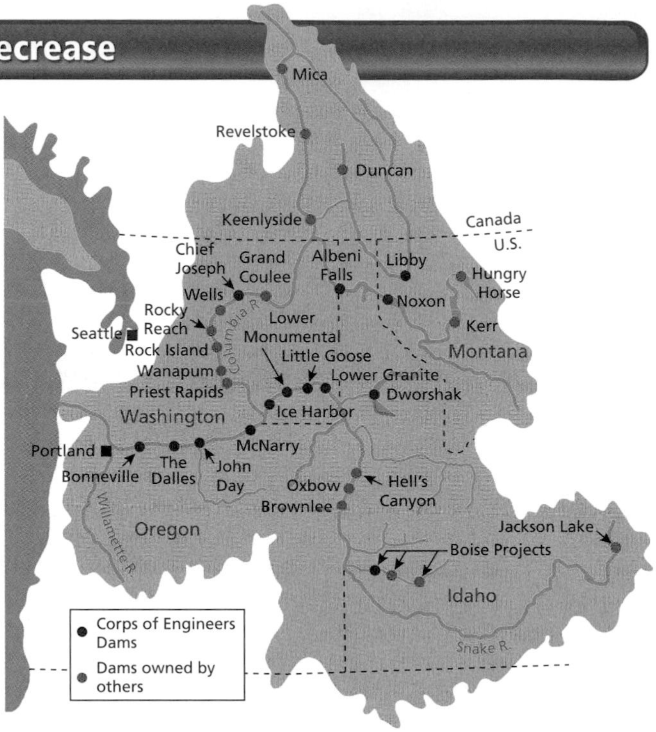

a. Copy and complete the table to show the number of young salmon that made it through the dams.

Dam	0	1	2	3	4	5	6	7	8
Salmon	1000	880	774						

88% of $1000 = 0.88 \cdot 1000$
$= 880$

88% of $880 = 0.88 \cdot 880$
$= 774.4$
≈ 774

b. Display the data in a bar graph.

c. By what percent did the number of young salmon decrease when passing through each dam?

COMMON CORE

Percents

In this lesson, you will
- find percents of increase.
- find percents of decrease.

Learning Standard
7.RP.3

2 ACTIVITY: Percent of Increase

Math Practice 1

Consider Similar Problems

How is this activity similar to the previous activity?

Work with a partner. In 2013, the population of a city was 18,000 people.

a. An organization projects that the population will increase by 2% each year for the next 7 years. Copy and complete the table to find the populations of the city for 2014 through 2020. Then display the data in a bar graph.

For 2014:

$$2\% \text{ of } 18{,}000 = 0.02 \cdot 18{,}000$$
$$= 360$$

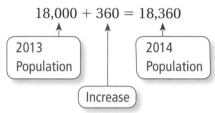

$$18{,}000 + 360 = 18{,}360$$

2013 Population — Increase — 2014 Population

Year	Population
2013	18,000
2014	18,360
2015	
2016	
2017	
2018	
2019	
2020	

b. Another organization projects that the population will increase by 3% each year for the next 7 years. Repeat part (a) using this percent.

c. Which organization projects the larger populations? How many more people do they project for 2020?

What Is Your Answer?

3. **IN YOUR OWN WORDS** What is a percent of decrease? What is a percent of increase?

4. Describe real-life examples of a percent of decrease and a percent of increase.

Practice Use what you learned about percent of increase and percent of decrease to complete Exercises 4–7 on page 680.

15.5 Lesson

Check It Out
Lesson Tutorials
BigIdeasMath.com

Key Vocabulary
percent of change, *p. 678*
percent of increase, *p. 678*
percent of decrease, *p. 678*
percent error, *p. 679*

A **percent of change** is the percent that a quantity changes from the original amount.

$$\text{percent of change} = \frac{\text{amount of change}}{\text{original amount}}$$

Key Idea

Percents of Increase and Decrease

When the original amount increases, the percent of change is called a **percent of increase**.

$$\text{percent of increase} = \frac{\text{new amount} - \text{original amount}}{\text{original amount}}$$

When the original amount decreases, the percent of change is called a **percent of decrease**.

$$\text{percent of decrease} = \frac{\text{original amount} - \text{new amount}}{\text{original amount}}$$

EXAMPLE 1 Finding a Percent of Increase

The table shows the numbers of hours you spent online last weekend. What is the percent of change in your online time from Saturday to Sunday?

Day	Hours Online
Saturday	2
Sunday	4.5

The number of hours on Sunday is greater than the number of hours on Saturday. So, the percent of change is a percent of increase.

$$\text{percent of increase} = \frac{\text{new amount} - \text{original amount}}{\text{original amount}}$$

$= \frac{4.5 - 2}{2}$ Substitute.

$= \frac{2.5}{2}$ Subtract.

$= 1.25$, or 125% Write as a percent.

So, your online time increased 125% from Saturday to Sunday.

On Your Own

Find the percent of change. Round to the nearest tenth of a percent if necessary.

1. 10 inches to 25 inches
2. 57 people to 65 people

EXAMPLE 2 Finding a Percent of Decrease

The bar graph shows a softball player's home run totals. What was the percent of change from 2012 to 2013?

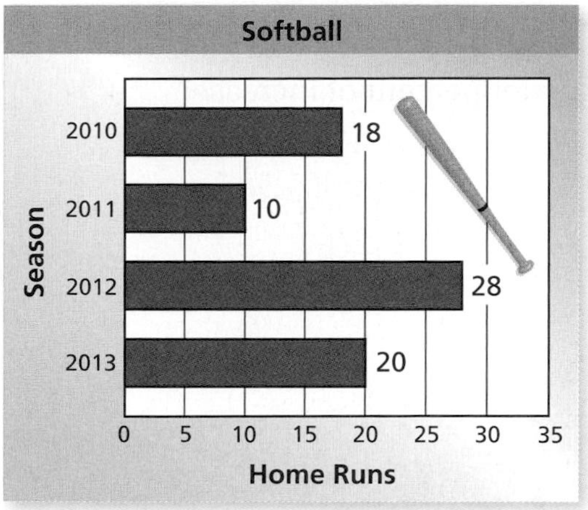

The number of home runs decreased from 2012 to 2013. So, the percent of change is a percent of decrease.

$$\text{percent of decrease} = \frac{\text{original amount} - \text{new amount}}{\text{original amount}}$$

$$= \frac{28 - 20}{28}$$ Substitute.

$$= \frac{8}{28}$$ Subtract.

$$\approx 0.286, \text{ or } 28.6\%$$ Write as a percent.

∴ So, the number of home runs decreased about 28.6%.

Key Idea

Percent Error

A **percent error** is the percent that an estimated quantity differs from the actual amount.

$$\text{percent error} = \frac{\text{amount of error}}{\text{actual amount}}$$

Study Tip

The amount of error is always positive.

EXAMPLE 3 Finding a Percent Error

You estimate that the length of your classroom is 16 feet. The actual length is 21 feet. Find the percent error.

The amount of error is $21 - 16 = 5$ feet.

$$\text{percent error} = \frac{\text{amount of error}}{\text{actual amount}}$$ Write percent error equation.

$$= \frac{5}{21}$$ Substitute.

$$\approx 0.238, \text{ or } 23.8\%$$ Write as a percent.

∴ The percent error is about 23.8%.

On Your Own

Now You're Ready Exercises 8–15 and 18

3. In Example 2, what was the percent of change from 2010 to 2011?
4. **WHAT IF?** In Example 3, your friend estimates that the length of the classroom is 23 feet. Who has the greater percent error? Explain.

15.5 Exercises

Check It Out
Help with Homework
BigIdeasMath.com

Vocabulary and Concept Check

1. **VOCABULARY** How do you know whether a percent of change is a *percent of increase* or a *percent of decrease*?

2. **NUMBER SENSE** Without calculating, which has a greater percent of increase?
   - 5 bonus points on a 50-point exam
   - 5 bonus points on a 100-point exam

3. **WRITING** What does it mean to have a 100% decrease?

Practice and Problem Solving

Find the new amount.

4. 8 meters increased by 25%
5. 15 liters increased by 60%
6. 50 points decreased by 26%
7. 25 penalties decreased by 32%

Identify the percent of change as an *increase* or a *decrease*. Then find the percent of change. Round to the nearest tenth of a percent if necessary.

1 2 8. 12 inches to 36 inches
9. 75 people to 25 people
10. 50 pounds to 35 pounds
11. 24 songs to 78 songs
12. 10 gallons to 24 gallons
13. 72 paper clips to 63 paper clips
14. 16 centimeters to 44.2 centimeters
15. 68 miles to 42.5 miles

16. **ERROR ANALYSIS** Describe and correct the error in finding the percent increase from 18 to 26.

$$\frac{26 - 18}{26} \approx 0.31 = 31\%$$

17. **VIDEO GAME** Last week, you finished Level 2 of a video game in 32 minutes. Today, you finish Level 2 in 28 minutes. What is your percent of change?

3 18. **PIG** You estimate that a baby pig weighs 20 pounds. The actual weight of the baby pig is 16 pounds. Find the percent error.

19. **CONCERT** You estimate that 200 people attended a school concert. The actual attendance was 240 people.

    a. Find the percent error.

    b. What other estimate gives the same percent error? Explain your reasoning.

Identify the percent of change as an *increase* or a *decrease*. Then find the percent of change. Round to the nearest tenth of a percent if necessary.

20. $\frac{1}{4}$ to $\frac{1}{2}$ **21.** $\frac{4}{5}$ to $\frac{3}{5}$ **22.** $\frac{3}{8}$ to $\frac{7}{8}$ **23.** $\frac{5}{4}$ to $\frac{3}{8}$

24. CRITICAL THINKING Explain why a change from 20 to 40 is a 100% increase, but a change from 40 to 20 is a 50% decrease.

25. POPULATION The table shows population data for a community.

Year	Population
2007	118,000
2013	138,000

a. What is the percent of change from 2007 to 2013?

b. Use this percent of change to predict the population in 2019.

26. GEOMETRY Suppose the length and the width of the sandbox are doubled.

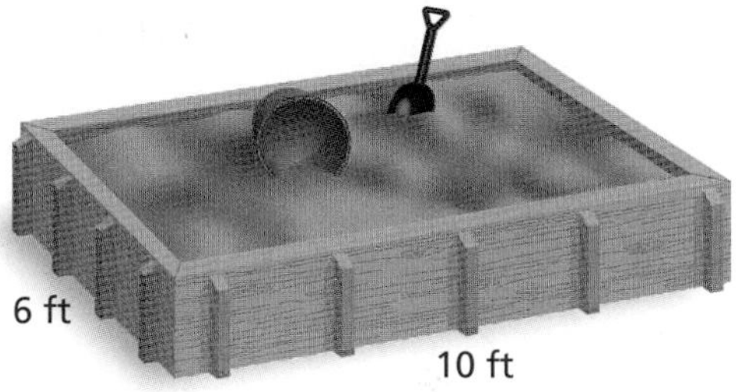

a. Find the percent of change in the perimeter.

b. Find the percent of change in the area.

27. CEREAL A cereal company fills boxes with 16 ounces of cereal. The acceptable percent error in filling a box is 2.5%. Find the least and the greatest acceptable weights.

28. PRECISION Find the percent of change from June to September in the time to run a mile.

29. CRITICAL THINKING A number increases by 10%, and then decreases by 10%. Will the result be *greater than*, *less than*, or *equal to* the original number? Explain.

30. DONATIONS Donations to an annual fundraiser are 15% greater this year than last year. Last year, donations were 10% greater than the year before. The amount raised this year is $10,120. How much was raised 2 years ago?

31. Reasoning Forty students are in the science club. Of those, 45% are girls. This percent increases to 56% after new girls join the club. How many new girls join?

Fair Game Review What you learned in previous grades & lessons

Write and solve an equation to answer the question. *(Section 15.4)*

32. What number is 25% of 64? **33.** 39.2 is what percent of 112?

34. 5 is 5% of what number? **35.** 18 is 32% of what number?

36. MULTIPLE CHOICE Which set of ratios does *not* form a proportion? *(Section 14.2)*

Ⓐ $\frac{1}{4}, \frac{6}{24}$ Ⓑ $\frac{4}{7}, \frac{7}{10}$ Ⓒ $\frac{16}{24}, \frac{2}{3}$ Ⓓ $\frac{36}{10}, \frac{18}{5}$

15.6 Discounts and Markups

Essential Question

How can you find discounts and selling prices?

1 ACTIVITY: Comparing Discounts

Work with a partner. The same pair of sneakers is on sale at three stores. Which one is the best buy? Explain.

a. Regular Price: $45

b. Regular Price: $49

c. Regular Price: $39

a.

b.

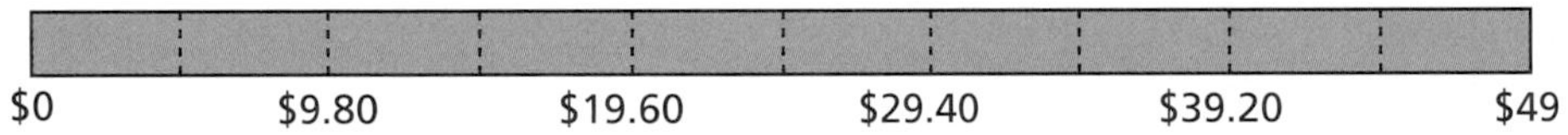

c.

2 ACTIVITY: Finding the Original Price

Work with a partner.

a. You buy a shirt that is on sale for 30% off. You pay $22.40. Your friend wants to know the original price of the shirt. Show how you can use the model below to find the original price.

b. Explain how you can use the percent proportion to find the original price.

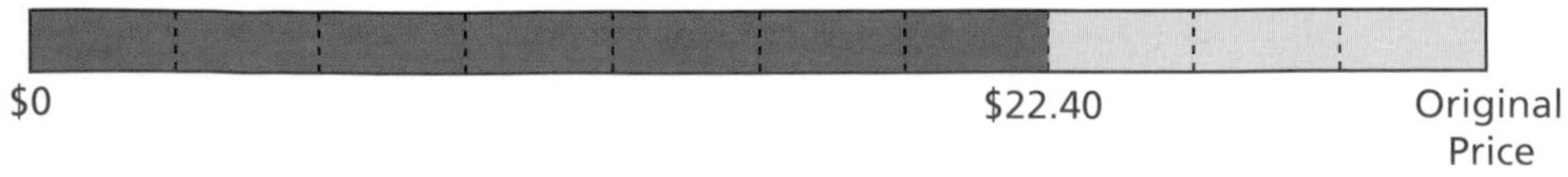

COMMON CORE

Percents

In this lesson, you will

- use percent of discounts to find prices of items.
- use percent of markups to find selling prices of items.

Learning Standard
7.RP.3

3 ACTIVITY: Finding Selling Prices

Math Practice 2

Make Sense of Quantities

What do the quantities represent? What is the relationship between the quantities?

You own a small jewelry store. You increase the price of the jewelry by 125%.

Work with a partner. Use a model to estimate the selling price of the jewelry. Then use a calculator to find the selling price.

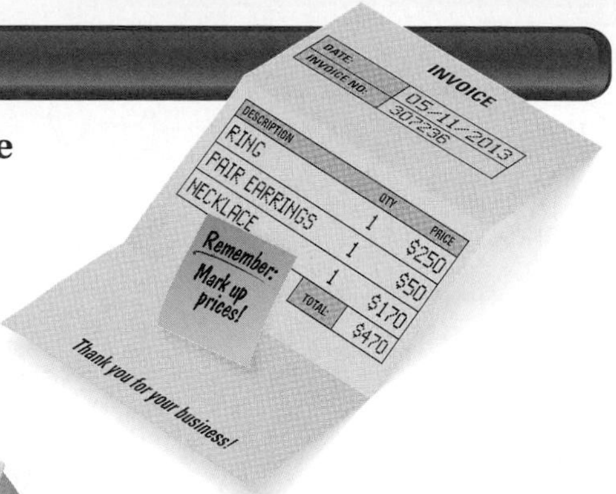

a. Your cost is $250.

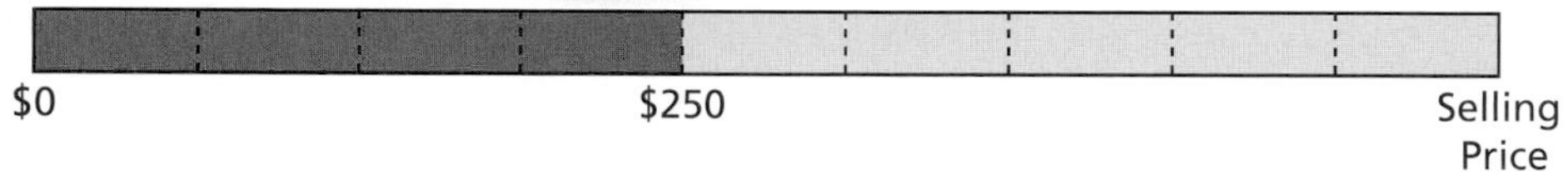

b. Your cost is $50.

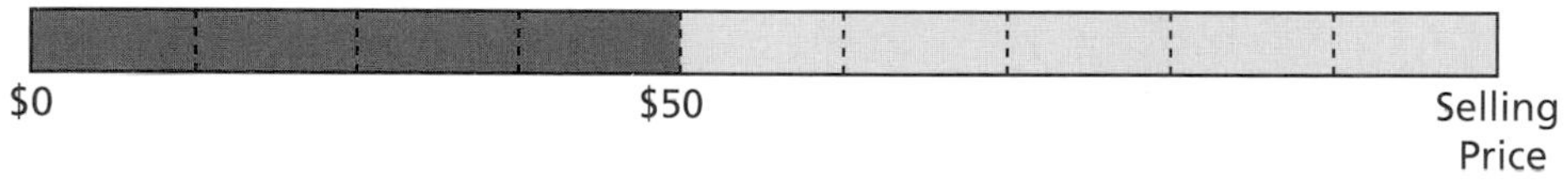

c. Your cost is $170.

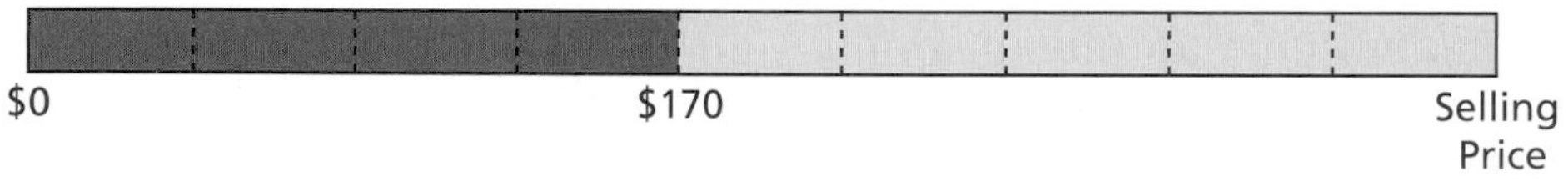

What Is Your Answer?

4. **IN YOUR OWN WORDS** How can you find discounts and selling prices? Give examples of each.

Practice Use what you learned about discounts to complete Exercises 4, 9, and 14 on page 686.

15.6 Lesson

Check It Out
Lesson Tutorials
BigIdeasMath.com

Key Vocabulary
discount, *p. 684*
markup, *p. 684*

Key Ideas

Discounts

A **discount** is a decrease in the original price of an item.

Markups

To make a profit, stores charge more than what they pay. The increase from what the store pays to the selling price is called a **markup**.

EXAMPLE 1 Finding a Sale Price

The original price of the shorts is $35. What is the sale price?

Method 1: First, find the discount. The discount is 25% of $35.

$a = p \cdot w$ Write percent equation.

$= 0.25 \cdot 35$ Substitute 0.25 for *p* and 35 for *w*.

$= 8.75$ Multiply.

Next, find the sale price.

sale price = original price − discount

$= 35 - 8.75$

$= 26.25$

So, the sale price is $26.25.

Method 2: First, find the percent of the original price.

$100\% - 25\% = 75\%$

Next, find the sale price.

sale price $= 75\%$ of $35

$= 0.75 \cdot 35$

$= 26.25$

So, the sale price is $26.25.

Check

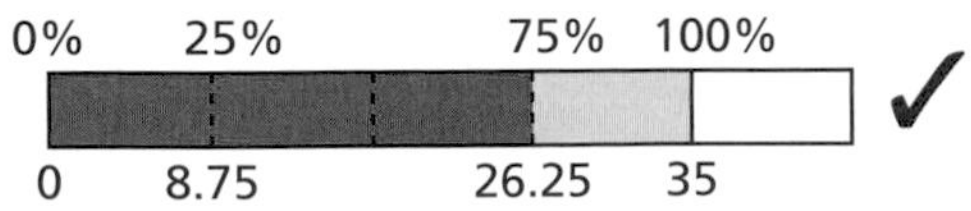

Study Tip

A 25% discount is the same as paying 75% of the original price.

On Your Own

Now You're Ready
Exercises 4–8

1. The original price of a skateboard is $50. The sale price includes a 20% discount. What is the sale price?

EXAMPLE 2 Finding an Original Price

What is the original price of the shoes?

The sale price is $100\% - 40\% = 60\%$ of the original price.

Answer the question: 33 is 60% of what number?

$a = p \cdot w$	Write percent equation.
$33 = 0.6 \cdot w$	Substitute 33 for *a* and 0.6 for *p*.
$55 = w$	Divide each side by 0.6.

∴ So, the original price of the shoes is $55.

Check

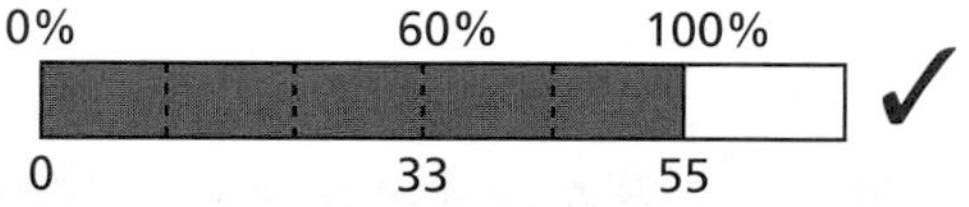

EXAMPLE 3 Finding a Selling Price

A store pays $70 for a bicycle. The percent of markup is 20%. What is the selling price?

Method 1: First, find the markup. The markup is 20% of $70.

$$a = p \cdot w$$
$$= 0.20 \cdot 70$$
$$= 14$$

Next, find the selling price.

selling price	=	cost to store	+	markup
	=	70	+	14
	= 84			

∴ So, the selling price is $84.

Method 2: Use a ratio table. The selling price is 120% of the cost to the store.

	Percent	Dollars	
	100%	$70	
÷ 5	20%	$14	÷ 5
× 6	120%	$84	× 6

∴ So, the selling price is $84.

Check

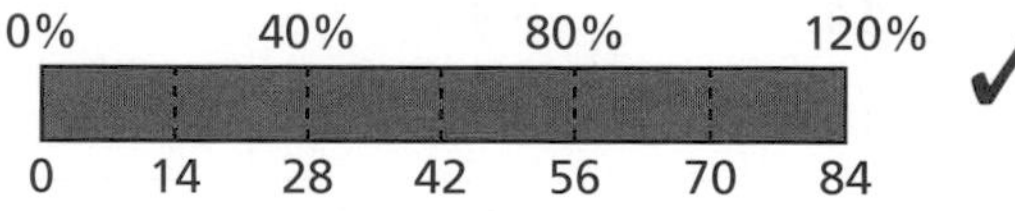

On Your Own

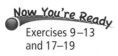

Now You're Ready
Exercises 9–13 and 17–19

2. The discount on a DVD is 50%. It is on sale for $10. What is the original price of the DVD?

3. A store pays $75 for an aquarium. The markup is 20%. What is the selling price?

15.6 Exercises

Check It Out
Help with Homework
BigIdeasMath.com

Vocabulary and Concept Check

1. **WRITING** Describe how to find the sale price of an item that has been discounted 25%.
2. **WRITING** Describe how to find the selling price of an item that has been marked up 110%.
3. **REASONING** Which would you rather pay? Explain your reasoning.

   a. 6% tax on a discounted price   or   6% tax on the original price

   b. 30% markup on a $30 shirt   or $30 markup on a $30 shirt

Practice and Problem Solving

Copy and complete the table.

	Original Price	Percent of Discount	Sale Price
1 **4.**	$80	20%	
5.	$42	15%	
6.	$120	80%	
7.	$112	32%	
8.	$69.80	60%	
2 **9.**		25%	$40
10.		5%	$57
11.		80%	$90
12.		64%	$72
13.		15%	$146.54
14.	$60		$45
15.	$82		$65.60
16.	$95		$61.75

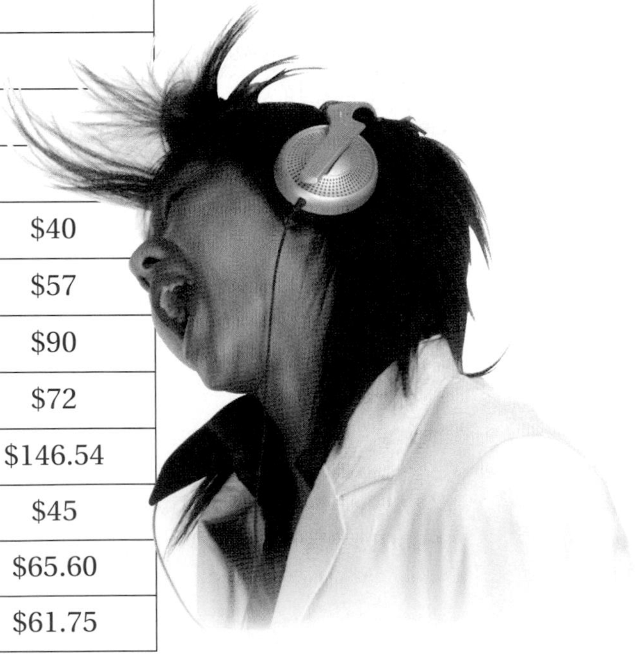

Find the selling price.

3 **17.** Cost to store: $50
Markup: 10%

18. Cost to store: $80
Markup: 60%

19. Cost to store: $140
Markup: 25%

20. **YOU BE THE TEACHER** The cost to a store for an MP3 player is \$60. The selling price is \$105. A classmate says that the markup is 175% because $\frac{\$105}{\$60} = 1.75$. Is your classmate correct? If not, explain how to find the correct percent of markup.

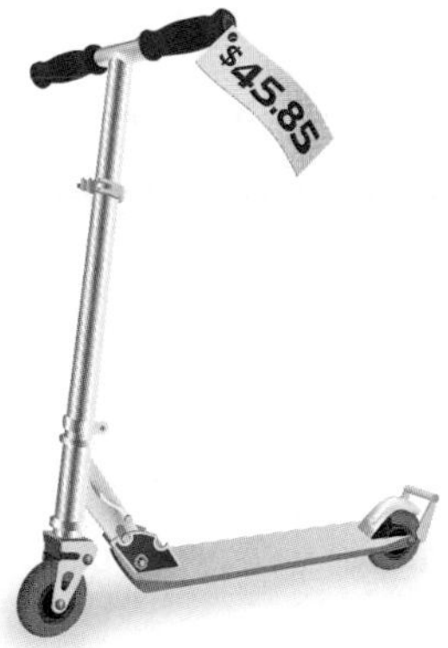

21. **SCOOTER** The scooter is on sale for 90% off the original price. Which of the methods can you use to find the sale price? Which method do you prefer? Explain.

Multiply \$45.85 by 0.9.	Multiply \$45.85 by 0.1.
Multiply \$45.85 by 0.9, then add to \$45.85.	Multiply \$45.85 by 0.9, then subtract from \$45.85.

22. **GAMING** You are shopping for a video game system.

Store	Cost to Store	Markup
A	\$162	40%
B	\$155	30%
C	\$160	25%

   a. At which store should you buy the system?
   b. Store A has a weekend sale. What discount must Store A offer for you to buy the system there?

23. **STEREO** A \$129.50 stereo is discounted 40%. The next month, the sale price is discounted 60%. Is the stereo now "free"? If not, what is the sale price?

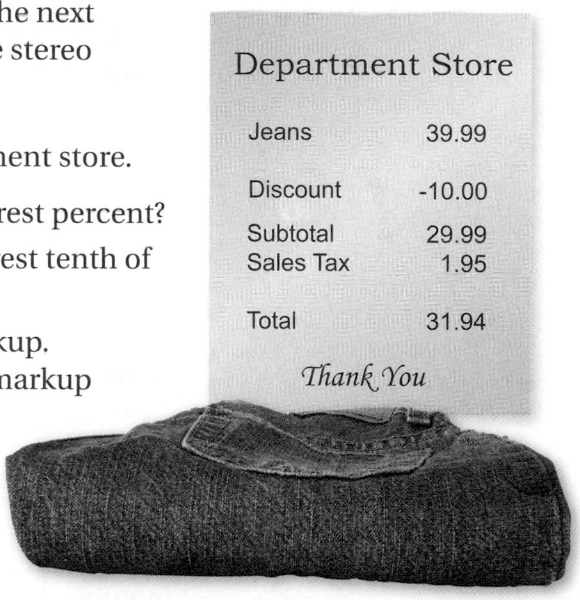

24. **CLOTHING** You buy a pair of jeans at a department store.
   a. What is the percent of discount to the nearest percent?
   b. What is the percent of sales tax to the nearest tenth of a percent?
   c. The price of the jeans includes a 60% markup. After the discount, what is the percent of markup to the nearest percent?

25. Critical Thinking You buy a bicycle helmet for \$22.26, which includes 6% sales tax. The helmet is discounted 30% off the selling price. What is the original price?

Fair Game Review What you learned in previous grades & lessons

Evaluate. *(Section 2.5)*

26. 2000(0.085)
27. 1500(0.04)(3)
28. 3200(0.045)(8)

29. **MULTIPLE CHOICE** Which measurement is greater than 1 meter? *(Section 5.7)*

Ⓐ 38 inches Ⓑ 1 yard Ⓒ 3.4 feet Ⓓ 98 centimeters

15.7 Simple Interest

Essential Question How can you find the amount of simple interest earned on a savings account? How can you find the amount of interest owed on a loan?

Simple interest is money earned on a savings account or an investment. It can also be money you pay for borrowing money.

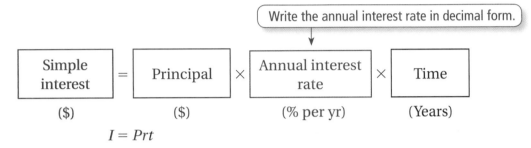

Write the annual interest rate in decimal form.

Simple interest		Principal		Annual interest rate		Time
($)	=	($)	×	(% per yr)	×	(Years)

$I = Prt$

1 ACTIVITY: Finding Simple Interest

Work with a partner. You put $100 in a savings account. The account earns 6% simple interest per year. (a) Find the interest earned and the balance at the end of 6 months. (b) Copy and complete the table. Then make a bar graph that shows how the balance grows in 6 months.

a. $I = Prt$ Write simple interest formula.

$=$ ______ Substitute values.

$=$ ______ Multiply.

∴ At the end of 6 months, you earn $____ in interest. So, your balance is $____.

b.

Time	Interest	Balance
0 month	$0	$100
1 month		
2 months		
3 months		
4 months		
5 months		
6 months		

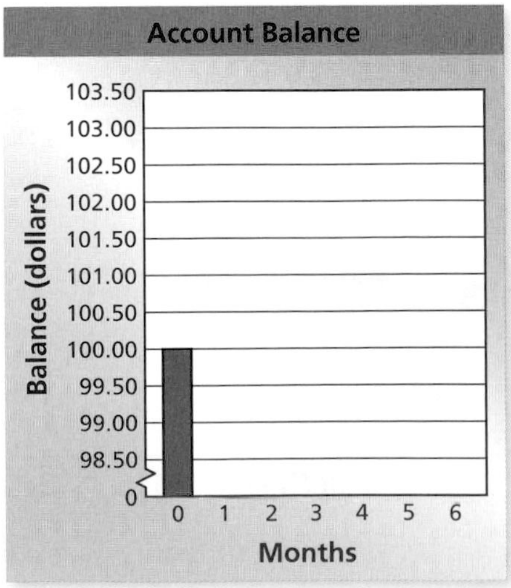

COMMON CORE

Percents

In this lesson, you will

- use the simple interest formula to find interest earned or paid, annual interest rates, and amounts paid on loans.

Learning Standard 7.RP.3

2 ACTIVITY: Financial Literacy

Work with a partner. Use the following information to write a report about credit cards. In the report, describe how a credit card works. Include examples that show the amount of interest paid each month on a credit card.

Math Practice 5

Use Other Resources

What resources can you use to find more information about credit cards?

U.S. Credit Card Data

- A typical household with credit card debt in the United States owes about $16,000 to credit card companies.
- A typical credit card interest rate is 14% to 16% per year. This is called the annual percentage rate.

3 ACTIVITY: The National Debt

Work with a partner. In 2012, the United States owed about $16 trillion in debt. The interest rate on the national debt is about 1% per year.

a. Write $16 trillion in decimal form. How many zeros does this number have?

b. How much interest does the United States pay each year on its national debt?

c. How much interest does the United States pay each day on its national debt?

d. The United States has a population of about 314 million people. Estimate the amount of interest that each person pays per year toward interest on the national debt.

What Is Your Answer?

4. **IN YOUR OWN WORDS** How can you find the amount of simple interest earned on a savings account? How can you find the amount of interest owed on a loan? Give examples with your answer.

Practice Use what you learned about simple interest to complete Exercises 4–7 on page 692.

15.7 Lesson

Check It Out
Lesson Tutorials
BigIdeasMath.com

Key Vocabulary
interest, *p. 690*
principal, *p. 690*
simple interest, *p. 690*

Interest is money paid or earned for the use of money. The **principal** is the amount of money borrowed or deposited.

Key Idea

Simple Interest

Words **Simple interest** is money paid or earned only on the principal.

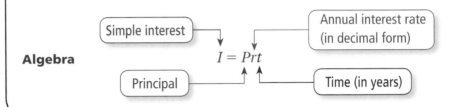

Algebra $I = Prt$

- I: Simple interest
- P: Principal
- r: Annual interest rate (in decimal form)
- t: Time (in years)

EXAMPLE 1 Finding Interest Earned

You put \$500 in a savings account. The account earns 3% simple interest per year. (a) What is the interest earned after 3 years? (b) What is the balance after 3 years?

a. $I = Prt$ — Write simple interest formula.

$= 500(0.03)(3)$ — Substitute 500 for P, 0.03 for r, and 3 for t.

$= 45$ — Multiply.

So, the interest earned is \$45 after 3 years.

b. To find the balance, add the interest to the principal.

So, the balance is \$500 + \$45 = \$545 after 3 years.

EXAMPLE 2 Finding an Annual Interest Rate

You put \$1000 in an account. The account earns \$100 simple interest in 4 years. What is the annual interest rate?

$I = Prt$ — Write simple interest formula.

$100 = 1000(r)(4)$ — Substitute 100 for I, 1000 for P, and 4 for t.

$100 = 4000r$ — Simplify.

$0.025 = r$ — Divide each side by 4000.

So, the annual interest rate of the account is 0.025, or 2.5%.

On Your Own

Now You're Ready
Exercises 4–16

1. In Example 1, what is the balance of the account after 9 months?
2. You put \$350 in an account. The account earns \$17.50 simple interest in 2.5 years. What is the annual interest rate?

EXAMPLE 3 Finding an Amount of Time

A bank offers three savings accounts. The simple interest rate is determined by the principal. How long does it take an account with a principal of \$800 to earn \$100 in interest?

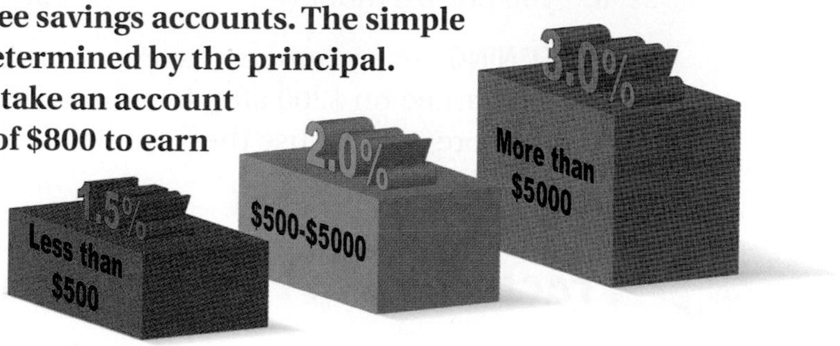

The pictogram shows that the interest rate for a principal of \$800 is 2%.

$I = Prt$	Write simple interest formula.
$100 = 800(0.02)(t)$	Substitute 100 for *I*, 800 for *P*, and 0.02 for *r*.
$100 = 16t$	Simplify.
$6.25 = t$	Divide each side by 16.

So, the account earns \$100 in interest in 6.25 years.

EXAMPLE 4 Finding an Amount Paid on a Loan

You borrow \$600 to buy a violin. The simple interest rate is 15%. You pay off the loan after 5 years. How much do you pay for the loan?

$I = Prt$	Write simple interest formula.
$= 600(0.15)(5)$	Substitute 600 for *P*, 0.15 for *r*, and 5 for *t*.
$= 450$	Multiply.

To find the amount you pay, add the interest to the loan amount.

So, you pay \$600 + \$450 = \$1050 for the loan.

On Your Own

Now You're Ready
Exercises 17–20 and 24–27

3. In Example 3, how long does it take an account with a principal of \$10,000 to earn \$750 in interest?
4. **WHAT IF?** In Example 4, you pay off the loan after 2 years. How much money do you save?

15.7 Exercises

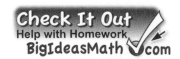

Check It Out
Help with Homework
BigIdeasMath.com

Vocabulary and Concept Check

1. **VOCABULARY** Define each variable in $I = Prt$.

2. **WRITING** In each situation, tell whether you would want a *higher* or *lower* interest rate. Explain your reasoning.

   a. you borrow money

   b. you open a savings account

3. **REASONING** An account earns 6% simple interest. You want to find the interest earned on \$200 after 8 months. What conversions do you need to make before you can use the formula $I = Prt$?

Practice and Problem Solving

An account earns simple interest. (a) Find the interest earned. (b) Find the balance of the account.

(1) **4.** \$600 at 5% for 2 years

5. \$1500 at 4% for 5 years

6. \$350 at 3% for 10 years

7. \$1800 at 6.5% for 30 months

8. \$700 at 8% for 6 years

9. \$1675 at 4.6% for 4 years

10. \$925 at 2% for 2.4 years

11. \$5200 at 7.36% for 54 months

12. ERROR ANALYSIS Describe and correct the error in finding the simple interest earned on \$500 at 6% for 18 months.

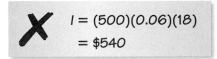

✗ $I = (500)(0.06)(18)$
$= \$540$

Find the annual interest rate.

(2) **13.** $I = \$24, P = \$400, t = 2$ years

14. $I = \$562.50, P = \$1500, t = 5$ years

15. $I = \$54, P = \$900, t = 18$ months

16. $I = \$160.67, P = \$2000, t = 8$ months

Find the amount of time.

(3) **17.** $I = \$30, P = \$500, r = 3\%$

18. $I = \$720, P = \$1000, r = 9\%$

19. $I = \$54, P = \$800, r = 4.5\%$

20. $I = \$450, P = \$2400, r = 7.5\%$

21. BANKING A savings account earns 5% simple interest per year. The principal is \$1200. What is the balance after 4 years?

22. SAVINGS You put \$400 in an account. The account earns \$18 simple interest in 9 months. What is the annual interest rate?

23. CD You put \$3000 in a CD (certificate of deposit) at the promotional rate. How long will it take to earn \$336 in interest?

Find the amount paid for the loan.

(4) **24.** \$1500 at 9% for 2 years

25. \$2000 at 12% for 3 years

26. \$2400 at 10.5% for 5 years

27. \$4800 at 9.9% for 4 years

Copy and complete the table.

	Principal	Interest Rate	Time	Simple Interest
28.	\$12,000	4.25%	5 years	
29.		6.5%	18 months	\$828.75
30.	\$15,500	8.75%		\$5425.00
31.	\$18,000		54 months	\$4252.50

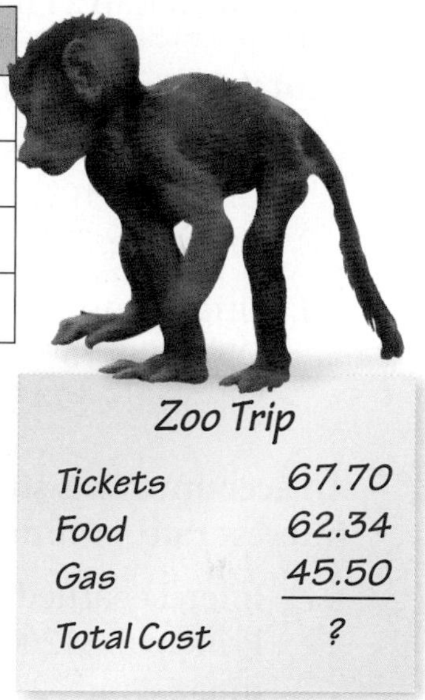

32. ZOO A family charges a trip to the zoo on a credit card. The simple interest rate is 12%. The charges are paid after 3 months. What is the total amount paid for the trip?

33. MONEY MARKET You deposit \$5000 in an account earning 7.5% simple interest. How long will it take for the balance of the account to be \$6500?

34. LOANS A music company offers a loan to buy a drum set for \$1500. What is the monthly payment?

35. REASONING How many years will it take for \$2000 to double at a simple interest rate of 8%? Explain how you found your answer.

36. PROBLEM SOLVING You have two loans, for 2 years each. The total interest for the two loans is \$138. On the first loan, you pay 7.5% simple interest on a principal of \$800. On the second loan, you pay 3% simple interest. What is the principal for the second loan?

37. Critical Thinking You put \$500 in an account that earns 4% annual interest. The interest earned each year is added to the principal to create a new principal. Find the total amount in your account after each year for 3 years.

Fair Game Review What you learned in previous grades & lessons

Solve the inequality. Graph the solution. *(Section 7.6)*

38. $x + 5 < 2$

39. $b - 2 \geq -1$

40. $w + 6 \leq -3$

41. MULTIPLE CHOICE What is the solution of $4x + 5 = -11$? *(Section 13.5)*

Ⓐ $x = -4$ Ⓑ $x = -1.5$ Ⓒ $x = 1.5$ Ⓓ $x = 4$

15.5–15.7 Quiz

Check It Out
Progress Check
BigIdeasMath.com

Identify the percent of change as an *increase* or a *decrease*. Then find the percent of change. Round to the nearest tenth of a percent if necessary. *(Section 15.5)*

1. 8 inches to 24 inches

2. 300 miles to 210 miles

Find the original price, discount, sale price, or selling price. *(Section 15.6)*

3. Original price: \$30
Discount: 10%
Sale price: ?

4. Original price: \$55
Discount: ?
Sale price: \$46.75

5. Original price: ?
Discount: 75%
Sale price: \$74.75

6. Cost to store: \$152
Markup: 50%
Selling price: ?

An account earns simple interest. Find the interest earned, principal, interest rate, or time. *(Section 15.7)*

7. Interest earned: ?
Principal: \$1200
Interest rate: 2%
Time: 5 years

8. Interest earned: \$25
Principal: \$500
Interest rate: 5%
Time: ?

9. Interest earned: \$76
Principal: \$800
Interest rate: ?
Time: 2 years

10. Interest earned: \$119.88
Principal: ?
Interest rate: 3.6%
Time: 3 years

11. HEIGHT You estimate that your friend is 50 inches tall. The actual height of your friend is 54 inches. Find the percent error. *(Section 15.5)*

12. DIGITAL CAMERA A digital camera costs \$230. The camera is on sale for 30% off, and you have a coupon for an additional 15% off the sale price. What is the final price? *(Section 15.6)*

13. WATER SKIS The original price of the water skis was \$200. What is the percent of discount? *(Section 15.6)*

SALE
\$150

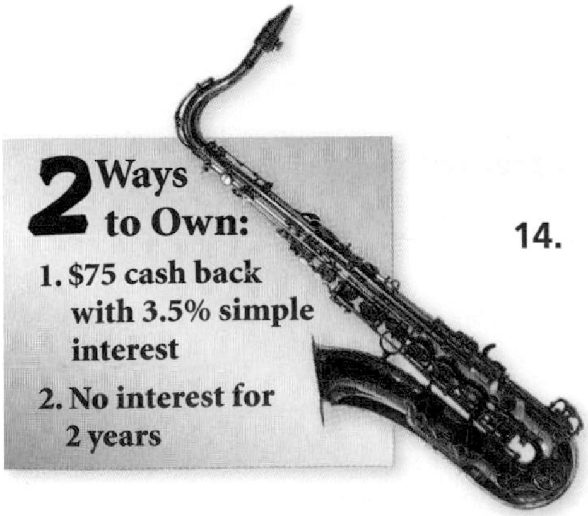

2 Ways to Own:
1. \$75 cash back with 3.5% simple interest
2. No interest for 2 years

14. SAXOPHONE A saxophone costs \$1200. A store offers two loan options. Which option saves more money if you pay the loan in 2 years? *(Section 15.7)*

15. LOAN You borrow \$200. The simple interest rate is 12%. You pay off the loan after 2 years. How much do you pay for the loan? *(Section 15.7)*

15 Chapter Review

Check It Out
Vocabulary Help
BigIdeasMath.com

Review Key Vocabulary

percent of change, *p. 678*
percent of increase, *p. 678*
percent of decrease, *p. 678*
percent error, *p. 679*
discount, *p. 684*
markup, *p. 684*
interest, *p. 690*
principal, *p. 690*
simple interest, *p. 690*

Review Examples and Exercises

15.1 Percents and Decimals *(pp. 650–655)*

a. Write 64% as a decimal.

64% = 64.% = 0.64

b. Write 0.023 as a percent.

0.023 = 0.023 = 2.3%

Exercises

Write the percent as a decimal. Use a model to check your answer.

1. 76%
2. 6%
3. 334%

Write the decimal as a percent. Use a model to check your answer.

4. 0.15
5. 1.24
6. 0.097

15.2 Comparing and Ordering Fractions, Decimals, and Percents *(pp. 656–661)*

Which is greater, $\frac{9}{10}$ or 88%?

Write $\frac{9}{10}$ as a percent: $\frac{9}{10} = \frac{90}{100} = 90\%$

∴ 88% is less than 90%. So, $\frac{9}{10}$ is the greater number.

Exercises

Tell which number is greater.

7. $\frac{1}{2}$, 52%
8. $\frac{12}{5}$, 245%
9. 0.46, 43%
10. 0.023, 22%

Use a number line to order the numbers from least to greatest.

11. $\frac{41}{50}$, 0.83, 80%
12. $\frac{9}{4}$, 220%, 2.15
13. 0.67, 66%, $\frac{2}{3}$
14. 0.88, $\frac{7}{8}$, 90%

15.3 The Percent Proportion (pp. 662–667)

a. What percent of 24 is 9?

$\frac{a}{w} = \frac{p}{100}$ Write the percent proportion.

$\frac{9}{24} = \frac{p}{100}$ Substitute 9 for *a* and 24 for *w*.

$100 \cdot \frac{9}{24} = 100 \cdot \frac{p}{100}$ Multiplication Property of Equality

$37.5 = p$ Simplify.

So, 37.5% of 24 is 9.

b. What number is 15% of 80?

$\frac{a}{w} = \frac{p}{100}$ Write the percent proportion.

$\frac{a}{80} = \frac{15}{100}$ Substitute 80 for *w* and 15 for *p*.

$80 \cdot \frac{a}{80} = 80 \cdot \frac{15}{100}$ Multiplication Property of Equality

$a = 12$ Simplify.

So, 12 is 15% of 80.

c. 120% of what number is 54?

$\frac{a}{w} = \frac{p}{100}$ Write the percent proportion.

$\frac{54}{w} = \frac{120}{100}$ Substitute 54 for *a* and 120 for *p*.

$54 \cdot 100 = w \cdot 120$ Cross Products Property

$5400 = 120w$ Multiply.

$45 = w$ Divide each side by 120.

So, 120% of 45 is 54.

Exercises

Write and solve a proportion to answer the question.

15. What percent of 60 is 18?

16. 40 is what percent of 32?

17. What number is 70% of 70?

18. $\frac{3}{4}$ is 75% of what number?

15.4 The Percent Equation (pp. 668–673)

a. What number is 72% of 25?

$a = p \cdot w$	Write percent equation.
$= 0.72 \cdot 25$	Substitute 0.72 for p and 25 for w.
$= 18$	Multiply.

So, 72% of 25 is 18.

b. 28 is what percent of 70?

$a = p \cdot w$	Write percent equation.
$28 = p \cdot 70$	Substitute 28 for a and 70 for w.
$\frac{28}{70} = \frac{p \cdot 70}{70}$	Division Property of Equality
$0.4 = p$	Simplify.

Because 0.4 equals 40%, 28 is 40% of 70.

c. 22.1 is 26% of what number?

$a = p \cdot w$	Write percent equation.
$22.1 = 0.26 \cdot w$	Substitute 22.1 for a and 0.26 for p.
$85 = w$	Divide each side by 0.26.

So, 22.1 is 26% of 85.

Exercises

Write and solve an equation to answer the question.

19. What number is 24% of 25?

20. 9 is what percent of 20?

21. 60.8 is what percent of 32?

22. 91 is 130% of what number?

23. 85% of what number is 10.2?

24. 83% of 20 is what number?

25. PARKING 15% of the school parking spaces are handicap spaces. The school has 18 handicap spaces. How many parking spaces are there?

26. FIELD TRIP Of the 25 students on a field trip, 16 students bring cameras. What percent of the students bring cameras?

15.5 Percents of Increase and Decrease (pp. 676–681)

The table shows the numbers of skim boarders at a beach on Saturday and Sunday. What was the percent of change in boarders from Saturday to Sunday?

The number of skim boarders on Sunday is less than the number of skim boarders on Saturday. So, the percent of change is a percent of decrease.

Day	Number of Skim Boarders
Saturday	12
Sunday	9

$$\text{percent of decrease} = \frac{\text{original amount} - \text{new amount}}{\text{original amount}}$$

$= \dfrac{12 - 9}{12}$ Substitute.

$= \dfrac{3}{12}$ Subtract.

$= 0.25 = 25\%$ Write as a percent.

So, the number of skim boarders decreased by 25% from Saturday to Sunday.

Exercises

Identify the percent of change as an *increase* or a *decrease*. Then find the percent of change. Round to the nearest tenth of a percent if necessary.

27. 6 yards to 36 yards

28. 120 meals to 52 meals

29. **MARBLES** You estimate that a jar contains 68 marbles. The actual number of marbles is 60. Find the percent error.

15.6 Discounts and Markups (pp. 682–687)

What is the original price of the tennis racquet?

The sale price is 100% − 30% = 70% of the original price.

Answer the question: 21 is 70% of what number?

$a = p \cdot w$ Write percent equation.

$21 = 0.7 \cdot w$ Substitute 21 for a and 0.7 for p.

$30 = w$ Divide each side by 0.7.

So, the original price of the tennis racquet is $30.

Exercises

Find the sale price or original price.

30. Original price: $50
Discount: 15%
Sale price: ?

31. Original price: ?
Discount: 20%
Sale price: $75

15.7 Simple Interest (pp. 688–693)

You put $200 in a savings account. The account earns 2% simple interest per year.

a. What is the interest earned after 4 years?

b. What is the balance after 4 years?

a. $I = Prt$ — Write simple interest formula.

$= 200(0.02)(4)$ — Substitute 200 for P, 0.02 for r, and 4 for t.

$= 16$ — Multiply.

∴ So, the interest earned is $16 after 4 years.

b. To find the balance, add the interest to the principal.

∴ So, the balance is $200 + $16 = $216 after 4 years.

You put $500 in an account. The account earns $55 simple interest in 5 years. What is the annual interest rate?

$I = Prt$ — Write simple interest formula.

$55 = 500(r)(5)$ — Substitute 55 for I, 500 for P, and 5 for t.

$55 = 2500r$ — Simplify.

$0.022 = r$ — Divide each side by 2500.

∴ So, the annual interest rate of the account is 0.022, or 2.2%.

Exercises

An account earns simple interest.

a. Find the interest earned.

b. Find the balance of the account.

32. $300 at 4% for 3 years

33. $2000 at 3.5% for 4 years

Find the annual simple interest rate.

34. $I = \$17$, $P = \$500$, $t = 2$ years

35. $I = \$426$, $P = \$1200$, $t = 5$ years

Find the amount of time.

36. $I = \$60$, $P = \$400$, $r = 5\%$

37. $I = \$237.90$, $P = \$1525$, $r = 2.6\%$

38. SAVINGS You put $100 in an account. The account earns $2 simple interest in 6 months. What is the annual interest rate?

15 Chapter Test

Check It Out
Test Practice
BigIdeasMath.com

Write the percent as a decimal.

1. 0.96% **2.** 65% **3.** 25.7%

Write the decimal as a percent.

4. 0.42 **5.** 7.88 **6.** 0.5854

Tell which number is greater.

7. $\frac{16}{25}$, 65% **8.** 56%, 5.6

Use a number line to order the numbers from least to greatest.

9. 85%, $\frac{15}{18}$, 0.84 **10.** 58.3%, 0.58, $\frac{7}{12}$

Answer the question.

11. What percent of 28 is 21?

12. 64 is what percent of 40?

13. What number is 80% of 45?

14. 0.8% of what number is 6?

Identify the percent of change as an *increase* or a *decrease*. Then find the percent of change. Round to the nearest tenth of a percent if necessary.

15. 4 strikeouts to 10 strikeouts

16. \$24 to \$18

Find the sale price or selling price.

17. Original price: \$15
Discount: 5%
Sale price: ?

18. Cost to store: \$5.50
Markup: 75%
Selling price: ?

An account earns simple interest. Find the interest earned or the principal.

19. Interest earned: ?
Principal: \$450
Interest rate: 6%
Time: 8 years

20. Interest earned: \$27
Principal: ?
Interest rate: 1.5%
Time: 2 years

21. BASKETBALL You, your cousin, and a friend each take the same number of free throws at a basketball hoop. Who made the most free throws?

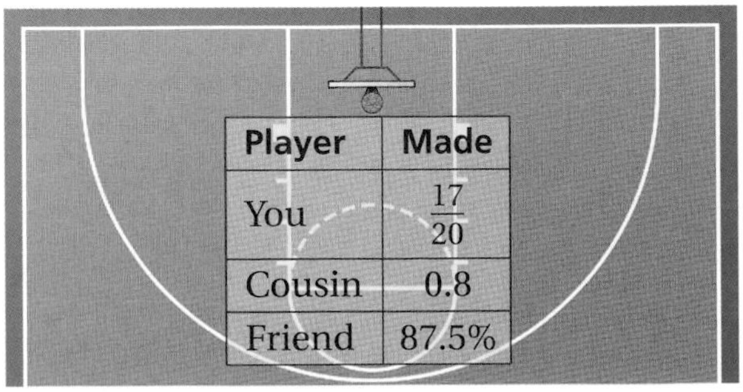

Player	Made
You	$\frac{17}{20}$
Cousin	0.8
Friend	87.5%

22. PARKING LOT You estimate that there are 66 cars in a parking lot. The actual number of cars is 75.

a. Find the percent error.

b. What other estimate gives the same percent error? Explain your reasoning.

23. INVESTMENT You put \$800 in an account that earns 4% simple interest. Find the total amount in your account after each year for 3 years.

15 Standards Assessment

Test-Taking Strategy
Read All Choices Before Answering

Which amount of increase in your catnip allowance do you want?
Ⓐ 50% Ⓑ 75% Ⓒ 98% Ⓓ 10%
I get it. C for catnip.

"Reading all choices before answering can really pay off!"

1. A movie theater offers 30% off the price of a movie ticket to students from your school. The regular price of a movie ticket is \$8.50. What is the discounted price that you would pay for a ticket? *(7.RP.3)*

   A. \$2.55 **C.** \$5.95

   B. \$5.50 **D.** \$8.20

2. You are comparing the prices of four boxes of cereal. Two of the boxes contain free extra cereal.

   - Box F costs \$3.59 and contains 16 ounces.
   - Box G costs \$3.79 and contains 16 ounces, plus an additional 10% for free.
   - Box H costs \$4.00 and contains 500 grams.
   - Box I costs \$4.69 and contains 500 grams, plus an additional 20% for free.

   Which box has the least unit cost? (1 ounce = 28.35 grams) *(7.RP.3)*

   F. Box F **H.** Box H

   G. Box G **I.** Box I

3. What value makes the equation $11 - 3x = -7$ true? *(7.EE.4a)*

4. Which proportion represents the problem below? *(7.RP.3)*

   "17% of a number is 43. What is the number?"

   A. $\frac{17}{43} = \frac{n}{100}$ **C.** $\frac{n}{43} = \frac{17}{100}$

   B. $\frac{n}{17} = \frac{43}{100}$ **D.** $\frac{43}{n} = \frac{17}{100}$

5. Which list of numbers is in order from least to greatest? *(7.EE.3)*

F. $0.8, \frac{5}{8}, 70\%, 0.09$

G. $0.09, \frac{5}{8}, 0.8, 70\%$

H. $\frac{5}{8}, 70\%, 0.8, 0.09$

I. $0.09, \frac{5}{8}, 70\%, 0.8$

6. What is the value of $\frac{9}{8} \div \left(-\frac{11}{4}\right)$? *(7.NS.2b)*

7. A pair of running shoes is on sale for 25% off the original price.

Which price is closest to the sale price of the running shoes? *(7.RP.3)*

A. $93

B. $99

C. $124

D. $149

8. What is the slope of the line? *(7.RP.2b)*

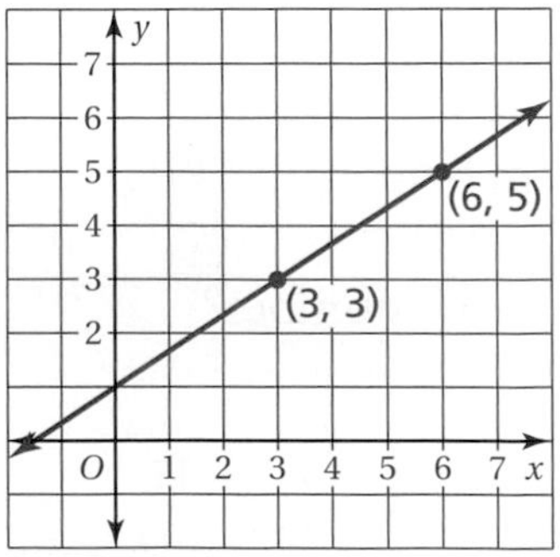

F. $\frac{2}{3}$

G. $\frac{3}{2}$

H. 2

I. 3

9. Brad solved the equation in the box shown.

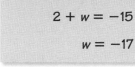

$$-3(2 + w) = -45$$
$$2 + w = -15$$
$$w = -17$$

What should Brad do to correct the error that he made? *(7.EE.4a)*

A. Multiply -45 by -3 to get $2 + w = 135$.

B. Add 3 to -45 to get $2 + w = -42$.

C. Add 2 to -15 to get $w = -13$.

D. Divide -45 by -3 to get 15.

10. You are comparing the costs of a certain model of ladder at a hardware store and at an online store. *(7.RP.3)*

Think Solve Explain

Part A What is the cost of the ladder at each of the stores? Show your work and explain your reasoning.

Part B Suppose that the hardware store is offering 10% off the price of the ladder and that the online store is offering free shipping and handling. Which store offers the better final cost? by how much? Show your work and explain your reasoning.

11. Which graph represents the inequality below? *(7.EE.4b)*

$$-5 - 3x \geq -11$$

F. 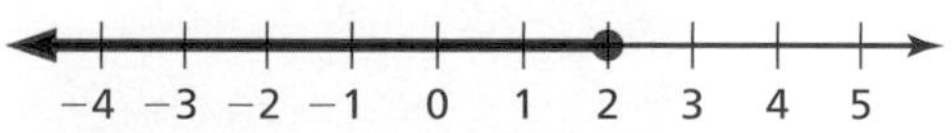

H. 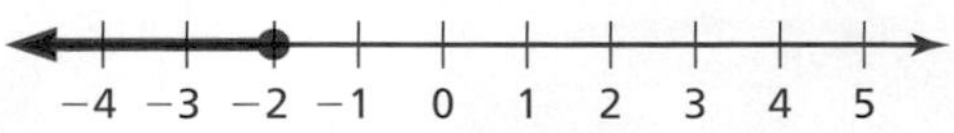

G. 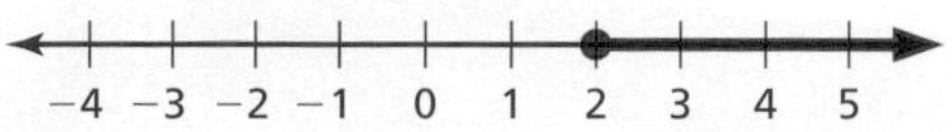

I. 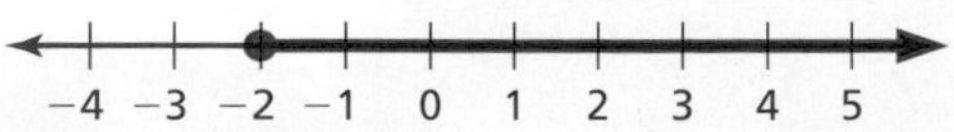

Appendix A
My Big Ideas Projects

A.1 Literature Project
Alice's Adventures in Wonderland

A.2 History Project
Mathematics in Ancient Egypt

My Big Ideas Projects

A.3 Art Project
Polyhedra in Art

A.4 Science Project
Why Does Ice Float?

A.1 Literature Project

Alice's Adventures in Wonderland

1 Getting Started

Lewis Carroll was a mathematician who also wrote literature books for children.

Essential Question How does mathematical problem solving influence a story plot?

Read *Alice's Adventures in Wonderland* by Lewis Carroll. In each chapter, rewrite one of the episodes so that it contains some of the math that you have studied this year.

Sample: You could rewrite Alice's shrinking adventure as follows.

Alice's normal height was 4′7″ or 55 inches. After drinking the bottle marked "Drink Me," Alice could feel herself shrinking more and more. After several minutes, Alice was only two-elevenths of her original height. Would Alice have to bend over to walk through a door that is only 1 foot high?

The illustrations on these two pages are from the 1907 edition of *Alice's Adventures in Wonderland*. They were drawn by Arthur Rackham (1867–1939), a book illustrator from England.

2 Things to Remember

- You can download each chapter of the book at *BigIdeasMath.com*.
- Add your own illustrations to your project.
- Try to include as many different math concepts as possible. Your goal is to include at least one concept from each of the chapters you studied this year.
- Organize your math stories in a folder, and think of a title for your report.

A.2 History Project

Mathematics in Ancient Egypt

1 Getting Started

The ancient Egyptian civilization in North Africa began around 3150 B.C. with the union of Upper and Lower Egypt under the first pharaoh. The rule of the pharaohs ended in 31 B.C. when the Romans conquered Egypt and made it a province.

The ancient Egyptians' achievements included a system of mathematics, glass technology, medicine, literature, irrigation and agricultural techniques, and art. Their mastery of surveying, quarrying, and construction techniques allowed them to build pyramids, temples, and obelisks.

Essential Question How is mathematical knowledge that was originally discovered by the ancient Egyptians used today?

Sample: To multiply two whole numbers, the ancient Egyptians repeatedly multiplied by 2. The "doubling lists" used to multiply 53 by 85 are shown. The lists stop when the last number in the first list is greater than the first number in the second list.

•	1	53
	2	106
•	4	212
	8	424
•	16	848
	32	1696
•	64	3392

Because $85 = 1 + 4 + 16 + 64$,
$85 \cdot 53 = 53 + 212 + 848 + 3392 = 4505$.

Ancient Papyrus

The Blue Sphinx

Ancient Egyptian Numbers

2 Things to Include

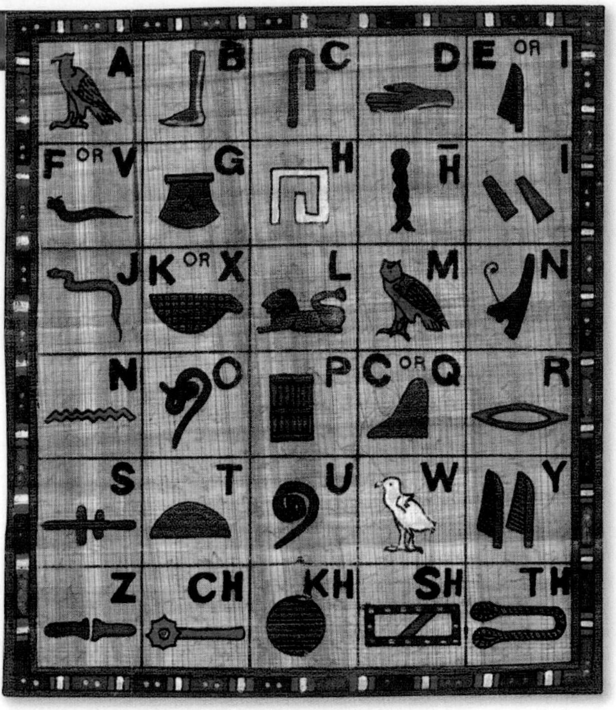

Ancient Egyptian Alphabet

- Explain how the ancient Egyptians multiplied two whole numbers. Give an example.
- Explain how the ancient Egyptians divided two whole numbers. Give an example.
- How did the ancient Egyptians write whole numbers? Give some examples.
- How did the ancient Egyptians write fractions? Give some examples.
- How did the ancient Egyptians use ratios to measure the height of a pyramid? Give an example.
- Write your name using ancient Egyptian hieroglyphics.
- Describe how ancient Egyptians used mathematics. How does this compare with the ways in which mathematics is used today?

3 Things to Remember

- Add your own illustrations to your project.
- Try to include as many different math concepts as possible. Your goal is to include at least one concept from each of the chapters you studied this year.
- Organize your report in a folder, and think of a title for your report.

What does this spell?

A.3 Art Project

Polyhedra in Art

1 Getting Started

Polyhedra is the plural of *polyhedron*. Polyhedra have been used in art for many centuries, in cultures all over the world.

Essential Question Do polyhedra influence the design of games and architecture?

Some of the most famous polyhedra are the five Platonic solids shown at the right.

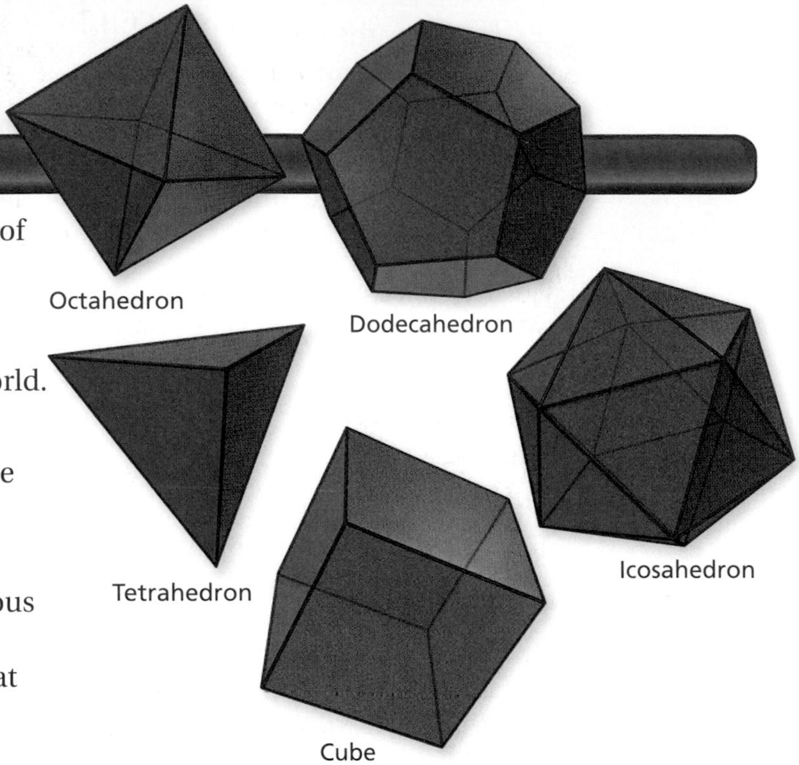

Mosaic by Paolo Uccello, 1430 A.D.

Sample: Three of the Platonic solids have faces that are equilateral triangles. One of these is the octahedron. Here is a net for an octahedron.

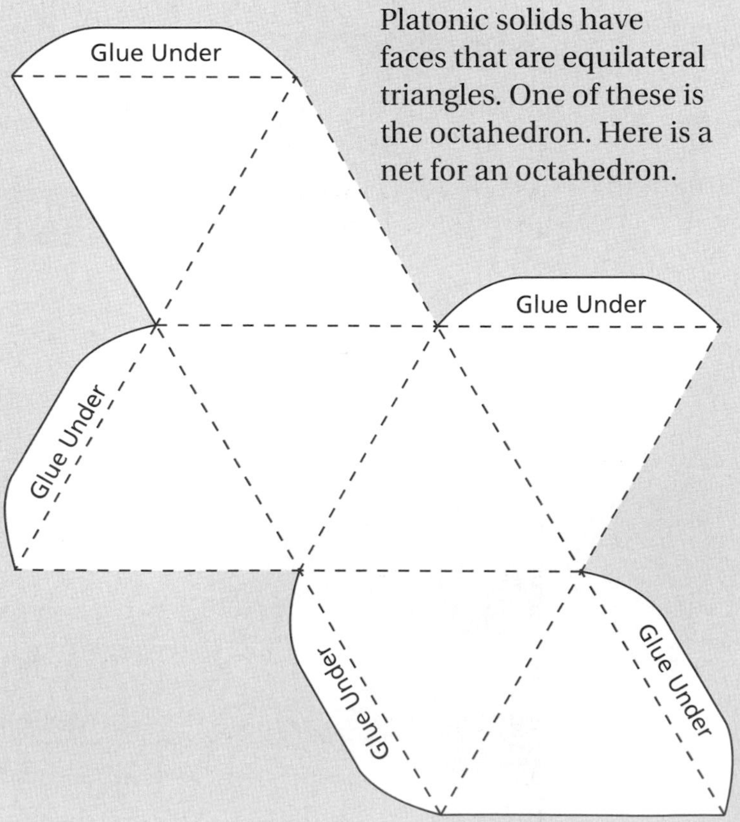

2 Things to Include

- Explain why the platonic solids are sometimes referred to as the cosmic figures.
- Draw a net for an icosahedron or a dodecahedron. Cut out the net and fold it to form the polyhedron.
- Describe the 13 polyhedra that are called Archimedean solids. What is the definition of this category of polyhedra? Draw a net for one of them. Then cut out the net and fold it to form the polyhedron.
- Find examples of polyhedra in games and architecture.

Faceted Cut Gem

Origami Polyhedron

3 Things to Remember

- Add your own illustrations or paper creations to your project.
- Organize your report in a folder, and think of a title for your report.

Concrete Tetrahedrons by Ocean

Bulatov Sculpture

Why Does Ice Float?

1 Getting Started

Among the planets in our solar system, Earth is clearly the "water planet." Water occurs on Earth's surface as a liquid, solid, and gas. Ocean waters cover about 70% of Earth's surface. Fresh water in lakes and rivers covers less than 1%. Thick sheets of ice permanently cover Earth's polar regions, and glaciers occur in its higher mountains. Water in the form of clouds covers about half of Earth's surface at any time.

Essential Question How does floating ice affect you and the world in which you live?

There are three parts to this question. First, you have to understand why things float in water. Make lists of things that float in water and things that do not float in water. How can you describe, scientifically, the difference between these two types of objects?

Second, you have to discover what is special about water. Almost all other elements and compounds have the property that their solid forms sink in their liquid forms.

Third, you have to research how water in solid form influences the environment and living organisms. You may want to explore ocean currents, weather, and various cycles including the water cycle.

Earth's North Polar Ice Cap

2 Things to Include

- Does vegetable cooking oil float in water? Does ice float in vegetable oil?
- Compare the density of water and the density of ice. How do these densities relate to the fact that ice floats in water?
- What percent of an iceberg is above the water? What percent is below?
- What environmental and biological systems would not work if ice sunk in the oceans instead of floating?
- Explain how the floating of ice helps keep the oceans warm and therefore helps aquatic life exist in Earth's polar oceans.

3 Things to Remember

- Add your own illustrations to your project.
- Try to include as many different math concepts as possible. Your goal is to include at least one concept from each of the chapters you studied this year.
- Organize your report in a folder, and think of a title for your report.

Floating Iceberg

Selected Answers

Section 1.1 Whole Number Operations
(pages 7–9)

1. addition

3. division

5. addition

7. **a.** dividend **b.** quotient **c.** divisor

9. $4785 - 3391$; 1394 people

11. 4785×2; 9570 people

13. 7081

15. 2462

17. 433

19. 6944

21. 31

23. 60

25. $47\frac{110}{173}$

27. The partial product 39 should be moved to the left so that the 3 is under the 2 and the 9 is under the 7. The answer should be 663.

29. multiplication

31. division

33. addition

35. 24 in.; 35 in.^2

37. 36 m; 80 m^2

39. You can use addition to check your answer by adding 93 to itself 6 times. You can use division to check your answer by dividing 558 by either 93 or 6.

41. no; If the remainder is greater than the divisor, then the quotient should be increased until the remainder is less than the divisor or equal to zero.

43. 46 tokens

45. **a.** \$424 **b.** $\frac{3}{4}$ qt, or $\frac{3}{16}$ gal

47 and 49.

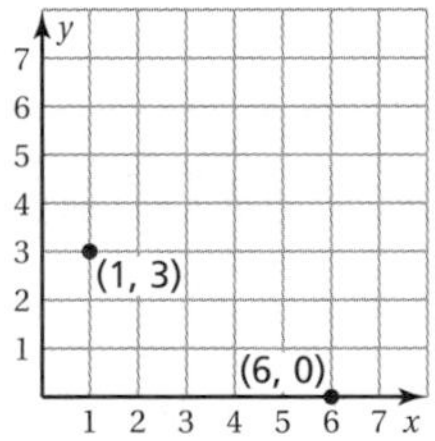

51. A

Section 1.2 Powers and Exponents
(pages 14 and 15)

1. An exponent indicates the number of times the base is used as a factor. A power is the entire expression (base and exponent). A power is a product of repeated factors.

3. $3 + 3 + 3 + 3 = 3(4)$ does not belong because it shows a product as a sum of repeated addends, whereas the other three show powers as products of repeated factors.

5. 13^2

7. 2^5

9. 8^4

11. 7^6

13. The base is written as the exponent and the exponent is written as the base.
$4 \cdot 4 \cdot 4 = 4^3$

15. 64

17. 196

19. 65,536

21. 1,419,857

23. 8,000,000 people

25. not a perfect square

27. perfect square

29. perfect square

31. not a perfect square

33. 40,000 cm^2

35. 8 squares

37.

Power	4^6	4^5	4^4	4^3	4^2	4^1
Value	4096	1024	256	64	16	4

As the exponent decreases, the value of the power is divided by 4. $4^0 = 1$

39. 13 blocks; add $7^2 - 6^2$ blocks; 19 blocks; add $10^2 - 9^2$ blocks; 39 blocks; add $20^2 - 19^2$ blocks

41. 165

43. 7

Section 1.3 Order of Operations
(pages 20 and 21)

1. Using the order of operations for $12 - 8 \div 2$, you divide 8 by 2 and then subtract the result from 12. Using the order of operations for $(12 - 8) \div 2$, you subtract 8 from 12 and then divide by 2.

3. 57

5. 2

7. 5

9. 24

11. 88

13. 2

15. Addition was performed before multiplication. $9 + 2 \times 3 = 9 + 6 = 15$

17. 8 pages

19. 25

21. 47

23. 8

25. 22 people

27. 1

29. \$34

31. \$23; Add the prices of the items you buy. Then subtract the amount of the gift card from the total.

33. **a.** $27 \div 3 + 5 \times 2 = 19$

b. *Sample answer:* $9^2 + 11 - 8 \times 4 \div 1 = 60$

c. $5 \times 6 - 15 + 9 = 24$

d. $14 \times 2 \div 7 - 3 + 9 = 10$

35. 6.1

37. 0.9

Section 1.4 Prime Factorization
(pages 28 and 29)

1. The prime factorization of a composite number is the number written as a product of its prime factors.

3. 6, 9 does not belong because it is a factor pair of 54 and the others are factor pairs of 56.

5. 3, 5, 9

7. None, 1709 is a prime number.

9. 1, 22; 2, 11

11. 1, 39; 3, 13

13. 1, 54; 2, 27; 3, 18; 6, 9

15. 1, 61

17. $5 \cdot 5$ or 5^2

19. $2 \cdot 13$

21. $2 \cdot 3 \cdot 3 \cdot 3$ or $2 \cdot 3^3$

23. $7 \cdot 11$

Selected Answers

Section 1.4 Prime Factorization *(continued)*

(pages 28 and 29)

25. 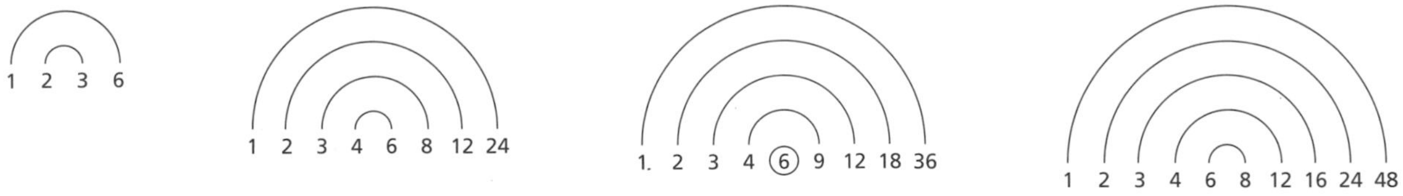

27. 1575

29. 4

31. 36

33. yes; 2 is a prime number because it only has 1 and itself as factors. The rest of the even whole numbers have 2 as a factor.

35. Use 36 objects to help you determine the possible group sizes.

37. cupcake table; Because 60 has more factors than 75, there are more rectangular arrangements.

39. 6 prisms; There are 6 unique arrangements of length, width, and height using the factors of 40.
(Note that $1 \times 1 \times 40$ names the same prism as $40 \times 1 \times 1$.);
$1 \times 1 \times 40$, $1 \times 2 \times 20$, $1 \times 4 \times 10$, $1 \times 5 \times 8$, $2 \times 2 \times 10$, $2 \times 4 \times 5$

41. 357

43. 1248

Section 1.5 Greatest Common Factor

(pages 34 and 35)

1. The GCF is the greatest factor that is shared by the two numbers.

3. What is the greatest prime factor of 24 and 32?; 2; 8

5. 2

7. 3

9. 1

11. 17

13. 15

15. 9

17. 1

19. 7 is the greatest common *prime* factor. The GCF is $2 \cdot 7 = 14$.

21. 23 packets

23. 7

25. 14

27. *Sample answer:* Prime factorization because it is tedious to find all the factors of large numbers.

29. always

31. 12; 6 red, 5 pink, and 4 yellow

33. **a.** Because 73 is a prime number and the GCF of the three numbers is 1.

b. 18; The GCF of 54 and 36 is 18. 18 divides evenly into 72 leaving one banana left over.

35. Commutative Property of Addition

37. Commutative Property of Multiplication

39. B

Section 1.6 Least Common Multiple
(pages 40 and 41)

1. The LCM of two numbers is the least of the multiples shared by the two numbers.

3. 21
5. 60
7. 12
9. 40
11. 36
13. 108
15. 66
17. 350
19. 15 days
21. D; This model represents multiples of 4 and 6 which have an LCM of 12. The other models represent multiples of 3 and 8, 8 and 12, and 6 and 8, which have an LCM of 24.
23. 165
25. 120
27. 1260
29. always
31. never
33. 300th caller
35. **a.**

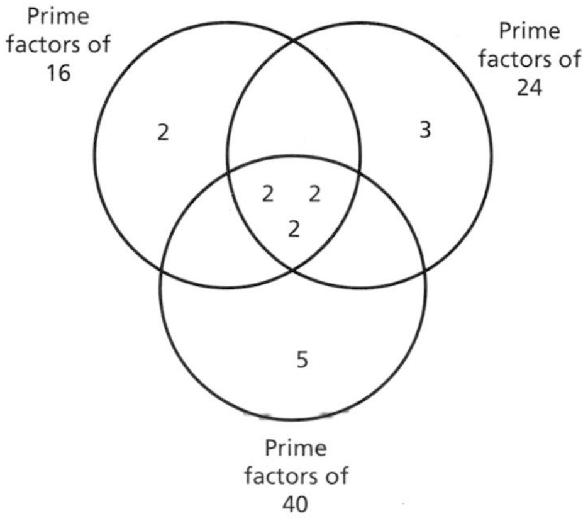

b. 240 **c.** 80; 120

37. 3^2
39. 17^5

Extension 1.6 Adding and Subtracting Fractions
(page 43)

1. $\frac{4}{24}, \frac{9}{24}$
3. $\frac{15}{36}, \frac{8}{36}$
5. $<$
7. $=$
9. $1\frac{5}{12}$
11. $\frac{17}{60}$
13. $5\frac{11}{18}$
15. $1\frac{1}{12}$
17. *Sample answer:* The LCD method uses numbers that are easier to work with, but there is extra work in finding the LCD. Using the other method, there are no preliminary steps for finding the LCD, but there may be more simplifying in the solution.

Selected Answers

Section 2.1 Multiplying Fractions
(pages 59–61)

1. Multiply numerators and multiply denominators, then simplify the fraction.

3. *Sample answer:* $3\frac{1}{2} \times 3\frac{1}{7} = 11$

5. $\frac{5}{16}$ **7.** $\frac{3}{28}$ **9.** $\frac{5}{8}$ **11.** $\frac{1}{3}$ **13.** $5\frac{1}{4}$

15. $\frac{16}{45}$ **17.** $\frac{5}{27}$ **19.** $1\frac{19}{30}$ **21.** $\frac{3}{10}$

23. $\frac{4}{7} > \left(\frac{9}{10} \times \frac{4}{7}\right)$; Because $\frac{9}{10} < 1$, the product will be less than $\frac{4}{7}$.

25. $\frac{5}{6} = \left(\frac{5}{6} \times \frac{7}{7}\right)$; Because $\frac{7}{7} = 1$, by the Multiplication Property of One, the two expressions are equal.

27. 2 **29.** 2 **31.** 2 **33.** $1\frac{1}{2}$

35. $1\frac{3}{14}$ **37.** $36\frac{2}{3}$ **39.** $6\frac{4}{9}$ **41.** $11\frac{3}{8}$

43. You must first rewrite the mixed number as improper fractions and then multiply.

$$2\frac{1}{2} \times 7\frac{4}{5} = \frac{5}{2} \times \frac{39}{5} = \frac{\overset{1}{\cancel{5}} \times 39}{2 \times \underset{1}{\cancel{5}}} = \frac{39}{2}\text{, or } 19\frac{1}{2}$$

45. **a.** 7 ft^2 **b.** $10\frac{1}{3}$ ft^2

47. $\frac{2}{15}$ **49.** $26\frac{2}{5}$ **51.** $\frac{9}{25}$ **53.** 462 $in.^2$

55. 4 mi

57. Which units of measure would make the calculations easier?

59. **a.** $\frac{3}{50}$ **b.** 45 people

61. $3^2 \cdot 5$ **63.** $2^2 \cdot 3 \cdot 5$

Section 2.2 Dividing Fractions
(pages 67–69)

1. *Sample answer:* $\frac{2}{5}, \frac{5}{2}$ **3.** B **5.** A

7. $\frac{1}{8}$ **9.** $\frac{5}{2}$ **11.** $\frac{1}{2}$ **13.** 16 **15.** $\frac{1}{14}$

17. $\frac{1}{3}$ **19.** 3 **21.** $\frac{2}{27}$ **23.** $\frac{27}{28}$ **25.** $20\frac{1}{4}$

27. You need to invert the second fraction before you multiply.

$\frac{4}{7} \div \frac{13}{28} = \frac{4}{7} \times \frac{28}{13} = \frac{4 \times \cancel{28}^{4}}{{}_{1}\cancel{7} \times 13} = \frac{16}{13}$, or $1\frac{3}{13}$

29. Round $\frac{2}{5}$ to $\frac{1}{2}$ and $\frac{8}{9}$ to 1. $\frac{1}{2} \div 1 = \frac{1}{2}$, which is not close to the incorrect answer of $\frac{20}{9}$.

31. $5\frac{5}{8}$ times

33. yes

35. yes

37. $\frac{1}{3}$

39. >; When you divide a number by a fraction less than 1, the quotient is greater than the number.

41. >; When you divide a number by a fraction less than 1, the quotient is greater than the number.

43. $\frac{1}{216}$

45. $1\frac{1}{6}$

47. 2

49. $\frac{3}{26}$

51. $\frac{2}{3}$

53. $2\frac{2}{5}$ hours, or 2 hours 24 minutes

55. **a.** $3\frac{3}{4}$ times **b.** $3\frac{1}{3}$ times **c.** $\frac{1}{4}$

57. **a.** $67\frac{1}{5}$ gal **b.** $8\frac{2}{5}$ gal **c.** $33\frac{3}{5}$ gal

59. 6

61. 5

Section 2.3 Dividing Mixed Numbers *(pages 74 and 75)*

1. $\frac{3}{22}$

3. sometimes; The reciprocal of $\frac{2}{2}$ is $\frac{2}{2}$, which is improper.

5. 3

7. $9\frac{3}{4}$

9. $3\frac{18}{19}$

11. $\frac{9}{10}$

13. $12\frac{1}{2}$

15. $1\frac{1}{5}$

17. $\frac{2}{7}$

19. $1\frac{5}{18}$

21. The mixed number $1\frac{2}{3}$ was not written as an improper fraction before inverting.

$$3\frac{1}{2} \div 1\frac{2}{3} = \frac{7}{2} \div \frac{5}{3} = \frac{7}{2} \times \frac{3}{5} = \frac{7 \times 3}{2 \times 5} = \frac{21}{10}, \text{ or } 2\frac{1}{10}$$

23. 14 hamburgers

25. no; The model shows $2\frac{1}{2} \div 1\frac{1}{6} = 2\frac{1}{7}$. There are 2 full groups of $1\frac{1}{6}$ plus one piece remaining, which represents $\frac{1}{7}$ of $1\frac{1}{6}$.

27. 4

29. $1\frac{1}{3}$

31. $5\frac{1}{6}$

33. $\frac{7}{54}$

35. $12\frac{1}{2}$

37. $\frac{22}{35}$

39. **a.** 6 ramps; *Sample answer:* The estimate is reasonable because $12\frac{1}{2}$ was rounded down.

b. 6 ramps; $1\frac{1}{4}$ feet left over

41. 0.43

43. 3.8

45. C

Section 2.4 Adding and Subtracting Decimals
(pages 82 and 83)

1. Estimating allows you to check that your answer is reasonable.

3. $1.15 + 0.43 = 1.58$ 5. 11.029 7. 22.899

9. 29.937 11. 1.46 13. 4.366 15. 2.644

17. Line up the decimal points before adding. Insert a 0 at the end of the second number so that both numbers have the same number of decimal places. $6.058 + 3.95 = 10.008$.

19. $8.30 21. 19.58 23. 10 25. 15.606

27. the decimal parts in the sum total 1; the decimal parts in the difference are exactly the same

29. 34.995 m 31. 4.816 AU 33. 20.189 AU 35. 6.85 units

37. $\frac{1}{4}$ 39. $\frac{1}{20}$

Section 2.5 Multiplying Decimals
(pages 89–91)

1. Place the decimal point so that there are two decimal places. $1.2 \times 2.4 = 2.88$

3. 8.722 5. 19.5750 7. 4 9. 3.15

11. 0.21 13. 33.6 15. 115.04 17. 21.45

19. 13.888 21. 2.4 23. 0.0342

25. The decimal is in the wrong place. $0.0045 \times 9 = 0.0405$

27. 30.06 lb 29. 3 mm; 9 mm 31. 0.024 33. 0.000072

35. 0.03 37. 0.000012 39. 109.74 41. 3.886

43. 13.7104 45. 51.3156 47. $3.24 49. $1284.78

51. $7.12 \times 8.22 \times 100 = 7.12 \times 822 = 5852.64$

53. 137 55. 23.112 57. 71.984 59. 36.225 61. 2; 3; 4

63. Each number is 0.1 times the previous number; 0.0015, 0.00015, 0.000015

65. Each number is 1.5 times the previous number; 25.3125, 37.96875, 56.953125

67. **a.** 190.06 mi

b. 91.29 mi

69. Which framing is thicker?

71. 5 73. 7

Section 2.6 Dividing Decimals
(pages 97–99)

1. $\begin{array}{r} 0.61 \\ 4\overline{)2.44} \end{array}$

3. $6.38 \div 11 = 0.58$

5. $47\overline{)136}$

7. $216\overline{)1850}$

9. 13

11. 9

13. 6.7

15. 1.3

17. 9.4

19. 3.2

21. 0.098

23. 3.57

25. A zero should be placed before the 8 in the quotient.

$$\begin{array}{r} 0.086 \\ 6\overline{)0.516} \\ \underline{48} \\ 36 \\ \underline{36} \\ 0 \end{array}$$

27. the 5-ounce bottle; The price per ounce is \$2.06 for the 5-ounce bottle and \$2.12 for the 4-ounce bottle.

29. 1.62

31. 10.12

33. 8.046

35. \$26.96

37. 9

39. 0.08

41. 400

43. 460

45. about 1.33

47. about 12.21

49. 113 tickets

51. 850 songs

53. =

55. $<$

57. about 5357 bees

59. When dividing, make sure your units cancel.

61. $1\frac{1}{6}$

63. $\frac{1}{20}$

65. B

Section 3.1 Algebraic Expressions
(pages 115–117)

1. $3(4) + 5$ does not belong because it is a numerical expression and the other three are algebraic expressions.

3. decrease; When you subtract greater and greater values from 20, you will have less and less left.

5. \$120

7. \$8

9. Terms: g, 12, $9g$
Coefficients: 1, 9
Constant: 12

11. Terms: $2m^2$, 15, $2p^2$
Coefficients: 2, 2
Constant: 15

13. Terms: $8x$, $\frac{x^2}{3}$
Coefficients: 8, $\frac{1}{3}$
Constant: none

15. **a.** Terms: 2ℓ, $2w$; Coefficients: 2, 2: Constant: none

b. The coefficient 2 of ℓ represents that there are 2 lengths on the rectangle. The coefficient 2 of w represents that there are 2 widths on the rectangle.

17. g^5

19. $5.2y^3$

21. $2.1xz^4$

23. $25d^2$

25. 9

27. 11

29. 10

31. 6

33. 5

35. 4

Section 3.1 Algebraic Expressions *(continued)*
(pages 115–117)

37. Multiplication should be done first, then addition.

$5m + 3 = 5 \cdot 8 + 3$

$= 40 + 3$

$= 43$

39. 34 mm; 118 mm

41.

x	2	4	8
$64 \div x$	32	16	8

43. 23 **45.** $2\frac{5}{6}$ **47.** 22

49. 46 **51.** 24

53. Start by drawing a visual image of moving 2000 feet in exactly 10 minutes.

55. 64 in.^{3} **57.** 512 **59.** 256

Section 3.2 Writing Expressions
(pages 122 and 123)

1. x take away 12; $x - 12$; $x + 12$ **3.** $8 - 5$ **5.** $28 \div 7$

7. $18 - 3$ **9.** $x - 13$ **11.** $18 \div a$

13. $7 + w$ or $w + 7$ **15.** $y + 4$ or $4 + y$ **17.** $2 \cdot z$ or $z \cdot 2$

19. The expression is not written in the correct order; $\frac{8}{y}$

21. **a.** $x \div 5$

b. *Sample answer:* If the total cost is \$30, then the cost per person is $x \div 5 = 30 \div 5 = \$6$. The result is reasonable.

23. *Sample answer:* The sum of n and 6; 6 more than a number n

25. *Sample answer:* A number b less than 15; 15 take away a number b

27. $\frac{y}{4} - 3$; 2 **29.** $8x + 6$; 46

31. **a.**

Game	1	2	3	4	5
Cost	\$5	\$8	\$11	\$14	\$17

b. $2 + 3g$

c. \$26

33. It might help to see the pattern if you make a table of the data in the bar graph.

35. $\frac{x}{4}$ **37.** 59 **39.** 140

Section 3.3 Properties of Addition and Multiplication
(pages 130 and 131)

1. *Sample answer:* $\frac{1}{5} + \frac{3}{5} = \frac{3}{5} + \frac{1}{5}$

$\frac{4}{5} = \frac{4}{5}$

3. *Sample answer:* $(5 \cdot x) \cdot 1 = 5 \cdot (x \cdot 1)$

$= 5x$

5. Comm. Prop. of Mult.

7. Assoc. Prop. of Mult.

9. Add. Prop. of Zero

11. The grouping of the numbers did not change. The statement illustrates the Commutative Property of Addition because the order of the addends changed.

13. $(14 + y) + 3 = (y + 14) + 3$ Comm. Prop. of Add.

$= y + (14 + 3)$ Assoc. Prop. of Add.

$= y + 17$ Add 14 and 3.

15. $7(9w) = (7 \cdot 9)w$ Assoc. Prop. of Mult.

$= 63w$ Multiply 7 and 9.

17. $(0 + a) + 8 = a + 8$ Add. Prop. of Zero

19. $(18.6 \cdot d) \cdot 1 = 18.6 \cdot (d \cdot 1)$ Assoc. Prop. of Mult.

$= 18.6d$ Mult. Prop. of One

21. $(2.4 + 4n) + 9 = (4n + 2.4) + 9$ Comm. Prop. of Add.

$= 4n + (2.4 + 9)$ Assoc. Prop. of Add.

$= 4n + 11.4$ Add 2.4 and 9.

23. $z \cdot 0 \cdot 12 = (z \cdot 0) \cdot 12$ Assoc. Prop. of Mult.

$= 0 \cdot 12$ Mult. Prop. of Zero

$= 0$ Mult. Prop. of Zero

25. **a.** x represents the cost of a box of cookies. **b.** $120x$

27. $7 + (x + 5) = x + 12$

29. $(7 \cdot 2) \cdot y$

31. $(17 + 6) + 2x$

33. $w \cdot 16$

35. 98

37. 90

39. 37 is already prime.

41. 3×7^2

43. B

Section 3.4 The Distributive Property
(pages 137–139)

1. *Sample answer:* You must distribute or give the number outside the parentheses to the numbers inside the parentheses.

3. $4 + (x \cdot 4)$ does not belong because it does not represent the Distributive Property.

5. 63

7. 516

9. 936

11. 504

13. $\frac{4}{7}$

15. $2\frac{1}{2}$

17. $3x + 12$

19. $6s - 54$

21. $96 + 8a$

23. $72 - 12k$

25. $63 + 9c$

27. $40g + 24$

29. $4x + 4y$

31. $7p + 7q + 63$

33. The 6 was not distributed to the 8 inside the parentheses; $6(y + 8) = 6y + 48$

35. $5(r + 15)$ and $5r + 5 \cdot 15$, because they are equivalent expressions.

Section 3.4 The Distributive Property *(continued)*

(pages 137–139)

37. $16(10 + x) = 160 + 16x$

39. $6x + 25$

41. $68 + 28k$

43. $19y + 5$

45. $3d + 1$

47. $5v$

49. $2.7w - 14.04$

51. $\frac{11}{4}z + \frac{3}{10}$

53. $7x + 12y$

55. $x = 8$

57. $x = 3$

59. Area: $8x + 64$
Perimeter: $2x + 32$

61. Area: $9x + 108$
Perimeter: $2x + 42$

63. **a.** 6.2 **b.** 14

Sample answer: The preferred method is not the same for both expressions.
For part (a), evaluating inside the parentheses first requires easier and less calculations.
For part (b), using the Distributive Property will eliminate the fractions.

65. $7(x + 3) + 8 \cdot x + 3 \cdot x + 8 - 9 = 2(9x + 10)$

67. 34.006

69. 0.387

Extension 3.4 Factoring Expressions

(page 141)

1. $7(1 + 2)$

3. $6(3 - 2)$

5. $12(5 - 3)$

7. $28(3 + 1)$

9. $2(x + 5)$

11. $13(2x - 1)$

13. $9(4x + 1)$

15. $5(2x - 5y)$

17. yes; yes; Because a and b are divisible by c, you can factor c out of each expression. Because c is a factor of the sum and the difference, each expression is divisible by c.

19. $(x + 4)$ feet

Section 4.1 Areas of Parallelograms

(pages 156 and 157)

1. The area of a polygon is the amount of surface it covers. The perimeter of a polygon is the distance around the polygon.

3. 18 ft^2

5. 187 km^2

7. 243 in.^2

9. 15 m was used for the height instead of 13 m.
$A = 8(13) = 104\ \text{m}^2$

11. 12 units^2

13. 24 units^2

15. 72 m^2

17. What shape could have an area of 128 square feet?
What shape could have an area of s^2 square feet?

19. 287 in.^2

21. n^2bh where b represents the base and h represents the height of the original parallelogram, or n^2A where A represents the area of the original parallelogram.

23. 1640

25. 118

Section 4.2 Areas of Triangles
(pages 162 and 163)

1. yes; To find the area of the triangle, you must also know the height of the triangle. That is, the perpendicular distance from the base to the opposite vertex.

3. 6 cm^2

5. 1620 $in.^2$

7. 1125 cm^2

9. The side length of 13 meters was used instead of the height.

$A = \frac{1}{2}(10)(12) = 60 \text{ m}^2$

11. 324 cm^2

13. 90 mi^2

15. *Sample answer:*

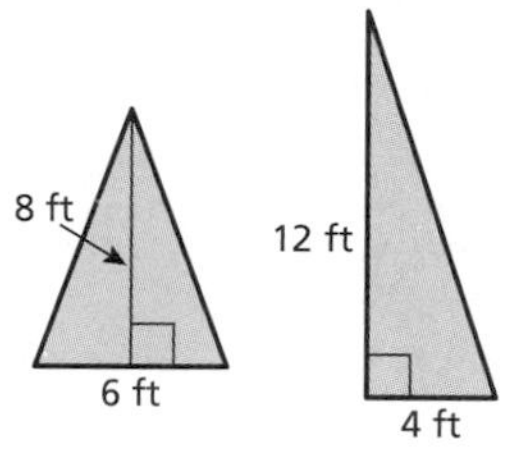

17. x^2 times greater

19. 4 times greater

21. Mult. Prop. of One

23. Assoc. Prop. of Add.

Section 4.3 Areas of Trapezoids
(pages 170 and 171)

1. bases: 4 ft and 7 ft; height: 15 ft

3. $2\ell + 2w$; This is an expression for the perimeter of a rectangle. The other three are expressions for area (triangle, rectangle, and trapezoid).

5. 24 $units^2$

7. 28 $in.^2$

9. 105 ft^2

11. 8 $units^2$

13. 12 $units^2$

15. 60 $in.^2$

17. 78 mi^2

19. 18 ft

21. Use the strategy *Solve a Simpler Problem* by assigning values to the variables.

23 and 25.

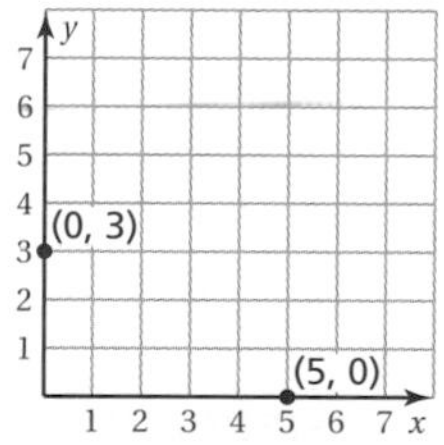

27. C

Extension 4.3 Areas of Composite Figures
(page 173)

1. 36 $units^2$

3. 20 $units^2$

5. $126\frac{1}{2}$ cm^2

7. 3400 yd^2

Sample answer: Separate the figure into a trapezoid and a square.

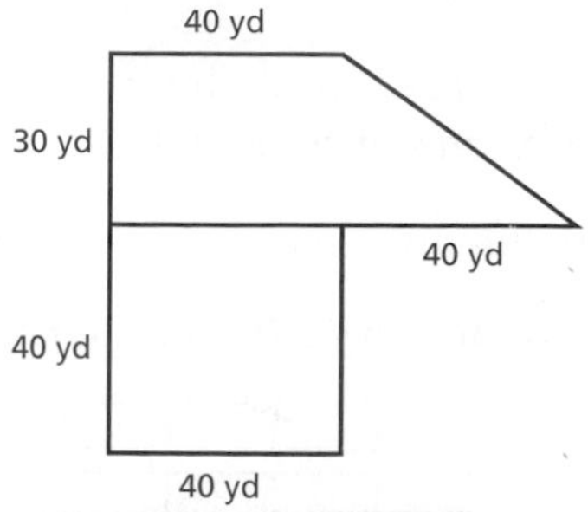

Section 4.4 Polygons in the Coordinate Plane
(pages 178 and 179)

1. Plot the points that represent the vertices of the polygon and connect the points in order.

3.

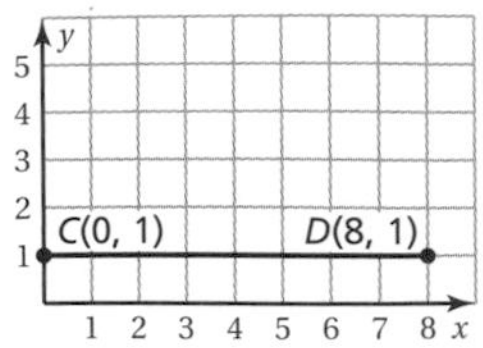

Length of CD is 8 units.

5.

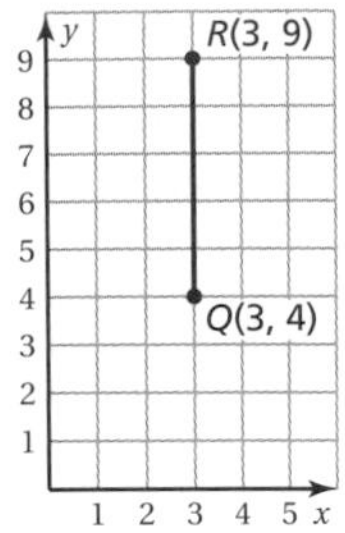

Length of QR is 5 units.

7.

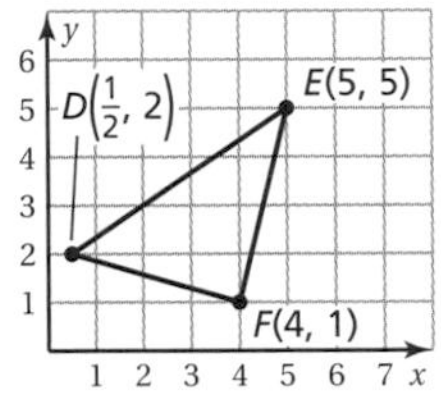

9.

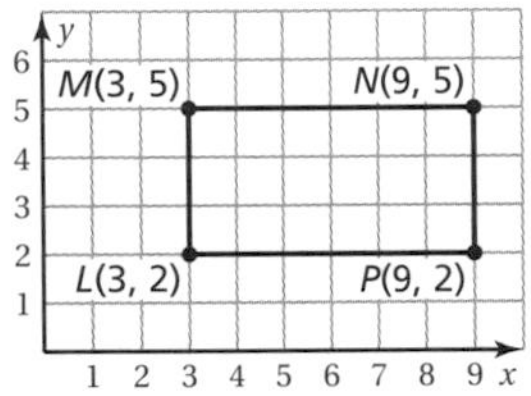

11.

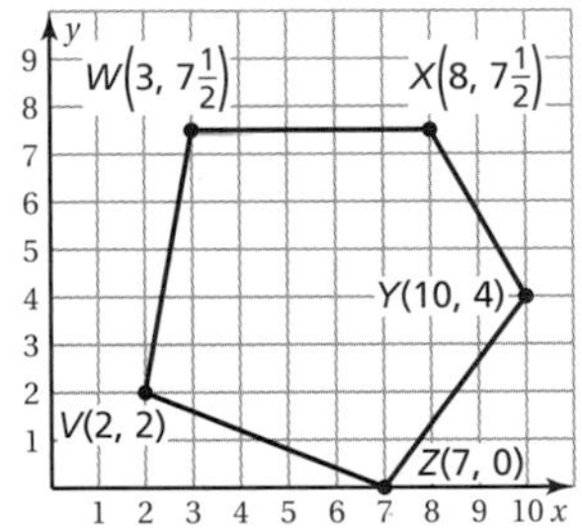

13. 24 units; 36 units^2

15. 28 units; 45 units^2

17. **a.** square

b. 28 ft; 49 ft^2

19. *Sample answer:*

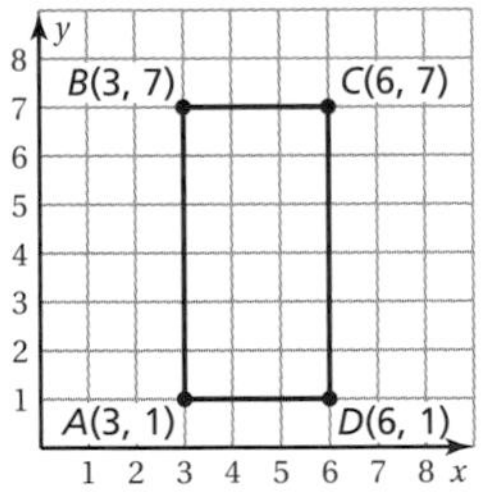

21. *Sample answer:*

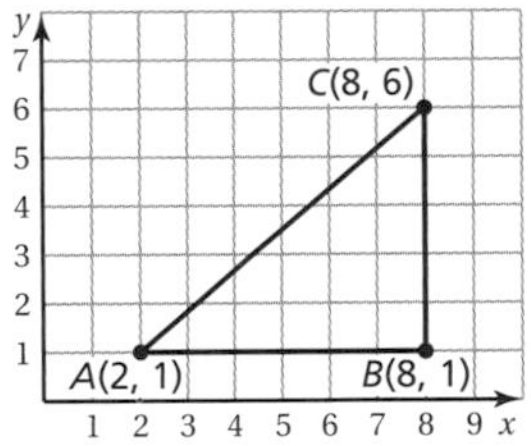

23. 27 miles; There are only two ways to go from station P to station L. Traveling from station P to N to M to L is 27 miles. Traveling from station P to J to K to L is 33 miles.

25. 2.5 times larger **27.** 2 **29.** $\frac{5}{16}$ **31.** D

Section 5.1 Ratios
(pages 194 and 195)

1. consonants; For every 5 vowels there are 7 consonants, so consonants outnumber the vowels.

3. 2 out of every 5; This ratio is 2 : 3, all other ratios are 2 : 5.

5.

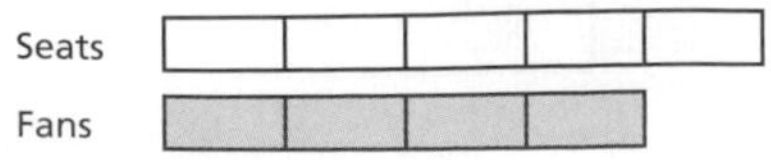

7. 6 to 4, or 6 : 4; For every 6 basketballs, there are 4 soccer balls.

9. 3 to 7, or 3 : 7; For every 3 shirts, . there are 7 pants

11. 8 to 15, or 8 : 15; 8 out of 15 movies are comedies.

13. 15 to 3, or 15 : 3; Out of 15 movies, 3 are dramas.

15. 9 h

17. 12 : 16

19. 6 black pieces; The ratio of black to red is 3 : 5, so each part is $16 \div 8 = 2$. So, there are $3 \cdot 2 = 6$ black pieces and $5 \cdot 2 = 10$ red pieces.

21. It may be helpful to organize your results in a table.

23. 4 pints of soda water, 8 pints of fruit punch concentrate, 20 pints of ginger ale; Yes; *Sample answer:* There is twice as much fruit punch as soda water (as in the original ratio). There is 5 times as much ginger ale as soda water (as in the original ratio).

25. 4.6

27. 2.53

29. B

Section 5.2 Ratio Tables
(pages 201–203)

1. Two ratios are equivalent if they can be written as the same ratio.

3. 12 : 15 does not belong because all other ratios are equivalent.

5. The ratio of ladybugs to bees can be described by 12 : 4, 6 : 2, or 3 : 1.

7.

Violins	8	24
Cellos	3	9

8 : 3 and 24 : 9

9.

Burgers	3	6	9
Hot Dogs	5	10	15

3 : 5, 6 : 10, and 9 : 15

11.

Forks	16	8	48
Spoons	10	5	30

16 : 10, 8 : 5, and 48 : 30

13.

You	3	6	9	12
Friend	4	8	12	16

16 tickets

15.

First	100	10	60
Second	60	6	36

$60

17. Adding the same number, 5 in this case, to each part of the ratio does not create equivalent ratios. You can add corresponding parts of equivalent ratios to create new equivalent ratios.

Sample answer:

A	3	6	9
B	7	14	21

19. 28 basketballs

21. Add the corresponding quantities of Recipes B and D to create Recipe E.

23. Subtract the corresponding quantities of Recipe B from Recipe C to create Recipe A.

25. *Sample answer:* Add the corresponding quantities of Recipes B and F to create a batch with 11 servings.

Section 5.2 Ratio Tables *(continued)* *(pages 201–203)*

27. $A = 16, B = 28$

29. $A = 65, B = 40$

31. 32

33. 36 bugs

35. $27(2 + 1)$

37. $14(3x + 2y)$

Section 5.3 Rates *(pages 208 and 209)*

1. *Sample answer:* You walk at a rate of 2 blocks per minute, so you walk 12 blocks in 6 minutes.

3. *Sample answer:* 45 words for every 30 minutes

5. *Sample answer:* 4 inches for every 12 years

7. \$7 per week

9. 45 miles per hour

11. 140 kilobytes per second

13. 72 miles per gallon

15. 100 times per second

17. \$20

19. equivalent

21. not equivalent

23. **a.** 6 min

b.

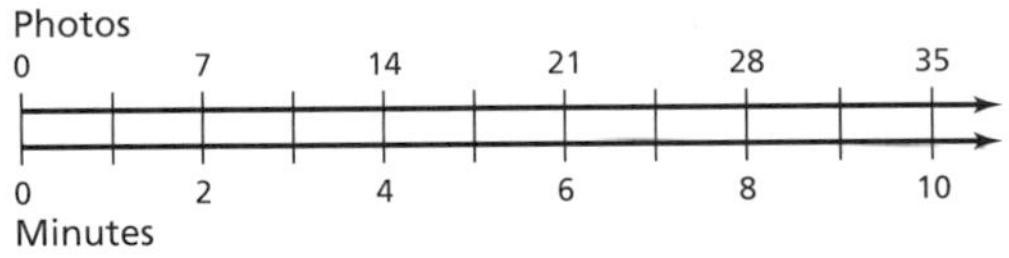

10 min

25. It may be helpful to organize your results in a table.

27. 1.2 h

29. *Sample answer:* $\frac{10}{12}, \frac{20}{24}$

31. *Sample answer:* $\frac{8}{18}, \frac{12}{27}$

Section 5.4 Comparing and Graphing Ratios *(pages 214 and 215)*

1. Use ratio tables to write equivalent ratios where one part from each ratio has the same numerical value. Compare the other part.

3. A

5. B

7. A

9. B

11. the first recipe

13.

Zoo	
People	**Cost (dollars)**
4	60
8	120
12	180
16	240

Museum	
People	**Cost (dollars)**
5	95
10	190
15	285
20	380

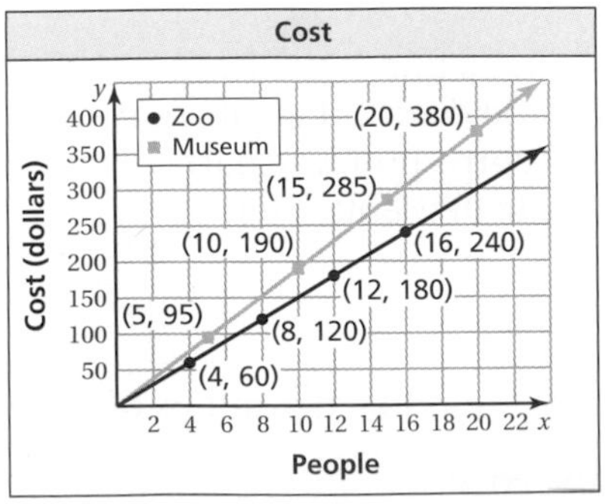

Both graphs begin at (0, 0). The graph for the museum is steeper, so the cost to attend the museum is greater than the cost to attend the zoo.

15. Begin by using double number lines to represent the situations.

17. a. *Sample answer:*

Old Mixture	**Peanuts**	7	14	21	28	35	42	49	56	63	70	77	84	91
	Almonds	4	8	12	16	20	24	28	32	36	40	44	48	52
	Total	11	22	33	44	55	66	77	88	99	110	121	132	**143**

New Mixture	**Peanuts**	8	16	24	32	40	48	56	64	72	80	88
	Almonds	5	10	15	20	25	30	35	40	45	50	55
	Total	13	26	39	52	65	78	91	104	117	130	**143**

143 nuts

b.

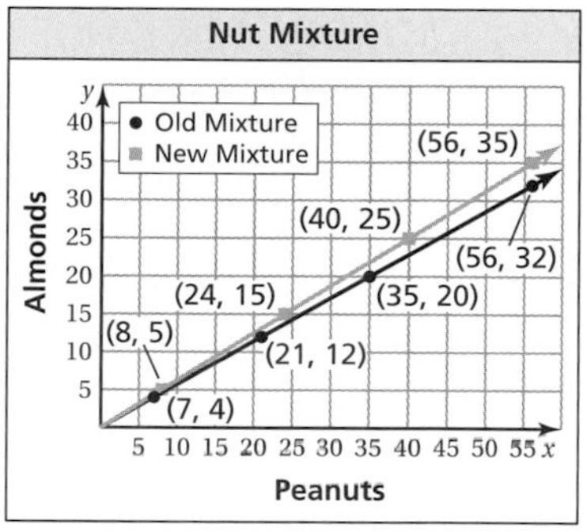

Both graphs begin at (0, 0). The graph for the new mixture is slightly steeper, so it has a greater concentration of almonds.

c. more; Almonds cost more and there are more of them in the new mixture than in the old mixture.

19. 16

21. 34 R109

Section 5.5

Percents
(pages 222 and 223)

1. You can shade 42 out of 100 squares to model 42%.

3. *Sample answer:* $\frac{3}{20}, \frac{23}{100}, \frac{1}{8}$

5.

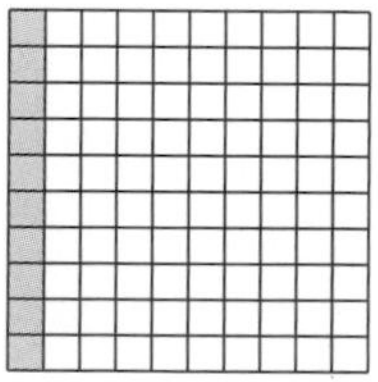

7.

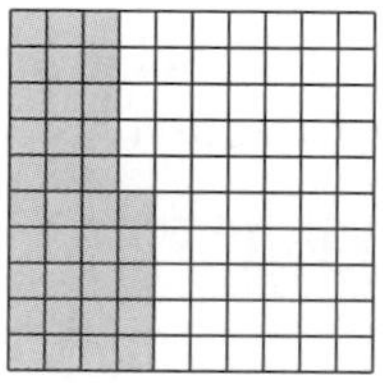

9. $\frac{9}{10}$

11. $\frac{7}{100}$

13. $\frac{79}{100}$

15. $1\frac{22}{25}$

17. $2\frac{6}{25}$

19. $\frac{1}{250}$

21. 10%

23. 55%

25. 54%

27. 185%

29. The decimal point should not have been added to the percent expression.

$$\frac{14}{25} = \frac{14 \times 4}{25 \times 4} = \frac{56}{100} = 56\%$$

31. $\frac{5}{4}$; No, you have more than you need.

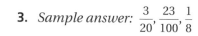

Section 5.5 Percents *(continued)*

(pages 222 and 223)

33. 81.25%

35. 82.5%

37. Organize the percents and fractions in a table. What operation should you use to compare Illinois to Hawaii?

39. You can shade half of one of the squares.

41. $\frac{1}{2}$

43. 16

45. D

Section 5.6 Solving Percent Problems

(pages 229–231)

1. Twenty percent of what number is 30?; 6; 150

3–21. Explanations will vary.

3. 12 **5.** 35 **7.** 9 **9.** 12.5 **11.** 21

13. 20.25 **15.** 24 **17.** 14 **19.** 84 **21.** 94.5

23. The percent was not written as a fraction before multiplying; $40\% \times 75 = \frac{2}{5} \times 75 = 30$

25. 35.2 in.

27–35. Explanations will vary.

27. 140 **29.** 84 **31.** 80 **33.** 25 **35.** 20

37. The percent should be written as a fraction before dividing; $5 \div 20\% = 5 \div \frac{1}{5} = 25$

39. **a.** 50 students **b.** 18 students

41. 21 cars

43. =

45. >

47. 48 min

49. **a.** 432 in.^{2}

b. 37.5%; Because the length is doubled, the width of the rectangle is now half of 75% of its length, or 37.5%.

51. *Sample answer:* Because 30% of n is equal to 2 times 15% of n and 45% of n is equal to 3 times 15% of n, you can write 30% of $n = 2 \times 12 = 24$ and 45% of $n = 3 \times 12 = 36$.

53. 97.2%

55. 16.5

57. 26.28

Section 5.7 Converting Measures

(pages 236 and 237)

1. yes; Because 1 centimeter is equal to 10 millimeters, the conversion factor equals 1.

3. Find the number of inches in 5 cm; 5 cm ≈ 1.97 in.; 5 in. = 12.7 cm

5. person weighing 75 kg; 75 kg ≈ 166.67 lb and 166.67 lb > 110 lb

7. 1.5 **9.** 12.63 **11.** 1.22 **13.** 0.19 **15.** 37.78 **17.** 14.49

19. **a.** about 60.67 **b.** about 8.04 km

21. $<$ **23.** $>$ **25.** $>$ **27.** 1320 **29.** 111.8 **31.** 0.001

33. When using conversion factors, make sure your units cancel.

35. about 669,600,000 mi/h

37. 30 **39.** 18 **41.** C

Section 6.1 Integers
(pages 252 and 253)

1. 8, −9, 22

3. *Sample answer:* below, under, lose

5.
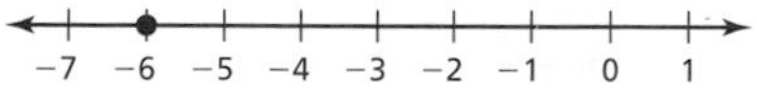

7.

9. 37,500 **11.** −56 **13.** 5 **15.** −318

17.
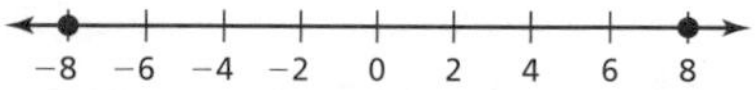

19.
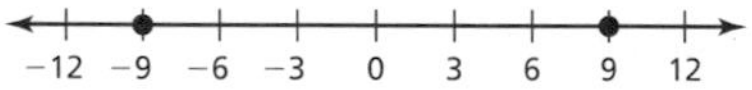

21.
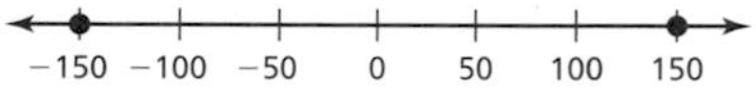

23.
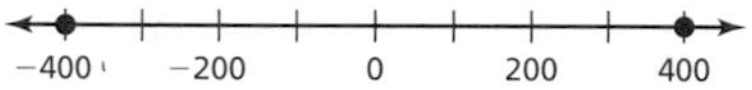

25.
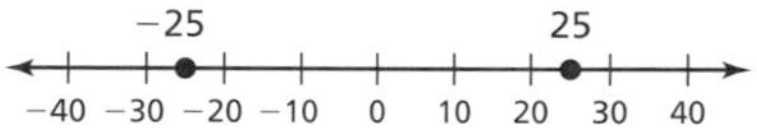

27. −8 **29.** 18

31. **a.** *Sample answer:* Choosing 8, the opposite is −8.

b. *Sample answer:* 8

c. The opposite of the opposite of an integer is the integer.; Yes; *Sample answer:*

Case 1:
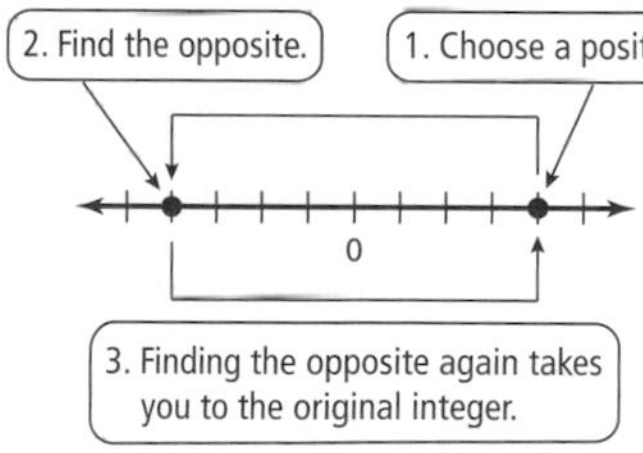

Case 2:
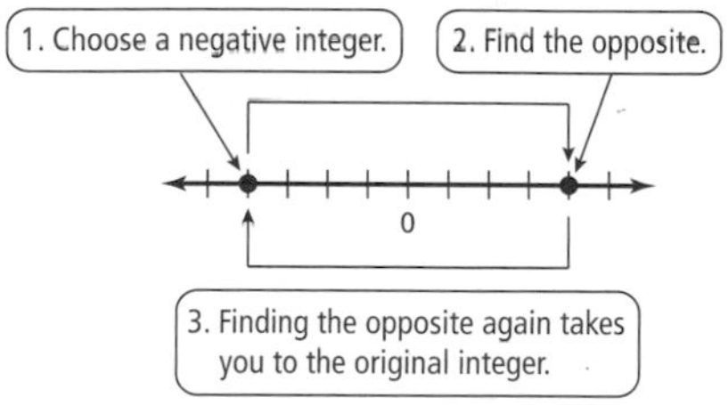

Case 3: Choose 0. The opposite of 0 is 0, so the opposite of the opposite of 0 is 0.

d. $-(-(-6))$ is the opposite of the opposite of -6; $-(-(-6)) = -6$

33. $\frac{3}{8}, \frac{1}{2}, \frac{3}{4}, \frac{7}{8}$ **35.** B

Section 6.2 Comparing and Ordering Integers
(pages 258 and 259)

1. On a number line, numbers to the left are less than numbers to the right. Numbers to the right are greater than numbers to the left.

3. The value of a is less than the value of b because a is to the left of b.

5. $<$ **7.** $>$ **9.** $>$ **11.** $>$

Selected Answers

Section 6.2 Comparing and Ordering Integers *(continued)*
(pages 258 and 259)

13. The explanation about where the integers are located on a number line is incorrect; $-7 < -3$; So, -7 is to the left of -3 on a number line.

15. $-4, -3, -2, 1, 2$

17. $-7, -4, 2, 3, 6$

19. $-20, -10, -5, 15, 25$

21. oxygen

23. always; The opposite of a positive integer is a negative integer. Positive integers are greater than negative integers.

25. **a.** Florida, Louisiana, Arkansas, Tennessee, California

b. California, Louisiana, Florida, Arkansas, Tennessee

c. An elevation of 0 feet represents sea level.

27. no; In order for the median to be below 0°F, at least 6 of the temperatures must be below 0°F.

29.

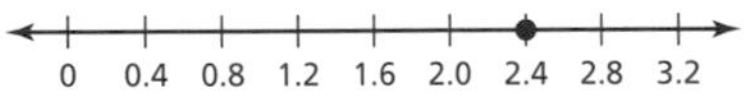

31.

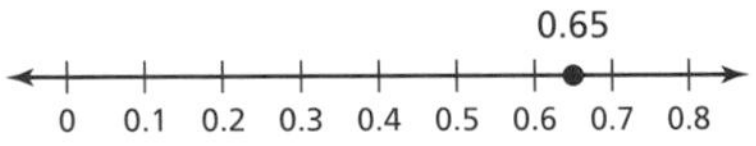

33. B

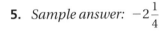

Section 6.3 Fractions and Decimals on the Number Line
(pages 264 and 265)

1. a

3. -2.6

5. *Sample answer:* $-2\frac{1}{4}$

7.

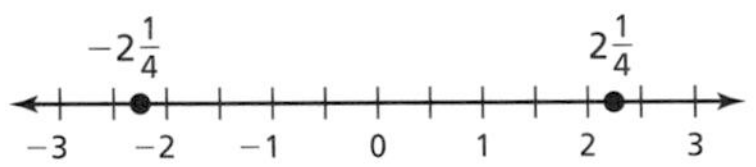

9.

11. $<$

13. $<$

15. $>$

17. $>$

19. the larger sand dollar

21. $-1, -\frac{3}{4}, -\frac{5}{8}, -\frac{1}{20}, 0$

23. $-5, -4.9, -4.35, -4.3, -4$

25. Write the numbers as decimals instead of finding a common denominator.

27. 1, 2, and any integer less than -3

29.

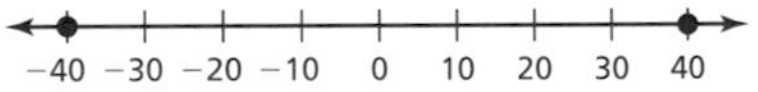

31.

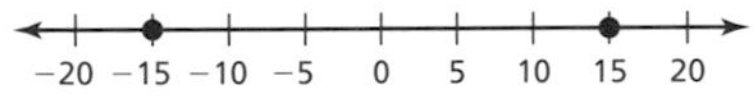

Section 6.4 Absolute Value
(pages 272 and 273)

1. Find the distance between the number and 0 on a number line.

3. What integer is 3 units to the left of 0?; -3; 3

5.

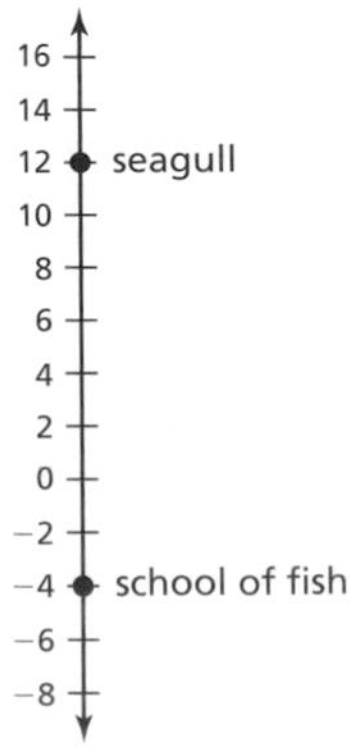

Seagull

7. 2

9. 8.35

11. $3\frac{2}{5}$

13. 14.06

15. $-10, 10$ **17.** $<$ **19.** $>$ **21.** $<$

23. Scientist B

25. A; You owe more than \$25, so debt > \$25.

27. $-2, 0, |-1|, |4|, 5$

29. $-11, 0, |3|, |-6|, 9, 10$

31. 0

33. -1

35. sometimes; If the number is negative then its absolute value is greater, but if the number is positive or zero then it is equal to its absolute value.

37. never; The absolute value of a positive number is the number itself.

39. *Sample answer:* $x = -2, y = -3$

41. 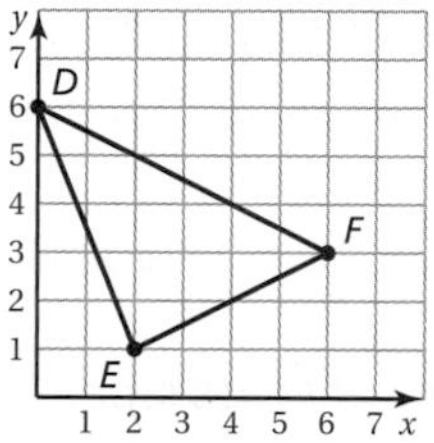

43. 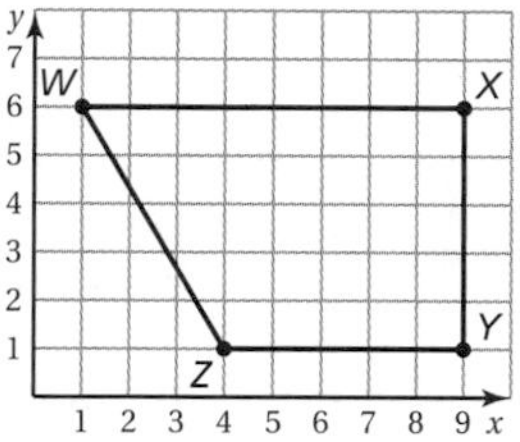

Section 6.5 The Coordinate Plane

(pages 279–281)

1. 4

3. $(2, -3)$; $(2, -3)$ is in Quadrant IV. The other three points are in Quadrant II.

5. $(3, 1)$ **7.** $(-2, 4)$ **9.** $(2, -2)$ **11.** $(-4, 2)$ **13.** $(4, 0)$

15–21. See graph below.

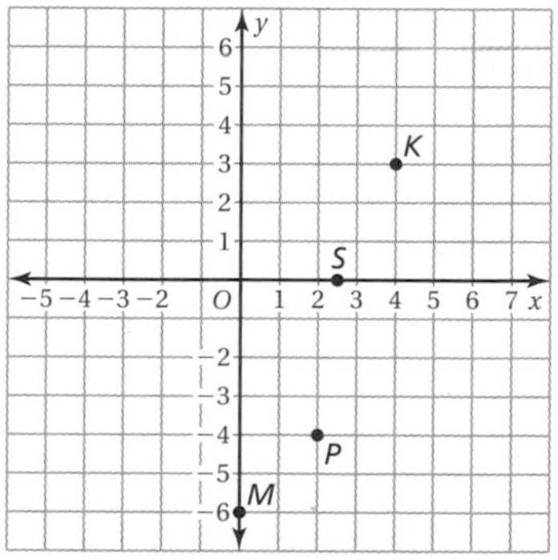

15. Quadrant I

17. y-axis

19. Quadrant IV

21. x-axis

23. The numbers are reversed. To plot (4, 5), start at (0, 0) and move 4 units right and 5 units up.

25. 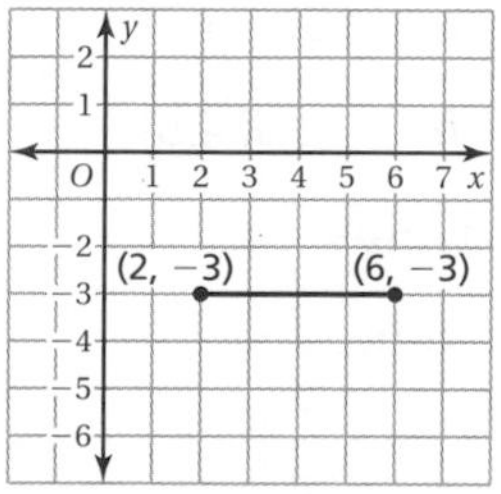

4

27. 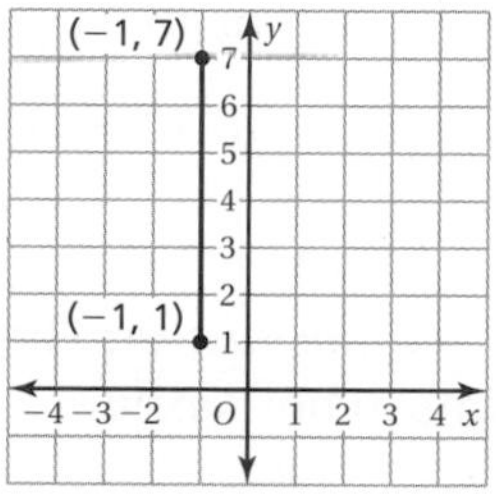

6

29. 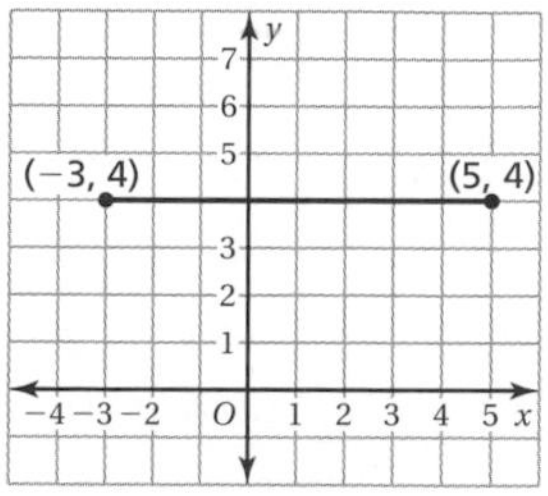

8

31. $(-2, 1)$

Section 6.5 The Coordinate Plane *(continued)* *(pages 279–281)*

33.

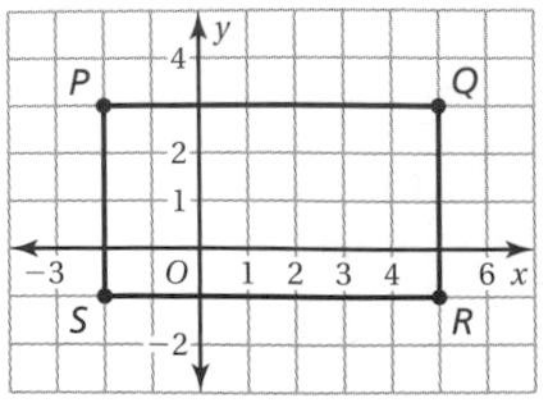

22 units; 28 units^2

35. **a.** about 142,000

b. 2011 and 2012

c. about 10,000

37. **a.**

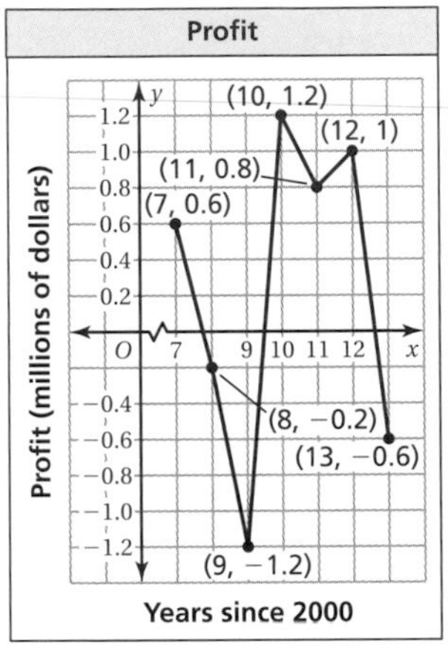

b. *Sample answer:* There are four years where the profit is positive and three years where the profit is negative. The profit decreased from 2007 to 2009. The profit increased the most from 2009 to 2010.

c. \$1.6 million

d. Because the x-axis represents the number of years since 2000, then 0 represents 2000. So, you could graph the profits for 1990 to 2006 by using x-values from -10 to 6.

39. Quadrant III

41. Quadrant I or Quadrant IV

43. origin

45. never; All points in Quadrant III have negative y-coordinates.

47. Reptiles

49. no; Quadrants III or IV

51. Because the rain forest is in Quadrant IV, the x-coordinate of the point will be positive and the y-coordinate of the point will be negative.

53. *Sample answer:* $(-6, 3)$, $(-2, 3)$, $(-2, -9)$, $(2, -9)$

55. $y - 4$

57. $x + 9$

59. C

Extension 6.5 Reflecting Points in the Coordinate Plane *(page 283)*

1. **a.** $(3, -2)$
b. $(-3, 2)$

3. **a.** $(-5, 6)$
b. $(5, -6)$

5. **a.** $(0, 1)$
b. $(0, -1)$

7. **a.** $(2.5, -4.5)$
b. $(-2.5, 4.5)$

9. $(-4, -5)$

11. $(2, 2)$

13. $(3, 9)$; 18 units

15. **a.** $(-4, -5)$; $(1, -7)$; $(2, 2)$; $(-6.5, 10.5)$; yes; *Sample answer:* The order of the reflections does not matter. You are still reflecting the points in both axes.

b. *Sample answer:* Use the opposite of each coordinate.

Section 7.1 Writing Equations in One Variable *(pages 298 and 299)*

1. An equation has an equal sign and an expression does not.

3. *Sample answer:* A number n subtracted from 28 is 5.

5. What was the high temperature if it was 4° less than 62°F? 58°F

7. $y - 9 = 8$ **9.** $w \div 5 = 6$ **11.** $5 = \frac{1}{4}c$ **13.** $n - 9 = 27$

15. $6042 = 1780 + a$ **17.** $16 = 3x$ **19.** $326 = 12(14) + 6(5) + 16x$

21. It might be helpful to organize the given information visually.

23. 13 **25.** 28 **27.** B

Section 7.2 Solving Equations Using Addition or Subtraction *(pages 305–307)*

1. Substitute your solution back into the original equation and see if you obtain a true statement.

3. subtraction **5.** so that x is by itself; so that the two sides remain equal

7. yes **9.** no **11.** yes **13.** $t = 1$

15. What number plus 5 equals 12?; $a = 7$ **17.** 20 is what number minus 6?; $d = 26$

19. $z = 16$ **21.** $p = 3$ **23.** $h = 34$ **25.** $q = 11$ **27.** $x = \frac{7}{30}$ **29.** $a = 11.8$

31. They must apply the same operations to both sides.

$$\begin{array}{rcl} 34 & = & y - 12 \\ +\,12 & & +\,12 \\ \hline 46 & = & y \end{array}$$

33. $x - 8 = 16$; 24th floor

35. Subtract 3 from each side. (Subtraction Property of Equality); Subtract 3 from 3. (Subtract.); Add x and 0. (Addition Property of Zero)

37. $k + 7 = 34$; $k = 27$ **39.** $93 = g + 58$; $g = 35$

41. $y = 15$ **43.** $v = 28$ **45.** $d = 54$

47. $x + 34 + 34 + 16 = 132$; 48 in. **49.** Addition is commutative.

51. Begin by writing the characteristics of each problem.

53. $x + 15.50 + 8.75 = 66.55$; \$42.30

55. **a.** \$5.25

b. no; You have \$5.25 left and it costs \$9.75 to ride each ride once.

57. 96 **59.** 5 **61.** C

Section 7.3 Solving Equations Using Multiplication or Division *(pages 312 and 313)*

1. 12

3. $\frac{4x}{4} = \frac{24}{4}$

5. $8 \cdot 3 = (n \div 3) \cdot 3$

7. $s = 70$

9. $x = 24$

11. $a = 4$

13. $y = 10$

15. $x = 15$

17. $d = 78$

19. $c = 66$

21. $n = 2.56$

23. They should have multiplied by 4.

$$x \div 4 = 28$$
$$(x \div 4) \cdot 4 = 28 \cdot 4$$
$$x = 112$$

25. $3x = 45$; 15 items

27. 9 units

29. 8 units

31. 20 cards

33. $x = 6$; Because $5x$ is on both sides of the equation, $3x$ must be equal to 18 so that the equation is true.

35. length: 20 in.; width: 5 in.

37. $\frac{t}{3} = 7$

Section 7.4 Writing Equations in Two Variables *(pages 319–321)*

1. *Sample answer:* An independent variable can change freely. A dependent variable depends on the independent variable.

3. $n = 4n - 6$; This one is not an equation in two variables.

5. $A = 9h$, where A is the area in square feet and h is the height in feet; A depends on h.

7. yes

9. no

11. yes

13. w is independent and A is dependent.

15. p is independent and t is dependent.

17. $270

19 and 21. Sample answers are given.

	Independent Variable	Dependent Variable
19.	The speed you are pedaling a bike	Time it takes to stop your bike
21.	The number of years of education	The amount of money you earn

23. *Sample answer:* $c = 25m + 35$ where m is the number of months and c is the total cost of the gym membership.

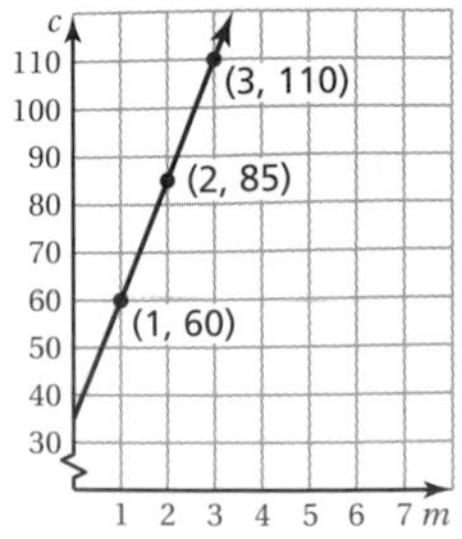

25. Methods to create the graph will vary.

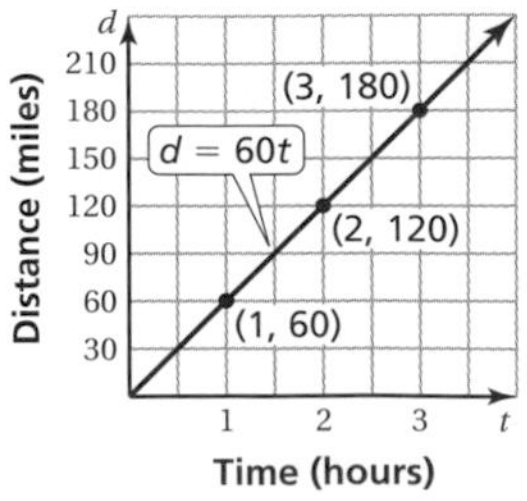

27. $d = \frac{5}{6}t$

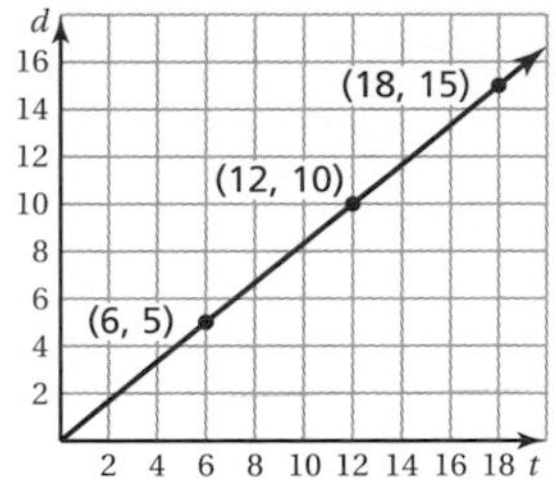

29. $d = 240t$

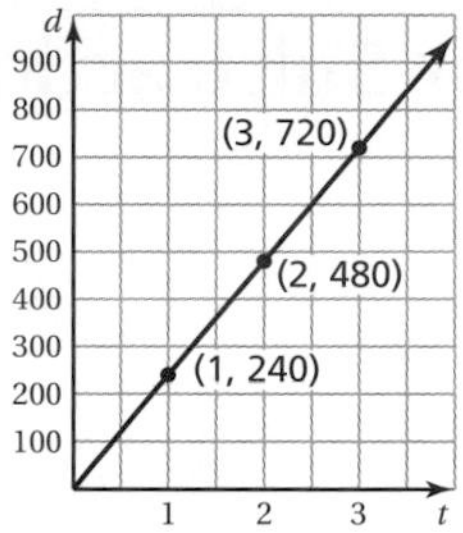

31. 1

33. no; By definition, the independent variable can change freely.

35. 50 city blocks

37. Methods to create the graph will vary.

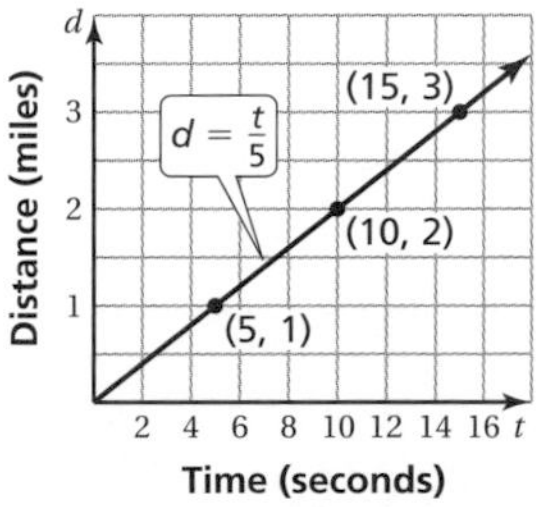

39. **a.** no; It does not make sense to draw a line between the points to show the solutions because you cannot sell part of a ticket.

b. $c = 10n$

41. 80%

43. 68%

Section 7.5 Writing and Graphing Inequalities

(pages 329–331)

1. Both phrases refer to numbers that are greater than a given number. The difference is that "greater than or equal to" includes the number itself, whereas "greater than" does not.

3. The graph of $x \leq 6$ has a closed circle at 6. The graph of $x < 6$ has an open circle at 6.

5. $k < 10$

7. $z < \frac{3}{4}$

9. $1 + y \leq -13$

11. yes

13. yes

15. no

17. B

19. D

21. $x < 1$; A number x is less than 1.

23. $x \geq -4$; A number x is at least -4.

25. 

27. 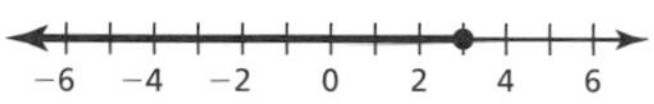

29. 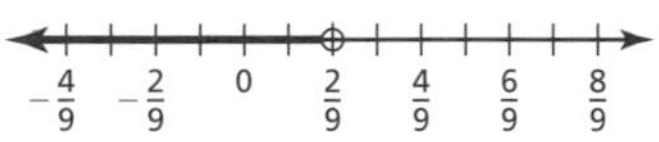

31. 

33. 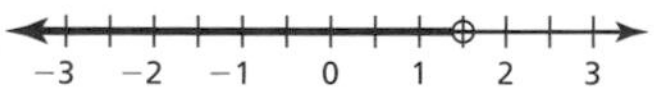

35. 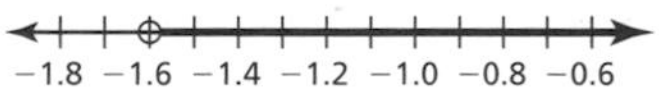

37. $x \geq 1$ means that 1 is also a solution, so a closed circle should be used.

39. **a.** $b \leq 3$; 

b. $\ell \geq 18$; 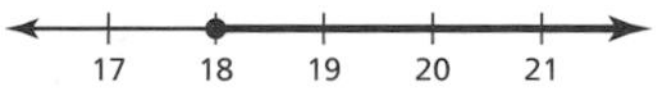

41. The cost of the necklace and another item should be less than or equal to \$33.

43. sometimes; The only time this is not true is if $x = 5$.

45. $p \leq 375$

47. $x = 9$

49. $x = 28$

51. D

Selected Answers

Section 7.6 Solving Inequalities Using Addition or Subtraction (pages 336 and 337)

1. *Sample answer:* $x + 7 \geq 143$

3. By solving the inequality to obtain $x \leq 1$, the graph has a closed circle at 1 and an arrow pointing in the negative direction.

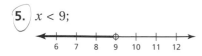

5. $x < 9$;

7. $5 \geq y$;

9. $6 > x$;
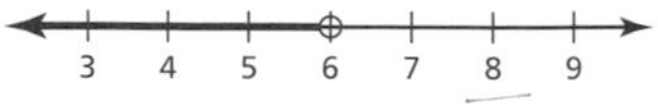

11. $y < 106$;
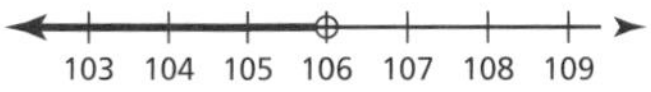

13. $3 < x$;
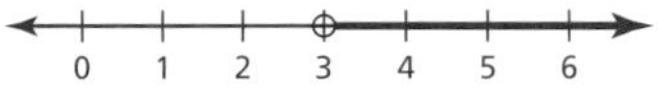

15. $\frac{1}{4} \leq n$;

17. To solve the inequality, 9 should be added to both sides, not subtracted.

$$\begin{array}{r} 28 \geq t - 9 \\ +9 \quad\quad +9 \\ \hline 37 \geq t \end{array}$$

19. $x + 18.99 \leq 24$; $x \leq \$5.01$

21. $x - 3 > 15$; $x > 18$

23. $11 > s$;

25. $34{,}280 + d + 1000 > 36{,}480$; $d > 1200$ dragonflies

27. The estimate for running a mile should be greater than 4 minutes, because the world record is under 4 minutes.

29. $t = 48$

31. $x = 9$

33. C

Section 7.7 Solving Inequalities Using Multiplication or Division (pages 342 and 343)

1. The solution of $2x \geq 10$ includes the solution of $2x = 10$, $x = 5$, and all other x values that are greater than 5.

3. Div. Prop. of Ineq.

5. *Sample answer:* $\frac{x}{2} \geq 4$, $2x \geq 16$

7. $n > 12$;
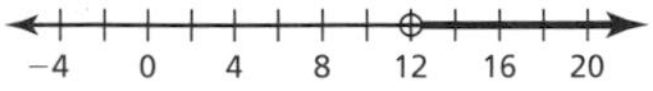

9. $c \geq 99$;
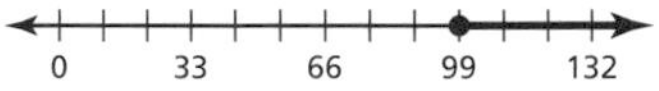

11. $x \geq 5$;

13. $p \leq 6$;
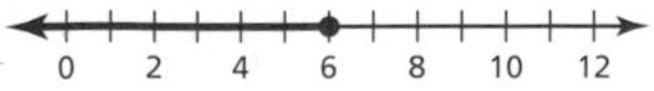

15. $x < 15$;

17. $v \leq 81$;

19. $w \leq 32$;

21. $x \geq 48$;
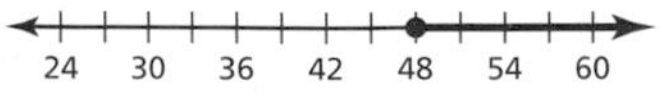

23. $8x < 168$; $x < 21$ ft

25. $8n < 72$; $n < 9$

27. $225 \geq 12w$; $18.75 \geq w$

29.
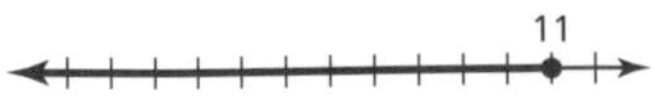

31. $80x > 2 \cdot 272$; $x > 6.8$ yards per play

33. *Sample answer:* the number of gallons of milk you can buy with \$20; the length of a park that has an area of at least 500 square feet

35. yes; $a > b$ and $x > y$

37. yes; $a > b$ and $x > y$

39. rectangle

41. parallelogram

Section 8.1 Three-Dimensional Figures

(pages 358 and 359)

1. false; It has two triangular faces.

3. true

5. false; Some are perpendicular and some are neither (skew).

7. front: 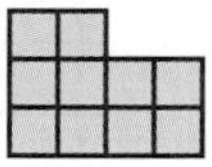 10 cubes

side: 

top: 

9. front: 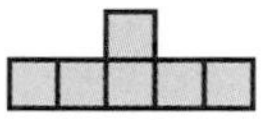 9 cubes

side: 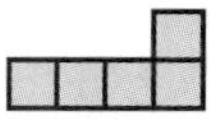

top: 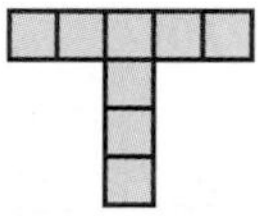

11. 10 faces, 24 edges, and 16 vertices

13. 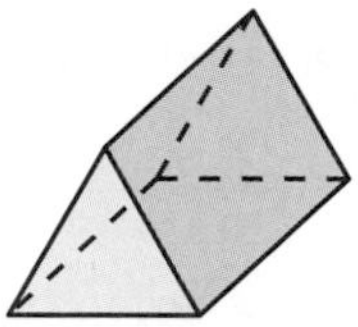

15. 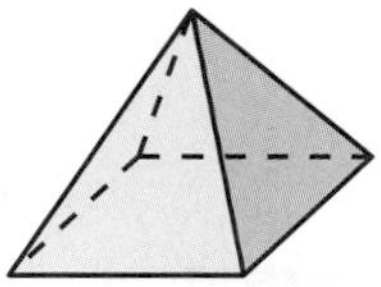

17. front: 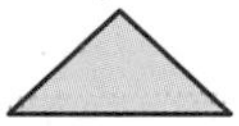

side: 

top: 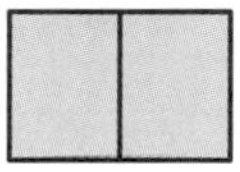

19. front: 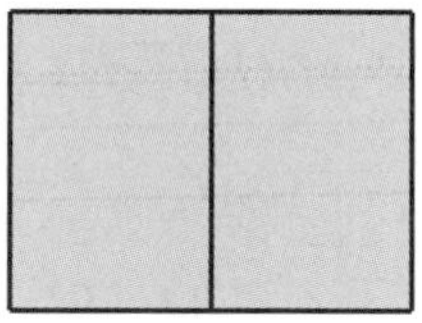

side:  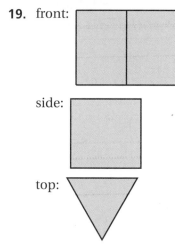

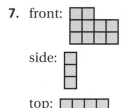

top: 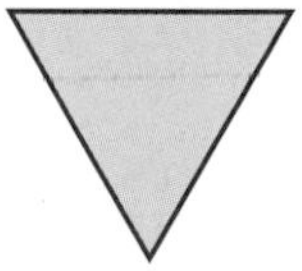

21. front: 

side: 

top: 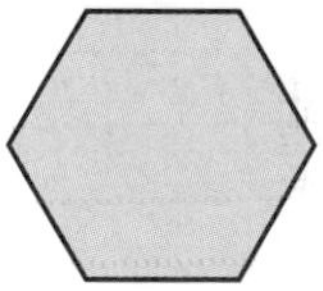

23. 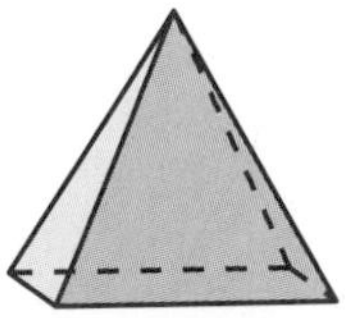

25. 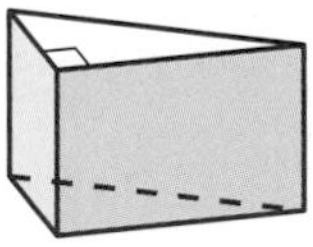

27. *Answer should include, but is not limited to:* an original drawing of a house; a description of any solids that make up any part of the house

29. *Sample answer:*

a. 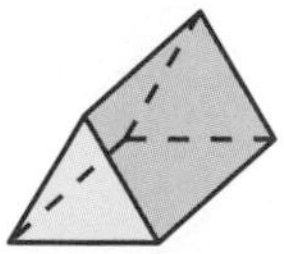

Triangular prism
6 vertices
9 edges

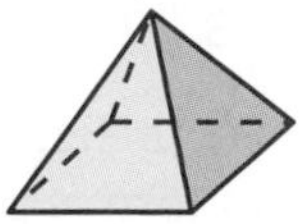

Square pyramid
5 vertices
8 edges

b. More than one solid can have the same number of faces, so knowing the number of edges and vertices can help you draw the intended solid.

31. 12 cm^2

33. D

Section 8.2 Surface Areas of Prisms

(pages 364 and 365)

1. Find the sum of the areas of the faces.

3. 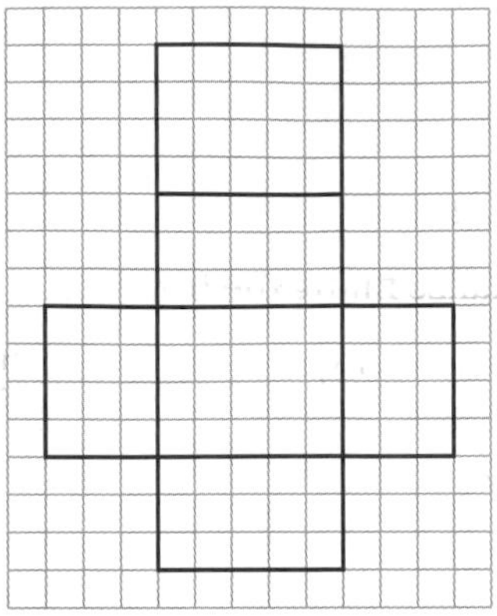 94 units^2

5. 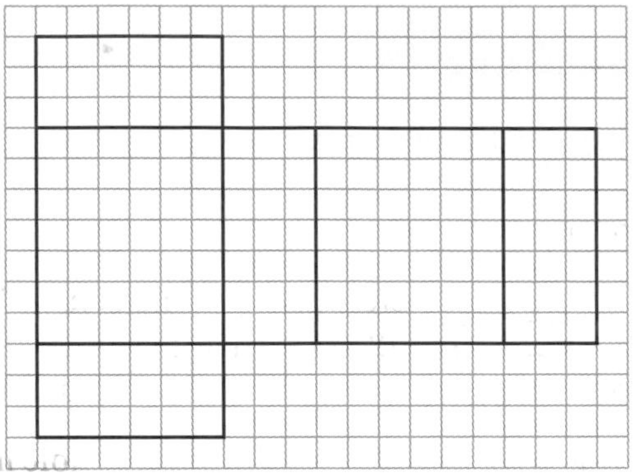 162 units^2

7. 198 cm^2 **9.** 17.6 ft^2 **11.** 57.1 mm^2

13. 136 ft^2 **15.** Draw diagrams of the given information.

17. 364 ft^2 **19.** 165 ft^2 **21.** C

Section 8.3 Surface Areas of Pyramids

(pages 372 and 373)

1. Find the sum of the areas of the faces.

3. 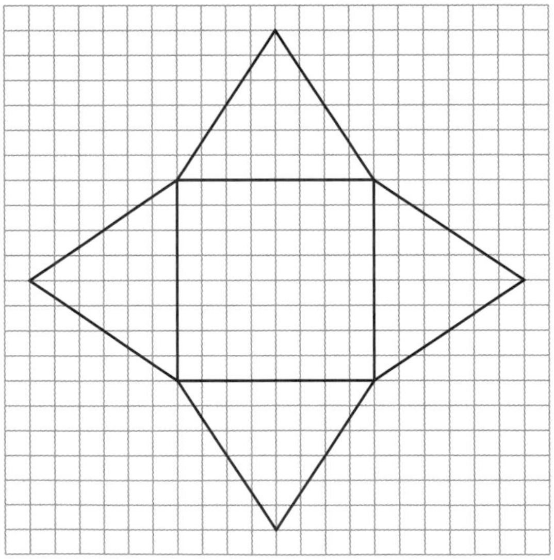

160 units^2

5. 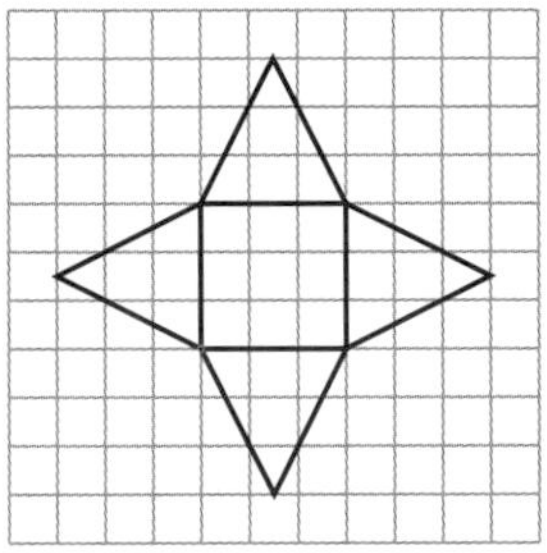

27 units^2

7. 172.8 yd^2 **9.** 224.4 ft^2 **11.** 55 m^2 **13.** 21,274.4 ft^2 **15.** 4

17. no; You can place the four triangles on top of the square and it covers the entire square. But when you lift up the triangles, they do not touch. So, they do not form a pyramid.

19.

Apples	10	5	30
Oranges	4	2	12

10 : 4, 5 : 2, 30 : 12

Section 8.4 Volumes of Rectangular Prisms
(pages 378 and 379)

1. The volume of an object is the amount of space it occupies. The surface area of an object is the sum of the areas of all of its faces.

3. How much does it take to cover the rectangular prism?; 310 cm^2; 350 cm^3

5. $1\frac{5}{16}$ cm^3

7. $\frac{15}{16}$ m^3

9. $12\frac{1}{2}$ m^3

11. $220.5 = 7 \cdot w \cdot 7$; 4.5 cm

13. Use unit cubes to visualize filling the fish tank.

15. 1728 1-inch cubes; There are 1728 1-inch cubes in a cube with a side length of 1 foot. The area of the cube with a side length of 1 foot is 1 cubic foot, or 1728 cubic inches. So, 1 cubic foot is equal to 1728 cubic inches. You can use the conversion factors $\frac{1728 \text{ in.}^3}{1 \text{ ft}^3}$ and $\frac{1 \text{ ft}^3}{1728 \text{ in.}^3}$ to convert between cubic inches and cubic feet.

17. 1152 cm^3

19. yes

21. no

Section 9.1 Introduction to Statistics
(pages 394 and 395)

1. A statistical question is one for which you do not expect to get a single answer. Instead, you expect a variety of answers, and you are interested in the distribution and tendency of those answers. *Sample answer:* How old are the teachers in middle school?

3. yes; There are many different answers.

5. *Sample answer:* 2 pets; no

7. 100 senators; yes

9. not statistical; There is only one answer.

11. statistical; There are many different answers.

13.

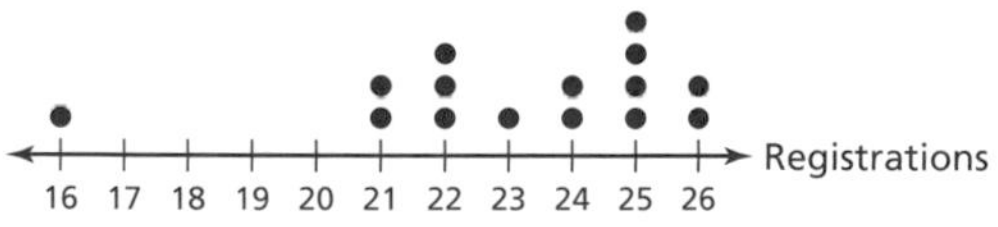

Most of the registrations are in a cluster from 21 to 26. The peak is 25. There is a gap between 16 and 21.

15.

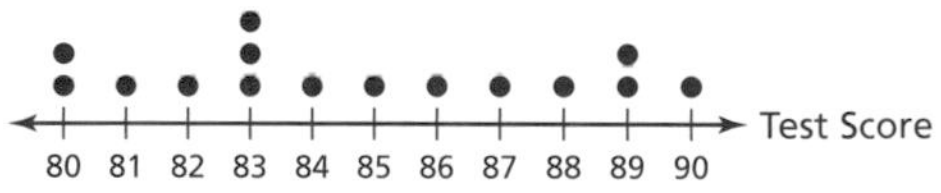

The test scores are spread out pretty evenly with no clusters or gaps. The peak is 83.

17. **a.** 21 earthworms
    b. *Sample answer:* Use a centimeter ruler. The units are centimeters.
    c. *Sample answer:* "What is the length of an earthworm?"; The lengths are spread out pretty evenly from 15 centimeters to 28 centimeters.

19. *Sample answer:* Anemometer; miles per hour

21. *Sample answer:* Richter scale; magnitude

23. *Sample answer:* 65 mi/h; Most of the data cluster around 65, and 65 miles per hour is a common speed limit.

25. Does changing the order of the bars in the bar graph affect the distribution?

27. no

29. D

Selected Answers

Section 9.2 Mean
(pages 400 and 401)

1. Add the data values then divide by the number of data values.

3. yes; Because of the variability of the answers to a statistical question, the mean gives an average of the answers. That way, you can use only one value, the mean, to answer the statistical question.

5. 1 movie seen this week; Find the total number of movies and divide by the number of people.

7. 3 brothers and sisters

9. 16 visits

11. **a.** yes; There will be variability in the lengths of the commercial breaks.

    b. 3.45 minutes

13. *Sample answer:* 20, 21, 21, 21, 21, 22; 20, 20.5, 20.5, 21.5, 21.5, 22

15. 3.9 inches; No, neither team has a height that is much shorter or taller than the other heights. So, you can say that the Tigers are taller than the Dolphins on average.

17. There are 5 different allowance amounts and 24 students. Each amount is used more than once.

19. 9

21. 18.5

23. B

Section 9.3 Measures of Center
(pages 407–409)

1. *Sample answer:* 1, 2, 3, 4, 5, 6

3. outlier; The other three are measures of center.

5. 5.5

7. median: 7; mode: 3

9. median: 92.5; mode: 94

11. median: 17; mode: 12

13. The data were not ordered from least to greatest; The median is 55.
    49, 50, 51, 55, 58, 59, 63

15. singing

17. mean: 35.875; median: 44; mode: 48
    Sample answer: The median is probably best, because it is close to most of the data. The mean is less than most of the data and the mode is the greatest value.

19. mean: 12; median: 8; mode: 2
    Sample answer: The median is the best measure, because the mean is greater than most of the data and the mode is the least value.

21.

With Outlier	*Without Outlier*
mean: 48.5	mean: 53
median: 53	median: 54
mode: none	mode: none

The outlier reduces the median slightly, but reduces the mean more. There is no mode with or without the outlier.

23. mean: 7.61; median: 7.42; no mode

25. **a.** mean: 94°F; median: 91°F; mode: 91°F
Sample answer: Both the median and mode are the best measures for the data, because both are very close to most of the values.

b. mean: 77°F; median: 77°F; modes: 77°F and 78°F
Sample answer: Both the mean and median are the best measures for the data, because there are two modes.

27. 10 hours; Using the mean as the average, you would need to work 12 hours. Using the median as the average, you would need to work 10 or more hours. Using the mode as the average, you would need to work 10 hours. So, the minimum number of hours is 10 and you can use the median or mode to justify your answer.

29. Ordering the data makes it easier to find the median and the mode.

31. **a.** mean: \$1794; median: \$1790; mode: \$1940

b. mean: \$1883.70; median: \$1879.50; mode: \$2037
The mean, median, and mode all increased by 5%.

c. annual salaries: \$23,280, \$19,920, \$22,320, \$25,200, \$20,640, \$18,480, \$21,120, \$23,280, \$21,840, \$19,200; mean: \$21,528; median: \$21,480; mode: \$23,280; They are 12 times the mean, median, and mode of the monthly salary.

33. 13

35. 119

37. D

Section 9.4 Measures of Variation
(pages 416 and 417)

1. A measure of center represents the center of a data set, but a measure of variation describes the distribution of a data set.

3. What is the range of the data?; 20; 12

5. median = 81.5; median of lower half = 67; median of upper half = 92; The data is spread out.

7. 23

9. 7.3

11. median = 37; $Q_1 = 33.5$; $Q_3 = 40.5$; IQR = 7

13. median = 133.5; $Q_1 = 128$; $Q_3 = 139$; IQR = 11

15. range = $21\frac{3}{4}$ ft; The distances traveled by the paper airplane vary by no more than $21\frac{3}{4}$ feet; IQR = 11 ft; The middle half of the distances traveled by the paper airplane vary by no more than 11 feet.

17. Exercise 11: 54
Exercise 12: none
Exercise 13: 106 and 158
Exercise 14: 38

19. **a.** range = 172 points; IQR = 42 points

b. The outlier is 193 points; range = 101; IQR = 34; range

Section 9.4 Measures of Variation *(continued)*
(pages 416 and 417)

21. **a.** Show A: mean = 20, median = 19.5, range = 13, IQR = 5
Show B: mean = 21, median = 20.5, range = 23, IQR = 6

The mean ages for the shows, 20 and 21, and the median ages for the shows, 19.5 and 20.5, are about the same. The interquartile ranges of the ages for the shows, 5 and 6, are about the same. The range of the ages for Show A is 13 years and the range for Show B is 23 years. So, the ages for Show B are more spread out.

b. Show A: The mean of the ages decreases a small amount, from 20 to $19\frac{8}{9}$. The median of the ages decreases from 19.5 to 18. The range of the ages stays at 13. The interquartile ranges of the ages increase from 5 to 6.5. Some of these values do not change by a large amount because 21 is towards the middle of the data set.

Show B: The mean of the ages decreases from 21 to $19\frac{1}{3}$. The median of the ages decreases a small amount, from 20.5 to 20. The range of the ages decreases a large amount, from 23 to 12. The interquartile ranges of the ages increase a small amount, from 6 to 6.5. Some of these values change by a large amount because 36 is an outlier of the data set.

23. 11

25. D

Section 9.5 Mean Absolute Deviation
(pages 422 and 423)

1. All the values in the data set are the same.

3. 2.8 years

5. 4.4; The prices differ from the mean price by an average of $4.40.

7. 4.9; The capacities differ from the mean capacity by an average of 4.9 thousand, or 4900 people.

9. When calculating the mean absolute deviation, you need to divide by 6, not 5. Even though the distance from the mean of one of the values (38) is 0, it is still included in the calculation.

$$\text{mean absolute deviation} = \frac{3 + 2 + 0 + 6 + 4 + 3}{6} = 3$$

So, the values differ from the mean by an average of 3.0.

11. The MAD of the five most-expensive dishes is 3.6. The MAD of the five least-expensive dishes is 1.76. The MAD of the five least-expensive dishes is much less than the MAD of the five most-expensive dishes. So, the data for the five least-expensive dishes is closer together compared to the five most-expensive dishes.

13. **a.** mean: 8.25; median: 8.5; mode: 5
range: 13; IQR: 5.5; MAD: 3

b. no; Using the interquartile range, 21 is inside the outlier boundaries.
mean: 9; median: 9; mode: 5
range: 19; IQR: 6.5; MAD: 3.5

The range is most affected by including this value. The mode stays the same. The mean, median, IQR, and MAD all increased slightly.

15. monthly amounts of water used in a home; *Sample answer:* The amount of rainfall that falls in a city during a month usually ranges from 0 to 5–6 inches. The monthly amounts of water used in a home are much greater numbers that will have more variation from month to month.

17. **a.** 50%; 87.5%; 2 and 15

    b. *Sample answer:* A good portion of a data set is within one MAD of the mean and most of the data set is within 2 MADs of the mean. As you get more and more MADs away from the mean, the percent increases because more and more data are included in the interval.

19. mean: 1.6; median: 1.7; modes: 1.2, 1.7

Section 10.1 Stem-and-Leaf Plots
(pages 438 and 439)

1. 3 is the stem; 4 is the leaf

3. From the leaves, you can see where most of the data lies and whether there are many values that are low or high.

5. **Hours Online**

Stem	Leaf
0	0 2 6 8
1	2 2 4 5 7 8
2	1 4

Key: 2 | 1 = 21 hours

7. **Points Scored**

Stem	Leaf
3	8
4	2 2 3 3 5
5	0 1 6 8 8
6	
7	0 1 1 5

Key: 3 | 8 = 38 points

9. **Minutes in Line**

Stem	Leaf
1	6 9
2	0 2 6 7 9
3	1 1 6 8
4	0

Key: 4 | 0 = 4.0 minutes

11. **Weights**

Stem	Leaf
0	8
1	2 5 7 8
2	4 4
3	1

Key: 2 | 4 = 24 pounds

Most of the weights are in the middle.

13. mean: 56.6; median: 53; modes: 41, 43, 63; range: 56; IQR = 20

15. 97; It increases the mean.

17. **a.** 6.4; The daily high temperatures differ from the mean daily high temperature by an average of 6.4 degrees.

    b. Because the mean absolute deviation increases, most of the data values for the rest of the month must be further from the mean than 6.4 degrees. So, most of the data values for the rest of the month are either less than 71.6 degrees or greater than 84.4 degrees.

19. 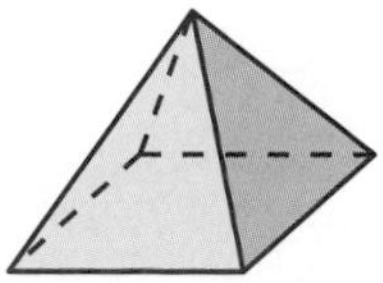

21. B

Selected Answers

Section 10.2 Histograms
(pages 445–447)

1. The *Test Scores* graph is a histogram because the number of students (frequency) achieving the test scores are shown in intervals of the same size (20).

3. No bar is shown on that interval.

5. *Sample answer:*

Interval	Tally	Total
20–29	I	1
30–39	III	3
40–49	~~IIII~~ II	7
50–59	IIII	4

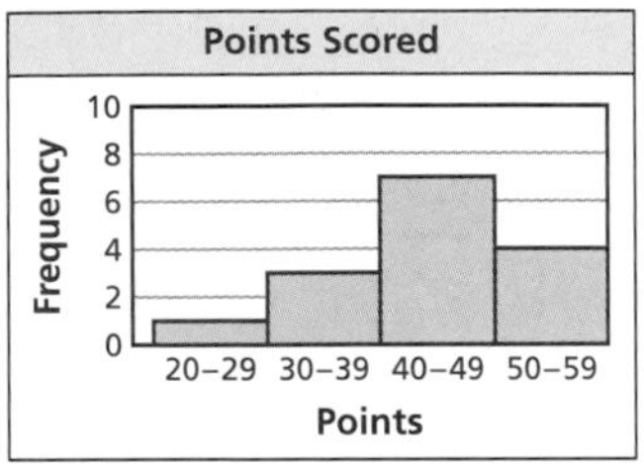

7.

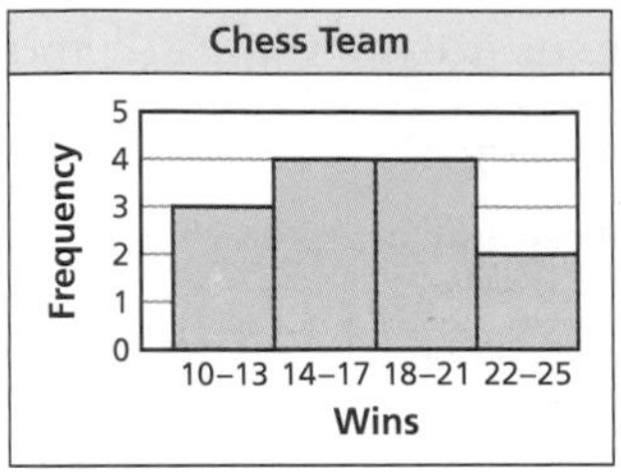

9. There should not be space between the bars of the histogram.

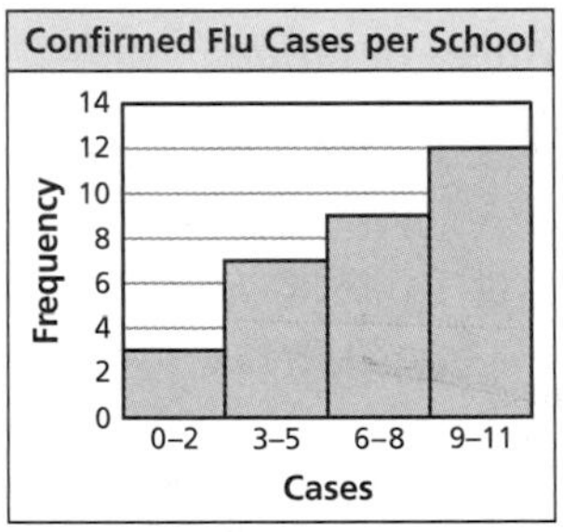

11. The frequency is the number of songs not the percent of songs. The statement should be "12 of the songs took 5–8 seconds to download."

13. Pennsylvania; You can see from the intervals and frequencies that Pennsylvania counties are greater in area, which makes up for it having fewer counties.

15. **a.** yes; The stem-and-leaf plot shows that 10 pounds is a data value.

b. no; Both displays show that 11 residents produced between 20 and 29 pounds of garbage.

17. Begin by ordering the data.

19. 45

21. 22.4

23. D

Section 10.3 Shapes of Distributions
(pages 454 and 455)

1. The shape of a skewed distribution will have a tail on one side. The shape of a symmetric distribution is when the data on the left are a mirror image of the data on the right.

3.

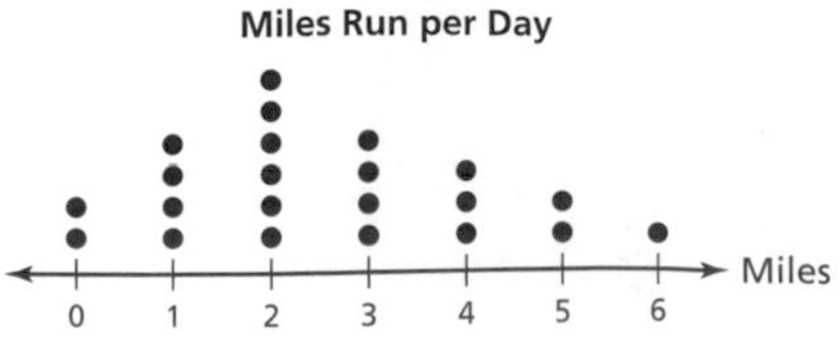

skewed

5. skewed left

7. skewed right

9.

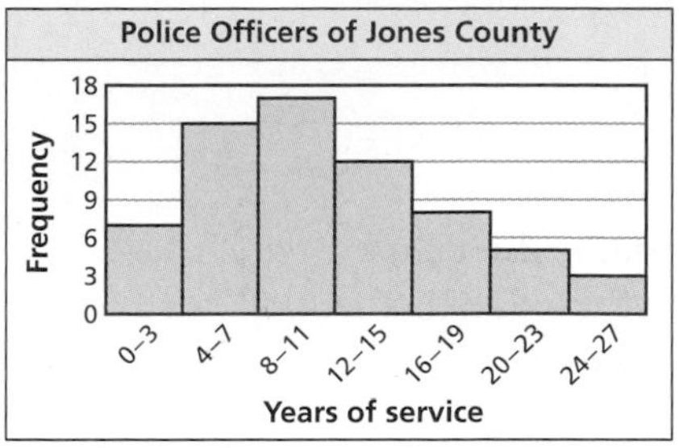

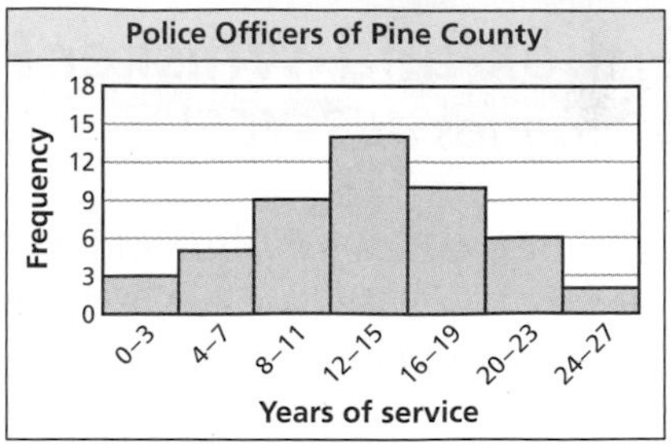

skewed right

symmetric

Jones County; The distribution of Jones County is skewed right, so most of the data values are on the left.

11. no; Distributions can have any shape.

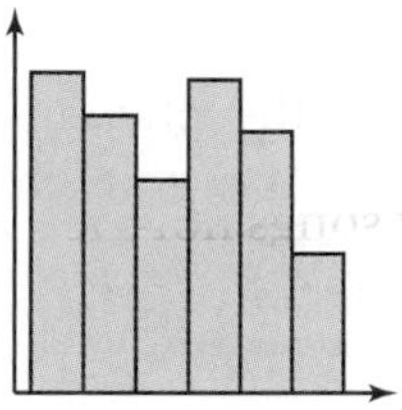

13. **a.**

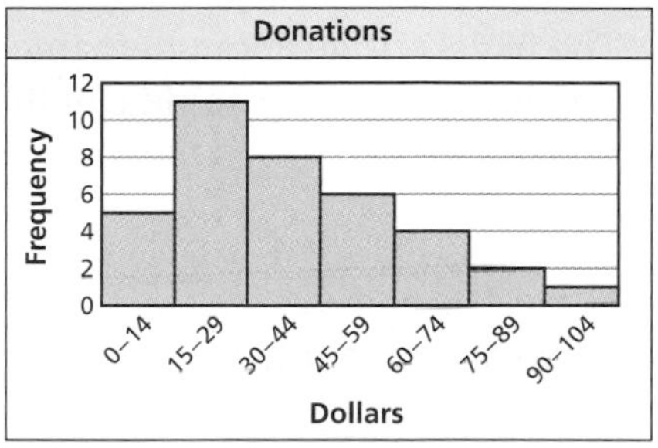

skewed right

b.

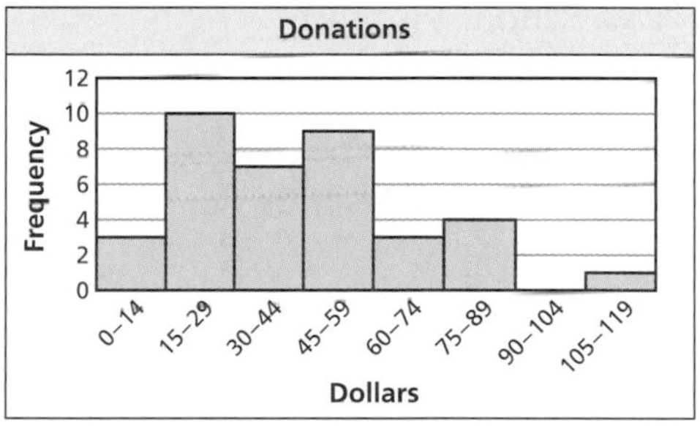

Both distributions are skewed right. The original donation distribution is more skewed right than the distribution when the increases are added to the donations. Some of the data values moved into different intervals when $5 is added to each donation, which is why the distributions are not exactly the same.

15. median = 70; $Q_1 = 65.5$; $Q_3 = 75$; IQR = 9.5

17. A

Extension 10.3 Choosing Appropriate Measures
(page 457)

1. median and interquartile range; median: $32; IQR: $6

3. no; You do not know the actual values in the data set. You can approximate the mean and MAD but your answers will not be exact.

5. *Sample answer*:

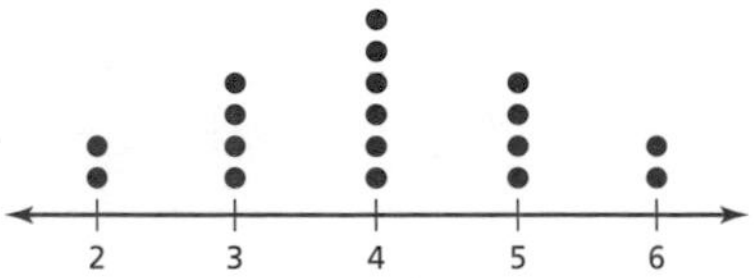

Section 10.4 Box-and-Whisker Plots
(pages 463–465)

1. Order the data. The first number is the *least value* and the last number is the *greatest value.* The middle value is the *median.* The middle value of the lower half of the data is the *first quartile.* The middle value of the upper half of the data is the *third quartile.*

3. Is the distribution skewed right?; yes; no

5. 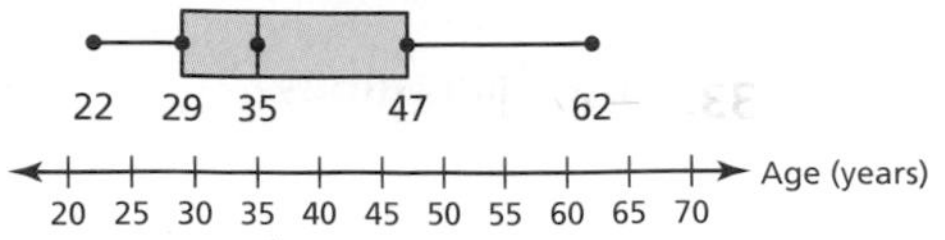

7. 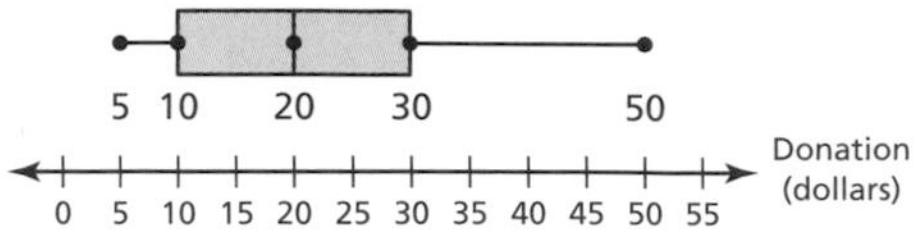

9. The data should be ordered before finding the five-number summary.

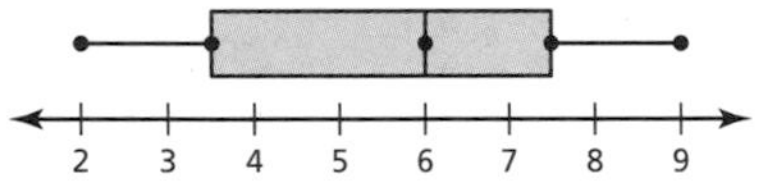

11. **a.** about $\frac{1}{2}$

b. The right whisker is longer than the left whisker. So the data are more spread out above the third quartile than below the first quartile.

c. 150; The middle half of the data varies by no more than 150 gallons.

13. skewed left; The left whisker is longer than the right whisker, and most of the data are on the right.

15. symmetric; The whiskers are about the same length, and the median is in the middle of the box.

17. **a.** School 1 is skewed left and School 2 is skewed right.

b. School 2; The range for School 2 is a half hour greater than the range for School 1. Also, the IQR of School 2 is greater than the IQR of School 1.

c. School 1; School 1 has more data on the left than School 2. So, School 1 is more likely to have recess before lunch.

19. 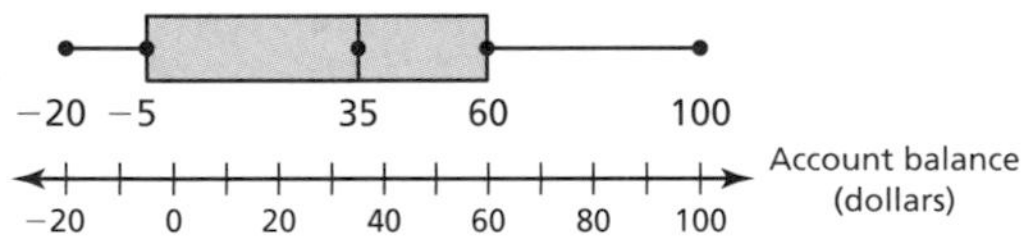

21. Use the median to describe the center and the interquartile range to describe the variation.

23. When the least value and the first quartile are equal, there is no whisker on the left. When the greatest value and the third quartile are equal, there is no whisker on the right.

25. **a.** Team 1; There is less variability in the data.

b. 24 games

c. Team 1; In 75% of the games, Team 1 scored 6 runs or more. However, Team 2 scored 6 runs or less in 75% of the games.

d. Team 1; *Sample answer:* By looking at the shapes of the distributions, you can see that the majority of the data for Team 1 is greater than the majority of the data for Team 2.

27. >

29. >

Section 11.1 Integers and Absolute Value
(pages 480 and 481)

1. 9, -1, 15 **3.** -6; All of the other expressions are equal to 6.

5. 6 **7.** 10 **9.** 13 **11.** 12 **13.** 8 **15.** 18

17. 45 **19.** 125 **21.** $|-4| < 7$ **23.** $|-4| > -6$ **25.** $|5| = |-5|$

27. Because $|-5| = 5$, the statement is incorrect; $|-5| > 4$

29. -8, 5 **31.** -7, -6, $|5|$, $|-6|$, 8 **33.** -17, $|-11|$, $|20|$, 21, $|-34|$

35. -4

37. **a.** MATE; 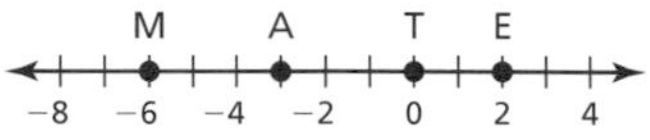

b. TEAM; 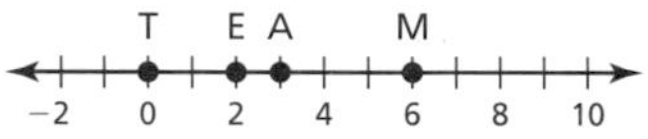

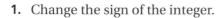

39. $n \geq 0$ **41.** The number closer to 0 is the greater integer.

43. **a.** Player 3 **b.** Player 2 **c.** Player 1

45. false; The absolute value of zero is zero, which is neither positive nor negative.

47. 144 **49.** 3170

Section 11.2 Adding Integers
(pages 486 and 487)

1. Change the sign of the integer. **3.** positive; 20 has the greater absolute value and is positive.

5. negative; The common sign is a negative sign. **7.** false; A positive integer and its absolute value are equal, not opposites.

9. -10 **11.** 7 **13.** 0 **15.** 10

17. -4 **19.** -11 **21.** -4 **23.** -34

25. -10 and -10 are not opposites; $-10 + (-10) = -20$ **27.** \$48

29. Use the Associative Property to add 13 and -13 first; -8

31. *Sample answer:* Use the Commutative Property to switch the last two terms; -12

33. *Sample answer:* Use the Commutative Property to switch the last two terms; 11

35. -27 **37.** 21 **39.** -85

41. *Sample answer:* $-26 + 1$; $-12 + (-13)$ **43.** -3

45. $d = -10$ **47.** $m = -7$

49. Find the number in each row or column that already has two numbers in it before guessing.

51. 8 **53.** 183

Section 11.3 Subtracting Integers
(pages 492 and 493)

1. You add the integer's opposite.
3. What is 3 less than -2?; -5; 5
5. C
7. B
9. 13
11. -5
13. -10
15. 3
17. 17
19. 1
21. -22
23. -20
25. $-3 - 9$
27. 6
29. 9
31. 8
33. $m = 14$
35. $c = 15$
37. 2
39. 3
41. *Sample answer:* $x = -2, y = -1;\ x = -3, y = -2$
43. sometimes; It's positive only if the first integer is greater.
45. always; It's always positive because the first integer is always greater.
47. all values of a and b
49. when a and b have the same sign and $|a| \geq |b|$ or $b = 0$
51. -45
53. 468
55. 2378

Section 11.4 Multiplying Integers
(pages 500 and 501)

1. **a.** They are the same. **b.** They are different.
3. negative; different signs
5. negative; different signs
7. false; The product of the first two negative integers is positive. The product of the positive result and the third negative integer is negative.
9. -21
11. 12
13. 27
15. 12
17. 0
19. -30
21. 78
23. 121
25. $-240{,}000$
27. 54
29. -700
31. 0
33. -1
35. -36
37. 54
39. The answer should be negative; $-10^2 = -(10 \cdot 10) = -100$
41. 32
43. -7500, 37,500
45. -12
47. **a.**

Month	Price of Skates
June	165 = \$165
July	165 + (−12) = \$153
August	165 + 2(−12) = \$141
September	165 + 3(−12) = \$129

b. The price drops \$12 every month.

c. no; yes; In August you have \$135 but the cost is \$141. In September you have \$153 and the cost is only \$129.

49. 3
51. 14
53. D

Section 11.5 Dividing Integers
(pages 506 and 507)

1. They have the same sign; They have different signs; The dividend is zero.

3. *Sample answer:* $-4, 2$

5. negative

7. negative

9. -3

11. 6

13. 0

15. -6

17. 7

19. -10

21. undefined

23. 12

25. The quotient should be 0; $0 \div (-5) = 0$

27. 15 pages

29. -8

31. 65

33. 5

35. 4

37. -400 ft/min

39. 5

41. *Sample answer:* $-20, -15, -10, -5, 0$; Start with -10, then pair -15 with -5 and -20 with 0.

43. 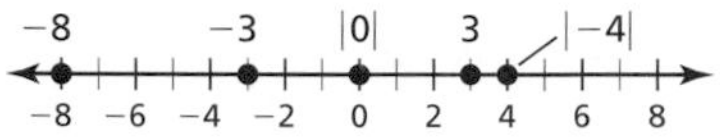

45. B

Section 12.1 Rational Numbers
(pages 522 and 523)

1. no; The denominator cannot be 0.

3. rational numbers, integers

5. rational numbers, integers, whole numbers

7. repeating

9. terminating

11. 0.875

13. $-0.\overline{7}$

15. $1.8\overline{3}$

17. $-5.58\overline{3}$

19. The bar should be over both digits to the right of the decimal point; $-\frac{7}{11} = -0.\overline{63}$

21. $\frac{9}{20}$

23. $-\frac{39}{125}$

25. $-1\frac{16}{25}$

27. $-12\frac{81}{200}$

29. $-2.5, -1.1, -\frac{4}{5}, 0.8, \frac{9}{5}$

31. $-\frac{9}{4}, -0.75, -\frac{6}{10}, \frac{5}{3}, 2.1$

33. $-2.4, -2.25, -\frac{11}{5}, \frac{15}{10}, 1.6$

35. spotted turtle

37. $-1.82 < -1.81$

39. $-4\frac{6}{10} > -4.65$

41. $-2\frac{13}{16} < -2\frac{11}{14}$

43. Michelle

45. no; The base of the skating pool is at -10 feet, which is deeper than $-9\frac{5}{6}$ feet.

47. **a.** when a is negative
    b. when a and b have the same sign, $a \neq 0 \neq b$

49. $\frac{7}{30}$

51. 21.15

Section 12.2 Adding Rational Numbers
(pages 528 and 529)

1. Because $|-8.46| > |5.31|$, subtract $|5.31|$ from $|-8.46|$ and the sign is negative.

3. What is the distance between -4.5 and 3.5?; 8; -1

5. $-1\frac{4}{5}$ **7.** $-\frac{5}{14}$ **9.** $-\frac{7}{12}$ **11.** 1.844

13. The decimals are not lined up correctly; Line up the decimals; -3.95

15. $-1\frac{5}{12}$ **17.** $1\frac{5}{12}$ **19.** $\frac{1}{2}$ **21.** $-9\frac{3}{4}$

23. The sum is an integer when the sum of the fractional parts of the numbers adds up to an integer.

25. less than; The water level for the three-month period compared to the normal level is $-1\frac{7}{16}$.

27. no; This is only true when a and b have the same sign.

29. Commutative Property of Addition; 7 **31.** Associative Property of Addition; $1\frac{1}{8}$

33. A

Section 12.3 Subtracting Rational Numbers
(pages 536 and 537)

1. Instead of subtracting, add the opposite of $\frac{3}{5}$, $-\frac{3}{5}$. Then, add $\left|-\frac{4}{5}\right|$ and $\left|-\frac{3}{5}\right|$, and the sign is negative.

3. $1\frac{1}{2}$ **5.** -3.5 **7.** $-18\frac{13}{24}$ **9.** -2.6

11. 14.963 **13.** $3\frac{1}{4}$ **15.** $3\frac{1}{3}$

17. **a.** 410.7 feet **b.** 136.9 feet per hour **19.** 1.2

21. The difference is an integer when (1) the decimals have the same sign and the digits to the right of the decimal point are the same, or (2) the decimals have different signs and the sum of the decimal parts of the numbers add up to 1.

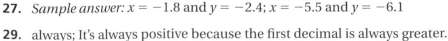

23. $-1\frac{7}{8}$ miles **25.** Subtract the least number from the greatest number.

27. *Sample answer:* $x = -1.8$ and $y = -2.4$; $x = -5.5$ and $y = -6.1$

29. always; It's always positive because the first decimal is always greater.

31. 35.88 **33.** $8\frac{2}{3}$ **35.** C

Section 12.4 Multiplying and Dividing Rational Numbers
(pages 542 and 543)

1. The same rules for signs of integers are applied to rational numbers.

3. positive **5.** negative **7.** $-\frac{4}{5}$ **9.** 0.25

11. $-\frac{2}{3}$ **13.** $-\frac{1}{100}$ **15.** $2\frac{5}{14}$ **17.** $3.\overline{63}$

19. -6 **21.** -2.5875 **23.** $-\frac{4}{9}$ **25.** 9

27. 0.025 **29.** $8\frac{1}{4}$

31. The answer should be negative; $-2.2 \times 3.7 = -8.14$

33. $-66°$ **35.** -19.59 **37.** -22.667 **39.** $-5\frac{11}{24}$

41. *Sample answer:* $-\frac{9}{10}, \frac{2}{3}$ **43.** $3\frac{5}{8}$ gal **45.** -1.28 sec

47. -1.5 **49.** $4\frac{1}{2}$ **51.** D

Section 13.1 Algebraic Expressions
(pages 558 and 559)

1. Terms of an expression are separated by addition. Rewrite the expression as $3y + (-4) + (-5y)$. The terms in the expression are $3y$, -4, and $-5y$.

3. no; The like terms $3x$ and $2x$ should be combined.

$$3x + 2x - 4 = (3 + 2)x - 4$$
$$= 5x - 4$$

5. Terms: t, 8, $3t$; Like terms: t and $3t$ **7.** Terms: $2n$, $-n$, -4, $7n$; Like terms: $2n$, $-n$, and $7n$

9. Terms: $1.4y$, 5, -4.2, $-5y^2$, z; Like terms: 5 and -4.2

11. $2x^2$ is not a like term because x is squared. The like terms are $3x$ and $9x$.

13. $11x + 2$ **15.** $-2.3v - 5$ **17.** $3 - \frac{1}{2}y$ **19.** $-p - 30$

21. $10.2x$; The weight carried by each hiker is 10.2 pounds. **23.** yes; Both expressions simplify to $11x^2 + 3y$. **25.** $(9 + 3x)$ ft^2

27. *Sample answer:*

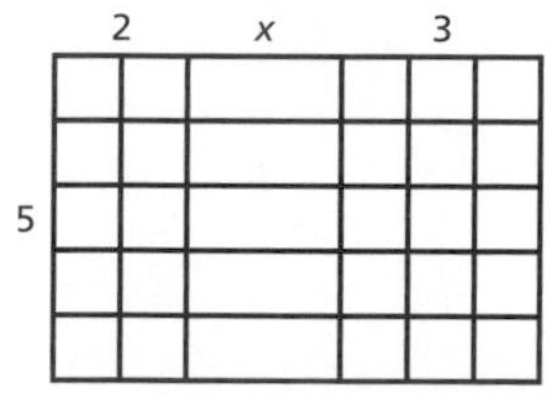

$5x + 25$

29. When you subtract the two red strips, you subtract their intersection twice. So, you need to add it back into the expression once.

31. 0.52 m, 0.545 m, 0.55 m, 0.6 m, 0.65 m

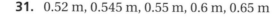

Section 13.2 Adding and Subtracting Linear Expressions *(pages 564 and 565)*

1. not linear; An exponent of a variable is not equal to 1.

3. not linear; An exponent of a variable is not equal to 1.

5. What is x more than $3x - 1$?; $4x - 1$; $2x - 1$

7. *Sample answer:* $(2x + 7) - (2x - 4) = 11$

9. $2b + 9$

11. $6x - 18$

13. 17

15. $m + 1$

17. $55w + 145$

19. $-3g - 4$

21. $-12y + 20$

23. $-2c$

25. The -3 was not distributed to both terms inside the parentheses.

$$\begin{aligned}(4m + 9) - 3(2m - 5) &= 4m + 9 - 6m + 15\\ &= 4m - 6m + 9 + 15\\ &= -2m + 24\end{aligned}$$

27. no; If the variable terms are opposites, the sum is a numerical expression.

29. $0.25x + 0.15$

31. $|x - 3|$, or equivalently $|-x + 3|$; 0; 6

33. $\frac{2}{5}$

35. D

Extension 13.2 Factoring Expressions *(page 567)*

1. $3(3 + 7)$

3. $2(4x + 1)$

5. $4(5z - 2)$

7. $4(9a + 4b)$

9. $\frac{1}{3}(b - 1)$

11. $2.2(x + 2)$

13. $-\frac{1}{2}(x - 12)$

15. $(3x - 8)$ ft

17. *Sample answer:* $2x - 1$ and x, $2x$ and $x - 1$

Section 13.3 Solving Equations Using Addition or Subtraction *(pages 574 and 575)*

1. Subtraction Property of Equality

3. No, $m = -8$ not -2 in the first equation.

5. $a = 19$

7. $k = -20$

9. $c = 3.6$

11. $q = -\frac{1}{6}$

13. $g = -10$

15. $y = -2.08$

17. $q = -\frac{7}{18}$

19. $w = -1\frac{13}{24}$

21. The 8 should have been subtracted rather than added.

$$\begin{aligned}x + 8 &= 10\\ -8 \quad & \;\; -8\\ \hline x &= 2\end{aligned}$$

23. $c + 10 = 3$; $c = -7$

25. $p - 6 = -14$; $p = -8$

27. $p + 2.54 = 1.38$; $-\$1.16$ million

29. $x + 8 = 12$; 4 cm

31. $x + 22.7 = 34.6$; 11.9 ft

33. Because your first jump is higher, your second jump went a farther distance than your first jump.

35. $m + 30.3 + 40.8 = 180$; 108.9°

37. -9

39. 6, -6

41. -56

43. -9

45. B

Section 13.4 Solving Equations Using Multiplication or Division *(pages 580 and 581)*

1. Multiplication is the inverse operation of division, so it can undo division.

3. dividing by 5

5. multiplying by -8

7. $h = 5$

9. $n = -14$

11. $m = -2$

13. $x = -8$

15. $p = -8$

17. $n = 8$

19. $g = -16$

21. $f = 6\frac{3}{4}$

23. They should divide by -4.2.

$$-4.2x = 21$$

$$\frac{-4.2x}{-4.2} = \frac{21}{-4.2}$$

$$x = -5$$

25. $\frac{2}{5}x = \frac{3}{20}$; $x = \frac{3}{8}$

27. $\frac{x}{-1.5} = 21$; $x = -31.5$

29. $\frac{x}{30} = 12\frac{3}{5}$; 378 ft

31 and 33. Sample answers are given.

31. **a.** $-2x = 4.4$ **b.** $\frac{x}{1.1} = -2$

33. **a.** $4x = -5$ **b.** $\frac{x}{5} = -\frac{1}{4}$

35. $-1.26n = -10.08$; 8 days

37. -50 ft

39. $-5, 5$

41. -7

43. 12

45. B

Section 13.5 Solving Two-Step Equations *(pages 586 and 587)*

1. Eliminate the constants on the side with the variable. Then solve for the variable using either division or multiplication.

3. D

5. A

7. $b = -3$

9. $t = -4$

11. $g = 4.22$

13. $p = 3\frac{1}{2}$

15. $h = -3.5$

17. $y = -6.4$

19. Each side should be divided by -3, not 3.

$$-3x + 2 = -7$$

$$-3x = -9$$

$$\frac{-3x}{-3} = \frac{-9}{-3}$$

$$x = 3$$

Section 13.5 Solving Two-Step Equations *(continued)* *(pages 586 and 587)*

21. $a = 1\frac{1}{3}$ **23.** $b = 13\frac{1}{2}$ **25.** $v = -\frac{1}{30}$

27. $2.5 + 2.25x = 9.25$; 3 games

29. $v = -5$ **31.** $d = -12$ **33.** $m = -9$

35. *Sample answer:* You travel halfway up a ladder. Then you climb down two feet and are 8 feet above the ground. How long is the ladder? $x = 20$

37. the initial fee

39. Find the number of insects remaining and then find the number of insects you caught.

41. decrease the length by 10 cm; $2(25 + x) + 2(12) = 54$

43. $-6\frac{2}{3}$ **45.** 6.2

Section 14.1 Ratios and Rates *(pages 603–605)*

1. It has a denominator of 1.

3. *Sample answer:* A basketball player runs 10 feet down the court in 2 seconds.

5. \$0.10 per fluid ounce **7.** \$72 **9.** 870 MB

11. $\frac{5}{9}$ **13.** $\frac{7}{3}$ **15.** $\frac{4}{3}$

17. 60 miles per hour **19.** \$2.40 per pound **21.** 54 words per minute

23. 4.5 servings per package **25.** 4.8 MB per minute **27.** 280 square feet per hour

29. no; Although the relative number of boys and girls are the same, the two ratios are inverses.

31. \$51

33. 2 cups of juice concentrate, 16 cups of water

35. **a.** rest: 72 beats per minute
running: 150 beats per minute
b. 234 beats

37. Try searching for "fire hydrant colors."

39. **a.** you; $\frac{1}{3}$ mile per hour faster
b. $3\frac{1}{2}$ hours
c. you; $1\frac{1}{6}$ miles

41. <

43. B

Section 14.2 Proportions
(pages 610 and 611)

1. Both ratios are equal.

3. *Sample answer:* $\frac{6}{10}, \frac{12}{20}$

5. yes

7. no

9. yes

11. no

13. no

15. yes

17. no

19. yes

21. yes; Both can do 45 sit-ups per minute.

23. yes

25. yes

27. yes; The ratio of height to base for both triangles is $\frac{4}{5}$.

29. Organize the information by using a table.

31. no; The ratios are not equivalent; $\frac{13}{19} \neq \frac{14}{20} \neq \frac{15}{21}$ etc.

33. -13

35. -18

37. D

Extension 14.2 Graphing Proportional Relationships
(pages 612 and 613)

1. 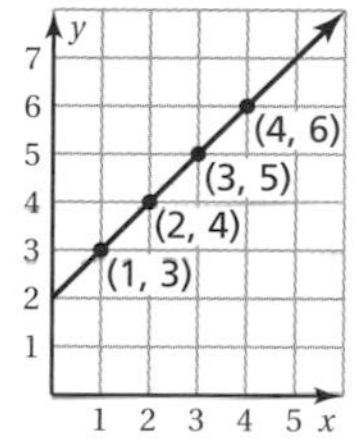

no

3. (0, 0): You earn \$0 for working 0 hours.

   (1, 15): You earn \$15 for working 1 hour; unit rate: $\frac{\$15}{1\text{ h}}$

   (4, 60): You earn \$60 for working 4 hours; unit rate: $\frac{\$60}{4\text{ h}} = \frac{\$15}{1\text{ h}}$

5. yes; 5 ft/h

7. $y = \frac{4}{3}$

Section 14.3 Writing Proportions
(pages 618 and 619)

1. You can use the columns or the rows of the table to write a proportion.

3. *Sample answer:* $\frac{x}{12} = \frac{5}{6}$; $x = 10$

5. $\frac{x}{50} = \frac{78}{100}$

7. $\frac{x}{150} = \frac{96}{100}$

9. $\frac{n\text{ winners}}{85\text{ entries}} = \frac{34\text{ winners}}{170\text{ entries}}$

11. $\frac{100\text{ meters}}{x\text{ seconds}} = \frac{200\text{ meters}}{22.4\text{ seconds}}$

13. $\frac{\$24}{3\text{ shirts}} = \frac{c}{7\text{ shirts}}$

15. $\frac{5\text{ 7th grade swimmers}}{16\text{ swimmers}} = \frac{s\text{ 7th grade swimmers}}{80\text{ swimmers}}$

17. $y = 16$

19. $c = 24$

21. $g = 14$

Section 14.3 Writing Proportions *(continued)*
(pages 618 and 619)

23. $\frac{1}{200} = \frac{19.5}{x}$; Dimensions for the model are in the numerators and the corresponding dimensions for the actual space shuttle are in the denominators.

25. Draw a diagram of the given information.

27. $x = 9$

29. $x = 140$

Section 14.4 Solving Proportions
(pages 626 and 627)

1. mental math; Multiplication Property of Equality; Cross Products Property

3. yes; Both cross products give the equation $3x = 60$.

5. $h = 80$

7. $n = 15$

9. $y = 7\frac{1}{3}$

11. $k = 5.6$

13. $n = 10$

15. $d = 5.76$

17. $m = 20$

19. $d = 15$

21. $k = 5.4$

23. 108 pens

25. $x = 1.5$

27. $k = 4$

29. $\frac{2.5}{x} = \frac{1}{0.26}$; about 0.65

31. true; Both cross products give the equation $3a = 2b$.

33. 15.5 lb

35. no; The relationship is not proportional. It should take more people less time to build the swing set.

37. 4 bags

39 and 41.

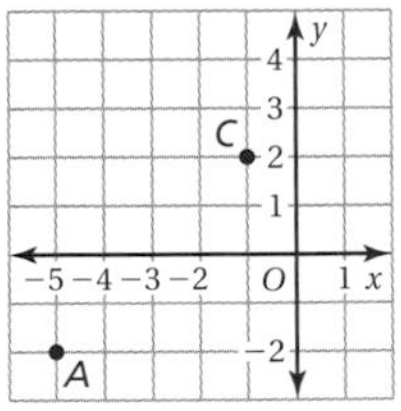

43. D

Section 14.5 Slope
(pages 632 and 633)

1. yes; Slope is the rate of change of a line.

3. 5; A ramp with a slope of 5 increases 5 units vertically for every 1 unit horizontally. A ramp with a slope of $\frac{1}{5}$ increases 1 unit vertically for every 5 units horizontally.

5. $\frac{3}{2}$

7. 1

9. $\frac{4}{5}$

11. 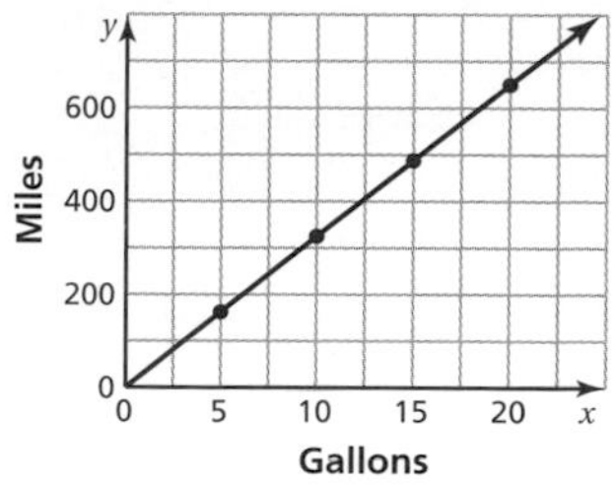

slope = 32.5;
32.5 miles per gallon

13. 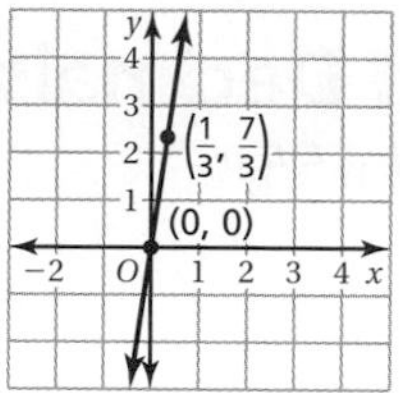

slope = 7

15. 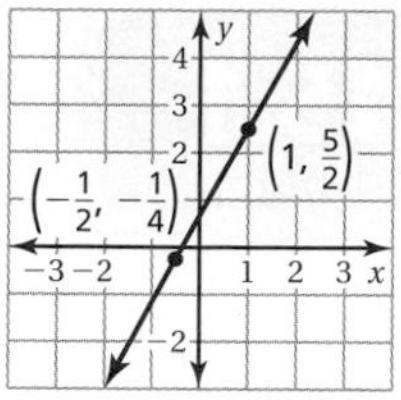

slope $= \frac{11}{6}$

17. a. 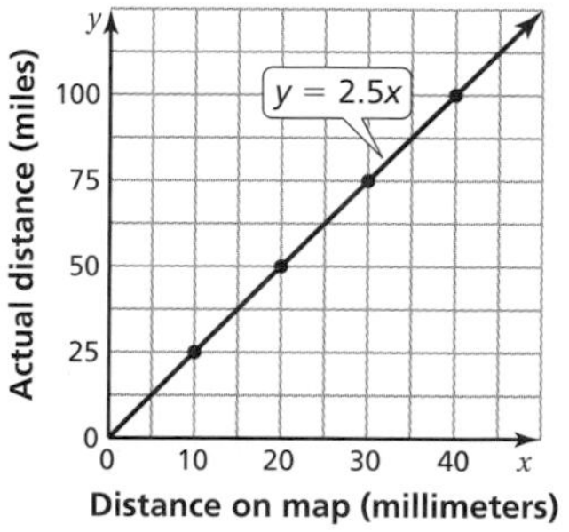

b. 2.5; Every millimeter represents 2.5 miles.

c. 120 mi

d. 90 mm

19. $y = 6$

21. $-\frac{3}{5}$

23. C

Section 14.6 Direct Variation

(pages 638 and 639)

1. $y = kx$, where k is a number and $k \neq 0$.

3. Is the graph of the relationship a line?; yes; no

5. 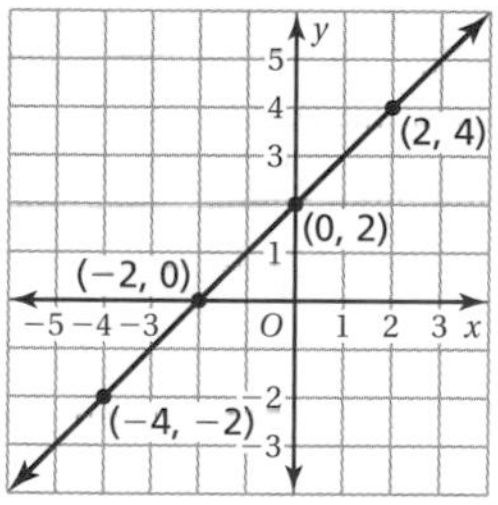

no; The line does not pass through the origin.

7. no; The line does not pass through the origin.

9. yes; The line passes through the origin; $k = \frac{2}{3}$

11. yes; The equation can be written as $y = kx$; $k = \frac{5}{2}$

13. no; The equation cannot be written as $y = kx$.

15. yes; The equation can be written as $y = kx$; $k = \frac{1}{2}$

17. no; The equation cannot be written as $y = kx$.

19. 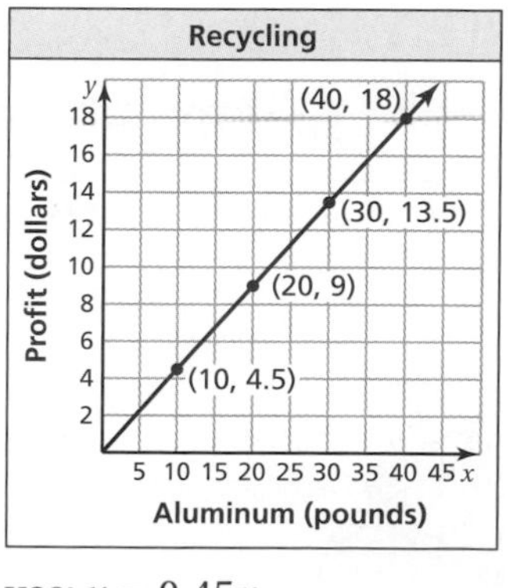

yes; $y = 0.45x$

21. $k = \frac{5}{3}$; $y = \frac{5}{3}x$

23. $y = 2.54x$

25. When $x = 0$, $y = 0$. So, the graph of a proportional relationship always passes through the origin.

27. no

29. Every graph of direct variation is a line; however, not all lines show direct variation because the line must pass through the origin.

31. 0.5625

33. 0.96

Section 15.1 Percents and Decimals
(pages 654 and 655)

1. B

3. C

5. *Sample answer:* 0.11, 0.13, 0.19

7. 0.78

9. 0.185

11. 0.33

13. 0.4763

15. 1.66

17. 0.0006

19. 74%

21. 89%

23. 99%

25. 48.7%

27. 368%

29. 3.71%

31. The decimal point was moved in the wrong direction. 0.86 = 0.86 = 86%

33. 34%

35. $\frac{9}{25} = 0.36$

37. $\frac{203}{1250} = 0.1624$

39. 40%

41. **a.** $16.\overline{6}\%$ or $16\frac{2}{3}\%$ **b.** $\frac{5}{6}$

43. $\frac{31}{100}$

45. $4\frac{8}{25}$

47. $-1.6n - 1$

49. $-b - \frac{3}{2}$

Section 15.2 Comparing and Ordering Fractions, Decimals, and Percents *(pages 660 and 661)*

1.

Fraction	Decimal	Percent
$\frac{18}{25}$	0.72	72%
$\frac{17}{20}$	0.85	85%
$\frac{13}{50}$	0.26	26%
$\frac{31}{50}$	0.62	62%
$\frac{9}{20}$	0.45	45%

3. 0.04; 0.04 = 4%, but 40%, $\frac{2}{5}$, and 0.4 are all equal to 40%.

5. 20%

7. $\frac{13}{25}$

9. 76%

11. 0.12

13. 140%

15. 80%

17.

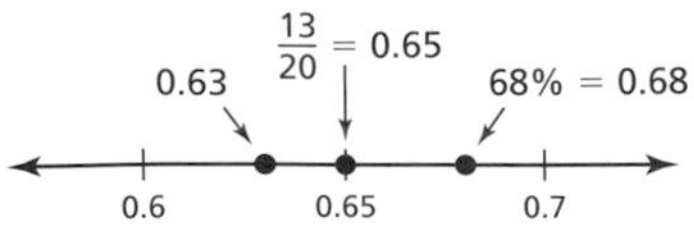

19.

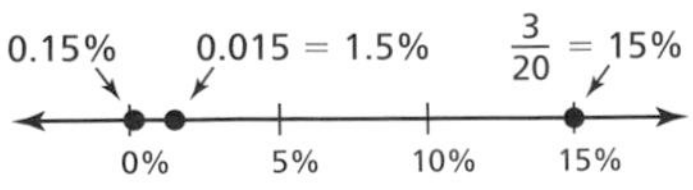

21.

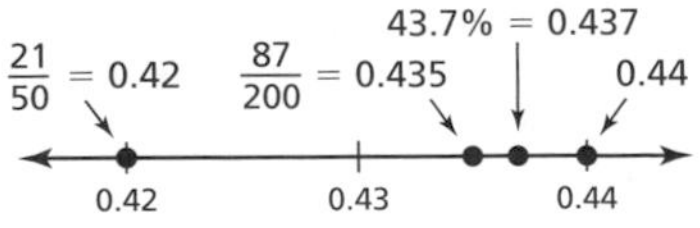

23. Russia, Brazil, United States, India

25. 21%, $0.2\overline{1}$, $\frac{11}{50}$, $\frac{2}{9}$

27. *D*

29. *C*

31. Write the numbers as percents or decimals to make the ordering easier.

33. yes

35. yes

Section 15.3 The Percent Proportion
(pages 666 and 667)

1. The percent proportion is $\frac{a}{w} = \frac{p}{100}$ where a is part of the whole w, and p%, or $\frac{p}{100}$, is the percent.

3. $\frac{a}{40} = \frac{60}{100}$; $a = 24$

5. 19.2

7. 50

9. about 38.5%

11. $\frac{12}{25} = \frac{p}{100}$; $p = 48$

13. $\frac{9}{w} = \frac{25}{100}$; $w = 36$

15. $\frac{a}{124} = \frac{75}{100}$; $a = 93$

17. $\frac{a}{40} = \frac{0.4}{100}$; $a = 0.16$

19. 34 represents the part, not the whole.

$$\frac{a}{w} = \frac{p}{100}$$

$$\frac{34}{w} = \frac{40}{100}$$

$$w = 85$$

21. $6000

23. $\frac{14.2}{w} = \frac{35.5}{100}$; $w = 40$

25. $\frac{a}{\frac{7}{8}} = \frac{25}{100}$; $a = \frac{7}{32}$

27. $8.40

29. **a.** a scale along the vertical axis

b. 6.25%; *Sample answer:* Although you do not know the actual number of votes, you can visualize each bar as a model with the horizontal lines breaking the data into equal parts. The sum of all the parts is 16. Greg has the least parts with 1, which is 100% ÷ 16 = 6.25%.

c. 31 votes

31. **a.** 62.5% **b.** $52x$

33. −0.6

35. B

Section 15.4 The Percent Equation
(pages 672 and 673)

1. A part of the whole is equal to a percent times the whole.

3. 55 is 20% of what number?; 275; 11

5. 37.5%

7. 84

9. 64

11. $45 = p \cdot 60$; 75%

13. $0.008 \cdot 150$; 1.2

15. $12 = 0.005 \cdot w$; 2400

17. $102 = 1.2 \cdot w$; 85

19. 30 represents the part of the whole.

$$30 = 0.6 \cdot w$$

$$50 = w$$

21. $5400

23. 26 years old

25. 56 signers

Selected Answers

Section 15.4 The Percent Equation *(continued)* *(pages 672 and 673)*

27. If the percent is less than 100%, the percent of a number is less than the number; 50% of 80 is 40; If the percent is equal to 100%, the percent of a number will equal the number; 100% of 80 is 80; If the percent is greater than 100%, the percent of a number is greater than the number; 150% of 80 is 120.

29. Remember when writing a proportion that either the units are the same on each side of the proportion, or the numerators have the same units and the denominators have the same units.

31. 92%

33. 0.88

35. 0.36

Section 15.5 Percents of Increase and Decrease *(pages 680 and 681)*

1. If the original amount decreases, the percent of change is a percent of decrease. If the original amount increases, the percent of change is a percent of increase.

3. The new amount is now 0.

5. 24 L

7. 17 penalties

9. decrease; 66.7%

11. increase; 225%

13. decrease; 12.5%

15. decrease; 37.5%

17. 12.5% decrease

19. **a.** about 16.7%

b. 280 people; To get the same percent error, the amount of error needs to be the same. Because your estimate was 40 people below the actual attendance, an estimate of 40 people above the actual attendance will give the same percent error.

21. decrease; 25%

23. decrease; 70%

25. **a.** about 16.95% increase

b. 161,391 people

27. 15.6 ounces; 16.4 ounces

29. less than; *Sample answer:* Let x represent the number. A 10% increase is equal to $x + 0.1x$, or $1.1x$. A 10% decrease of this new number is equal to $1.1x - 0.1(1.1x)$, or $0.99x$. Because $0.99x < x$, the result is less than the original number.

31. 10 girls

33. $39.2 = p \cdot 112$; 35%

35. $18 = 0.32 \cdot w$; 56.25

Section 15.6 Discounts and Markups
(pages 686 and 687)

1. *Sample answer:* Multiply the original price by 100% − 25% = 75% to find the sale price.
3. **a.** 6% tax on a discounted price; The discounted price is less, so the tax is less.

   b. 30% markup on a \$30 shirt; 30% of \$30 is less than \$30.
5. \$35.70
7. \$76.16
9. \$53.33
11. \$450
13. \$172.40
15. 20%
17. \$55
19. \$175
21. "Multiply \$45.85 by 0.1" and "Multiply \$45.85 by 0.9, then subtract from \$45.85." Both will give the sale price of \$4.59. The first method is easier because it is only one step.
23. no; \$31.08
25. \$30
27. 180
29. C

Section 15.7 Simple Interest
(pages 692 and 693)

1. I = simple interest, P = principal, r = annual interest rate (in decimal form), t = time (in years)
3. You have to change 6% to a decimal and 8 months to a fraction of a year.
5. **a.** \$300 **b.** \$1800
7. **a.** \$292.50 **b.** \$2092.50
9. **a.** \$308.20 **b.** \$1983.20
11. **a.** \$1722.24 **b.** \$6922.24
13. 3%
15. 4%
17. 2 yr
19. 1.5 yr
21. \$1440
23. 2 yr
25. \$2720
27. \$6700.80
29. \$8500
31. 5.25%
33. 4 yr
35. 12.5 yr; Substitute \$2000 for P and I, 0.08 for r, and solve for t.
37. Year 1 = \$520; Year 2 = \$540.80; Year 3 = \$562.43
39. $b \geq 1$; 

41. A

Key Vocabulary Index

Mathematical terms are best understood when you see them used and defined *in context*. This index lists where you will find key vocabulary. A full glossary is available in your Record and Practice Journal and at *BigIdeasMath.com*.

Student Index

This student-friendly index will help you find vocabulary, key ideas, and concepts. It is easily accessible and designed to be a reference for you whether you are looking for a definition, real-life application, or help with avoiding common errors.

D

Student Index

L

M

Student Index

N

O

P

S

W

Common Core State Standards

Domain: Ratios and Proportional Relationships

Understand ratio concepts and use ratio reasoning to solve problems.

6.RP.1 Understand the concept of a ratio and use ratio language to describe a ratio relationship between two quantities.

6.RP.2 Understand the concept of a unit rate a/b associated with a ratio $a:b$ with $b \neq 0$, and use rate language in the context of a ratio relationship.

6.RP.3 Use ratio and rate reasoning to solve real-world and mathematical problems.

- **a.** Make tables of equivalent ratios relating quantities with whole-number measurements, find missing values in the tables, and plot the pairs of values on the coordinate plane. Use tables to compare ratios.
- **b.** Solve unit rate problems including those involving unit pricing and constant speed.
- **c.** Find a percent of a quantity as a rate per 100; solve problems involving finding the whole, given a part and the percent.
- **d.** Use ratio reasoning to convert measurement units; manipulate and transform units appropriately when multiplying or dividing quantities.

Analyze proportional relationships and use them to solve real-world and mathematical problems.

7.RP.1 Compute unit rates associated with ratios of fractions, including ratios of lengths, areas and other quantities measured in like or different units.

7.RP.2 Recognize and represent proportional relationships between quantities.

- **a.** Decide whether two quantities are in a proportional relationship.
- **b.** Identify the constant of proportionality (unit rate) in tables, graphs, equations, and diagrams, and verbal descriptions of proportional relationships.
- **c.** Represent proportional relationships by equations.
- **d.** Explain what a point (x, y) on the graph of a proportional relationship means in terms of the situation, with special attention to the points $(0, 0)$ and $(1, r)$ where r is the unit rate.

7.RP.3 Use proportional relationships to solve multistep ratio and percent problems.

Domain: The Number System

Apply and extend previous understandings of multiplication and division to divide fractions by fractions.

6.NS.1 Interpret and compute quotients of fractions, and solve word problems involving division of fractions by fractions.

Compute fluently with multi-digit numbers and find common factors and multiples.

6.NS.2 Fluently divide multi-digit numbers using the standard algorithm.

6.NS.3 Fluently add, subtract, multiply, and divide multi-digit decimals using the standard algorithm for each operation.

6.NS.4 Find the greatest common factor of two whole numbers less than or equal to 100 and the least common multiple of two whole numbers less than or equal to 12. Use the distributive property to express a sum of two whole numbers 1–100 with a common factor as a multiple of a sum of two whole numbers with no common factor.

Apply and extend previous understandings of numbers to the system of rational numbers.

6.NS.5 Understand that positive and negative numbers are used together to describe quantities having opposite directions or values; use positive and negative numbers to represent quantities in real-world contexts, explaining the meaning of 0 in each situation.

6.NS.6 Understand a rational number as a point on the number line. Extend number line diagrams and coordinate axes familiar from previous grades to represent points on the line and in the plane with negative number coordinates.

- **a.** Recognize opposite signs of numbers as indicating locations on opposite sides of 0 on the number line; recognize that the opposite of the opposite of a number is the number itself, and that 0 is its own opposite.
- **b.** Understand signs of numbers in ordered pairs as indicating locations in quadrants of the coordinate plane; recognize that when two ordered pairs differ only by signs, the locations of the points are related by reflections across one or both axes.
- **c.** Find and position integers and other rational numbers on a horizontal or vertical number line diagram; find and position pairs of integers and other rational numbers on a coordinate plane.

6.NS.7 Understand ordering and absolute value of rational numbers.

- **a.** Interpret statements of inequality as statements about the relative position of two numbers on a number line diagram.
- **b.** Write, interpret, and explain statements of order for rational numbers in real-world contexts.
- **c.** Understand the absolute value of a rational number as its distance from 0 on the number line; interpret absolute value as magnitude for a positive or negative quantity in a real-world situation.
- **d.** Distinguish comparisons of absolute value from statements about order.

6.NS.8 Solve real-world and mathematical problems by graphing points in all four quadrants of the coordinate plane. Include use of coordinates and absolute value to find distance between points with the same first coordinate or the same second coordinate.

Apply and extend previous understandings of operations with fractions to add, subtract, multiply, and divide rational numbers.

7.NS.1 Apply and extend previous understandings of addition and subtraction to add and subtract rational numbers; represent addition and subtraction on a horizontal or vertical number line diagram.

- **a.** Describe situations in which opposite quantities combine to make 0.
- **b.** Understand $p + q$ as the number located a distance $|q|$ from p, in the positive or negative direction depending on whether q is positive or negative. Show that a number and its opposite have the sum of 0 (are additive inverses). Interpret sums of rational numbers by describing real-world contexts.
- **c.** Understand subtraction of rational numbers as adding the additive inverse, $p - q = p + (-q)$. Show that the distance between two rational numbers on the number line is the absolute value of their difference, and apply this principle in real-world contexts.
- **d.** Apply properties of operations as strategies to add and subtract rational numbers.

7.NS.2 Apply and extend previous understandings of multiplication and division and of fractions to multiply and divide rational numbers.

- **a.** Understand that multiplication is extended from fractions to rational numbers by requiring that operations continue to satisfy the properties of operations, particularly the distributive property, leading to products such as $(-1)(-1) = 1$ and the rules for multiplying signed numbers. Interpret products of rational numbers by describing real-world contexts.
- **b.** Understand that integers can be divided, provided that the divisor is not zero, and every quotient of integers (with non-zero divisor) is a rational number. If p and q are integers, then $-(p/q) = (-p)/q = p/(-q)$. Interpret quotients of rational numbers by describing real-world contexts.

c. Apply properties of operations as strategies to multiply and divide rational numbers.

d. Convert a rational number to a decimal using long division; know that the decimal form of a rational number terminates in 0s or eventually repeats.

7.NS.3 Solve real-world and mathematical problems involving the four operations with rational numbers.

Domain: Expressions and Equations

Apply and extend previous understandings of arithmetic to algebraic expressions.

6.EE.1 Write and evaluate numerical expressions involving whole-number exponents.

6.EE.2 Write, read, and evaluate expressions in which letters stand for numbers.

a. Write expressions that record operations with numbers and with letters standing for numbers.

b. Identify parts of an expression using mathematical terms (sum, term, product, factor, quotient, coefficient); view one or more parts of an expression as a single entity.

c. Evaluate expressions at specific values of their variables. Include expressions that arise from formulas used in real-world problems. Perform arithmetic operations, including those involving whole-number exponents, in the conventional order when there are no parentheses to specify a particular order (Order of Operations).

6.EE.3 Apply the properties of operations to general equivalent expressions.

6.EE.4 Identify when two expressions are equivalent.

Reason about and solve one-variable equations and inequalities.

6.EE.5 Understand solving an equation or inequality as a process of answering a question: which values from a specified set, if any, make the equation or inequality true? Use substitution to determine whether a given number in a specified set makes an equation or inequality true.

6.EE.6 Use variables to represent numbers and write expressions when solving a real-world or mathematical problem; understand that a variable can represent an unknown number, or, depending on the purpose at hand, any number in a specified set.

6.EE.7 Solve real-world and mathematical problems by writing and solving equations of the form $x + p = q$ and $px = q$ for cases in which p, q, and x are all nonnegative rational numbers.

6.EE.8 Write an inequality of the form $x > c$ or $x < c$ to represent a constraint or condition in a real-world or mathematical problem. Recognize that inequalities of the form $x > c$ or $x < c$ have infinitely many solutions; represent solutions of such inequalities on number line diagrams.

Represent and analyze quantitative relationships between dependent and independent variables.

6.EE.9 Use variables to represent two quantities in a real-world problem that change in relationship to one another; write an equation to express one quantity thought of as the dependent variable, in terms of the other quantity, thought of as the independent variable. Analyze the relationship between the dependent and independent variables using graphs and tables, and relate these to the equation.

Use properties of operations to generate equivalent expressions.

7.EE.1 Apply properties of operations as strategies to add, subtract, factor, and expand linear expressions with rational coefficients.

7.EE.2 Understand that rewriting an expression in different forms in a problem context can shed light on the problem and how the quantities in it are related.

Solve real-life and mathematical problems using numerical and algebraic expressions and equations.

7.EE.3 Solve multi-step real-life and mathematical problems posed with positive and negative rational numbers in any form (whole numbers, fractions, and decimal), using tools strategically. Apply properties of operations to calculate with numbers in any form; convert between forms as appropriate; and assess the reasonableness of answers using mental computation and estimation strategies.

7.EE.4 Use variables to represent quantities in a real-world or mathematical problem, and construct simple equations and inequalities to solve problems by reasoning about the quantities.

a. Solve word problems leading to equations of the form $px + q = r$ and $p(x + q) = r$, where p, q, and r are specific rational numbers. Solve equations of these forms fluently. Compare an algebraic solution to an arithmetic solution, identifying the sequence of the operations used in each approach.

Domain: Geometry

Solve real-world and mathematical problems involving area, surface area, and volume.

6.G.1 Find the area of right triangles, other triangles, special quadrilaterals, and polygons by composing into rectangles or decomposing into triangles and other shapes; apply these techniques in the context of solving real-world and mathematical problems.

6.G.2 Find the volume of a right rectangular prism with fractional edge lengths by packing it with unit cubes of the appropriate unit fraction edge lengths, and show that the volume is the same as would be found by multiplying the edge length of the prism. Apply the formulas $V = \ell wh$ and $V = bh$ to find volumes of right rectangular prisms with fractional edge lengths in the context of solving real-world and mathematical problems.

6.G.3 Draw polygons in the coordinate plane given coordinates for the vertices; use coordinates to find the length of a side joining points with the same first coordinate or the same second coordinate. Apply these techniques in the context of solving real-world and mathematical problems.

6.G.4 Represent three-dimensional figures using nets made up of rectangles and triangles, and use the nets to find the surface area of these figures. Apply these techniques in the context of solving real-world and mathematical problems.

Domain: Statistics and Probability

Develop understanding of statistical variability.

6.SP.1 Recognize a statistical question as one that anticipates variability in the data related to the question and accounts for it in the answers.

6.SP.2 Understand that a set of data collected to answer a statistical question has a distribution which can be described by its center, spread, and overall shape.

6.SP.3 Recognize that a measure of center for a numerical data set summarizes all of its values with a single number, while a measure of variation describes how its values vary with a single number.

Summarize and describe distributions.

6.SP.4 Display numerical data in plots on a number line, including dot plots, histograms, and box plots.

6.SP.5 Summarize numerical data sets in relation to their context, such as by:

a. Reporting the number of observations.

b. Describing the nature of the attribute under investigation, including how it was measured and its units of measurement.

c. Giving quantitative measures of center (median and/or mean) and variability (interquartile range and/or mean absolute deviation), as well as describing any overall pattern and any striking deviations from the overall pattern with reference to the context in which the data were gathered.

d. Relating the choice of measures of center and variability to the shape of the data distribution and the context in which the data were gathered.

Photo Credits

Cover
Pavelk/Shutterstock.com, © Dwight Smith | Dreamstime.com, valdis torms /Shutterstock.com

Front matter
i Pavelk/Shutterstock.com, © Dwight Smith | Dreamstime.com, valdis torms /Shutterstock.com; **iv** Big Ideas Learning, LLC; **viii** *top* ©iStockphoto.com/ALEAIMAGE, ©iStockphoto.com/Ann Marie Kurtz; *bottom* ©iStockphoto.com/Jane norton; **ix** *top* ©iStockphoto.com/ALEAIMAGE, ©iStockphoto.com/Ann Marie Kurtz; *bottom* wavebreakmedia ltd/Shutterstock.com; **x** *top* stephan kerkhofs/Shutterstock.com, Cigdem Sean Cooper/Shutterstock.com, ©iStockphoto.com/Andreas Gradin; *bottom* Odua Images/Shutterstock.com; **xi** *top* ©iStockphoto.com/sumnersgraphicsinc, ©iStockphoto.com/Ann Marie Kurtz; *bottom* James Flint/Shutterstock.com; **xii** *top* ©iStockphoto.com/Lisa Thornberg, ©iStockphoto.com/Ann Marie Kurtz; *bottom* william casey/Shutterstock.com; **xiii** *top* ©iStockphoto.com/Jonathan Larsen; *bottom* Edyta Pawlowska/Shutterstock.com; **xiv** *top* Varina and Jay Patel/Shutterstock.com, ©iStockphoto.com/Ann Marie Kurtz; *bottom* PETER CLOSE/Shutterstock.com; **xv** *top* ©iStockphoto/Michael Flippo, ©iStockphoto.com/Ann Marie Kurtz; *bottom* ©iStockphoto.com/ranplett; **xvi** *top* Chiyacat/Shutterstock.com, Zoom Team/Shutterstock.com; *bottom* ©iStockphoto.com/Noraznen Azit; **xvii** *top* ©iStockphoto.com/Alistair Cotton; *bottom* ©iStockphoto.com/Thomas Perkins; **xviii** *top* ©iStockphoto.com/ALEAIMAGE, ©iStockphoto.com/Ann Marie Kurtz; *bottom* Evok20/Shutterstock.com; **xix** *top* Chiyacat/Shutterstock.com, Zoom Team/Shutterstock.com; *bottom* Sinisa Bobic/Shutterstock.com; **xx** *top* ©iStockphoto.com/ALEAIMAGE, ©iStockphoto.com/Ann Marie Kurtz; *bottom* Artpose Adam Borkowski/Shutterstock.com; **xxi** *top* Varina and Jay Patel/Shutterstock.com, ©iStockphoto.com/Ann Marie Kurtz; *bottom* ©iStockphoto.com/Aldo Murillo; **xxii** *top* Kasiap/Shutterstock.com, ©iStockphoto.com/Ann Marie Kurtz; *bottom* ©iStockphoto.com/kate_sept2004; **xxiii** *top* ©iStockphoto.com/Björn Kindler; *bottom* ©iStockphoto.com/Ferli Achirulli; **xxiv** Ljupco Smokovski/Shutterstock.com

Chapter 1
1 ©iStockphoto.com/ALEAIMAGE, ©iStockphoto.com/Ann Marie Kurtz; **2** *top right* S.Dashkevych/Shutterstock.com; *bottom right* auremar/Shutterstock.com; **5** yxm2008/Shutterstock.com; **6** ©iStockphoto.com/Santino Ambrogio; **8** Jiang Dao Hua/Shutterstock.com; **9** *top right* Africa Studio/Shutterstock.com; *center left* Podriv Ustoev/Shutterstock.com; **10** Danomyte/Shutterstock.com; **11** Nicole Gordine/Shutterstock.com; **13** vonDUCK/Shutterstock.com; **15** ©iStockphoto.com/WestLight; **20** ©iStockphoto.com/clu; **21** ©iStockphoto.com/mladn61; **23** *center right* Wendy Nero/Shutterstock.com; *bottom right* ©iStockphoto.com/craftvision; **24** James Hoenstine/Shutterstock.com; **26** farbeffekte/Shutterstock.com; **29** Michael Mitchell/Shutterstock.com; **33** jmatzick/Shutterstock.com; **34** Glenda M. Powers/Shutterstock.com; **35** Africa Studio/Shutterstock.com; **39** Alinochka/Shutterstock.com; **40** Orla/Shutterstock.com; **41** yuyangc/Shutterstock.com; **44** *center left* Mike Flippo/Shutterstock.com; *bottom right* Nitr/Shutterstock.com; **48** beboy/Shutterstock.com

Chapter 2
52 ©iStockphoto.com/ALEAIMAGE, ©iStockphoto.com/Ann Marie Kurtz; **59** *center left* fivespots/Shutterstock.com; *center right* bluehand/Shutterstock.com; **68** tele52/Shutterstock.com, lithian/Shutterstock.com, Galina Barskaya/Shutterstock.com; **69** *center left* ©iStockphoto.com/Michael Plumb; *center right* g215/Shutterstock.com; **73** ©iStockphoto.com/Karin Lau; **81** Vacclav/Shutterstock.com; **82** Peredniankina/Shutterstock.com; **83** SSSCCC/Shutterstock.com; **89** Luba V Nel/Shutterstock.com; **90** KENCKOphotography/Shutterstock.com; **97** ©iStockphoto.com/bananahuman; **99** ©iStockphoto.com/Janis Litavnieks; **104** ©iStockphoto.com/graham heywood

Chapter 3
108 stephan kerkhofs/Shutterstock.com, Cigdem Sean Cooper/Shutterstock.com, ©iStockphoto.com/Andreas Gradin; **121** Kateryna Larina/Shutterstock.com; **125** Vivid Pixels/Shutterstock.com; **138** Pal Teravagimov/Shutterstock.com; **141** Inhabitant/Shutterstock.com; **142** Fotofermer/Shutterstock.com; **143** Helder Almeida/Shutterstock.com; **146** *center right* Aptyp_koK/Shutterstock.com; *bottom left* Andy Cash/Shutterstock.com

Chapter 4
150 ©iStockphoto.com/sumnersgraphicsinc, ©iStockphoto.com/Ann Marie Kurtz; **162** Terrance Emerson/Shutterstock.com; **163** ©iStockphoto.com/bamse009; **171** Kharidehal Abhirama Ashwin/Shutterstock.com; **177** Christian Musat/Shutterstock.com; **179** Brandon Seidel/Shutterstock.com; **184** *top right* ©iStockphoto.com/AlexMax; *bottom right* U.S. Geological Survey

Chapter 5
188 ©iStockphoto.com/Lisa Thornberg, ©iStockphoto.com/Ann Marie Kurtz; **190** Elnur/Shutterstock.com; **192** Vladimir Wrangel/Shutterstock.com; **193** Lepas/Shutterstock.com; **196** *top* foxie/Shutterstock.com; *center right* Constantinos/Shutterstock.com; **197** Danny Smythe/Shutterstock.com; **199** South 12th Photography/Shutterstock.com; **200** Petr Malyshev/Shutterstock.com; **202** ©iStockphoto.com/otakumania; **203** Anneka/Shutterstock.com; **205** vita khorzhevska/Shutterstock.com; **206** © HO/Reuters/Corbis; **207** Alex Staroseltsev/Shutterstock.com; **208** *top right* ©iStockphoto.com/Gord Horne; *bottom right* leonid_tit/Shutterstock.com; **209** *top left* ©iStockphoto.com/ljpat; *bottom right* ©iStockphoto.com/Birgitte Magnus; **210** Mike Flippo/Shutterstock.com; **212** *top right* ultimathule/Shutterstock.com; *top left* Christopher Kolaczan/Shutterstock.com; **213** ©iStockphoto.com/Bart Wolczyk; **214** Catalin Petolea/Shutterstock.com, Kaspri/Shutterstock.com; **215** R. Gino Santa Maria/Shutterstock.com; **225** Spasiblo/Shutterstock.com; **230** Thank You/Shutterstock.com; **236** Derek Wong / CC-BY-SA-3.0; **237** ©iStockphoto.com/Paul Tessier; **238** *bottom left* Kitch Bain/Shutterstock.com; *bottom* ©iStockphoto.com/Ermin Gutenberger; **242** Joe Gough/Shutterstock.com

Chapter 6
246 ©iStockphoto.com/Jonathan Larsen; **251** ©iStockphoto.com/Egor Mopanko; **253** haveseen/Shutterstock.com; **254** *top right* NASA/Kim Shiflett; *bottom* NASA; **255** NASA; **259** ©iStockphoto.com/Andrew Penner; **260** NASA; **261** *left* Andrey Armyagov/Shutterstock.com; *center* ©iStockphoto.com/Island Effects; *right* U.S. Navy photo by Photographer's Mate 2nd Class Prince Hughes III; **264** ©iStockphoto.com/jclegg, ©iStockphoto.com/spxChrome, ©iStockphoto.com/Laura Eisenberg; **265** A'lya/Shutterstock.com; **267** *top right* NicolasMcComber/Shutterstock.com; *bottom* ©iStockphoto.com/Zuki; **269** *left* ©iStockphoto.com/james steidl; *right* U.S. Navy photo by Photographers Mate 2nd Class Michael Sandberg; **271** Shane W Thompson/Shutterstock.com; **278** Christophe Testi/Shutterstock.com

Chapter 7

292 Varina and Jay Patel/Shutterstock.com, ©iStockphoto.com/Ann Marie Kurtz; **299** *top right* ©iStockphoto.com/Joop Snijder, ©iStockphoto.com/Michael MacFadden, ©iStockphoto.com/Steve Goodwin; *center left* ©iStockphoto.com/Kenneth C. Zirkel; **304** jocic/Shutterstock.com, Marko Poplasen/Shutterstock.com; **306** *top left* ©iStockphoto.com/Jeremy Wee; *top right* ©iStockphoto.com/Jan Will; **307** ©iStockphoto.com/Keith Reicher; **309** ©iStockphoto.com/Leo Blanchette; **312** ©iStockphoto.com/Christopher Futcher; **313** ©iStockphoto.com/Eric Isselée; **314** mangostock/Shutterstock.com; **315** llya Andriyanov/Shutterstock.com; **316** violetkaipa/Shutterstock.com; **318** kokandr/Shutterstock.com; **319** discpicture/Shutterstock.com; **320** ©iStockphoto.com/Mutlu Kurtbas; **324** *first* ©iStockphoto.com/Studio-Annika; *second* ©iStockphoto.com/nicholas belton; *third* ©iStockphoto.com/Robert Dant; **328** NASA/Johns Hopkins University Applied Physics Laboratory; **330** ©iStockphoto.com/George Peters; **331** ©iStockphoto.com/Anthony Ladd; **332** *top right* CLS Design/Shutterstock.com; *bottom right* Liquid Productions, LLC / Shutterstock.com; **337** *top left* ©iStockphoto.com/o-che; *center left* ©iStockphoto.com/sunygraphics; **338** *top right* Dennis Owusu-Ansah/Shutterstock.com; *bottom right* Julia Zakharova /Shutterstock.com; **341** Nathan Till/Shutterstock.com; **342** ©iStockphoto.com/rami ben ami; **343** Karin Hildebrand Lau/Shutterstock.com; **344** ©iStockphoto.com/Algimantas Balezentis; **348** ©iStockphoto.com/Jani Bryson

Chapter 8

352 ©iStockphoto.com/Michael Flippo, ©iStockphoto.com/Ann Marie Kurtz; **358** *Exercise 17* ©iStockphoto.com/Rich Koele; *Exercise 21* design56/Shutterstock.com; **359** *top right* ©iStockphoto.com/Hedda Gjerpen; *center* ©iStockphoto.com/rzdeb; **363** Niki Crucillo/Shutterstock.com; **370** PeterG/Shutterstock.com; **372** Itana/Shutterstock.com; **373** Tupungato/Shutterstock.com; **377** *top left* ©iStockphoto.com/William Britten; *center left* Denis Barbulat/Shutterstock.com; **379** *center left* ©iStockphoto.com/Jill Chen; *center right* ©iStockphoto.com/LongHa2006

Chapter 9

388 Chiyacat/Shutterstock.com, Zoom Team/Shutterstock.com; **390** LeventeGyori/Shutterstock.com; **391** *center left and right* Nattika/Shutterstock.com; **392** Eric Isselée/Shutterstock.com; **393** *top right* Iznogood/Shutterstock.com; *bottom left* AISPIX by Image Source/Shutterstock.com; **395** Laralova/Shutterstock.com; **396** Rob Byron/Shutterstock.com; **397** Denis Vrublevski/Shutterstock.com; **399** ©iStockphoto.com/Eric Isselée; **402** *top right* Hein Nouwens/Shutterstock.com; *bottom* ivelly/Shutterstock.com; **403** ©iStockphoto.com/Andrew Rich; **409** *top right* ©iStockphoto.com/suemack; *top left* ©iStockphoto.com/muratkoc; **415** Charlie Hutton/Shutterstock.com; **416** Talvi/Shutterstock.com; **417** ©iStockphoto.com/Jason Lugo; **420** Danny Smythe/Shutterstock.com; **421** *top left* Ganko/Shutterstock.com; *top right* Mark Herreid/Shutterstock.com; **422** alarich/Shutterstock.com; **423** tab62/Shutterstock.com; **424** Racheal Grazias/Shutterstock.com; **426** Jan Martin Will/Shutterstock.com; **427** Kitch Bain/Shutterstock.com

Chapter 10

432 ©iStockphoto.com/Alistair Cotton; **434** Elzbieta Szpak/Shutterstock.com; **435** ©CORBIS; **437** ©iStockphoto.com/Pekka Nikonen; **438** ©iStockphoto.com/Mehmet Salih Guler; **442** stockshoppe/Shutterstock.com; **443** *top right* ©iStockphoto.com/susaro; *bottom left* Arman Zhenikeyev/Shutterstock.com; **444** Tomasz Trojanowski/Shutterstock.com; **447** ©iStockphoto.com/Eric Isselée; **449** ©iStockphoto.com/vincent chien chow chine; **450** Sergey Mironov/Shutterstock.com; **451** fresher/Shutterstock.com; **453** mmaxer/Shutterstock.com; **457** Lightspring/Shutterstock.com; **459** *first* windu/Shutterstock.com; *second* motorolka/Shutterstock.com; *third* Preto Perola/Shutterstock.com; *fourth* nikkytok/Shutterstock.com; *center right* Apollofoto/Shutterstock.com; **460** ©iStockphoto.com/rusm; **461** Sebastian Knight/Shutterstock.com; **462** Garbuzov/Shutterstock.com; **464** *center left* Ffooter/Shutterstock.com; *bottom right* zhuda/Shutterstock.com; **465** Rob Marmion/Shutterstock.com; **469** Nikola Bilic/Shutterstock.com; **470** Monkey Business Images/Shutterstock.com

Chapter 11

475 ©iStockphoto.com/ALEAIMAGE, ©iStockphoto.com/Ann Marie Kurtz; **495** ©iStockphoto.com/RonTech2000; **501** Dmitry Melnikov/Shutterstock.com; **508** *center right* ©iStockphoto.com/Rich Legg; *bottom left* CLFProductions/Shutterstock.com; **511** Liem Bahneman/Shutterstock.com; **512** *center right* ©iStockphoto.com/susaro; *bottom left* © Leonard J. DeFrancisci / Wikimedia Commons / CC-BY-SA-3.0 / GFDL

Chapter 12

516 Chiyacat/Shutterstock.com, Zoom Team/Shutterstock.com; **518** ©iStockphoto.com/Shantell; **527** ultrapro/Shutterstock.com; **528** margouillat photo/Shutterstock.com; **529** Heide Hellebrand/Shutterstock.com; **531** ©iStockphoto.com/Jason Lugo; **547** Pinosub/Shutterstock.com; **548** EdBockStock/Shutterstock.com; **550** Laborant/Shutterstock.com

Chapter 13

552 ©iStockphoto.com/ALEAIMAGE, ©iStockphoto.com/Ann Marie Kurtz; **558** *bottom left* photo25th/Shutterstock.com; *bottom center* ©iStockphoto.com/Don Nichols; **563** Andrew Burgess/Shutterstock.com; **564** Suzanne Tucker/Shutterstock.com; **565** ©iStockphoto.com/Vadim Ponomarenko; **571** MarcelClemens/Shutterstock.com; **575** ©iStockphoto.com/fotoVoyager; **577** *top right* John Kropewnicki/Shutterstock.com; *center left* Yuri Bathan (yuri10b)/Shutterstock.com; *bottom right* Dim Dimich/Shutterstock.com; **595** wacpan/Shutterstock.com

Chapter 14

596 Varina and Jay Patel/Shutterstock.com, ©iStockphoto.com/Ann Marie Kurtz; **599** qingqing/Shutterstock.com; **600** the808/Shutterstock.com; **604** Sergey Peterman/Shutterstock.com; **605** VikaSuh/Shutterstock.com; **610** ©iStockphoto.com/Kemter; **611** ©iStockphoto.com/VikaValter; **619** NASA/Carla Thomas; **622** Jean Tompson; **634** Baldwin Online: Children's Literature Project at www.mainlesson.com; **635** ©iStockphoto.com/Brian Pamphilon; **637** John Kasawa/Shutterstock.com; **640** ©iStockphoto.com/Uyen Le; **646** Peter zljlstra/Shutterstock.com

Chapter 15

648 Kasiap/Shutterstock.com, ©iStockphoto.com/Ann Marie Kurtz; **661** ©iStockphoto.com/Eric Isselée; **666** Image courtesy the President's Challenge, a program of the President's Council on Fitness, Sports and Nutrition; **677** Rob Byron/Shutterstock.com; **678** ©iStockphoto.com/NuStock; **680** ©iStockphoto.com/ARENA Creative; **685** ©iStockphoto.com/amriphoto; **686** ©iStockphoto.com/Albert Smirnov; **687** ©iStockphoto.com/Lori Sparkia; **691** ©iStockphoto.com/anne de Haas; **693** *top right* Big Ideas Learning, LLC; *center left* ©iStockphoto.com/Rui Matos; **694** ©iStockphoto.com/Michael Fernahl; **697** AISPIX by Image Source/Shutterstock.com; **698** ©iStockphoto.com/ted johns

Appendix A

A0 *background* ©iStockphoto.com/Björn Kindler; *top left and bottom* ©iStockphoto.com/Ralf Hettler; **A1** *top right* Emmer, Michele, ed., The Visual Mind: Art and Mathematics, Plate 2, © 1993 Massachusetts Institute of Technology, by permission of The MIT Press.; *bottom left* ©iStockphoto.com/Andrew Cribb; *bottom right* ©iStockphoto.com/Liz Leyden; **A4** *top right* ©iStockphoto.com/Ralf Hettler; *center right* ©iStockphoto.com/Ragnarocks; *bottom left* ©iStockphoto.com/Linda Steward; *bottom right* ©iStockphoto.com/rackermann; **A5** *top and bottom right* ©iStockphoto.com/rackermann; *bottom left* ©iStockphoto.com/Ralf Hettler; **A6** *top right* Stannered, DTR; *bottom left* Emmer, Michele, ed., The Visual Mind: Art and Mathematics, Plate 2, © 1993 Massachusetts Institute of Technology, by permission of The MIT Press.; **A7** *top right* Elena Borodynkina/Shutterstock.com; *center right* ©iStockphoto.com/Andrew Cribb; *bottom left* ©iStockphoto.com/smokyme; *bottom right* Sculpture by Vladimir Bulatov; **A8** *top right* ©iStockphoto.com/Dan Van Oss; *bottom left* ©iStockphoto.com/Tomasz Tulik; **A9** *top right* ©iStockphoto.com/timoph; *bottom left* ©iStockphoto.com/Liz Leyden; *bottom right* ©iStockphoto.com/Pauline S Mills

Cartoon illustrations Tyler Stout

Common Core State Standards

Kindergarten

Counting and Cardinality	– Count to 100 by Ones and Tens; Compare Numbers
Operations and Algebraic Thinking	– Understand and Model Addition and Subtraction
Number and Operations in Base Ten	– Work with Numbers 11–19 to Gain Foundations for Place Value
Measurement and Data	– Describe and Compare Measurable Attributes; Classify Objects into Categories
Geometry	– Identify and Describe Shapes

Grade 1

Operations and Algebraic Thinking	– Represent and Solve Addition and Subtraction Problems
Number and Operations in Base Ten	– Understand Place Value for Two-Digit Numbers; Use Place Value and Properties to Add and Subtract
Measurement and Data	– Measure Lengths Indirectly; Write and Tell Time; Represent and Interpret Data
Geometry	– Draw Shapes; Partition Circles and Rectangles into Two and Four Equal Shares

Grade 2

Operations and Algebraic Thinking	– Solve One- and Two-Step Problems Involving Addition and Subtraction; Build a Foundation for Multiplication
Number and Operations in Base Ten	– Understand Place Value for Three-Digit Numbers; Use Place Value and Properties to Add and Subtract
Measurement and Data	– Measure and Estimate Lengths in Standard Units; Work with Time and Money
Geometry	– Draw and Identify Shapes; Partition Circles and Rectangles into Two, Three, and Four Equal Shares

Grade 3

Operations and Algebraic Thinking	– Represent and Solve Problems Involving Multiplication and Division; Solve Two-Step Problems Involving Four Operations
Number and Operations in Base Ten	– Round Whole Numbers; Add, Subtract, and Multiply Multi-Digit Whole Numbers
Number and Operations—Fractions	– Understand Fractions as Numbers
Measurement and Data	– Solve Time, Liquid Volume, and Mass Problems; Understand Perimeter and Area
Geometry	– Reason with Shapes and Their Attributes

Grade 4

Operations and Algebraic Thinking	– Use the Four Operations with Whole Numbers to Solve Problems; Understand Factors and Multiples
Number and Operations in Base Ten	– Generalize Place Value Understanding; Perform Multi-Digit Arithmetic
Number and Operations—Fractions	– Build Fractions from Unit Fractions; Understand Decimal Notation for Fractions
Measurement and Data	– Convert Measurements; Understand and Measure Angles
Geometry	– Draw and Identify Lines and Angles; Classify Shapes

Grade 5

Operations and Algebraic Thinking	– Write and Interpret Numerical Expressions
Number and Operations in Base Ten	– Perform Operations with Multi-Digit Numbers and Decimals to Hundredths
Number and Operations—Fractions	– Add, Subtract, Multiply, and Divide Fractions
Measurement and Data	– Convert Measurements within a Measurement System; Understand Volume
Geometry	– Graph Points in the First Quadrant of the Coordinate Plane; Classify Two-Dimensional Figures

Mathematics Reference Sheet

Conversions

U.S. Customary
1 foot = 12 inches
1 yard = 3 feet
1 mile = 5280 feet
1 acre ≈ 43,560 square feet
1 cup = 8 fluid ounces
1 pint = 2 cups
1 quart = 2 pints
1 gallon = 4 quarts
1 gallon = 231 cubic inches
1 pound = 16 ounces
1 ton = 2000 pounds
1 cubic foot ≈ 7.5 gallons

U.S. Customary to Metric
1 inch = 2.54 centimeters
1 foot ≈ 0.3 meter
1 mile ≈ 1.61 kilometers
1 quart ≈ 0.95 liter
1 gallon ≈ 3.79 liters
1 cup ≈ 237 milliliters
1 pound ≈ 0.45 kilogram
1 ounce ≈ 28.3 grams
1 gallon ≈ 3785 cubic centimeters

Time
1 minute = 60 seconds
1 hour = 60 minutes
1 hour = 3600 seconds
1 year = 52 weeks

Temperature
$C = \frac{5}{9}(F - 32)$

$F = \frac{9}{5}C + 32$

Metric
1 centimeter = 10 millimeters
1 meter = 100 centimeters
1 kilometer = 1000 meters
1 liter = 1000 milliliters
1 kiloliter = 1000 liters
1 milliliter = 1 cubic centimeter
1 liter = 1000 cubic centimeters
1 cubic millimeter = 0.001 milliliter
1 gram = 1000 milligrams
1 kilogram = 1000 grams

Metric to U.S. Customary
1 centimeter ≈ 0.39 inch
1 meter ≈ 3.28 feet
1 kilometer ≈ 0.62 mile
1 liter ≈ 1.06 quarts
1 liter ≈ 0.26 gallon
1 kilogram ≈ 2.2 pounds
1 gram ≈ 0.035 ounce
1 cubic meter ≈ 264 gallons

Number Properties

Commutative Properties of Addition and Multiplication

$a + b = b + a$

$a \cdot b = b \cdot a$

Associative Properties of Addition and Multiplication

$(a + b) + c = a + (b + c)$

$(a \cdot b) \cdot c = a \cdot (b \cdot c)$

Addition Property of Zero

$a + 0 = a$

Multiplication Properties of Zero and One

$a \cdot 0 = 0$

$a \cdot 1 = a$

Distributive Property:

$a(b + c) = ab + ac$

$a(b - c) = ab - ac$

Properties of Equality

Addition Property of Equality

If $a = b$, then $a + c = b + c$.

Subtraction Property of Equality

If $a = b$, then $a - c = b - c$.

Multiplication Property of Equality

If $a = b$, then $a \cdot c = b \cdot c$.

Multiplicative Inverse Property

$n \cdot \frac{1}{n} = \frac{1}{n} \cdot n = 1, n \neq 0$

Division Property of Equality

If $a = b$, then $a \div c = b \div c, c \neq 0$.

Properties of Inequality

Addition Property of Inequality

If $a > b$, then $a + c > b + c$.

Subtraction Property of Inequality

If $a > b$, then $a - c > b - c$.

Multiplication Property of Inequality

If $a > b$ and c is positive, then $a \cdot c > b \cdot c$.

Division Property of Inequality

If $a > b$ and c is positive, then $a \div c > b \div c$.

Perimeter and Area

Square	Rectangle	Parallelogram	Triangle	Trapezoid
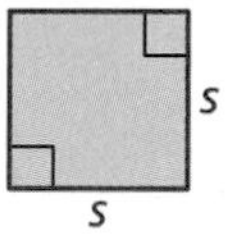	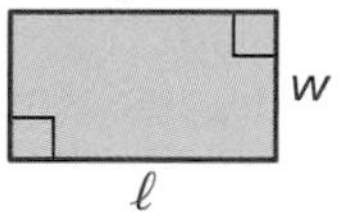	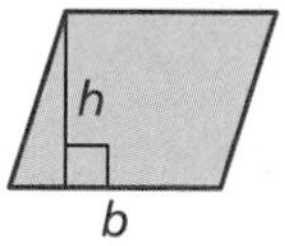	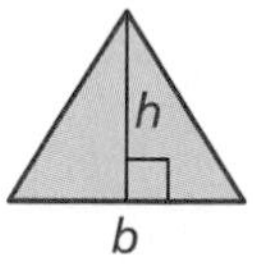	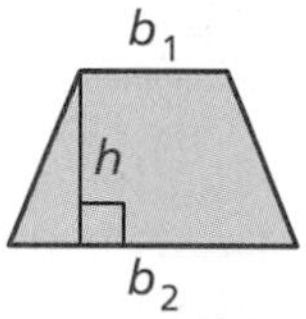
$P = 4s$ $A = s^2$	$P = 2\ell + 2w$ $A = \ell w$	$A = bh$	$A = \frac{1}{2}bh$	$A = \frac{1}{2}h(b_1 + b_2)$

Surface Area

Prism

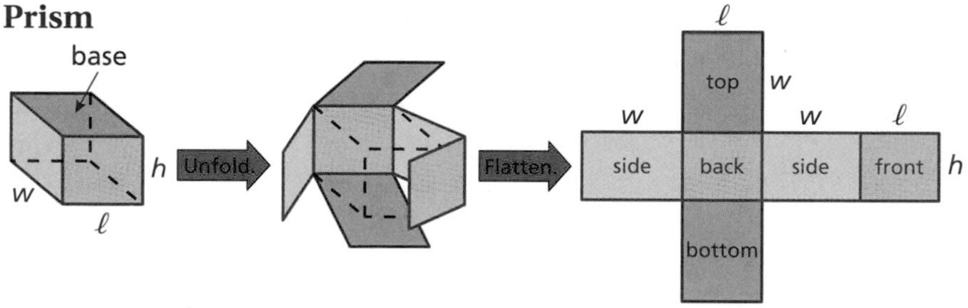

$S =$ areas of bases + areas of lateral faces

Pyramid

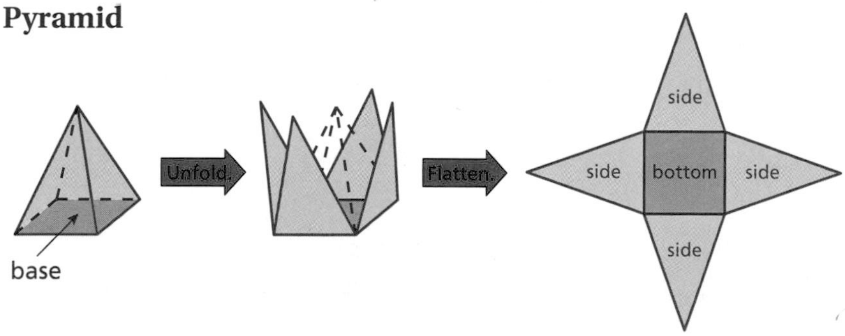

$S =$ area of base + areas of lateral faces

Volume of a Rectangular Prism

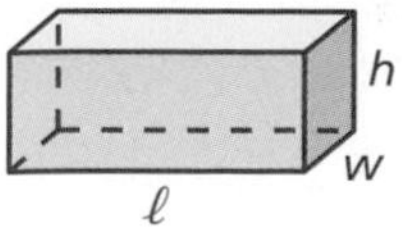

$V = Bh = \ell wh$

Simple Interest

Simple interest formula

$I = Prt$

The Coordinate Plane

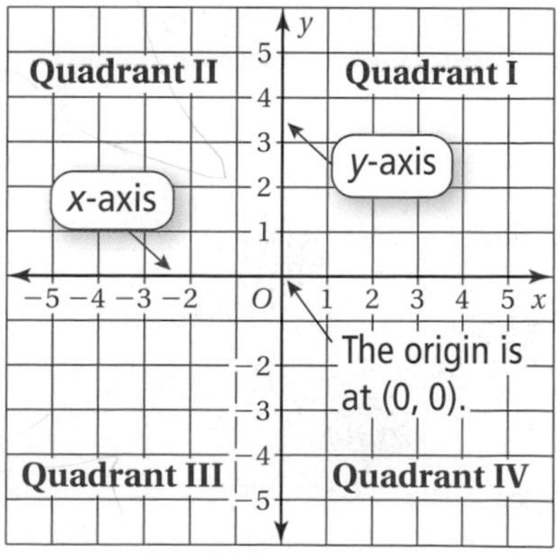